P9-BIB-998

BRIEF CALCULUS

with Applications

ALTERNATE FOURTH EDITION

FREE REVIEW COPY FREE REVIEW COPY

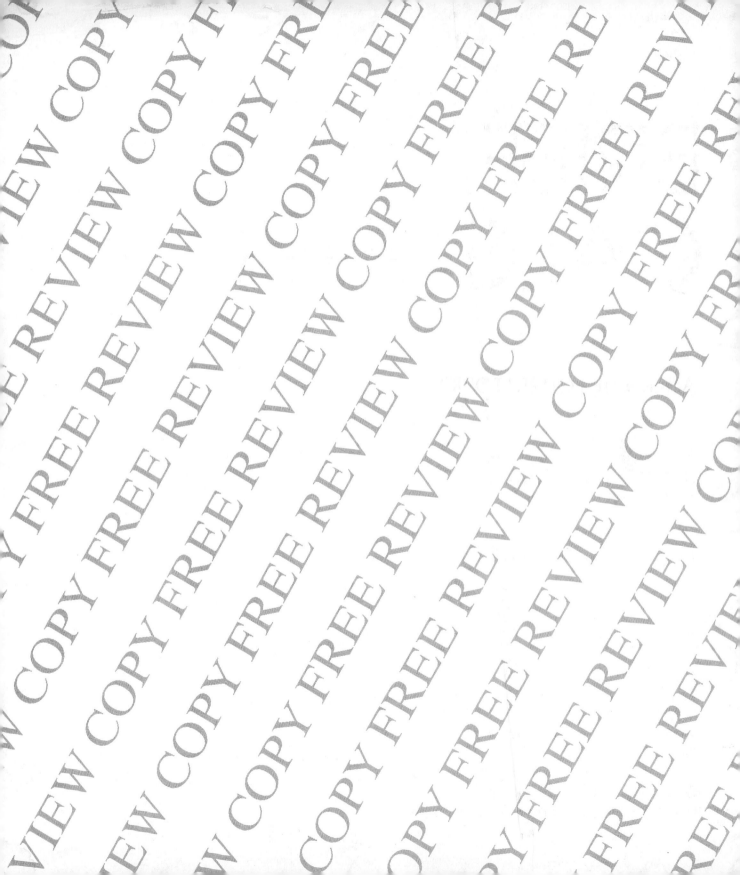

BRIEF CALCULUS

with Applications

ALTERNATE FOURTH EDITION

ROLAND E. LARSON
The Pennsylvania State University
The Behrend College

ROBERT P. HOSTETLER
The Pennsylvania State University
The Behrend College

BRUCE H. EDWARDS
University of Florida

Case Studies Contributed by
James A. Kurre
Kenneth T. Louie
The Pennsylvania State University
The Behrend College

CUSTOM PUBLISHING AVAILABLE

D. C. Heath and Company
Lexington, Massachusetts   Toronto

REVIEW COPY FREE REVIEW COPY FREE

Address editorial correspondence to:

D. C. Heath and Company
125 Spring Street
Lexington, MA 02173

Acquisitions Editors: Ann Marie Jones, Charlie Hartford
Managing Editor: Cathy Cantin
Developmental Editor: Emily Keaton
Production Editor: Karen Carter
Marketing Manager: Christine Hoag
Designer: Jan Shapiro
Art Editor: Gary Crespo
Production Coordinator: Lisa Merrill
Photo Researchers: Melina Freedman, Donna Simons
Composition: Meridian Creative Group
Technical Art: Folium, Inc.
Cover: Tony Stone Images/Ed Honowitz

Trademark Acknowledgments: TI is a registered trademark of Texas Instruments Incorporated. Casio is a registered trademark of Casio, Inc. Sharp is a registered trademark of Sharp Electronics Corporation. *Derive* is a registered trademark of Soft Warehouse, Inc. *Maple* is a registered trademark of Waterloo Maple Software, Inc. *Mathematica* is a registered trademark of Wolfram Research, Inc.

We have included examples and exercises that use real-life data as well as technology output from a variety of software. This would not have been possible without the help of many people and organizations. Our wholehearted thanks goes to all for their time and effort.

Copyright © 1995 by D. C. Heath and Company.

Previous editions copyright © 1991, 1987, 1983 by D. C. Heath and Company.

All rights reserved. No part of this publication may be reproduced or transmitted in any form or by any means, electronic or mechanical, including photocopy, recording, or any information storage or retrieval system, without permission in writing from the publisher.

Published simultaneously in Canada.

Printed in the United States of America.

International Standard Book Number: 0-669-35163-6

Library of Congress Catalog Number: 94-71598

10 9 8 7 6 5 4 3 2 1

Preface

Brief Calculus with Applications, Fourth Edition, is designed for use in a beginning calculus course for students in business, management, economics, and the social and life sciences. In writing this edition, we were guided by two primary objectives. For students, our objective was to write in a precise and readable manner with the basic concepts, techniques, and applications of calculus clearly defined and demonstrated. For instructors, our objective was to create a comprehensive teaching instrument that uses proven pedagogical techniques, freeing instructors to make the most efficient use of classroom time.

Changes in the Fourth Edition

In the Fourth Edition, all textual elements were considered for revision, and many new examples, exercises, and applications were added to the text. The following are the major changes in the Fourth Edition.

Technology   Recognizing that graphing calculators and computer algebra systems are becoming increasingly available, we have provided the opportunity to incorporate their use throughout the Fourth Edition. Technology has been incorporated in the form of special notes in the expository text, technology-dedicated examples, and problems in the exercise sets that either require or can be solved with the use of technology. Technology is used as a means of discovery (see Discovery, page 26, and Exercises 37–40, Section 2.5), to reinforce concepts (see Technology Note, page 88, and Exercises 39 and 40, Section 3.5), and as an efficient problem-solving tool (see Technology Note, page 364, and Exercises 58 and 59, Section 2.2). In addition to addressing the benefits of technology, the text also points out the drawbacks of its use. For example, see the discussions given in the Technology Notes on pages 288 and 328.

Problem Solving   Helping students learn to be efficient and creative problem-solvers is the primary goal of any mathematics course. While the mechanics of calculus are stressed, the text has increased emphasis on the concepts of calculus. The range of exercises is much broader than in previous editions. Although the text still contains an abundance of exercises that are designed to develop skills, it now has many other types of exercises as well. Some exercises have multiple parts, leading students through mathematical exploration (Exercise 54, Section 2.6, and Exercises 33 and 34, Section 4.2); some exercises ask for interpretation (Exercise 60, Section 2.4, and Exercise 75, Section 7.4); and many exercises ask students to consider solving problems by combining numerical, analytical, and graphical methods (Exercises 51 and 52, Section 1.6, and Exercises 59 and 60, Section 2.1).

Real Applications and Real Data   In writing the Fourth Edition, we have included many new and interesting real-life applications and have tried, whenever possible, to use real rather than contrived data, as indicated by source lines. Applications are incorporated through the use of regular textual examples, through extended applications that introduce special topics, and in the exercise sets throughout the text. The applications come from a variety of fields, with an emphasis on applications in business, economics, and the life sciences. An index of applications follows the Table of Contents on page xxv.

Case Studies   To allow students the opportunity to use mathematics as a tool in decision making, case studies are included throughout the text. These feature a wide variety of real businesses and offer both consumer- and business-oriented decision-making opportunities. Each case study incorporates the mathematical skills presented in the chapter in an analysis of real data that was collected in a situation typical of the featured business. The case studies conclude with a set of questions that require the student to place him- or herself in a similar situation and to decide on a suitable course of action based on mathematical analysis.

Communicating Mathematics   In addition to problem solving, a way to reinforce students' understanding of calculus concepts is to give them the opportunity to communicate mathematical ideas to others; therefore, in the Fourth Edition, the development of communication skills has been emphasized. We have included many exercises that ask for descriptions, discussions, justifications, and conjectures. For example, see Exercise 59, Section 3.3, and Exercise 71, Section 5.1. Other opportunities for writing include essay questions (see Exercises 47 and 48, Section 5.5), discussion questions (see page 457), and case studies (see pages 360–361).

Improved Coverage   Many users of the Third Edition preferred to cover Exponential and Logarithmic Functions (previously Chapter 5) before covering Integration and Its Applications (previously Chapter 4). To accommodate this preference in topic order, these two chapters have been transposed and rewritten for the Fourth Edition. In addition, approaches to several topics have been enhanced. For instance, curve sketching in Section 3.7 now includes an expanded list of guidelines for sketching the graph of a function. New material on logistics growth has been added to Section 6.3. At the request of reviewers, L'Hôpital's Rule has been added as a new topic in the Trigonometry chapter.

Additional Appendices   Several new appendices in the Fourth Edition focus on utilizing technology to the best advantage. These include an introduction to the use of graphing utilities, translations of programs appearing in the text for additional graphing calculator models, and an introduction to the use of various computer algebra systems.

How to Study Calculus   Pages xxix and xxx of the front matter contain a summary of hints for how to study calculus effectively. Approaches to absorbing the material, checking homework, preparing for exams, and taking exams are discussed. This practice of presenting study strategies to students is carried throughout the text in the form of Study Tips for each chapter.

Accuracy   Throughout the entire revision process for the Fourth Edition, special care was given to creating a text that is accurate. We thank all of those who helped us improve the accuracy of the text.

Custom Publishing Option   Because topical coverage differs from school to school, we offer this edition with custom publication options. Any or all of the following chapters can be bound for your text: A Precalculus Review, Chapters 1–7, Probability and Calculus, Differential Equations, Series and Taylor Polynomials, and Trigonometric Functions.

Features of the Fourth Edition

The Fourth Edition has many new and improved design features that enhance the relevance of the text and make it an even more effective teaching and learning tool. These are shown on the following pages.

Features of the Fourth Edition

CHAPTER OPENERS

Each chapter opens with a portrait and brief discussion of a famous mathematician who is connected with or responsible for a key topic in that chapter. Also featured on the chapter openers are important formulas or notation presented within the chapter, as well as a list of section topics.

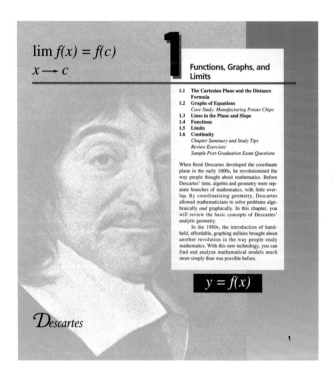

$$\lim_{x \to c} f(x) = f(c)$$

1

Functions, Graphs, and Limits

1.1 The Cartesian Plane and the Distance Formula
1.2 Graphs of Equations
    Case Study: Manufacturing Potato Chips
1.3 Lines in the Plane and Slope
1.4 Functions
1.5 Limits
1.6 Continuity
    Chapter Summary and Study Tips
    Review Exercises
    Sample Post-Graduation Exam Questions

When René Descartes developed the coordinate plane in the early 1600s, he revolutionized the way people thought about mathematics. Before Descartes' time, algebra and geometry were separate branches of mathematics, with little overlap. By coordinatizing geometry, Descartes allowed mathematicians to solve problems algebraically *and* graphically. In this chapter, you will review the basic concepts of Descartes' analytic geometry.

In the 1980s, the introduction of handheld, affordable, graphing utilities brought about another revolution in the way people study mathematics. With this new technology, you can find and analyze mathematical models much more simply than was possible before.

$$y = f(x)$$

Descartes

1

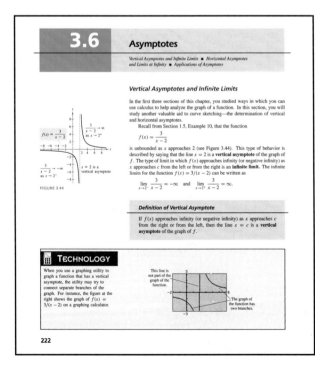

3.6 **Asymptotes**

Vertical Asymptotes and Infinite Limits ■ *Horizontal Asymptotes and Limits at Infinity* ■ *Applications of Asymptotes*

Vertical Asymptotes and Infinite Limits

In the first three sections of this chapter, you studied ways in which you can use calculus to help analyze the graph of a function. In this section, you will study another valuable aid to curve sketching—the determination of vertical and horizontal asymptotes.

Recall from Section 1.5, Example 10, that the function

$$f(x) = \frac{3}{x - 2}$$

is unbounded as x approaches 2 (see Figure 3.44). This type of behavior is described by saying that the line $x = 2$ is a **vertical asymptote** of the graph of f. The type of limit in which $f(x)$ approaches infinity (or negative infinity) as x approaches c from the left or from the right is an **infinite limit**. The infinite limits for the function $f(x) = 3/(x - 2)$ can be written as

$$\lim_{x \to 2^-} \frac{3}{x - 2} = -\infty \quad \text{and} \quad \lim_{x \to 2^+} \frac{3}{x - 2} = \infty.$$

FIGURE 3.44

Definition of Vertical Asymptote

If $f(x)$ approaches infinity (or negative infinity) as x approaches c from the right or from the left, then the line $x = c$ is a **vertical asymptote** of the graph of f.

TECHNOLOGY

When you use a graphing utility to graph a function that has a vertical asymptote, the utility may try to connect separate branches of the graph. For instance, the figure at the right shows the graph of $f(x) = 3/(x - 2)$ on a graphing calculator.

222

SECTION OUTLINES

Each text section begins with a list of subsection topics. This outline can be used for easy reference and can help students study and synthesize the material in the section more efficiently.

TECHNOLOGY NOTES

Throughout the text, special notes teach students how to enhance their learning of calculus by utilizing technology. A graphing utility or a computer algebra system is used as a problem-solving tool to execute complicated computations, to visualize theoretical concepts, to discover approaches, and to verify the results of solution methods. Through technology notes, the text offers guidance on which tasks are best suited to the use of technology, indicates which commands and programs are to be used with certain utilities, and presents opportunities to develop an awareness of the possible misuses of technology.

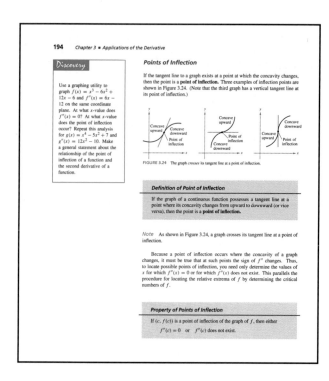

DISCOVERY

Students are encouraged to explore mathematical concepts and discover mathematical relationships through the use of these investigations. Carefully posed questions, often in conjunction with instructions for using a graphing utility, help students develop an intuitive understanding of theoretical topics.

DEFINITIONS AND THEOREMS

All definitions and theorems, as well as important results and formulas, are highlighted in blue for emphasis and easy reference.

EXAMPLES

To increase the usefulness of the text as a study tool, a wide variety of examples has been included to illustrate important mathematical concepts, problem-solving approaches, and computational techniques. An effort has been made to present not only theoretical examples but also those utilizing technology and those demonstrating real-life applications, often with actual data. Each example is titled for easy reference, and examples of real-life applications are labeled as such. Most examples in the Fourth Edition now include side comments in color that clarify the steps of the solution.

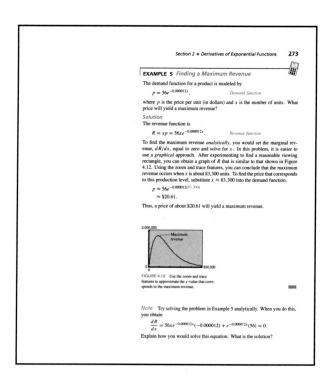

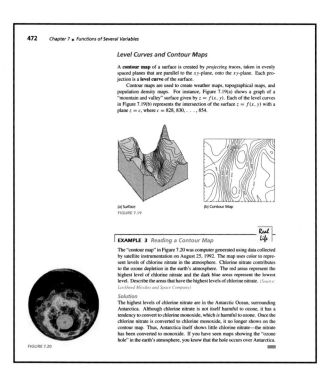

472 Chapter 7 ▪ Functions of Several Variables

Level Curves and Contour Maps

A **contour map** of a surface is created by *projecting* traces, taken in evenly spaced planes that are parallel to the xy-plane, onto the xy-plane. Each projection is a **level curve** of the surface.

Contour maps are used to create weather maps, topographical maps, and population density maps. For instance, Figure 7.19(a) shows a graph of a "mountain and valley" surface given by $z = f(x, y)$. Each of the level curves in Figure 7.19(b) represents the intersection of the surface $z = f(x, y)$ with a plane $z = c$, where $c = 828, 830, \ldots, 854$.

(a) Surface
(b) Contour Map
FIGURE 7.19

EXAMPLE 3 *Reading a Contour Map*

The "contour map" in Figure 7.20 was computer generated using data collected by satellite instrumentation on August 25, 1992. The map uses color to represent levels of chlorine nitrate in the atmosphere. Chlorine nitrate contributes to the ozone depletion in the earth's atmosphere. The red areas represent the highest level of chlorine nitrate and the dark blue areas represent the lowest level. Describe the areas that have the highest levels of chlorine nitrate. *(Source: Lockheed Missiles and Space Company)*

Solution
The highest levels of chlorine nitrate are in the Antarctic Ocean, surrounding Antarctica. Although chlorine nitrate is not itself harmful to ozone, it has a tendency to convert to chlorine monoxide, which *is* harmful to ozone. Once the chlorine nitrate is converted to chlorine monoxide, it no longer shows on the contour map. Thus, Antarctica itself shows little chlorine nitrate—the nitrate has been converted to monoxide. If you have seen maps showing the "ozone hole" in the earth's atmosphere, you know that the hole occurs over Antarctica.

FIGURE 7.20

GRAPHICS
The Fourth Edition has over 1200 text and exercise figures and over 700 figures in the Answers to Odd-Numbered Exercises. Computer-generated for accuracy and designed with full color, this art program will help students interpret graphs and better visualize mathematical concepts. This is particularly true in the presentation of complex, three-dimensional material, for which the art was drawn in true perspective. Some figures were generated by the commercial software Mathematica.

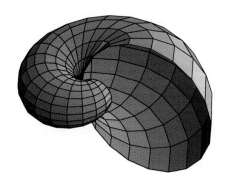

CONNECTIONS TO REAL LIFE
Color photographs help to reinforce the connection between calculus and its application to real-life situations.

DISCUSSION PROBLEMS
Each section in the text, except those in the Precalculus Review, closes with a discussion problem designed for use in class or assignment to small groups. These problems encourage thinking, reasoning, and writing about calculus and emphasize synthesis or further exploration of the concepts presented in the section.

410 Chapter 6 ▪ Techniques of Integration

EXAMPLE 6 *Modeling a Population*

The state game commission releases 100 deer into a game refuge. During the first 5 years, the population increases to 432 deer. The commission believes that the population can be modeled by logistics growth with a limit of 2000 deer. Write the logistics growth model for this population. Then use the model to create a table showing the size of the deer population over the next 30 years.

Solution
Let y represent the number of deer in year t. Assuming a logistics growth model means that the rate of change in the population is proportional to both y and $(2000 - y)$. That is

$$\frac{dy}{dt} = ky(2000 - y).$$

The solution of this equation is

$$y = \frac{2000}{1 + be^{-kt}}.$$

Using the fact that $y = 100$ when $t = 0$, you can solve for b.

$$100 = \frac{2000}{1 + be^{-k(0)}} \quad \longrightarrow \quad b = 19$$

Then, using the fact that $y = 432$ when $t = 5$, you can solve for k.

$$432 = \frac{2000}{1 + 19e^{-k(5)}} \quad \longrightarrow \quad k \approx 0.33106$$

Thus, the logistics growth model for the population is

$$y = \frac{2000}{1 + 19e^{-0.33106t}}. \qquad \textit{Logistics growth model}$$

The population, in 5-year intervals, is shown in Table 6.1.

TABLE 6.1

Time, t	0	5	10	15	20	25	30
Population, y	100	432	1181	1766	1951	1990	1998

In the United States, the National Park System has more than 330 protected areas for wildlife, and the National Wildlife Refuge System includes more than 400 refuges. The deer shown in this photo is a white-tailed buck.

Discussion Problem *Logistics Growth*

Analyze the graph of the logistics growth function in Example 6. During which years is the *rate of growth* of the herd increasing? During which years is the *rate of growth* of the herd decreasing? How would these answers change if, instead of a limit of 2000 deer, the game refuge could contain a limit of 3000 deer?

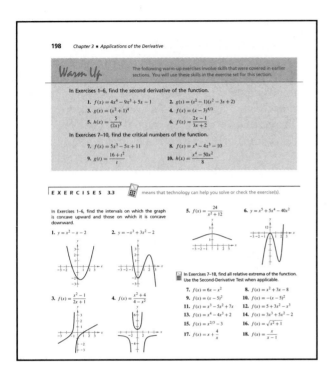

WARM-UP EXERCISES

Beginning with Chapter 1, the exercise set in each section is preceded by a set of ten warm-up exercises to help students review skills that are needed in the main exercise set. All warm-up exercises are answered in the back of the text.

RESEARCH PROJECTS

Appearing throughout the text are research projects that are keyed to business capsules. Using a library or other reference source, students are required first to collect real data or information about businesses similar to those described in the business capsules and then to analyze or synthesize this researched information to answer a posed question. Research projects are suitable for individual or group assignments.

BUSINESS CAPSULES

Nearly every chapter includes business capsules describing business situations that are connected to the mathematical concepts covered in the chapter.

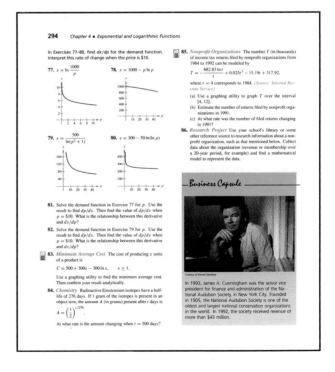

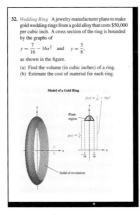

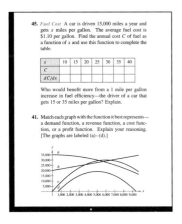

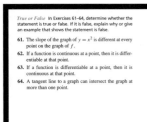

VARIETY OF PROBLEMS

The text offers a rich variety of problems to accommodate a wide range of teaching and learning styles: computational, conceptual, interpretational, matching (graphs to functions), true/false, essay, multipart, exploratory, and applied problems. An assortment of applications from a variety of disciplines is also included in the exercise sets. To emphasize the relevance of calculus to real-world situations, an effort has been made to incorporate real product and business names in addition to real data, as indicated by source lines. Problems are often labeled to indicate the subject of the application (as in "Fuel Cost" and "Wedding Ring") or to identify a special type of question (as in "True or False" and "Research Project") for easy reference.

EXERCISES

The approximately 5700 exercises in the Fourth Edition are designed to build competence, skill, and understanding. Each exercise set is carefully graded in difficulty to allow students to gain confidence as they progress. All odd-numbered exercises are solved in detail in the Student Solutions Guide, with answers appearing in the back of this text.

The symbol ▦ identifies all exercises in which technology can assist in solving or checking the exercise or in which students are specifically asked to use a graphing utility or computer algebra system.

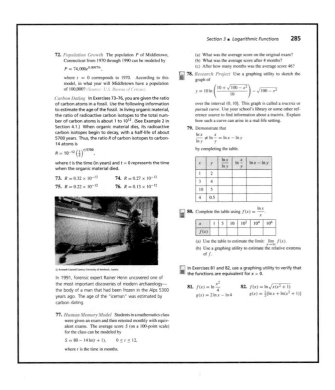

CHAPTER SUMMARIES AND STUDY TIPS

Chapter summaries have been completely revised for the Fourth Edition to list the concepts taught in the chapter and point out the Review Exercises that test these concepts. Following each chapter summary is a short list of study hints for handling topics or situations specific to the chapter.

REVIEW EXERCISES

A set of Review Exercises at chapter's end gives students an opportunity for additional practice and review. Beginning in Chapter 1, the Review Exercises offer computational, conceptual, and applied problems covering a wide range of topics. Answers to the odd-numbered Review Exercises are given in the back of the text.

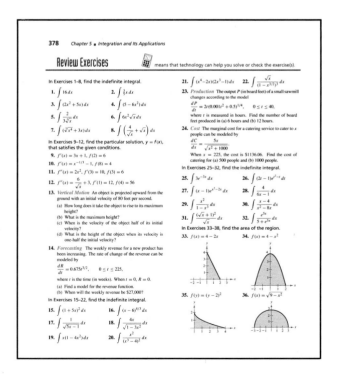

SAMPLE POST-GRADUATION EXAM QUESTIONS

Chapters 1–7 conclude with a full page of sample questions taken from certified public accountant exams, graduate management admissions tests, graduate records exams, actuarial exams, or college level academic skills tests to emphasize the variety of situations in which calculus problems arise. The answers to all Sample Post-Graduation Exam Questions are given in the back of the text.

CASE STUDIES

Chapters 1–7 include case studies that show how a real person or business uses mathematics to help make an informed consumer or business decision. Each case study gives a short profile of the featured business and concludes with a set of questions titled "What Would You Do?" which gives students the opportunity to make decisions based on mathematical reasoning. These case studies provide opportunities for individual or group work.

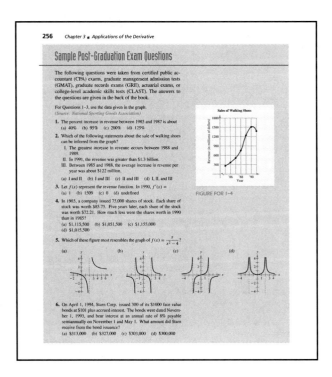

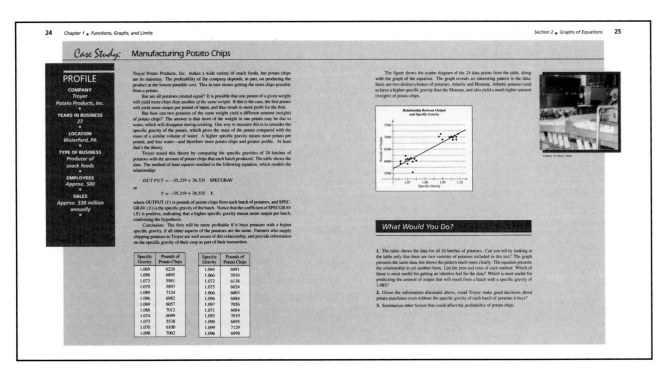

Supplements

Brief Calculus with Applications, Fourth Edition, by Larson, Hostetler, and Edwards is accompanied by a comprehensive supplements package with ancillaries for students, for instructors, and for classroom resources. Most items are keyed directly to the book.

PRINTED RESOURCES

For the student:

Student Solutions Guide by Dianna L. Zook, Indiana University and Purdue University at Fort Wayne, and Bruce H. Edwards, University of Florida
• Step-by-step solutions to all odd-numbered text exercises

The Algebra of Calculus by Eric J. Braude
• Review of algebra, trigonometry, and analytical geometry required for calculus
• Over 200 examples with solutions
• Pretests and exercises

Graphing Technology Keystroke Guide by Benjamin N. Levy
• Keystroke-level commands and instructions for use with the TI-81, TI-82, TI-85, Casio fx-7700 G, Sharp EL 9200-9300, and Hewlett Packard 48G-48GX
• Numerous screen displays
• Examples and technology tips

Study Guide and Workbook by Ronnie Khuri, University of Florida
• Numerous additional examples and exercises with answers
• Practice tests with answers

For the instructor:

Instructor's Guide
• Notes to the new teacher
• Chapter summaries
• Ready-made chapter tests, some of which require technology
• Answers to Discussion Problems, Warm-up Exercises, and Post-Graduation Exam Questions
• Teaching strategies

Solutions to Even-Numbered Exercises by Dianna L. Zook, Indiana University and Purdue University at Fort Wayne, and Bruce H. Edwards, University of Florida
• Step-by-step solutions to even-numbered text exercises

Test Item File
• Printed test bank
• Approximately 2000 test items
• Multiple-choice, open-ended, and writing questions coded by level of difficulty
• Technology-required test items coded for easy reference

Transparency Package
• Color transparencies

SOFTWARE RESOURCES

Brief Calculus TUTOR by Timothy R. Larson
• An IBM-PC-format "electronic study guide" for the text
• Examples with step-by-step solutions
• Additional exercises with diagnostic feedback
• An on-line glossary of key concepts
• Post-tests reviewing newly learned concepts

Computerized Testing
• Test-generating software for both IBM and Macintosh computers
• Approximately 2000 test items
• Also available as a printed test bank

Derive
• Symbolic algebra system
• Available to adopters at a discount

BestGrapher by George Best
• Function grapher for both IBM and Macintosh computers
• Simultaneous display of equation, graph, and table of values
• Zoom and print features

For information on ordering supplements, please call D. C. Heath's Customer Service toll free at 1-800-334-3284.

Acknowledgments

We would like to thank the many people who have helped us at various stages of this project during the past 14 years. Their encouragement, criticisms, and suggestions have been invaluable to us.

D.C. Heath would like to express our appreciation to the following people who offered their comments on pre-publication of the Fourth Edition: Yvonne Sandoval-Brown, Pima Community College; Mary Chabot, Mt. San Antonio College; Karen Hay, Mesa Community College; Arlene Jesky, Rose State College; Melvin Lax, California State University-Long Beach; Kevin McDonald, Mt. San Antonio College; Mike Nasab, Long Beach City College; Rita Richards, Scottsdale Community College; Jane Y. Smith, University of Florida; Carol G. Williams, Pepperdine University; Carlton Woods, Auburn University at Montgomery.

Fourth edition survey respondents: We would also like to thank the over 75 professors who took time to respond to the *Brief Calculus* Survey. We appreciate your comments.

Case Studies: Special thanks to James A. Kurre and Kenneth T. Louie, Professors of Economics, The Behrend College, The Pennsylvania State University, for creating the Case Studies.

Our thanks to David E. Heyd, The Behrend College, The Pennsylvania State University, for his contributions to this project.

Third Edition Reviewers: Carol Achs, Mesa Community College; John Chuchel, University of California, Davis; William Conway, University of Arizona; Ronnie Khuri, University of Florida; Peter J. Livorsi, Oakton Community College; Samuel A. Lynch, Southwest Missouri State University; Philip R. Montgomery, University of Kansas; Karla Neal, Louisiana State University; James Osterburg, University of Cincinnati; Jan E. Wynn, Brigham Young University.

First and Second Edition Reviewers: Miriam E. Connellan, Marquette University; Bruce H. Edwards, University of Florida; Roger A. Engle, Clarion University of Pennsylvania; William C. Huffman, Loyola University of Chicago; James A. Kurre, The Pennsylvania State University; Norbert Lerner, State University of New York at Cortland; Earl H. McKinney, Ball State University; Stephen B. Rodi, Austin Community College; DeWitt L. Sumners, Florida State University; Jonathan Wilkin, Northern Virginia Community College; Melvin R. Woodard, Indiana University of Pennsylvania; Robert A. Yawin, Springfield Technical Community College.

A special thanks to all the people at D. C. Heath and Company who worked with us in the development of the Fourth Edition, especially Ann Marie Jones and Charlie Hartford, Mathematics Acquisitions Editors; Cathy Cantin, Managing Editor; Emily Keaton, Developmental Editor; Karen Carter, Production Editor; Christine Hoag, Marketing Manager; Jan Shapiro, Designer; Gary Crespo, Art Editor; Carolyn Johnson, Editorial Associate; Lisa Merrill, Production Manager; and Melina Freedman and Donna Simons, Photo Researchers.

We would also like to thank the staff at Larson Texts, Inc., who assisted with proofreading the manuscript; preparing and proofreading the art package; and checking and typesetting the supplements.

On a personal level, we are grateful to our wives, Deanna Gilbert Larson, Eloise Hostetler, and Consuelo Edwards, for their love, patience, and support. Also, a special thanks goes to R. Scott O'Neil.

If you have suggestions for improving the text, please feel free to write to us. Over the past two decades we have received many useful comments from both instructors and students, and we value these comments very much.

Roland E. Larson
Robert P. Hostetler
Bruce H. Edwards

Contents

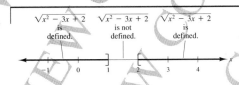

A Precalculus Review   0-1

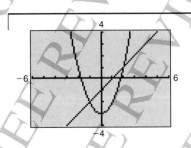

Functions, Graphs, and Limits   1

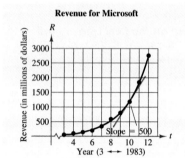

Revenue for Microsoft

Differentiation   81

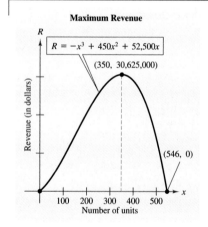

Maximum Revenue

Applications of the Derivative   171

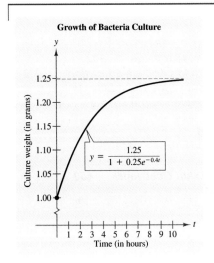

Growth of Bacteria Culture

$$y = \frac{1.25}{1 + 0.25e^{-0.4t}}$$

Exponential and Logarithmic Functions   *257*

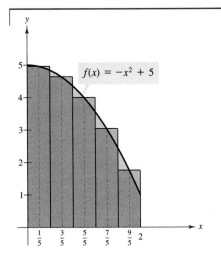

$f(x) = -x^2 + 5$

Integration and Its Applications   *313*

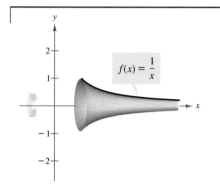

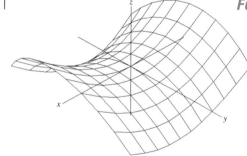

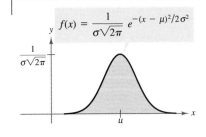

$$f(x) = \frac{1}{\sigma\sqrt{2\pi}} e^{-(x-\mu)^2/2\sigma^2}$$

Applications

Applications in Business and Economics

Life Sciences

Social and Behavioral Sciences

Physical Sciences

How to Study Calculus

Studying Mathematics   When studying mathematics—and calculus—the material learned on one day is built upon material learned previously. It is necessary to keep up with course work day by day.

Making a Plan   Make your own course plan right now! Determine the number of hours per week you need to spend on calculus. A general guideline is to study two to four hours for every hour spent in class.

Absorbing the Material   The following practices will help you absorb the concepts of calculus:

a) Before attending class, read the portion of the text to be covered, paying special attention to boxed definitions, theorems, formulas, and summaries. Going to class prepared will enable you to benefit much more from your instructor's presentation.

b) Attend every class and arrive on time. If you have to miss a class, get the notes from another student or get help from your school's tutorial service. Learn the information that was taught in the missed class before attending the next class.

c) Take notes in class, focusing on the instructor's cues indicating important material. Then, as soon after class as possible, read through your notes, adding any explanations necessary to make your notes understandable to you.

Doing the Homework   Learning calculus is like acquiring any other skill. Practice, in the form of homework exercises, is essential to developing your understanding of calculus. Doing the homework while the concepts are still fresh in your mind increases your chances of retaining the information.

Finding a Study Partner   When you get stuck on a problem, it may help to try to work with someone else. Group work is an effective study technique. Even if you feel that you are giving more help than you are getting, you will find that teaching others is an excellent way to better understand calculus concepts.

Building a Math Library   Build a library of books that can help you with this course and future math courses. Consider using the Student Solutions Guide that accompanies this text. Also, after you finish this course, we suggest that you keep the text. It will be a valuable reference book in whatever discipline you are pursuing, not only for use in future classes but also for preparation for post-graduation examinations such as the Graduate Records Exam or the Graduate Management Admissions Test. Computer software and manuals on technology will also be valuable additions to your mathematics library.

Keeping Up with the Work   Don't let yourself fall behind in the course. If you think you are having trouble, seek help immediately. Ask your instructor or teaching assistant, attend your school's tutorial service, talk with your study partner, use additional study aids—do something. If you are having trouble with the material in one chapter of your calculus text, there is a good chance that you will also have trouble with subsequent chapters.

Getting Stuck   Anyone who has taken a math course has had the following experience: you are working on a problem and cannot solve it, or you have solved it but your answer does not agree with the answer given in the back of the text. One approach that often helps when you are stuck is to sketch a graph, either by hand or with a graphing utility. Not only will the sketch give you insight into whatever function you are facing, but it can also clarify such items as limits of integration or whether your answer should be positive or negative. Other approaches include asking for help, taking a break to clear your thoughts, sleeping on it, reworking the problem, comparing notes with a study partner, and rereading your notes or the section in the text. Don't get frustrated or spend too much time on a single problem.

Keeping Your Skills Sharp   Preceding each exercise set in the text is a short set of Warm-Up Exercises. These exercises will help you review skills that you learned in previous sections and retain them. These sets are designed to take only a few minutes to solve. We suggest that you work the entire warm-up set before you start the main exercise set. (All of the Warm-Up Exercises are answered at the end of the text.)

Checking Your Work   Learning to check your work is an important part of learning calculus and is especially useful in testing situations. One easy way to check your work is to ask yourself "Does my answer make sense?" or "Is this the answer I expected to get?" in the context of a specific problem. This simple step will help you determine whether or not your answer has the correct sign (positive or negative) or the correct magnitude. Other ways to check your work include studying in a group and then comparing answers and using technology or a different solution method to check your solution. Check the Study Tips at the ends of the chapters for more hints on checking problems specific to each chapter.

Preparing for Exams   If you have kept up with the work and followed the suggestions given here, you will be nearly ready for the exam. We have included several features that should help in your final preparation. Review the Chapter Summary, work the Review Exercises, and, if you have the Student Solutions Guide, take the practice tests offered there. These should point out areas in which you need more work. For additional practice in certain areas, you can always rework the assigned homework exercises or work the exercises not previously assigned. Analyze your work to locate possible test-taking errors.

Taking Exams   Most instructors suggest that you do not study right up to the minute you take a test. This practice tends to make people anxious. The best cure for anxiety during tests is good preparation. Once the test has begun, read the directions carefully and try to work at a steady pace. Rushing tends to cause people to make careless errors. Remember, there is no rule that says you must tackle the problems in the order given. You may solve them in whatever order with which you feel most comfortable. Often, approaching first the ones you feel are easiest will boost your confidence. If you finish early, take a few moments to check your work.

Learning from Mistakes   When you get an exam back, be sure to go over any errors that you have made. If you don't understand something that was marked as incorrect on your exam, be sure to ask your instructor about it or talk with a study partner. Understanding your errors will prevent you from repeating the same conceptual or systematic errors.

BRIEF CALCULUS

with Applications

ALTERNATE FOURTH EDITION

$$x = \frac{-b \pm \sqrt{b^2 - 4ac}}{2a}$$

0

A Precalculus Review

Mathematics evolved over the past few thousand years in many stages. All of the early stages were centered around the use of mathematics to answer questions about real life. This is much like the way we wrote *Brief Calculus with Applications*. We centered the concepts around the real-life use of mathematics.

The mathematics that you will study in this text is a combination of the work of literally thousands of people whose work took place independently in many different parts of the world: Africa, Asia, Europe, North America, and South America. For instance, the principles of geometry, as cataloged by Euclid (c. 365?–300 B.C.), and the principles of algebra, as developed in India and the Middle East, were known and used by many other civilizations throughout the world.

$$ax^2 + bx + c = 0$$

Euclid

The Bettmann Archive

The Real Line and Order

The Real Line ■ *Order and Intervals on the Real Line* ■
Solving Inequalities ■ *Applications*

The Real Line

Real numbers can be represented with a coordinate system called the **real line** (or x-axis), as shown in Figure 0.1. The **positive direction** (to the right) is denoted by an arrowhead and indicates the direction of increasing values of x. The real number corresponding to a particular point on the real line is called the **coordinate** of the point. As shown in Figure 0.1, it is customary to label those points whose coordinates are integers.

The point on the real line corresponding to zero is called the **origin.** Numbers to the right of the origin are **positive,** and numbers to the left of the origin are **negative.** The term **nonnegative** describes a number that is either positive or zero.

The importance of the real line is that it provides you with a conceptually perfect picture of the real numbers. That is, each point on the real line corresponds to one and only one real number, and each real number corresponds to one and only one point on the real line. This type of relationship is called a **one-to-one correspondence** and is illustrated in Figure 0.2.

Each of the four points in Figure 0.2 corresponds to a real number that can be expressed as the ratio of two integers. (Note that $1.85 = \frac{37}{20}$ and $-2.6 = -\frac{13}{5}$.) Such numbers are called **rational.** Rational numbers have either terminating or infinitely repeating decimal representations.

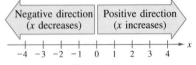

Every point on the real line corresponds to a real number.

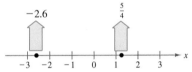

Every real number corresponds to a point on the real line.

FIGURE 0.2

Terminating Decimals	*Infinitely Repeating Decimals*
$\frac{2}{5} = 0.4$	$\frac{1}{3} = 0.333\ldots = 0.\overline{3}^{*}$
$\frac{7}{8} = 0.875$	$\frac{12}{7} = 1.714285714285\ldots = 1.\overline{714285}$

Real numbers that are not rational are called **irrational,** and they cannot be represented as the ratio of two integers (or as terminating or infinitely repeating decimals). To represent an irrational number, one usually resorts to a decimal approximation. Some irrational numbers occur so frequently in applications that mathematicians have invented special symbols to represent them. For example, the symbols $\sqrt{2}$, π, and e represent irrational numbers whose decimal approximations are as follows. (See Figure 0.3.)

$$\sqrt{2} \approx 1.4142135623$$
$$\pi \approx 3.1415926535$$
$$e \approx 2.7182818284$$

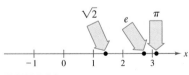

FIGURE 0.3

*The bar indicates which digit or digits repeat.

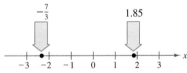

FIGURE 0.1   The Real Line

Order and Intervals on the Real Line

One important property of the real numbers is that they are **ordered:** 0 is less than 1, -3 is less than -2.5, π is less than $\frac{22}{7}$, and so on. You can visualize this property on the real line by observing that a is less than b if and only if a lies to the left of b. Symbolically, "a is less than b" is denoted by the inequality

$$a < b.$$

For example, the inequality $\frac{3}{4} < 1$ follows from the fact that $\frac{3}{4}$ lies to the left of 1 on the real line, as shown in Figure 0.4.

When three real numbers a, x, and b are ordered such that $a < x$ and $x < b$, we say that x is **between** a and b and write

$$a < x < b. \qquad\qquad x \text{ is between } a \text{ and } b.$$

$\frac{3}{4}$ lies to the left of 1, so $\frac{3}{4} < 1$

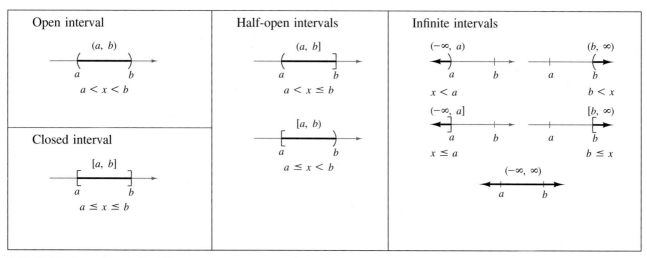

FIGURE 0.4

The set of *all* real numbers between a and b is called the **open interval** between a and b and is denoted by (a, b). An interval of the form (a, b) does not contain the "endpoints" a and b. Intervals that include their endpoints are called **closed** and are denoted by $[a, b]$. Intervals of the form $[a, b)$ and $(a, b]$ are called **half-open intervals.** Figure 0.5 shows the nine types of intervals on the real line.

Open interval	Half-open intervals	Infinite intervals
(a, b) $a < x < b$	$(a, b]$ $a < x \le b$	$(-\infty, a)$ $x < a$ $\quad$ (b, ∞) $b < x$
	$[a, b)$ $a \le x < b$	$(-\infty, a]$ $x \le a$ $\quad$ $[b, \infty)$ $b \le x$
Closed interval		
$[a, b]$ $a \le x \le b$		$(-\infty, \infty)$

FIGURE 0.5   Intervals on the Real Line

Note   Note that a square bracket is used to denote "less than or equal to" ($\le$). Furthermore, the symbols ∞ and $-\infty$ denote positive and negative infinity. These symbols do not denote real numbers; they merely let you describe unbounded conditions more concisely. For instance, the interval $[b, \infty)$ is unbounded to the right because it includes *all* real numbers that are greater than or equal to b.

Solving Inequalities

In calculus, you are frequently required to "solve inequalities" involving variable expressions such as $3x - 4 < 5$. The number a is a **solution** of an inequality if the inequality is true when a is substituted for x. The set of all values of x that satisfy an equality is called the **solution set** of the inequality. The following properties are useful for solving inequalities. (Similar properties are obtained if $<$ is replaced by $\leq$ and $>$ is replaced by $\geq$.)

Properties of Inequalities

1. Transitive property: $a < b$ and $b < c \Longrightarrow a < c$
2. Adding inequalities: $a < b$ and $c < d \Longrightarrow a + c < b + d$
3. Multiplying by a (positive) constant: $a < b \Longrightarrow ac < bc, \quad c > 0$
4. Multiplying by a (negative) constant: $a < b \Longrightarrow ac > bc, \quad c < 0$
5. Adding a constant: $a < b \Longrightarrow a + c < b + c$
6. Subtracting a constant: $a < b \Longrightarrow a - c < b - c$

Note   Note that you *reverse the inequality* when you multiply by a negative number. For example, if $x < 3$, then $-4x > -12$. This principle also applies to division by a negative number. Thus, if $-2x > 4$, then $x < -2$.

EXAMPLE 1 *Solving an Inequality*

Find the solution set of the inequality $3x - 4 < 5$.

Solution

$$3x - 4 < 5 \qquad \text{\textit{Original inequality}}$$
$$3x - 4 + 4 < 5 + 4 \qquad \text{\textit{Add 4 to each side.}}$$
$$3x < 9 \qquad \text{\textit{Simplify.}}$$
$$\frac{1}{3}(3x) < \frac{1}{3}(9) \qquad \text{\textit{Multiply each side by } } \tfrac{1}{3}.$$
$$x < 3 \qquad \text{\textit{Simplify.}}$$

Thus, the solution set is the interval $(-\infty, 3)$, as shown in Figure 0.6.   ▬

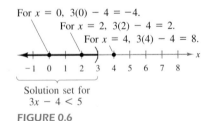

For $x = 0$, $3(0) - 4 = -4$.
For $x = 2$, $3(2) - 4 = 2$.
For $x = 4$, $3(4) - 4 = 8$.

Solution set for
$3x - 4 < 5$

FIGURE 0.6

Note   In Example 1, all five inequalities listed as steps in the solution have the same solution set, and they are called **equivalent inequalities.**

Once you have solved an inequality, it is a good idea to check some x-values in your solution interval to see whether they satisfy the original inequality. You might also check some values outside your solution interval to verify that they do *not* satisfy the inequality. For example, Figure 0.6 shows that when $x = 0$ or $x = 2$ the inequality is satisfied, but when $x = 4$ the inequality is not satisfied.

The inequality in Example 1 involves a first-degree polynomial. To solve inequalities involving polynomials of higher degree, you can use the fact that a polynomial can change signs *only* at its real zeros (the real numbers that make the polynomial zero). Between two consecutive real zeros, a polynomial must be entirely positive or entirely negative. This means that when the real zeros of a polynomial are put in order, they divide the real line into **test intervals** in which the polynomial has no sign changes. That is, if a polynomial has the factored form

$$(x - r_1)(x - r_2), \ldots, (x - r_n), \qquad r_1 < r_2 < r_3 < \cdots < r_n,$$

then the test intervals are

$$(-\infty, r_1), \quad (r_1, r_2), \quad \ldots, \quad (r_{n-1}, r_n), \quad \text{and} \quad (r_n, \infty).$$

For example, the polynomial

$$x^2 - x - 6 = (x - 3)(x + 2)$$

can change signs only at $x = -2$ and $x = 3$. To determine the sign of the polynomial in the intervals $(-\infty, -2)$, $(-2, 3)$, and $(3, \infty)$, you need to test only *one value* from each interval.

EXAMPLE 2 *Solving a Polynomial Inequality*

Find the solution set of the inequality $x^2 < x + 6$.

Solution

$$\begin{aligned} x^2 &< x + 6 && \textit{Original inequality} \\ x^2 - x - 6 &< 0 && \textit{Polynomial form} \\ (x - 3)(x + 2) &< 0 && \textit{Factor.} \end{aligned}$$

Thus, the polynomial $x^2 - x - 6$ has $x = -2$ and $x = 3$ as its zeros. You can solve the inequality by testing the sign of the polynomial in each of the following intervals.

$$x < -2, \qquad -2 < x < 3, \qquad x > 3$$

To test an interval, choose a representative number in the interval and compute the sign of each factor. For example, for any $x < -2$, both of the factors $(x - 3)$ and $(x + 2)$ are negative. Consequently, the product (of two negative numbers) is positive, and the inequality is *not* satisfied in the interval

$$x < -2.$$

We suggest the testing format shown in Figure 0.7. Because the inequality is satisfied only by the center test interval, you conclude that the solution set is given by the interval

$$-2 < x < 3. \qquad\qquad \textit{Solution set}$$

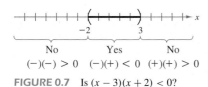

No   Yes   No
$(-)(-) > 0$ $(-)(+) < 0$ $(+)(+) > 0$

FIGURE 0.7 Is $(x - 3)(x + 2) < 0$?

Applications

Inequalities are frequently used to describe conditions that occur in business and science. For instance, the inequality

$$118 \leq W \leq 138$$

describes the recommended weight W for a woman whose height is 5 feet 6 inches. Example 3 shows how an inequality can be used to describe the production level of a manufacturing plant.

Real Life

EXAMPLE 3 *Production Levels*

In addition to fixed overhead costs of $500 per day, the cost of producing x units of a certain item is $2.50 per unit. During the month of August, the total cost of production varied from a high of $1325 to a low of $1200 per day. Find the high and low *production levels* during the month.

Solution

Because it costs $2.50 to produce one unit, it costs $2.5x$ to produce x units. Furthermore, because the fixed cost per day is $500, the total daily cost of producing x units is

$$C = 2.5x + 500.$$

Now, because the cost ranged from $1200 to $1325, you can write the following.

$$1200 \leq 2.5x + 500 \leq 1325$$
$$1200 - 500 \leq 2.5x + 500 - 500 \leq 1325 - 500$$
$$700 \leq 2.5x \leq 825$$
$$\frac{700}{2.5} \leq \frac{2.5x}{2.5} \leq \frac{825}{2.5}$$
$$280 \leq x \leq 330$$

Thus, the daily production levels during the month varied between a low of 280 units and a high of 330 units, as pictured in Figure 0.8.

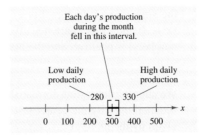

FIGURE 0.8

EXERCISES 0.1

In Exercises 1–10, determine whether the real number is rational or irrational.

1. 0.7

2. -3678

3. $\dfrac{3\pi}{2}$

4. $3\sqrt{2} - 1$

5. $4.3451\overline{451}$

6. $\dfrac{22}{7}$

7. $\sqrt[3]{64}$

8. $0.81778\overline{177}$

9. $\sqrt[3]{60}$

10. $2e$

In Exercises 11–14, determine whether the given value of x satisfies the inequality.

11. $5x - 12 > 0$
  (a) $x = 3$
  (b) $x = -3$
  (c) $x = \frac{5}{2}$
  (d) $x = \frac{3}{2}$

12. $x + 1 < \dfrac{2x}{3}$
  (a) $x = 0$
  (b) $x = 4$
  (c) $x = -4$
  (d) $x = -3$

13. $0 < \dfrac{x - 2}{4} < 2$
  (a) $x = 4$
  (b) $x = 10$
  (c) $x = 0$
  (d) $x = \frac{7}{2}$

14. $-1 < \dfrac{3 - x}{2} \le 1$
  (a) $x = 0$
  (b) $x = \sqrt{5}$
  (c) $x = 1$
  (d) $x = 5$

In Exercises 15–32, solve the inequality and sketch the graph of the solution on the real number line.

15. $x - 5 \ge 7$

16. $2x > 3$

17. $4x + 1 < 2x$

18. $2x + 7 < 3$

19. $2x - 1 \ge 0$

20. $3x + 1 \ge 2x + 2$

21. $4 - 2x < 3x - 1$

22. $x - 4 \le 2x + 1$

23. $-4 < 2x - 3 < 4$

24. $0 \le x + 3 < 5$

25. $\dfrac{3}{4} > x + 1 > \dfrac{1}{4}$

26. $-1 < -\dfrac{x}{3} < 1$

27. $\dfrac{x}{2} + \dfrac{x}{3} > 5$

28. $\dfrac{x}{2} - \dfrac{x}{3} > 5$

29. $x^2 \le 3 - 2x$

30. $x^2 - x \le 0$

31. $x^2 + x - 1 \le 5$

32. $2x^2 + 1 < 9x - 3$

33. *Simple Interest* P dollars is invested at a (simple) interest rate of r. After t years, the balance in the account is given by

$$A = P + Prt,$$

where the interest rate r is expressed in decimal form. In order for an investment of \$1000 to grow to *more than* \$1250 in 2 years, what must the interest rate be?

34. *Sales* A doughnut shop at a shopping mall sells a dozen doughnuts for \$2.95. Beyond the fixed costs (for rent, utilities, and insurance) of \$150 per day, it costs \$1.45 for enough materials (flour, sugar, etc.) and labor to produce each dozen doughnuts. If the daily profit *varies between* \$50 and \$200, between what levels (in dozens) do the daily sales vary?

35. *Profit* The revenue for selling x units of a product is

$$R = 115.95x,$$

and the cost of producing x units is

$$C = 95x + 750.$$

In order to obtain a profit, the revenue must be *greater than* the cost. For what values of x will this product return a profit?

36. *Operating Cost* A utility company has a fleet of vans for which the annual operating cost per van is

$$C = 0.32m + 2300,$$

where m is the number of miles traveled by a van in a year. What number of miles will yield an annual operating cost per van that is *less than* \$10,000?

37. *Area* A square region is to have an area of *at least* 500 square meters. What must the length of the sides of the region be?

38. *Area* An isosceles right triangle is to have an area of *at least* 32 square feet. What must the length of each side be?

In Exercises 39 and 40, determine whether each statement is true or false, given $a < b$.

39. (a) $-2a < -2b$
  (b) $a + 2 < b + 2$
  (c) $6a < 6b$
  (d) $\dfrac{1}{a} < \dfrac{1}{b}$

40. (a) $a - 4 < b - 4$
  (b) $4 - a < 4 - b$
  (c) $-3b < -3a$
  (d) $\dfrac{a}{4} < \dfrac{b}{4}$

0.2 Absolute Value and Distance on the Real Line

Absolute Value of a Real Number ∎ *Distance on the Real Line* ∎
Intervals Defined by Absolute Value ∎ *Applications*

Absolute Value of a Real Number

Definition of Absolute Value

The **absolute value** of a real number a is
$$|a| = \begin{cases} a, & \text{if } a \geq 0 \\ -a, & \text{if } a < 0. \end{cases}$$

At first glance it may appear from this definition that the absolute value of a real number can be negative, but this is not possible. For example, let $a = -3$. Then, because $-3 < 0$, you have

$$|a| = |-3| = -(-3) = 3.$$

The following properties are useful for working with absolute values.

Properties of Absolute Value

1. Multiplication: $|ab| = |a||b|$

2. Division: $\left|\dfrac{a}{b}\right| = \dfrac{|a|}{|b|}, \quad b \neq 0$

3. Power: $|a^n| = |a|^n$

4. Square root: $\sqrt{a^2} = |a|$

Be sure you understand the fourth property in this list. A common error in algebra is to imagine that by squaring a number and then taking the square root, we come back to the original number. But this is true only if the original number is nonnegative. For instance, if $a = 2$, then

$$\sqrt{2^2} = \sqrt{4} = 2,$$

but if $a = -2$, then

$$\sqrt{(-2)^2} = \sqrt{4} = 2.$$

The reason for this is that (by definition) the square root symbol $\sqrt{}$ denotes only the nonnegative root.

Distance on the Real Line

Consider two distinct points on the real line, as shown in Figure 0.9.

Directed distance
from *a* to *b*:

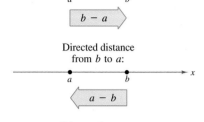

1. The **directed distance from *a* to *b*** is $b - a$.
2. The **directed distance from *b* to *a*** is $a - b$.
3. The **distance between *a* and *b*** is $|a - b|$ or $|b - a|$.

Directed distance
from *b* to *a*:

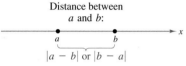

In Figure 0.9, note that because *b* is to the right of *a*, the directed distance from *a* to *b* (moving to the right) is positive. Moreover, because *a* is to the left of *b*, the directed distance from *b* to *a* (moving to the left) is negative. The distance *between* two points on the real line can never be negative.

Distance between
a and *b*:

$|a - b|$ or $|b - a|$

FIGURE 0.9

Distance Between Two Points on the Real Line

The distance d between points x_1 and x_2 on the real line is given by

$$d = |x_2 - x_1| = \sqrt{(x_2 - x_1)^2}.$$

Note that the order of subtraction with x_1 and x_2 does not matter because

$$|x_2 - x_1| = |x_1 - x_2| \quad \text{and} \quad (x_2 - x_1)^2 = (x_1 - x_2)^2.$$

EXAMPLE 1 *Finding Distance on the Real Line*

Determine the distance between -3 and 4 on the real line. What is the directed distance from -3 to 4? From 4 to -3?

Solution
The distance between -3 and 4 is given by

$$|4 - (-3)| = |7| = 7 \quad \text{or} \quad |-3 - 4| = |-7| = 7.$$

The directed distance from -3 to 4 is

$$4 - (-3) = 7.$$

The directed distance from 4 to -3 is

$$-3 - 4 = -7.$$

(See Figure 0.10.)

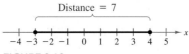

Distance = 7

FIGURE 0.10

Intervals Defined by Absolute Value

EXAMPLE 2 *Defining an Interval on the Real Line*

Find the interval on the real line that contains all numbers that lie no more than two units from 3.

Solution

Let x be any point in this interval. You need to find all x such that the distance between x and 3 is less than or equal to 2. This implies that

$$|x - 3| \leq 2.$$

Requiring the absolute value of $x - 3$ to be less than or equal to 2 means that $x - 3$ must lie between -2 and 2. Hence, you can write

$$-2 \leq x - 3 \leq 2.$$

Solving this pair of inequalities, you have

$$-2 + 3 \leq x - 3 + 3 \leq 2 + 3$$
$$1 \leq \quad x \quad \leq 5. \qquad \textit{Solution set}$$

Therefore, the interval is [1, 5], as shown in Figure 0.11.

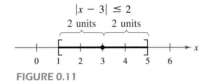

FIGURE 0.11

Two Basic Types of Inequalities Involving Absolute Value

Let a and d be real numbers, where $d > 0$.

$$|x - a| \leq d \text{ if and only if } a - d \leq x \leq a + d.$$
$$|x - a| \geq d \text{ if and only if } x \leq a - d \text{ or } a + d \leq x.$$

Inequality	Interpretation	Graph
$\|x - a\| \leq d$	All numbers x whose distance from a is less than or equal to d.	
$\|x - a\| \geq d$	All numbers x whose distance from a is greater than or equal to d.	

Note Be sure you see that inequalities of the form $|x - a| \geq d$ have solution sets consisting of two intervals. To describe the two intervals without using absolute values, we must use *two* separate inequalities, connected by an "or" to indicate union.

Applications

Real
Life

EXAMPLE 3 *Quality Control*

A large manufacturer hired a quality control firm to determine the reliability of a certain product. Using statistical methods, the firm determined that the manufacturer could expect $0.35\% \pm 0.17\%$ of the units to be defective. If the manufacturer offers a money-back guarantee on this product, how much should be budgeted to cover the refunds on 100,000 units? (Assume that the retail price is \$8.95.)

Solution

Let r represent the percent of defective units (written in decimal form). You know that r will differ from 0.0035 by at most 0.0017.

$$0.0035 - 0.0017 \leq r \leq 0.0035 + 0.0017$$

$$0.0018 \leq r \leq 0.0052 \qquad \text{Figure 0.12(a)}$$

Now, letting x be the number of defective units out of 100,000, it follows that $x = 100,000r$ and you have

$$0.0018(100,000) \leq 100,000r \leq 0.0052(100,000)$$

$$180 \leq \quad x \quad \leq 520. \qquad \text{Figure 0.12(b)}$$

Finally, letting C be the cost of refunds, you have $C = 8.95x$. Thus, the total cost of refunds for 100,000 units should fall within the interval given by

$$180(8.95) \leq 8.95x \leq 520(8.95)$$

$$\$1611 \leq \quad C \quad \leq \$4654. \qquad \text{Figure 0.12(c)}$$ ▬

(a) Percent of defective units

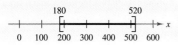

(b) Number of defective units

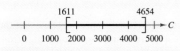

(c) Cost of refunds

FIGURE 0.12

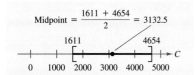

FIGURE 0.13

In Example 3, the manufacturer should expect to spend between \$1611 and \$4654 for refunds. Of course, the safer budget figure for refunds would be the higher of these estimates. However, from a statistical point of view, the most representative estimate would be the average of these two extremes. Graphically, the average of two numbers is the **midpoint** of the interval with the two numbers as endpoints, as shown in Figure 0.13.

Midpoint of an Interval

The **midpoint** of the interval with endpoints a and b is found by taking the average of the endpoints.

$$\text{Midpoint} = \frac{a+b}{2}$$

EXERCISES 0.2

In Exercises 1–4, find (a) the directed distance from a to b, (b) the directed distance from b to a, and (c) the distance between a and b.

1. $a = 126, b = 75$

2. $a = -126, b = -75$

3. $a = 9.34, b = -5.65$

4. $a = \frac{16}{5}, b = \frac{112}{75}$

In Exercises 5–8, find the midpoint of the given interval.

5. $[7, 21]$

6. $[8.6, 11.4]$

7. $[-6.85, 9.35]$

8. $[-4.6, -1.3]$

In Exercises 9–22, solve the inequality and sketch the graph of the solution on the real line.

9. $|x| < 5$

10. $|2x| < 6$

11. $\left|\dfrac{x}{2}\right| > 3$

12. $|5x| > 10$

13. $|x + 2| < 5$

14. $|3x + 1| \geq 4$

15. $\left|\dfrac{x - 3}{2}\right| \geq 5$

16. $|2x + 1| < 5$

17. $|10 - x| > 4$

18. $|25 - x| \geq 20$

19. $|9 - 2x| < 1$

20. $\left|1 - \dfrac{2x}{3}\right| < 1$

21. $|x - a| \leq b$

22. $|2x - a| \geq b$

In Exercises 23–34, use absolute values to describe the given interval (or pair of intervals) on the real line.

23. $[-2, 2]$

24. $(-3, 3)$

25. $(-\infty, -2) \cup (2, \infty)$

26. $(-\infty, -3] \cup [3, \infty)$

27. $[2, 6]$

28. $(-7, -1)$

29. $(-\infty, 0) \cup (4, \infty)$

30. $(-\infty, 20) \cup (24, \infty)$

31. All numbers *less than* two units from 4.

32. All numbers *more than* six units from 3.

33. y is *at most* two units from a.

34. y is *less than* h units from c.

35. *Statistics* The heights h of two-thirds of the members of a certain population satisfy the inequality

$$\left|\frac{h - 68.5}{2.7}\right| \leq 1,$$

where h is measured in inches. Determine the interval on the real line in which these heights lie.

36. *Statistics* To determine whether a coin is fair, an experimenter tosses it 100 times and records the number of heads x. Statistical theory states that the coin should be declared unfair if

$$\left|\frac{x - 50}{5}\right| \geq 1.645.$$

For what values of x will the coin be declared unfair?

37. *Production* The estimated daily production x at a refinery is given by

$$|x - 200,000| \leq 125,000,$$

where x is measured in barrels of oil. Determine the high and low production levels.

38. *Stock Price* A stock market analyst predicts that over the next year the price p of a certain stock will not change from its current price of $\$33\frac{1}{8}$ by more than $\$2$. Use absolute values to write this prediction as an inequality.

Budget Variance In Exercises 39–42, (a) use absolute value notation to represent the two intervals in which expenses must lie if they are to be within $\$500$ and within 5% of the specified budget amount, and (b) using the more stringent constraint, determine whether the given expense is at variance with the budget restriction.

Item	Budget	Expense
39. Utilities	$4,750.00	$5,116.37
40. Insurance	$15,000.00	$14,695.00
41. Maintenance	$20,000.00	$22,718.35
42. Taxes	$7,500.00	$8,691.00

Exponents and Radicals

Expressions Involving Exponents or Radicals ■ *Operations with Exponents* ■ *Domain of an Algebraic Expression*

Expressions Involving Exponents or Radicals

Note If n is even, then the principal nth root is positive. For example, $\sqrt{4} = +2$ and $\sqrt[4]{81} = +3$.

Properties of Exponents

1. Whole number exponents: $x^n = \underbrace{x \cdot x \cdot x \cdots x}_{n \text{ factors}}$

2. Zero exponent: $x^0 = 1, \quad x \neq 0$

3. Negative exponents: $x^{-n} = \dfrac{1}{x^n}, \quad x \neq 0$

4. Radicals (principal nth root): $\sqrt[n]{x} = a \implies x = a^n$

5. Rational exponents $(1/n)$: $x^{1/n} = \sqrt[n]{x}$

6. Rational exponents (m/n): $x^{m/n} = (x^{1/n})^m = \left(\sqrt[n]{x} \right)^m$

   $x^{m/n} = (x^m)^{1/n} = \sqrt[n]{x^m}$

7. Special convention (square root): $\sqrt[2]{x} = \sqrt{x}$

EXAMPLE 1 *Evaluating Expressions*

	Expression	*x-Value*	*Substitution*
a.	$y = -2x^2$	$x = 4$	$y = -2(4^2) = -2(16) = -32$
b.	$y = 3x^{-3}$	$x = -1$	$y = 3(-1)^{-3} = \dfrac{3}{(-1)^3} = \dfrac{3}{-1} = -3$
c.	$y = (-x)^2$	$x = \dfrac{1}{2}$	$y = \left(-\dfrac{1}{2}\right)^2 = \dfrac{1}{4}$
d.	$y = \dfrac{2}{x^{-2}}$	$x = 3$	$y = \dfrac{2}{3^{-2}} = 2(3^2) = 18$

EXAMPLE 2 *Evaluating Expressions*

	Expression	*x-Value*	*Substitution*
a.	$y = 2x^{1/2}$	$x = 4$	$y = 2\sqrt{4} = 2(2) = 4$
b.	$y = \sqrt[3]{x^2}$	$x = 8$	$y = 8^{2/3} = (8^{1/3})^2 = 2^2 = 4$

Operations with Exponents

Operations with Exponents

1. **Multiplying like bases:** $x^n x^m = x^{n+m}$ *Add exponents.*

2. **Dividing like bases:** $\dfrac{x^n}{x^m} = x^{n-m}$ *Subtract exponents.*

3. **Removing parentheses:** $(xy)^n = x^n y^n$

$$\left(\frac{x}{y}\right)^n = \frac{x^n}{y^n}$$

$$(x^n)^m = x^{nm}$$

4. **Special conventions:** $-x^n = -(x^n),\quad -x^n \neq (-x)^n$

$$cx^n = c(x^n),\quad cx^n \neq (cx)^n$$

$$x^{n^m} = x^{(n^m)},\quad x^{n^m} \neq (x^n)^m$$

EXAMPLE 3 *Simplifying Expressions with Exponents*

Simplify the following expressions.

a. $2x^2(x^3)$ **b.** $(3x)^2 \sqrt[3]{x}$ **c.** $\dfrac{3x^2}{\left(\sqrt{x}\right)^3}$

d. $\dfrac{5x^4}{(x^2)^3}$ **e.** $x^{-1}(2x^2)$ **f.** $\dfrac{-\sqrt{x}}{5x^{-1}}$

Solution

a. $2x^2(x^3) = 2x^{2+3} = 2x^5$

b. $(3x)^2 \sqrt[3]{x} = 9x^2 x^{1/3} = 9x^{2+(1/3)} = 9x^{7/3}$

c. $\dfrac{3x^2}{\left(\sqrt{x}\right)^3} = \dfrac{3x^2}{(x^{1/2})^3} = 3\left(\dfrac{x^2}{x^{3/2}}\right) = 3x^{2-(3/2)} = 3x^{1/2}$

d. $\dfrac{5x^4}{(x^2)^3} = \dfrac{5x^4}{x^6} = \dfrac{5}{x^{6-4}} = \dfrac{5}{x^2}$

e. $x^{-1}(2x^2) = 2x^{-1}x^2 = 2x^{2-1} = 2x$

f. $\dfrac{-\sqrt{x}}{5x^{-1}} = -\dfrac{1}{5}\left(\dfrac{x^{1/2}}{x^{-1}}\right) = -\dfrac{1}{5}x^{(1/2)+1} = -\dfrac{1}{5}x^{3/2}$

Note in Example 3 that one characteristic of simplified expressions is the absence of negative exponents. Another characteristic of simplified expressions is that we usually prefer to write sums and differences in *factored form.* To do this, you can use the **distributive property.**

$$abx^n + acx^{n+m} = ax^n(b + cx^m)$$

Study the next example carefully to be sure that you understand the concepts involved in the factoring process.

EXAMPLE 4 *Simplifying by Factoring*

a. $2x^2 - x^3 = x^2(2 - x)$

b. $2x^3 + x^2 = x^2(2x + 1)$

c. $2x^{1/2} + 4x^{5/2} = 2x^{1/2}(1 + 2x^2)$

d. $2x^{-1/2} + 3x^{5/2} = x^{-1/2}(2 + 3x^3) = \dfrac{2 + 3x^3}{\sqrt{x}}$ ▬

Many algebraic expressions obtained in calculus occur in unsimplified form. For instance, the two expressions shown in the following example are the result of an operation in calculus called **differentiation.** [The first is the derivative of $2(x + 1)^{3/2}(2x - 3)^{5/2}$, and the second is the derivative of $2(x + 1)^{1/2}(2x - 3)^{5/2}$.]

EXAMPLE 5 *Simplifying by Factoring*

Simplify the following expressions by factoring.

a. $3(x + 1)^{1/2}(2x - 3)^{5/2} + 10(x + 1)^{3/2}(2x - 3)^{3/2}$

b. $(x + 1)^{-1/2}(2x - 3)^{5/2} + 10(x + 1)^{1/2}(2x - 3)^{3/2}$

Solution

a. $3(x + 1)^{1/2}(2x - 3)^{5/2} + 10(x + 1)^{3/2}(2x - 3)^{3/2}$

$$= (x + 1)^{1/2}(2x - 3)^{3/2}[3(2x - 3) + 10(x + 1)]$$
$$= (x + 1)^{1/2}(2x - 3)^{3/2}(6x - 9 + 10x + 10)$$
$$= (x + 1)^{1/2}(2x - 3)^{3/2}(16x + 1)$$

b. $(x + 1)^{-1/2}(2x - 3)^{5/2} + 10(x + 1)^{1/2}(2x - 3)^{3/2}$

$$= (x + 1)^{-1/2}(2x - 3)^{3/2}[(2x - 3) + 10(x + 1)]$$
$$= (x + 1)^{-1/2}(2x - 3)^{3/2}(2x - 3 + 10x + 10)$$
$$= (x + 1)^{-1/2}(2x - 3)^{3/2}(12x + 7)$$
$$= \dfrac{(2x - 3)^{3/2}(12x + 7)}{(x + 1)^{1/2}}$$ ▬

Example 6 shows some additional types of expressions that can occur in calculus. [The expression in part c is an antiderivative of $(x+1)^{2/3}(2x+3)$, and the expression in part d is the derivative of $(x+2)^3/(x-1)^3$.]

EXAMPLE 6   *Factors Involving Quotients*

Simplify the following expressions by factoring.

a. $\dfrac{3x^2+x^4}{2x}$

b. $\dfrac{\sqrt{x}+x^{3/2}}{x}$

c. $\dfrac{3}{5}(x+1)^{5/3}+\dfrac{3}{4}(x+1)^{8/3}$

d. $\dfrac{3(x+2)^2(x-1)^3-3(x+2)^3(x-1)^2}{[(x-1)^3]^2}$

Solution

a. $\dfrac{3x^2+x^4}{2x}=\dfrac{x^2(3+x^2)}{2x}=\dfrac{x^{2-1}(3+x^2)}{2}=\dfrac{x(3+x^2)}{2}$

b. $\dfrac{\sqrt{x}+x^{3/2}}{x}=\dfrac{x^{1/2}(1+x)}{x}=\dfrac{1+x}{x^{1-(1/2)}}=\dfrac{1+x}{\sqrt{x}}$

c. $\dfrac{3}{5}(x+1)^{5/3}+\dfrac{3}{4}(x+1)^{8/3}=\dfrac{12}{20}(x+1)^{5/3}+\dfrac{15}{20}(x+1)^{8/3}$

$=\dfrac{3}{20}(x+1)^{5/3}[4+5(x+1)]$

$=\dfrac{3}{20}(x+1)^{5/3}(4+5x+5)$

$=\dfrac{3}{20}(x+1)^{5/3}(5x+9)$

d. $\dfrac{3(x+2)^2(x-1)^3-3(x+2)^3(x-1)^2}{[(x-1)^3]^2}$

$=\dfrac{3(x+2)^2(x-1)^2[(x-1)-(x+2)]}{(x-1)^6}$

$=\dfrac{3(x+2)^2(x-1-x-2)}{(x-1)^{6-2}}$

$=\dfrac{-9(x+2)^2}{(x-1)^4}$

Domain of an Algebraic Expression

When working with algebraic expressions involving x, we face the potential difficulty of substituting a value of x for which the expression is not defined (does not produce a real number). For example, the expression $\sqrt{2x+3}$ is *not defined* when $x = -2$ because $\sqrt{2(-2)+3}$ is not a real number.

The set of all values for which an expression is defined is called its **domain.** Thus, the domain of $\sqrt{2x+3}$ is the set of all values of x such that $\sqrt{2x+3}$ is a real number. In order for $\sqrt{2x+3}$ to represent a real number, it is necessary that $2x+3 \geq 0$. In other words, $\sqrt{2x+3}$ is defined only for those values of x that lie in the interval $\left[-\frac{3}{2}, \infty\right)$, as shown in Figure 0.14.

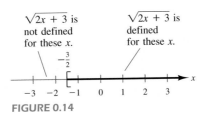

FIGURE 0.14

EXAMPLE 7 *Finding the Domain of an Expression*

Find the domain of each of the following expressions.

a. $\sqrt{3x-2}$

b. $\dfrac{1}{\sqrt{3x-2}}$

c. $\sqrt[3]{9x+1}$

Solution

a. The domain of $\sqrt{3x-2}$ consists of all x such that

$$3x - 2 \geq 0,$$

which implies that $x \geq \frac{2}{3}$. Therefore, the domain is $\left[\frac{2}{3}, \infty\right)$.

b. The domain of $1/\sqrt{3x-2}$ is the same as the domain of $\sqrt{3x-2}$, except that $1/\sqrt{3x-2}$ is not defined when $3x - 2 = 0$. Because this occurs when $x = \frac{2}{3}$, we conclude that the domain is $\left(\frac{2}{3}, \infty\right)$.

c. Because $\sqrt[3]{9x+1}$ is defined for *all* real numbers, we conclude that its domain is $(-\infty, \infty)$.

EXERCISES 0.3

In Exercises 1–20, evaluate the expression for the indicated value of x.

Expression	x-value	Expression	x-value
1. $-3x^3$	$x = 2$	2. $\dfrac{x^2}{2}$	$x = 6$
3. $4x^{-3}$	$x = 2$	4. $7x^{-2}$	$x = 4$
5. $\dfrac{1 + x^{-1}}{x^{-1}}$	$x = 2$	6. $x - 4x^{-2}$	$x = 3$
7. $3x^2 - 4x^3$	$x = -2$	8. $5(-x)^3$	$x = 3$
9. $6x^0 - (6x)^0$	$x = 10$	10. $\dfrac{1}{(-x)^{-3}}$	$x = 4$
11. $\sqrt[3]{x^2}$	$x = 27$	12. $\sqrt{x^3}$	$x = \frac{1}{9}$
13. $x^{-1/2}$	$x = 4$	14. $x^{-3/4}$	$x = 16$
15. $x^{-2/5}$	$x = -32$	16. $(x^{2/3})^3$	$x = 10$
17. $500x^{60}$	$x = 1.01$	18. $\dfrac{10{,}000}{x^{120}}$	$x = 1.075$
19. $\sqrt[3]{x}$	$x = -154$	20. $\sqrt[6]{x}$	$x = 325$

In Exercises 21–34, simplify the expression.

21. $5x^4(x^2)$

22. $(8x^4)(2x^3)$

23. $6y^2(2y^4)^2$

24. $z^{-3}(3z^4)$

25. $10(x^2)^2$

26. $(4x^3)^2$

27. $\dfrac{7x^2}{x^{-3}}$

28. $\dfrac{r^4}{r^6}$

29. $\dfrac{12(x + y)^3}{9(x + y)}$

30. $\left(\dfrac{12s^2}{9s}\right)^3$

31. $\dfrac{3x\sqrt{x}}{x^{1/2}}$

32. $\left(\sqrt[3]{x^2}\right)^3$

33. $\left(\dfrac{\sqrt{2}\sqrt{x^3}}{\sqrt{x}}\right)^4$

34. $(2x^2yz^5)^0$

In Exercises 35–40, simplify by removing all possible factors from the radical.

35. (a) $\sqrt{8}$ (b) $\sqrt{18}$

36. (a) $\sqrt[3]{\frac{16}{27}}$ (b) $\sqrt[3]{\frac{24}{125}}$

37. (a) $\sqrt[3]{16x^5}$ (b) $\sqrt[4]{32x^4z^5}$

38. (a) $\sqrt[4]{(3x^2y^3)^4}$ (b) $\sqrt[3]{54x^7}$

39. (a) $\sqrt{75x^2y^{-4}}$ (b) $\sqrt[5]{5(x - y)^3}$

40. (a) $\sqrt[5]{96b^6c^3}$ (b) $\sqrt{72(x + 1)^4}$

In Exercises 41–50, find the domain of the given expression.

41. $\sqrt{x - 1}$

42. $\sqrt{5 - 2x}$

43. $\sqrt{x^2 + 3}$

44. $\sqrt[5]{1 - x}$

45. $\dfrac{1}{\sqrt[3]{x - 1}}$

46. $\dfrac{1}{\sqrt{x + 4}}$

47. $\dfrac{1}{\sqrt[4]{2x - 6}}$

48. $\dfrac{\sqrt{x - 1}}{x + 1}$

49. $\sqrt{x - 1} + \sqrt{5 - x}$

50. $\dfrac{1}{\sqrt{2x + 3}} + \sqrt{6 - 4x}$

Compound Interest In Exercises 51 and 52, a certificate of deposit has a principal of P and an annual percentage rate of r (expressed as a decimal) compounded n times per year. Use the compound interest formula

$$A = P\left(1 + \frac{r}{n}\right)^N$$

to find the balance after N compoundings.

51. $P = \$10{,}000,\quad r = 9.5\%,\quad n = 12,\quad N = 120$

52. $P = \$7000,\quad r = 8\%,\quad n = 365,\quad N = 1000$

53. *Installment Loan* A 5-year auto loan for $P = \$7000$ with an annual percentage rate of $r = 0.14$ requires a monthly payment of $M = \$162.88$. Payments have been made on the loan for 2 years ($N = 24$ monthly payments). Determine the balance B due on the loan if that balance is given by the formula

$$B = \left(1 + \frac{r}{12}\right)^N \left(P - \frac{12M}{r}\right) + \frac{12M}{r}.$$

54. *Annuity* A balance A, after n annual payments of P dollars have been made into an annuity earning an annual percentage rate of r compounded annually, is given by

$$A = P(1 + r) + P(1 + r)^2 + \cdots + P(1 + r)^n.$$

Rewrite this formula by completing the following factorization: $A = P(1 + r)(\quad)$.

Factoring Polynomials

Factorization Techniques ▪ *Factoring Polynomials of Degree Three or More* ▪ *The Rational Zero Theorem*

Factorization Techniques

The Fundamental Theorem of Algebra states that every nth-degree polynomial

$$a_n x^n + a_{n-1} x^{n-1} + \cdots + a_1 x + a_0$$

has precisely n **zeros.** (The zeros may be repeated or imaginary.) The problem of finding the zeros of a polynomial is equivalent to the problem of factoring the polynomial into linear factors.

Special Products and Factorization Techniques

Quadratic Formula

$$ax^2 + bx + c = 0 \implies x = \frac{-b \pm \sqrt{b^2 - 4ac}}{2a}$$

Example

$$x^2 + 3x - 1 = 0 \implies x = \frac{-3 \pm \sqrt{13}}{2}$$

Special Products

$x^2 - a^2 = (x - a)(x + a)$
$x^3 - a^3 = (x - a)(x^2 + ax + a^2)$
$x^3 + a^3 = (x + a)(x^2 - ax + a^2)$
$x^4 - a^4 = (x - a)(x + a)(x^2 + a^2)$

Examples

$x^2 - 9 = (x - 3)(x + 3)$
$x^3 - 8 = (x - 2)(x^2 + 2x + 4)$
$x^3 + 64 = (x + 4)(x^2 - 4x + 16)$
$x^4 - 16 = (x - 2)(x + 2)(x^2 + 4)$

*Binomial Theorem**

$(x + a)^2 = x^2 + 2ax + a^2$
$(x - a)^2 = x^2 - 2ax + a^2$
$(x + a)^3 = x^3 + 3ax^2 + 3a^2x + a^3$
$(x - a)^3 = x^3 - 3ax^2 + 3a^2x - a^3$
$(x + a)^4 = x^4 + 4ax^3 + 6a^2x^2 + 4a^3x + a^4$
$(x - a)^4 = x^4 - 4ax^3 + 6a^2x^2 - 4a^3x + a^4$

Examples

$(x + 3)^2 = x^2 + 6x + 9$
$(x^2 - 5)^2 = x^4 - 10x^2 + 25$
$(x + 2)^3 = x^3 + 6x^2 + 12x + 8$
$(x - 1)^3 = x^3 - 3x^2 + 3x - 1$
$(x + 2)^4 = x^4 + 8x^3 + 24x^2 + 32x + 16$
$(x - 4)^4 = x^4 - 16x^3 + 96x^2 - 256x + 256$

$$(x + a)^n = x^n + nax^{n-1} + \frac{n(n-1)}{2!}a^2x^{n-2} + \frac{n(n-1)(n-2)}{3!}a^3x^{n-3} + \cdots + na^{n-1}x + a^n$$

$$(x - a)^n = x^n - nax^{n-1} + \frac{n(n-1)}{2!}a^2x^{n-2} - \frac{n(n-1)(n-2)}{3!}a^3x^{n-3} + \cdots \pm na^{n-1}x \mp a^n$$

Factoring by Grouping

$acx^3 + adx^2 + bcx + bd = ax^2(cx + d) + b(cx + d)$
$\qquad\qquad\qquad\qquad\quad = (ax^2 + b)(cx + d)$

Example

$3x^3 - 2x^2 - 6x + 4 = x^2(3x - 2) - 2(3x - 2)$
$\qquad\qquad\qquad\qquad = (x^2 - 2)(3x - 2)$

*The *factorial* symbol ! is defined as follows $0! = 1$, $1! = 1$, $2! = 2 \cdot 1 = 2$, $3! = 3 \cdot 2 \cdot 1 = 6$, $4! = 4 \cdot 3 \cdot 2 \cdot 1 = 24$, and so on.

EXAMPLE 1 *Applying the Quadratic Formula*

Use the Quadratic Formula to find all real zeros of the following polynomials.

a. $4x^2 + 6x + 1$ **b.** $x^2 + 6x + 9$ **c.** $2x^2 - 6x + 5$

Solution

a. Using $a = 4$, $b = 6$, and $c = 1$, you can write

$$x = \frac{-b \pm \sqrt{b^2 - 4ac}}{2a}$$

$$= \frac{-6 \pm \sqrt{36 - 16}}{8}$$

$$= \frac{-6 \pm \sqrt{20}}{8}$$

$$= \frac{-6 \pm 2\sqrt{5}}{8}$$

$$= \frac{2\left(-3 \pm \sqrt{5}\right)}{2(4)}$$

$$= \frac{-3 \pm \sqrt{5}}{4}.$$

Thus, there are two real zeros:

$$x = \frac{-3 - \sqrt{5}}{4} \approx -1.309 \quad \text{and} \quad x = \frac{-3 + \sqrt{5}}{4} \approx -0.191.$$

b. In this case, $a = 1$, $b = 6$, and $c = 9$, and the Quadratic Formula yields

$$x = \frac{-b \pm \sqrt{b^2 - 4ac}}{2a} = \frac{-6 \pm \sqrt{36 - 36}}{2} = -\frac{6}{2} = -3.$$

Thus, there is one (repeated) real zero: $x = -3$.

c. For this quadratic equation, $a = 2$, $b = -6$, and $c = 5$. Thus,

$$x = \frac{-b \pm \sqrt{b^2 - 4ac}}{2a} = \frac{6 \pm \sqrt{36 - 40}}{4} = \frac{6 \pm \sqrt{-4}}{4}.$$

Because $\sqrt{-4}$ is imaginary, there are no real zeros.   ▬

In Example 1, the zeros in part a are irrational, and the zeros in part c are imaginary. In both of these cases the quadratic is said to be **irreducible** because it cannot be factored into linear factors with rational coefficients. The next example shows how to find the zeros associated with *reducible* quadratics. In this example, factoring is used to find the zeros of each quadratic. Try using the Quadratic Formula to obtain the same zeros.

EXAMPLE 2 *Factoring Quadratics*

Find the zeros of the following quadratic polynomials.

a. $x^2 - 5x + 6$ **b.** $x^2 - 5x - 6$ **c.** $2x^2 + 5x - 3$

Solution

a. Because

$$x^2 - 5x + 6 = (x - 2)(x - 3),$$

the zeros are $x = 2$ and $x = 3$.

b. Because

$$x^2 - 5x - 6 = (x + 1)(x - 6),$$

the zeros are $x = -1$ and $x = 6$.

c. Because

$$2x^2 + 5x - 3 = (2x - 1)(x + 3),$$

the zeros are $x = \frac{1}{2}$ and $x = -3$.

EXAMPLE 3 *Finding the Domain of a Radical Expression*

Find the domain of $\sqrt{x^2 - 3x + 2}$.

Solution

Because

$$x^2 - 3x + 2 = (x - 1)(x - 2),$$

you know that the zeros of the quadratic are $x = 1$ and $x = 2$. Thus, you need to test the sign of the quadratic in the three intervals $(-\infty, \ 1), (1, \ 2),$ and $(2, \ \infty)$, as shown in Figure 0.15. After testing each of these intervals, you can see that the quadratic is negative in the center interval and positive in the outer two intervals. Moreover, because the quadratic is zero when $x = 1$ and $x = 2$, you can conclude that the domain of $\sqrt{x^2 - 3x + 2}$ is

$(-\infty, \ 1] \cup [2, \ \infty).$ *Domain*

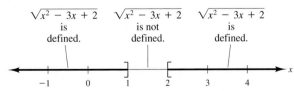

FIGURE 0.15

Factoring Polynomials of Degree Three or More

It can be difficult to find the zeros of polynomials of degree three or more. However, if one of the zeros of a polynomial is known, then you can use that zero to reduce the degree of the polynomial. For example, if you know that $x = 2$ is a zero of $x^3 - 4x^2 + 5x - 2$, then you know that $(x - 2)$ is a factor, and you can use long division to factor the polynomial as follows.

$$x^3 - 4x^2 + 5x - 2 = (x - 2)(x^2 - 2x + 1)$$
$$= (x - 2)(x - 1)(x - 1)$$

As an alternative to long division, many people prefer to use **synthetic division** to reduce the degree of a polynomial.

Synthetic Division for a Cubic Polynomial

Given: $x = x_1$ is a zero of $ax^3 + bx^2 + cx + d$.

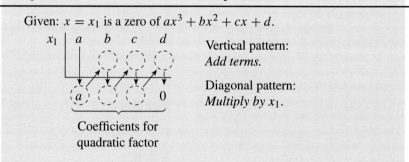

Vertical pattern:
Add terms.

Diagonal pattern:
Multiply by x_1.

Coefficients for quadratic factor

For example, performing synthetic division on the polynomial $x^3 - 4x^2 + 5x - 2$ using the given zero, $x = 2$, produces the following.

$$
\begin{array}{r|rrrr}
2 & 1 & -4 & 5 & -2 \\
  &   & 2 & -4 & 2 \\
\hline
  & 1 & -2 & 1 & 0
\end{array}
$$

$$(x - 2)(x^2 - 2x + 1) = x^3 - 4x^2 + 5x - 2$$

When you use synthetic division, remember to take *all* coefficients into account—*even if some of them are zero.* For instance, if you know that $x = -2$ is a zero of $x^3 + 3x + 14$, you can apply synthetic division as follows.

$$
\begin{array}{r|rrrr}
-2 & 1 & 0 & 3 & 14 \\
   &   & -2 & 4 & -14 \\
\hline
   & 1 & -2 & 7 & 0
\end{array}
$$

$$(x + 2)(x^2 - 2x + 7) = x^3 + 3x + 14$$

The Rational Zero Theorem

There is a systematic way to find the *rational* zeros of a polynomial. You can use the **Rational Zero Theorem** (also called the Rational Root Theorem).

Rational Zero Theorem

If a polynomial

$$a_n x^n + a_{n-1} x^{n-1} + \cdots + a_1 x + a_0$$

has integer coefficients, then every *rational* zero is of the form $x = p/q$, where p is a factor of a_0, and q is a factor of a_n.

EXAMPLE 4   *Using the Rational Zero Theorem*

Find all real zeros of $2x^3 + 3x^2 - 8x + 3$.

Solution

$$2x^3 + 3x^2 - 8x + 3$$

Factors of constant term: $\pm 1, \pm 3$

Factors of leading coefficient: $\pm 1, \pm 2$

The possible rational zeros are the factors of the constant term divided by the factors of the leading coefficient.

$$1, \ -1, \ 3, \ -3, \ \frac{1}{2}, \ -\frac{1}{2}, \ \frac{3}{2}, \ -\frac{3}{2}$$

By testing these possible zeros, you can see that $x = 1$ works.

$$2(1)^3 + 3(1)^2 - 8(1) + 3 = 2 + 3 - 8 + 3 = 0$$

Now, by synthetic division you have the following.

$$
\begin{array}{r|rrrr}
1 & 2 & 3 & -8 & 3 \\
  &   & 2 & 5 & -3 \\
\hline
  & 2 & 5 & -3 & 0
\end{array}
$$

$$(x - 1)(2x^2 + 5x - 3) = 2x^3 + 3x^2 - 8x + 3$$

Finally, by factoring the quadratic, $2x^2 + 5x - 3 = (2x - 1)(x + 3)$, you have

$$2x^3 + 3x^2 - 8x + 3 = (x - 1)(2x - 1)(x + 3)$$

and you can conclude that the zeros are $x = 1$, $x = \frac{1}{2}$, and $x = -3$.   ■

EXERCISES 0.4

 means that technology can help you solve or check the exercise(s).

In Exercises 1–6, use the Quadratic Formula to find all real zeros of the second-degree polynomial.

1. $6x^2 - x - 1$ **2.** $8x^2 - 2x - 1$

3. $4x^2 - 12x + 9$ **4.** $9x^2 + 12x + 4$

5. $y^2 + 4y + 1$ **6.** $x^2 + 6x - 1$

In Exercises 7–16, write the second-degree polynomial as the product of two linear factors.

7. $x^2 - 4x + 4$ **8.** $x^2 + 10x + 25$

9. $4x^2 + 4x + 1$ **10.** $9x^2 - 12x + 4$

11. $x^2 + x - 2$ **12.** $2x^2 - x - 1$

13. $3x^2 - 5x + 2$ **14.** $x^2 - xy - 2y^2$

15. $x^2 - 4xy + 4y^2$ **16.** $a^2b^2 - 2abc + c^2$

In Exercises 17–30, completely factor the polynomial.

17. $81 - y^4$ **18.** $x^4 - 16$

19. $x^3 - 8$ **20.** $y^3 - 64$

21. $y^3 + 64$ **22.** $z^3 + 125$

23. $x^3 - 27$ **24.** $(x - a)^3 + b^3$

25. $x^3 - 4x^2 - x + 4$ **26.** $x^3 - x^2 - x + 1$

27. $2x^3 - 3x^2 + 4x - 6$ **28.** $x^3 - 5x^2 - 5x + 25$

29. $2x^3 - 4x^2 - x + 2$ **30.** $x^3 - 7x^2 - 4x + 28$

In Exercises 31–48, find all real solutions of the equation.

31. $x^2 - 5x = 0$ **32.** $2x^2 - 3x = 0$

33. $x^2 - 9 = 0$ **34.** $x^2 - 25 = 0$

35. $x^2 - 3 = 0$ **36.** $x^2 - 8 = 0$

37. $(x - 3)^2 - 9 = 0$ **38.** $(x + 1)^2 - 8 = 0$

39. $x^2 + x - 2 = 0$ **40.** $x^2 + 5x + 6 = 0$

41. $x^2 - 5x + 6 = 0$ **42.** $x^2 + x - 20 = 0$

43. $x^3 + 64 = 0$ **44.** $x^3 - 216 = 0$

45. $x^4 - 16 = 0$ **46.** $x^4 - 625 = 0$

47. $x^3 - x^2 - 4x + 4 = 0$ **48.** $2x^3 + x^2 + 6x + 3 = 0$

In Exercises 49–52, find the interval (or intervals) on which the given expression is defined.

49. $\sqrt{x^2 - 7x + 12}$ **50.** $\sqrt{x^2 - 4}$

51. $\sqrt{4 - x^2}$ **52.** $\sqrt{144 - 9x^2}$

In Exercises 53–56, use synthetic division to complete the indicated factorization.

53. $x^3 + 8 = (x + 2)(\quad)$

54. $x^3 - 2x^2 - x + 2 = (x + 1)(\quad)$

55. $2x^3 - x^2 - 2x + 1 = (x - 1)(\quad)$

56. $x^4 - 16x^3 + 96x^2 - 256x + 256 = (x - 4)(\quad)$

In Exercises 57–64, use the Rational Zero Theorem as an aid in finding all real solutions of the given equation.

57. $x^3 - x^2 - x + 1 = 0$

58. $x^3 - x^2 - 4x + 4 = 0$

59. $x^3 - 6x^2 + 11x - 6 = 0$

60. $x^3 + 2x^2 - 5x - 6 = 0$

61. $4x^3 - 4x^2 - x + 1 = 0$

62. $18x^3 - 9x^2 - 8x + 4 = 0$

63. $x^3 - 3x^2 - 3x - 4 = 0$

64. $4x^3 - 6x^2 + 2x + 3 = 0$

65. *Average Cost* The minimum average cost of producing x units of a certain product occurs when the production level is set at the (positive) solution of

$$0.0003x^2 - 1200 = 0.$$

Determine this production level.

66. *Profit* The profit P from sales is given by

$$P = -200x^2 + 2000x - 3800,$$

where x is the number of units sold per day (in hundreds). Determine the interval for x so that the profit will be greater than 1000.

Fractions and Rationalization

Operations with Fractions ■ *Expressions Involving Radicals* ■
Rationalization Techniques

Operations with Fractions

In the final section of this chapter we review operations involving fractional expressions such as

$$\frac{2}{x}, \quad \frac{x^2 + 2x - 4}{x + 6}, \quad \text{and} \quad \frac{1}{\sqrt{x^2 + 1}}.$$

The first two expressions have polynomials as both numerator and denominator and are called **rational expressions.** A rational expression is **proper** if the degree of the numerator is less than the degree of the denominator. For example, $x/(x^2 + 1)$ is proper. If the degree of the numerator is greater than or equal to the degree of the denominator, then the rational expression is improper. For example, $x^2/(x^2 + 1)$ is improper.

Operations with Fractions

1. Add fractions (find a common denominator):

$$\frac{a}{b} + \frac{c}{d} = \frac{a}{b}\left(\frac{d}{d}\right) + \frac{c}{d}\left(\frac{b}{b}\right) = \frac{ad}{bd} + \frac{bc}{bd} = \frac{ad + bc}{bd}$$

2. Subtract fractions (find a common denominator):

$$\frac{a}{b} - \frac{c}{d} = \frac{a}{b}\left(\frac{d}{d}\right) - \frac{c}{d}\left(\frac{b}{b}\right) = \frac{ad}{bd} - \frac{bc}{bd} = \frac{ad - bc}{bd}$$

3. Multiply fractions:

$$\left(\frac{a}{b}\right)\left(\frac{c}{d}\right) = \frac{ac}{bd}$$

4. Divide fractions (invert and multiply):

$$\frac{a/b}{c/d} = \left(\frac{a}{b}\right)\left(\frac{d}{c}\right) = \frac{ad}{bc}, \qquad \frac{a/b}{c} = \frac{a/b}{c/1} = \left(\frac{a}{b}\right)\left(\frac{1}{c}\right) = \frac{a}{bc}$$

5. Cancel:

$$\frac{\cancel{a}b}{\cancel{a}c} = \frac{b}{c}, \qquad \frac{ab + ac}{ad} = \frac{\cancel{a}(b + c)}{\cancel{a}d} = \frac{b + c}{d}$$

EXAMPLE 1 *Adding and Subtracting Rational Expressions*

Perform the indicated operations and simplify.

a. $x + \dfrac{1}{x}$ **b.** $\dfrac{1}{x+1} - \dfrac{2}{2x-1}$

Solution

a. $x + \dfrac{1}{x} = \dfrac{x^2}{x} + \dfrac{1}{x}$

$\qquad = \dfrac{x^2 + 1}{x}$

b. $\dfrac{1}{x+1} - \dfrac{2}{2x-1} = \dfrac{(2x-1)}{(x+1)(2x-1)} - \dfrac{2(x+1)}{(x+1)(2x-1)}$

$\qquad = \dfrac{2x - 1 - 2x - 2}{2x^2 + x - 1}$

$\qquad = \dfrac{-3}{2x^2 + x - 1}$

In adding (or subtracting) fractions whose denominators have no common factors, it is convenient to use the following pattern.

$$\frac{a}{b} + \frac{c}{d} = \frac{a}{b} \diagdown \!\!\!\!\diagup \frac{c}{d} = \frac{ad + bc}{bd}$$

For instance, in part b of Example 1, you could have used this pattern as follows.

$$\frac{1}{x+1} - \frac{2}{2x-1} = \frac{(2x-1) - 2(x+1)}{(x+1)(2x-1)}$$

$$= \frac{2x - 1 - 2x - 2}{(x+1)(2x-1)}$$

$$= \frac{-3}{2x^2 + x - 1}$$

In Example 1, the denominators of the rational expressions have no common factors. When the denominators do have common factors, it is best to find the least common denominator before adding or subtracting. For instance, when adding $1/x$ and $2/x^2$, you can recognize that the least common denominator is x^2 and write

$$\frac{1}{x} + \frac{2}{x^2} = \frac{x}{x^2} + \frac{2}{x^2}$$

$$= \frac{x + 2}{x^2}.$$

This is further demonstrated in Example 2.

EXAMPLE 2 *Adding and Subtracting Rational Expressions*

Perform the indicated operations and simplify.

a. $\dfrac{x}{x^2-1} + \dfrac{3}{x+1}$ **b.** $\dfrac{1}{2(x^2+2x)} - \dfrac{1}{4x}$

Solution

a. Because $x^2 - 1 = (x+1)(x-1)$, the least common denominator is $x^2 - 1$.

$$\frac{x}{x^2-1} + \frac{3}{x+1} = \frac{x}{(x-1)(x+1)} + \frac{3}{x+1}$$

$$= \frac{x}{(x-1)(x+1)} + \frac{3(x-1)}{(x-1)(x+1)}$$

$$= \frac{x+3x-3}{(x-1)(x+1)}$$

$$= \frac{4x-3}{x^2-1}$$

b. In this case, the least common denominator is $4x(x+2)$.

$$\frac{1}{2(x^2+2x)} - \frac{1}{4x} = \frac{1}{2x(x+2)} - \frac{1}{2(2x)}$$

$$= \frac{2}{2(2x)(x+2)} - \frac{x+2}{2(2x)(x+2)}$$

$$= \frac{2-x-2}{4x(x+2)}$$

$$= \frac{-\cancel{x}}{4\cancel{x}(x+2)} \qquad \text{Cancel } x.$$

$$= \frac{-1}{4(x+2)}$$

To add more than two fractions you must find a denominator that is common to all the fractions. For instance, to add $\frac{1}{2}$, $\frac{1}{3}$, and $\frac{1}{5}$, use a (least) common denominator of 30 and write

$$\frac{1}{2} + \frac{1}{3} + \frac{1}{5} = \frac{15}{30} + \frac{10}{30} + \frac{6}{30}$$

$$= \frac{31}{30}.$$

To add more than two rational expressions, use a similar procedure, as demonstrated in Example 3. (Expressions such as those shown in this example are used in calculus to perform an integration technique called integration by partial fractions.)

EXAMPLE 3 *Adding More Than Two Rational Expressions*

Perform the indicated addition of rational expressions.

a. $\dfrac{A}{x+2} + \dfrac{B}{x-3} + \dfrac{C}{x+4}$

b. $\dfrac{A}{x+2} + \dfrac{B}{(x+2)^2} + \dfrac{C}{x-1}$

Solution

a. The least common denominator is $(x+2)(x-3)(x+4)$.

$$\frac{A}{x+2} + \frac{B}{x-3} + \frac{C}{x+4}$$

$$= \frac{A(x-3)(x+4) + B(x+2)(x+4) + C(x+2)(x-3)}{(x+2)(x-3)(x+4)}$$

$$= \frac{A(x^2 + x - 12) + B(x^2 + 6x + 8) + C(x^2 - x - 6)}{(x+2)(x-3)(x+4)}$$

$$= \frac{Ax^2 + Bx^2 + Cx^2 + Ax + 6Bx - Cx - 12A + 8B - 6C}{(x+2)(x-3)(x+4)}$$

$$= \frac{(A+B+C)x^2 + (A+6B-C)x + (-12A+8B-6C)}{(x+2)(x-3)(x+4)}$$

b. Here the least common denominator is $(x+2)^2(x-1)$.

$$\frac{A}{x+2} + \frac{B}{(x+2)^2} + \frac{C}{x-1}$$

$$= \frac{A(x+2)(x-1) + B(x-1) + C(x+2)^2}{(x+2)^2(x-1)}$$

$$= \frac{A(x^2 + x - 2) + B(x-1) + C(x^2 + 4x + 4)}{(x+2)^2(x-1)}$$

$$= \frac{Ax^2 + Cx^2 + Ax + Bx + 4Cx - 2A - B + 4C}{(x+2)^2(x-1)}$$

$$= \frac{(A+C)x^2 + (A+B+4C)x + (-2A-B+4C)}{(x+2)^2(x-1)}$$

Expressions Involving Radicals

In calculus, the operation of differentiation tends to produce "messy" expressions when applied to fractional expressions. This is especially true when the fractional expression involves radicals. When differentiation is used, it is important to be able to simplify these expressions so that you can obtain more manageable forms. All of the expressions in Examples 4 and 5 are the results of differentiation. In each case, note how much *simpler* the simplified form is than the original form.

EXAMPLE 4 *Simplifying an Expression with Radicals*

Simplify the following expressions.

a. $\dfrac{\sqrt{x+1} - \dfrac{x}{2\sqrt{x+1}}}{x+1}$

b. $\left(\dfrac{1}{x+\sqrt{x^2+1}}\right)\left(1 + \dfrac{2x}{2\sqrt{x^2+1}}\right)$

Solution

a.
$$\dfrac{\sqrt{x+1} - \dfrac{x}{2\sqrt{x+1}}}{x+1} = \dfrac{\dfrac{2(x+1)}{2\sqrt{x+1}} - \dfrac{x}{2\sqrt{x+1}}}{x+1}$$

$$= \dfrac{\dfrac{2x+2-x}{2\sqrt{x+1}}}{x+1}$$

$$= \dfrac{x+2}{2\sqrt{x+1}}\left(\dfrac{1}{x+1}\right)$$

$$= \dfrac{x+2}{2(x+1)^{3/2}}$$

b.
$$\left(\dfrac{1}{x+\sqrt{x^2+1}}\right)\left(1 + \dfrac{2x}{2\sqrt{x^2+1}}\right)$$

$$= \left(\dfrac{1}{x+\sqrt{x^2+1}}\right)\left(1 + \dfrac{x}{\sqrt{x^2+1}}\right)$$

$$= \left(\dfrac{1}{x+\sqrt{x^2+1}}\right)\left(\dfrac{\sqrt{x^2+1}}{\sqrt{x^2+1}} + \dfrac{x}{\sqrt{x^2+1}}\right)$$

$$= \left(\dfrac{1}{x+\sqrt{x^2+1}}\right)\left(\dfrac{x+\sqrt{x^2+1}}{\sqrt{x^2+1}}\right)$$

$$= \dfrac{1}{\sqrt{x^2+1}}$$

EXAMPLE 5 *Simplifying an Expression with Radicals*

Simplify the following expression.

$$\frac{-x\left(\dfrac{2x}{2\sqrt{x^2+1}}\right)+\sqrt{x^2+1}}{x^2}+\left(\frac{1}{x+\sqrt{x^2+1}}\right)\left(1+\frac{2x}{2\sqrt{x^2+1}}\right)$$

Solution

From part b of Example 4, you already know that the second part of this sum simplifies to $1/\sqrt{x^2+1}$. The first part simplifies as follows.

$$\frac{-x\left(\dfrac{2x}{2\sqrt{x^2+1}}\right)+\sqrt{x^2+1}}{x^2}=\frac{-x^2}{x^2\sqrt{x^2+1}}+\frac{\sqrt{x^2+1}}{x^2}$$

$$=\frac{-x^2}{x^2\sqrt{x^2+1}}+\frac{x^2+1}{x^2\sqrt{x^2+1}}$$

$$=\frac{-x^2+x^2+1}{x^2\sqrt{x^2+1}}$$

$$=\frac{1}{x^2\sqrt{x^2+1}}$$

Therefore, the sum is

$$\frac{-x\left(\dfrac{2x}{2\sqrt{x^2+1}}\right)+\sqrt{x^2+1}}{x^2}+\left(\frac{1}{x+\sqrt{x^2+1}}\right)\left(1+\frac{2x}{2\sqrt{x^2+1}}\right)$$

$$=\frac{1}{x^2\sqrt{x^2+1}}+\frac{1}{\sqrt{x^2+1}}$$

$$=\frac{1}{x^2\sqrt{x^2+1}}+\frac{x^2}{x^2\sqrt{x^2+1}}$$

$$=\frac{x^2+1}{x^2\sqrt{x^2+1}}$$

$$=\frac{\sqrt{x^2+1}}{x^2}.$$

Note To check that the simplified expression in Example 5 is equivalent to the original expression, try substituting values of x into each expression. For instance, when you substitute $x=1$ into each expression, you obtain $\sqrt{2}$.

Rationalization Techniques

In working with quotients involving radicals, it is often convenient to move the radical expression from the denominator to the numerator, or vice versa. For example, we can move $\sqrt{2}$ from the denominator to the numerator in the following quotient by multiplying by $\sqrt{2}/\sqrt{2}$.

Radical in Denominator   *Rationalize*   *Radical in Numerator*

$$\frac{1}{\sqrt{2}} \quad \longrightarrow \quad \frac{1}{\sqrt{2}}\left(\frac{\sqrt{2}}{\sqrt{2}}\right) \longrightarrow \quad \frac{\sqrt{2}}{2}$$

This process is called **rationalizing the denominator.** A similar process is used to **rationalize the numerator.**

Rationalizing Techniques

1. If the denominator is $\sqrt{a}$, multiply by $\dfrac{\sqrt{a}}{\sqrt{a}}$.

Note   The success of the second and third rationalizing techniques stems from the following.

$$(\sqrt{a} - \sqrt{b})(\sqrt{a} + \sqrt{b}) = a - b$$

2. If the denominator is $\sqrt{a} - \sqrt{b}$, multiply by $\dfrac{\sqrt{a} + \sqrt{b}}{\sqrt{a} + \sqrt{b}}$.

3. If the denominator is $\sqrt{a} + \sqrt{b}$, multiply by $\dfrac{\sqrt{a} - \sqrt{b}}{\sqrt{a} - \sqrt{b}}$.

The same guidelines apply to rationalizing numerators.

EXAMPLE 6 *Rationalizing Denominators and Numerators*

a. $\dfrac{3}{\sqrt{12}} = \dfrac{3}{2\sqrt{3}} = \dfrac{3}{2\sqrt{3}}\left(\dfrac{\sqrt{3}}{\sqrt{3}}\right) = \dfrac{3\sqrt{3}}{2(3)} = \dfrac{\sqrt{3}}{2}$

b. $\dfrac{\sqrt{x+1}}{2} = \dfrac{\sqrt{x+1}}{2}\left(\dfrac{\sqrt{x+1}}{\sqrt{x+1}}\right) = \dfrac{x+1}{2\sqrt{x+1}}$

c. $\dfrac{1}{\sqrt{5}+\sqrt{2}} = \dfrac{1}{\sqrt{5}+\sqrt{2}}\left(\dfrac{\sqrt{5}-\sqrt{2}}{\sqrt{5}-\sqrt{2}}\right) = \dfrac{\sqrt{5}-\sqrt{2}}{5-2} = \dfrac{\sqrt{5}-\sqrt{2}}{3}$

d. $\dfrac{1}{\sqrt{x}-\sqrt{x+1}} = \dfrac{1}{\sqrt{x}-\sqrt{x+1}}\left(\dfrac{\sqrt{x}+\sqrt{x+1}}{\sqrt{x}+\sqrt{x+1}}\right)$

$$= \dfrac{\sqrt{x}+\sqrt{x+1}}{x-(x+1)} = -\sqrt{x}-\sqrt{x+1}$$

E X E R C I S E S   0.5

In Exercises 1–28, perform the indicated operations and simplify your answer.

1. $\dfrac{5}{x-1}+\dfrac{x}{x-1}$

2. $\dfrac{2x-1}{x+3}+\dfrac{1-x}{x+3}$

3. $\dfrac{2x}{x^2+2}-\dfrac{1-3x}{x^2+2}$

4. $\dfrac{5x+10}{2x-1}-\dfrac{2x+10}{2x-1}$

5. $\dfrac{4}{x}-\dfrac{3}{x^2}$

6. $\dfrac{5}{x-1}+\dfrac{3}{x}$

7. $\dfrac{2}{x+2}-\dfrac{1}{x-2}$

8. $\dfrac{x}{x^2+x-2}-\dfrac{1}{x+2}$

9. $\dfrac{5}{x-3}+\dfrac{3}{3-x}$

10. $\dfrac{x}{2-x}+\dfrac{2}{x-2}$

11. $\dfrac{A}{x-6}+\dfrac{B}{x+3}$

12. $\dfrac{Ax+B}{x^2+2}+\dfrac{C}{x-4}$

13. $-\dfrac{1}{x}+\dfrac{2}{x^2+1}$

14. $\dfrac{2}{x+1}+\dfrac{1-x}{x^2-2x+3}$

15. $\dfrac{-x}{(x+1)^{3/2}}+\dfrac{2}{(x+1)^{1/2}}$

16. $2\sqrt{x}(x-2)+\dfrac{(x-2)^2}{2\sqrt{x}}$

17. $\dfrac{2-t}{2\sqrt{1+t}}-\sqrt{1+t}$

18. $-\dfrac{\sqrt{x^2+1}}{x^2}+\dfrac{1}{\sqrt{x^2+1}}$

19. $\dfrac{1}{x^2-x-2}-\dfrac{x}{x^2-5x+6}$

20. $\dfrac{x-1}{x^2+5x+4}+\dfrac{2}{x^2-x-2}+\dfrac{10}{x^2+2x-8}$

21. $\dfrac{A}{x+1}+\dfrac{B}{(x+1)^2}+\dfrac{C}{x-2}$

22. $\dfrac{A}{x-5}+\dfrac{B}{x+5}+\dfrac{C}{(x+5)^2}$

23. $\left(2x\sqrt{x^2+1}-\dfrac{x^3}{\sqrt{x^2+1}}\right)\div(x^2+1)$

24. $\left(\sqrt{x^3+1}-\dfrac{3x^3}{2\sqrt{x^3+1}}\right)\div(x^3+1)$

25. $\dfrac{(x^2+2)^{1/2}-x^2(x^2+2)^{-1/2}}{x^2}$

26. $\dfrac{x(x+1)^{-1/2}-(x+1)^{1/2}}{x^2}$

27. $\dfrac{\dfrac{\sqrt{x+1}}{\sqrt{x}}-\dfrac{\sqrt{x}}{\sqrt{x+1}}}{2(x+1)}$

28. $\dfrac{\dfrac{2x^2}{3(x^2-1)^{2/3}}-(x^2-1)^{1/3}}{x^2}$

In Exercises 29–48, rationalize the numerator or denominator and simplify.

29. $\dfrac{3}{\sqrt{27}}$

30. $\dfrac{5}{\sqrt{10}}$

31. $\dfrac{\sqrt{2}}{3}$

32. $\dfrac{\sqrt{26}}{2}$

33. $\dfrac{x}{\sqrt{x-4}}$

34. $\dfrac{4y}{\sqrt{y+8}}$

35. $\dfrac{\sqrt{y^3}}{6y}$

36. $\dfrac{x\sqrt{x^2+4}}{3}$

37. $\dfrac{49(x-3)}{\sqrt{x^2-9}}$

38. $\dfrac{10(x+2)}{\sqrt{x^2-x-6}}$

39. $\dfrac{5}{\sqrt{14}-2}$

40. $\dfrac{13}{6+\sqrt{10}}$

41. $\dfrac{2x}{5-\sqrt{3}}$

42. $\dfrac{x}{\sqrt{2}+\sqrt{3}}$

43. $\dfrac{1}{\sqrt{6}+\sqrt{5}}$

44. $\dfrac{\sqrt{15}+3}{12}$

45. $\dfrac{\sqrt{3}-\sqrt{2}}{x}$

46. $\dfrac{x-1}{\sqrt{x}+x}$

47. $\dfrac{2x-\sqrt{4x-1}}{2x-1}$

48. $\dfrac{10}{\sqrt{x}+\sqrt{x+5}}$

49. *Installment Loan* Determine the monthly payment M for an installment loan of $P = \$10{,}000$ at an annual percentage rate of 14% ($r = 0.14$) for 5 years ($N = 60$ monthly payments) given the formula

$$M = P\left[\dfrac{r/12}{1-\left(\dfrac{1}{(r/12)+1}\right)^N}\right].$$

50. *Inventory* A retailer has determined that the cost C of ordering and storing x units of a product is

$$C = 6x + \dfrac{900{,}000}{x}.$$

(a) Write the expression for cost as a single fraction.
(b) Determine the cost for ordering and storing $x = 240$ units of this product.

$$\lim_{x \to c} f(x) = f(c)$$

1

Functions, Graphs, and Limits

When René Descartes developed the coordinate plane in the early 1600s, he revolutionized the way people thought about mathematics. Before Descartes' time, algebra and geometry were separate branches of mathematics, with little overlap. By coordinatizing geometry, Descartes allowed mathematicians to solve problems algebraically *and* graphically. In this chapter, you will review the basic concepts of Descartes' analytic geometry.

In the 1980s, the introduction of hand-held, affordable, graphing utilities brought about another revolution in the way people study mathematics. With this new technology, you can find and analyze mathematical models much more simply than was possible before.

$$y = f(x)$$

Descartes

1.1 The Cartesian Plane and the Distance Formula

The Cartesian Plane ■ *The Distance Formula* ■ *The Midpoint Formula* ■ *Extended Application: Computer Graphics*

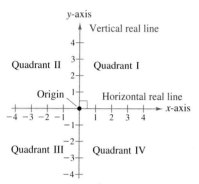

FIGURE 1.1   The Cartesian Plane

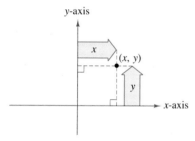

FIGURE 1.2

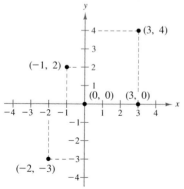

FIGURE 1.3

The Cartesian Plane

Just as you can represent real numbers by points on a real number line, you can represent ordered pairs of real numbers by points in a plane called the **rectangular coordinate system,** or the **Cartesian plane,** after the French mathematician René Descartes (1596–1650).

The Cartesian plane is formed by using two real lines intersecting at right angles, as shown in Figure 1.1. The horizontal real line is usually called the **x-axis,** and the vertical real line is usually called the **y-axis.** The point of intersection of these two axes is the **origin,** and the two axes divide the plane into four parts called **quadrants.**

Each point in the plane corresponds to an **ordered pair** (x, y) of real numbers x and y, called **coordinates** of the point. The **x-coordinate** represents the directed distance from the y-axis to the point, and the **y-coordinate** represents the directed distance from the x-axis to the point, as shown in Figure 1.2.

$$(x, y)$$

Directed distance from y-axis   Directed distance from x-axis

Note   The notation (x, y) denotes both a point in the plane and an open interval on the real line. The context will tell you which meaning is intended.

EXAMPLE 1 *Plotting Points in the Cartesian Plane*

Plot the points $(-1, 2)$, $(3, 4)$, $(0, 0)$, $(3, 0)$, and $(-2, -3)$.

Solution
To plot the point

$$(-1, 2),$$

x-coordinate y-coordinate

imagine a vertical line through -1 on the x-axis and a horizontal line through 2 on the y-axis. The intersection of these two lines is the point $(-1, 2)$. The other four points can be plotted in a similar way, and are shown in Figure 1.3.

The beauty of a rectangular coordinate system is that it allows you to visualize relationships between two variables. It would be difficult to overestimate the importance of Descartes's introduction of coordinates to the plane. Today, his ideas are in common use in virtually every scientific and business-related field. In Example 2, notice how much your intuition is enhanced by the use of a graphical presentation.

Real Life

EXAMPLE 2 *Sketching a Scatter Plot*

From 1981 through 1990, the amount A (in millions of dollars) spent on water skis in the United States is given in Table 1.1, where t represents the year. Sketch a scatter plot of the data. *(Source: National Sporting Goods Association)*

TABLE 1.1

t	1981	1982	1983	1984	1985	1986	1987	1988	1989	1990
A	123	106	133	146	125	132	148	160	96	96

Solution
To sketch a *scatter plot* of the data given in the table, you simply represent each pair of values by an ordered pair (t, A), and plot the resulting points, as shown in Figure 1.4. For instance, the first pair of values is represented by the ordered pair (1981, 123). Note that the break in the t-axis indicates that the numbers between 0 and 1981 have been omitted.

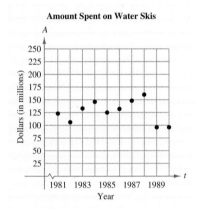

Amount Spent on Water Skis

FIGURE 1.4

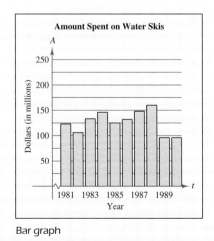

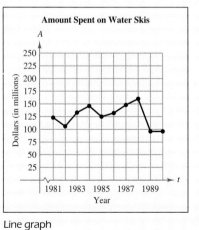

Bar graph   Line graph

TECHNOLOGY

The scatter plot in Example 2 is only one way to graphically represent the given data. Two other techniques are shown at right. The first is a *bar graph* and the second is a *line graph*. All three graphical representations were created with a computer. If you have access to computer graphing software, try using it to graphically represent the data given in Example 2.

Note In Example 2, you could have let $t = 1$ represent the year 1981. In that case, the horizontal axis would not have been broken, and the tick marks would have been labeled 1 through 10 (instead of 1981 through 1990).

The Distance Formula

Recall from the Pythagorean Theorem that, for a right triangle with hypotenuse of length c and sides of lengths a and b, you have

$$a^2 + b^2 = c^2, \qquad \textit{Pythagorean Theorem}$$

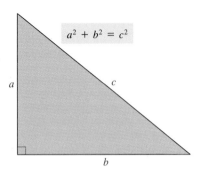

FIGURE 1.5   Pythagorean Theorem

as shown in Figure 1.5. (The converse is also true. That is, if $a^2 + b^2 = c^2$, then the triangle is a right triangle.)

Suppose you want to determine the distance d between two points (x_1, y_1) and (x_2, y_2) in the plane. With these two points, a right triangle can be formed, as shown in Figure 1.6. The length of the vertical side of the triangle is $|y_2 - y_1|$, and the length of the horizontal side is $|x_2 - x_1|$. By the Pythagorean Theorem, you can write

$$d^2 = |x_2 - x_1|^2 + |y_2 - y_1|^2$$
$$d = \sqrt{|x_2 - x_1|^2 + |y_2 - y_1|^2}$$
$$d = \sqrt{(x_2 - x_1)^2 + (y_2 - y_1)^2}.$$

This result is the **Distance Formula.**

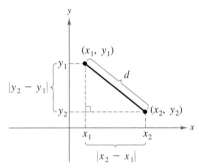

FIGURE 1.6   Distance Between Two Points

The Distance Formula

The distance d between the points (x_1, y_1) and (x_2, y_2) in the plane is

$$d = \sqrt{(x_2 - x_1)^2 + (y_2 - y_1)^2}.$$

EXAMPLE 3   *Finding a Distance*

Find the distance between the points $(-2, 1)$ and $(3, 4)$.

Solution
Let $(x_1, y_1) = (-2, 1)$ and $(x_2, y_2) = (3, 4)$. Then apply the Distance Formula as follows.

$$
\begin{aligned}
d &= \sqrt{(x_2 - x_1)^2 + (y_2 - y_1)^2} && \textit{Distance Formula}\\
&= \sqrt{[3 - (-2)]^2 + (4 - 1)^2} && \textit{Substitute for } x_1,\ y_1,\ x_2,\ \text{and } y_2.\\
&= \sqrt{(5)^2 + (3)^2} && \textit{Simplify.}\\
&= \sqrt{34}\\
&\approx 5.83 && \textit{Use a calculator.}
\end{aligned}
$$

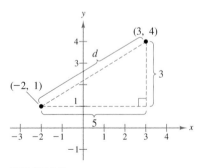

FIGURE 1.7

Note in Figure 1.7 that a distance of 5.83 looks about right.

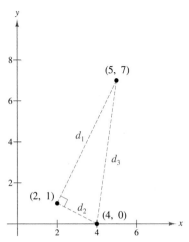

FIGURE 1.8

EXAMPLE 4 *Verifying a Right Triangle*

Use the Distance Formula to show that the points $(2, 1)$, $(4, 0)$, and $(5, 7)$ are vertices of a right triangle.

Solution

The three points are plotted in Figure 1.8. Using the Distance Formula, you can find the lengths of the three sides as follows.

$$d_1 = \sqrt{(5 - 2)^2 + (7 - 1)^2} = \sqrt{9 + 36} = \sqrt{45}$$
$$d_2 = \sqrt{(4 - 2)^2 + (0 - 1)^2} = \sqrt{4 + 1} = \sqrt{5}$$
$$d_3 = \sqrt{(5 - 4)^2 + (7 - 0)^2} = \sqrt{1 + 49} = \sqrt{50}$$

Because

$$d_1{}^2 + d_2{}^2 = 45 + 5 = 50 = d_3{}^2,$$

you can apply the converse of the Pythagorean Theorem to conclude that the triangle must be a right triangle. ▬

The figures provided with Examples 3 and 4 were not really essential to the solution. *Nevertheless*, we strongly recommend that you develop the habit of including sketches with your solutions—even if they are not required.

Real Life

EXAMPLE 5 *Finding the Length of a Pass*

In a football game, a quarterback throws a pass from the 5-yard line, 20 yards from the sideline. The pass is caught by a wide receiver on the 45-yard line, 50 yards from the same sideline, as shown in Figure 1.9. How long was the pass?

Solution

You can find the length of the pass by finding the distance between the points $(20, 5)$ and $(50, 45)$.

$$d = \sqrt{(50 - 20)^2 + (45 - 5)^2} \qquad \textit{Distance Formula}$$
$$= \sqrt{900 + 1600}$$
$$= 50 \qquad\qquad\qquad\qquad \textit{Simplify.}$$

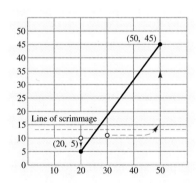

FIGURE 1.9

Thus, the pass was 50 yards long. ▬

Note In Example 5, the scale along the goal line does not normally appear on a football field. However, when you use coordinate geometry to solve real-life problems, you are free to place the coordinate system in any way that is convenient to the solution of the problem.

The Midpoint Formula

To find the **midpoint** of the line segment that joins two points in a coordinate plane, you can simply find the average values of the respective coordinates of the two endpoints.

The Midpoint Formula

The midpoint of the segment joining the points (x_1, y_1) and (x_2, y_2) is

$$\text{Midpoint} = \left(\frac{x_1 + x_2}{2}, \frac{y_1 + y_2}{2} \right).$$

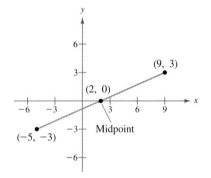

FIGURE 1.10

EXAMPLE 6 *Finding a Segment's Midpoint*

Find the midpoint of the line segment joining the points $(-5, -3)$ and $(9, 3)$, as shown in Figure 1.10.

Solution

Let $(x_1, y_1) = (-5, -3)$ and $(x_2, y_2) = (9, 3)$.

$$\text{Midpoint} = \left(\frac{x_1 + x_2}{2}, \frac{y_1 + y_2}{2} \right) \qquad \textit{Midpoint Formula}$$

$$= \left(\frac{-5 + 9}{2}, \frac{-3 + 3}{2} \right) \qquad \textit{Substitute for } x_1, y_1, x_2, \textit{ and } y_2.$$

$$= (2, 0) \qquad \textit{Simplify.}$$

Real Life

EXAMPLE 7 *Estimating Annual Sales*

Eastman Kodak had annual sales of $17.0 billion in 1988 and $20.2 billion in 1992. Without knowing any additional information, what would you estimate the 1990 sales to have been? *(Source: Eastman Kodak Company)*

Solution

One solution to the problem is to assume that sales followed a linear pattern. With this assumption, you can estimate the 1990 sales by finding the midpoint of the segment connecting the points $(1988, 17.0)$ and $(1992, 20.2)$.

$$\text{Midpoint} = \left(\frac{1988 + 1992}{2}, \frac{17.0 + 20.2}{2} \right) = (1990, 18.6)$$

Hence, you would estimate the 1990 sales to have been about $18.6 billion, as shown in Figure 1.11. (The actual 1990 sales were $18.9 billion.)

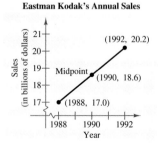

Eastman Kodak's Annual Sales

FIGURE 1.11

Extended Application: Computer Graphics

The photo at the right shows a scene from *Terminator 2*. The pilot's expression is mirrored on the head of the computer-animated polyalloy robot, which takes on human form after pouring itself through a hole in the helicopter window. This movie was the first box-office success to use extensive computer graphics. Much of computer graphics consists of transformations of points in two- and three-dimensional space.

Shooting Star

EXAMPLE 8 *Translating Points in the Plane*

Figure 1.12 shows the vertices of a parallelogram. Find the vertices of the parallelogram after it has been translated two units down and four units to the right.

Solution

To translate each vertex two units down, subtract 2 from each y-coordinate. To translate each vertex four units to the right, add 4 to each x-coordinate.

Original Point	Translated Point
$(1, 0)$	$(1 + 4, 0 - 2) = (5, -2)$
$(3, 2)$	$(3 + 4, 2 - 2) = (7, 0)$
$(3, 6)$	$(3 + 4, 6 - 2) = (7, 4)$
$(1, 4)$	$(1 + 4, 4 - 2) = (5, 2)$

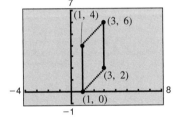

FIGURE 1.12

Discussion Problem Extending the Example

Example 8 shows how to translate points in a coordinate plane. How are the following transformed points related to the original points?

Original Point	Transformed Point
(x, y)	$(-x, y)$
(x, y)	$(x, -y)$
(x, y)	$(-x, -y)$

Warm Up

The following warm-up exercises involve skills that were covered in earlier sections. You will use these skills in the exercise set for this section.

In Exercises 1–6, simplify the expression.

1. $\sqrt{(3-6)^2 + [1-(-5)]^2}$

2. $\sqrt{(-2-0)^2 + [-7-(-3)]^2}$

3. $\dfrac{5+(-4)}{2}$

4. $\dfrac{-3+(-1)}{2}$

5. $\sqrt{27} + \sqrt{12}$

6. $\sqrt{8} - \sqrt{18}$

In Exercises 7–10, solve for *x* or *y*.

7. $\sqrt{(3-x)^2 + (7-4)^2} = \sqrt{45}$

8. $\sqrt{(6-2)^2 + (-2-y)^2} = \sqrt{52}$

9. $\dfrac{x+(-5)}{2} = 7$

10. $\dfrac{-7+y}{2} = -3$

EXERCISES 1.1

means that technology can help you solve or check the exercise(s).

In Exercises 1–4, (a) find the length of each side of the right triangle, and (b) show that these lengths satisfy the Pythagorean Theorem.

*1.

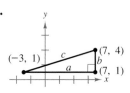

2.

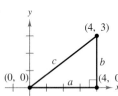

3.

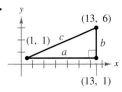

4.

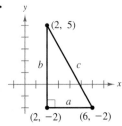

In Exercises 5–12, (a) plot the points, (b) find the distance between the points, and (c) find the midpoint of the line segment joining the points.

5. $(3, 1), (5, 5)$

6. $(-3, 2), (3, -2)$

7. $\left(\frac{1}{2}, 1\right), \left(-\frac{3}{2}, -5\right)$

8. $\left(\frac{2}{3}, -\frac{1}{3}\right), \left(\frac{5}{6}, 1\right)$

9. $(2, 2), (4, 14)$

10. $(-3, 7), (1, -1)$

11. $\left(1, \sqrt{3}\right), (-1, 1)$

12. $(-2, 0), \left(0, \sqrt{2}\right)$

In Exercises 13–16, show that the points form the vertices of the indicated figure. (A rhombus is a quadrilateral whose sides have the same length.)

Vertices	*Figure*
13. $(0, 1), (3, 7), (4, -1)$	Right triangle
14. $(1, -3), (3, 2), (-2, 4)$	Isosceles triangle
15. $(0, 0), (1, 2), (2, 1), (3, 3)$	Rhombus
16. $(0, 1), (3, 7), (4, 4), (1, -2)$	Parallelogram

** The *answers* to the odd-numbered exercises are given in the back of the text. *Worked-out solutions* to the odd-numbered exercises are given in the *Student Solutions Guide*.

In Exercises 17–20, use the Distance Formula to determine whether the points are collinear.

17. $(0, -4), (2, 0), (3, 2)$ **18.** $(0, 4), (7, -6), (-5, 11)$

19. $(-2, -6), (1, -3), (5, 2)$ **20.** $(-1, 1), (3, 3), (5, 5)$

In Exercises 21 and 22, find x so that the distance between the points is 5.

21. $(1, 0), (x, -4)$ **22.** $(2, -1), (x, 2)$

In Exercises 23 and 24, find y so that the distance between the points is 8.

23. $(0, 0), (3, y)$ **24.** $(5, 1), (5, y)$

25. Use the Midpoint Formula repeatedly to find the three points that divide the segment joining (x_1, y_1) and (x_2, y_2) into four equal parts.

26. Show that $\left(\frac{1}{3}[2x_1 + x_2], \frac{1}{3}[2y_1 + y_2]\right)$ is one of the points of trisection of the line segment joining (x_1, y_1) and (x_2, y_2). Then, find the second point of trisection by finding the midpoint of the segment joining

$$\left(\frac{1}{3}[2x_1 + x_2], \frac{1}{3}[2y_1 + y_2]\right) \text{ and } (x_2, y_2).$$

27. Use Exercise 25 to find the points that divide the line segment joining the given points into four equal parts.
(a) $(1, -2), (4, -1)$ (b) $(-2, -3), (0, 0)$

28. Use Exercise 26 to find the points of trisection of the line segment joining the given points.
(a) $(1, -2), (4, 1)$ (b) $(-2, -3), (0, 0)$

29. *Building Dimensions* The base and height of the trusses for the roof of a house are 32 feet and 5 feet, respectively (see figure).
(a) Find the distance from the eaves to the peak of the roof.
(b) The length of the house is 40 feet. Use the result of part (a) to find the number of square feet of roofing.

30. *Wire Length* A guy wire is stretched from a broadcasting tower at a point 200 feet above the ground to an anchor 125 feet from the base (see figure). How long is the wire?

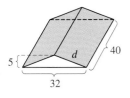

FIGURE FOR 29

FIGURE FOR 30

In Exercises 31 and 32, use a scatter plot, a bar graph, or a line graph to represent the data. Describe any trends that appear.

31. *Cable TV* For 1983–1992, the number (in millions) of cable television subscribers in the United States is given in the table. *(Source: Television Bureau of Advertising, Inc.)*

Year	1983	1984	1985	1986	1987
Subscribers	25.0	37.3	39.9	42.2	44.9

Year	1988	1989	1990	1991	1992
Subscribers	48.6	52.6	54.9	55.8	57.2

32. *Magazine Purchases* From 1981 through 1990, the amount (in billions of dollars) spent on magazines, newspapers, and sheet music in the United States is given in the table. *(Source: U.S. Bureau of Economic Analysis)*

Year	1981	1982	1983	1984	1985
Amount	11.0	11.4	12.0	12.7	13.2

Year	1986	1987	1988	1989	1990
Amount	13.9	15.4	17.0	18.5	19.2

Dow Jones Average In Exercises 33 and 34, use the figure showing the Dow Jones Industrial Average for common stocks. *(Source: Dow Jones, Inc.)*

33. Estimate the Dow Jones Average for the following dates.
(a) August 1990   (b) November 1990
(c) December 1991   (d) March 1992

34. Estimate the percent increase or decrease in the Dow Jones Industrial Average (a) from August 1990 to November 1990 and (b) from December 1991 to June 1992.

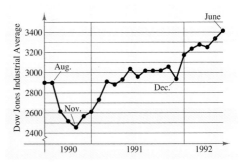

FIGURE FOR 33 AND 34

Housing Starts In Exercises 35 and 36, use the figure showing the number of housing starts (in millions) in the United States from 1990 through 1992. *(Source: U.S. Bureau of Census, Construction Reports)*

35. Estimate the number of housing starts for the following.
   (a) February 1990   (b) October 1990
   (c) January 1991   (d) April 1992

36. Estimate the percent increase or decrease in housing starts (a) from January 1990 to January 1991, and (b) from January 1991 to March 1992.

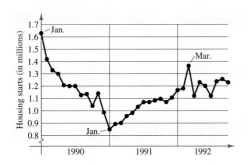

FIGURE FOR 35 AND 36

Research Project In Exercises 37 and 38, (a) use the Midpoint Formula to estimate the revenue and profit of the company in 1990. (b) Then use your school's library or some other reference source to find the actual revenue and profit for 1990. (c) Did the revenue and profit increase in a linear pattern from 1988 to 1992? Explain your reasoning. (d) What were the company's expenses during each of the given years? (e) How would you rate the company's growth from 1988 to 1992? *(Source: The Gillette Company and Helene Curtis)*

37. The Gillette Company

Year	1988	1990	1992
Revenue (millions of $)	3581.2	?	5162.8
Profit (millions of $)	268.5	?	513.4

38. Helene Curtis

Year	1988	1990	1992
Revenue (millions of $)	629.2	?	1167.8
Profit (millions of $)	14.4	?	22.1

Computer Graphics In Exercises 39 and 40, the red figure is translated to a new position in the plane to form the blue figure. (a) Find the vertices of the transformed figure. (b) Then use a graphing utility to draw both figures.

39.   **40.**

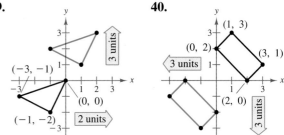

41. *Research Project* Use your school's library or some other reference source to collect data about a small business. Organize the data graphically. Then, interpret the data.

Graphs of Equations

The Graph of an Equation

In Section 1.1, you used a coordinate system to graphically represent the relationship between two quantities. There, the graphical picture consisted of a collection of points in a coordinate plane (see Example 2 in Section 1.1).

Frequently, a relationship between two quantities is expressed as an equation. For instance, degrees on the Fahrenheit scale are related to degrees on the Celsius scale by the equation $F = \frac{9}{5}C + 32$. In this section, you will study some basic procedures for sketching the graph of such an equation. The **graph** of an equation is the set of all points that are solutions of the equation.

EXAMPLE 1 *Sketching the Graph of an Equation*

Sketch the graph of $y = 7 - 3x$.

Solution

The simplest way to sketch the graph of an equation is the *point-plotting method*. With this method, you construct a table of values that consists of several solution points of the equation, as shown in Table 1.2. For instance, when $x = 0$,

$$y = 7 - 3(0) = 7,$$

which implies that $(0, 7)$ is a solution point of the graph.

TABLE 1.2

x	0	1	2	3	4
$y = 7 - 3x$	7	4	1	−2	−5

From the table, it follows that $(0, 7)$, $(1, 4)$, $(2, 1)$, $(3, -2)$, and $(4, -5)$ are solution points of the equation. After plotting these points, you can see that they appear to lie on a line, as shown in Figure 1.13. The graph of the equation is the line that passes through the five plotted points. ▬

Note Even though we refer to the sketch shown in Figure 1.13 as the graph of $y = 7 - 3x$, it actually represents only a *portion* of the graph. The entire graph is a line that would extend off the page.

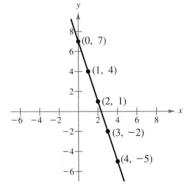

FIGURE 1.13 Solution Points for $y = 7 - 3x$

EXAMPLE 2 *Sketching the Graph of an Equation*

Sketch the graph of $y = x^2 - 2$.

Solution
Begin by constructing a table of values, as shown in Table 1.3.

TABLE 1.3

x	−2	−1	0	1	2	3
$y = x^2 - 2$	2	−1	−2	−1	2	7

Next, plot the points given in the table, as shown in Figure 1.14(a). Finally, connect the points by a smooth curve, as shown in Figure 1.14(b).

Note The graph shown in Example 2 is a **parabola.** The graph of any second-degree equation of the form

$$y = ax^2 + bx + c, \qquad a \neq 0$$

has a similar shape. If $a > 0$, the parabola opens upward, as in Figure 1.14(b), and if $a < 0$, the parabola opens downward.

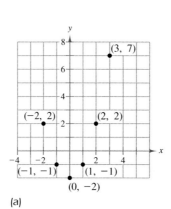

(a)

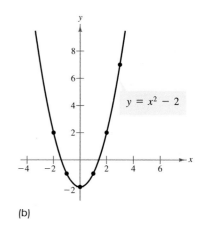

(b)

FIGURE 1.14

The point-plotting technique demonstrated in Examples 1 and 2 is easy to use, but it does have some shortcomings. With too few solution points, you can badly misrepresent the graph of a given equation. For instance, how would you connect the four points in Figure 1.15? Without further information, any one of the three graphs in Figure 1.16 would be reasonable.

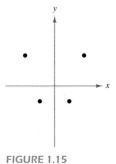

FIGURE 1.15

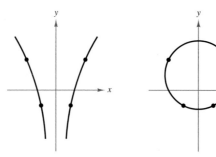

FIGURE 1.16

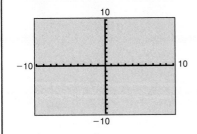

■ TECHNOLOGY

Creating a Viewing Rectangle

A **viewing rectangle** for a graph is a rectangular portion of the coordinate plane. A viewing rectangle is determined by six values: the minimum x-value, the maximum x-value, the x-scale, the minimum y-value, the maximum y-value, and the y-scale. When you enter these six values into a graphing utility, you are setting the **range** or **window.** Some graphing utilities have a standard viewing rectangle, as shown at the left.

By choosing different viewing rectangles for a graph, it is possible to obtain very different impressions of the graph's shape. For instance, below are four different viewing rectangles for the graph of

$$y = 0.1x^4 - x^3 + 2x^2.$$

Of these, the view shown in part (a) is the most complete.

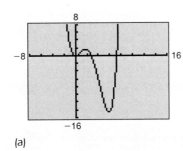

(a)

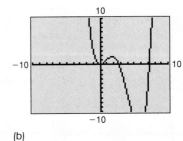

(b)

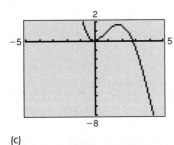

(c)

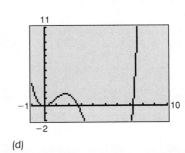

(d)

On most graphing utilities, the display screen is two-thirds as high as it is wide. On such screens, you can obtain a graph with a true geometric perspective by using a **square setting**—one in which

$$\frac{Y_{max} - Y_{min}}{X_{max} - X_{min}} = \frac{2}{3}.$$

One such setting is shown at the left. Notice that the x and y tick marks are equally spaced on a square setting, but not on a standard setting.

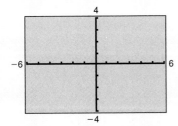

Intercepts of a Graph

It is often easy to determine the solution points that have zero as either the x-coordinate or the y-coordinate. These points are called **intercepts** because they are the points at which the graph intersects the x- or y-axis.

It is possible for a graph to have no intercepts or several intercepts, as shown in Figure 1.17.

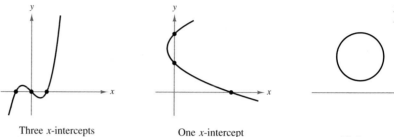

No x-intercept
One y-intercept

Three x-intercepts
One y-intercept

One x-intercept
Two y-intercepts

No intercepts

FIGURE 1.17

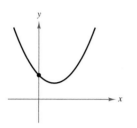

$y = x^3 - 4x$

$(-2, 0)$ $(0, 0)$ $(2, 0)$

FIGURE 1.18

Finding Intercepts

1. To find x-intercepts, let y be zero and solve the equation for x.
2. To find y-intercepts, let x be zero and solve the equation for y.

EXAMPLE 3 *Finding x- and y-Intercepts*

Find the x- and y-intercepts for the graphs of the following equations.

a. $y = x^3 - 4x$

b. $y^2 - 3 = x$

Solution

a. Let $y = 0$. Then $0 = x(x^2 - 4)$ has solutions $x = 0$ and $x = \pm 2$.

x-intercepts: $(0, 0)$, $(2, 0)$, $(-2, 0)$

Let $x = 0$. Then $y = 0$.

y-intercept: $(0, 0)$ *(See Figure 1.18.)*

b. Let $y = 0$. Then $-3 = x$.

x-intercept: $(-3, 0)$

Let $x = 0$. Then $y^2 - 3 = 0$ has solutions $y = \pm\sqrt{3}$.

y-intercepts: $\left(0, \sqrt{3}\right)$, $\left(0, -\sqrt{3}\right)$ *(See Figure 1.19.)*

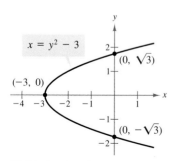

$x = y^2 - 3$

$(0, \sqrt{3})$

$(-3, 0)$

$(0, -\sqrt{3})$

FIGURE 1.19

TECHNOLOGY

Zooming in to Find Intercepts

You can use the **zoom** feature of a graphing utility to approximate the x-intercepts of a graph. Suppose you want to approximate the x-intercept(s) of the graph of $y = 2x^3 - 3x + 2$. Begin by graphing the equation, as shown below in part (a). From the viewing rectangle shown, the graph appears to have only one x-intercept. This intercept lies between -2 and -1. By zooming in on the intercept, you can improve the approximation, as shown in part (b). To three decimal places, the solution is $x \approx -1.476$.

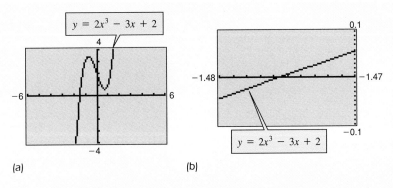

(a)   (b)

Here are some suggestions for using the zoom feature.

1. With each successive zoom-in, adjust the x-scale so that the viewing rectangle shows at least one tick mark on each side of the x-intercept.

2. The error in your approximation will be less than the distance between two scale marks.

3. The **trace** feature can usually be used to add one more decimal place of accuracy without changing the viewing rectangle.

Part (a) below shows the graph of $y = x^2 - 5x + 3$. Parts (b) and (c) show "zoom-in views" of the two intercepts. From these views, you can approximate the x-intercepts to be $x \approx 0.697$ and $x \approx 4.303$.

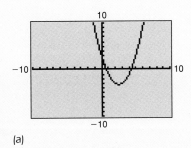

(a)

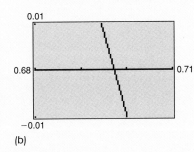

(b)

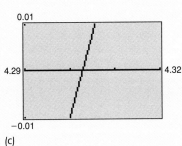

(c)

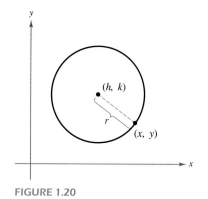

FIGURE 1.20

Circles

Throughout this course, you will learn to recognize several types of graphs from their equations. For instance, you should recognize that the graph of a second-degree equation of the form $y = ax^2 + bx + c$ is a parabola (see Example 2). Another easily recognized graph is that of a **circle.**

Consider the circle shown in Figure 1.20. A point (x, y) is on the circle if and only if its distance from the center (h, k) is r. By the Distance Formula,

$$\sqrt{(x - h)^2 + (y - k)^2} = r.$$

By squaring both sides of this equation, you obtain the **standard form of the equation of a circle.**

Standard Form of the Equation of a Circle

The point (x, y) lies on the circle of radius r and center (h, k) if and only if

$$(x - h)^2 + (y - k)^2 = r^2.$$

From this result, you can see that the standard form of the equation of a circle with its center at the origin, $(h, k) = (0, 0)$, is simply

$$x^2 + y^2 = r^2. \qquad \text{\textit{Circle with center at origin}}$$

EXAMPLE 4 *Finding the Equation of a Circle*

The point $(3, 4)$ lies on a circle whose center is at $(-1, 2)$, as shown in Figure 1.21. Find the standard form of the equation for this circle.

Solution
The radius of the circle is the distance between $(-1, 2)$ and $(3, 4)$.

$$r = \sqrt{[3 - (-1)]^2 + (4 - 2)^2} \qquad \text{\textit{Distance Formula}}$$
$$= \sqrt{16 + 4} \qquad \text{\textit{Simplify.}}$$
$$= \sqrt{20} \qquad \text{\textit{Radius}}$$

Using $(h, k) = (-1, 2)$, the standard form of the equation of the circle is

$$(x - h)^2 + (y - k)^2 = r^2$$
$$[x - (-1)]^2 + (y - 2)^2 = \left(\sqrt{20}\right)^2 \qquad \text{\textit{Substitute for h, k, and r.}}$$
$$(x + 1)^2 + (y - 2)^2 = 20. \qquad \text{\textit{Standard form}}$$

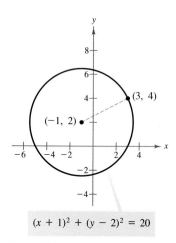

$(x + 1)^2 + (y - 2)^2 = 20$

FIGURE 1.21

The general form of the equation of a circle is

$$Ax^2 + Ay^2 + Dx + Ey + F = 0, \qquad A \neq 0 \qquad \text{\textit{General form}}$$

To change from general form to standard form, you can use a process called **completing the square,** as demonstrated in Example 5.

EXAMPLE 5 *Completing the Square*

Sketch the graph of the circle whose general equation is

$$4x^2 + 4y^2 + 20x - 16y + 37 = 0.$$

Solution

First divide by 4 so that the coefficients of x^2 and y^2 are both 1.

$$4x^2 + 4y^2 + 20x - 16y + 37 = 0 \qquad \text{\textit{General form}}$$

$$x^2 + y^2 + 5x - 4y + \tfrac{37}{4} = 0 \qquad \text{\textit{Divide by 4.}}$$

$$(x^2 + 5x + \quad) + (y^2 - 4y + \quad) = -\tfrac{37}{4} \qquad \text{\textit{Group terms.}}$$

$$\left(x^2 + 5x + \tfrac{25}{4}\right) + (y^2 - 4y + 4) = -\tfrac{37}{4} + \tfrac{25}{4} + 4 \qquad \begin{array}{l}\text{\textit{Complete the}}\\\text{\textit{square.}}\end{array}$$

$$\underbrace{\qquad}_{(\text{Half})^2} \qquad \underbrace{\qquad}_{(\text{Half})^2}$$

$$\left(x + \tfrac{5}{2}\right)^2 + (y - 2)^2 = 1 \qquad \text{\textit{Standard form}}$$

From the standard form, you can see that the circle is centered at $\left(-\tfrac{5}{2}, 2\right)$ and has a radius of 1, as shown in Figure 1.22.

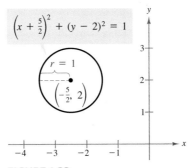

FIGURE 1.22

The general equation $Ax^2 + Ay^2 + Dx + Ey + F = 0$ may not always represent a circle. In fact, such an equation will have no solution points if the procedure of completing the square yields the impossible result

$$(x - h)^2 + (y - k)^2 = \text{negative number.} \qquad \text{\textit{No solution points}}$$

TECHNOLOGY

To graph a circle on a graphing utility, you can solve its equation for y and sketch the top and bottom halves of the circle separately. For instance, you can graph the circle $(x + 1)^2 + (y - 2)^2 = 20$ by graphing the following equations.

$$y = 2 + \sqrt{20 - (x + 1)^2}$$
$$y = 2 - \sqrt{20 - (x + 1)^2}$$

If you want the result to appear circular, you need to use a square setting, as shown below.

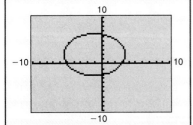

Standard setting

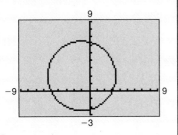

Square setting

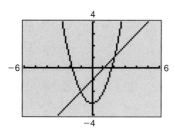

FIGURE 1.23

Points of Intersection

A **point of intersection** of two graphs is an ordered pair that is a solution point of both graphs. For instance, Figure 1.23 shows that the graphs of

$$y = x^2 - 3 \quad \text{and} \quad y = x - 1$$

have two points of intersection: $(2, 1)$ and $(-1, -2)$. To find the points analytically, set the two y-values equal to each other and solve the equation

$$x^2 - 3 = x - 1$$

for x.

A common business application that involves points of intersection is **break-even analysis.** The marketing of a new product typically requires an initial investment. When sufficient units have been sold so that the total revenue has offset the total cost, the sale of the product has reached the **break-even point.** The **total cost** of producing x units of a product is denoted by C, and the **total revenue** from the sale of x units of the product is denoted by R. Thus, you can find the break-even point by setting the cost C equal to the revenue R, and solving for x.

Real Life

EXAMPLE 6 *Finding a Break-Even Point*

A business manufactures a product at a cost of $0.65 per unit and sells the product for $1.20 per unit. The company's initial investment to produce the product was $10,000. How many units must the company sell to break even?

Solution
The total cost of producing x units of the product is given by

$$C = 0.65x + 10,000. \qquad \textit{Cost equation}$$

The total revenue from the sale of x units is given by

$$R = 1.2x. \qquad \textit{Revenue equation}$$

To find the break-even point, set the cost equal to the revenue and solve for x.

$R = C$	*Set revenue equal to cost.*
$1.2x = 0.65x + 10,000$	*Substitute for R and C.*
$0.55x = 10,000$	*Subtract $0.65x$ from both sides.*
$x = \dfrac{10,000}{0.55}$	*Divide both sides by 0.55.*
$x \approx 18,182$	*Use a calculator.*

Thus the company must sell 18,182 units before it breaks even. This result is shown graphically in Figure 1.24.

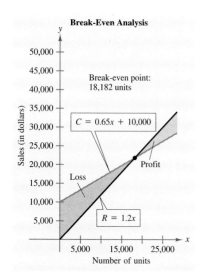

FIGURE 1.24

Mathematical Models

In this text, you will see many examples of the use of equations as **mathematical models** of real-life phenomena. In developing a mathematical model to represent actual data, you should strive for two (often conflicting) goals—accuracy and simplicity.

Real Life

EXAMPLE 7 *Using Mathematical Models*

Table 1.4 shows the annual sales, in millions of dollars, for K Mart Corporation and Wal-Mart Stores for 1985 through 1992. In the summer of 1993, the publication of *Value Line* listed projected 1993 sales for the companies as $41,000 million and $68,470 million. How do you think these projections were obtained? *(Source: K Mart Corporation and Wal-Mart Stores)*

TABLE 1.4

Year	1985	1986	1987	1988	1989	1990	1991	1992
K Mart	22,420	23,812	25,627	27,301	29,533	32,070	34,580	37,725
Wal-Mart	8,452	11,909	15,959	20,649	25,811	32,602	43,887	55,550

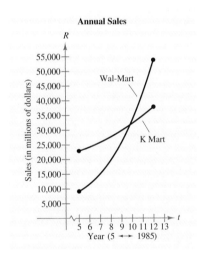

Annual Sales

Sales (in millions of dollars)

Year (5 ↔ 1985)

FIGURE 1.25

Solution

The projections were obtained by using past revenues to predict future revenues. The past revenues were modeled by equations that were found by a statistical procedure called least-squares regression analysis.

$$R = 131.8t^2 - 67.9t + 19,494.6, \qquad 5 \le t \le 12 \qquad \text{K Mart}$$

$$R = 28.72t^3 + 5263.8, \qquad 5 \le t \le 12 \qquad \text{Wal-Mart}$$

Using $t = 13$ to represent 1993, you can predict the 1993 revenues to be

$$R = 131.8(13)^2 - 67.9(13) + 19,494.6 = \$40,886 \qquad \text{K Mart}$$

$$R = 28.72(13)^3 + 5263.8 = \$68,362. \qquad \text{Wal-Mart}$$

These two projections are close to those projected by *Value Line*. The graphs of the two models are shown in Figure 1.25.

To test the accuracy of a model, you can compare the actual data with the values given by the model. For instance, Table 1.5 compares the actual K Mart revenues with those given by the model.

TABLE 1.5

Year	1985	1986	1987	1988	1989	1990	1991	1992
Actual	22,420	23,812	25,627	27,301	29,533	32,070	34,580	37,725
Model	22,450	23,832	25,478	27,387	29,559	31,996	34,696	37,659

Much of your study of calculus will center around the behavior of the graphs of mathematical models. Figure 1.26 shows the graphs of six basic algebraic equations. Familiarity with these graphs will help you in the creation and use of mathematical models.

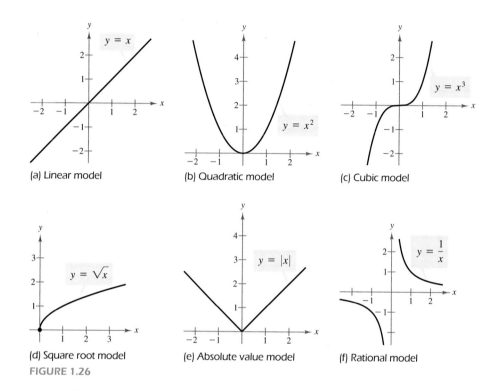

(a) Linear model

(b) Quadratic model

(c) Cubic model

(d) Square root model

(e) Absolute value model

(f) Rational model

FIGURE 1.26

Discussion Problem   Graphical, Numerical, and Analytical Solutions

Most problems in calculus can be solved in a variety of ways. Often, you can solve a problem graphically, numerically (using a table), and analytically. For instance, Example 6 compares a graphical and analytical approach to finding points of intersection.

In Example 7, suppose you were asked to find the time that Wal-Mart's revenues exceeded K Mart's revenues. Explain how to use *each* of the three approaches to answer the question. For this question, which approach do you think is best? Explain.

Suppose you answered the question and obtained $t = 9.66$. What date does this represent—August 1989 or August 1990? Explain.

Warm Up

The following warm-up exercises involve skills that were covered in earlier sections. You will use these skills in the exercise set for this section.

In Exercises 1–6, solve for *y*.

1. $5y - 12 = x$

2. $-y = 15 - x$

3. $x^3 y + 2y = 1$

4. $x^2 + x - y^2 - 6 = 0$

5. $(x - 2)^2 + (y + 1)^2 = 9$

6. $(x + 6)^2 + (y - 5)^2 = 81$

In Exercises 7–10, factor the expression.

7. $x^2 - 3x + 2$

8. $x^2 + 5x + 6$

9. $y^2 - 3y + \frac{9}{4}$

10. $y^2 - 7y + \frac{49}{4}$

E X E R C I S E S   1.2   means that technology can help you solve or check the exercise(s).

In Exercises 1–4, determine whether the points are solution points of the given equation.

1. $2x - y - 3 = 0$

  (a) $(1, 2)$ (b) $(1, -1)$ (c) $(4, 5)$

2. $x^2 + y^2 = 4$

  (a) $(1, -\sqrt{3})$ (b) $\left(\frac{1}{2}, -1\right)$ (c) $\left(\frac{3}{2}, \frac{7}{2}\right)$

3. $x^2 y + x^2 - 5y = 0$

  (a) $\left(0, \frac{1}{5}\right)$ (b) $(2, 4)$ (c) $(-2, -4)$

4. $x^2 - xy + 4y = 3$

  (a) $(0, 2)$ (b) $\left(-2, -\frac{1}{6}\right)$ (c) $(3, -6)$

In Exercises 5–10, match the equation with its graph. Use a graphing utility, set with a square setting, to confirm your result. [The graphs are labeled (a)–(f).]

5. $y = x - 2$

6. $y = -\frac{1}{2}x + 2$

7. $y = x^2 + 2x$

8. $y = \sqrt{9 - x^2}$

9. $y = |x| - 2$

10. $y = x^3 - x$

(a)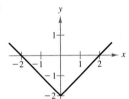

(b)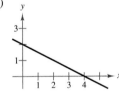

(c)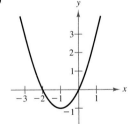

(d)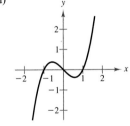

(e)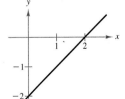

(f)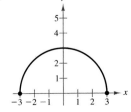

In Exercises 11–20, find the intercepts of the graph of the given equation.

11. $2x - y - 3 = 0$ **12.** $y = (x - 1)(x - 3)$

13. $y = x^2 + x - 2$ **14.** $y^2 = x^3 - 4x$

15. $y = x^2\sqrt{9 - x^2}$ **16.** $xy = 4$

17. $y = \dfrac{x^2 - 4}{x - 2}$ **18.** $y = \dfrac{x^2 + 3x}{(3x + 1)^2}$

19. $x^2y - x^2 + 4y = 0$ **20.** $y = x^2 + 1$

 In Exercises 21–38, sketch the graph of the equation and plot the intercepts.

21. $y = x$ **22.** $y = x + 3$

23. $y = -3x + 2$ **24.** $y = 2x - 3$

25. $y = -3 - x^2$ **26.** $y = x^2 + 3$

27. $y = x^3 + 2$ **28.** $y = 1 - x^3$

29. $y = (x + 2)^2$ **30.** $y = (x - 1)^2$

31. $y = \sqrt{x + 1}$ **32.** $y = -\sqrt[3]{x}$

33. $y = |x + 1|$ **34.** $y = -|x - 2|$

35. $y = \dfrac{1}{x - 3}$ **36.** $y = \dfrac{1}{x^2 + 1}$

37. $x = y^2 - 4$ **38.** $x = 4 - y^2$

In Exercises 39–46, write the general form of the equation of the circle.

39. Center: $(0, 0)$; radius: 3

40. Center: $(0, 0)$; radius: 5

41. Center: $(2, -1)$; radius: 4

42. Center: $(-4, 3)$; radius: $\frac{5}{8}$

43. Center: $(-1, 2)$; solution point: $(0, 0)$

44. Center: $(3, -2)$; solution point: $(-1, 1)$

45. Endpoints of a diameter: $(3, 3), (-3, 3)$

46. Endpoints of a diameter: $(-4, -1), (4, 1)$

 In Exercises 47–54, complete the square to write the equation of the circle in standard form. Then use a graphing utility to graph the circle.

47. $x^2 + y^2 - 2x + 6y + 6 = 0$

48. $x^2 + y^2 - 2x + 6y - 15 = 0$

49. $x^2 + y^2 + 4x + 6y - 3 = 0$

50. $3x^2 + 3y^2 - 6y - 1 = 0$

51. $2x^2 + 2y^2 - 2x - 2y - 3 = 0$

52. $4x^2 + 4y^2 - 4x + 2y - 1 = 0$

53. $16x^2 + 16y^2 + 16x + 40y - 7 = 0$

54. $x^2 + y^2 - 4x + 2y + 3 = 0$

 In Exercises 55–64, find the points of intersection (if any) of the graphs of the equations. Use a graphing utility to check your results.

55. $x + y = 2, \ 2x - y = 1$

56. $2x - 3y = 13, \ 5x + 3y = 1$

57. $x + y = 7, \ 3x - 2y = 11$

58. $x^2 + y^2 = 25, \ 2x + y = 10$

59. $x^2 + y^2 = 5, \ x - y = 1$

60. $x^2 + y = 4, \ 2x - y = 1$

61. $y = x^3, \ y = 2x$

62. $y = \sqrt{x}, \ y = x$

63. $y = x^4 - 2x^2 + 1, \ y = 1 - x^2$

64. $y = x^3 - 2x^2 + x - 1, \ y = -x^2 + 3x - 1$

 65. *Break-Even Analysis* You are setting up a part-time business with an initial investment of $5000. The unit cost of the product is $11.80, and the selling price is $19.30.

(a) Find equations for the total cost C and total revenue R for x units.

(b) Find the break-even point by finding the point of intersection of the cost and revenue equations.

(c) How many units would yield a profit of $100?

66. *Break-Even Analysis* A certain car model costs $14,500 with a gasoline engine and $15,450 with a diesel engine. The numbers of miles per gallon of fuel for cars with these two engines are 22 and 31, respectively. Assume that the price of each type of fuel is $1.389 per gallon.

(a) Show that the cost C_g of driving the gasoline-powered car x miles is

$$C_g = 14{,}500 + \frac{1.389x}{22}$$

and the cost C_d of driving the diesel model x miles is

$$C_d = 15{,}450 + \frac{1.389x}{31}.$$

(b) Find the break-even point. That is, find the mileage at which the diesel-powered car becomes more economical than the gasoline-powered car.

Break-Even Analysis In Exercises 67–70, find the sales necessary to break even for the given cost and revenue equations. (Round your answer up to the nearest whole unit.)

67. $C = 0.85x + 35{,}000$, $R = 1.55x$

68. $C = 6x + 500{,}000$, $R = 35x$

69. $C = 8650x + 250{,}000$, $R = 9950x$

70. $C = 5.5\sqrt{x} + 10{,}000$, $R = 3.29x$

71. *College Text Expenses* The amount of money y (in millions of dollars) spent on college textbooks in the United States is given in the table. (*Source: Book Industry Study Group, Inc.*)

Year	1982	1985	1987	1988	1989	1990
Expense	1388	1575	1803	1998	2144	2319

A mathematical model for the data is

$$y = 10.40t^2 - 6.52t + 1355.74,$$

where t represents the year, with $t = 0$ corresponding to 1980.

(a) Compare the actual expenses with those given by the model. How good is the model? Explain your reasoning.

(b) Use the model to predict the expenses in 1995.

72. *Farm Population* The farm population in the United States, as a percent of the total population, is given in the table. (*Source: U.S. Department of Agriculture*)

Year	1950	1960	1970	1975	1980
Percent	15.3	8.7	4.8	4.2	2.8

Year	1986	1987	1988	1989	1990
Percent	2.2	2.1	2.1	2.0	1.9

A mathematical model for the data is

$$y = \frac{15.28 - 0.24t}{1 + 0.05t},$$

where y represents the percent and t represents the year, with $t = 0$ corresponding to 1950.

(a) Compare the actual percent with that given by the model. How good is the model? Explain your reasoning.

(b) Use the model to predict the farm population as a percent of the total population in 1996.

(c) Discuss the validity of your prediction in part (b).

73. *Annual Salary* A mathematical model for the average annual salary y of a person in finance, insurance, or real estate is

$$y = 15{,}848.32 + 1519.23t + 291.82\sqrt{t},$$

where t represents the year, with $t = 0$ corresponding to 1980. (*Source: U.S. Bureau of Economic Analysis*)

Year	1980	1985	1990	1991	1994
Salary					

(a) Use the model to complete the table.

(b) This model was created using actual data from 1980 through 1991. How accurate do you think the model is in predicting the 1994 average salary? Explain your reasoning.

(c) What does this model predict the average salary to be in the year 2000? How valid do you think this prediction is?

74. Define the break-even point for a business marketing a new product. Give examples of a linear cost equation and a linear revenue equation for which the break-even point is 10,000 units.

 In Exercises 75–80, find the point(s) of intersection of the graphs of the equations.

75. $y = 2x - 5$, $y = -x + 4$

76. $y = x + 6$, $y = 2x - 10$

77. $y = x + 2$, $y = -x^2 + 4$

78. $y = 2x - 3$, $y = x^2 - 3$

79. $y = x^2 + 1$, $y = -x^2 + 9$

80. $y = x^2$, $y = x^2 - 2x + 4$

 In Exercises 81–88, use a graphing utility to graph the equation. Use the graphing utility to approximate the intercepts of the graph.

81. $y = 0.24x^2 + 1.32x + 5.36$

82. $y = -0.56x^2 - 5.34x + 6.25$

83. $y = -0.1x^3 + 1.3x^2 - 4.3$

84. $y = 0.2x^3 - 2.4x^2 + 0.4x$

85. $y = \sqrt{0.3x^2 - 4.3x + 5.7}$

86. $y = \sqrt{-1.21x^2 + 2.34x + 5.6}$

87. $y = \dfrac{0.2x^2 + 1}{0.1x + 2.4}$ **88.** $y = \dfrac{0.4x - 5.3}{0.4x^2 + 5.3}$

Case Study: ## Manufacturing Potato Chips

PROFILE

COMPANY
*Troyer
Potato Products, Inc.*
■

YEARS IN BUSINESS
27
■

LOCATION
Waterford, PA
■

TYPE OF BUSINESS
*Producer of
snack foods*
■

EMPLOYEES
Approx. 500
■

SALES
*Approx. $38 million
annually*
■

Troyer Potato Products, Inc. makes a wide variety of snack foods, but potato chips are its mainstay. The profitability of the company depends, in part, on producing the product at the lowest possible cost. This in turn means getting the most chips possible from a potato.

But are all potatoes created equal? It is possible that one potato of a given weight will yield more chips than another *of the same weight.* If this is the case, the first potato will yield more output per pound of input, and thus result in more profit for the firm.

But how can two potatoes of the same weight yield a different amount (weight) of potato chips? The answer is that more of the weight in one potato may be due to water, which will disappear during cooking. One way to measure this is to consider the specific gravity of the potato, which gives the mass of the potato compared with the mass of a similar volume of water. A higher specific gravity means more potato per pound, and less water—and therefore more potato chips and greater profits. At least that's the theory.

Troyer tested this theory by comparing the specific gravities of 24 batches of potatoes with the amount of potato chips that each batch produced. The table shows the data. The method of least squares resulted in the following equation, which models the relationship:

$$OUTPUT = -35,219 + 38,535 \cdot SPECGRAV$$

or

$$Y = -35,219 + 38,535 \cdot X,$$

where OUTPUT (Y) is pounds of potato chips from each batch of potatoes, and SPECGRAV (X) is the specific gravity of the batch. Notice that the coefficient of SPECGRAV (X) is positive, indicating that a higher specific gravity means more output per batch, confirming the hypothesis.

Conclusion: The firm will be more profitable if it buys potatoes with a higher specific gravity, if all other aspects of the potatoes are the same. Farmers who supply chipping potatoes to Troyer are well aware of this relationship, and provide information on the specific gravity of their crop as part of their transaction.

Specific Gravity	Pounds of Potato Chips	Specific Gravity	Pounds of Potato Chips
1.069	6226	1.094	6991
1.098	6895	1.066	5910
1.073	5981	1.072	6138
1.070	5893	1.075	6024
1.089	7124	1.066	6003
1.096	6982	1.096	6884
1.069	6057	1.097	7056
1.088	7012	1.071	6084
1.074	6099	1.095	7019
1.073	5538	1.090	6895
1.070	6100	1.099	7129
1.098	7002	1.096	6998

The figure shows the scatter diagram of the 24 data points from the table, along with the graph of the equation. The graph reveals an interesting pattern in the data: there are two distinct clusters of potatoes, Atlantic and Monona. Atlantic potatoes tend to have a higher specific gravity than the Monona, and also yield a much higher amount (weight) of potato chips.

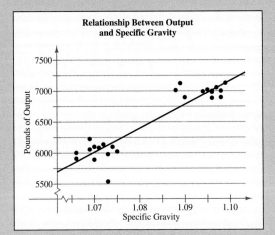

Courtesy of Troyer Farms

What Would You Do?

1. The table shows the data for all 24 batches of potatoes. Can you tell by looking at the table only that there are two varieties of potatoes included in this test? The graph presents the same data, but shows the pattern much more clearly. The equation presents the relationship in yet another form. List the pros and cons of each method. Which of these is most useful for getting an intuitive feel for the data? Which is most useful for predicting the amount of output that will result from a batch with a specific gravity of 1.083?

2. Given the information discussed above, could Troyer make good decisions about potato purchases even without the specific gravity of each batch of potatoes it buys?

3. Summarize other factors that could affect the profitability of potato chips.

Lines in the Plane and Slope

Using Slope ■ *Finding the Slope of a Line* ■ *Writing Linear Equations* ■
Parallel and Perpendicular Lines ■ *Extended Application: Linear Depreciation*

Using Slope

The simplest mathematical model for relating two variables is the **linear equation** $y = mx + b$. The equation is called *linear* because its graph is a line. (In this text, we use the term *line* to mean *straight line*.) By letting $x = 0$, you can see that the line crosses the y-axis at $y = b$, as shown in Figure 1.27. In other words, the y-intercept is $(0, b)$. The steepness or slope of the line is m.

$$y = mx + b$$

Slope ⤴ ⤴ y-Intercept

The **slope** of a line is the number of units the line rises (or falls) vertically for each unit of horizontal change from left to right, as shown in Figure 1.27.

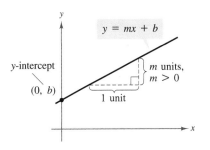

Positive slope, line rises.　　　　Negative slope, line falls.

FIGURE 1.27

A linear equation that is written in the form $y = mx + b$ is said to be written in **slope-intercept form.**

The Slope-Intercept Form of the Equation of a Line

The graph of the equation

$$y = mx + b$$

is a line whose slope is m and whose y-intercept is $(0, b)$.

Discovery

Use a graphing utility to compare the slopes of the lines $y = mx$, where $m = 0.5, 1, 2,$ and 4. Which line rises most quickly? Now, let $m = -0.5, -1, -2,$ and -4. Which line falls most quickly? Let $m = 0.01, 0.001,$ and 0.0001. What is the slope of a horizontal line? Use a square setting to obtain a true geometric perspective.

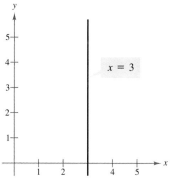

FIGURE 1.28   When the line is vertical, the slope is undefined.

Once you have determined the slope and the *y*-intercept of a line, it is a relatively simple matter to sketch its graph.

In the following example, note that none of the lines is vertical. A vertical line has an equation of the form

$$x = a. \qquad\qquad \textit{Vertical line}$$

Because such an equation cannot be written in the form $y = mx + b$, it follows that the slope of a vertical line is undefined, as indicated in Figure 1.28.

EXAMPLE 1   *Graphing a Linear Equation*

Sketch the graphs of the following linear equations.

a. $y = 2x + 1$ **b.** $y = 2$ **c.** $x + y = 2$

Solution

a. Because $b = 1$, the *y*-intercept is $(0, 1)$. Moreover, because the slope is $m = 2$, the line *rises* two units for each unit the line moves to the right, as shown in Figure 1.29(a).

b. By writing this equation in the form $y = (0)x + 2$, you can see that the *y*-intercept is $(0, 2)$ and the slope is zero. A zero slope implies that the line is horizontal—that is, it doesn't rise *or* fall, as shown in Figure 1.29(b).

c. By writing this equation in slope-intercept form

$$\begin{aligned} x + y &= 2 && \textit{Original equation} \\ y &= -x + 2 && \textit{Subtract x from both sides.} \\ y &= (-1)x + 2 && \textit{Slope-intercept form} \end{aligned}$$

you can see that the *y*-intercept is $(0, 2)$. Moreover, because the slope is $m = -1$, this line *falls* one unit for each unit the line moves to the right, as shown in Figure 1.29(c).

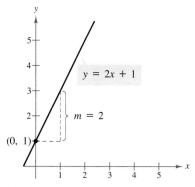

(a) When m is positive, the line rises.

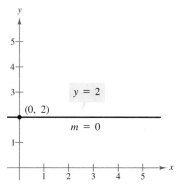

(b) When m is zero, the line is horizontal.

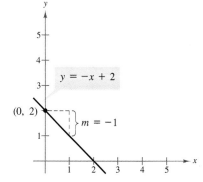

(c) When m is negative, the line falls.

FIGURE 1.29

In real-life problems, the slope of a line can be interpreted as either a *ratio* or a *rate*. If the x-axis and y-axis have the same unit of measure, then the slope has no units and is a **ratio.** If the x-axis and y-axis have different units of measure, then the slope is a **rate** or **rate of change.**

EXAMPLE 2 *Using Slope as a Ratio*

The maximum recommended slope of a wheelchair ramp is $\frac{1}{12}$. A business is installing a wheelchair ramp that rises 22 inches over a horizontal length of 24 feet. Is the ramp steeper than recommended? *(Source: American Disabilities Act Handbook)*

Solution
The horizontal length of the ramp is 12(24) or 288 inches, as shown in Figure 1.30. Thus the slope of the ramp is

$$\text{Slope} = \frac{\text{vertical change}}{\text{horizontal change}} = \frac{22 \text{ in.}}{288 \text{ in.}} \approx 0.076.$$

Thus, the slope is not steeper than recommended.

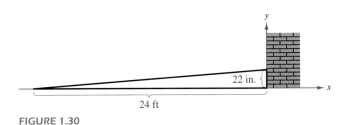

FIGURE 1.30

EXAMPLE 3 *Using Slope as a Rate of Change*

A manufacturing company determines that the total cost in dollars of producing x units of a product is

$$C = 25x + 3500. \qquad \textit{Cost equation}$$

Describe the practical significance of the y-intercept and slope of the line given by this equation.

Solution
The y-intercept $(0, 3500)$ tells you that the cost of producing zero units is $3500. This is the **fixed cost** of production—it includes costs that must be paid regardless of the number of units produced. The slope of $m = 25$ tells you that the cost of producing each unit is $25, as shown in Figure 1.31. Economists call the cost per unit the **marginal cost.** If the production increases by one unit, then the "margin" or extra amount of cost is $25.

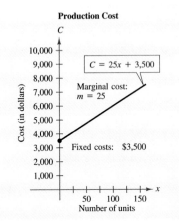

FIGURE 1.31

Finding the Slope of a Line

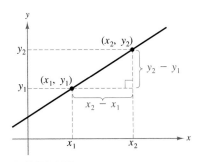

FIGURE 1.32

Given an equation of a line, you can find its slope by writing the equation in slope-intercept form. If you are not given an equation, you can still find the slope of a line. For instance, suppose you want to find the slope of the line passing through the points (x_1, y_1) and (x_2, y_2), as shown in Figure 1.32. As you move from left to right along this line, a change of $(y_2 - y_1)$ units in the vertical direction corresponds to a change of $(x_2 - x_1)$ units in the horizontal direction. These two changes are denoted by the symbols

$$\Delta y = y_2 - y_1 = \text{the change in } y$$

and

$$\Delta x = x_2 - x_1 = \text{the change in } x.$$

(Δ is the Greek capital letter delta, and the symbols Δy and Δx are read as "delta y" and "delta x.") The ratio of Δy to Δx represents the slope of the line that passes through the points (x_1, y_1) and (x_2, y_2).

$$\text{Slope} = \frac{\Delta y}{\Delta x} = \frac{y_2 - y_1}{x_2 - x_1}$$

Be sure you see that Δx represents a single number, not the product of two numbers (Δ and x). The same is true for Δy.

The Slope of a Line Passing Through Two Points

The **slope** m of the line passing through (x_1, y_1) and (x_2, y_2) is

$$m = \frac{\Delta y}{\Delta x} = \frac{y_2 - y_1}{x_2 - x_1}$$

where $x_1 \neq x_2$.

When this formula is used for slope, the *order of subtraction* is important. Given two points on a line, you are free to label either one of them as (x_1, y_1) and the other as (x_2, y_2). However, once you have done this, you must form the numerator and denominator using the same order of subtraction.

$$m = \frac{y_2 - y_1}{x_2 - x_1} \qquad m = \frac{y_1 - y_2}{x_1 - x_2} \qquad m = \frac{y_2 - y_1}{x_1 - x_2}$$

$$\text{Correct} \qquad\qquad \text{Correct} \qquad\qquad \text{Incorrect}$$

For instance, the slope of the line passing through the points $(3, 4)$ and $(5, 7)$ can be calculated as

$$m = \frac{7 - 4}{5 - 3} = \frac{3}{2} \quad \text{or} \quad m = \frac{4 - 7}{3 - 5} = \frac{-3}{-2} = \frac{3}{2}.$$

EXAMPLE 4 *Finding the Slope of a Line*

Find the slope of the line passing through the pairs of points.

a. $(-2, 0)$ and $(3, 1)$

b. $(-1, 2)$ and $(2, 2)$

c. $(0, 4)$ and $(1, -1)$

d. $(3, 4)$ and $(3, 1)$

Solution

a. Letting $(x_1, y_1) = (-2, 0)$ and $(x_2, y_2) = (3, 1)$, you obtain a slope of

$$m = \frac{y_2 - y_1}{x_2 - x_1} = \frac{1 - 0}{3 - (-2)} = \frac{1}{5}.$$ ⟵ Difference in y-values
⟵ Difference in x-values

b. The slope of the line passing through $(-1, 2)$ and $(2, 2)$ is

$$m = \frac{2 - 2}{2 - (-1)} = \frac{0}{3} = 0.$$

c. The slope of the line passing through $(0, 4)$ and $(1, -1)$ is

$$m = \frac{-1 - 4}{1 - 0} = \frac{-5}{1} = -5.$$

d. The slope of the vertical line passing through $(3, 4)$ and $(3, 1)$ is not defined because division by zero is undefined. (See Figure 1.33.)

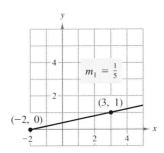

(a) Positive slope; line rises.

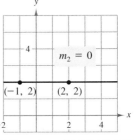

(b) Zero slope; line is horizontal.

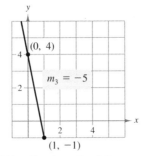

(c) Negative slope; line falls.

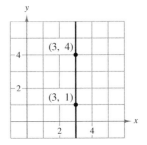

(d) Vertical line; undefined slope.

FIGURE 1.33

Writing Linear Equations

If (x_1, y_1) is a point lying on a line of slope m and (x, y) is *any other* point on the line, then

$$\frac{y - y_1}{x - x_1} = m.$$

This equation, involving the variables x and y, can be rewritten in the form $y - y_1 = m(x - x_1)$, which is the **point-slope form** of the equation of a line.

Point-Slope Form of the Equation of a Line

The equation of the line with slope m passing through the point (x_1, y_1) is

$$y - y_1 = m(x - x_1).$$

The point-slope form is most useful for *finding* the equation of a line. You should remember this formula—it is used repeatedly throughout the text.

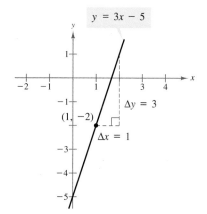

$y = 3x - 5$

$\Delta y = 3$

$(1, -2)$

$\Delta x = 1$

FIGURE 1.34

EXAMPLE 5   *Using the Point-Slope Form*

Find the equation of the line that has a slope of 3 and passes through the point $(1, -2)$, as shown in Figure 1.34.

Solution

Use the point-slope form with $m = 3$ and $(x_1, y_1) = (1, -2)$.

$$
\begin{aligned}
y - y_1 &= m(x - x_1) && \textit{Point-slope form} \\
y - (-2) &= 3(x - 1) && \textit{Substitute for } m, x_1, \textit{ and } y_1. \\
y + 2 &= 3x - 3 && \textit{Simplify.} \\
y &= 3x - 5 && \textit{Slope-intercept form}
\end{aligned}
$$

The point-slope form can be used to find an equation of the line passing through points (x_1, y_1) and (x_2, y_2). To do this, first find the slope of the line

$$m = \frac{y_2 - y_1}{x_2 - x_1}, \qquad x_1 \neq x_2,$$

and then use the point-slope form to obtain the equation

$$y - y_1 = \frac{y_2 - y_1}{x_2 - x_1}(x - x_1). \qquad \textit{Two-point form}$$

This is sometimes called the **two-point form** of the equation of a line.

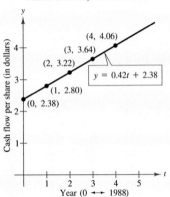

Bausch & Lomb, Inc. Cash Flow

Cash flow per share (in dollars)

(4, 4.06)
(3, 3.64)
(2, 3.22)
$y = 0.42t + 2.38$
(1, 2.80)
(0, 2.38)

Year (0 ←→ 1988)

FIGURE 1.35

EXAMPLE 6 *Predicting Cash Flow Per Share*

Real Life

The cash flow per share for Bausch & Lomb, Inc. was $2.38 in 1988 and $2.80 in 1989. Using only this information, write a linear equation that gives the cash flow per share in terms of the year. *(Source: Bausch & Lomb, Inc.)*

Solution

Let $t = 0$ represent 1988. Then the two given values are represented by the ordered pairs $(0, 2.38)$ and $(1, 2.80)$. The slope of the line passing through these points is

$$m = \frac{2.80 - 2.38}{1 - 0} = 0.42.$$

Using the point-slope form, you can find the equation that relates the cash flow C and the year t to be

$$y = 0.42t + 2.38.$$

You can use this model to predict future cash flows. For instance, it predicts the cash flows in 1990, 1991, and 1992 to be $3.22, $3.64, and $4.06, as shown in Figure 1.35. (In this case, the predictions are quite good—the actual cash flows in 1990, 1991, and 1992 were $3.38, $3.65, and $4.16.) ▄

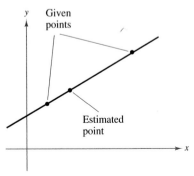

(a) Linear Extrapolation

Given points

Estimated point

(b) Linear Interpolation

FIGURE 1.36

The prediction method illustrated in Example 6 is called **linear extrapolation.** Note in Figure 1.36(a) that an extrapolated point does not lie between the given points. When the estimated point lies between two given points, as shown in Figure 1.36(b), the procedure is called **linear interpolation.**

Because the slope of a vertical line is not defined, its equation cannot be written in slope-intercept form. However, every line has an equation that can be written in the **general form**

$$Ax + By + C = 0, \qquad \textit{General form}$$

where A and B are not both zero. For instance, the vertical line given by $x = a$ can be represented by the general form $x - a = 0$. The five most common forms of equations of lines are summarized below.

Equations of Lines

1. General form: $Ax + By + C = 0$
2. Vertical line: $x = a$
3. Horizontal line: $y = b$
4. Slope-intercept form: $y = mx + b$
5. Point-slope form: $y - y_1 = m(x - x_1)$

Parallel and Perpendicular Lines

The slope of a line is a convenient tool for determining whether two lines are parallel or perpendicular.

TECHNOLOGY

On a graphing utility, lines will not appear to have the correct slope unless you use a viewing rectangle that has a "square setting." For instance, try graphing the lines in Example 7 using the standard setting $-10 \le x \le 10$ and $-10 \le y \le 10$. Then reset the viewing rectangle with the square setting $-9 \le x \le 9$ and $-6 \le y \le 6$. On which setting do the lines $y = \frac{2}{3}x - \frac{5}{3}$ and $y = -\frac{3}{2}x + 2$ appear perpendicular?

Parallel and Perpendicular Lines

1. Two distinct nonvertical lines are **parallel** if and only if their slopes are equal. That is, $m_1 = m_2$.
2. Two nonvertical lines are **perpendicular** if and only if their slopes are negative reciprocals of each other. That is, $m_1 = -1/m_2$.

EXAMPLE 7 *Finding Parallel and Perpendicular Lines*

Find an equation of the line that passes through the point $(2, -1)$ and is

a. parallel to the line $2x - 3y = 5$
b. perpendicular to the line $2x - 3y = 5$.

Solution
By writing the given equation in slope-intercept form

$$2x - 3y = 5 \qquad \text{\textit{Given equation}}$$
$$-3y = -2x + 5 \qquad \text{\textit{Subtract 2x from both sides.}}$$
$$y = \tfrac{2}{3}x - \tfrac{5}{3} \qquad \text{\textit{Slope-intercept form}}$$

you can see that it has a slope of $m = \frac{2}{3}$, as shown in Figure 1.37.

a. Any line parallel to the given line must also have a slope of $\frac{2}{3}$. Thus, the line through $(2, -1)$ that is parallel to the given line has the following equation.

$$y - (-1) = \tfrac{2}{3}(x - 2) \qquad \text{\textit{Point-slope form}}$$
$$3(y + 1) = 2(x - 2) \qquad \text{\textit{Multiply both sides by 3.}}$$
$$3y + 3 = 2x - 4 \qquad \text{\textit{Distribute.}}$$
$$2x - 3y = 7 \qquad \text{\textit{Simplify.}}$$
$$y = \tfrac{2}{3}x - \tfrac{7}{3} \qquad \text{\textit{Slope-intercept form}}$$

b. Any line perpendicular to the given line must have a slope of $-1/(2/3)$ or $-\frac{3}{2}$. Therefore, the line through $(2, -1)$ that is perpendicular to the given line has the following equation.

$$y - (-1) = -\tfrac{3}{2}(x - 2) \qquad \text{\textit{Point-slope form}}$$
$$2(y + 1) = -3(x - 2) \qquad \text{\textit{Multiply both sides by 2.}}$$
$$3x + 2y = 4 \qquad \text{\textit{General form}}$$
$$y = -\tfrac{3}{2}x + 2 \qquad \text{\textit{Slope-intercept form}}$$

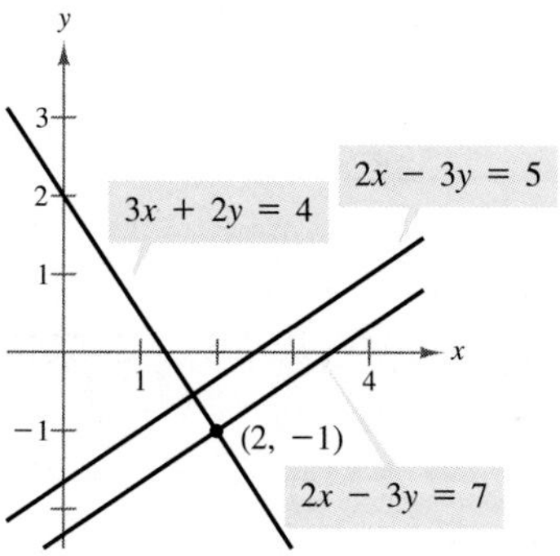

FIGURE 1.37

Extended Application: Linear Depreciation

Most business expenses can be deducted the same year they occur. One exception to this is the cost of property that has a useful life of more than one year, such as buildings, cars, or equipment. Such costs must be **depreciated** over the useful life of the property. If the *same amount* is depreciated each year, the procedure is called **linear depreciation.**

Real Life

EXAMPLE 8 *Depreciating Equipment*

Your company has purchased a $12,000 machine that has a useful life of 8 years. The salvage value at the end of 8 years is $2000. Write a linear equation that describes the *nondepreciated* value of the machine each year.

Solution

Let V represent the value of the machine at the end of year t. You can represent the initial value of the machine by the ordered pair (0, 12,000) and the salvage value of the machine by the ordered pair (8, 2000). The slope of the line is

$$m = \frac{2000 - 12,000}{8 - 0} = -\$1250,$$

which represents the annual depreciation in *dollars per year.* Using the point-slope form, you can write the equation of the line as follows.

$$V - 12,000 = -1250(t - 0) \qquad \textit{Point-slope form}$$

$$V = -1250t + 12,000 \qquad \textit{Slope-intercept form}$$

Table 1.6 shows the nondepreciated value of the machine at the end of each year.

TABLE 1.6

t	0	1	2	3	4	5	6	7	8
V	12,000	10,750	9500	8250	7000	5750	4500	3250	2000

The graph of this equation is shown in Figure 1.38. ■

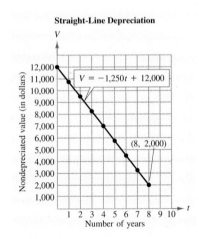

Straight-Line Depreciation

$V = -1,250t + 12,000$

(8, 2,000)

Nondepreciated value (in dollars)

Number of years

FIGURE 1.38

Discussion Problem **Comparing Different Types of Depreciation**

The Internal Revenue Service allows businesses to choose different types of depreciation. Another type is

$$\textit{Uniform Declining Balances: } V = 12,000 \left(\frac{n - 1.605}{n} \right)^{t}, \quad n = 8.$$

Construct a table that compares this type of depreciation with linear depreciation. What are the advantages of each type?

Warm Up

The following warm-up exercises involve skills that were covered in earlier sections. You will use these skills in the exercise set for this section.

In Exercises 1 and 2, simplify the expression.

1. $\dfrac{5 - (-2)}{-3 - 4}$

2. $\dfrac{-7 - (-0)}{4 - 1}$

3. Evaluate $-\dfrac{1}{m}$ when $m = -3$.

4. Evaluate $-\dfrac{1}{m}$ when $m = \dfrac{6}{7}$.

In Exercises 5–10, solve for y in terms of x.

5. $-4x + y = 7$

6. $3x - y = 7$

7. $y - 2 = 3(x - 4)$

8. $y - (-5) = -1[x - (-2)]$

9. $y - (-3) = \dfrac{4 - (-3)}{2 - 1}(x - 2)$

10. $y - 1 = \dfrac{-3 - 1}{-7 - (-1)}[x - (-1)]$

E X E R C I S E S   1.3   means that technology can help you solve or check the exercise(s).

In Exercises 1–6, estimate the slope of the line.

1.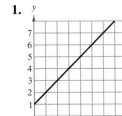

2.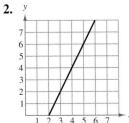

5.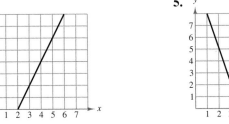

6.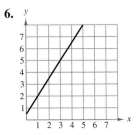

3.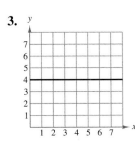

4.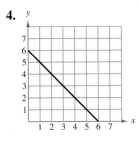

In Exercises 7–14, plot the points and find the slope of the line passing through the points.

7. $(3, -4), (5, 2)$

8. $(-2, 1), (4, -3)$

9. $\left(\dfrac{1}{2}, 2\right), (6, 2)$

10. $\left(-\dfrac{3}{2}, -5\right), \left(\dfrac{5}{6}, 4\right)$

11. $(-8, -3), (-8, -5)$

12. $(2, 1), (2, 5)$

13. $(1, 2), (-2, 2)$

14. $\left(\dfrac{7}{8}, \dfrac{3}{4}\right), \left(\dfrac{5}{4}, -\dfrac{1}{4}\right)$

In Exercises 15–22, use the point on the line and slope of the line to find three additional points on the line. (There are many correct solutions.)

	Point	Slope		Point	Slope
15.	$(2, 1)$	$m = 0$	**16.**	$(-3, 4)$	m undefined
17.	$(6, -4)$	$m = -2$	**18.**	$(-2, -2)$	$m = 2$
19.	$(1, 7)$	$m = -3$	**20.**	$(10, -6)$	$m = -1$
21.	$(-8, 1)$	m undefined	**22.**	$(-3, -1)$	$m = 0$

In Exercises 23–28, find the slope and y-intercept (if possible) of the line.

23. $x + 5y = 20$ **24.** $2x + y = 40$

25. $7x - 5y = 15$ **26.** $6x - 5y = 15$

27. $x = 4$ **28.** $y = -1$

 In Exercises 29–36, write an equation of the line that passes through the points. Then use the equation to sketch the line.

29. $(4, 3), (0, -5)$ **30.** $(-3, -4), (1, 4)$

31. $(0, 0), (-1, 3)$ **32.** $(-3, 6), (1, 2)$

33. $(2, 3), (2, -2)$ **34.** $(6, 1), (10, 1)$

35. $(1, -2), (3, -2)$ **36.** $\left(\frac{7}{8}, \frac{3}{4}\right), \left(\frac{5}{4}, -\frac{1}{4}\right)$

 In Exercises 37–46, write an equation of the indicated line. Then use the equation to sketch the line.

	Point	Slope		Point	Slope
37.	$(0, 3)$	$m = \frac{3}{4}$	**38.**	$(-1, 2)$	m undefined
39.	$(0, 0)$	$m = \frac{2}{3}$	**40.**	$(-1, -4)$	$m = \frac{1}{4}$
41.	$(-2, 7)$	$m = -3$	**42.**	$(-2, 4)$	$m = -\frac{3}{5}$
43.	$(0, 2)$	$m = 4$	**44.**	$\left(0, -\frac{2}{3}\right)$	$m = \frac{1}{6}$
45.	$\left(0, \frac{2}{3}\right)$	$m = \frac{3}{4}$	**46.**	$(0, 4)$	$m = 0$

In Exercises 47 and 48, explain how to use the concept of slope to determine whether the three points are collinear. Then explain how to use the Distance Formula to determine whether the points are collinear.

47. $(-2, 1), (-1, 0), (2, -2)$

48. $(0, 4), (7, -6), (-5, 11)$

49. Write an equation of the vertical line with x-intercept at 3.

50. Write an equation of the horizontal line through $(0, -5)$.

 In Exercises 51–58, write an equation of the line through the given point (a) parallel to the given line and (b) perpendicular to the given line. Then graph all three equations on the same set of coordinate axes.

	Point	Line		Point	Line
51.	$(-3, 2)$	$x + y = 7$	**52.**	$(2, 1)$	$4x - 2y = 3$
53.	$(-6, 4)$	$3x + 4y = 7$	**54.**	$\left(\frac{7}{8}, \frac{3}{4}\right)$	$5x + 3y = 0$
55.	$(-1, 0)$	$y = -3$	**56.**	$(2, 5)$	$x = 4$
57.	$(1, 1)$	$-2x + 3y = -3$	**58.**	$(12, -3)$	$x + 4y = 4$

In Exercises 59–66, sketch the graph of the equation.

59. $y = -2$ **60.** $x = 4$

61. $2x - y - 3 = 0$ **62.** $x + 2y + 6 = 0$

63. $y = -2x + 1$ **64.** $y - 1 = 3(x + 4)$

65. $y + 2 = -4(x + 1)$ **66.** $4x + 5y = 20$

67. *Temperature Conversion* Write a linear equation that expresses the relationship between the temperature in degrees Celsius, C, and degrees Fahrenheit, F. Use the fact that water freezes at 0° Celsius (32° Fahrenheit) and boils at 100° Celsius (212° Fahrenheit).

68. *Temperature Conversion* Use the result of Exercise 67 to complete the table.

C	$-10°$	$10°$			$177°$
F	$0°$			$68°$	$90°$	

69. *Reimbursed Expenses* A company reimburses its sales representatives $100 per day for lodging and meals, plus $0.25 per mile driven. Write a linear equation giving the daily cost C in terms of x, the number of miles driven.

70. *Union Negotiation* You are on a negotiating panel in a union hearing for a large corporation. The union is asking for a base pay of $8.75 per hour *plus* an additional piecework rate of $0.80 per unit produced. The corporation is offering a base pay of $6.35 per hour *plus* a piecework rate of $1.15.

(a) Write a linear equation for the hourly wages W in terms of x, the number of units produced per hour, for each pay schedule.
(b) Use a graphing utility to graph each linear equation and find the point of intersection.
(c) Interpret the meaning of the point of intersection of the graphs. How would you use this information to advise the corporation and the union?

71. *Annual Salary* Your annual salary was $26,300 in 1992 and $29,700 in 1994. Assume your salary can be modeled by a linear equation.

(a) Write a linear equation giving your salary S in terms of the year t where $t = 0$ corresponds to the year 1992.

(b) Use the linear model to predict your salary in 1997.

(c) Does a linear model assume that your salary increases by the same *amount* each year or the same *percent* each year? Explain.

(d) If you assume your salary can be modeled by the equation $S = 26,300(1.0627)^t$, would it be increasing by the same *amount* each year or the same *percent*? Would you prefer this model over the linear one? Explain.

72. *Linear Depreciation* A company constructs a warehouse for $825,000. The warehouse has an estimated useful life of 25 years, after which its value is expected to be $75,000. Write a linear equation giving the value y of the warehouse during its 25 years of useful life. (Let t represent the time in years.)

73. *Linear Depreciation* A small business purchases a piece of equipment for $1025. After 5 years the equipment will be outdated, having no value.

(a) Write a linear equation giving the value of the equipment in terms of the time t, $0 \le t \le 5$.

(b) Use a graphing utility to graph the equation.

(c) Move the cursor along the graph and estimate (to two decimal place accuracy) the value of the equipment when $t = 3$.

(d) Move the cursor along the graph and estimate (to two decimal place accuracy) the time when the value of the equipment will be $600.

74. *Apartment Rental* A real estate office handles an apartment complex with 50 units. When the rent is $380 per month, all 50 units are occupied. When the rent is $425, however, the average number of occupied units drops to 47. Assume that the relationship between the monthly rent p and the demand x is linear. (The term *demand* refers to the number of occupied units.)

(a) Write a linear equation expressing x in terms of p.

(b) *Linear Extrapolation* Predict the number of occupied units when the rent is set at $455.

(c) *Linear Interpolation* Predict the number of occupied units when the rent is set at $395.

75. *Profit* You are a contractor and have purchased a piece of equipment for $26,500. The equipment costs an average of $5.25 per hour for fuel and maintenance, and the operator is paid $9.50 per hour.

(a) Write a linear equation giving the total cost C of operating the equipment for t hours.

(b) You charge your customers $25 per hour of machine use. Write an equation for the revenue R derived from t hours of use.

(c) Use the formula for profit, $P = R - C$, to write an equation for the profit derived from t hours of use.

(d) Find the number of hours you must operate the equipment before you break even.

76. *Research Project* Personal income (in billions of dollars) in the United States was 3380 in 1985 and 4834 in 1991. Assume that the relationship between the personal income Y and the time t (in years) is linear. Let $t = 0$ correspond to 1985. *(Source: U.S. Bureau of Economic Analysis)*

(a) Write a linear model for the data.

(b) *Linear Interpolation* Estimate the personal income in 1988.

(c) *Linear Extrapolation* Estimate the personal income in 1992.

(d) Use your school's library or some other reference source to find the actual personal income in 1988 and 1992. How close were your estimates?

 In Exercises 77–82, use a graphing utility to graph the cost function. Determine the maximum production level x, given that the cost C cannot exceed $100,000.

77. $C = 23,500 + 3100x$ **78.** $C = 30,000 + 575x$

79. $C = 18,375 + 1150x$ **80.** $C = 24,900 + 1785x$

81. $C = 75,500 + 89x$ **82.** $C = 83,620 + 67x$

83. *Sales Commission* As a salesperson, you receive a monthly salary of $2000, plus a commission of 7% of sales. You are offered a new job at $2300 per month, plus a commission of 5% of sales.

(a) Write a linear equation for your current monthly wage W in terms of your monthly sales S.

(b) Write a linear equation for the monthly wage W of your job offer in terms of the monthly sales S.

(c) Use a graphing utility to graph both equations on the same viewing rectangle. Find the point of intersection. What does it signify?

(d) You think you can sell $20,000 per month. Should you change jobs? Explain.

Functions

Functions ■ *The Graph of a Function* ■ *Function Notation* ■
Combinations of Functions ■ *Inverse Functions*

Functions

In many common relationships between two variables, the value of one of the variables depends on the value of the other variable. For example, the sales tax on an item depends on its selling price, the distance an object moves in a given amount of time depends on its speed, the price of mailing a package with an overnight delivery service depends on the package's weight, and the area of a circle depends on its radius.

Consider the relationship between the area of a circle and its radius. This relationship can be expressed by the equation

$$A = \pi r^2.$$

In this equation, the value of A depends on the choice of r. Because of this, A is the **dependent variable** and r is the **independent variable.**

Most of the relationships that you will study in this course have the property that for a given value of the independent variable, there corresponds exactly one value of the dependent variable. Such a relationship is a **function.**

Definition of a Function

A **function** is a relationship between two variables such that to each value of the independent variable there corresponds exactly one value of the dependent variable.

The **domain** of the function is the set of all values of the independent variable for which the function is defined. The **range** of the function is the set of all values taken on by the dependent variable.

In Figure 1.39, notice that you can think of a function as a machine that inputs values of the independent variable and outputs values of the dependent variable.

Although functions can be described by various means such as tables, graphs, or diagrams, they are most often specified by formulas or equations. For instance, the equation $y = 4x^2 + 3$ describes y as a function of x. For this function, x is the independent variable and y is the dependent variable.

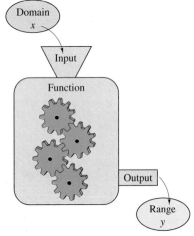

FIGURE 1.39

▦ TECHNOLOGY

The procedure used in Example 1, isolating the dependent variable on the left side, is also useful for graphing equations with a graphing utility. In fact, the standard graphing program on most graphing utilities is a "function grapher." To graph equations in which y is not a function of x, such as a circle, you usually have to enter two or more equations into the graphing utility.

EXAMPLE 1 *Deciding Whether Relations Are Functions*

Which of the following equations define y as a function of x?

a. $x + y = 1$ **b.** $x^2 + y^2 = 1$
c. $x^2 + y = 1$ **d.** $x + y^2 = 1$

Solution
To decide whether an equation defines a function, it is helpful to isolate the dependent variable on the left side. For instance, to decide whether the equation $x + y = 1$ defines y as a function of x, write the equation in the form

$$y = 1 - x.$$

From this form, you can see that for any value of x, there is exactly one value of y. Thus, y is a function of x.

Original Equation	Rewritten Equation	Test: Is y a function of x?
a. $x + y = 1$	$y = 1 - x$	Yes, each value of x determines exactly one value of y.
b. $x^2 + y^2 = 1$	$y = \pm\sqrt{1 - x^2}$	No, some values of x determine two values of y.
c. $x^2 + y = 1$	$y = 1 - x^2$	Yes, each value of x determines exactly one value of y.
d. $x + y^2 = 1$	$y = \pm\sqrt{1 - x}$	No, some values of x determine two values of y.

Note that the equations that assign two values ($\pm$) to the dependent variable for a given value of the independent variable do not define functions of x. For instance, in part b, when $x = 0$, the equation

$$y = \pm\sqrt{1 - x^2}$$

indicates that $y = +1$ or $y = -1$. Figure 1.40 shows the graphs of the four equations.

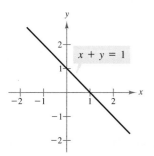

(a)

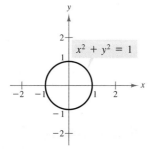

(b)

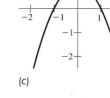

(c)

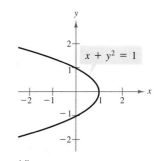

(d)

FIGURE 1.40

The Graph of a Function

When the graph of a function is sketched, the standard convention is to let the horizontal axis represent the independent variable. When this convention is used, the test described in Example 1 has a nice graphical interpretation called the **vertical line test.** This test states that if every vertical line intersects the graph of an equation at most once, then the equation defines y as a function of x. For instance, in Figure 1.40, the graphs in parts (a) and (c) pass the vertical line test, but those in parts (b) and (d) do not.

The domain of a function may be described explicitly, or it may be *implied* by an equation used to define the function. For example, the function given by $y = 1/(x^2 - 4)$ has an implied domain that consists of all real x except $x = \pm 2$. These two values are excluded from the domain because division by zero is undefined.

Another type of implied domain is that used to avoid even roots of negative numbers, as indicated in Example 2.

EXAMPLE 2 *Finding the Domain and Range of a Function*

Find the domain and range of the following functions.

a. $y = \sqrt{x - 1}$ **b.** $y = \begin{cases} 1 - x, & x < 1 \\ \sqrt{x - 1}, & x \geq 1 \end{cases}$

Solution

a. Because $\sqrt{x - 1}$ is not defined for $x - 1 < 0$ (that is, for $x < 1$), it follows that the domain of the function is the interval $x \geq 1$ or $[1, \infty)$. To find the range, observe that $\sqrt{x - 1}$ is never negative. Moreover, as x takes on the various values in the domain, y takes on all nonnegative values. Thus, the range is the interval $y \geq 0$ or $[0, \infty)$. The graph of the function, shown in Figure 1.41(a), confirms these results.

b. Because this **compound** function is defined for $x < 1$ *and* for $x \geq 1$, the domain is the entire set of real numbers. When $x \geq 1$, the function behaves as in part a. For $x < 1$, the value of $1 - x$ is positive, and therefore the range of the function is $y \geq 0$ or $[0, \infty)$, as shown in Figure 1.41(b). ▪

A function is **one-to-one** if to each value of the dependent variable in the range there corresponds exactly one value of the independent variable. For instance, the function in Example 2a is one-to-one, whereas the function in Example 2b is not one-to-one.

Geometrically, a function is one-to-one if every horizontal line intersects the graph of the function at most once. This geometrical interpretation is the **horizontal line test** for one-to-one functions. Thus, a graph that represents a one-to-one function must satisfy *both* the vertical line test and the horizontal line test.

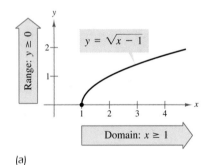

(a)

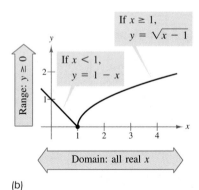

(b)

FIGURE 1.41

Function Notation

When using an equation to define a function, you generally isolate the dependent variable on the left. For instance, writing the equation $x + 2y = 1$ as

$$y = \frac{1-x}{2}$$

indicates that y is the dependent variable. In **function notation,** this equation has the form

$$f(x) = \frac{1-x}{2}. \qquad \qquad \textit{Function notation}$$

The independent variable is x, and the name of the function is "f." The symbol $f(x)$ is read as "f of x," and it denotes the value of the dependent variable. For instance, the value of f when $x = 3$ is

$$f(3) = \frac{1-(3)}{2} = \frac{-2}{2} = -1.$$

The value $f(3)$ is called a **function value,** and it lies in the range of f. This means that the point $(3, f(3))$ lies on the graph of f. One of the advantages of function notation is that it allows you to be less wordy. For instance, instead of asking "What is the value of y when $x = 3$?" you can ask "What is $f(3)$?"

EXAMPLE 3 *Evaluating a Function*

Find the value of the function

$$f(x) = 2x^2 - 4x + 1$$

when x is -1, 0, and 2. Is f one-to-one?

Solution

When $x = -1$, the value of f is

$$f(-1) = 2(-1)^2 - 4(-1) + 1 = 2 + 4 + 1 = 7.$$

When $x = 0$, the value of f is

$$f(0) = 2(0)^2 - 4(0) + 1 = 0 - 0 + 1 = 1.$$

When $x = 2$, the value of f is

$$f(2) = 2(2)^2 - 4(2) + 1 = 8 - 8 + 1 = 1.$$

Because two different values of x yield the same value of $f(x)$, the function is *not* one-to-one, as shown in Figure 1.42. ▬

Note You can use the horizontal line test to determine whether the function in Example 3 is one-to-one. Because the line $y = 1$ intersects the graph of the function twice, the function is *not* one-to-one.

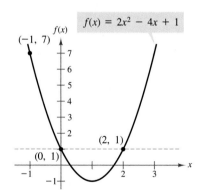

$f(x) = 2x^2 - 4x + 1$

$(-1, 7)$

$(2, 1)$

$(0, 1)$

FIGURE 1.42

⊞ **TECHNOLOGY**

Most graphing utilities can be programmed to evaluate functions. The program depends on the calculator used. Here is a sample program for the *TI-81* or *TI-82*. See Appendix D for other calculator models.

```
Prgm: EVALUATE
:Lbl 1
:Disp ''ENTER X''
:Input X
:Disp Y₁
:Goto 1
```

To use this program, enter a function in Y_1. Then run the program—it will allow you to evaluate the function at several values of x.

Example 3 suggests that the role of the variable x in the equation

$$f(x) = 2x^2 - 4x + 1$$

is simply that of a placeholder. Informally, f could be defined by the equation

$$f() = 2()^2 - 4() + 1.$$

To evaluate $f(-2)$, simply place -2 in each set of parentheses.

$$f(-2) = 2(-2)^2 - 4(-2) + 1 = 8 + 8 + 1 = 17$$

The ratio in part b of the next example is called a *difference quotient*. In Section 2.1, you will see that it has special significance in calculus.

EXAMPLE 4 *Evaluating a Function*

For the function $f(x) = x^2 - 4x + 7$, evaluate the following.

a. $f(x + \Delta x)$ **b.** $\dfrac{f(x + \Delta x) - f(x)}{\Delta x}$

Solution

a. To evaluate f at $x + \Delta x$, substitute $x + \Delta x$ into each set of parentheses, as follows.

$$f(x + \Delta x) = (x + \Delta x)^2 - 4(x + \Delta x) + 7$$
$$= x^2 + 2x\Delta x + (\Delta x)^2 - 4x - 4\Delta x + 7$$

b. Using the result of part a, you can write the following.

$$\frac{f(x + \Delta x) - f(x)}{\Delta x}$$
$$= \frac{[(x + \Delta x)^2 - 4(x + \Delta x) + 7] - [x^2 - 4x + 7]}{\Delta x}$$
$$= \frac{x^2 + 2x\Delta x + (\Delta x)^2 - 4x - 4\Delta x + 7 - x^2 + 4x - 7}{\Delta x}$$
$$= \frac{2x\Delta x + (\Delta x)^2 - 4\Delta x}{\Delta x}$$
$$= 2x + \Delta x - 4, \qquad \Delta x \neq 0$$ ▬

Although f is often used as a convenient function name and x as the independent variable, you can use other symbols. For instance, the following equations all define the same function.

$$f(x) = x^2 - 4x + 7$$
$$f(t) = t^2 - 4t + 7$$
$$g(s) = s^2 - 4s + 7$$

Combinations of Functions

Two functions can be combined in various ways to create new functions. For instance, if $f(x) = 2x - 3$ and $g(x) = x^2 + 1$, you can form the following functions.

$$f(x) + g(x) = (2x - 3) + (x^2 + 1) = x^2 + 2x - 2 \qquad \textit{Sum}$$

$$f(x) - g(x) = (2x - 3) - (x^2 + 1) = -x^2 + 2x - 4 \qquad \textit{Difference}$$

$$f(x)g(x) = (2x - 3)(x^2 + 1) = 2x^3 - 3x^2 + 2x - 3 \qquad \textit{Product}$$

$$\frac{f(x)}{g(x)} = \frac{2x - 3}{x^2 + 1} \qquad \textit{Quotient}$$

You can combine two functions in yet another way called the **composition**. The resulting function is a **composite function.**

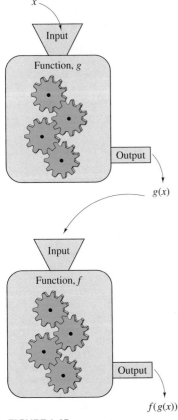

x

Input

Function, *g*

Output

$g(x)$

Input

Function, *f*

Output

$f(g(x))$

FIGURE 1.43

Definition of Composite Function

The function given by $(f \circ g)(x) = f(g(x))$ is the **composite** of f with g. The **domain** of $(f \circ g)$ is the set of all x in the domain of g such that $g(x)$ is in the domain of f, as indicated in Figure 1.43.

The composite of f with g may not be equal to the composite of g with f, as illustrated in the following example.

EXAMPLE 5 *Forming Composite Functions*

Given $f(x) = 2x - 3$ and $g(x) = x^2 + 1$, find the following.

a. $f(g(x))$ **b.** $g(f(x))$

Solution

a. The composite of f with g is given by

$$\begin{aligned} f(g(x)) &= 2(g(x)) - 3 &&\textit{Evaluate } f \textit{ at } g(x). \\ &= 2(x^2 + 1) - 3 &&\textit{Substitute } x^2 + 1 \textit{ for } g(x). \\ &= 2x^2 - 1. &&\textit{Simplify.} \end{aligned}$$

b. The composite of g with f is given by

$$\begin{aligned} g(f(x)) &= (f(x))^2 + 1 &&\textit{Evaluate } g \textit{ at } f(x). \\ &= (2x - 3)^2 + 1 &&\textit{Substitute } 2x - 3 \textit{ for } f(x). \\ &= 4x^2 - 12x + 10. &&\textit{Simplify.} \end{aligned}$$

Inverse Functions

Informally, the inverse of a function f is another function g that "undoes" what f has done.

$$x \xrightarrow{\quad f \quad} f(x) \xrightarrow{\quad g \quad} g(f(x)) = x$$

Definition of Inverse Function

Two functions, f and g, are **inverses** of each other if

$$f(g(x)) = x$$

for each x in the domain of g and

$$g(f(x)) = x$$

for each x in the domain of f. The function g is denoted by f^{-1}, which is read as "f inverse." For f and g to be inverses of each other, the range of g must be equal to the domain of f, and vice versa.

Note   Don't be confused by the "exponential notation" for inverse functions. Whenever we write $f^{-1}(x)$, we will *always* be referring to the inverse of the function f, not to the reciprocal of f.

EXAMPLE 6 *Finding Inverse Functions*

Several functions and their inverses are shown below. In each case, note that the inverse function "undoes" the original function. For instance, to undo multiplication by 2, you should divide by 2.

a. $f(x) = 2x$ $\qquad$ $f^{-1}(x) = \frac{1}{2}x$

b. $f(x) = \frac{1}{3}x$ $\qquad$ $f^{-1}(x) = 3x$

c. $f(x) = x + 4$ $\qquad$ $f^{-1}(x) = x - 4$

d. $f(x) = 2x - 5$ $\qquad$ $f^{-1}(x) = \frac{1}{2}(x + 5)$

e. $f(x) = x^3$ $\qquad$ $f^{-1}(x) = \sqrt[3]{x}$

f. $f(x) = \dfrac{1}{x}$ $\qquad$ $f^{-1}(x) = \dfrac{1}{x}$

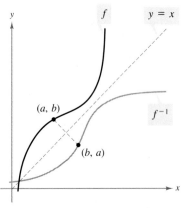

FIGURE 1.44   The graph of f^{-1} is a reflection of the graph of f in the line $y = x$.

The graphs of f and f^{-1} are mirror images of each other (with respect to the line $y = x$), as shown in Figure 1.44. Try using a graphing utility to confirm this for each of the functions given in Example 6.

The functions in Example 6 are simple enough so that their inverses can be found by inspection. The next example demonstrates a strategy for finding the inverses of more complicated functions.

EXAMPLE 7 *Finding the Inverse of a Function*

Find the inverse of the function $f(x) = \sqrt{2x - 3}$.

Solution

Begin by substituting y for $f(x)$. Then, interchange x and y and solve for y.

$$f(x) = \sqrt{2x - 3} \qquad\qquad \textit{Original function}$$

$$y = \sqrt{2x - 3} \qquad\qquad \textit{Substitute } y \textit{ for } f(x).$$

$$x = \sqrt{2y - 3} \qquad\qquad \textit{Interchange } x \textit{ and } y.$$

$$x^2 = 2y - 3 \qquad\qquad \textit{Square both sides.}$$

$$x^2 + 3 = 2y \qquad\qquad \textit{Add 3 to both sides.}$$

$$\frac{x^2 + 3}{2} = y \qquad\qquad \textit{Divide both sides by 2.}$$

Thus, the inverse function has the form

$$f^{-1}(\rule{0.8cm}{0.4pt}) = \frac{(\rule{0.8cm}{0.4pt})^2 + 3}{2}.$$

Using x as the independent variable, you can write

$$f^{-1}(x) = \frac{x^2 + 3}{2}, \qquad x \geq 0.$$

In Figure 1.45, note that the domain of f^{-1} coincides with the range of f. ■

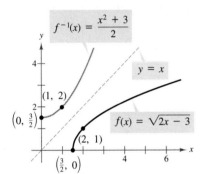

$f^{-1}(x) = \dfrac{x^2 + 3}{2}$

$y = x$

$(1, 2)$

$\left(0, \frac{3}{2}\right)$

$f(x) = \sqrt{2x - 3}$

$(2, 1)$

$\left(\frac{3}{2}, 0\right)$

FIGURE 1.45

TECHNOLOGY

A graphing utility can help you check that the graphs of f and f^{-1} are reflections of each other in the line $y = x$. To do this, graph $y = f(x)$, $y = f^{-1}(x)$, and $y = x$ on the same viewing rectangle, using a *square setting*.

After you have found the inverse of a function, you should check your results. You can check your results *graphically* by observing that the graphs of f and f^{-1} are reflections of each other in the line $y = x$. You can check your results *algebraically* by evaluating $f(f^{-1}(x))$ and $f^{-1}(f(x))$—both should be equal to x.

Check That $f(f^{-1}(x)) = x$

$$f(f^{-1}(x)) = f\left(\frac{x^2 + 3}{2}\right)$$

$$= \sqrt{2\left(\frac{x^2 + 3}{2}\right) - 3}$$

$$= \sqrt{x^2}$$

$$= x, \qquad x \geq 0$$

Check That $f^{-1}(f(x)) = x$

$$f^{-1}(f(x)) = f^{-1}\left(\sqrt{2x - 3}\right)$$

$$= \frac{(\sqrt{2x - 3})^2 + 3}{2}$$

$$= \frac{2x}{2}$$

$$= x, \qquad x \geq \frac{3}{2}$$

Not every function possesses an inverse. In fact, for a function to have an inverse, it must be one-to-one.

EXAMPLE 8 *A Function That Has No Inverse*

Show that the function

$$f(x) = x^2 - 1$$

has no inverse. (Assume that the domain of f is the set of all real numbers.)

Solution

Begin by sketching the graph of f, as shown in Figure 1.46. Note that

$$f(2) = (2)^2 - 1 = 3$$

and

$$f(-2) = (-2)^2 - 1 = 3.$$

Thus f does not pass the horizontal line test, which implies that it is not one-to-one, and therefore has no inverse. The same conclusion can by obtained by trying to find the inverse of f.

$f(x) = x^2 - 1$	*Original function*
$y = x^2 - 1$	*Substitute y for $f(x)$.*
$x = y^2 - 1$	*Interchange x and y.*
$x + 1 = y^2$	*Add 1 to both sides.*
$\pm\sqrt{x + 1} = y$	*Take square root of both sides.*

The last equation does not define y as a function of x, and thus f has no inverse. ■

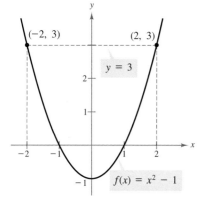

FIGURE 1.46 f is not one-to-one and has no inverse.

Discussion Problem   Comparing Composition Orders

You are buying an automobile whose price is $14,500. Which of the following options would you choose? Explain.

a. You are given a factory rebate of $2000, followed by a dealer discount of 10%.

b. You are given a dealer discount of 10%, followed by a factory rebate of $2000.

Let $f(x) = x - 2000$ and let $g(x) = 0.9x$. Which option is represented by the composite $f(g(x))$? Which is represented by the composite $g(f(x))$?

Warm Up

The following warm-up exercises involve skills that were covered in earlier sections. You will use these skills in the exercise set for this section.

In Exercises 1–6, simplify the expression.

1. $5(-1)^2 - 6(-1) + 9$

2. $(-2)^3 + 7(-2)^2 - 10$

3. $(x - 2)^2 + 5x - 10$

4. $(3 - x) + (x + 3)^3$

5. $\dfrac{1}{1 - (1 - x)}$

6. $1 + \dfrac{x - 1}{x}$

In Exercises 7–10, solve for *y* in terms of *x*.

7. $2x + y - 6 = 11$

8. $5y - 6x^2 - 1 = 0$

9. $(y - 3)^2 = 5 + (x + 1)^2$

10. $y^2 - 4x^2 = 2$

E X E R C I S E S   1.4   means that technology can help you solve or check the exercise(s).

In Exercises 1–6, evaluate the function at the specified values of the independent variable. Simplify the result.

1. $f(x) = 2x - 3$
  (a) $f(0)$ (b) $f(-3)$
  (c) $f(x - 1)$ (d) $f(1 + \Delta x)$

2. $f(x) = x^2 - 2x + 2$
  (a) $f\left(\frac{1}{2}\right)$ (b) $f(-1)$
  (c) $f(c)$ (d) $f(x + \Delta x)$

3. $g(x) = \frac{1}{2}x^2$
  (a) $g(-2)$ (b) $g(6)$
  (c) $g(c)$ (d) $g(x + \Delta x)$

4. $g(x) = \dfrac{1}{x}$
  (a) $g(2)$ (b) $g\left(\frac{1}{4}\right)$
  (c) $g(x + \Delta x)$ (d) $g(x + \Delta x) - g(x)$

5. $f(x) = \dfrac{|x|}{x}$
  (a) $f(2)$ (b) $f(-2)$
  (c) $f(x^2)$ (d) $f(x - 1)$

6. $f(x) = |x| + 4$
  (a) $f(2)$ (b) $f(-2)$
  (c) $f(x^2)$ (d) $f(x + \Delta x) - f(x)$

In Exercises 7–12, evaluate the difference quotient and simplify the result.

7. $f(x) = 3x - 1$
  $\dfrac{f(x + \Delta x) - f(x)}{\Delta x}$

8. $f(x) = 5 - 4x$
  $\dfrac{f(x + \Delta x) - f(x)}{\Delta x}$

9. $h(x) = x^2 - x + 1$
  $\dfrac{h(2 + \Delta x) - h(2)}{\Delta x}$

10. $h(x) = x^3$
  $\dfrac{h(x + \Delta x) - h(x)}{\Delta x}$

11. $f(x) = x^3 - x$
  $\dfrac{f(x + \Delta x) - f(x)}{\Delta x}$

12. $f(x) = \dfrac{1}{\sqrt{x - 1}}$
  $\dfrac{f(x) - f(2)}{x - 2}$

In Exercises 13–20, decide whether the equation defines *y* as a function of *x*.

13. $x^2 + y^2 = 4$

14. $3x - 2y + 5 = 0$

15. $\frac{1}{2}x - 6y = -3$

16. $x + y^2 = 4$

17. $x^2 + y = 4$

18. $x^2 + y^2 - 2x - 4y + 1 = 0$

19. $y^2 = x^2 - 1$

20. $x^2y - x^2 + 4y = 0$

In Exercises 21–26, find the domain and range of the function. Use interval notation to write your result.

21. $f(x) = 4 - 2x$

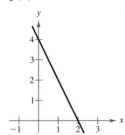

22. $f(x) = x^3$

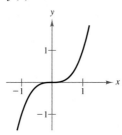

23. $f(x) = \sqrt{2x - 3}$

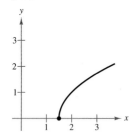

24. $f(x) = 4 - x^2$

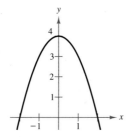

25. $f(x) = \dfrac{1}{|x|}$

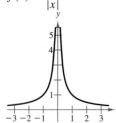

26. $f(x) = |x - 2|$

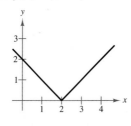

In Exercises 27–32, use a graphing utility to graph the function. Then determine the domain and range of the function.

27. $f(x) = \sqrt{9 - x^2}$

28. $f(x) = 5x^3 + 6x^2 - 1$

29. $f(x) = \dfrac{|x|}{x}$

30. $f(x) = \dfrac{1}{\sqrt{x}}$

31. $f(x) = \dfrac{x - 2}{x + 4}$

32. $f(x) = \dfrac{x^2}{1 - x}$

In Exercises 33–38, use the vertical line test to determine whether *y* is a function of *x*.

33. $y = x^2$

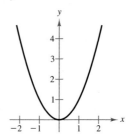

34. $y = x^3 - 1$

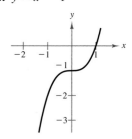

35. $x^2 + y^2 = 9$

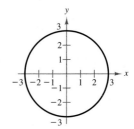

36. $x - xy + y + 1 = 0$

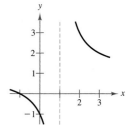

37. $x^2 = xy - 1$

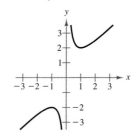

38. $x = |y|$

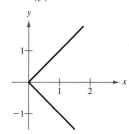

In Exercises 39–46, find (a) $f(x) + g(x)$, (b) $f(x) \cdot g(x)$, and (c) $f(x)/g(x)$. Also, find (d) $f(g(x))$ and (e) $g(f(x))$ if defined.

39. $f(x) = x + 1$ $g(x) = x - 1$

40. $f(x) = 2x - 5$ $g(x) = 1 - x$

41. $f(x) = 2x^2$ $g(x) = x^2 - 1$

42. $f(x) = 2x - 5$ $g(x) = 5$

43. $f(x) = x^2 + 5$ $g(x) = \sqrt{1 - x}$

44. $f(x) = \sqrt{x^2 - 4}$ $g(x) = \dfrac{x^2}{x^2 + 1}$

45. $f(x) = \dfrac{1}{x}$ $g(x) = \dfrac{1}{x^2}$

46. $f(x) = \dfrac{x}{x+1}$ $g(x) = x^3$

In Exercises 47–52, show that f and g are inverse functions by showing that $f(g(x)) = x$ and $g(f(x)) = x$. Then sketch the graphs of f and g on the same coordinate axes.

47. $f(x) = x^3$ $g(x) = \sqrt[3]{x}$

48. $f(x) = \dfrac{1}{x}$ $g(x) = \dfrac{1}{x}$

49. $f(x) = 5x + 1$ $g(x) = \dfrac{x - 1}{5}$

50. $f(x) = 9 - x^2$, $x \geq 0$ $g(x) = \sqrt{9 - x}$, $x \leq 9$

51. $f(x) = \sqrt{x - 4}$ $g(x) = x^2 + 4$, $x \geq 0$

52. $f(x) = 1 - x^3$ $g(x) = \sqrt[3]{1 - x}$

In Exercises 53–60, find the inverse of f. Then sketch the graphs of f and f^{-1} on the same coordinate axes.

53. $f(x) = 2x - 3$ **54.** $f(x) = 6 - 3x$

55. $f(x) = x^5$ **56.** $f(x) = x^3 + 1$

57. $f(x) = \sqrt{9 - x^2}$, $0 \leq x \leq 3$

58. $f(x) = \sqrt{x^2 - 4}$, $x \geq 2$

59. $f(x) = x^{2/3}$, $x \geq 0$

60. $f(x) = x^{3/5}$

In Exercises 61–66, graph the function. Then use the horizontal line test to decide whether the function is one-to-one. If it is, find its inverse.

61. $f(x) = 3 - 7x$ **62.** $f(x) = \sqrt{x - 2}$

63. $f(x) = x^2$ **64.** $f(x) = x^4$

65. $f(x) = |x - 2|$ **66.** $f(x) = 3$

67. Given $f(x) = \sqrt{x}$ and $g(x) = x^2 - 1$, find the composite functions.
   (a) $f(g(1))$ (b) $g(f(1))$
   (c) $g(f(0))$ (d) $f(g(-4))$
   (e) $f(g(x))$ (f) $g(f(x))$

68. Given $f(x) = 1/x$ and $g(x) = x^2 - 1$, find the composite functions.
   (a) $f(g(2))$ (b) $g(f(2))$
   (c) $f\left(g\left(1/\sqrt{2}\right)\right)$ (d) $g\left(f\left(1/\sqrt{2}\right)\right)$
   (e) $f(g(x))$ (f) $g(f(x))$

In Exercises 69–72, select a function from (a) $f(x) = cx$, (b) $g(x) = cx^2$, (c) $h(x) = c\sqrt{|x|}$, or (d) $r(x) = c/x$ and determine the value of the constant c so that the function fits the data in the table.

69.

x	-4	-1	0	1	4
y	-32	-2	0	-2	-32

70.

x	-4	-1	0	1	4
y	-1	$-\frac{1}{4}$	0	$\frac{1}{4}$	1

71.

x	-4	-1	0	1	4
y	-8	-32	undef.	32	8

72.

x	-4	-1	0	1	4
y	6	3	0	3	6

73. Use the graph of $f(x) = \sqrt{x}$ below to sketch the graph of each function.
   (a) $y = \sqrt{x} + 2$
   (b) $y = -\sqrt{x}$
   (c) $y = \sqrt{x - 2}$
   (d) $y = \sqrt{x + 3}$
   (e) $y = \sqrt{x - 4}$
   (f) $y = 2\sqrt{x}$

$f(x) = \sqrt{x}$

74. Use the graph of $f(x) = x^2$ to find a formula for each of the functions whose graphs are shown.
   (a)

$(-3, 0)$

   (b)

$(0, 3)$

75. *Real Estate* Express the value V of a real estate firm in terms of x, the number of acres of property owned. Each acre is valued at \$2500 and other company assets total \$750,000.

76. *Cost* The inventor of a new game believes that the variable cost for producing the game is $0.95 per unit. The fixed cost is $6000.

(a) Express the total cost C as a function of x, the number of games sold.

(b) Find a formula for the average cost per unit $\overline{C} = C/x$.

(c) The selling price for each game is $1.69. How many units must be sold before the average cost per unit falls below the selling price?

77. *Demand* The demand function for a commodity is

$$p = \frac{14.75}{1 + 0.01x}, \qquad x \geq 0,$$

where p is the price per unit and x is the number of units sold.

(a) Find x as a function of p.

(b) Find the number of units sold when the price is $10.

78. *Cost* A power station is on one side of a river that is $\frac{1}{2}$ mile wide. A factory is 3 miles downstream on the other side of the river (see figure). It costs $10/ft to run the power lines on land and $15/ft to run them underwater. Express the cost C of running the line from the power station to the factory as a function of x.

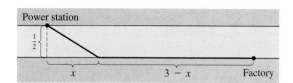

79. *Market Equilibrium* The supply function for a product relates the number of units x that producers are willing to supply for a given price per unit p. The supply and demand functions for a market are

$$p = \tfrac{2}{5}x + 4 \qquad\qquad \textit{Supply}$$
$$p = -\tfrac{16}{25}x + 30. \qquad \textit{Demand}$$

(a) Use a graphing utility to graph the supply and demand functions on the same viewing rectangle.

(b) Use the trace feature of the graphing utility to find the *equilibrium point* for the market. That is, find the point of intersection of the two graphs.

(c) For what values of x does the demand exceed the supply?

(d) For what values of x does the supply exceed the demand?

80. *Profit* A radio manufacturer charges $90 per unit for units that cost $60 to produce. To encourage large orders from distributors, the manufacturer will reduce the price by $0.01 per unit for each unit in excess of 100 units. (For example, an order of 101 units would have a price of $89.99 per unit, and an order of 102 units would have a price of $89.98 per unit.) This price reduction is discontinued when the price per unit drops to $75.

(a) Express the price per unit p as a function of the order size x.

(b) Express the profit P as a function of the order size x.

81. *Cost, Revenue, and Profit* A company invests $98,000 for equipment to produce a new product. Each unit of the product costs $12.30 and is sold for $17.98. Let x be the number of units produced and sold.

(a) Write the total cost C as a function of x.

(b) Write the revenue R as a function of x.

(c) Write the profit P as a function of x.

82. *Charter Bus Fares* For groups of 80 or more people, a charter bus company determines the rate r (in dollars per person) according to the formula

$$r = 8 - 0.05(n - 80), \qquad n \geq 80,$$

where n is the number of people.

(a) Express the revenue R for the bus company as a function of n.

(b) Complete the following table.

n	90	100	110	120	130	140	150
R							

(c) Criticize the formula for the rate. Would you use this formula? Explain your reasoning.

In Exercises 83–88, use a graphing utility to graph the function. Then use the zoom and trace features to find the zeros of the function. Is the function one-to-one?

83. $f(x) = x\sqrt{9 - x^2}$

84. $f(x) = 2\left(3x^2 - \dfrac{6}{x}\right)$

85. $g(t) = \dfrac{t + 3}{1 - t}$

86. $h(x) = 6x^3 - 12x^2 + 4$

87. $f(x) = \dfrac{4 - x^2}{x}$

88. $g(x) = \left|\tfrac{1}{2}x^2 - 4\right|$

Limits

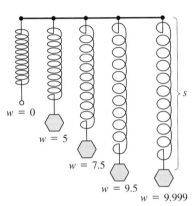

FIGURE 1.52 What is the limit of *s* as *w* approaches 10 lb?

The Limit of a Function

In everyday language, we refer to a speed limit, a wrestler's weight limit, the limit of one's endurance, or stretching a spring to its limit. These phrases all suggest that a limit is a bound, which on some occasions may not be reached but on other occasions may be reached or exceeded.

   Consider a spring that will break only if a weight of 10 or more pounds is attached. To determine how far the spring will stretch without breaking, you could attach increasingly heavier weights and measure the spring length *s* for each weight *w*, as shown in Figure 1.47. If the spring length approaches a value of *L*, then we say that "the limit of *s* as *w* approaches 10 is *L*." A mathematical limit is much like the limit of a spring. The notation for a limit is

$$\lim_{x \to c} f(x) = L,$$

which is read as "the limit of $f(x)$ as x approaches c is L."

EXAMPLE 1 *Finding a Limit*

Find the limit: $\lim\limits_{x \to 1} (x^2 + 1)$.

Solution
Let $f(x) = x^2 + 1$. From the graph of f in Figure 1.48, it appears that $f(x)$ approaches 2 as x approaches 1 from either side, and you can write

$$\lim_{x \to 1} (x^2 + 1) = 2.$$

Table 1.7 yields the same conclusion. Notice that as x gets closer and closer to 1, $f(x)$ gets closer and closer to 2.

TABLE 1.7

	x approaches 1				*x* approaches 1		
x	0.900	0.990	0.999	1.000	1.001	1.010	1.100
$f(x)$	1.810	1.980	1.998	2.000	2.002	2.020	2.200

f(x) approaches 2   *f(x)* approaches 1

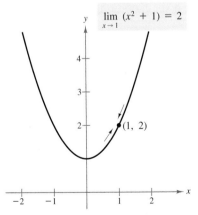

$$\lim_{x \to 1} (x^2 + 1) = 2$$

FIGURE 1.48

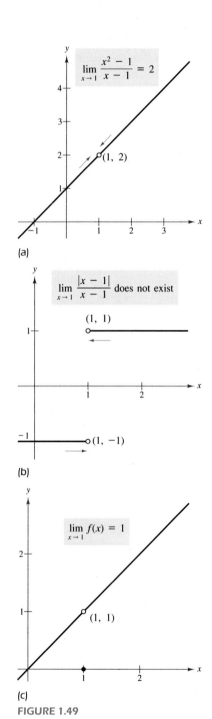

(a)

(b)

(c)

FIGURE 1.49

EXAMPLE 2 *Finding Limits Graphically and Numerically*

Find the limit: $\lim\limits_{x\to 1} f(x)$.

a. $f(x) = \dfrac{x^2 - 1}{x - 1}$ **b.** $f(x) = \dfrac{|x - 1|}{x - 1}$ **c.** $f(x) = \begin{cases} x, & x \neq 1 \\ 0, & x = 1 \end{cases}$

Solution

a. From the graph of f, in Figure 1.49(a), it appears that $f(x)$ approaches 2 as x approaches 1 from either side. This conclusion is reinforced by Table 1.8. Be sure you see that *it does not matter that $f(x)$ is undefined when $x = 1$.* The limit depends only on values of $f(x)$ near 1, not at 1.

TABLE 1.8

	x approaches 1				x approaches 1		
x	0.900	0.990	0.999	1.000	1.001	1.010	1.100
$f(x)$	1.900	1.990	1.999	?	2.001	2.010	2.100

f(x) approaches 2   f(x) approaches 1

b. From the graph of f, in Figure 1.49(b), you can see that $f(x) = -1$ for all values to the left of $x = 1$ and $f(x) = 1$ for all values to the right of $x = 1$. Thus, $f(x)$ is approaching a different value from the left of $x = 1$ than it is from the right of $x = 1$. In such situations, we say that the *limit does not exist.* This conclusion is reinforced by Table 1.9.

TABLE 1.9

	x approaches 1				x approaches 1		
x	0.900	0.990	0.999	1.000	1.001	1.010	1.100
$f(x)$	-1.000	-1.000	-1.000	?	1.000	1.000	1.000

f(x) approaches –1   f(x) approaches 1

c. From the graph of f, in Figure 1.49(c), it appears that $f(x)$ approaches 1 as x approaches 1 from either side. This conclusion is reinforced by Table 1.10. It does not matter that $f(1) = 0$. The limit depends only on values of $f(x)$ near 1, not at 1.

TABLE 1.10

	x approaches 1				x approaches 1		
x	0.900	0.990	0.999	1.000	1.001	1.010	1.100
$f(x)$	0.900	0.990	0.999	0.000	1.001	1.010	1.100

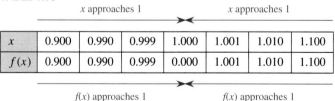

f(x) approaches 1   f(x) approaches 1

TECHNOLOGY

Try using a graphing utility to determine the following limit.

$$\lim_{x \to 1} \frac{x^3 + 4x - 5}{x - 1}$$

You can do this by graphing

$$f(x) = \frac{x^3 + 4x - 5}{x - 1}$$

and zooming in near $x = 1$. From the graph, what does the limit appear to be?

There are three important ideas to learn from Examples 1 and 2.

1. Saying that the limit of $f(x)$ approaches L as x approaches c means that the value of $f(x)$ may be made *arbitrarily close* to the number L by choosing x closer and closer to c.

2. For a limit to exist, you must allow x to approach c from *either side* of c. If $f(x)$ approaches a different number as x approaches c from the left than it does when x approaches c from the right, then the limit *does not exist*. (See Example 2b.)

3. The value of $f(x)$ when $x = c$ has no bearing on the existence or nonexistence of the limit of $f(x)$ as x approaches c. For instance, in Example 2a, the limit of $f(x)$ exists as x approaches 1 even though the function f is not defined at $x = 1$.

Definition of the Limit of a Function

If $f(x)$ becomes arbitrarily close to a single number L as x approaches c from either side, then

$$\lim_{x \to c} f(x) = L$$

which is read as "the **limit** of $f(x)$ as x approaches c is L."

Properties of Limits

Many times the limit of $f(x)$ as x approaches c is simply $f(c)$, as illustrated in Example 1. Whenever the limit of $f(x)$ as x approaches c is

$$\lim_{x \to c} f(x) = f(c), \qquad \textit{Direct substitution}$$

we say that the limit can be evaluated by **direct substitution.** (In the next section, you will learn that a function that has this property is *continuous at c.*) It is important that you learn to recognize the types of functions that have this property. Some basic ones are given in the following list.

Properties of Limits

Let b and c be real numbers, and let n be a positive integer.

1. $\lim\limits_{x \to c} b = b$ 2. $\lim\limits_{x \to c} x = c$

3. $\lim\limits_{x \to c} x^n = c^n$ 4. $\lim\limits_{x \to c} \sqrt[n]{x} = \sqrt[n]{c}$

In Property 4, if n is even, then c must be positive.

By combining the properties of limits with the following rules for operating with limits, you can find limits for a wide variety of algebraic functions.

▦ TECHNOLOGY

Symbolic algebra utilities, such as *Derive, Mathematica,* and *Maple*, are capable of evaluating limits. For instance, *Derive* can evaluate the limit in Example 3 as follows.

1: $x^2 + 2x - 3$ *Author*

2: $\lim\limits_{x \to 2} (x^2 + 2x - 3)$ *Limit*

3: 5   *Simplify.*

Operations with Limits

Let b and c be real numbers and let n be a positive integer. If the limits of $f(x)$ and $g(x)$ exist as x approaches c, then the following operations are valid.

1. Constant multiple: $\lim\limits_{x \to c} [bf(x)] = b\left[\lim\limits_{x \to c} f(x)\right]$

2. Addition: $\lim\limits_{x \to c} [f(x) \pm g(x)] = \lim\limits_{x \to c} f(x) \pm \lim\limits_{x \to c} g(x)$

3. Multiplication: $\lim\limits_{x \to c} [f(x) \cdot g(x)] = \left[\lim\limits_{x \to c} f(x)\right]\left[\lim\limits_{x \to c} g(x)\right]$

4. Division: $\lim\limits_{x \to c} \dfrac{f(x)}{g(x)} = \dfrac{\lim\limits_{x \to c} f(x)}{\lim\limits_{x \to c} g(x)}$, $\lim\limits_{x \to c} g(x) \neq 0$

5. Power: $\lim\limits_{x \to c} [f(x)]^n = \left[\lim\limits_{x \to c} f(x)\right]^n$

6. Radical: $\lim\limits_{x \to c} \sqrt[n]{f(x)} = \sqrt[n]{\lim\limits_{x \to c} f(x)}$

In Property 6, if n is even and $\lim\limits_{x \to c} f(x) = L$, then L must be positive.

Discovery

Use a graphing utility to graph $Y_1 = 1/x^2$. Does Y_1 approach a limit as x approaches 0? Evaluate $Y_1 = 1/x^2$ at several positive and negative values of x near 0 to confirm your answer. Does $\lim\limits_{x \to 1} 1/x^2$ exist?

EXAMPLE 3   *Evaluating the Limit of a Polynomial*

$$\lim_{x \to 2} (x^2 + 2x - 3) = \lim_{x \to 2} x^2 + \lim_{x \to 2} 2x - \lim_{x \to 2} 3 \qquad \textit{Addition Property}$$
$$= 2^2 + 2(2) - 3 \qquad \textit{Direct substitution}$$
$$= 4 + 4 - 3 \qquad \textit{Simplify.}$$
$$= 5$$

Example 3 is an illustration of the following important result, which states that the limit of a polynomial can be evaluated by direct substitution.

The Limit of a Polynomial Function

If p is a polynomial function and c is any real number, then

$$\lim_{x \to c} p(x) = p(c).$$

Techniques for Evaluating Limits

Many techniques for evaluating limits are based on the following important theorem. Basically, the theorem states that if two functions agree at all but a single point c, then they have identical limit behavior at $x = c$.

The Replacement Theorem

Let c be a real number and $f(x) = g(x)$ for all $x \neq c$. If the limit of $g(x)$ exists as $x \to c$, then the limit of $f(x)$ also exists and

$$\lim_{x \to c} f(x) = \lim_{x \to c} g(x).$$

To apply the Replacement Theorem, you can use a result from algebra which states that for a polynomial function p, $p(c) = 0$ if and only if $(x - c)$ is a factor of $p(x)$. This concept is demonstrated in Example 4.

EXAMPLE 4 *Evaluating a Limit*

Evaluate the limit: $\displaystyle\lim_{x \to 1} \frac{x^3 - 1}{x - 1}$.

Solution

Note that the numerator and denominator are zero when $x = 1$. This implies that $(x - 1)$ is a factor of both, and you can cancel this common factor.

$$\frac{x^3 - 1}{x - 1} = \frac{(x - 1)(x^2 + x + 1)}{x - 1} \qquad \text{\textit{Factor numerator.}}$$

$$= \frac{\cancel{(x - 1)}(x^2 + x + 1)}{\cancel{x - 1}} \qquad \text{\textit{Cancel common factor.}}$$

$$= x^2 + x + 1, \qquad x \neq 1 \qquad \text{\textit{Simplify.}}$$

Thus, the rational function $(x^3 - 1)/(x - 1)$ and the polynomial function $x^2 + x + 1$ agree for all values of x other than $x = 1$, and you can apply the Replacement Theorem.

$$\lim_{x \to 1} \frac{x^3 - 1}{x - 1} = \lim_{x \to 1} (x^2 + x + 1) = 1^2 + 1 + 1 = 3$$

Figure 1.50 illustrates this result graphically. Note that the two graphs are identical except that the graph of g contains the point $(1, 3)$, whereas this point is missing on the graph of f. (In Figure 1.50, the missing point is denoted by an open dot.)

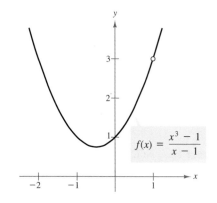

$$f(x) = \frac{x^3 - 1}{x - 1}$$

$$g(x) = x^2 + x + 1$$

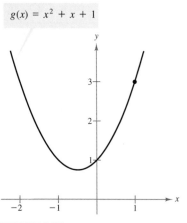

FIGURE 1.50

The technique used to evaluate the limit in Example 4 is called the **cancellation technique.** This technique is further demonstrated in the next example.

EXAMPLE 5 *Using the Cancellation Technique*

Evaluate the limit: $\lim\limits_{x \to -3} \dfrac{x^2 + x - 6}{x + 3}$.

Solution

Direct substitution fails because both the numerator and the denominator are zero when $x = -3$.

$$\lim_{x \to -3} \frac{x^2 + x - 6}{x + 3} \quad \begin{aligned} &\longleftarrow \lim_{x \to -3} (x^2 + x - 6) = 0 \\ &\longleftarrow \lim_{x \to -3} (x + 3) = 0 \end{aligned}$$

However, because the limits of both the numerator and denominator are zero, you know that they have a *common factor* of $(x + 3)$. Thus, for all $x \neq -3$, you can cancel this factor to obtain the following.

$$
\begin{aligned}
\lim_{x \to -3} \frac{x^2 + x - 6}{x + 3} &= \lim_{x \to -3} \frac{(x - 2)(x + 3)}{x + 3} && \textit{Factor numerator.} \\
&= \lim_{x \to -3} \frac{(x - 2)\cancel{(x + 3)}}{\cancel{x + 3}} && \textit{Cancel common factor.} \\
&= \lim_{x \to -3} (x - 2) && \textit{Simplify.} \\
&= -5 && \textit{Direct substitution}
\end{aligned}
$$

This result is shown graphically in Figure 1.51. Note that the graph of f coincides with the graph of $g(x) = x - 2$, except that the graph of f has a hole at $(-3, -5)$.   ■

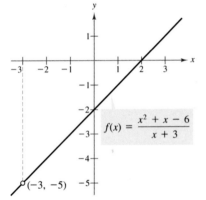

$$f(x) = \frac{x^2 + x - 6}{x + 3}$$

$(-3, -5)$

FIGURE 1.51 f is undefined when $x = -3$.

EXAMPLE 6 *Evaluating a Limit*

Evaluate the limit: $\lim\limits_{x \to -2} \dfrac{x^2 + x - 2}{x - 2}$.

Solution

Because the limit of the denominator is not zero, you can use the limit of a quotient (Property 4 of Operations with Limits) to obtain

$$
\begin{aligned}
\lim_{x \to -2} \frac{x^2 + x - 2}{x - 2} &= \lim_{x \to -2} \frac{(-2)^2 - 2 - 2}{-2 - 2} && \textit{Direct substitution} \\
&= \frac{0}{-4} && \textit{Simplify.} \\
&= 0. && \textit{The limit is 0.}
\end{aligned}
$$

■

One-Sided Limits

In Example 2b, you saw that one way in which a limit can fail to exist is when a function approaches a different value from the left of c than it approaches from the right of c. This type of behavior can be described more concisely with the concept of a **one-sided limit.**

$$\lim_{x \to c^-} f(x) = L \qquad\qquad \textit{Limit from the left}$$

$$\lim_{x \to c^+} f(x) = L \qquad\qquad \textit{Limit from the right}$$

The first of these two limits is read as "the limit of $f(x)$ as x approaches c from the left is L." The second is read as "the limit of $f(x)$ as x approaches c from the right is L."

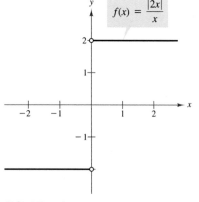

$$f(x) = \frac{|2x|}{x}$$

FIGURE 1.52

EXAMPLE 7 *Evaluating One-Sided Limits*

Find the limit as $x \to 0$ from the left and the limit as $x \to 0$ from the right for the function

$$f(x) = \frac{|2x|}{x}.$$

Solution

From the graph of f, shown in Figure 1.52, you can see that $f(x) = -2$ for all $x < 0$. Therefore, the limit from the left is

$$\lim_{x \to 0^-} \frac{|2x|}{x} = -2. \qquad\qquad \textit{Limit from the left}$$

Because $f(x) = 2$ for all $x > 0$, the limit from the right is

$$\lim_{x \to 0^+} \frac{|2x|}{x} = 2. \qquad\qquad \textit{Limit from the right}$$ ▬

In Example 7, note that the function approaches different limits from the left and from the right. In such cases, the limit of $f(x)$ as $x \to c$ does not exist. For the limit of a function to exist as $x \to c$, it must be true that *both* one-sided limits exist and are equal.

Existence of a Limit

If f is a function and c and L are real numbers, then

$$\lim_{x \to c} f(x) = L$$

if and only if both the left and right limits are equal to L.

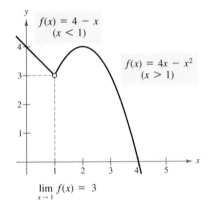

$f(x) = 4 - x$
$(x < 1)$

$f(x) = 4x - x^2$
$(x > 1)$

$$\lim_{x \to 1} f(x) = 3$$

FIGURE 1.53

EXAMPLE 8 *Evaluating One-Sided Limits*

Find the limit of $f(x)$ as x approaches 1.

$$f(x) = \begin{cases} 4 - x, & x < 1 \\ 4x - x^2, & x > 1 \end{cases}$$

Solution

Remember that you are concerned about the value of f near $x = 1$ rather than at $x = 1$. Thus, for $x < 1$, $f(x)$ is given by $4 - x$, and you can use direct substitution to obtain

$$\lim_{x \to 1^-} f(x) = \lim_{x \to 1^-} (4 - x) = 4 - 1 = 3.$$

For $x > 1$, $f(x)$ is given by $4x - x^2$, and you can use direct substitution to obtain

$$\lim_{x \to 1^+} f(x) = \lim_{x \to 1^+} (4x - x^2) = 4 - 1 = 3.$$

Because both one-sided limits exist and are equal to 3, it follows that

$$\lim_{x \to 1} f(x) = 3.$$

The graph in Figure 1.53 confirms this conclusion.

Real Life

EXAMPLE 9 *Comparing One-Sided Limits*

An overnight delivery service charges $8 for the first pound and $2 for each additional pound. Let x represent the weight of a parcel and let $f(x)$ represent the shipping cost.

$$f(x) = \begin{cases} 8, & 0 < x \le 1 \\ 10, & 1 < x \le 2 \\ 12, & 2 < x \le 3 \end{cases}$$

Show that the limit of $f(x)$ as $x \to 2$ does not exist.

Solution

The graph of f is shown in Figure 1.54. The limit of $f(x)$ as x approaches 2 from the left is

$$\lim_{x \to 2^-} f(x) = 10,$$

whereas the limit of $f(x)$ as x approaches 2 from the right is

$$\lim_{x \to 2^+} f(x) = 12.$$

Because these one-sided limits are not equal, the limit of $f(x)$ as $x \to 2$ does not exist.

Delivery Service Rates

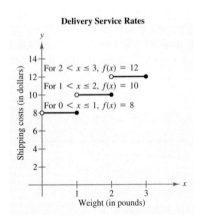

For $2 < x \le 3$, $f(x) = 12$

For $1 < x \le 2$, $f(x) = 10$

For $0 < x \le 1$, $f(x) = 8$

Shipping costs (in dollars)

Weight (in pounds)

FIGURE 1.54

Unbounded Behavior

Example 9 illustrates a limit that fails to exist because the limits from the left and right differ. Another important way in which a limit can fail to exist is when $f(x)$ increases or decreases without bound as x approaches c.

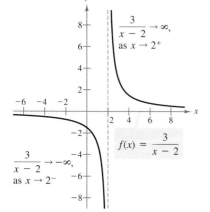

$\dfrac{3}{x-2} \to \infty,$ as $x \to 2^+$

$f(x) = \dfrac{3}{x-2}$

$\dfrac{3}{x-2} \to -\infty,$ as $x \to 2^-$

FIGURE 1.55

EXAMPLE 10 *An Unbounded Function*

Evaluate the limit (if possible): $\displaystyle\lim_{x \to 2} \frac{3}{x-2}$.

Solution

From Figure 1.55, you can see that $f(x)$ decreases without bound as x approaches 2 from the left and $f(x)$ increases without bound as x approaches 2 from the right. Symbolically, you can write this as

$$\lim_{x \to 2^-} \frac{3}{x-2} = -\infty$$

and

$$\lim_{x \to 2^+} \frac{3}{x-2} = \infty.$$

Because f is unbounded as x approaches 2, the limit does not exist. ■

Note   The equal sign in the statement

$$\lim_{x \to c^+} f(x) = \infty$$

does not mean that the limit exists! On the contrary, it tells you how the limit *fails to exist* by denoting the unbounded behavior of $f(x)$ as x approaches c.

Discussion Problem   *Evaluating a Limit*

Consider the following limit.

$$\lim_{x \to 0} \frac{\sqrt{x+1} - 1}{x}$$

a. Approximate the limit graphically, using a graphing utility.
b. Approximate the limit numerically, by constructing a table.
c. Evaluate the limit analytically, using the following equation.

$$\frac{\sqrt{x+1} - 1}{x} = \left(\frac{\sqrt{x+1} - 1}{x} \right) \left(\frac{\sqrt{x+1} + 1}{\sqrt{x+1} + 1} \right) = \frac{1}{\sqrt{x+1} + 1}$$

Which method do you prefer? Explain.

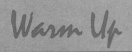

The following warm-up exercises involve skills that were covered in earlier sections. You will use these skills in the exercise set for this section.

In Exercises 1–4, evaluate the expression and simplify.

1. $f(x) = x^2 - 3x + 3$ (a) $f(-1)$ (b) $f(c)$ (c) $f(x + h)$

2. $f(x) = \begin{cases} 2x - 2, & x < 1 \\ 3x + 1, & x \geq 1 \end{cases}$ (a) $f(-1)$ (b) $f(3)$ (c) $f(t^2 + 1)$

3. $f(x) = x^2 - 2x + 2$ $\dfrac{f(1 + h) - f(1)}{h}$

4. $f(x) = 4x$ $\dfrac{f(2 + h) - f(2)}{h}$

In Exercises 5–8, find the domain and range of the function and sketch its graph.

5. $h(x) = -\dfrac{5}{x}$ 6. $g(x) = \sqrt{25 - x^2}$

7. $f(x) = |x - 3|$ 8. $f(x) = \dfrac{|x|}{x}$

In Exercises 9 and 10, determine whether y is a function of x.

9. $9x^2 + 4y^2 = 49$ 10. $2x^2y + 8x = 7y$

E X E R C I S E S   1.5   means that technology can help you solve or check the exercise(s).

In Exercises 1–8, complete the table and estimate the limit.

1. $\lim\limits_{x \to 2} (5x + 4)$

x	1.9	1.99	1.999	2	2.001	2.01	2.1
$f(x)$	?	?	?	?	?	?	?

2. $\lim\limits_{x \to 2} \dfrac{x - 2}{x^2 - x - 2}$

x	1.9	1.99	1.999	2	2.001	2.01	2.1
$f(x)$	?	?	?	?	?	?	?

3. $\lim\limits_{x \to 2} \dfrac{x - 2}{x^2 - 4}$

x	1.9	1.99	1.999	2	2.001	2.01	2.1
$f(x)$	?	?	?	?	?	?	?

4. $\lim\limits_{x \to 2} \dfrac{x^5 - 32}{x - 2}$

x	1.9	1.99	1.999	2	2.001	2.01	2.1
$f(x)$	?	?	?	?	?	?	?

5. $\lim\limits_{x \to 0} \dfrac{\sqrt{x + 3} - \sqrt{3}}{x}$

x	-0.1	-0.01	-0.001	0	0.001	0.01	0.1
$f(x)$	?	?	?	?	?	?	?

6. $\lim\limits_{x \to 0} \dfrac{\sqrt{x + 2} - \sqrt{2}}{x}$

x	-0.1	-0.01	-0.001	0	0.001	0.01	0.1
$f(x)$	?	?	?	?	?	?	?

7. $\lim\limits_{x \to 2^-} \dfrac{2-x}{\sqrt{4-x^2}}$

x	1.5	1.9	1.99	1.999	2
$f(x)$	?	?	?	?	?

8. $\lim\limits_{x \to 0^+} \dfrac{[1/(2+x)] - (1/2)}{2x}$

x	0.5	0.1	0.01	0.001	0
$f(x)$	?	?	?	?	?

In Exercises 9–12, use the graph to visually determine the limit.

9.

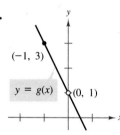

(a) $\lim\limits_{x \to 0} f(x)$

(b) $\lim\limits_{x \to -1} f(x)$

10.

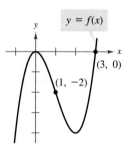

(a) $\lim\limits_{x \to 1} f(x)$

(b) $\lim\limits_{x \to 3} f(x)$

11.

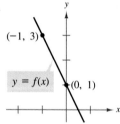

(a) $\lim\limits_{x \to 0} g(x)$

(b) $\lim\limits_{x \to -1} g(x)$

12.

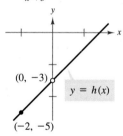

(a) $\lim\limits_{x \to -2} h(x)$

(b) $\lim\limits_{x \to 0} h(x)$

In Exercises 13 and 14, find the limit of (a) $f(x) + g(x)$, (b) $f(x)g(x)$, and (c) $f(x)/g(x)$ as x approaches c.

13. $\lim\limits_{x \to c} f(x) = 3$
$\lim\limits_{x \to c} g(x) = 9$

14. $\lim\limits_{x \to c} f(x) = \frac{3}{2}$
$\lim\limits_{x \to c} g(x) = \frac{1}{2}$

In Exercises 15–20, use the graph to visually determine the limit (if it exists).

(a) $\lim\limits_{x \to c^+} f(x)$ (b) $\lim\limits_{x \to c^-} f(x)$ (c) $\lim\limits_{x \to c} f(x)$

15.

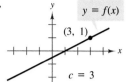

16.

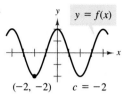

17.

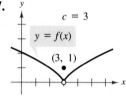

18.

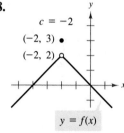

19.

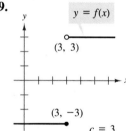

20.

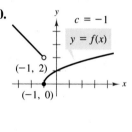

In Exercises 21–32, find the limit.

21. $\lim\limits_{x \to 2} x^2$

22. $\lim\limits_{x \to -3} (3x + 2)$

23. $\lim\limits_{x \to 0} (2x - 3)$

24. $\lim\limits_{x \to 1} (1 - x^2)$

25. $\lim\limits_{x \to 2} (-x^2 + x - 2)$

26. $\lim\limits_{x \to 1} (3x^3 - 2x^2 + 4)$

27. $\lim\limits_{x \to 3} \sqrt{x + 1}$

28. $\lim\limits_{x \to 4} \sqrt[3]{x + 4}$

29. $\lim\limits_{x \to 2} -\dfrac{2}{x}$

30. $\lim\limits_{x \to -3} \dfrac{2}{x + 2}$

31. $\lim\limits_{x \to 3} \dfrac{3x - 5}{x + 3}$

32. $\lim\limits_{x \to -2} \dfrac{3x + 1}{2 - x}$

In Exercises 33–54, find the limit (if it exists).

33. $\lim\limits_{x \to -1} \dfrac{x^2 - 1}{x + 1}$

34. $\lim\limits_{x \to -1} \dfrac{2x^2 - x - 3}{x + 1}$

35. $\lim\limits_{x \to 2} \dfrac{x - 2}{x^2 - 4x + 4}$

36. $\lim\limits_{x \to 2} \dfrac{2 - x}{x^2 - 4}$

37. $\lim\limits_{t \to 5} \dfrac{t - 5}{t^2 - 25}$

38. $\lim\limits_{t \to 1} \dfrac{t^2 + t - 2}{t^2 - 1}$

39. $\lim\limits_{x \to -2} \dfrac{x^3 + 8}{x + 2}$

40. $\lim\limits_{x \to 1} \dfrac{x^3 - 1}{x - 1}$

41. $\lim\limits_{x \to 0} \dfrac{|x|}{x}$.

42. $\lim\limits_{x \to 2} \dfrac{|x - 2|}{x - 2}$

43. $\lim\limits_{x \to 3} f(x)$, where $f(x) = \begin{cases} \frac{1}{3}x - 2, & x \le 3 \\ -2x + 5, & x > 3 \end{cases}$

44. $\lim\limits_{s \to 1} f(s)$, where $f(s) = \begin{cases} s, & s \le 1 \\ 1 - s, & s > 1 \end{cases}$

45. $\lim\limits_{x \to 1} f(x)$, where $f(x) = \begin{cases} x^3 + 1, & x < 1 \\ x + 1, & x \ge 1 \end{cases}$

46. $\lim\limits_{x \to 2} f(x)$, where $f(x) = \begin{cases} x^2 - 4x + 6, & x < 2 \\ -x^2 + 4x - 2, & x \ge 2 \end{cases}$

47. $\lim\limits_{\Delta x \to 0} \dfrac{2(x + \Delta x) - 2x}{\Delta x}$

48. $\lim\limits_{\Delta x \to 0} \dfrac{(1 + \Delta x)^3 - 1}{\Delta x}$

49. $\lim\limits_{\Delta x \to 0} \dfrac{(x + \Delta x)^3 - x^3}{\Delta x}$

50. $\lim\limits_{\Delta x \to 0} \dfrac{\sqrt{x + \Delta x} - \sqrt{x}}{\Delta x}$

51. $\lim\limits_{\Delta t \to 0} \dfrac{(t + \Delta t)^2 - 5(t + \Delta t) - (t^2 - 5t)}{\Delta t}$

52. $\lim\limits_{\Delta x \to 0} \dfrac{(x + \Delta x)^2 - 2(x + \Delta x) + 1 - (x^2 - 2x + 1)}{\Delta x}$

53. $\lim\limits_{x \to 0} \dfrac{\sqrt{2 + x} - \sqrt{2}}{x}$

54. $\lim\limits_{x \to 4} \dfrac{\sqrt{x} - 2}{x - 4}$

In Exercises 55–60, find the one-sided limit (if it exists).

55. $\lim\limits_{x \to 1^-} \dfrac{2}{x^2 - 1}$

56. $\lim\limits_{x \to 1^+} \dfrac{5}{1 - x}$

57. $\lim\limits_{x \to -2^-} \dfrac{1}{x + 2}$

58. $\lim\limits_{x \to 0^-} \dfrac{x + 1}{x}$

59. $\lim\limits_{x \to 5^+} \dfrac{2}{(x - 5)^2}$

60. $\lim\limits_{x \to 3^+} \dfrac{x}{x^2 - 2x - 3}$

61. Find $\lim\limits_{x \to 0} f(x)$, given
$$4 - x^2 \le f(x) \le 4 + x^2, \quad \text{for all } x.$$

62. The limit of $f(x) = (1 + x)^{1/x}$ is a natural base for many business applications.

$$\lim\limits_{x \to 0} (1 + x)^{1/x} = e \approx 2.718$$

(a) Show the reasonableness of this limit by completing the table.

x	−0.01	−0.001	−0.0001	0	0.0001	0.001	0.01
$f(x)$	?	?	?	?	?	?	?

(b) Use a graphing utility to graph f and to confirm the answer in part (a).

(c) Find the domain and range of the function.

 In Exercises 63–66, use a graphing utility to estimate the limit (if it exists).

63. $\lim\limits_{x \to 2} \dfrac{x^2 - 5x + 6}{x^2 - 4x + 4}$

64. $\lim\limits_{x \to 1} \dfrac{x^2 + 6x - 7}{x^3 - x^2 + 2x - 2}$

65. $\lim\limits_{x \to -4} \dfrac{x^3 + 4x^2 + x + 4}{2x^2 + 7x - 4}$

66. $\lim\limits_{x \to -2} \dfrac{4x^3 + 7x^2 + x + 6}{3x^2 - x - 14}$

67. *Compound Interest* You deposit $1000 into an account that is compounded quarterly at an annual rate of r (in decimal form). The balance A after 10 years is

$$A = 1000 \left(1 + \dfrac{r}{4}\right)^{40}.$$

Does the limit of A exist as the interest rate approaches 6%? If so, what is the limit?

68. *Compound Interest* Consider a certificate of deposit that pays 10% (annual percentage rate) on an initial deposit of $500. The balance A after 10 years is

$$A = 500(1 + 0.1x)^{10/x},$$

where x is the length of the compounding period (in years).

(a) Use a graphing utility to graph A, where $0 \le x \le 1$.

(b) Use the zoom and trace features to estimate the balance for quarterly compounding and daily compounding.

(c) Use the zoom and trace features to estimate

$$\lim\limits_{x \to 0^+} A.$$

What do you think this limit represents? Explain your reasoning.

1.6

Continuity

Continuity ■ *Continuity on a Closed Interval* ■ *The Greatest Integer Function* ■ *Extended Application: Compound Interest*

Continuity

In mathematics, the term "continuous" has much the same meaning as it does in everyday use. To say that a function is continuous at $x = c$ means that there is no interruption in the graph of f at c. The graph of f is unbroken at c, and there are no holes, jumps, or gaps. As simple as this concept may seem, its precise definition eluded mathematicians for many years. In fact, it was not until the early 1800s that a precise definition was finally developed.

Before looking at this definition, consider the function whose graph is shown in Figure 1.56. This figure identifies three values of x at which the function f is not continuous.

1. At $x = c_1$, $f(c_1)$ is not defined.

2. At $x = c_2$, $\lim_{x \to c_2} f(x)$ does not exist.

3. At $x = c_3$, $f(c_3) \neq \lim_{x \to c_3} f(x)$.

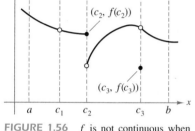
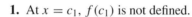

FIGURE 1.56 f is not continuous when $x = c_1, c_2, c_3$.

At all other points in the interval (a, b), the graph of f is uninterrupted, which implies that the function f is continuous at all other points in the interval (a, b).

Definition of Continuity

Let c be a number in the interval (a, b), and let f be a function whose domain contains the interval (a, b). The function f is **continuous at the point** c if the following conditions are true.

1. $f(c)$ is defined.

2. $\lim_{x \to c} f(x)$ exists.

3. $\lim_{x \to c} f(x) = f(c)$.

If f is continuous at every point in the interval (a, b), then it is **continuous on the interval** (a, b).

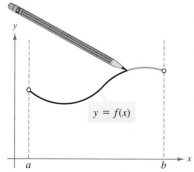

FIGURE 1.57   On the interval (a, b), the graph of f can be traced with a pencil.

Roughly, you can say that a function is continuous on an interval if its graph on the interval can be traced using a pencil and paper without lifting the pencil from the paper, as shown in Figure 1.57.

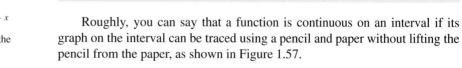

Most graphing utilities can draw graphs in two different modes: connected mode and dot mode. The connected mode works well as long as the function is continuous on the entire interval represented by the viewing rectangle. If, however, the function is not continuous at one or more x-values in the viewing rectangle, then the connected mode may try to "connect" parts of the graphs that should not be connected. For instance, try graphing the function $y_1 = (x + 3)/(x - 2)$ on the viewing rectangle $-9 \le x \le 9$ and $-6 \le y \le 6$. Do you notice any problems?

In Section 1.5, you studied several types of functions that meet the three conditions for continuity. Specifically, if *direct substitution* can be used to evaluate the limit of a function at c, then the function is continuous at c. Two types of functions that have this property are polynomial functions and rational functions.

Continuity of Polynomial and Rational Functions

1. A polynomial function is continuous at every real number.
2. A rational function is continuous at every number in its domain.

EXAMPLE 1 Determining Continuity of a Function

Discuss the continuity of the following functions.

a. $f(x) = x^2 - 2x + 3$ **b.** $f(x) = x^3 - x$

Solution

Each of these functions is a *polynomial function*. Therefore, each is continuous on the entire real line, as indicated in Figure 1.58.

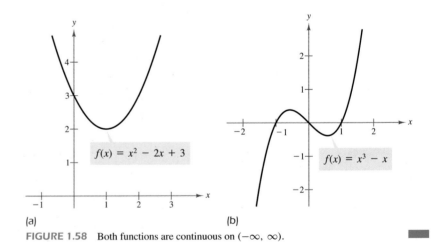

(a)   (b)

FIGURE 1.58   Both functions are continuous on $(-\infty, \infty)$. ▬

Polynomial functions are one of the most important types of functions used in calculus. Be sure you see from Example 1 that the graph of a polynomial function is continuous on the entire real line, and therefore has no holes, jumps, or gaps. Rational functions, on the other hand, need not be continuous on the entire real line, as shown in Example 2.

EXAMPLE 2 *Determining Continuity of a Function*

Discuss the continuity of the following functions.

a. $f(x) = \dfrac{1}{x}$ **b.** $f(x) = \dfrac{x^2 - 1}{x - 1}$ **c.** $f(x) = \dfrac{1}{x^2 + 1}$

Solution

Each of these functions is a rational function and is therefore continuous at every number in its domain.

a. The domain of $f(x) = 1/x$ consists of all real numbers other than $x = 0$. Therefore, this function is continuous on the intervals $(-\infty, 0)$ and $(0, \infty)$. (See Figure 1.59a.)

b. The domain of $f(x) = (x^2 - 1)/(x - 1)$ consists of all real numbers other than $x = 1$. Therefore, this function is continuous on the intervals $(-\infty, 1)$ and $(1, \infty)$. (See Figure 1.59b.)

c. The domain of $f(x) = 1/(x^2 + 1)$ consists of all real numbers. Therefore, this function is continuous on the entire real line. (See Figure 1.59c.)

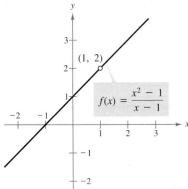
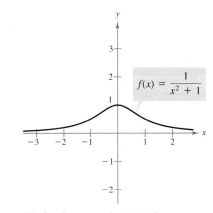

(a) Continuous on $(-\infty, 0)$ and $(0, \infty)$ (b) Continuous on $(-\infty, 1)$ and $(1, \infty)$ (c) Continuous on $(-\infty, \infty)$

FIGURE 1.59

If a function is continuous at every number in an open interval except c, then c is a **discontinuity** of the function. Discontinuities fall into two categories: removable and nonremovable. The number c is a **removable** discontinuity of f if the graph of f has a hole at the point (c, L) where L is the limit of $f(x)$ as x approaches c. For instance, the function in Example 2b has a removable discontinuity at $(1, 2)$. To remove the discontinuity, all you need to do is redefine the function so that $f(1) = 2$.

A discontinuity at $x = c$ is **nonremovable** if the function cannot be made continuous at $x = c$ by defining or redefining the function at $x = c$. For instance, the function in Example 2a has a nonremovable discontinuity at $x = 0$.

Continuity on a Closed Interval

The intervals discussed in Examples 1 and 2 are open. To discuss continuity on a closed interval, you can use the concept of one-sided limits, as defined in Section 1.5.

Definition of Continuity on a Closed Interval

Let f be defined on a closed interval $[a, b]$. If f is continuous on the open interval (a, b) and

$$\lim_{x \to a^+} f(x) = f(a) \quad \text{and} \quad \lim_{x \to b^-} f(x) = f(b),$$

then f is **continuous on the closed interval** $[a, b]$. Moreover, f is **continuous from the right** at a and **continuous from the left** at b.

Similar definitions can be made to cover continuity on half-open intervals of the form $(a, b]$ and $[a, b)$, or on infinite intervals. For example, the function

$$f(x) = \sqrt{x}$$

is continuous on the infinite interval $[0, \infty)$.

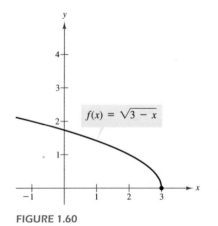

FIGURE 1.60

EXAMPLE 3   *Examining Continuity at an Endpoint*

Discuss the continuity of

$$f(x) = \sqrt{3 - x}.$$

Solution

Notice that the domain of f is the set $(-\infty, 3]$. Moreover, f is continuous from the left at $x = 3$ because

$$\lim_{x \to 3^-} f(x) = \lim_{x \to 3^-} \sqrt{3 - x}$$
$$= 0$$
$$= f(3).$$

For all $x < 3$, the function f satisfies the three conditions for continuity. Thus, you can conclude that f is continuous on the interval $(-\infty, 3]$, as shown in Figure 1.60. ■

Note   When working with radical functions of the form $f(x) = \sqrt{g(x)}$, remember that the domain of f coincides with the solution of $g(x) \geq 0$.

EXAMPLE 4 *Examining Continuity on a Closed Interval*

Discuss the continuity of

$$g(x) = \begin{cases} 5 - x, & -1 \le x \le 2 \\ x^2 - 1, & 2 < x \le 3. \end{cases}$$

Solution

The polynomial functions $5 - x$ and $x^2 - 1$ are continuous on the intervals $[-1, 2)$ and $(2, 3]$, respectively. Thus, to conclude that g is continuous on the entire interval $[-1, 3]$, you need only check the behavior of g when $x = 2$. You can do this by taking the one-sided limits when $x = 2$.

$$\lim_{x \to 2^-} g(x) = \lim_{x \to 2^-} (5 - x) = 3 \qquad \text{\textit{Limit from the left}}$$

and

$$\lim_{x \to 2^+} g(x) = \lim_{x \to 2^+} (x^2 - 1) = 3 \qquad \text{\textit{Limit from the right}}$$

Because these two limits are equal,

$$\lim_{x \to 2} g(x) = g(2) = 3.$$

Thus, g is continuous at $x = 2$ and, consequently, it is continuous on the entire interval $[-1, 3]$. The graph of g is shown in Figure 1.61. ■

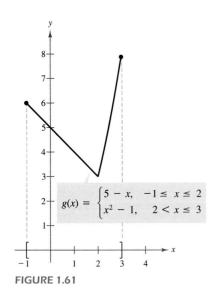

$$g(x) = \begin{cases} 5 - x, & -1 \le x \le 2 \\ x^2 - 1, & 2 < x \le 3 \end{cases}$$

FIGURE 1.61

The Greatest Integer Function

Many functions that are used in business applications are **step functions.** For instance, the function in Example 9 in Section 1.5 is a step function. The **greatest integer function** is another example of a step function. This function is denoted by

$$[\![x]\!] = \text{greatest integer less than or equal to } x.$$

For example,

$$[\![-2.1]\!] = \text{greatest integer less than or equal to } -2.1 = -3$$
$$[\![-2]\!] = \text{greatest integer less than or equal to } -2 = -2$$
$$[\![1.5]\!] = \text{greatest integer less than or equal to } 1.5 = 1.$$

Note that the graph of the greatest integer function (Figure 1.62) jumps up one unit at each integer. This implies that the function is not continuous at each integer.

In real-life applications, the domain of the greatest integer function is often restricted to nonnegative values of x. In such cases this function serves the purpose of **truncating** the decimal portion of x. For example, 1.345 is truncated to 1 and 3.57 is truncated to 3. That is,

$$[\![1.345]\!] = 1 \quad \text{and} \quad [\![3.57]\!] = 3.$$

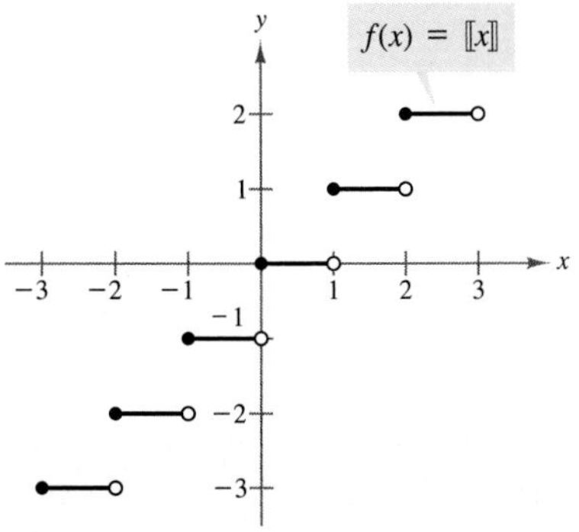

$f(x) = [\![x]\!]$

FIGURE 1.62   Greatest Integer Function

Courtesy of R. R. Donnelley

R. R. Donnelley & Sons Company is the world's largest commercial printer. It prints and binds a major share of the national publications in the United States, including *Time*, *Newsweek*, *TV Guide*, and this book. In 1993, printing and binding of books accounted for 11% of Donnelley's business. The other part of its business came from printing and binding of catalogs, magazines, directories, and other types of publications. In the photo, an employee is aligning a roll of paper on a printing press in one of Donnelley's plants.

Real Life

EXAMPLE 5 *Modeling a Cost Function*

A bookbinding company produces 10,000 books in an 8-hour shift. The fixed costs *per shift* amount to $5000, and the unit cost *per book* is $3. Using the greatest integer function, you can write the cost of producing x books as

$$C = 5000 \left(1 + \left[\!\left[\frac{x-1}{10,000} \right]\!\right] \right) + 3x.$$

Sketch the graph of this cost function.

Solution

Note that during the first 8-hour shift

$$\left[\!\left[\frac{x-1}{10,000} \right]\!\right] = 0, \quad 1 \le x \le 10,000,$$

which implies

$$C = 5000 \left(1 + \left[\!\left[\frac{x-1}{10,000} \right]\!\right] \right) + 3x = 5000 + 3x.$$

During the second 8-hour shift

$$\left[\!\left[\frac{x-1}{10,000} \right]\!\right] = 1, \quad 10,001 \le x \le 20,000,$$

which implies

$$C = 5000 \left(1 + \left[\!\left[\frac{x-1}{10,000} \right]\!\right] \right) + 3x = 10,000 + 3x.$$

The graph of C is shown in Figure 1.63. Note the graph's discontinuities.

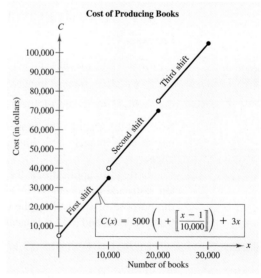

FIGURE 1.63

▥ TECHNOLOGY

Step Functions and Compound Functions

To graph a step function or compound function with a graphing utility, you must be familiar with the utility's programming language. For instance, the *TI-81*, *TI-82*, and *TI-85* have two different "integer truncation" functions. One is IPart(x), and it yields the truncated integer part of x. For example, IPart(-1.2) $= -1$ and IPart(3.4) $= 3$. The other function is Int(x), which is the greatest integer function. The graphs of these two functions are shown below. When graphing a step function, you should set your graphing utility to dot mode.

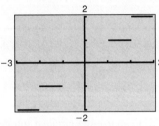

Graph of $f(x) = \text{IPart}(x)$ Graph of $f(x) = \text{Int}(x)$

On the *TI-81*, *TI-82*, or *TI-85*, you can graph a compound function such as

$$f(x) = \begin{cases} x^2 - 4, & x \le 2 \\ -x + 2, & 2 < x \end{cases}$$

by entering

$$Y_1 = (X^2 - 4)(X \le 2) + (-X + 2)(2 < X).$$

The graph of this function is shown below.

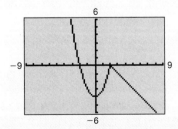

Extended Application: Compound Interest

Banks and other financial institutions differ on how interest is paid to an account. If the interest is added to the account so that future interest is paid on previously earned interest, then the interest is said to be **compounded.** Suppose, for example, that you deposited $10,000 in an account that pays 6% interest, compounded quarterly. Because the 6% is the annual interest rate, the quarterly rate is $\frac{1}{4}(0.06) = 0.015$ or 1.5%. The balances during the first five quarters are shown below.

Quarter	Balance
1st	$10,000.00
2nd	$10,000.00 + (0.015)(10,000.00) = \$10,150.00$
3rd	$10,150.00 + (0.015)(10,150.00) = \$10,302.25$
4th	$10,302.25 + (0.015)(10,302.25) = \$10,456.78$
5th	$10,456.78 + (0.015)(10,456.78) = \$10,613.63$

Quarterly Compounding

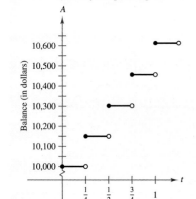

FIGURE 1.64

Real Life

EXAMPLE 6 *Graphing Compound Interest*

Sketch the graph of the balance in the account described above.

Solution

Let A represent the balance in the account and let t represent the time, in years. You can use the greatest integer function to represent the balance, as follows.

$$A = 10,000(1 + 0.015)^{[\![4t]\!]}$$

From the graph shown in Figure 1.64, notice that the function has a discontinuity at each quarter.   ■

Discussion Problem Compound Interest

If P dollars is deposited into an account, compounded n times per year, with an annual rate of r (in decimal form), then the balance A after t years is given by

$$A = P\left(1 + \frac{r}{n}\right)^{nt}.$$

Sketch the graph of each of the following. Which function is continuous? Describe the differences in policy between a bank that uses the first formula and a bank that uses the second formula.

a. $A = P\left(1 + \dfrac{r}{n}\right)^{nt}$ **b.** $A = P\left(1 + \dfrac{r}{n}\right)^{[\![nt]\!]}$

Warm Up

The following warm-up exercises involve skills that were covered in earlier sections. You will use these skills in the exercise set for this section.

In Exercises 1–4, simplify the expression.

1. $\dfrac{x^2 + 6x + 8}{x^2 - 6x - 16}$

2. $\dfrac{x^2 - 5x - 6}{x^2 - 9x + 18}$

3. $\dfrac{2x^2 - 2x - 12}{4x^2 - 24x + 36}$

4. $\dfrac{x^3 - 16x}{x^3 + 2x^2 - 8x}$

In Exercises 5–8, solve for x.

5. $x^2 + 7x = 0$

6. $x^2 + 4x - 5 = 0$

7. $3x^2 + 8x + 4 = 0$

8. $x^3 + 5x^2 - 24x = 0$

In Exercises 9 and 10, find the limit.

9. $\lim\limits_{x \to 3} (2x^2 - 3x + 4)$

10. $\lim\limits_{x \to -2} (3x^3 - 8x + 7)$

EXERCISES 1.6

 means that technology can help you solve or check the exercise(s).

In Exercises 1–4, decide whether the function is continuous on the entire real line. Explain your reasoning.

1. $5x^3 - x^2 + 2$

2. $(x^2 - 1)^3$

3. $\dfrac{1}{x^2 - 4}$

4. $\dfrac{1}{4 + x^2}$

In Exercises 5–32, find the intervals on which the function is continuous.

5. $f(x) = -\dfrac{x^3}{2}$

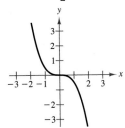

6. $f(x) = \dfrac{x^2 - 1}{x}$

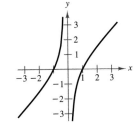

7. $f(x) = \dfrac{x^2 - 1}{x + 1}$

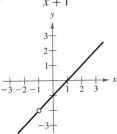

8. $f(x) = \dfrac{1}{x^2 - 4}$

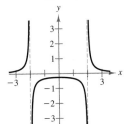

9. $f(x) = x^2 - 2x + 1$

10. $f(x) = \dfrac{1}{x^2 + 1}$

11. $f(x) = \dfrac{1}{x - 1}$

12. $f(x) = \dfrac{x}{x^2 - 1}$

13. $f(x) = \dfrac{x}{x^2 + 1}$

14. $f(x) = \dfrac{x - 3}{x^2 - 9}$

15. $f(x) = \dfrac{x - 5}{x^2 - 9x + 20}$

16. $f(x) = \dfrac{x - 1}{x^2 + x - 2}$

17. $f(x) = \begin{cases} -x, & x < 1 \\ 1, & x = 1 \\ x, & x > 1 \end{cases}$

18. $f(x) = \dfrac{[\![x]\!]}{2} + x$

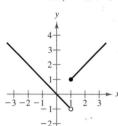

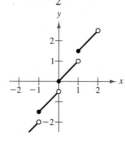

19. $f(x) = \begin{cases} x, & x \le 1 \\ x^2, & x > 1 \end{cases}$

20. $f(x) = \begin{cases} -2x + 3, & x < 1 \\ x^2, & x \ge 1 \end{cases}$

21. $f(x) = \begin{cases} \frac{1}{2}x + 1, & x \le 2 \\ 3 - x, & x > 2 \end{cases}$

22. $f(x) = \begin{cases} -2x, & x \le 2 \\ x^2 - 4x + 1, & x > 2 \end{cases}$

23. $f(x) = \begin{cases} 3 + x, & x \le 2 \\ x^2 + 1, & x > 2 \end{cases}$

24. $f(x) = \begin{cases} |x - 2| + 3, & x < 0 \\ x + 5, & x \ge 0 \end{cases}$

25. $f(x) = \dfrac{|x + 1|}{x + 1}$

26. $f(x) = \dfrac{|4 - x|}{4 - x}$

27. $f(x) = [\![x - 1]\!]$

28. $f(x) = x - [\![x]\!]$

29. $f(x) = [\![x]\!] - 2$

30. $f(x) = \frac{1}{2}[\![x]\!]$

31. $h(x) = f(g(x)), \ f(x) = \dfrac{1}{\sqrt{x}}, \ g(x) = x - 1, x > 1$

32. $h(x) = f(g(x)), \ f(x) = \dfrac{1}{x - 1}, \ g(x) = x^2 + 5$

In Exercises 33–36, discuss the continuity of the function on the closed interval. If there are any discontinuities, determine whether they are removable.

     Function   *Interval*

33. $f(x) = x^2 - 4x - 5$ $[-1, 5]$

34. $f(x) = \dfrac{5}{x^2 + 1}$ $[-2, 2]$

     Function   *Interval*

35. $f(x) = \dfrac{1}{x - 2}$ $[1, 4]$

36. $f(x) = \dfrac{x}{x^2 - 4x + 3}$ $[0, 4]$

In Exercises 37–42, sketch the graph of the function and describe the intervals on which the function is continuous.

37. $f(x) = \dfrac{x^2 - 16}{x - 4}$

38. $f(x) = \dfrac{2x^2 + x}{x}$

39. $f(x) = \dfrac{x^3 + x}{x}$

40. $f(x) = \dfrac{x^4 - 1}{x^2 - 1}$

41. $f(x) = \begin{cases} x^2 + 1, & x < 0 \\ x - 1, & x \ge 0 \end{cases}$

42. $f(x) = x - 2[\![x]\!]$

43. Determine the constant a so that the function is continuous on the entire real line.

$$f(x) = \begin{cases} x^3, & x \le 2 \\ ax^2, & x > 2 \end{cases}$$

44. Determine the constants a and b so that the function is continuous on the entire real line.

$$f(x) = \begin{cases} 2, & x \le -1 \\ ax + b, & -1 < x < 3 \\ -2, & x \ge 3 \end{cases}$$

In Exercises 45 and 46, use a graphing utility to graph the function. Then use the graph to determine any x-values at which the function is not continuous.

45. $h(x) = \dfrac{1}{x^2 - x - 2}$

46. $f(x) = \begin{cases} 2x - 4, & x \le 3 \\ x^2 - 2x, & x > 3 \end{cases}$

In Exercises 47–50, find the interval(s) on which the function is continuous.

47. $f(x) = \dfrac{x}{x^2 + 1}$

48. $f(x) = x\sqrt{x + 3}$

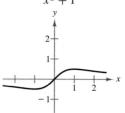

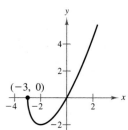

49. $f(x) = \frac{1}{2} [\![2x]\!]$

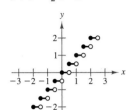

50. $f(x) = \dfrac{x+1}{\sqrt{x}}$

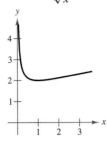

Essay In Exercises 51 and 52, use a graphing utility to graph the function on the interval $[-4, 4]$. Does the graph of the function appear continuous on this interval? *Is the* function continuous on $[-4, 4]$? Write a short paragraph about the importance of examining a function analytically as well as graphically.

51. $f(x) = \dfrac{x^2 + x}{x}$

52. $f(x) = \dfrac{x^3 - 8}{x - 2}$

 53. *Compound Interest* A deposit of $7500 is made into an account that pays 6% compounded quarterly. The amount A in the account after t years is

$$A = 7500(1.015)^{[\![4t]\!]}, \qquad t \geq 0.$$

(a) Sketch the graph of A. Is the graph continuous? Explain.

(b) What is the balance after 7 years?

 54. *Environmental Cost* The cost C (in millions of dollars) of removing x percent of the pollutants emitted from the smokestack of a factory can be modeled by

$$C = \dfrac{2x}{100 - x}.$$

(a) What is the implied domain of C? Explain.

(b) Graph the cost function. Is the function continuous on its domain? Explain.

(c) Find the cost of removing 75% of the pollutants from the smokestack.

 55. *Telephone Rates* A dial-direct long distance call between two cities costs $1.04 for the first 2 minutes and $0.36 for each additional minute or fraction thereof.

(a) Use the greatest integer function to write the cost C of a call in terms of the time t (in minutes). Graph the cost function and discuss its continuity.

(b) Find the cost of a 9-minute call.

 56. *Salary Contract* A union contract guarantees a 9% increase yearly for 5 years. For a current salary of $28,500, the salary for the next 5 years is given by

$$S = 28,500(1.09)^{[\![t]\!]},$$

where $t = 0$ represents the present year.

(a) Graph the salary function and discuss its continuity.

(b) Find the salary during the fifth year (when $t = 5$).

57. *Inventory Management* The number of units in inventory in a small company is

$$N = 25 \left(2 \left[\!\!\left[\dfrac{t+2}{2} \right]\!\!\right] - t \right), \qquad 0 \leq t \leq 12,$$

where the real number t is the time in months.

(a) Graph this function and discuss its continuity.

(b) How often must the company replenish its inventory?

58. *Owning a Franchise* You have purchased a franchise similar to that described below. You have determined a linear model for your revenue as a function of time. Is the model a continuous function? Would your actual revenue be a continuous function of time? Explain.

Business Capsule

Dan Bryant for "5 Keys to Successful Franchise Ownership" by Ingrid Sturgis in *Black Enterprise* May 1993, Vol. 23, No. 10, pages 77-82.

Twins Rodney and Roger Wagner used an inheritance of $25,000 and a small business loan of $85,000 to purchase a franchise for "I Can't Believe It's Yogurt" in 1988. By 1993, they had expanded to three outlets in the Dallas, Texas area.

Chapter Summary and Study Tips

After studying this chapter, you should have acquired the following skills. The exercise numbers are keyed to the Review Exercises that begin on page 76. Answers to odd-numbered Review Exercises are given in the back of the text.

▪ Plot points in a coordinate plane and read data presented graphically. (*Section 1.1*)   Review Exercises 1–4

▪ Find the distance between two points in a coordinate plane. (*Section 1.1*)   Review Exercises 5–8

$$d = \sqrt{(x_2 - x_1)^2 + (y_2 - y_1)^2}$$

▪ Find the midpoint of the segment connecting two points. (*Section 1.1*)   Review Exercises 9–12

$$\text{Midpoint} = \left(\frac{x_1 + x_2}{2}, \frac{y_1 + y_2}{2} \right)$$

▪ Interpret real-life data that is presented graphically. (*Section 1.1*)   Review Exercises 13, 14

▪ Translate points in a coordinate plane. (*Section 1.1*)   Review Exercises 15, 16

▪ Sketch the graph of an equation by hand *and* using a graphing utility. (*Section 1.2*)   Review Exercises 17–24

▪ Find the *x*- and *y*-intercepts of the graph of an equation algebraically *and* graphically using a graphing utility. (*Section 1.2*)   Review Exercises 25, 26

▪ Write the standard equation of a circle, given the center and a point on the circle. (*Section 1.2*)   Review Exercises 27, 28

$$(x - h)^2 + (y - k)^2 = r^2$$

▪ Convert an equation of a circle from general form to standard form by completing the square. (*Section 1.2*)   Review Exercises 29, 30

▪ Find the points of intersection of two graphs algebraically *and* graphically using a graphing utility. (*Section 1.2*)   Review Exercises 31–34

▪ Find the break-even point for a business. (*Section 1.2*)   Review Exercises 35, 36

The break-even point occurs when the revenue R is equal to the cost C.

▪ Use the slope-intercept form of a linear equation to sketch its graph. (*Section 1.3*)   Review Exercises 37–42

$$y = mx + b$$

▪ Find the slope of the line passing through two points. (*Section 1.3*)   Review Exercises 43–46

$$m = \frac{y_2 - y_1}{x_2 - x_1}$$

▪ Use the point-slope form to write an equation of a line. (*Section 1.3*)   Review Exercises 47, 48

$$y - y_1 = m(x - x_1)$$

---

*Several student study aids are available with this text. The *Student Solutions Guide* includes detailed solutions to all odd-numbered exercises, as well as practice chapter tests with answers. The *Graphics Calculator Guide* offers instructions on the use of a variety of graphing calculators and computer graphing software. The *Brief Calculus TUTOR* includes additional examples for selected exercises in the text.

■ Find equations of parallel and perpendicular lines. (*Section 1.3*)   Review Exercises 49, 50

   Parallel lines: $m_1 = m_2$ Perpendicular lines: $m_1 = -\dfrac{1}{m_2}$

■ Use linear equations to solve real-life problems such as predicting future sales or   Review Exercises 51, 52
   creating a linear depreciation schedule. (*Section 1.3*)

■ Use the vertical line test to decide whether an equation defines a function. (*Section   Review Exercises 53–56
   1.4*)

■ Use function notation to evaluate functions. (*Section 1.4*)   Review Exercises 57, 58

■ Find the domain and range of a function. (*Section 1.4*)   Review Exercises 59–64

■ Combine functions to form other functions. (*Section 1.4*)   Review Exercises 65, 66

■ Use the horizontal line test to determine whether a function has an inverse. If it   Review Exercises 67–70
   does, find the inverse. (*Section 1.4*)

■ Determine whether a limit exists. If it does, approximate the limit graphically and   Review Exercises 71–88
   numerically using a table. (*Section 1.5*)

■ Evaluate limits analytically and evaluate one-sided limits. (*Section 1.5*)   Review Exercises 89–94

■ Determine whether a function is continuous at a point, in an open interval, and in a   Review Exercises 95–104
   closed interval. (*Section 1.6*)

■ Create mathematical models using the greatest integer function and other step func-   Review Exercises 105, 106
   tions. (*Section 1.6*)

On pages xxix–xxx of the preface, we included a feature called *How to Study Calculus*. If you have not already read this, we encourage you to do so now. Here are some other tips that can help you succeed in this course.

■ Use a Graphing Utility   A graphing calculator or graphing software for a computer can help you in this course in two important ways. As an *exploratory device*,
   a graphing utility allows you to learn concepts by allowing you to compare graphs
   of equations. For instance, sketching the graphs of $y = x^2$, $y = x^2 + 1$, and
   $y = x^2 - 1$ helps confirm that adding (or subtracting) a constant from a function
   shifts the graph of the function vertically. As a *problem-solving tool*, a graphing
   utility frees you of some of the drudgery of sketching complicated graphs by hand.
   The time that you save can be spent using mathematics to solve real-life problems.

■ Use the Warm-Up Exercises   Each exercise set in this text begins with a set of
   10 warm-up exercises. We urge you to begin each homework session by quickly
   working all 10 warm-up exercises (all are answered in the back of the text). The
   "old" skills covered in the warm-up exercises are needed to master the "new" skills
   in the section-exercise set. The warm-up exercises remind you that mathematics is
   cumulative—to be successful in this course, you must retain "old" skills.

■ Use the Additional Study Aids   The additional study aids were prepared specifically to help you master the concepts discussed in the text. They are the *Student
   Solutions Guide*, the *Brief Calculus TUTOR*, and the *Graphics Calculator Guide*.

Review Exercises

 means that technology can help you solve or check the exercise(s).

In Exercises 1–4, match the data with the real-life situation that it represents. [Graphs are labeled (a)–(d).]

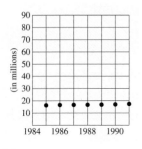

(a)

(b)

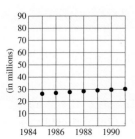

(c)

(d)

1. Population of Texas

2. Population of California

3. Number of U.S. Business Failures

4. IBM Revenues

In Exercises 5–8, find the distance between the two points.

5. $(0, 0), (5, 2)$ **6.** $(1, 2), (4, 3)$

7. $(-1, 3), (-4, 6)$ **8.** $(6, 8), (-3, 7)$

In Exercises 9–12, find the midpoint of the segment connecting the two points.

9. $(5, 6), (9, 2)$ **10.** $(0, 0), (-4, 8)$

11. $(-10, 4), (-6, 8)$ **12.** $(7, -9), (-3, 5)$

In Exercises 13 and 14, use the graph below, which gives the revenues, costs, and profits for *Kimberly-Clark Corporation* from 1987 through 1992. *(Source: Kimberly-Clark Corporation)*

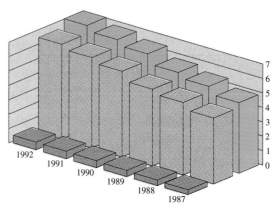

FIGURE FOR 13 AND 14

13. Which bars on the graph represent the revenues? Which represent the costs? Which represent the profits? Explain your reasoning. Write an equation that relates the revenue R, cost C, and profit P.

14. Estimate the revenue, cost, and profit for Kimberly-Clark Corporation in 1992. Do you think these numbers are in billions of dollars, millions of dollars, or thousands of dollars? Explain your reasoning. (Kimberly-Clark is the maker of *Kleenex, Huggies Disposable Diapers*, and other paper products.)

15. Translate the triangle whose vertices are $(1, 3)$, $(2, 4)$, and $(5, 6)$ three units to the right and four units up. Find the coordinates of the translated vertices.

16. Translate the quadrilateral whose vertices are $(-2, 4)$, $(5, 0)$, $(2, 6)$, and $(-1, 3)$ four units to the left and three units down. Find the coordinates of the translated vertices.

In Exercises 17–24, sketch the graph of the equation.

17. $y = 4x - 12$ **18.** $y = 4 - 3x$

19. $y = x^2 + 5$ **20.** $y = 1 - x^2$

21. $y = x^2 + 5x + 6$ **22.** $y = |2x - 3|$

23. $y = x^3 + 4$ **24.** $y = \sqrt{2x}$

In Exercises 25 and 26, find the intercepts of the graph of the equation algebraically. Check your results with a graphing utility.

25. $4x + y + 3 = 0$

26. $y = (x-1)^3 + 2(x-1)^2$

In Exercises 27 and 28, write the standard form of the circle.

27. Center: $(0, 0)$;
   Solution point: $\left(2, \sqrt{5}\right)$

28. Center: $(2, -1)$;
   Solution point: $(-1, 7)$

In Exercises 29 and 30, complete the square to write the equation of the circle in standard form. Determine the radius and center of the circle. Then sketch the circle.

29. $x^2 + y^2 - 6x + 8y = 0$

30. $x^2 + y^2 + 10x + 4y - 7 = 0$

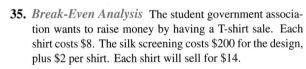 In Exercises 31–34, find the points of intersection of the graphs algebraically. Then use a graphing utility to check your results.

31. $x+y = 2$, $2x-y = 1$

32. $x^2 + y^2 = 5$, $x - y = 1$

33. $y = x^3$, $y = x$

34. $y = \sqrt{x}$, $y = x$

35. *Break-Even Analysis* The student government association wants to raise money by having a T-shirt sale. Each shirt costs \$8. The silk screening costs \$200 for the design, plus \$2 per shirt. Each shirt will sell for \$14.

(a) Find equations for the total cost C and the total revenue R for selling x shirts.

(b) Find the break-even point.

36. *Break-Even Analysis* You are starting a part-time business. You make an initial investment of \$6000. The unit cost of the product is \$6.50, and the selling price is \$13.90.

(a) Find equations for the total cost C and the total revenue R for selling x units of the product.

(b) Find the break-even point.

 In Exercises 37–42, find the slope and *y*-intercept (if possible) of the linear equation. Then sketch the graph of the equation.

37. $3x + y = -2$

38. $-\frac{1}{3}x + \frac{5}{6}y = 1$

39. $y = -\frac{5}{3}$

40. $x = -3$

41. $-2x - 5y - 5 = 0$

42. $3.2x - 0.8y + 5.6 = 0$

In Exercises 43–46, find the slope of the line passing through the two points.

43. $(0, 0)$, $(7, 6)$

44. $(-1, 5)$, $(-5, 7)$

45. $(10, 17)$, $(-11, -3)$

46. $(-11, -3)$, $(-1, -3)$

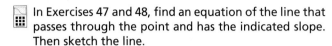 In Exercises 47 and 48, find an equation of the line that passes through the point and has the indicated slope. Then sketch the line.

47. Point: $(3, -1)$; Slope: $m = -2$

48. Point: $(-3, -3)$; Slope: $m = \frac{1}{2}$

In Exercises 49 and 50, find the general form of the equation of the line passing through the point and satisfying the given condition.

49. Point: $(-3, 6)$

(a) Slope is $\frac{7}{8}$.

(b) Parallel to the line $4x + 2y = 7$.

(c) Passes through the origin.

(d) Perpendicular to the line $3x - 2y = 2$.

50. Point: $(1, -3)$

(a) Parallel to the x-axis.

(b) Perpendicular to the x-axis.

(c) Parallel to the line $-4x + 5y = -3$.

(d) Perpendicular to the line $5x - 2y = 3$.

 51. *Demand* When a wholesaler sold a product at \$32 per unit, sales were 750 units per week. After a price increase of \$5 per unit, however, the sales dropped to 700 units per week.

(a) Write the quantity demanded x as a linear function of the price p.

(b) *Linear Interpolation* Predict the number of units sold at a price of \$34.50 per unit.

(c) *Linear Extrapolation* Predict the number of units sold at a price of \$42.00 per unit.

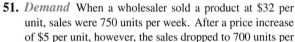 **52.** *Linear Depreciation* A small business purchases a typesetting system for \$117,000. After 9 years, the system will be obsolete and have no value.

(a) Write a linear equation giving the value v of the system in terms of the time t.

(b) Use a graphing utility to graph the function.

(c) Use a graphing utility to estimate the value of the system after 4 years.

(d) Use a graphing utility to estimate the time when the system's value will be \$84,000.

In Exercises 53–56, use the vertical line test to determine whether y is a function of x.

53. $y = -x^2 + 2$

54. $x^2 + y^2 = 4$

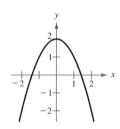

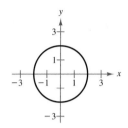

55. $y^2 - \frac{1}{4}x^2 = 4$

56. $y = |x + 4|$

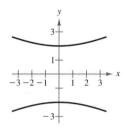

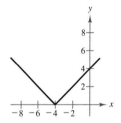

57. Given $f(x) = 3x + 4$, find the following.

(a) $f(1)$ (b) $f(x + 1)$ (c) $f(2 + \Delta x)$

58. Given $f(x) = x^2 + 4x + 3$, find the following.

(a) $f(0)$ (b) $f(x - 1)$ (c) $f(x + \Delta x) - f(x)$

In Exercises 59–64, use a graphing utility to graph the function. Then find the domain and range of the function.

59. $f(x) = x^2 + 3x + 2$

60. $f(x) = 2$

61. $f(x) = \sqrt{x + 1}$

62. $f(x) = \dfrac{x - 3}{x^2 + x - 12}$

63. $f(x) = -|x| + 3$

64. $f(x) = -\frac{12}{13}x - \frac{7}{8}$

In Exercises 65 and 66, use f and g to find the following.

(a) $f(x) + g(x)$ (b) $f(x) - g(x)$ (c) $f(x)g(x)$

(d) $\dfrac{f(x)}{g(x)}$ (e) $f(g(x))$ (f) $g(f(x))$

65. $f(x) = 1 + x^2$, $g(x) = 2x - 1$

66. $f(x) = 2x - 3$, $g(x) = \sqrt{x + 1}$

In Exercises 67–70, find the inverse of f (if it exists).

67. $f(x) = \frac{3}{2}x$

68. $f(x) = |x + 1|$

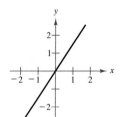

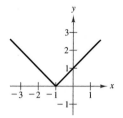

69. $f(x) = -x^2 + \frac{1}{2}$

70. $f(x) = x^3 - 1$

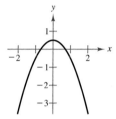

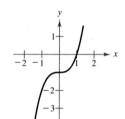

In Exercises 71–86, find the limit (if it exists).

71. $\lim\limits_{x \to 2} (5x - 3)$

72. $\lim\limits_{x \to 2} (2x + 9)$

73. $\lim\limits_{x \to 2} (5x - 3)(2x + 3)$

74. $\lim\limits_{x \to 2} \dfrac{5x - 3}{2x + 9}$

75. $\lim\limits_{t \to 3} \dfrac{t^2 + 1}{t}$

76. $\lim\limits_{t \to 1} \dfrac{t + 1}{t - 2}$

77. $\lim\limits_{t \to 0} \dfrac{t^2 + 1}{t}$

78. $\lim\limits_{t \to 2} \dfrac{t + 1}{t - 2}$

79. $\lim\limits_{x \to -2} \dfrac{x + 2}{x^2 - 4}$

80. $\lim\limits_{x \to 3^-} \dfrac{x^2 - 9}{x - 3}$

81. $\lim\limits_{x \to 0^+} \left(x - \dfrac{1}{x}\right)$

82. $\lim\limits_{x \to 1/2} \dfrac{2x - 1}{6x - 3}$

83. $\lim\limits_{x \to 0} \dfrac{[1/(x - 2)] - 1}{x}$

84. $\lim\limits_{s \to 0} \dfrac{(1/\sqrt{1 + s}) - 1}{s}$

85. $\lim\limits_{\Delta x \to 0} \dfrac{(x + \Delta x)^3 - (x + \Delta x) - (x^3 - x)}{\Delta x}$

86. $\lim\limits_{\Delta x \to 0} \dfrac{1 - (x + \Delta x)^2 - (1 - x^2)}{\Delta x}$

In Exercises 87 and 88, use a table to estimate the limit.

87. $\lim\limits_{x \to 1^+} \dfrac{\sqrt{2x + 1} - \sqrt{3}}{x - 1}$

88. $\lim\limits_{x \to 1^+} \dfrac{1 - \sqrt[3]{x}}{x - 1}$

In Exercises 89–94, determine whether the statement is true or false.

89. $\lim\limits_{x \to 0} \dfrac{|x|}{x} = 1$

90. $\lim\limits_{x \to 0} x^3 = 0$

91. $\lim\limits_{x \to 0} \sqrt{x} = 0$

92. $\lim\limits_{x \to 0} \sqrt[3]{x} = 0$

93. $\lim\limits_{x \to 2} f(x) = 3$, $f(x) = \begin{cases} 3, & x \le 2 \\ 0, & x > 2 \end{cases}$

94. $\lim\limits_{x \to 3} f(x) = 1$, $f(x) = \begin{cases} x - 2, & x \le 3 \\ -x^2 + 8x - 14, & x > 3 \end{cases}$

 In Exercises 95–102, find the intervals on which the function is continuous.

95. $f(x) = \dfrac{1}{(x+4)^2}$

96. $f(x) = \dfrac{x+2}{x}$

97. $f(x) = \dfrac{3}{x+1}$

98. $f(x) = \dfrac{x+1}{2x+2}$

99. $f(x) = [\![x+3]\!]$

100. $f(x) = \dfrac{3x^2 - x - 2}{x - 1}$

101. $f(x) = \begin{cases} x, & x \le 0 \\ x+1, & x > 0 \end{cases}$

102. $f(x) = \begin{cases} x, & x \le 0 \\ x^2, & x > 0 \end{cases}$

In Exercises 103 and 104, find *a* so that *f* is continuous on $(-\infty, \infty)$.

103. $f(x) = \begin{cases} -x+1, & x \le 3 \\ ax - 8, & x > 3 \end{cases}$

104. $f(x) = \begin{cases} x+1, & x < 1 \\ 2x + a, & x \ge 1 \end{cases}$

 105. *National Debt* The table lists the national debt D (in billions of dollars) for selected years. A mathematical model for the national debt is

$$D = 0.2t^3 + 2.74t^2 + 8.47t + 358.78,$$

where $t = 0$ represents 1970. *(Source: U.S. Office of Management and Budget)*

t	0	5	10	15	20	21
D	381	542	909	1817	3206	3599

(a) Use a graphing utility to graph the model.
(b) Create a table that compares the values given by the model with the actual data.
(c) Use the model to estimate the national debt in 1995.

 106. *Recycling* A recycling center pays $0.25 for each pound of aluminum cans. Twenty-four aluminum cans weigh 1 pound. A mathematical model for the amount paid A by the recycling center is

$$A = \frac{1}{4}\left[\!\left[\frac{x}{24}\right]\!\right],$$

where x is the number of cans.

(a) Sketch the graph of the function and then discuss its continuity.
(b) How much does a recycling center pay out for 1500 cans?

Sample Post-Graduation Exam Questions

The following questions were taken from certified public accountant (CPA) exams, graduate management admission tests (GMAT), graduate records exams (GRE), actuarial exams, or college-level academic skills tests (CLAST). The answers to the questions are given in the back of the book.

In Questions 1–5, use the data given in the graph at the right. *(Source: U.S. Bureau of the Census)*

1. The total labor force in 1985 was about *y* million with *y* equal to

(a) 65   (b) 81   (c) 100   (d) 108   (e) 115

2. In 1975, the percent of women in the labor force who were married was about

(a) 19   (b) 32   (c) 45   (d) 60   (e) 82

3. What was the first year when more than 50 million women were in the labor force?

(a) 1976   (b) 1984   (c) 1985   (d) 1988   (e) 1991

4. Between 1974 and 1991, the number of women in the labor force

(a) increased by about 25%   (b) increased by about 50%
(c) increased by about 100%   (d) increased by about 150%
(e) increased by about 200%

5. Which of the following statements about the labor force can be inferred from the graphs?

 I. Between 1974 and 1991, there were no years when more than 15 million widowed, divorced, or separated women were in the labor force.

 II. In every year between 1974 and 1991, the number of single women in the labor force increased.

 III. In every year between 1975 and 1991, women made up at least $\frac{2}{5}$ of the total labor force.

(a) I only   (b) II only   (c) I and II only   (d) I and III only
(e) I, II, and III

6. What is the length of the line segment connecting $(1, 3)$ to $(-1, 5)$?

(a) $\sqrt{3}$ (b) 2   (c) $2\sqrt{2}$ (d) 4   (e) 8

7. The interest charged on a loan is *p* dollars per $1000 for the first month and *q* dollars per $1000 for each month after the first month. How much interest will be charged during the first 3 months on a loan of $10,000?

(a) $30p$ (b) $30q$ (c) $p+2q$ (d) $20p+10q$ (e) $10p+20q$

8. If $x + y > 5$ and $x - y > 3$, which of the following describes the *x* solutions?

(a) $x > 3$ (b) $x > 4$ (c) $x > 5$ (d) $x < 5$ (e) $x < 3$

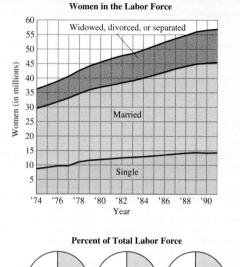

Women in the Labor Force

Percent of Total Labor Force

1975   1985   1991

FIGURE FOR 1–5

$$\lim_{\Delta x \to 0} \frac{\Delta y}{\Delta x} = \frac{dy}{dx}$$

Differentiation

Calculus is the mathematics of change. When Isaac Newton and other 17th century mathematicians developed calculus, they were concerned with describing changes in moving objects, such as the planets. As their investigations deepened, their unresolved questions centered around four problems: the tangent line problem, the velocity and acceleration problem, the minimum-maximum problem, and the area problem.

In what turned out to be the most famous mathematical discovery in history, Newton realized that all four problems were intimately connected and could be solved by calculating *derivatives* and *antiderivatives*.

$$y' = f'(x)$$

Newton

Aip Niels Bohr Library, W. F. Meggars Collection

2.1 The Derivative and the Slope of a Graph

Tangent Line to a Graph ■ *Slope of a Graph* ■ *Slope and the Limit Process* ■ *The Derivative of a Function* ■ *Differentiability and Continuity*

Tangent Line to a Graph

Calculus is a branch of mathematics that studies rates of change of functions. In this course, you will learn that rates of change have many applications in real life. In Section 1.3, you learned how the slope of a line indicates the rate at which a line rises or falls. For a line, this rate (or slope) is the same at every point on the line. For graphs other than lines, the rate at which the graph rises or falls changes from point to point. For instance, in Figure 2.1, the parabola is rising more quickly at the point (x_1, y_1) than it is at the point (x_2, y_2). At the vertex (x_3, y_3), the graph levels off, and at the point (x_4, y_4), the graph is falling.

To determine the rate at which a graph rises or falls at a *single point*, you can find the slope of the tangent line at the point. In simple terms, the **tangent line** to the graph of a function f at a point $P(x_1, y_1)$ is the line that best approximates the graph at that point, as shown in Figure 2.1. Figure 2.2 shows other examples of tangent lines.

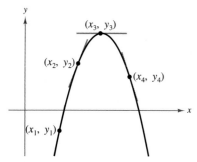

FIGURE 2.1 The slope of a graph changes from one point to another.

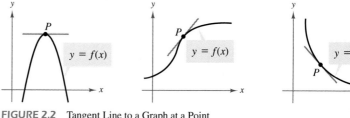

FIGURE 2.2 Tangent Line to a Graph at a Point

When Isaac Newton (1642–1727) was working on the "tangent line problem," he realized that it is difficult to define precisely what is meant by a tangent to a general curve. From geometry, you know that a line is tangent to a circle if the line intersects the circle at only one point, as shown in Figure 2.3. Tangent lines to noncircular graphs, however, can intersect the graph at more than one point. For instance, in the second graph in Figure 2.2, if the tangent line were extended, it would intersect the graph at a point other than the point of tangency. In this section, you will see how the notion of a limit can be used to define a general tangent line.

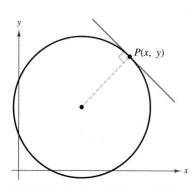

FIGURE 2.3 Tangent Line to a Circle

Slope of a Graph

Because a tangent line approximates the graph at a point, the problem of finding the slope of a graph at a point becomes one of finding the slope of the tangent line at the point.

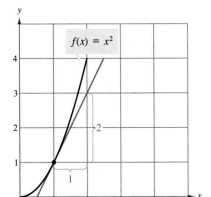

FIGURE 2.4

EXAMPLE 1   *Approximating the Slope of a Graph*

Use the graph in Figure 2.4 to approximate the slope of the graph of $f(x) = x^2$ at the point $(1, 1)$.

Solution

From the graph of $f(x) = x^2$, you can see that the tangent line at $(1, 1)$ rises approximately two units for each unit change in x. Thus, the slope of the tangent line at $(1, 1)$ is given by

$$\text{Slope} = \frac{\text{change in } y}{\text{change in } x} \approx \frac{2}{1} = 2.$$

Because the tangent line at the point $(1, 1)$ has a slope of about 2, you can conclude that the graph has a slope of about 2 at the point $(1, 1)$.   ▬

Note   When visually approximating the slope of a graph, note that the scales on the horizontal and vertical axes may differ. When this happens (as it frequently does in applications), the slope of the tangent line is distorted, and you must be careful to account for the difference in scales.

Real Life

EXAMPLE 2   *Interpreting Slope*

Figure 2.5 graphically depicts the average daily temperature (in degrees Fahrenheit) in Duluth, Minnesota. Estimate the slope of this graph at the indicated point and give a physical interpretation of the result. *(Source: National Oceanic and Atmospheric Administration)*

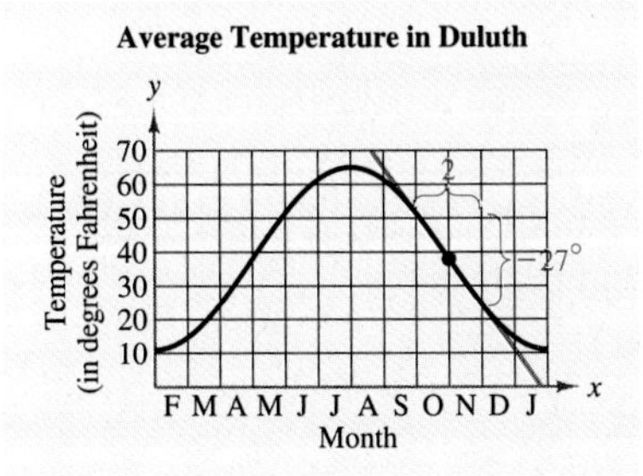

Solution

From the graph, you can see that the tangent line at the given point falls approximately 27 units for each two-unit change in x. Thus, you can estimate the slope at the given point to be

$$\text{Slope} = \frac{\text{change in } y}{\text{change in } x} \approx \frac{-27}{2} = -13.5 \text{ degrees per month.}$$

This means that you can expect the average daily temperatures in November to be about 13.5 degrees *lower* than the corresponding temperatures in October.   ▬

Average Temperature in Duluth

FIGURE 2.5

Slope and the Limit Process

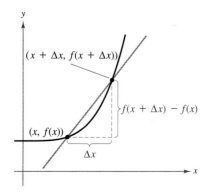

FIGURE 2.6   The Secant Line Through $(x, f(x))$ and $(x + \Delta x, f(x + \Delta x))$

In Examples 1 and 2, you approximated the slope of a graph at a point by making a careful graph and then "eyeballing" the tangent line at the point of tangency. A more precise method of approximating tangent lines makes use of a **secant line** through the point of tangency and a second point on the graph, as shown in Figure 2.6. If $(x, f(x))$ is the point of tangency and $(x + \Delta x, f(x + \Delta x))$ is a second point on the graph of f, then the slope of the secant line through the two points is

$$m_{\text{sec}} = \frac{f(x + \Delta x) - f(x)}{\Delta x}. \qquad \textit{Slope of secant line}$$

The right side of this equation is called the **difference quotient.** The denominator Δx is the **change in x,** and the numerator is the **change in y.** The beauty of this procedure is that you obtain better and better approximations of the slope of the tangent line by choosing the second point closer and closer to the point of tangency, as shown in Figure 2.7.

Using the limit process, you can find the *exact* slope of the tangent line at $(x, f(x))$.

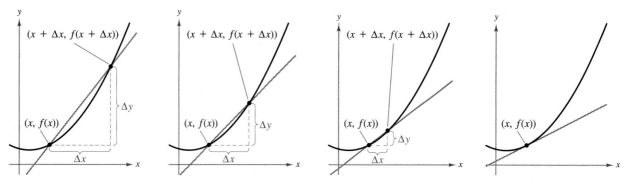

FIGURE 2.7   As Δx approaches 0, the secant lines approach the tangent line.

Note Δx is used as a variable to represent the change in x in the definition of the slope of a graph. Other variables may also be used. For instance, this definition is sometimes written as

$$m = \lim_{h \to 0} \frac{f(x + h) - f(x)}{h}.$$

Definition of the Slope of a Graph

The **slope** m of the graph of f at the point $(x, f(x))$ is equal to the slope of its tangent line at $(x, f(x))$, and is given by

$$m = \lim_{\Delta x \to 0} m_{\text{sec}} = \lim_{\Delta x \to 0} \frac{f(x + \Delta x) - f(x)}{\Delta x}$$

provided this limit exists.

EXAMPLE 3 *Finding Slope by the Limit Process*

Find the slope of the graph of $f(x) = x^2$ at the point $(-2, 4)$.

Solution

Begin by finding an expression that represents the slope of a secant line at the point $(-2, 4)$.

$$m_{\text{sec}} = \frac{f(-2 + \Delta x) - f(-2)}{\Delta x}$$ *Set up difference quotient.*

$$= \frac{(-2 + \Delta x)^2 - (-2)^2}{\Delta x}$$ *Use $f(x) = x^2$.*

$$= \frac{4 - 4\Delta x + (\Delta x)^2 - 4}{\Delta x}$$ *Expand terms.*

$$= \frac{-4\Delta x + (\Delta x)^2}{\Delta x}$$ *Simplify.*

$$= \frac{\cancel{\Delta x}(-4 + \Delta x)}{\cancel{\Delta x}}$$ *Factor and cancel.*

$$= -4 + \Delta x, \qquad \Delta x \neq 0$$ *Simplify.*

Next, take the limit of m_{sec} as $\Delta x \to 0$.

$$m = \lim_{\Delta x \to 0} m_{\text{sec}} = \lim_{\Delta x \to 0} (-4 + \Delta x) = -4$$

Thus, the graph of f has a slope of -4 at the point $(-2, 4)$, as shown in Figure 2.8.   ▬

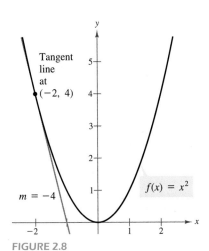

FIGURE 2.8

EXAMPLE 4 *Finding the Slope of a Graph*

Find the slope of $f(x) = -2x + 4$.

Solution

You know from your study of linear functions that the line given by $f(x) = -2x + 4$ has a slope of -2, as shown in Figure 2.9. This conclusion is consistent with the limit definition of slope.

$$m = \lim_{\Delta x \to 0} \frac{f(x + \Delta x) - f(x)}{\Delta x}$$

$$= \lim_{\Delta x \to 0} \frac{[-2(x + \Delta x) + 4] - [-2x + 4]}{\Delta x}$$

$$= \lim_{\Delta x \to 0} \frac{-2x - 2\Delta x + 4 + 2x - 4}{\Delta x}$$

$$= \lim_{\Delta x \to 0} \frac{-2\cancel{\Delta x}}{\cancel{\Delta x}}$$

$$= -2$$ ▬

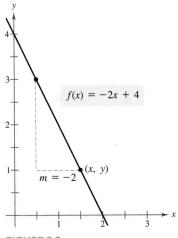

FIGURE 2.9

Discovery

Use a graphing utility to graph the function $Y_1 = x^2 + 1$ and the three lines $Y_2 = 3x - 1$, $Y_3 = 4x - 3$, and $Y_4 = 5x - 5$. Which of these lines appears to be tangent to Y_1 at the point $(2, 5)$? Confirm your answer by showing that the graphs of Y_1 and its tangent line have only one point of intersection, whereas the graphs of Y_1 and the other lines each have two points of intersection.

It is important that you see the distinction between the ways the difference quotients were set up in Examples 3 and 4. In Example 3, you were finding the slope of a graph at a specific point $(c, f(c))$. To find the slope, you can use the following form of a difference quotient.

$$m = \lim_{\Delta x \to 0} \frac{f(c + \Delta x) - f(c)}{\Delta x} \qquad \text{Slope at specific point}$$

In Example 4, however, you were finding a formula for the slope at *any* point on the graph. In such cases, you should use x, rather than c, in the difference quotient.

$$m = \lim_{\Delta x \to 0} \frac{f(x + \Delta x) - f(x)}{\Delta x} \qquad \text{Formula for slope}$$

Except for linear functions, this form will always produce a function of x, which can then be evaluated to find the slope at any desired point.

EXAMPLE 5 *Finding a Formula for the Slope of a Graph*

Find a formula for the slope of the graph of $f(x) = x^2 + 1$. What is the slope at the points $(-1, 2)$ and $(2, 5)$?

Solution

$$\begin{aligned} m_{\text{sec}} &= \frac{f(x + \Delta x) - f(x)}{\Delta x} && \text{Set up difference quotient.} \\ &= \frac{[(x + \Delta x)^2 + 1] - [x^2 + 1]}{\Delta x} && \text{Use } f(x) = x^2 + 1. \\ &= \frac{x^2 + 2x\Delta x + (\Delta x)^2 + 1 - x^2 - 1}{\Delta x} && \text{Expand terms.} \\ &= \frac{2x\Delta x + (\Delta x)^2}{\Delta x} && \text{Simplify.} \\ &= \frac{\Delta x(2x + \Delta x)}{\Delta x} && \text{Factor and cancel.} \\ &= 2x + \Delta x, \qquad \Delta x \neq 0 && \text{Simplify.} \end{aligned}$$

Next, take the limit of m_{sec} as $\Delta x \to 0$.

$$\begin{aligned} m &= \lim_{\Delta x \to 0} m_{\text{sec}} \\ &= \lim_{\Delta x \to 0} (2x + \Delta x) \\ &= 2x \end{aligned}$$

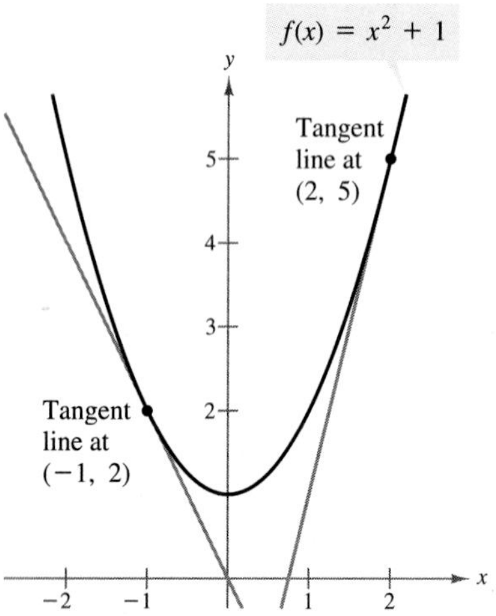

Using the formula $m = 2x$, you can find the slope at the specified points. At $(-1, 2)$, the slope is $m = 2(-1) = -2$, and at $(2, 5)$, the slope is $m = 2(2) = 4$. The graph of f is shown in Figure 2.10.

$f(x) = x^2 + 1$

Tangent line at $(2, 5)$

Tangent line at $(-1, 2)$

FIGURE 2.10

The Derivative of a Function

In Example 5, you started with the function $f(x) = x^2 + 1$, and used the limit process to derive another function, $m = 2x$, that represents the slope of the graph of f at the point $(x, f(x))$. This derived function is called the **derivative** of f at x. It is denoted by $f'(x)$, which is read as "f prime of x."

Note   The notation dy/dx is read "the derivative of y with respect to x," and using limit notation, you can write

$$\frac{dy}{dx} = \lim_{\Delta x \to 0} \frac{\Delta y}{\Delta x}$$

$$= \lim_{\Delta x \to 0} \frac{f(x + \Delta x) - f(x)}{\Delta x}$$

$$= f'(x).$$

Definition of the Derivative

The **derivative of f at x** is given by

$$f'(x) = \lim_{\Delta x \to 0} \frac{f(x + \Delta x) - f(x)}{\Delta x}$$

provided this limit exists. A function is **differentiable** at x if its derivative exists at x. The process of finding derivatives is called **differentiation.**

In addition to $f'(x)$, other notations can be used to denote the derivative of $y = f(x)$. The most common are

$$\frac{dy}{dx}, \quad y', \quad \frac{d}{dx}[f(x)], \quad \text{and} \quad D_x[y].$$

EXAMPLE 6   *Finding a Derivative*

Find the derivative of $f(x) = 3x^2 - 2x$.

Solution

$$f'(x) = \lim_{\Delta x \to 0} \frac{f(x + \Delta x) - f(x)}{\Delta x}$$

$$= \lim_{\Delta x \to 0} \frac{[3(x + \Delta x)^2 - 2(x + \Delta x)] - [3x^2 - 2x]}{\Delta x}$$

$$= \lim_{\Delta x \to 0} \frac{3x^2 + 6x\Delta x + 3(\Delta x)^2 - 2x - 2\Delta x - 3x^2 + 2x}{\Delta x}$$

$$= \lim_{\Delta x \to 0} \frac{6x\Delta x + 3(\Delta x)^2 - 2\Delta x}{\Delta x}$$

$$= \lim_{\Delta x \to 0} \frac{\cancel{\Delta x}(6x + 3\Delta x - 2)}{\cancel{\Delta x}}$$

$$= \lim_{\Delta x \to 0} (6x + 3\Delta x - 2)$$

$$= 6x - 2$$

Thus, the derivative of $f(x) = 3x^2 - 2x$ is $f'(x) = 6x - 2$.   ▬

In many applications, it is convenient to use a variable other than x as the independent variable. Example 7 shows a function that uses t as the independent variable.

⊞ **TECHNOLOGY**

You can use a graphing utility to confirm the result given in Example 7. One way to do this is to choose a point on the graph of $y = 2/t$, such as $(1, 2)$, and find the equation of the tangent line at that point. Using the derivative found in the example, you know that the slope of the tangent line when $t = 1$ is $m = -2$. This means that the tangent line at the point $(1, 2)$ is

$y - 2 = -2(t - 1)$ or
$y = -2t + 4$.

By graphing $y = 2/t$ and $y = -2t + 4$ on the same screen, as shown below, you can confirm that the line is tangent to the graph at the point $(1, 2)$.

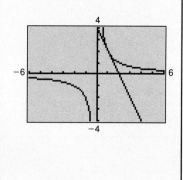

EXAMPLE 7   *Finding a Derivative*

Find the derivative of y with respect to t for the function

$$y = \frac{2}{t}.$$

Solution

Consider $y = f(t)$, and use the limit process as follows.

$$\frac{dy}{dt} = \lim_{\Delta t \to 0} \frac{f(t + \Delta t) - f(t)}{\Delta t} \qquad \textit{Set up difference quotient.}$$

$$= \lim_{\Delta t \to 0} \frac{\left(\dfrac{2}{t + \Delta t} - \dfrac{2}{t} \right)}{\Delta t} \qquad \textit{Use } f(t) = 2/t.$$

$$= \lim_{\Delta t \to 0} \frac{\left(\dfrac{2t - 2t - 2\Delta t}{t(t + \Delta t)} \right)}{\Delta t} \qquad \textit{Expand terms.}$$

$$= \lim_{\Delta t \to 0} \frac{-2\Delta t}{t(\Delta t)(t + \Delta t)} \qquad \textit{Factor and cancel.}$$

$$= \lim_{\Delta t \to 0} \frac{-2}{t(t + \Delta t)} \qquad \textit{Simplify.}$$

$$= -\frac{2}{t^2} \qquad \textit{Evaluate the limit.}$$

Thus, the derivative of y with respect to t is

$$\frac{dy}{dt} = -\frac{2}{t^2}.$$

Remember that the derivative of a function gives you a formula for finding the slope of the tangent at any point on the function's graph. For example, the slope of the tangent line to the graph of f at the point $(1, 2)$ is given by

$$f'(1) = -\frac{2}{1^2} = -2.$$

To find the slope of the graph at other points, substitute the x-coordinate of the point into the derivative, as shown below.

Point	x-Coordinate	Slope
$(2, 1)$	$x = 1$	$m = f'(2) = -\dfrac{2}{2^2} = -\dfrac{1}{2}$
$(-2, -1)$	$x = 1$	$m = f'(-2) = -\dfrac{2}{(-2)^2} = -\dfrac{1}{2}$

Differentiability and Continuity

Not every function is differentiable. Figure 2.11 shows some common situations in which a function will not be differentiable at a point—vertical tangent lines, discontinuities, and sharp turns in the graph. Each of the functions shown in Figure 2.11 is differentiable at every value of x *except $x = 0$*.

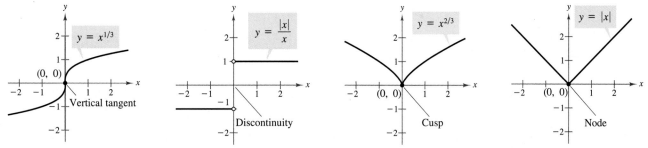

FIGURE 2.11   Functions That Are Not Differentiable at $(0, 0)$

From Figure 2.11, you can see that continuity is not a strong enough condition to guarantee differentiability. [All but one of the functions are continuous at $(0, 0)$, but none is differentiable there.] On the other hand, if a function is differentiable at a point, then it must be continuous. This important result is stated in the following theorem.

Differentiability Implies Continuity

If a function is differentiable at $x = c$, then it is continuous at $x = c$.

Discussion Problem   A Graphing Utility Experiment

Use a graphing utility to sketch the graph of $f(x) = x^3 - 3x$. On the same screen, graph the following lines. Which of the lines appear to be tangent lines to the graph of f? If the line appears to be tangent to the graph, state the point of tangency. (*Hint:* Use the zoom feature of your graphing utility.)

a. $y = 2$ **b.** $y = x - 3$ **c.** $y = -3x$ **d.** $y = -x + 1$

Explain how you can verify your conclusion analytically.

$\mathcal{Warm\ Up}$ The following warm-up exercises involve skills that were covered in earlier sections. You will use these skills in the exercise set for this section.

In Exercises 1 and 2, find an equation of the line containing P and Q.

1. $P(2, 1),\ Q(2, 4)$

2. $P(2, 2),\ Q(-5, 2)$

In Exercises 3–6, evaluate the limit.

3. $\displaystyle\lim_{\Delta x \to 0} \frac{2x\Delta x + (\Delta x)^2}{\Delta x}$

4. $\displaystyle\lim_{\Delta x \to 0} \frac{3x^2\Delta x + 3x(\Delta x)^2 + (\Delta x)^3}{\Delta x}$

5. $\displaystyle\lim_{\Delta x \to 0} \frac{1}{x(x + \Delta x)}$

6. $\displaystyle\lim_{\Delta x \to 0} \frac{(x + \Delta x)^2 - x^2}{\Delta x}$

In Exercises 7–10, find the domain of the function.

7. $f(x) = \dfrac{1}{x - 1}$

8. $f(x) = \dfrac{1}{5}x^3 - 2x^2 + \dfrac{1}{3}x - 1$

9. $f(x) = \dfrac{6x}{x^3 + x}$

10. $f(x) = \dfrac{x^2 - 2x - 24}{x^2 + x - 12}$

E X E R C I S E S   2.1   means that technology can help you solve or check the exercise(s).

In Exercises 1–4, trace the graph and sketch the tangent lines at (x_1, y_1) and (x_2, y_2).

In Exercises 5–10, estimate the slope of the graph at the point (x, y). (Each square on the grid is 1 unit by 1 unit.)

1.

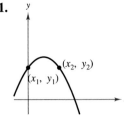

2.

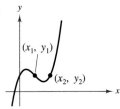

5.

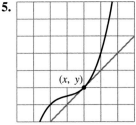

6.

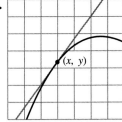

3.

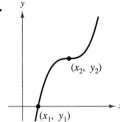

4.

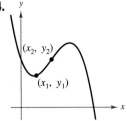

7.

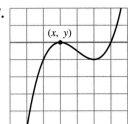

8.

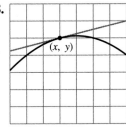

9.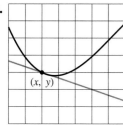

10.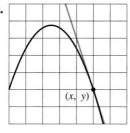

11. *Revenue* The graph represents the revenue R (in millions of dollars) for Dairy Queen from 1989 through 1993. Estimate the slope of the graph in 1990 and in 1992. (*Source: International Dairy Queen*)

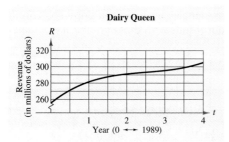

12. *Revenue* The graph represents the revenue R (in billions of dollars) for Chrysler Corporation from 1989 through 1993. Estimate the slope of the graph in 1990 and in 1992. (*Source: Chrysler Corporation*)

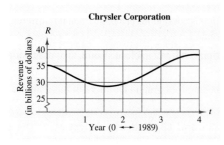

In Exercises 13–26, use the limit definition to find the derivative of the function.

13. $f(x) = 3$

14. $f(x) = -4$

15. $f(x) = -5x + 3$

16. $f(x) = \frac{1}{2}x$

17. $f(x) = x^2$

18. $f(x) = 1 - x^2$

19. $f(x) = 3x^2 - 5x - 2$

20. $f(x) = x - x^2$

21. $h(t) = \sqrt{t - 1}$

22. $f(x) = \sqrt{2x}$

23. $f(t) = t^3 - 12t$

24. $f(t) = t^3 + t^2$

25. $f(x) = \dfrac{1}{x^2}$

26. $g(s) = \dfrac{1}{s - 1}$

In Exercises 27–38, find the slope of the tangent line to the graph of f at the indicated point.

Function	Point of Tangency
27. $f(x) = 6 - 2x$	$(2, 2)$
28. $f(x) = 2x + 4$	$(1, 6)$
29. $f(x) = -x$	$(0, 0)$
30. $f(x) = 6$	$(-2, 6)$
31. $f(x) = x^2 - 2$	$(2, 2)$
32. $f(x) = x^2 + 2x + 1$	$(-3, 4)$
33. $f(x) = x^3$	$(2, 8)$
34. $f(x) = x^3 + 3$	$(-2, -5)$
35. $f(x) = \sqrt{x + 1}$	$(3, 2)$
36. $f(x) = \dfrac{1}{x + 1}$	$(0, 1)$
37. $f(x) = x^3 + 2x$	$(1, 3)$
38. $f(x) = \sqrt{2x - 2}$	$(9, 4)$

In Exercises 39–44, find an equation of the tangent line to the graph of f at the indicated point. Then verify your result by sketching the graph of f and the tangent line.

Function	Point of Tangency
39. $f(x) = \frac{1}{2}x^2$	$(2, 2)$
40. $f(x) = -x^2$	$(-1, -1)$
41. $f(x) = (x - 1)^2$	$(-2, 9)$
42. $f(x) = 2x^2 - 1$	$(0, -1)$
43. $f(x) = \sqrt{x} + 1$	$(4, 3)$
44. $f(x) = \dfrac{1}{x}$	$(1, 1)$

In Exercises 45–48, find an equation of the line that is tangent to the graph of f *and* parallel to the given line.

Function	Line
45. $f(x) = -\frac{1}{4}x^2$	$x + y = 0$
46. $f(x) = x^2 + 1$	$2x + y = 0$
47. $f(x) = -\frac{1}{2}x^3$	$6x + y + 4 = 0$
48. $f(x) = \dfrac{1}{\sqrt{x}}$	$x + 2y - 6 = 0$

In Exercises 49–58, describe the *x*-values at which the function is differentiable. Explain your reasoning.

49. $y = |x + 3|$

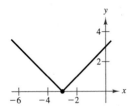

50. $y = |x^2 - 9|$

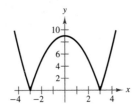

51. $y = \dfrac{1}{x + 1}$

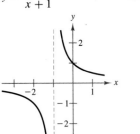

52. $y = \dfrac{2x}{x - 1}$

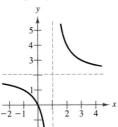

53. $y = (x - 3)^{2/3}$

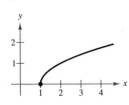

54. $y = x^{2/5}$

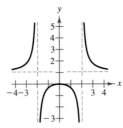

55. $y = \sqrt{x - 1}$

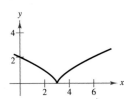

56. $y = \dfrac{x^2}{x^2 - 4}$

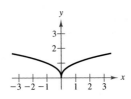

57. $y = \begin{cases} x^3 + 3, & x < 0 \\ x^3 - 3, & x \geq 0 \end{cases}$

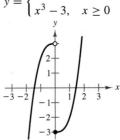

58. $y = \begin{cases} x^2, & x \leq 1 \\ -x^2, & x > 1 \end{cases}$

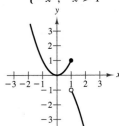

In Exercises 59 and 60, use a graphing utility to graph *f* over the interval $[-2, 2]$. Then complete the table by *graphically* estimating the slope of the graph at the indicated points. Finally, evaluate the slope *analytically* and compare your results with those graphically obtained.

x	-2	$-\frac{3}{2}$	-1	$-\frac{1}{2}$	0	$\frac{1}{2}$	1	$\frac{3}{2}$	2
$f(x)$									
$f'(x)$									

59. $f(x) = \frac{1}{4}x^3$

60. $f(x) = \frac{1}{2}x^2$

True or False In Exercises 61–64, determine whether the statement is true or false. If it is false, explain why or give an example that shows the statement is false.

61. The slope of the graph of $y = x^2$ is different at every point on the graph of f.

62. If a function is continuous at a point, then it is differentiable at that point.

63. If a function is differentiable at a point, then it is continuous at that point.

64. A tangent line to a graph can intersect the graph at more than one point.

In Exercises 65 and 66, find the derivative of the given function *f*. Then graph *f* and its derivative on the same set of coordinate axes. What does the *x*-intercept of the derivative indicate about the graph of *f*?

65. $f(x) = x^2 - 4x$

66. $f(x) = 2 + 6x - x^2$

2.2 Some Rules for Differentiation

The Constant Rule ▪ *The Power Rule* ▪ *The Constant Multiple Rule* ▪
The Sum and Difference Rules ▪ *Extended Application: Increasing Sales*

The Constant Rule

In Section 2.1, you found derivatives by the limit process. This process is tedious, even for simple functions, and fortunately there are rules that greatly simplify differentiation. These rules allow you to calculate derivatives without the *direct* use of limits.

The Constant Rule

The derivative of a constant function is zero. That is,

$$\frac{d}{dx}[c] = 0, \qquad c \text{ is a constant.}$$

Proof Let $f(x) = c$. Then, by the limit definition of the derivative, you can write

$$f'(x) = \lim_{\Delta x \to 0} \frac{f(x + \Delta x) - f(x)}{\Delta x} = \lim_{\Delta x \to 0} \frac{c - c}{\Delta x} = \lim_{\Delta x \to 0} 0 = 0.$$

Therefore,

$$\frac{d}{dx}[c] = 0.$$

Note Note in Figure 2.12 that the Constant Rule is equivalent to saying that the slope of a horizontal line is zero.

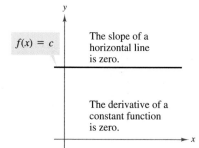

$f(x) = c$

The slope of a horizontal line is zero.

The derivative of a constant function is zero.

FIGURE 2.12

EXAMPLE 1 *Finding Derivatives of Constant Functions*

a. $\dfrac{d}{dx}[7] = 0$

b. If $f(x) = 0$, then $f'(x) = 0$.

c. If $y = 2$, then $\dfrac{dy}{dx} = 0$.

d. If $g(t) = -\dfrac{3}{2}$, then $g'(t) = 0$.

93

The Power Rule

The binomial expansion process is used to prove the Power Rule.

$$(x + \Delta x)^2 = x^2 + 2x\Delta x + (\Delta x)^2$$

$$(x + \Delta x)^3 = x^3 + 3x^2\Delta x + 3x(\Delta x)^2 + (\Delta x)^3$$

$$(x + \Delta x)^n = x^n + nx^{n-1}\Delta x + \underbrace{\frac{n(n-1)x^{n-2}}{2}(\Delta x)^2 + \cdots + (\Delta x)^n}_{(\Delta x)^2 \text{ is a factor of these terms.}}$$

The (Simple) Power Rule

$$\frac{d}{dx}[x^n] = nx^{n-1}, \qquad n \text{ is any real number.}$$

Proof We prove only the case in which n is a positive integer. Let $f(x) = x^n$. Using the binomial expansion, you can write

$$f'(x) = \lim_{\Delta x \to 0} \frac{f(x + \Delta x) - f(x)}{\Delta x}$$

$$= \lim_{\Delta x \to 0} \frac{(x + \Delta x)^n - x^n}{\Delta x}$$

$$= \lim_{\Delta x \to 0} \frac{x^n + nx^{n-1}\Delta x + \dfrac{n(n-1)x^{n-2}}{2}(\Delta x)^2 + \cdots + (\Delta x)^n - x^n}{\Delta x}$$

$$= \lim_{\Delta x \to 0} \left[nx^{n-1} + \frac{n(n-1)x^{n-2}}{2}(\Delta x) + \cdots + (\Delta x)^{n-1} \right]$$

$$= nx^{n-1} + 0 + \cdots + 0$$

$$= nx^{n-1}.$$

For the Power Rule, the case in which $n = 1$ is worth remembering as a separate differentiation rule. That is,

$$\frac{d}{dx}[x] = 1. \qquad \qquad \textit{The derivative of x is 1.}$$

This rule is consistent with the fact that the slope of the line given by $y = x$ is 1, as shown in Figure 2.13.

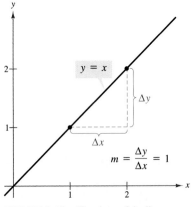

FIGURE 2.13 The slope of the line $y = x$ is 1.

EXAMPLE 2 *Applying the Power Rule*

Function	*Derivative*
a. $f(x) = x^3$	$f'(x) = 3x^2$
b. $y = \dfrac{1}{x^2} = x^{-2}$	$\dfrac{dy}{dx} = (-2)x^{-3} = -\dfrac{2}{x^3}$
c. $g(t) = t$	$g'(t) = 1$
d. $R = x^4$	$\dfrac{dR}{dx} = 4x^3$

In Example 2b, note that *before* differentiating, you should rewrite $1/x^2$ as x^{-2}. Rewriting is the first step in *many* differentiation problems.

Function:	Rewrite:	Differentiate:	Simplify:
$y = \dfrac{1}{x^2}$	$y = x^{-2}$	$\dfrac{dy}{dx} = (-2)x^{-3}$	$\dfrac{dy}{dx} = -\dfrac{2}{x^3}$

Remember that the derivative of a function f is another function that gives the slope of the graph of f at any point at which f is differentiable. Thus, you can use the derivative to find slopes, as shown in Example 3.

EXAMPLE 3 *Finding the Slope of a Graph*

Find the slope of the graph of

$$f(x) = x^2 \qquad \text{\textit{Original function}}$$

when $x = -2, -1, 0, 1,$ and 2.

Solution

Begin by using the Power Rule to find the derivative of f.

$$f'(x) = 2x \qquad \text{\textit{Derivative}}$$

You can use the derivative to find the slope of the graph of f, as follows.

x-Value	*Slope of Graph of f*
$x = -2$	$m = f'(-2) = 2(-2) = -4$
$x = -1$	$m = f'(-1) = 2(-1) = -2$
$x = 0$	$m = f'(0) = 2(0) = 0$
$x = 1$	$m = f'(1) = 2(1) = 2$
$x = 2$	$m = f'(2) = 2(2) = 4$

The graph of f is shown in Figure 2.14.

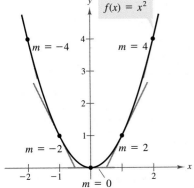

FIGURE 2.14

The Constant Multiple Rule

To prove the Constant Multiple Rule, the following property of limits is used.

$$\lim_{x \to a} cg(x) = c \left[\lim_{x \to a} g(x) \right]$$

The Constant Multiple Rule

If f is a differentiable function of x, and c is a real number, then

$$\frac{d}{dx}[cf(x)] = cf'(x), \qquad c \text{ is a constant.}$$

Proof Apply the definition of the derivative to produce

$$\frac{d}{dx}[cf(x)] = \lim_{\Delta x \to 0} \frac{cf(x + \Delta x) - cf(x)}{\Delta x}$$

$$= \lim_{\Delta x \to 0} c \left[\frac{f(x + \Delta x) - f(x)}{\Delta x} \right]$$

$$= c \left[\lim_{\Delta x \to 0} \frac{f(x + \Delta x) - f(x)}{\Delta x} \right]$$

$$= cf'(x). \qquad \blacksquare$$

Informally, the Constant Multiple Rule states that constants can be factored out of the differentiation process.

$$\frac{d}{dx}[cf(x)] = c\frac{d}{dx}[f(x)] = cf'(x)$$

The usefulness of this rule is often overlooked, especially when the constant appears in the denominator, as follows.

$$\frac{d}{dx}\left[\frac{f(x)}{c}\right] = \frac{d}{dx}\left[\frac{1}{c}f(x)\right] = \frac{1}{c}\left[\frac{d}{dx}[f(x)]\right] = \frac{1}{c}f'(x)$$

To use the Constant Multiple Rule efficiently, look for constants that can be factored out *before* differentiating. For example,

$$\frac{d}{dx}[5x^2] = 5\frac{d}{dx}[x^2] = 5(2x) = 10x$$

and

$$\frac{d}{dx}\left[\frac{x^2}{5}\right] = \frac{1}{5}\left[\frac{d}{dx}[x^2]\right] = \frac{1}{5}(2x) = \frac{2}{5}x.$$

▦ TECHNOLOGY

If you have access to a symbolic differentiation utility such as *Derive*, *Mathematica*, or *Maple*, try using it to confirm the derivatives shown in this section. For instance, if you use *Derive* to differentiate the function in Example 4b, you will obtain the following.

```
Author: (4t^2)/5

1:  4t²
    ──
     5

Calculus: Differentiate

2:  d  4t²
    ── ──
    dt  5

Simplify

3:  8t
    ──
     5
```

EXAMPLE 4 *Using the Power and Constant Multiple Rules*

Differentiate the following functions.

a. $y = 2x^{1/2}$ **b.** $f(t) = \dfrac{4t^2}{5}$

Solution

a. Using the Constant Multiple Rule and the Power Rule, you can write

$$\frac{dy}{dx} = \frac{d}{dx}[2x^{1/2}] = 2\underbrace{\frac{d}{dx}[x^{1/2}]}_{\text{Constant Multiple Rule}} = 2\underbrace{\left(\frac{1}{2}x^{-1/2}\right)}_{\text{Power Rule}} = x^{-1/2} = \frac{1}{\sqrt{x}}.$$

b. Begin by rewriting $f(t)$ as

$$f(t) = \frac{4t^2}{5} = \frac{4}{5}t^2.$$

Then, use the Constant Multiple Rule and the Power Rule to obtain

$$f'(t) = \frac{d}{dt}\left[\frac{4}{5}t^2\right] = \frac{4}{5}\left[\frac{d}{dt}(t^2)\right] = \frac{4}{5}(2t) = \frac{8}{5}t.$$

■

You may find it helpful to combine the Constant Multiple Rule and the Power Rule into one combined rule.

$$\frac{d}{dx}[cx^n] = cnx^{n-1}, \qquad n \text{ is a real number, } c \text{ is a constant.}$$

For instance, in Example 4b, you can apply this combined rule to obtain

$$\frac{d}{dx}\left[\frac{4}{5}t^2\right] = \left(\frac{4}{5}\right)(2)(t) = \frac{8}{5}t.$$

The three functions in the next example are simple, yet errors are frequently made in differentiating functions involving a constant multiple of the first power of x. Keep in mind that

$$\frac{d}{dx}[cx] = c, \qquad c \text{ is a constant.}$$

EXAMPLE 5 *Applying the Constant Multiple Rule*

Original Function	*Derivative*
a. $y = -\dfrac{3x}{2}$	$y' = -\dfrac{3}{2}$
b. $y = 3\pi x$	$y' = 3\pi$
c. $y = -\dfrac{x}{2}$	$y' = -\dfrac{1}{2}$

■

TECHNOLOGY

Most graphing utilities have a built-in program to approximate the derivative of a function at a specific point. Here is a sample program for the TI-81 that lets you conveniently evaluate the derivative of a function at several points x.

```
Prgm0: DERIV
:Lbl 1
:Disp ''ENTER X''
:Input X
:NDeriv(Y1,.001)⟶D
:Disp ''DERIVATIVE''
:Disp D
:Goto 1
:End
```

To use this program, enter the function in Y_1. Then run the program—it will allow you to approximate the derivative of the function at several values of x.

Parentheses can play an important role in the use of the Constant Multiple Rule and the Power Rule. In Example 6, be sure you understand the mathematical conventions involving the use of parentheses.

EXAMPLE 6 *Using Parentheses When Differentiating*

Find the derivative of each function.

a. $y = \dfrac{5}{2x^3}$ **b.** $y = \dfrac{5}{(2x)^3}$ **c.** $y = \dfrac{7}{3x^{-2}}$ **d.** $y = \dfrac{7}{(3x)^{-2}}$

Solution

	Function	*Rewrite*	*Differentiate*	*Simplify*
a.	$y = \dfrac{5}{2x^3}$	$y = \dfrac{5}{2}(x^{-3})$	$y' = \dfrac{5}{2}(-3x^{-4})$	$y' = -\dfrac{15}{2x^4}$
b.	$y = \dfrac{5}{(2x)^3}$	$y = \dfrac{5}{8}(x^{-3})$	$y' = \dfrac{5}{8}(-3x^{-4})$	$y' = -\dfrac{15}{8x^4}$
c.	$y = \dfrac{7}{3x^{-2}}$	$y = \dfrac{7}{3}(x^2)$	$y' = \dfrac{7}{3}(2x)$	$y' = \dfrac{14x}{3}$
d.	$y = \dfrac{7}{(3x)^{-2}}$	$y = 63(x^2)$	$y' = 63(2x)$	$y' = 126x$

When differentiating functions involving radicals, you should rewrite the function with rational exponents. For instance, you should rewrite $y = \sqrt[3]{x}$ as $y = x^{1/3}$, and you should rewrite

$$y = \dfrac{1}{\sqrt[3]{x^4}} \quad \text{as} \quad y = x^{-4/3}.$$

EXAMPLE 7 *Differentiating Radical Functions*

Find the derivative of each function.

a. $y = \sqrt{x}$ **b.** $y = \dfrac{1}{2\sqrt[3]{x^2}}$ **c.** $y = \sqrt{2x}$

Solution

	Function	*Rewrite*	*Differentiate*	*Simplify*
a.	$y = \sqrt{x}$	$y = x^{1/2}$	$y' = \left(\dfrac{1}{2}\right)x^{-1/2}$	$y' = \dfrac{1}{2\sqrt{x}}$
b.	$y = \dfrac{1}{2\sqrt[3]{x^2}}$	$y = \dfrac{1}{2}x^{-2/3}$	$y' = \dfrac{1}{2}\left(-\dfrac{2}{3}\right)x^{-5/3}$	$y' = -\dfrac{1}{3x^{5/3}}$
c.	$y = \sqrt{2x}$	$y = \sqrt{2}(x^{1/2})$	$y' = \sqrt{2}\left(\dfrac{1}{2}\right)x^{-1/2}$	$y' = \dfrac{1}{\sqrt{2x}}$

The Sum and Difference Rules

The next two rules are ones that you might expect to be true, and you may have used them without thinking about it. For instance, if you were asked to differentiate $y = 3x + 2x^3$, you would probably write

$$y' = 3 + 6x^2$$

without questioning your answer. The validity of differentiating a sum term by term is given by the Sum and Difference Rules.

The Sum and Difference Rules

The derivative of the sum (or difference) of two differentiable functions is the sum (or difference) of their derivatives.

$$\frac{d}{dx}[f(x) + g(x)] = f'(x) + g'(x) \qquad \text{Sum Rule}$$

$$\frac{d}{dx}[f(x) - g(x)] = f'(x) - g'(x) \qquad \text{Difference Rule}$$

Proof Let $h(x) = f(x) + g(x)$. Then, you can prove the Sum Rule as follows.

$$h'(x) = \lim_{\Delta x \to 0} \frac{h(x + \Delta x) - h(x)}{\Delta x}$$

$$= \lim_{\Delta x \to 0} \frac{f(x + \Delta x) + g(x + \Delta x) - f(x) - g(x)}{\Delta x}$$

$$= \lim_{\Delta x \to 0} \frac{f(x + \Delta x) - f(x) + g(x + \Delta x) - g(x)}{\Delta x}$$

$$= \lim_{\Delta x \to 0} \left[\frac{f(x + \Delta x) - f(x)}{\Delta x} + \frac{g(x + \Delta x) - g(x)}{\Delta x} \right]$$

$$= \lim_{\Delta x \to 0} \frac{f(x + \Delta x) - f(x)}{\Delta x} + \lim_{\Delta x \to 0} \frac{g(x + \Delta x) - g(x)}{\Delta x}$$

$$= f'(x) + g'(x)$$

Thus,

$$\frac{d}{dx}[f(x) + g(x)] = f'(x) + g'(x).$$

The Difference Rule can be proved in a similar manner.   ▬

The Sum and Difference Rules can be extended to the sum or difference of any finite number of functions. For instance, if $y = f(x) + g(x) + h(x)$, then $y' = f'(x) + g'(x) + h'(x)$.

With the four differentiation rules listed in this section, you can differentiate *any* polynomial function.

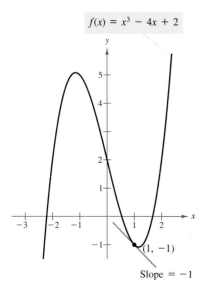

FIGURE 2.15

EXAMPLE 8 *Using the Sum and Difference Rules*

Find the slope of the graph of $f(x) = x^3 - 4x + 2$ at the point $(1, -1)$.

Solution
The derivative of $f(x)$ is

$$f'(x) = 3x^2 - 4.$$

Thus, the slope of the graph of f at $(1, -1)$ is

$$\text{Slope} = f'(1) = 3(1)^2 - 4 = -1,$$

as shown in Figure 2.15.

Example 8 illustrates the use of the derivative for determining the shape of a graph. A rough sketch of the graph of $f(x) = x^3 - 4x + 2$ might lead you to think that the point $(1, -1)$ is a minimum point of the graph. After finding the slope at this point to be -1, however, you can conclude that the minimum point (where the slope is 0) is farther to the right. (You will study techniques for finding minimum and maximum points in Section 3.2.)

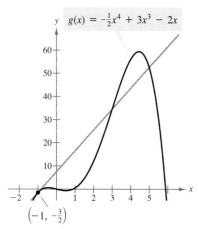

FIGURE 2.16

EXAMPLE 9 *Using the Sum and Difference Rules*

Find an equation of the tangent line to the graph of

$$g(x) = -\tfrac{1}{2}x^4 + 3x^3 - 2x$$

at the point $\left(-1, -\tfrac{3}{2}\right)$.

Solution
The derivative of $g(x)$ is

$$g'(x) = -2x^3 + 9x^2 - 2$$

which implies that the slope of the graph at the point $\left(-1, -\tfrac{3}{2}\right)$ is

$$\begin{aligned}
\text{Slope} = g'(-1) &= -2(-1)^3 + 9(-1)^2 - 2 \\
&= 2 + 9 - 2 \\
&= 9,
\end{aligned}$$

as shown in Figure 2.16. Using the point-slope form, you can write the equation of the tangent line at $\left(-1, -\tfrac{3}{2}\right)$ as follows.

$$y - \left(-\tfrac{3}{2}\right) = 9[x - (-1)] \qquad \textit{Point-slope form}$$

$$y = 9x + \tfrac{15}{2} \qquad \textit{Equation of tangent line}$$

Extended Application: Increasing Sales

Real Life

EXAMPLE 10 *Modeling Sales*

From 1983 through 1992, the revenue R (in millions of dollars) for Microsoft Corporation can be modeled by

$$R = 0.76t^4 - 16.77t^3 + 153.05t^2 - 571.84t + 789.84,$$

where $t = 3$ corresponds to 1983. At what rate was Microsoft's revenue changing in 1990? *(Source: Microsoft Corporation)*

Solution

One way to answer this question is to find the derivative of the revenue model with respect to time.

$$\frac{dR}{dt} = 3.04t^3 - 50.31t^2 + 306.1t - 571.84$$

In 1990 (when $t = 10$), the rate of change of the revenue with respect to the time is given by

$$\frac{dR}{dt} = 3.04(10)^3 - 50.31(10)^2 + 306.1(10) - 571.84$$

$$\approx 500.$$

Because R is measured in millions of dollars and t is measured in years, it follows that the derivative dR/dt is measured in millions of dollars per year. Thus, at the end of 1990, Microsoft's revenues were increasing at a rate of about \$500 million per year, as shown in Figure 2.17. ▬

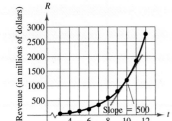

Revenue for Microsoft

FIGURE 2.17

Discussion Problem **Units for Rates of Change**

In Example 10, the units for R are millions of dollars and the units for dR/dt are millions of dollars per year. State the units for the derivatives of each of the following models.

a. A population model which gives the population P (in millions of people) of the United States in terms of the year, t. What are the units for dP/dt?

b. A position model which gives the height s (in feet) of an object in terms of the time t (in seconds). What are the units for ds/dt?

c. A demand model which gives the price per unit p (in dollars) of a product in terms of the number of units x sold. What are the units for dp/dx?

Warm Up

The following warm-up exercises involve skills that were covered in earlier sections. You will use these skills in the exercise set for this section.

In Exercises 1 and 2, evaluate each expression when $x = 2$.

1. (a) $2x^2$ (b) $(2x)^2$ (c) $2x^{-2}$ **2.** (a) $\dfrac{1}{(3x)^2}$ (b) $\dfrac{1}{4x^3}$ (c) $\dfrac{(2x)^{-3}}{4x^{-2}}$

In Exercises 3–6, simplify the expression.

3. $4(3)x^3 + 2(2)x$

4. $\frac{1}{2}(3)x^2 - \frac{3}{2}x^{1/2}$

5. $\left(\frac{1}{4}\right)x^{-3/4}$

6. $\frac{1}{3}(3)x^2 - 2\left(\frac{1}{2}\right)x^{-1/2} + \frac{1}{3}x^{-2/3}$

In Exercises 7–10, solve the equation.

7. $3x^2 + 2x = 0$

8. $x^3 - x = 0$

9. $x^2 + 8x - 20 = 0$

10. $x^2 - 10x - 24 = 0$

E X E R C I S E S   2.2   means that technology can help you solve or check the exercise(s).

In Exercises 1–4, find the slope of the tangent line to $y = x^n$ at the point $(1, 1)$.

1. (a) $y = x^2$ (b) $y = x^{1/2}$

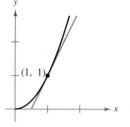

3. (a) $y = x^{-1}$ (b) $y = x^{-1/3}$

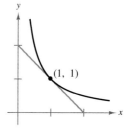   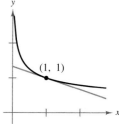

2. (a) $y = x^{3/2}$ (b) $y = x^3$

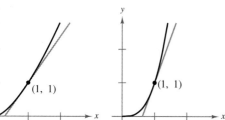

4. (a) $y = x^{-1/2}$ (b) $y = x^{-2}$

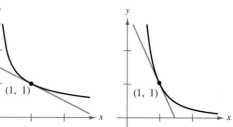

In Exercises 5–20, use the differentiation rules in this section to find the derivative of the function.

5. $y = 3$

6. $f(x) = -2$

7. $f(x) = x + 1$

8. $g(x) = 3x - 1$

9. $g(x) = x^2 + 4$

10. $y = t^2 + 2t - 3$

11. $f(t) = -3t^2 + 2t - 4$

12. $y = x^3 - 9$

13. $s(t) = t^3 - 2t + 4$

14. $y = 2x^3 - x^2 + 3x - 1$

15. $y = 4t^{4/3}$

16. $h(x) = x^{5/2}$

17. $f(x) = 4\sqrt{x}$

18. $g(x) = 4\sqrt[3]{x} + 2$

19. $y = 4x^{-2} + 2x^2$

20. $s(t) = 4t^{-1} + 1$

In Exercises 21–26, use Example 6 as a model to find the derivative.

Function	Rewrite	Derivative	Simplify
21. $y = \dfrac{1}{4x^3}$			
22. $y = \dfrac{2}{3x^2}$			
23. $y = \dfrac{1}{(4x)^3}$			
24. $y = \dfrac{\pi}{(3x)^2}$			
25. $y = \dfrac{\sqrt{x}}{x}$			
26. $y = \dfrac{4}{x^{-3}}$			

In Exercises 27–32, find the value of the derivative of the function at the indicated point.

Function	Point
27. $f(x) = \dfrac{1}{x}$	$(1, 1)$
28. $f(x) = -\frac{1}{2} + \frac{7}{5}x^3$	$\left(0, -\frac{1}{2}\right)$
29. $f(t) = 4 - \dfrac{4}{3t}$	$\left(\frac{1}{2}, \frac{4}{3}\right)$
30. $y = 3x\left(x^2 - \dfrac{2}{x}\right)$	$(2, 18)$
31. $y = (2x + 1)^2$	$(0, 1)$
32. $f(x) = 3(5 - x)^2$	$(5, 0)$

In Exercises 33–44, find $f'(x)$.

33. $f(x) = x^2 - \dfrac{4}{x}$

34. $f(x) = x^2 - 3x - 3x^{-2} + 5x^{-3}$

35. $f(x) = x^2 - 2x - \dfrac{2}{x^4}$

36. $f(x) = x^2 + 4x + \dfrac{1}{x}$

37. $f(x) = \dfrac{2x^3 - 4x^2 + 3}{x^2}$

38. $f(x) = \dfrac{2x^2 - 3x + 1}{x}$

39. $f(x) = x(x^2 + 1)$

40. $f(x) = (x^2 + 2x)(x + 1)$

41. $f(x) = x^{4/5}$

42. $f(x) = x^{1/3} - 1$

43. $f(x) = \sqrt[3]{x} + \sqrt[5]{x}$

44. $f(x) = \dfrac{1}{\sqrt[3]{x^2}}$

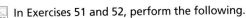 In Exercises 45 and 46, find an equation of the tangent line to the graph of the function at the indicated point.

Function	Point
45. $y = -2x^4 + 5x^2 - 3$	$(1, 0)$
46. $y = x^3 + x$	$(-1, -2)$

In Exercises 47–50, determine the point(s), if any, at which the graph of the function has a horizontal tangent line.

47. $y = -x^4 + 3x^2 - 1$

48. $y = \frac{1}{2}x^2 + 5x$

49. $y = x^3 + x$

50. $y = x^2 + 1$

In Exercises 51 and 52, perform the following.

(a) Sketch the graphs of f, g, and h on the same set of coordinate axes.

(b) Find $f'(1)$, $g'(1)$, and $h'(1)$.

(c) Sketch the graph of the tangent lines to each graph when $x = 1$.

51. $f(x) = x^3$
  $g(x) = x^3 + 3$
  $h(x) = x^3 - 2$

52. $f(x) = \sqrt{x}$
  $g(x) = \sqrt{x} + 4$
  $h(x) = \sqrt{x} - 2$

53. Use the Constant Rule, the Constant Multiple Rule, and the Sum Rule to find $h'(1)$ given that $f'(1) = 3$.

(a) $h(x) = f(x) - 2$

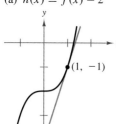

(b) $h(x) = 2f(x)$

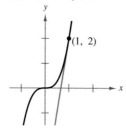

(c) $h(x) = -f(x)$

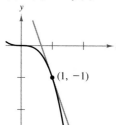

(d) $h(x) = -1 + 2f(x)$

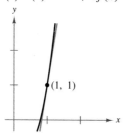

54. *Revenue* The revenue R (in millions of dollars) for Dairy Queen from 1989 through 1993 can be modeled by

$$R = 1.83t^3 - 13.7t^2 + 37.97t + 255.4,$$

where $t = 0$ represents 1989. *(Source: International Dairy Queen)*

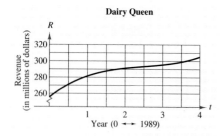

(a) Find the slope of the graph in 1990 ($t = 1$) and in 1992 ($t = 3$).

(b) Compare your results with those obtained in Exercise 11 in Section 2.1.

(c) What are the units for the slope of the graph? Interpret the slope of the graph in the context of the problem.

55. *Revenue* The revenue R (in billions of dollars) for Chrysler Corporation from 1989 through 1993 can be modeled by

$$R = -6.1t^3 + 23.6t^{5/2} - 22.6t^2 + 35.2,$$

where $t = 0$ represents 1989. *(Source: Chrysler Corp.)*

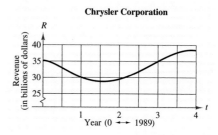

(a) Find the slope of the graph in 1990 ($t = 1$) and in 1992 ($t = 3$).

(b) Compare your results with those obtained in Exercise 12 in Section 2.1.

(c) What are the units for the slope of the graph? Interpret the slope of the graph in the context of the problem.

56. *Cost* The variable cost for manufacturing an electrical component is $7.75 per unit, and the fixed cost is $500. Write the cost C as a function of x, the number of units produced. Show that the derivative of this cost function is constant and equal to the variable cost.

57. *Cost* A college club raises funds by selling candy bars for $1.00 each. The club pays $0.60 for each candy bar and has annual fixed costs of $250. Write the profit P as a function of x, the number of candy bars sold. Show that the derivative of the profit function is a constant and that it is equal to the profit on each candy bar sold.

In Exercises 58 and 59, use a graphing utility to graph f and f' over the given interval. Determine any points at which the graph of f has horizontal tangents.

Function	*Interval*
58. $f(x) = 4.1x^3 - 12x^2 + 2.5x$	$[0, 3]$
59. $f(x) = x^3 - 1.4x^2 - 0.96x + 1.44$	$[-2, 2]$

Average Rate of Change ■ *Instantaneous Rate of Change and Velocity* ■ *Rates of Change in Economics: Marginals*

Average Rate of Change

In Sections 2.1 and 2.2, you studied the two primary applications of derivatives.

1. **Slope** The derivative of f is a function that gives the slope of the graph of f at a point $(x, f(x))$.
2. **Rate of Change** The derivative of f is a function that gives the rate of change of $f(x)$ with respect to x at the point $(x, f(x))$.

In this section, you will see that there are many real-life applications of rates of change. A few are velocity, acceleration, population growth rates, unemployment rates, production rates, and water flow rates. Although rates of change often involve change with respect to time, you can investigate the rate of change of one variable with respect to any other related variable.

When determining the rate of change of one variable with respect to another, you must be careful to distinguish between *average* and *instantaneous* rates of change. The distinction between these two rates of change is comparable to the distinction between the slope of the secant line through two points on a graph and the slope of the tangent line at one point on the graph.

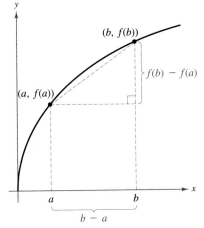

FIGURE 2.18

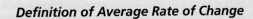

Definition of Average Rate of Change

If $y = f(x)$, then the **average rate of change** of y with respect to x on the interval $[a, b]$ is

$$\text{Average rate of change} = \frac{f(b) - f(a)}{b - a} = \frac{\Delta y}{\Delta x}.$$

Note that $f(a)$ is the value of the function at the *left* endpoint of the interval, $f(b)$ is the value of the function at the *right* endpoint of the interval, and $b - a$ is the width of the interval, as shown in Figure 2.18.

Note In real-life problems, it is important to list the units of measure for a rate of change. The units for $\Delta y/\Delta x$ are "y-units" per "x-units." For example, if y is measured in miles and x is measured in hours, then $\Delta y/\Delta x$ is measured in *miles per hour*.

Real Life

EXAMPLE 1 *Drug Concentration*

The concentration C (in milligrams per milliliter) of a drug in a patient's blood-stream is monitored over 10-minute intervals for 2 hours, where t is measured in minutes, as shown in Table 2.1. Find the average rates of change over the following intervals.

a. [0, 10]   **b.** [0, 20]   **c.** [100, 110]

Note In Example 1, the average rate of change is positive when the concentration increases and negative when the concentration decreases, as shown in Figure 2.19.

TABLE 2.1

t	0	10	20	30	40	50	60	70	80	90	100	110	120
C	0	2	17	37	55	73	89	103	111	113	113	103	68

Solution

a. For the interval [0, 10], the average rate of change is

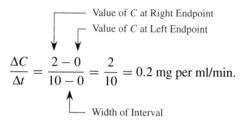

Value of C at Right Endpoint
Value of C at Left Endpoint

$$\frac{\Delta C}{\Delta t} = \frac{2 - 0}{10 - 0} = \frac{2}{10} = 0.2 \text{ mg per ml/min.}$$

Width of Interval

b. For the interval [0, 20], the average rate of change is

$$\frac{\Delta C}{\Delta t} = \frac{17 - 0}{20 - 0} = \frac{17}{20} = 0.85 \text{ mg per ml/min.}$$

c. For the interval [100, 110], the average rate of change is

$$\frac{\Delta C}{\Delta t} = \frac{103 - 113}{110 - 100} = \frac{-10}{10} = -1 \text{ mg per ml/min.}$$

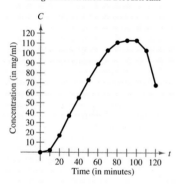

Drug Concentration in Bloodstream

Concentration (in mg/ml)

Time (in minutes)

FIGURE 2.19

The rates of change in Example 1 are in milligrams per milliliter per minute because the concentration is measured in milligrams per milliliter and the time is measured in minutes.

Concentration is measured in milligrams per milliliter.

Rate of change is measured in milligrams per milliliter per minute.

$$\frac{\Delta C}{\Delta t} = \frac{2 - 0}{10 - 0} = \frac{2}{10} = 0.2 \text{ mg per ml/min.}$$

Time is measured in minutes.

A common application of an average rate of change is to find the **average velocity** of an object that is moving in a straight line. That is,

$$\text{Average velocity} = \frac{\text{change in distance}}{\text{change in time}}.$$

This formula is demonstrated in Example 2.

Real Life

EXAMPLE 2 *Finding an Average Velocity*

If a free-falling object is dropped from a height of 100 feet, and *air resistance is neglected*, the height h (in feet) of the object at time t (in seconds) is given by

$$h = -16t^2 + 100,$$

as indicated in Figure 2.20. Find the average velocity of the object over the following intervals.

a. [1, 2]   **b.** [1, 1.5]   **c.** [1, 1.1]

Solution

You can use the position equation $h = -16t^2 + 100$ to determine the heights at $t = 1$, $t = 1.1$, $t = 1.5$, and $t = 2$, as shown in Table 2.2.

TABLE 2.2

t (in seconds)	0	1	1.1	1.5	2
h (in feet)	100	84	80.64	64	36

a. For the interval [1, 2], the object falls from a height of 84 feet to a height of 36 feet. Thus, the average velocity is

$$\frac{\Delta h}{\Delta t} = \frac{36 - 84}{2 - 1} = \frac{-48}{1} = -48 \text{ ft/sec.}$$

b. For the interval [1, 1.5], the average velocity is

$$\frac{\Delta h}{\Delta t} = \frac{64 - 84}{1.5 - 1} = \frac{-20}{0.5} = -40 \text{ ft/sec.}$$

c. For the interval [1, 1.1], the average velocity is

$$\frac{\Delta h}{\Delta t} = \frac{80.64 - 84}{1.1 - 1} = \frac{-3.36}{0.1} = -33.6 \text{ ft/sec.}$$

Note In Example 2, the average velocities are negative because the object is moving downward.

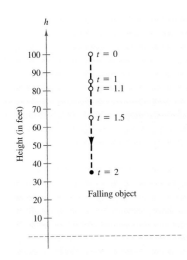

FIGURE 2.20
Some falling objects have considerable air resistance. Other falling objects have negligible air resistance. When modeling a falling-body problem, you must decide whether to account for air resistance or neglect it.

Instantaneous Rate of Change and Velocity

Suppose in Example 2 you wanted to find the rate of change of h at the instant $t = 1$ second. Such a rate is called an **instantaneous rate of change.** You can approximate the instantaneous rate of change at $t = 1$ by calculating the average rate of change over smaller and smaller intervals of the form $[1, 1 + \Delta t]$, as shown in Table 2.3. From the table, it seems reasonable to conclude that the instantaneous rate of change of the height when $t = 1$ is -32 feet per second.

TABLE 2.3

Δt approaches 0

Δt	1	0.5	0.1	0.01	0.001	0.0001	0
$\dfrac{\Delta h}{\Delta t}$	-48	-40	-33.6	-32.16	-32.016	-32.0016	-32

$\dfrac{\Delta h}{\Delta t}$ approaches -32

Note The limit in this definition is the same as the limit in the definition of the derivative of f at x. This is the second major interpretation of the derivative — as an *instantaneous rate of change in one variable with respect to another.* Recall that the first interpretation of the derivative is as the slope of the graph of f at x.

Definition of Instantaneous Rate of Change

The **instantaneous rate of change** (or simply **rate of change**) of $y = f(x)$ at x is the limit of the average rate of change on the interval $[x, x + \Delta x]$, as Δx approaches 0.

$$\lim_{\Delta x \to 0} \frac{\Delta y}{\Delta x} = \lim_{\Delta x \to 0} \frac{f(x + \Delta x) - f(x)}{\Delta x}$$

If y is a length and x is time, then the rate of change is a **velocity.**

Real Life

EXAMPLE 3 *Finding a Rate of Change*

Find the velocity of the object in Example 2 when $t = 1$.

Solution
From Example 2, you know that the height of the falling object is given by

$h = -16t^2 + 100.$　　　　　*Position function*

By taking the derivative of this position function, you obtain the velocity function.

$h'(t) = -32t$　　　　　*Velocity function*

The velocity function gives the velocity at *any* time. Thus, when $t = 1$, the velocity is $h'(1) = -32(1) = -32$ feet per second.

The general **position function** for a free-falling object, neglecting air resistance, is

$$h = -16t^2 + v_0t + h_0,$$ *Position function*

where h is the height (in feet), t is the time (in seconds), v_0 is the initial velocity (in feet per second), and h_0 is the initial height (in feet). Remember that the model assumes that positive velocities indicate upward motion and negative velocities indicate downward motion. The derivative $h' = -32t + v_0$ is the **velocity function.** The absolute value of the velocity is the **speed** of the object.

Real Life

EXAMPLE 4 *Finding the Velocity of a Diver*

At time $t = 0$, a diver jumps from a diving board that is 32 feet high, as shown in Figure 2.21. Because the diver's initial velocity is 16 feet per second, his position function is

$$h = -16t^2 + 16t + 32.$$ *Position function*

a. When does the diver hit the water?

b. What is the diver's velocity at impact?

Solution

a. To find the time at which the diver hits the water, let $h = 0$ and solve for t.

$$-16t^2 + 16t + 32 = 0$$ *Set h equal to 0.*

$$-16(t^2 - t - 2) = 0$$ *Factor out common factor.*

$$-16(t + 1)(t - 2) = 0$$ *Factor trinomial.*

$$t = -1 \text{ or } t = 2$$ *Solve for t.*

The solution $t = -1$ doesn't make sense in the problem, so you can conclude that the diver hits the water when $t = 2$ seconds.

b. The velocity at time t is given by the derivative

$$h' = -32t + 16.$$ *Velocity function*

The velocity at time $t = 2$ is $h' = -32(2) + 16 = -48$ feet per second.

In Example 4, note that the diver's initial velocity is $v_0 = 16$ feet per second (upward) and his initial height is $h_0 = 32$ feet.

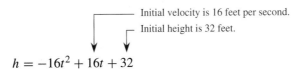

Initial velocity is 16 feet per second.

Initial height is 32 feet.

$$h = -16t^2 + 16t + 32$$

Discovery

Graph the polynomial function $h = -16t^2 + 16t + 32$ from Example 4 on the domain $0 \le t \le 2$. What is the maximum value of h? What is the derivative of h at this maximum point? In general, discuss how the derivative can be used to find the maximum or minimum values of a function.

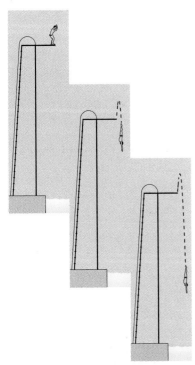

FIGURE 2.21

Rates of Change in Economics: Marginals

Another important use of rates of change is in the field of economics. Econo-mists refer to *marginal profit*, *marginal revenue*, and *marginal cost* as the rates of change of the profit, revenue, and cost with respect to the number x of units produced or sold. An equation that relates these three quantities is

$$P = R - C,$$

where P, R, and C represent the following quantities.

$$P = \text{total profit}, \quad R = \text{total revenue}, \quad \text{and} \quad C = \text{total cost}$$

The derivatives of these quantities are called the **marginal profit, marginal revenue,** and **marginal cost,** respectively.

$$\frac{dP}{dx} = \text{marginal profit}$$

$$\frac{dR}{dx} = \text{marginal revenue}$$

$$\frac{dC}{dx} = \text{marginal cost}$$

In many business and economics problems, the number of units produced or sold is restricted to positive integer values, as indicated in Figure 2.22(a). (Of course, it could happen that a sale involves half or quarter units, but it is hard to conceive of a sale involving $\sqrt{2}$ units.) The variable that denotes such units is called a **discrete variable.** To analyze a function of a discrete variable x, you can temporarily assume that x is a **continuous variable** and is able to take on any real value in a given interval, as indicated in Figure 2.22(b). Then, you can use the methods of calculus to find the x-value that corresponds to the marginal revenue, maximum profit, minimum cost, or whatever is called for. Finally, you should round the solution to the nearest sensible x-value—cents, dollars, units, or days, depending on the context of the problem.

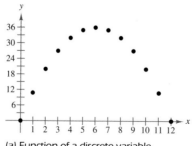

(a) Function of a discrete variable

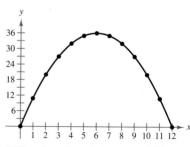

(b) Function of a continuous variable

FIGURE 2.22

Real Life

EXAMPLE 5 *Finding the Marginal Profit*

The profit derived from selling x units of an item is given by

$$P = 0.0002x^3 + 10x.$$

a. Find the marginal profit for a production level of 50 units.

b. Compare this with the actual gain in profit obtained by increasing the production level from 50 to 51 units.

Solution

a. Because the profit is $P = 0.0002x^3 + 10x$, the marginal profit is given by the derivative

$$\frac{dP}{dx} = 0.0006x^2 + 10.$$

When $x = 50$, the marginal profit is

$$\frac{dP}{dx} = 0.0006(50)^2 + 10 \qquad \textit{Substitute 50 for x.}$$

$$= 1.5 + 10$$

$$= \$11.50 \text{ per unit.} \qquad \textit{Marginal profit for } x = 50$$

b. For $x = 50$, the actual profit is

$$P = (0.0002)(50)^3 + 10(50) \qquad \textit{Substitute 50 for x.}$$

$$= 25 + 500$$

$$= \$525.00, \qquad \textit{Actual profit for } x = 50$$

and for $x = 51$, the actual profit is

$$P = (0.0002)(51)^3 + 10(51) \qquad \textit{Substitute 51 for x.}$$

$$= 26.53 + 510$$

$$= \$536.53. \qquad \textit{Actual profit for } x = 51$$

Thus, the additional profit obtained by increasing the production level from 50 to 51 units is

$$536.53 - 525.00 = \$11.53. \qquad \textit{Extra profit for one unit}$$

Note that the actual profit increase of \$11.53 (when x increases from 50 to 51 units) can be approximated by the marginal profit of \$11.50 per unit (when $x = 50$), as shown in Figure 2.23. ■

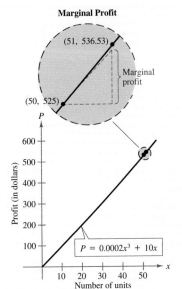

FIGURE 2.23

Note The reason the marginal profit gives a good approximation of the actual change in profit is that the graph of P is nearly straight over the interval $50 \leq x \leq 51$. You will study more about the use of marginals to approximate actual changes in Section 3.8.

The profit function in Example 5 is unusual in that the profit continues to increase as long as the number of units sold increases. In practice, it is more common to encounter situations in which sales can be increased only by lowering the price per item. Such reductions in price will ultimately cause the profit to decline.

The number of units x that consumers are willing to purchase at a given price per unit p is given by the **demand function**

$$p = f(x). \qquad\qquad \text{\textit{Demand function}}$$

The total revenue R is then related to the price per unit and the quantity demanded (or sold) by the equation

$$R = xp. \qquad\qquad \text{\textit{Revenue function}}$$

Real Life

EXAMPLE 6 *Finding a Demand Function*

A business sells 2000 items per month at a price of $10 each. It is estimated that monthly sales will increase 250 units for each $0.25 reduction in price. Use this information to find the demand function and total revenue function.

Solution

From the given estimate, x increases 250 units each time p drops $0.25 from the original cost of $10. This is described by the equation

$$x = 2000 + 250\left(\frac{10 - p}{0.25}\right)$$
$$= 2000 + 10{,}000 - 1000p$$
$$= 12{,}000 - 1000p.$$

Solving for p in terms of x produces

$$p = 12 - \frac{x}{1000}. \qquad\qquad \text{\textit{Demand function}}$$

This, in turn, implies that the revenue function is

$$R = xp \qquad\qquad \text{\textit{Formula for revenue}}$$
$$= x\left(12 - \frac{x}{1000}\right)$$
$$= 12x - \frac{x^2}{1000}. \qquad\qquad \text{\textit{Revenue function}}$$

The graph of the demand function is shown in Figure 2.24. Notice that the price decreases as the quantity demanded increases.

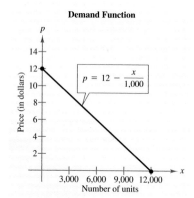

Demand Function

$p = 12 - \dfrac{x}{1{,}000}$

Price (in dollars)

3,000 6,000 9,000 12,000
Number of units

FIGURE 2.24

TECHNOLOGY

Modeling a Demand Function

To model a demand function, you need data that indicates how many units of a product will sell at a given price. As you might imagine, this type of data is not easy to obtain for a new product. After a product has been on the market awhile, however, its sales history can provide the necessary data.

As an example, consider the two bar graphs shown below. From these graphs, you can see that from 1981 through 1989, the price of prerecorded video cassettes dropped from an average price of about $65 to an average price of about $15. During that time, the number of units sold increased from about 1 million to about 89 million. *(Source: Paul Kagan Associates, Inc.)*

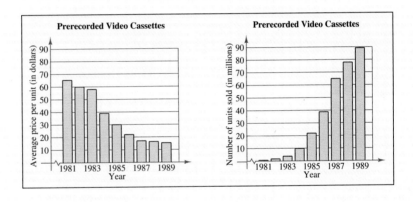

The information in the two bar graphs is combined in Table 2.4, where p represents the price (in dollars) and x represents the units sold (in millions).

TABLE 2.4

t	1981	1982	1983	1984	1985	1986	1987	1988	1989
x	0.72	1.72	3.76	9.80	21.94	38.78	64.90	77.90	89.20
p	65.28	59.88	57.98	38.98	29.99	22.00	17.00	16.34	15.36

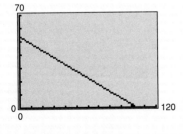

By entering the ordered pairs (x, p) into a calculator or computer that has a linear regression program, you can find the following linear model for the demand for prerecorded video cassettes: $p = -0.52x + 53.70$. A graph of this demand function is shown at the left.

Demand Function

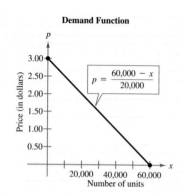

$$p = \frac{60,000 - x}{20,000}$$

Price (in dollars): 3.00, 2.50, 2.00, 1.50, 1.00, 0.50

Number of units: 20,000   40,000   60,000

FIGURE 2.25   As the price decreases, more hamburgers are sold.

Real Life

EXAMPLE 7 *Finding the Marginal Revenue*

A fast-food restaurant has determined that the monthly demand for their hamburgers is given by

$$p = \frac{60,000 - x}{20,000}.$$

Figure 2.25 shows that as the price decreases, the quantity demanded increases. Table 2.5 shows the demand for hamburgers at various prices.

TABLE 2.5

x	60,000	50,000	40,000	30,000	20,000	10,000	0
p	\$0.00	\$0.50	\$1.00	\$1.50	\$2.00	\$2.50	\$3.00

Find the increase in revenue per hamburger for monthly sales of 20,000 hamburgers. In other words, find the marginal revenue when $x = 20,000$.

Solution

Because the demand is given by

$$p = \frac{60,000 - x}{20,000},$$

the revenue is given by $R = xp$, and you have

$$R = xp = x\left(\frac{60,000 - x}{20,000}\right)$$

$$= \frac{1}{20,000}(60,000x - x^2).$$

By differentiating, you can find the marginal revenue to be

$$\frac{dR}{dx} = \frac{1}{20,000}(60,000 - 2x).$$

Thus, when $x = 20,000$, the marginal revenue is

$$\frac{dR}{dx} = \frac{1}{20,000}[60,000 - 2(20,000)]$$

$$= \frac{20,000}{20,000}$$

$$= \$1 \text{ per unit.}$$

Note   Writing a demand function in the form $p = f(x)$ is a convention used in economics. From a consumer's point of view, it might seem more reasonable to think that the quantity demanded is a function of the price. Mathematically, however, the two points of view are equivalent because a typical demand function is one-to-one and therefore possesses an inverse. For instance, in Example 7, you could write the demand function as $x = 60,000 - 20,000p$.

Real
Life

EXAMPLE 8 *Finding the Marginal Profit*

Suppose in Example 7 that the cost of producing x hamburgers is

$$C = 5000 + 0.56x, \qquad 0 \le x \le 50{,}000.$$

Find the profit and the marginal profit for the following production levels.

a. $x = 20{,}000$ **b.** $x = 24{,}400$ **c.** $x = 30{,}000$

Solution

From Example 7, you know that the total revenue from selling x units is

$$R = \frac{1}{20{,}000}(60{,}000x - x^2).$$

Because the total profit is given by $P = R - C$, you have

$$P = \frac{1}{20{,}000}(60{,}000x - x^2) - (5000 + 0.56x)$$

$$= 3x - \frac{x^2}{20{,}000} - 5000 - 0.56x$$

$$= 2.44x - \frac{x^2}{20{,}000} - 5000.$$

(See Figure 2.26.) Thus, the marginal profit is

$$\frac{dP}{dx} = 2.44 - \frac{x}{10{,}000}.$$

Using these formulas, you can compute the profit and marginal profit.

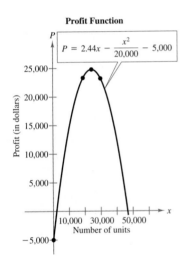

Profit Function

$$P = 2.44x - \frac{x^2}{20{,}000} - 5{,}000$$

Profit (in dollars)

Number of units

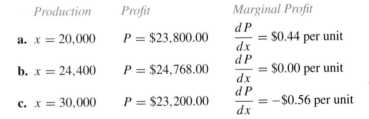

Production	Profit	Marginal Profit
a. $x = 20{,}000$	$P = \$23{,}800.00$	$\dfrac{dP}{dx} = \$0.44$ per unit
b. $x = 24{,}400$	$P = \$24{,}768.00$	$\dfrac{dP}{dx} = \$0.00$ per unit
c. $x = 30{,}000$	$P = \$23{,}200.00$	$\dfrac{dP}{dx} = -\$0.56$ per unit

FIGURE 2.26

Discussion Problem   *Interpreting Marginal Profit*

In Example 8 and Figure 2.26, notice that when 20,000 hamburgers are
sold the marginal profit is positive, when 24,400 are sold, the marginal
profit is zero, and when 30,000 are sold, the marginal profit is negative.
What does this mean? If you owned this business, how much would
you charge for hamburgers? Explain your reasoning.

Warm Up

The following warm-up exercises involve skills that were covered in earlier sections. You will use these skills in the exercise set for this section.

In Exercises 1 and 2, evaluate the expression.

1. $\dfrac{-63-(-105)}{21-7}$

2. $\dfrac{-37-54}{16-3}$

In Exercises 3–10, find the derivative of the function.

3. $y = 4x^2 - 2x + 7$

4. $y = -3t^3 + 2t^2 - 8$

5. $s = -16t^2 + 24t + 30$

6. $y = -16x^2 + 54x + 70$

7. $A = \frac{1}{10}(-2r^3 + 3r^2 + 5r)$

8. $y = \frac{1}{9}(6x^3 - 18x^2 + 63x - 15)$

9. $y = 12x - \dfrac{x^2}{5000}$

10. $y = 138 + 74x - \dfrac{x^3}{10{,}000}$

E X E R C I S E S   2.3

means that technology can help you solve or check the exercise(s).

1. *Research and Development* The graph below shows the amount A (in billions of constant 1982 dollars) spent on research and development in the United States from 1960 through 1990. Approximate the average rate of change of A during the following periods. (*Source: National Science Foundation*)

(a) 1960–1968 (b) 1968–1975
(c) 1980–1985 (d) 1960–1990

2. *Trade Deficit* The graph below shows the value I (in billions of dollars) of goods imported to the United States and the value E (in billions of dollars) of goods exported from the United States from 1970 through 1991. Approximate the indicated average rate of change. (*Source: U.S. Bureau of Census*)

(a) Imports: 1970–1980 (b) Exports: 1970–1980
(c) Imports: 1981–1985 (d) Exports: 1981–1985

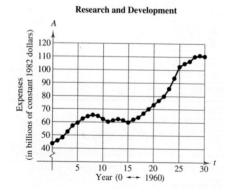

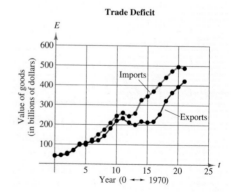

In Exercises 3–12, sketch the graph of the function and find its average rate of change on the interval. Compare this with the instantaneous rates of change at the endpoints of the interval.

Function	*Interval*
3. $f(t) = 2t + 7$	$[1, 2]$
4. $f(t) = 3t - 1$	$\left[0, \frac{1}{3}\right]$
5. $h(x) = 1 - x^2$	$[0, 1]$
6. $h(x) = x^2 - 4$	$[-2, 2]$
7. $f(t) = t^2 + 2$	$[1, 1.1]$
8. $f(x) = x^2 - 6x - 1$	$[-1, 3]$
9. $f(x) = \dfrac{1}{x}$	$[1, 4]$
10. $f(x) = x^{-1/2}$	$[1, 4]$
11. $g(x) = 4\sqrt{x}$	$[1, 9]$
12. $g(x) = x^3 - 1$	$[-1, 1]$

13. *Drug Effectiveness* The effectiveness E (on a scale from 0 to 1) of a pain-killing drug t hours after entering the bloodstream is given by

$$E = \tfrac{1}{27}(9t + 3t^2 - t^3), \qquad 0 \le t \le 4.5.$$

Find the average rate of change of E on the indicated intervals and compare this with the instantaneous rates of change at the endpoints of the intervals.

(a) $[0, 1]$ (b) $[1, 2]$ (c) $[2, 3]$ (d) $[3, 4]$

14. *Wind Chill* At $0°$ Celsius, the heat loss H (in kilocalories per square meter per hour) from a person's body can be modeled by

$$H = 33(10\sqrt{v} - v + 10.45),$$

where v is the wind speed (in meters per second). Find the rate of change of H when (a) $v = 2$ and (b) $v = 5$.

15. *Velocity* The height s (in feet) at time t (in seconds) of a silver dollar dropped from the top of the Washington Monument is

$$s = -16t^2 + 555.$$

(a) Find the average velocity on the interval $[2, 3]$.
(b) Find the instantaneous velocity when $t = 2$ and when $t = 3$.
(c) How long will it take the dollar to hit the ground?
(d) Find the velocity of the dollar when it hits the ground.

16. *Velocity* The height s (in feet) of an object fired straight up from ground level with an initial velocity of 200 feet per second is given by

$$s = -16t^2 + 200t,$$

where t is the time (in seconds).

(a) How fast is the object moving after one second?
(b) During which interval of time is the speed decreasing?
(c) During which interval of time is the speed increasing?

17. *Velocity* The position s (in feet) of an accelerating car is

$$s = 10t^{3/2}, \qquad 0 \le t \le 10,$$

where t is the time (in seconds). Find the velocity of the car at the following times.

(a) $t = 0$ (b) $t = 1$ (c) $t = 4$ (d) $t = 9$

18. *Velocity* Derive the equation for the velocity of an object whose position s (in meters) is $s(t) = t^2 + 2t$, where t is the time (in seconds).

Marginal Cost In Exercises 19–22, find the marginal cost for producing x units. (The cost is measured in dollars.)

19. $C = 4500 + 1.47x$

20. $C = 104{,}000 + 7200x$

21. $C = 55{,}000 + 470x - 0.25x^2, \quad 0 \le x \le 940$

22. $C = 100\left(9 + 3\sqrt{x}\right)$

Marginal Revenue In Exercises 23–26, find the marginal revenue for producing x units. (The revenue is measured in dollars.)

23. $R = 50x - 0.5x^2$

24. $R = 30x - x^2$

25. $R = -6x^3 + 8x^2 + 200x$

26. $R = 50(20x - x^{3/2})$

Marginal Profit In Exercises 27–30, find the marginal profit for producing x units. (The profit is measured in dollars.)

27. $P = -2x^2 + 72x - 145$

28. $P = -0.25x^2 + 2000x - 1{,}250{,}000$

29. $P = -0.00025x^2 + 12.2x - 25{,}000$

30. $P = -0.5x^3 + 30x^2 - 164.25x - 1000$

31. *Marginal Revenue* The revenue (in dollars) from producing x units of a product is

$$R = 125x - 0.002x^2.$$

(a) Find the additional revenue when production is increased from 15,000 units to 15,001 units.
(b) Find the marginal revenue when $x = 15,000$.
(c) Compare the results of parts (a) and (b).

32. *Marginal Revenue* The revenue (in dollars) from renting x apartments can be modeled by

$$R = 2x(900 + 32x - x^2).$$

(a) Find the additional revenue when the number of rentals is increased from 14 to 15.
(b) Find the marginal revenue when $x = 14$.
(c) Compare the results of parts (a) and (b).

33. *Marginal Cost* The cost (in dollars) of producing x units of a product is given by

$$C = 3.6\sqrt{x} + 500.$$

(a) Find the additional cost when the production increases from nine to ten units.
(b) Find the marginal cost when $x = 9$.
(c) Compare the results of parts (a) and (b).

34. *Marginal Profit* The profit (in dollars) from selling x units of a product is

$$P = -0.1x^2 + 2x - 100.$$

(a) Find the additional profit when the sales increase from seven to eight units.
(b) Find the marginal profit when $x = 7$.
(c) Compare the results of parts (a) and (b).

35. *Marginal Profit* The profit (in dollars) for selling x units of a product can be modeled by

$$P = 2400 - 403.4x + 32x^2 - 0.664x^3, \qquad 10 \le x \le 25.$$

Find the marginal profit for the following sales.

(a) $x = 10$ (b) $x = 20$ (c) $x = 23$ (d) $x = 25$

36. *Marginal Profit* The profit (in dollars) for selling x units of a product is given by

$$P = 36,000 + 2048\sqrt{x} - \frac{1}{8x^2}, \qquad 150 \le x \le 275.$$

Find the marginal profit for the following sales.

(a) $x = 150$ (b) $x = 175$ (c) $x = 200$
(d) $x = 225$ (e) $x = 250$ (f) $x = 275$

37. *Profit* The monthly demand function and cost function for x newspapers at a newsstand are

$$p = 5 - 0.001x \quad \text{and} \quad C = 35 + 1.5x.$$

(a) Find the monthly revenue as a function of x.
(b) Find the monthly profit as a function of x.
(c) Complete the table.

x	600	1200	1800	2400	3000
dR/dx					
dP/dx					
P					

38. *Profit* The demand function and cost function for x units of a product are

$$p = 25\left(20 - \sqrt{x}\right) \quad \text{and} \quad C = 42.50 + 1.25x.$$

(a) Find the revenue as a function of x.
(b) Find the profit as a function of x.
(c) Complete the table.

x	40	80	120	160	200
dR/dx					
dP/dx					
P					

39. *Marginal Profit* When the admission price to a baseball game was $6 per ticket, 36,000 tickets were sold. When the price was raised to $7, only 33,000 tickets were sold. Assume that the demand function is linear and that the variable and fixed costs for the ballpark owners are $0.20 and $85,000, respectively.

(a) Find the profit P as a function of x, the number of tickets sold.
(b) Find the marginal profit when 18,000 tickets are sold.

40. *Marginal Profit* In Exercise 39, suppose ticket sales decreased to 30,000 when the price increased to $7. How does this change the answers to Exercise 39?

41. *Marginal Cost* The cost C of producing x units is modeled by

$$C = v(x) + k,$$

where v represents the variable cost and k represents the fixed cost. Show that the marginal cost is independent of the fixed cost.

42. *Profit* The demand function for a product is

$$p = \frac{50}{\sqrt{x}}, \quad 1 \le x \le 8000,$$

and the cost function is

$$C = 0.5x + 500, \quad 0 \le x \le 8000.$$

Find the marginal profit for the following sales.

(a) $x = 900$ (b) $x = 1600$
(c) $x = 2500$ (d) $x = 3600$

If you were in charge of setting the price for this product, what price would you set? Explain your reasoning.

43. *Inventory Management* The annual inventory cost for a manufacturer is

$$C = \frac{1,008,000}{Q} + 6.3Q,$$

where Q is the order size when the inventory is replenished. Find the change in annual cost when Q is increased from 350 to 351, and compare this with the instantaneous rate of change when $Q = 350$.

44. *Center of Population* Since 1790, the center of population of the United States has been gradually moving westward. Use the figure below to estimate the rate (in miles per year) at which the center of population was moving *westward* during the given period. *(Source: U.S. Bureau of Census)*

(a) From 1790 to 1900
(b) From 1900 to 1990
(c) From 1790 to 1990

45. *Fuel Cost* A car is driven 15,000 miles a year and gets x miles per gallon. The average fuel cost is $1.10 per gallon. Find the annual cost C of fuel as a function of x and use this function to complete the table.

x	10	15	20	25	30	35	40
C							
dC/dx							

Who would benefit more from a 1 mile per gallon increase in fuel efficiency—the driver of a car that gets 15 or 35 miles per gallon? Explain.

46. Use a graphing utility to sketch the graph of f and f' for

$$f(x) = \frac{4}{x}, \quad 0 < x \le 5.$$

(a) Complete the table.

x	$\frac{1}{8}$	$\frac{1}{4}$	$\frac{1}{2}$	1	2	3	4	5
$f(x)$								
$f'(x)$								

(b) Find the average rate of change of f over the intervals determined by consecutive x-values in the table.

In Exercises 47 and 48, use a graphing utility to sketch the graphs of f and f'. Then determine the points (if any) at which f has a horizontal tangent.

47. $f(x) = \frac{1}{4}x^3, \quad -2 \le x \le 2$
48. $f(x) = x^4 - 12x^3 + 52x^2 - 96x + 64, \quad 1 \le x \le 5$

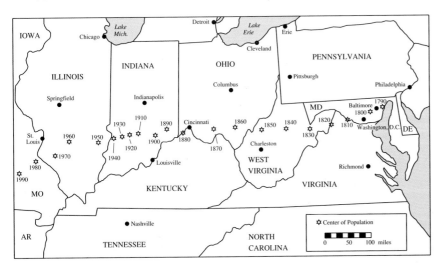

FIGURE FOR 44

Case Study: Shipping Costs to the Far East

PROFILE

COMPANY
*Hosford
International, Inc.*

▪

YEARS IN BUSINESS
30

▪

LOCATION
Erie, PA

▪

TYPE OF BUSINESS
*Customs brokers and
international freight
forwarders*

▪

EMPLOYEES
8

▪

Hosford International, Inc. of Erie, Pennsylvania is an international freight forwarding company that provides services to firms that export and import products worldwide. The company assists its clients in preparing documentation such as letters of credit, which spell out the terms of payment, as well as in arranging the scheduling and transportation of freight shipments using the most cost-efficient methods of shipment.

Actual shipping costs will, of course, vary according to such factors as the mode of transportation, the distance over which the cargo must be shipped, and the actual freight line chosen to transport the cargo. Approximately half of the overseas shipments of Hosford International's clients are transported by air, while the remainder is transported by ship.

In addition, some ocean freight lines base their shipping fees on how much space (out of a standard-sized full container) is taken up by the cargo. A full container's capacity is generally 1800 cubic feet, or in the language of the trade, 45 "measures," where one "measure" is equal to 40 cubic feet. One particular ocean freight shipping company, Maersk Line, has quoted the following costs as of March 1994 for shipping machinery from Baltimore to Hong Kong:

$332 per measure (i.e., 40 cubic feet) of cargo, or

$4690 for a full container of 45 measures (i.e., 1800 cubic feet).

A company with a small volume of machinery to ship might choose to pay the per-measure cost, but a company with a much larger freight volume could lower its average shipping costs per measure by renting a full container. For example, the average shipping cost per measure of cargo remains fixed at $332 up to the point where the cargo load is 14.13 measures ($4690/$332), but begins to decline thereafter, reaching a minimum of $104.22 per measure at a full load of 45 measures. Thus, for all cargo loads ranging from 14.13 measures up to a full load of 45 measures, the average shipping cost per measure of cargo is given by the equation

$$Y = 4690X^{-1},$$

where:

Y = average shipping cost (in dollars) per measure of cargo

X = volume of cargo (in number of measures).

Note: freight charges are also assessed on fractions of measures.

The diagram shows the graph of this function over the relevant domain. For all cargo volumes over 14.13 measures, up to a full load, average shipping costs decline, but at a diminishing rate. The *rate* at which average shipping costs decline as the volume of cargo increases is obtained by taking the derivative of the average cost function with respect to cargo volume:

$$\frac{dY}{dX} = -4690X^{-2}.$$

Such patterns in freight shipping costs may have practical implications for a company that is in the process of shipping a final product overseas. As illustrated in the graph, knowing the rate at which average shipping costs decline with cargo volume will help in making the right decisions.

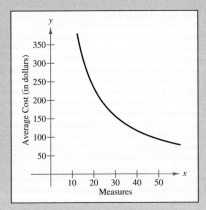

What Would You Do?

1. If transportation costs were the only consideration in shipping your final product overseas, how would you decide whether to transport partial shipments of the product as it rolls off the assembly line or to wait until more of the product is produced in order to increase cargo volume? Summarize the issues relevant to such a decision.

2. Suppose you have more than enough cargo to fill one full load, but not enough to fill two loads. How would you use the cost functions discussed above to decide whether to make one shipment first or ship all of your goods at the same time?

3. Of course, transport costs must be considered in light of many other factors and constraints that a company faces. What other considerations do you think are important in deciding when and how much to ship?

4. Suppose that a shipping line charges the following rates for shipping freight from Baltimore to London, England:

   $261 per measure of cargo
   $5079 per full container of 45 measures

(a) Formulate a general guideline for when to choose the per-measure cost or the full measure cost, and justify with calculations.

(b) Construct a function that reflects the average shipping cost per measure for this situation, and sketch a graph of the function.

(c) At which shipment level is it more beneficial in terms of cost to increase the shipment size by one measure – at 25 measures or at 35 measures? Explain your reasoning.

The Product and Quotient Rules

The Product Rule ■ *The Quotient Rule* ■ *Simplifying Derivatives* ■
Extended Application: Blood Pressure

The Product Rule

In Section 2.2 you saw that the derivative of a sum or difference of two functions is simply the sum or difference of their derivatives. The rules for the derivative of a product or quotient of two functions are not as simple.

The Product Rule

The derivative of the product of two differentiable functions is equal to the first function times the derivative of the second plus the second function times the derivative of the first.

$$\frac{d}{dx}[f(x)g(x)] = f(x)g'(x) + g(x)f'(x)$$

Note Rather than trying to remember the formula for the Product Rule, we suggest that you remember its verbal statement: *the first function times the derivative of the second plus the second function times the derivative of the first.*

Proof Some mathematical proofs, such as the proof of the Sum Rule, are straightforward. Others involve clever steps that may appear unmotivated. The following proof involves such a step—adding and subtracting the same quantity. (This step is shown in color.) Let $F(x) = f(x)g(x)$.

$$F'(x) = \lim_{\Delta x \to 0} \frac{F(x + \Delta x) - F(x)}{\Delta x}$$

$$= \lim_{\Delta x \to 0} \frac{f(x + \Delta x)g(x + \Delta x) - f(x)g(x)}{\Delta x}$$

$$= \lim_{\Delta x \to 0} \frac{f(x + \Delta x)g(x + \Delta x) - f(x + \Delta x)g(x) + f(x + \Delta x)g(x) - f(x)g(x)}{\Delta x}$$

$$= \lim_{\Delta x \to 0} \left[f(x + \Delta x)\frac{g(x + \Delta x) - g(x)}{\Delta x} + g(x)\frac{f(x + \Delta x) - f(x)}{\Delta x} \right]$$

$$= \lim_{\Delta x \to 0} f(x + \Delta x)\frac{g(x + \Delta x) - g(x)}{\Delta x} + \lim_{\Delta x \to 0} g(x)\frac{f(x + \Delta x) - f(x)}{\Delta x}$$

$$= \left[\lim_{\Delta x \to 0} f(x + \Delta x) \right]\left[\lim_{\Delta x \to 0} \frac{g(x + \Delta x) - g(x)}{\Delta x} \right]$$

$$+ \left[\lim_{\Delta x \to 0} g(x) \right]\left[\lim_{\Delta x \to 0} \frac{f(x + \Delta x) - f(x)}{\Delta x} \right]$$

$$= f(x)g'(x) + g(x)f'(x)$$

EXAMPLE 1 *Finding the Derivative of a Product*

Find the derivative of $y = (3x - 2x^2)(5 + 4x)$.

Solution

Using the Product Rule, you can write

$$\frac{dy}{dx} = \overbrace{(3x - 2x^2)}^{\text{(First)}}\overbrace{\frac{d}{dx}[5 + 4x]}^{\left(\begin{array}{c}\text{Derivative}\\\text{of second}\end{array}\right)} + \overbrace{(5 + 4x)}^{\text{(Second)}}\overbrace{\frac{d}{dx}[3x - 2x^2]}^{\left(\begin{array}{c}\text{Derivative}\\\text{of first}\end{array}\right)}$$

$$= (3x - 2x^2)(4) + (5 + 4x)(3 - 4x)$$

$$= (12x - 8x^2) + (15 - 8x - 16x^2)$$

$$= 15 + 4x - 24x^2.$$

■

Note   In general, the derivative of the product of two functions is not equal to the product of the derivatives of the two functions. To see this, compare the product of the derivatives of $f(x) = 3x - 2x^2$ and $g(x) = 5 + 4x$ with the derivative found in Example 1.

In the next example, notice that the first step in differentiating is *rewriting the original function.*

EXAMPLE 2 *Finding the Derivative of a Product*

Find the derivative of

$$f(x) = \left(\frac{1}{x} + 1\right)(x - 1). \qquad \text{\textit{Original function}}$$

Solution

Rewrite the function. Then use the Product Rule to find the derivative.

$$f(x) = (x^{-1} + 1)(x - 1). \qquad \text{\textit{Rewrite function.}}$$

$$f'(x) = (x^{-1} + 1)\frac{d}{dx}[x - 1] + (x - 1)\frac{d}{dx}[x^{-1} + 1] \quad \text{\textit{Product Rule}}$$

$$= (x^{-1} + 1)(1) + (x - 1)(-x^{-2})$$

$$= \frac{1}{x} + 1 - \frac{x - 1}{x^2}$$

$$= \frac{x + x^2 - x + 1}{x^2}$$

$$= \frac{x^2 + 1}{x^2} \qquad \text{\textit{Simplify.}}$$

■

▦ TECHNOLOGY

If you have access to a symbolic differentiation utility, such as *Derive*, *Maple*, or *Mathematica*, try using it to confirm several of the derivatives in this section. The form of the derivative can depend on how you use software. For instance, when we used *Derive* to differentiate the function in Example 2, we obtained the same form listed in the example. But when we used *Derive* to first expand the function, and then differentiate, we obtained

$$f'(x) = 1 + \frac{1}{x^2}.$$

You now have two differentiation rules that deal with products—the Constant Multiple Rule and the Product Rule. The difference between these two rules is that the Constant Multiple Rule deals with the product of a constant and a variable quantity,

Constant   Variable Quantity

$$F(x) = c\,f(x),$$ *Use Constant Multiple Rule.*

whereas the Product Rule deals with the product of two variable quantities,

Variable Quantity   Variable Quantity

$$F(x) = f(x)g(x).$$ *Use Product Rule.*

The next example compares these two rules.

EXAMPLE 3 *Comparing Differentiation Rules*

Find the derivatives of the functions.

a. $y = 2x(x^2 + 3x)$

b. $y = 2(x^2 + 3x)$

Solution

a. By the Product Rule,

$$\frac{dy}{dx} = (2x)\frac{d}{dx}[x^2 + 3x] + (x^2 + 3x)\frac{d}{dx}[2x] \quad \text{\textit{Product Rule}}$$

$$= (2x)(2x + 3) + (x^2 + 3x)(2)$$

$$= 4x^2 + 6x + 2x^2 + 6x$$

$$= 6x^2 + 12x.$$

b. By the Constant Multiple Rule,

$$\frac{dy}{dx} = 2\frac{d}{dx}[x^2 + 3x] \quad\quad\quad\quad\quad \text{\textit{Constant Multiple Rule}}$$

$$= 2(2x + 3)$$

$$= 4x + 6.$$ ▬

The Product Rule can be extended to products that have more than two factors. For example, if f, g, and h, are differentiable functions of x, then

$$\frac{d}{dx}[f(x)g(x)h(x)] = f'(x)g(x)h(x) + f(x)g'(x)h(x) + f(x)g(x)h'(x).$$

The Quotient Rule

In Section 2.2, you saw that by using the Constant Rule, Power Rule, Constant Multiple Rule, and Sum and Difference Rules, you were able to differentiate *any* polynomial function. By combining these rules with the Quotient Rule, you can now differentiate *any* rational function.

Note From this differentiation rule, you can see that the derivative of a quotient is not, in general, the quotient of the derivatives. That is,

$$\frac{d}{dx}\left[\frac{f(x)}{g(x)}\right] \neq \frac{f'(x)}{g'(x)}.$$

The Quotient Rule

The derivative of the quotient of two differentiable functions is equal to the denominator times the derivative of the numerator minus the numerator times the derivative of the denominator, all divided by the square of the denominator.

$$\frac{d}{dx}\left[\frac{f(x)}{g(x)}\right] = \frac{g(x)f'(x) - f(x)g'(x)}{[g(x)]^2}, \qquad g(x) \neq 0$$

Proof Let $F(x) = f(x)/g(x)$. As in the proof of the Product Rule, a key step in this proof is adding and subtracting the same quantity.

$$F'(x) = \lim_{\Delta x \to 0} \frac{F(x + \Delta x) - F(x)}{\Delta x}$$

$$= \lim_{\Delta x \to 0} \frac{\dfrac{f(x + \Delta x)}{g(x + \Delta x)} - \dfrac{f(x)}{g(x)}}{\Delta x}$$

$$= \lim_{\Delta x \to 0} \frac{g(x)f(x + \Delta x) - f(x)g(x + \Delta x)}{\Delta x g(x)g(x + \Delta x)}$$

$$= \lim_{\Delta x \to 0} \frac{g(x)f(x + \Delta x) - f(x)g(x) + f(x)g(x) - f(x)g(x + \Delta x)}{\Delta x g(x)g(x + \Delta x)}$$

$$= \frac{\displaystyle\lim_{\Delta x \to 0} \frac{g(x)[f(x + \Delta x) - f(x)]}{\Delta x} - \lim_{\Delta x \to 0} \frac{f(x)[g(x + \Delta x) - g(x)]}{\Delta x}}{\displaystyle\lim_{\Delta x \to 0} [g(x)g(x + \Delta x)]}$$

$$= \frac{g(x)\left[\displaystyle\lim_{\Delta x \to 0} \dfrac{f(x + \Delta x) - f(x)}{\Delta x}\right] - f(x)\left[\displaystyle\lim_{\Delta x \to 0} \dfrac{g(x + \Delta x) - g(x)}{\Delta x}\right]}{\displaystyle\lim_{\Delta x \to 0} [g(x)g(x + \Delta x)]}$$

$$= \frac{g(x)f'(x) - f(x)g'(x)}{[g(x)]^2}.$$

Note As suggested for the Product Rule, we recommend that you remember the verbal statement of the Quotient Rule rather than trying to remember the formula for the rule.

EXAMPLE 4 *Finding the Derivative of a Quotient*

Find the derivative of

$$y = \frac{x-1}{2x+3}.$$

Solution

Apply the Quotient Rule, as follows.

$$\frac{dy}{dx} = \frac{(2x+3)\dfrac{d}{dx}[x-1] - (x-1)\dfrac{d}{dx}[2x+3]}{(2x+3)^2}$$

$$= \frac{(2x+3)(1) - (x-1)(2)}{(2x+3)^2}$$

$$= \frac{2x+3-2x+2}{(2x+3)^2}$$

$$= \frac{5}{(2x+3)^2}$$

EXAMPLE 5 *Finding the Derivative of a Quotient*

Find the derivative of

$$y = \frac{2x^2 - 4x + 3}{2 - 3x}.$$

Solution

Apply the Quotient Rule, as follows.

$$\frac{dy}{dx} = \frac{(2-3x)\dfrac{d}{dx}[2x^2 - 4x + 3] - (2x^2 - 4x + 3)\dfrac{d}{dx}[2 - 3x]}{(2-3x)^2}$$

$$= \frac{(2-3x)(4x-4) - (2x^2 - 4x + 3)(-3)}{(2-3x)^2}$$

$$= \frac{-12x^2 + 20x - 8 - (-6x^2 + 12x - 9)}{(2-3x)^2}$$

$$= \frac{-12x^2 + 20x - 8 + 6x^2 - 12x + 9}{(2-3x)^2}$$

$$= \frac{-6x^2 + 8x + 1}{(2-3x)^2}$$

Note When applying the Quotient Rule, we suggest that you enclose all factors and derivatives in parentheses. Also, pay special attention to the subtraction required in the numerator.

EXAMPLE 6 *Finding the Derivative of a Quotient*

Find the derivative of

$$y = \frac{3 - (1/x)}{x + 5}.$$

Solution

Begin by rewriting the original function. Then apply the Quotient Rule and simplify the result.

$$y = \frac{3 - (1/x)}{x + 5} \qquad \text{\textit{Original function}}$$

$$= \frac{3x - 1}{x(x + 5)} \qquad \begin{array}{l} \textit{Multiply numerator and} \\ \textit{denominator by } x. \end{array}$$

$$= \frac{3x - 1}{x^2 + 5x} \qquad \text{\textit{Rewritten function}}$$

$$\frac{dy}{dx} = \frac{(x^2 + 5x)(3) - (3x - 1)(2x + 5)}{(x^2 + 5x)^2} \qquad \text{\textit{Apply Quotient Rule.}}$$

$$= \frac{(3x^2 + 15x) - (6x^2 + 13x - 5)}{(x^2 + 5x)^2}$$

$$= \frac{-3x^2 + 2x + 5}{(x^2 + 5x)^2} \qquad \text{\textit{Simplify.}} \qquad \blacksquare$$

Not every quotient needs to be differentiated by the Quotient Rule. For instance, each of the quotients in the next example can be considered as the product of a constant and a function of x. In such cases, the Constant Multiple Rule is more efficient than the Quotient Rule.

EXAMPLE 7 *Rewriting Before Differentiating*

Original Function	*Rewrite*	*Differentiate*	*Simplify*
a. $y = \dfrac{x^2 + 3x}{6}$	$y = \dfrac{1}{6}(x^2 + 3x)$	$y' = \dfrac{1}{6}(2x + 3)$	$y' = \dfrac{1}{3}x + \dfrac{1}{2}$
b. $y = \dfrac{5x^4}{8}$	$y = \dfrac{5}{8}x^4$	$y' = \dfrac{5}{8}(4x^3)$	$y' = \dfrac{5}{2}x^3$
c. $y = \dfrac{-3(3x - 2x^2)}{7x}$	$y = -\dfrac{3}{7}(3 - 2x)$	$y' = -\dfrac{3}{7}(-2)$	$y' = \dfrac{6}{7}$
d. $y = \dfrac{9}{5x^2}$	$y = \dfrac{9}{5}(x^{-2})$	$y' = \dfrac{9}{5}(-2x^{-3})$	$y' = -\dfrac{18}{5x^3}$

$\blacksquare$

Note   To see the efficiency of using the Constant Multiple Rule in Example 7, try using the Quotient Rule to find the derivatives of the four functions.

Simplifying Derivatives

EXAMPLE 8 *Combining the Product and Quotient Rules*

Find the derivative of
$$y = \frac{(1 - 2x)(3x + 2)}{5x - 4}.$$

Solution

This function contains a product within a quotient. You could first multiply the factors in the numerator and then apply the Quotient Rule. However, to gain practice in using the Product Rule within the Quotient Rule, try differentiating as follows.

$$y' = \frac{(5x - 4)\dfrac{d}{dx}[(1 - 2x)(3x + 2)] - (1 - 2x)(3x + 2)\dfrac{d}{dx}[5x - 4]}{(5x - 4)^2}$$

$$= \frac{(5x - 4)[(1 - 2x)(3) + (3x + 2)(-2)] - (1 - 2x)(3x + 2)(5)}{(5x - 4)^2}$$

$$= \frac{(5x - 4)(-12x - 1) - (1 - 2x)(15x + 10)}{(5x - 4)^2}$$

$$= \frac{(-60x^2 + 43x + 4) - (-30x^2 - 5x + 10)}{(5x - 4)^2}$$

$$= \frac{-30x^2 + 48x - 6}{(5x - 4)^2}$$

In the examples in this section, much of the work in obtaining the final form of the derivative occurs *after* the differentiation. As summarized in the following list, direct application of differentiation rules often yields results that are not in simplified form. Note that two characteristics of simplified form are the absence of negative exponents and the combining of like terms.

	$f'(x)$ *After Differentiating*	$f'(x)$ *After Simplifying*
Example 1	$(3x - 2x^2)(4) + (5 + 4x)(3 - 4x)$	$15 + 4x - 24x^2$
Example 2	$(x^{-1} + 1)(1) + (x - 1)(-x^{-2})$	$\dfrac{x^2 + 1}{x^2}$
Example 5	$\dfrac{(2 - 3x)(4x - 4) - (2x^2 - 4x + 3)(-3)}{(2 - 3x)^2}$	$\dfrac{-6x^2 + 8x + 1}{(2 - 3x)^2}$
Example 8	$\dfrac{(5x - 4)[(1 - 2x)(3) + (3x + 2)(-2)] - (1 - 2x)(3x + 2)(5)}{(5x - 4)^2}$	$\dfrac{-30x^2 + 48x - 6}{(5x - 4)^2}$

Extended Application: Blood Pressure

Real Life

EXAMPLE 9 *Rate of Change of Blood Pressure*

As blood moves from the heart through the major arteries out to the capillaries and back through the veins, the systolic pressure continuously drops. (See Figure 2.27.) Consider a person whose blood pressure P (in millimeters of mercury) is given by

$$P = \frac{25t^2 + 125}{t^2 + 1}, \qquad 0 \le t \le 10,$$

where t is measured in seconds. At what rate is the blood pressure changing 5 seconds after blood leaves the heart?

Solution

Begin by applying the Quotient Rule.

$$\frac{dP}{dt} = \frac{(t^2 + 1)(50t) - (25t^2 + 125)(2t)}{(t^2 + 1)^2} \qquad \textit{Quotient Rule}$$

$$= \frac{50t^3 + 50t - 50t^3 - 250t}{(t^2 + 1)^2}$$

$$= -\frac{200t}{(t^2 + 1)^2} \qquad \textit{Simplify.}$$

When $t = 5$, the rate of change is

$$\frac{dP}{dt} = -\frac{200(5)}{26^2}$$

$$\approx -1.48 \text{ mm/sec.}$$

Therefore, the pressure is *dropping* at a rate of 1.48 millimeters per second when $t = 5$ seconds. ■

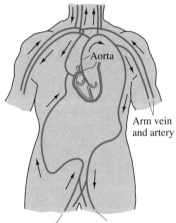

Aorta

Arm vein and artery

Leg vein and artery

FIGURE 2.27

Discussion Problem   Testing Differentiation Rules

What is wrong with the following differentiations?

$$f(x) = (x^3)(x^4)$$
$$f'(x) = (3x^2)(4x^3)$$

$$g(x) = \frac{x^3}{x^4}$$
$$g'(x) = \frac{3x^2}{4x^3}$$

If you were teaching a calculus class, how could you convince your class that these derivatives are incorrect?

Warm Up

The following warm-up exercises involve skills that were covered in earlier sections. You will use these skills in the exercise set for this section.

In Exercises 1–6, simplify the expression.

1. $(x^2 + 1)(2) + (2x + 7)(2x)$

2. $(2x - x^3)(8x) + (4x^2)(2 - 3x^2)$

3. $\dfrac{(2x + 7)(5) - (5x + 6)(2)}{(2x + 7)^2}$

4. $\dfrac{(x^2 - 4)(2x + 1) - (x^2 + x)(2x)}{(x^2 - 4)^2}$

5. $(x^{-1} + x)(2) + (2x - 3)(-x^{-2} + 1)$

6. $\dfrac{(1 - x^{-1})(1) - (x - 4)(x^{-2})}{(1 - x^{-1})^2}$

In Exercises 7–10, find $f'(2)$.

7. $f(x) = 3x^2 - x + 4$

8. $f(x) = -x^3 + x^2 + 8x$

9. $f(x) = \dfrac{1}{x}$

10. $f(x) = x^2 - \dfrac{1}{x^2}$

E X E R C I S E S   2.4   means that technology can help you solve or check the exercise(s).

In Exercises 1–10, find the value of the derivative of the function at the indicated point.

Function	Point
1. $f(x) = x^2(3x^3 - 1)$	$(1, 2)$
2. $f(x) = (x^2 + 1)(2x + 5)$	$(-1, 6)$
3. $f(x) = \frac{1}{3}(2x^3 - 4)$	$\left(0, -\frac{4}{3}\right)$
4. $f(x) = \frac{1}{7}(5 - 6x^2)$	$\left(1, -\frac{1}{7}\right)$
5. $g(x) = (x^2 - 4x + 3)(x - 2)$	$(4, 6)$
6. $g(x) = (x^2 - 2x + 1)(x^3 - 1)$	$(1, 0)$
7. $h(x) = \dfrac{x}{x - 5}$	$(6, 6)$
8. $h(x) = \dfrac{x^2}{x + 3}$	$\left(-1, \frac{1}{2}\right)$
9. $f(t) = \dfrac{2t^2 - 3}{3t}$	$\left(2, \frac{5}{6}\right)$
10. $f(x) = \dfrac{x + 1}{x - 1}$	$(2, 3)$

In Exercises 11–18, find the derivative of the function. Use Example 7 as a model.

Function	Rewrite	Derivative	Simplify
11. $y = \dfrac{x^2 + 2x}{x}$			
12. $y = \dfrac{4x^{3/2}}{x}$			
13. $y = \dfrac{7}{3x^3}$			
14. $y = \dfrac{4}{5x^2}$			
15. $y = \dfrac{4x^2 - 3x}{8\sqrt{x}}$			
16. $y = \dfrac{x^2 - 4}{x + 2}$			
17. $y = \dfrac{x^2 - 4x + 3}{x - 1}$			
18. $y = \dfrac{3x^2 - 4x}{6x}$			

In Exercises 19–40, differentiate the function.

19. $f(x) = (x^3 - 3x)(2x^2 + 3x + 5)$

20. $f(x) = (x - 1)(x^2 - 3x + 2)$

21. $h(t) = (t^5 - 1)(4t^2 - 7t - 3)$

22. $g(t) = (2t^2 - 3)(4 - t^2 - t^4)$

23. $h(p) = (p^3 - 2)^2$ **24.** $h(x) = (x^2 - 1)^2$

25. $f(x) = \sqrt[3]{x}\left(\sqrt{x} + 3\right)$ **26.** $f(x) = \sqrt[3]{x}(x + 1)$

27. $f(x) = \dfrac{3x - 2}{2x - 3}$ **28.** $f(x) = \dfrac{x^3 + 3x + 2}{x^2 - 1}$

29. $f(x) = \dfrac{3 - 2x - x^2}{x^2 - 1}$ **30.** $f(x) = x^3 \left(\dfrac{x - 2}{x + 1}\right)$

31. $f(x) = (x^5 - 3x)\left(\dfrac{1}{x^2}\right)$ **32.** $f(x) = x\left(1 - \dfrac{2}{x + 1}\right)$

33. $h(t) = \dfrac{t + 2}{t^2 + 5t + 6}$ **34.** $g(s) = \dfrac{5}{s^2 + 2s + 2}$

35. $f(x) = \dfrac{x + 1}{\sqrt{x}}$ **36.** $f(x) = \dfrac{\sqrt{x}}{x + 1}$

37. $g(x) = \left(\dfrac{x - 3}{x + 4}\right)(x^2 + 2x + 1)$

38. $f(x) = \left(\dfrac{x^2 - x - 3}{x^2 + 1}\right)(x^2 + x + 1)$

39. $f(x) = (3x^3 + 4x)(x - 5)(x + 1)$

40. $f(x) = (x^2 - x)(x^2 + 1)(x^2 + x + 1)$

In Exercises 41–44, find an equation of the tangent line to the graph of the function at the indicated point.

Function	*Point*
41. $f(x) = \dfrac{2x}{x + 2}$	$(2, 1)$
42. $f(x) = (x - 1)(x^2 - 2)$	$(0, 2)$
43. $f(x) = (x^3 - 3x + 1)(x + 2)$	$(1, -3)$
44. $f(x) = \dfrac{x - 1}{x + 1}$	$\left(2, \dfrac{1}{3}\right)$

In Exercises 45–48, find the point(s), if any, at which the graph of *f* has a horizontal tangent.

45. $f(x) = \dfrac{x^2}{x - 1}$ **46.** $f(x) = \dfrac{x^2}{x^2 + 1}$

47. $f(x) = \dfrac{x^4}{x^3 + 1}$ **48.** $f(x) = \dfrac{x^4 + 3}{x^2 + 1}$

In Exercises 49–52, use a graphing utility to graph *f* and *f'* on the interval $[-2, 2]$.

49. $f(x) = x(x + 1)$

50. $f(x) = x^2(x + 1)$

51. $f(x) = x(x + 1)(x - 1)$

52. $f(x) = x^2(x + 1)(x - 1)$

Demand In Exercises 53 and 54, use the demand function to find the rate of change in the demand *x* for the given price *p*.

53. $x = 275\left(1 - \dfrac{3p}{5p + 1}\right)$, $p = \$4$

54. $x = 300 - p - \dfrac{2p}{p + 1}$, $p = \$3$

55. *Oxygen Level* The model
$$f(t) = \frac{t^2 - t + 1}{t^2 + 1}$$

measures the percent of the normal level of oxygen in a pond, where *t* is the time (in weeks) after organic waste is dumped into the pond. Find the rate of change of *f* with respect to *t* when (a) $t = 0.5$, (b) $t = 2$, and (c) $t = 8$.

56. *Refrigeration* The temperature *T* of food placed in a refrigerator is modeled by
$$T = 10\left(\frac{4t^2 + 16t + 75}{t^2 + 4t + 10}\right),$$

where *t* is the time (in hours). What is the initial temperature of the food? Find the rate of change of *T* with respect to *t* when (a) $t = 1$, (b) $t = 3$, (c) $t = 5$, and (d) $t = 10$.

57. *Population Growth* A population of bacteria is introduced into a culture. The number of bacteria *P* can be modeled by
$$P = 500\left(1 + \frac{4t}{50 + t^2}\right),$$

where *t* is the time (in hours). Find the rate of change of the population when $t = 2$.

58. *Quality Control* The percent *P* of defective parts produced by a new employee *t* days after the employee starts work can be modeled by
$$P = \frac{t + 1750}{50(t + 2)}.$$

Find the rate of change of *P* when (a) $t = 1$ and (b) $t = 10$.

59. *Profit* The demand x for a product is inversely proportional to the square of the price for $x \geq 5$.

   (a) The price is $1000, the demand is 16 units. Find the demand function.

   (b) The product costs $250 per unit and the fixed cost is $10,000. Find the cost function.

   (c) Find the profit function and graph it. From the graph, what price would you set for this product? Explain your reasoning.

60. *Profit* You are managing a store and have been adjusting the price of an item. You have found that you make a profit of $50 when 10 units are sold, $60 when 12 units are sold, and $65 when 14 units are sold.

   (a) Fit this data to the following quadratic model.

$$P = ax^2 + bx + c$$

   (b) Use a graphing utility to graph P.

   (c) Find the point on the graph at which the marginal profit is zero. Interpret this point in the context of the problem.

61. *Demand Function* Given $f(x) = x^2 + 1$, which of the following would most likely represent a demand function?

   (a) $p = f(x)$ (b) $p = xf(x)$

   (c) $p = \dfrac{f(x)}{x}$ (d) $p = \dfrac{1}{f(x)}$

Explain your reasoning. Use a graph of each function as part of your explanation.

62. *Cost* The cost of producing x units of a product is

$$C = x^3 - 15x^2 + 87x - 73, \quad 4 \leq x \leq 9.$$

   (a) Use a graphing utility to graph the marginal cost function and the average cost function, C/x, on the same set of axes.

   (b) Find the point of intersection of the graphs of dC/dx and C/x. What significance, if any, does this point have?

63. *Inventory Replenishment* The ordering and transportation cost C (in thousands of dollars) of the components used in manufacturing a product is

$$C = 100 \left(\frac{200}{x^2} + \frac{x}{x + 30} \right), \quad 1 \leq x,$$

where x is the order size (in hundreds). Find the rate of change of C with respect to x for the following order sizes.

   (a) $x = 10$ (b) $x = 15$ (c) $x = 20$

What do these rates of change imply about increasing the size of an order?

64. *Sales Analysis* The monthly sale of memberships M at a newly built fitness center is modeled by

$$M(t) = \frac{300t}{t^2 + 1} + 8,$$

where t is the number of months that the center has been open.

   (a) Find $M'(t)$.

   (b) Find $M(3)$ and $M'(3)$ and interpret the results.

   (c) Find $M(24)$ and $M'(24)$ and interpret the results.

65. *Research Project* Use your school's library or some other reference source to find information about a company that performs market research. Write a short paper about the company. (One such company is described below.)

Business Capsule

Hispanic Market Connections, Inc.

Isabel Valdes is the founder of Hispanic Market Connections Inc. The database compiled by her company was considered so useful that the company won an award from *American Demographics* magazine as one of America's "Best 100 Sources of Marketing Information." In 1993, Isabel Valdes, who has two Master's degrees from Stanford University, co-authored a 450-page book entitled *Handbook on Marketing to Hispanics*. Of her book, Valdes says that "it offers a lot more than the traditional demographic tables and statistics."

The Chain Rule ■ *The General Power Rule* ■ *Simplification Techniques* ■
Extended Application: Revenue Per Share ■ *Summary of Differentiation Rules*

The Chain Rule

In this section, you will study one of the most powerful rules of differential calculus—the **Chain Rule.** This differentiation rule deals with composite functions and adds versatility to the rules presented in Sections 2.2 and 2.4. For example, compare the following functions. Those on the left can be differentiated without the Chain Rule, whereas those on the right are best done with the Chain Rule.

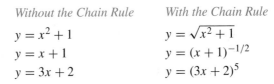

Without the Chain Rule

$$y = x^2 + 1$$
$$y = x + 1$$
$$y = 3x + 2$$

With the Chain Rule

$$y = \sqrt{x^2 + 1}$$
$$y = (x + 1)^{-1/2}$$
$$y = (3x + 2)^5$$

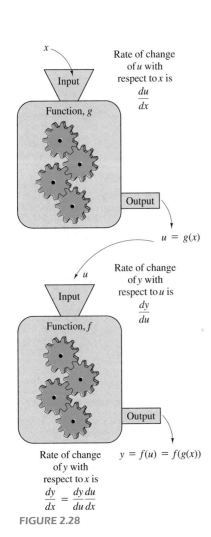

Rate of change of u with respect to x is $\dfrac{du}{dx}$

Input

Function, g

Output

$u = g(x)$

u

Rate of change of y with respect to u is $\dfrac{dy}{du}$

Input

Function, f

Output

$y = f(u) = f(g(x))$

Rate of change of y with respect to x is $\dfrac{dy}{dx} = \dfrac{dy}{du}\dfrac{du}{dx}$

FIGURE 2.28

The Chain Rule

If $y = f(u)$ is a differentiable function of u, and $u = g(x)$ is a differentiable function of x, then $y = f(g(x))$ is a differentiable function of x, and

$$\frac{dy}{dx} = \frac{dy}{du}\frac{du}{dx}$$

or, equivalently,

$$\frac{d}{dx}[f(g(x))] = f'(g(x))g'(x).$$

Basically, the Chain Rule states that if y changes dy/du times as fast as u, and u changes du/dx times as fast as x, then y changes

$$\frac{dy}{du}\frac{du}{dx}$$

times as fast as x, as illustrated in Figure 2.28. One advantage of the dy/dx notation for derivatives is that it helps you remember differentiation rules, such as the Chain Rule. For instance, in the formula $dy/dx = (dy/du)(du/dx)$, you can imagine that the du's cancel.

When applying the Chain Rule, it helps to think of the composite function $y = f(g(x))$ or $y = f(u)$ as having two parts—an *inside* and an *outside*—as illustrated below.

$$
\begin{array}{c}
\text{Inside} \\
\downarrow \quad \downarrow \\
y = f(g(x)) = f(u) \\
\uparrow \qquad \uparrow \\
\text{Outside}
\end{array}
$$

The Chain Rule tells you that the derivative of $y = f(u)$ is the derivative of the outer function (at the inner function u) *times* the derivative of the inner function. That is,

$$y' = f'(u) \cdot u'.$$

EXAMPLE 1 *Decomposing Composite Functions*

Write each function as the composition of two functions.

a. $y = \dfrac{1}{x+1}$ **b.** $y = \sqrt{3x^2 - x + 1}$

Solution

There is more than one correct way to decompose each function. One way for each is shown below.

	$y = f(g(x))$	$u = g(x)$ *(inside)*	$y = f(u)$ *(outside)*
a.	$y = \dfrac{1}{x+1}$	$u = x+1$	$y = \dfrac{1}{u}$
b.	$y = \sqrt{3x^2 - x + 1}$	$u = 3x^2 - x + 1$	$y = \sqrt{u}$

EXAMPLE 2 *Using the Chain Rule*

Note Try checking the result of Example 2 by expanding the function to obtain

$$y = x^6 + 3x^4 + 3x^2 + 1$$

and differentiating. Do you obtain the same answer?

Find the derivative of $y = (x^2 + 1)^3$.

Solution

To apply the Chain Rule, you need to identify the inside function u.

$$
y = \overbrace{(x^2 + 1)}^{u}{}^{3} = u^3
$$

By the Chain Rule, you can write the derivative as follows.

$$
\frac{dy}{dx} = \overbrace{3(x^2 + 1)^2}^{\frac{dy}{du}}\overbrace{(2x)}^{\frac{du}{dx}}
$$

$$
= 6x(x^2 + 1)^2
$$

The General Power Rule

The function in Example 2 illustrates one of the most common types of composite functions—a power function of the form

$$y = [u(x)]^n.$$

The rule for differentiating such functions is called the **General Power Rule,** and it is a special case of the Chain Rule.

The General Power Rule

If $y = [u(x)]^n$, where u is a differentiable function of x and n is a real number, then

$$\frac{dy}{dx} = n[u(x)]^{n-1}\frac{du}{dx}$$

or, equivalently,

$$\frac{d}{dx}[u^n] = nu^{n-1}u'.$$

Proof Apply the Chain Rule and the Simple Power Rule as follows.

$$\frac{dy}{dx} = \frac{dy}{du}\frac{du}{dx}$$

$$= \frac{d}{du}[u^n]\frac{du}{dx}$$

$$= nu^{n-1}\frac{du}{dx}$$

■

TECHNOLOGY

In you have access to a symbolic differentiation utility, such as *Derive*, *Mathematica*, or *Maple*, try using it to confirm the result of Example 3. For instance, when we used *Maple* to differentiate the function in Example 3, we obtained

$$f'(x) = 3(3x - 2x^2)^2(3 - 4x).$$

EXAMPLE 3 *Using the General Power Rule*

Find the derivative of $f(x) = (3x - 2x^2)^3$.

Solution

The inside function is $u = 3x - 2x^2$. Thus, by the General Power Rule,

$$f'(x) = \overbrace{3}^{n}\overbrace{(3x - 2x^2)^2}^{u^{n-1}}\overbrace{\frac{d}{dx}[3x - 2x^2]}^{u'}$$

$$= 3(3x - 2x^2)^2(3 - 4x)$$

$$= (9 - 12x)(3x - 2x^2)^2.$$

■

EXAMPLE 4 *Rewriting Before Differentiating*

Find the derivative of

$$y = \sqrt[3]{(x^2 + 2)^2}. \qquad\qquad \textit{Original function}$$

Solution

Begin by rewriting the function in rational exponent form.

$$y = (x^2 + 2)^{2/3} \qquad\qquad \textit{Rewrite original function.}$$

Then, using the inside function, $u = x^2 + 2$, apply the General Power Rule.

$$\frac{dy}{dx} = \overbrace{\frac{2}{3}}^{n}\overbrace{(x^2 + 2)^{-1/3}}^{u^{n-1}}\overbrace{(2x)}^{u'} \qquad\qquad \textit{Apply General Power Rule.}$$

$$= \frac{4x(x^2 + 2)^{-1/3}}{3}$$

$$= \frac{4x}{3\sqrt[3]{x^2 + 2}} \qquad\qquad \textit{Simplify.}$$

The derivative of a quotient can sometimes be found more easily with the General Power Rule than with the Quotient Rule. This is especially true when the numerator is a constant, as illustrated in the next example.

EXAMPLE 5 *Finding the Derivative of a Quotient*

Find the derivative of the functions.

a. $y = \dfrac{3}{x^2 + 1}$ **b.** $y = \dfrac{3}{(x + 1)^2}$

Solution

a. Begin by rewriting the function as

$$y = 3(x^2 + 1)^{-1}. \qquad\qquad \textit{Rewrite original function.}$$

Then apply the General Power Rule to obtain

$$\frac{dy}{dx} = -3(x^2 + 1)^{-2}(2x) = -\frac{6x}{(x^2 + 1)^2}. \qquad \textit{Apply General Power Rule.}$$

b. Begin by rewriting the function as

$$y = 3(x + 1)^{-2}. \qquad\qquad \textit{Rewrite original function.}$$

Then apply the General Power Rule to obtain

$$\frac{dy}{dx} = -6(x + 1)^{-3}(1) = -\frac{6}{(x + 1)^3}. \qquad \textit{Apply General Power Rule.}$$

Simplification Techniques

Throughout this chapter we have emphasized writing derivatives in simplified form. The reason for this is that most applications of derivatives require a simplified form. The next two examples illustrate some useful simplification techniques.

EXAMPLE 6 *Simplifying by Factoring Out Least Powers*

Note In Example 6, note that you subtract exponents when factoring. That is, when $(1 - x^2)^{-1/2}$ is factored out of $(1 - x^2)^{1/2}$, the *remaining* factor has an exponent of $\frac{1}{2} - \left(-\frac{1}{2}\right)$ or 1. Thus,

$$(1 - x^2)^{1/2} = (1 - x^2)^{-1/2}(1 - x^2)^1.$$

$$y = x^2\sqrt{1 - x^2} \qquad \text{\textit{Original function}}$$

$$= x^2(1 - x^2)^{1/2} \qquad \text{\textit{Rewrite function.}}$$

$$y' = x^2\frac{d}{dx}[(1 - x^2)^{1/2}] + (1 - x^2)^{1/2}\frac{d}{dx}[x^2] \qquad \text{\textit{Product Rule}}$$

$$= x^2\left[\frac{1}{2}(1 - x^2)^{-1/2}(-2x)\right] + (1 - x^2)^{1/2}(2x) \qquad \text{\textit{Power Rule}}$$

$$= -x^3(1 - x^2)^{-1/2} + 2x(1 - x^2)^{1/2}$$

$$= x(1 - x^2)^{-1/2}[-x^2(1) + 2(1 - x^2)] \qquad \text{\textit{Factor.}}$$

$$= x(1 - x^2)^{-1/2}(2 - 3x^2)$$

$$= \frac{x(2 - 3x^2)}{\sqrt{1 - x^2}} \qquad \text{\textit{Simplify.}}$$

EXAMPLE 7 *Differentiating a Quotient Raised to a Power*

Note In Example 7, try to find $f'(x)$ by applying the Quotient Rule to

$$f(x) = \frac{(3x - 1)^2}{(x^2 + 3)^2}.$$

Which method do you prefer?

Find the derivative of

$$f(x) = \left(\frac{3x - 1}{x^2 + 3}\right)^2.$$

Solution

$$f'(x) = \overset{n}{2}\left(\overset{u^{n-1}}{\overbrace{\frac{3x - 1}{x^2 + 3}}}\right)\overset{u'}{\overbrace{\frac{d}{dx}\left[\frac{3x - 1}{x^2 + 3}\right]}}$$

$$= \left[\frac{2(3x - 1)}{x^2 + 3}\right]\left[\frac{(x^2 + 3)(3) - (3x - 1)(2x)}{(x^2 + 3)^2}\right]$$

$$= \frac{2(3x - 1)(3x^2 + 9 - 6x^2 + 2x)}{(x^2 + 3)^3}$$

$$= \frac{2(3x - 1)(-3x^2 + 2x + 9)}{(x^2 + 3)^3}$$

Extended Application: Revenue Per Share

Real Life

EXAMPLE 8 *Finding Rates of Change*

From 1979 through 1993, the revenue per share r (in dollars) for MCI can be modeled by

$$r = (0.288t + 0.909)^2, \qquad -1 \le t \le 13,$$

where $t = 0$ corresponds to 1980. Use the model to approximate the rate of change in the revenue per share in 1984, 1988, and 1992. If you had been an MCI stockholder from 1979 through 1993, would you have been satisfied with the performance of this stock? *(Source: MCI Communications Corporation)*

Solution

The rate of change in r is given by the derivative, dr/dt. You can use the General Power Rule to find the derivative.

$$\frac{dr}{dt} = 2(0.288t + 0.909)^1(0.288) = 0.576(0.288t + 0.909)$$

In 1984, the revenue per share was increasing at a rate of

$$\frac{dr}{dt} = 0.576[0.288(4) + 0.909] \approx \$1.19 \text{ per share per year.} \qquad 1984$$

In 1988, the revenue per share was increasing at a rate of

$$\frac{dr}{dt} = 0.576[0.288(8) + 0.909] \approx \$1.85 \text{ per share per year.} \qquad 1988$$

In 1992, the revenue per share was increasing at a rate of

$$\frac{dr}{dt} = 0.576[0.288(12) + 0.909] \approx \$2.51 \text{ per share per year.} \qquad 1992$$

The graph of the revenue per share function is shown in Figure 2.29. For most investors, the performance of MCI stock would be considered to be very good.

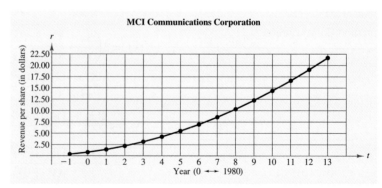

MCI Communications Corporation

FIGURE 2.29

The Stock Market/© Ted Horowitz

After AT&T was required to split up its national telephone system, several new telecommunications companies started and prospered. The most successful of these new companies is MCI. In 1993, it had 2107 million circuit-miles of microwave and fiber optics in its network. The photo shows an example of fiber optics.

Note   Example 8 deals with *revenues* per share, which is not the same as *dividends* per share.

Summary of Differentiation Rules

You now have all the rules you need to differentiate *any* algebraic function. For your convenience, they are summarized below.

Summary of Differentiation Rules

1. Constant Rule	$\dfrac{d}{dx}[c] = 0, \quad c$ is a constant.
2. Constant Multiple Rule	$\dfrac{d}{dx}[cu] = c\dfrac{du}{dx}, \quad c$ is a constant.
3. Sum and Difference Rules	$\dfrac{d}{dx}[u \pm v] = \dfrac{du}{dx} \pm \dfrac{dv}{dx}$
4. Product Rule	$\dfrac{d}{dx}[uv] = u\dfrac{dv}{dx} + v\dfrac{du}{dx}$
5. Quotient Rule	$\dfrac{d}{dx}\left[\dfrac{u}{v}\right] = \dfrac{v\dfrac{du}{dx} - u\dfrac{dv}{dx}}{v^2}$
6. Power Rules	$\dfrac{d}{dx}[x^n] = nx^{n-1}$
	$\dfrac{d}{dx}[u^n] = nu^{n-1}\dfrac{du}{dx}$
7. Chain Rule	$\dfrac{dy}{dx} = \dfrac{dy}{du}\dfrac{du}{dx}$

Discussion Problem Comparing Power Rules

You now have two power rules for differentiation.

$$\dfrac{d}{dx}[x^n] = nx^{n-1} \qquad\qquad \textit{Simple Power Rule}$$

$$\dfrac{d}{dx}[u^n] = nu^{n-1}\dfrac{du}{dx} \qquad\qquad \textit{General Power Rule}$$

Explain how you can tell which rule to use. Then state whether you would use the simple or general rule for each of the following.

a. $y = x^4$ **b.** $y = (x^2 + 1)$ **c.** $y = (x^2 + 1)^2$

d. $y = \dfrac{1}{x^4}$ **e.** $y = \dfrac{1}{x - 1}$ **f.** $y = \dfrac{1}{\sqrt{2x - 1}}$

Warm Up

The following warm-up exercises involve skills that were covered in earlier sections. You will use these skills in the exercise set for this section.

In Exercises 1–6, rewrite the expression with rational exponents.

1. $\sqrt[5]{(1 - 5x)^2}$

2. $\sqrt[4]{(2x - 1)^3}$

3. $\dfrac{1}{\sqrt{4x^2 + 1}}$

4. $\dfrac{1}{\sqrt[3]{x - 6}}$

5. $\dfrac{\sqrt{x}}{\sqrt[3]{1 - 2x}}$

6. $\dfrac{\sqrt{(3 - 7x)^3}}{2x}$

In Exercises 7–10, factor the expression.

7. $3x^3 - 6x^2 + 5x - 10$

8. $5x\sqrt{x} - x - 5\sqrt{x} + 1$

9. $4(x^2 + 1)^2 - x(x^2 + 1)^3$

10. $-x^5 + 3x^3 + x^2 - 3$

E X E R C I S E S   2.5   means that technology can help you solve or check the exercise(s).

In Exercises 1–8, identify the inside function, $u = g(x)$, and the outside function, $y = f(u)$.

$y = f(g(x))$	$u = g(x)$	$y = f(u)$
1. $y = (6x - 5)^4$		
2. $y = (x^2 - 2x + 3)^3$		
3. $y = (4 - x^2)^{-1}$		
4. $y = (x^2 + 1)^{4/3}$		
5. $y = \sqrt{5x - 2}$		
6. $y = \sqrt{9 - x^2}$		
7. $y = (3x + 1)^{-1}$		
8. $y = (x + 1)^{-1/2}$		

In Exercises 9–12, match the function with the rule that you would use to *most efficiently* find the derivative.

(a) Simple Power Rule   (b) Constant Rule

(c) General Power Rule   (d) Quotient Rule

9. $f(x) = \dfrac{2}{1 - x^3}$

10. $f(x) = \dfrac{2x}{1 - x^3}$

11. $f(x) = \sqrt[3]{8^2}$

12. $f(x) = \sqrt[3]{x^2}$

In Exercises 13–34, use the General Power Rule to find the derivative of the function.

13. $y = (2x - 7)^3$

14. $y = (3x^2 + 1)^4$

15. $f(x) = 2(x^2 - 1)^3$

16. $g(x) = 3(9x - 4)^4$

17. $g(x) = (4 - 2x)^3$

18. $h(t) = (1 - t^2)^4$

19. $h(x) = (6x - x^3)^2$

20. $f(x) = (4x - x^2)^3$

21. $f(x) = (x^2 - 9)^{2/3}$

22. $f(t) = (9t + 2)^{2/3}$

23. $f(t) = \sqrt{t + 1}$

24. $g(x) = \sqrt{2x + 3}$

25. $s(t) = \sqrt{2t^2 + 5t + 2}$

26. $y = \sqrt[3]{3x^3 + 4x}$

27. $y = \sqrt[3]{9x^2 + 4}$

28. $g(x) = \sqrt{x^2 - 2x + 1}$

29. $y = 2\sqrt{4 - x^2}$

30. $f(x) = -3\sqrt[4]{2 - 9x}$

31. $f(x) = (25 + x^2)^{-1/2}$

32. $g(x) = (6 + 3x^2)^{-2/3}$

33. $h(x) = (4 - x^3)^{-4/3}$

34. $f(x) = (4 - 3x)^{-5/2}$

In Exercises 35 and 36, find an equation of the tangent line to the graph of f at the point $(2, f(2))$.

35. $f(x) = \sqrt{4x^2 - 7}$

36. $f(x) = x\sqrt{x^2 + 5}$

In Exercises 37–40, use a symbolic differentiation utility to find the derivative of the function. Graph the function and its derivative on the same viewing rectangle. Describe the behavior of the function when the derivative is zero.

37. $f(x) = \dfrac{\sqrt{x+1}}{x^2+1}$

38. $f(x) = \sqrt{\dfrac{2x}{x+1}}$

39. $f(x) = \sqrt{\dfrac{x+1}{x}}$

40. $f(x) = \sqrt{x}(2 - x^2)$

In Exercises 41–66, find the derivative of the function.

41. $y = \dfrac{1}{x-2}$

42. $s(t) = \dfrac{1}{t^2+3t-1}$

43. $f(t) = \left(\dfrac{6}{3-t}\right)^2$

44. $y = -\dfrac{4}{(t+2)^2}$

45. $f(x) = \dfrac{3}{x^3-4}$

46. $f(x) = \dfrac{1}{(x^2-3x)^2}$

47. $y = \dfrac{1}{\sqrt{x+2}}$

48. $g(t) = \dfrac{1}{t^2-2}$

49. $g(x) = \dfrac{3}{\sqrt[3]{x^3-1}}$

50. $s(x) = \dfrac{1}{\sqrt{x^2-3x+4}}$

51. $y = x\sqrt{2x+3}$

52. $y = t\sqrt{t+1}$

53. $y = t^2\sqrt{t-2}$

54. $y = \dfrac{x}{\sqrt{25+x^2}}$

55. $f(x) = x^3(x-4)^2$

56. $f(x) = x(3x-9)^3$

57. $f(t) = (t^2-9)\sqrt{t+2}$

58. $y = \sqrt{x}(x-2)^2$

59. $f(x) = \sqrt{\dfrac{3-2x}{4x}}$

60. $y = \sqrt{\dfrac{2x}{x+1}}$

61. $g(t) = \dfrac{3t^2}{\sqrt{t^2+2t-1}}$

62. $f(x) = \dfrac{x+1}{2x-3}$

63. $f(x) = \dfrac{\sqrt[3]{x^3+1}}{x}$

64. $y = \sqrt{x-1} + \sqrt{x+1}$

65. $y = \left(\dfrac{6-5x}{x^2-1}\right)^2$

66. $y = \left(\dfrac{4x^2}{3-x}\right)^3$

67. *Compound Interest* You deposit $1000 in an account with an annual rate of r (in decimal form) compounded monthly. At the end of 5 years, the balance is

$$A = 1000\left(1 + \dfrac{r}{12}\right)^{60}.$$

Find the rate of change of A with respect to r when (a) $r = 0.08$, (b) $r = 0.10$, and (c) $r = 0.12$.

68. *Cost* An assembly plant purchases electric motors to install in one of its products. Management estimates that the cost per motor over the next 5 years is

$$C = 9(1.68t + 12)^{4/3},$$

where t is the time in years. Complete the following table. What can you conclude?

t	1	2	3	4	5
dC/dt					

69. *Population Growth* The number N of bacteria in a culture after t days is modeled by

$$N = 400\left[1 - \dfrac{3}{(t^2+2)^2}\right].$$

Complete the following table. What can you conclude?

t	0	1	2	3	4
dN/dt					

70. *Depreciation* The value V of a machine t years after it is purchased is inversely proportional to the square root of $t + 1$. The initial value of the machine is $10,000.

(a) Write V as a function of t.
(b) Find the rate of depreciation when $t = 1$.
(c) Find the rate of depreciation when $t = 3$.

71. *Depreciation* Repeat Exercise 70 given that the value of the machine t years after it is purchased is inversely proportional to the cube root of $t + 1$.

72. *Credit-Card Rate* The average annual rate r (in percent form) for commercial bank credit cards from 1980 through 1991 can be modeled by

$$r = \dfrac{-0.05t^5 + 1.43t^4 - 12.55t^3 + 32.5t^2 + 31.2t + 1730}{100},$$

where $t = 0$ represents 1980. *(Source: Federal Reserve Bulletin)*

(a) Find the derivative of this model. Which differentiation rules did you use?
(b) Use a graphing utility to graph the derivative. Use the interval $0 \le t \le 11$.
(c) Use the trace feature to find the years during which the finance rate was changing the most.
(d) Use the trace feature to find the years during which the finance rate was changing the least.

Higher-Order Derivatives

Second, Third, and Higher-Order Derivatives ■
Acceleration

Second, Third, and Higher-Order Derivatives

The derivative of f' is the **second derivative** of f and is denoted by f''.

$$\frac{d}{dx}[f'(x)] = f''(x) \qquad \textit{Second derivative}$$

The derivative of f'' is the **third derivative** of f and is denoted by f'''.

$$\frac{d}{dx}[f''(x)] = f'''(x) \qquad \textit{Third derivative}$$

By continuing this process, you obtain **higher-order derivatives** of f.

Note In the context of higher-order derivatives, the "standard" derivative f' is often called the **first derivative** of f.

EXAMPLE 1 *Finding Higher-Order Derivatives*

$f(x) = 2x^4 - 3x^2$	*Original function*
$f'(x) = 8x^3 - 6x$	*First derivative*
$f''(x) = 24x^2 - 6$	*Second derivative*
$f'''(x) = 48x$	*Third derivative*
$f^{(4)}(x) = 48$	*Fourth derivative*
$f^{(5)}(x) = 0$	*Fifth derivative*

Notation for Higher-Order Derivatives

1. 1st derivative: $\quad y', \quad f'(x), \quad \dfrac{dy}{dx}, \quad \dfrac{d}{dx}[f(x)], \quad D_x[y]$

2. 2nd derivative: $\quad y'', \quad f''(x), \quad \dfrac{d^2y}{dx^2}, \quad \dfrac{d^2}{dx^2}[f(x)], \quad D_x{}^2[y]$

3. 3rd derivative: $\quad y''', \quad f'''(x), \quad \dfrac{d^3y}{dx^3}, \quad \dfrac{d^3}{dx^3}[f(x)], \quad D_x{}^3[y]$

4. 4th derivative: $\quad y^{(4)}, \quad f^{(4)}(x), \quad \dfrac{d^4y}{dx^4}, \quad \dfrac{d^4}{dx^4}[f(x)], \quad D_x{}^4[y]$

5. *n*th derivative: $\quad y^{(n)}, \quad f^{(n)}(x), \quad \dfrac{d^ny}{dx^n}, \quad \dfrac{d^n}{dx^n}[f(x)], \quad D_x{}^n[y]$

EXAMPLE 2 *Finding Higher-Order Derivatives*

Find the value of $g'''(2)$ for the function

$$g(t) = -t^4 + 2t^3 + t + 4. \qquad \textit{Original function}$$

Solution

Begin by differentiating three times.

$$g'(t) = -4t^3 + 6t^2 + 1 \qquad \textit{First derivative}$$

$$g''(t) = -12t^2 + 12t \qquad \textit{Second derivative}$$

$$g'''(t) = -24t + 12 \qquad \textit{Third derivative}$$

Then, evaluate the third derivative of g at $t = 2$.

$$g'''(2) = -24(2) + 12 = -36 \qquad \textit{Value of third derivative}$$

■

Examples 1 and 2 show how to find higher-order derivatives of *polynomial* functions. Note that with each successive differentiation, the degree of the polynomial drops by one. Eventually, higher-order derivatives of polynomial functions degenerate to a constant function. Specifically, the nth-order derivative of an nth-degree polynomial function

$$f(x) = a_n x^n + a_{n-1} x^{n-1} + \cdots + a_1 x + a_0$$

is the constant function

$$f^{(n)}(x) = n! a_n,$$

where $n! = 1 \cdot 2 \cdot 3 \cdots n$. Each derivative of order higher than n is the zero function. Polynomial functions are the *only* functions with this characteristic. For other functions, successive differentiation never produces a constant function.

EXAMPLE 3 *Finding Higher-Order Derivatives*

$$y = x^{-1} = \frac{1}{x} \qquad \textit{Original function}$$

$$y' = (-1)x^{-2} = -\frac{1}{x^2} \qquad \textit{First derivative}$$

$$y'' = (-1)(-2)x^{-3} = \frac{2}{x^3} \qquad \textit{Second derivative}$$

$$y''' = (-1)(-2)(-3)x^{-4} = -\frac{6}{x^4} \qquad \textit{Third derivative}$$

$$y^{(4)} = (-1)(-2)(-3)(-4)x^{-5} = \frac{24}{x^5} \qquad \textit{Fourth derivative}$$

■

TECHNOLOGY

Higher-order derivatives of nonpolynomial functions can be difficult to find by hand. If you have access to a symbolic differentiation utility such as *Derive*, *Maple*, or *Mathematica*, try using it to find higher-order derivatives. For instance, if you use *Derive* to find the third derivative of $y = 1/(x^2 + 1)$, you will obtain the following.

1: $\dfrac{1}{x^2 + 1}$

2: $\left[\dfrac{d}{dx}\right]^3 \dfrac{1}{x^2 + 3}$

3: $\dfrac{24x(1 - x^2)}{(x^2 + 1)^2}$

Acceleration

In Section 2.3, you saw that the velocity of an object moving in a straight path is given by the derivative of its position function. In other words, the rate of change of the position with respect to time is defined to be the velocity. In a similar way, the rate of change of the velocity with respect to time is defined to be the **acceleration** of the object.

Note   Acceleration is measured in units of length per unit of time squared. For instance, if the velocity is measured in feet per second, then the acceleration is measured in "feet per second squared," or more formally as "feet per second per second."

$$s = f(t) \qquad\qquad \text{\textit{Position function}}$$

$$\frac{ds}{dt} = f'(t) \qquad\qquad \text{\textit{Velocity function}}$$

$$\frac{d^2s}{dt^2} = f''(t) \qquad\qquad \text{\textit{Acceleration function}}$$

To find the position, velocity, or acceleration at a particular time t, substitute the given value of t into the appropriate function, as illustrated in Example 4.

Real Life

EXAMPLE 4 *Finding Acceleration*

A ball is thrown into the air from the top of a 160-foot cliff, as shown in Figure 2.30. The initial velocity of the ball is 48 feet per second, which implies that the position function is

$$s = -16t^2 + 48t + 160,$$

where the time t is measured in seconds. Find the height, the velocity, and the acceleration of the ball when $t = 3$.

Solution
Begin by differentiating to find the velocity and acceleration functions.

$$s = -16t^2 + 48t + 160 \qquad\qquad \text{\textit{Position function}}$$

$$\frac{ds}{dt} = -32t + 48 \qquad\qquad \text{\textit{Velocity function}}$$

$$\frac{d^2s}{dt^2} = -32 \qquad\qquad \text{\textit{Acceleration function}}$$

To find the height, velocity, and acceleration when $t = 3$, substitute $t = 3$ into each of the above functions.

$$s = -16(3)^2 + 48(3) + 160 = 160 \text{ ft} \qquad \text{\textit{Position when } } t = 3$$

$$\frac{ds}{dt} = -32(3) + 48 = -48 \text{ ft/sec} \qquad \text{\textit{Velocity when } } t = 3$$

$$\frac{d^2s}{dt^2} = -32 \text{ ft/sec}^2 \qquad \text{\textit{Acceleration when } } t = 3$$

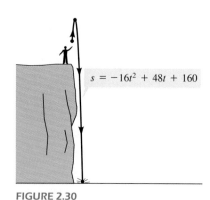

$$s = -16t^2 + 48t + 160$$

FIGURE 2.30

In Example 4, notice that the acceleration of the ball is -32 ft/sec^2 at any time t. This constant acceleration is due to the gravitational force of the earth and is called the **acceleration due to gravity.** Note that the negative value indicates that the ball is being pulled *down*—toward the earth.

Although the acceleration exerted on a falling object is relatively constant near the earth's surface, it varies greatly throughout our solar system. Large planets exert a much greater gravitational pull than do small planets or moons. The next example describes the motion of a free-falling object on the moon.

Real Life

EXAMPLE 5 *Finding Acceleration on the Moon*

An astronaut standing on the surface of the moon throws a rock into the air. The height s (in feet) of the rock is given by

$$s = -\frac{27}{10}t^2 + 27t + 6,$$

where t is measured in seconds. How does the acceleration due to gravity on the moon compare to that on the earth?

Solution

$$s = -\frac{27}{10}t^2 + 27t + 6 \qquad \textit{Position function}$$

$$\frac{ds}{dt} = -\frac{27}{5}t + 27 \qquad \textit{Velocity function}$$

$$\frac{d^2s}{dt^2} = -\frac{27}{5} \qquad \textit{Acceleration function}$$

Thus, the acceleration at any time is $-\frac{27}{5} = -5.4$ ft/sec^2—about one-sixth of the acceleration due to gravity on the earth.

The position function described in Example 5 neglects air resistance, which is appropriate because the moon has no atmosphere—and *no air resistance.* This means that the position function for any free-falling object on the moon is given by

$$s = -\frac{27}{10}t^2 + v_0 t + h_0,$$

where s is the height (in feet), t is the time (in seconds), v_0 is the initial velocity, and h_0 is the initial height. For instance, the rock in Example 5 was thrown upward with an initial velocity of 27 feet per second and had an initial height of 6 feet. This position function is valid for all objects—heavy ones such as hammers and light ones such as feathers.

In 1971, astronaut David R. Scott demonstrated the lack of atmosphere on the moon by dropping a hammer and feather from the same height. Both took exactly the same time to fall to the ground. If they were dropped from a height of 6 feet, how long did each take to hit the ground?

NASA

The acceleration due to gravity on the surface of the moon is only about one-sixth that exerted by the earth. Thus, if you were on the moon and threw an object into the air, it would rise to a greater height than it would on the earth's surface.

This drawing depicts the first moon landing on July 20, 1969. The event was witnessed live by millions of observers on the earth via satellite television. Can you see what is wrong with the drawing?

Real Life

EXAMPLE 6 *Finding Velocity and Acceleration*

The velocity v (in feet per second) of a certain automobile starting from rest is

$$v = \frac{80t}{t+5},$$ *Velocity function*

where t is the time (in seconds). The position of the automobile is shown in Figure 2.31. Find the velocity and acceleration of the automobile at 10-second intervals from $t = 0$ to $t = 60$.

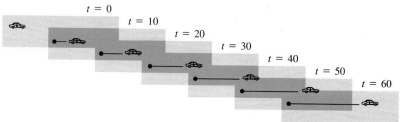

$t = 0$

$t = 10$

$t = 20$

$t = 30$

$t = 40$

$t = 50$

$t = 60$

FIGURE 2.31

Solution

To find the acceleration function, differentiate the velocity function.

$$\frac{dv}{dt} = \frac{(t+5)(80) - (80t)(1)}{(t+5)^2} = \frac{400}{(t+5)^2}$$ *Acceleration function*

TABLE 2.6

t (seconds)	0	10	20	30	40	50	60
v (ft/sec)	0	53.3	64.0	68.6	71.1	72.7	73.8
$\dfrac{dv}{dt}$ (ft/sec^2)	16	1.78	0.64	0.33	0.20	0.13	0.09

In Table 2.6, note that the acceleration approaches zero as the velocity levels off. This observation should agree with your experience—when riding in an accelerating automobile, you do not feel the velocity, but you do feel the acceleration. In other words, you feel changes in velocity. ■

Discussion Problem   Acceleration Due to Gravity

Newton's law of universal gravitation states that the gravitational attraction of two objects is directly proportional to their masses and inversely proportional to the square of the distance between their centers. The earth has a mass of 5.979×10^{24} kilograms and a radius of 6371 kilometers. The moon has a mass of 7.354×10^{22} kilograms and a radius of 1738 kilometers. Find the ratio of the earth's gravity to the moon's gravity.

Warm Up

The following warm-up exercises involve skills that were covered in earlier sections. You will use these skills in the exercise set for this section.

In Exercises 1–4, solve the equation.

1. $-16t^2 + 24t = 0$

2. $-16t^2 + 80t + 224 = 0$

3. $-16t^2 + 128t + 320 = 0$

4. $-16t^2 + 9t + 1440 = 0$

In Exercises 5–8, find dy/dx.

5. $y = x^2(2x + 7)$

6. $y = (x^2 + 3x)(2x^2 - 5)$

7. $y = \dfrac{x^2}{2x + 7}$

8. $y = \dfrac{x^2 + 3x}{2x^2 - 5}$

In Exercises 9 and 10, find the domain and range of f.

9. $f(x) = x^2 - 4$

10. $f(x) = \sqrt{x - 7}$

E X E R C I S E S 2.6   means that technology can help you solve or check the exercise(s).

In Exercises 1–16, find the second derivative of the function.

1. $f(x) = 5 - 4x$

2. $f(x) = 3x - 1$

3. $f(x) = x^2 + 7x - 4$

4. $f(x) = 3x^2 + 4x$

5. $g(t) = \frac{1}{3}t^3 - 4t^2 + 2t$

6. $g(t) = 5t^4 + 10t^2 + 3$

7. $g(t) = t^{-1/3}$

8. $f(t) = \dfrac{3}{4t^2}$

9. $f(x) = 4(x^2 - 1)^2$

10. $f(x) = 3(2 - x^2)^3$

11. $f(x) = x\sqrt[3]{x}$

12. $f(x) = \sqrt{9 - x^2}$

13. $f(x) = \dfrac{x + 1}{x - 1}$

14. $g(t) = -\dfrac{4}{(t + 2)^2}$

15. $y = x^2(x^2 + 4x + 8)$

16. $h(s) = s^3(s^2 - 2s + 1)$

In Exercises 17–24, find the third derivative of the function.

17. $f(x) = x^5 - 3x^4$

18. $f(x) = x^4 - 2x^3$

19. $f(x) = 5x(x + 4)^3$

20. $f(x) = (x - 1)^2$

21. $f(x) = \dfrac{3}{(4x)^2}$

22. $f(x) = \dfrac{1}{x}$

23. $f(x) = \sqrt{4 - x}$

24. $f(t) = \sqrt{2t + 3}$

In Exercises 25–30, find the indicated derivative.

Given	*Derivative*
25. $f'(x) = 2x^2$	$f''(x)$
26. $f''(x) = 20x^3 - 36x^2$	$f'''(x)$
27. $f''(x) = \dfrac{2x - 2}{x}$	$f'''(x)$
28. $f'''(x) = 2\sqrt{x - 1}$	$f^{(4)}(x)$
29. $f^{(4)}(x) = (x + 1)^2$	$f^{(6)}(x)$
30. $f(x) = x^3 - 2x$	$f''(x)$

In Exercises 31–38, find the second derivative and solve the equation $f''(x) = 0$.

31. $f(x) = x^3 - 9x^2 + 27x - 27$

32. $f(x) = 3x^3 - 9x + 1$

33. $f(x) = (x + 3)(x - 4)(x + 5)$

34. $f(x) = x^4 - 8x^3 + 18x^2 - 16x + 2$

35. $f(x) = 3x^4 - 18x^2$

36. $f(x) = x\sqrt{4 - x^2}$

37. $f(x) = \dfrac{x}{x^2 + 3}$

38. $f(x) = \dfrac{x}{x^2 + 1}$

39. *Velocity and Acceleration* A ball is propelled straight up from ground level with an initial velocity of 144 feet per second.

(a) Write the position function of the ball.

(b) Write the velocity and acceleration functions.

(c) When is the ball at its highest point? How high is this point?

(d) How fast is the ball traveling when it hits the ground? How is this speed related to its initial velocity?

40. *Velocity and Acceleration* A brick becomes dislodged from the top of the Empire State Building (at a height of 1250 feet) and falls to the sidewalk below.

(a) Write the position function of the brick.

(b) Write the velocity and acceleration functions.

(c) How long does it take the brick to hit the sidewalk?

(d) How fast is the brick traveling when it hits the sidewalk?

41. *Velocity and Acceleration* The velocity (in feet per second) of an automobile starting from rest is modeled by

$$\frac{ds}{dt} = \frac{90t}{t + 10}.$$

Complete the table showing the velocity and acceleration at 10-second intervals during the first minute of travel. What can you conclude?

t	0	10	20	30	40	50	60
$\dfrac{ds}{dt}$							
$\dfrac{d^2s}{dt^2}$							

42. *Stopping Distance* A car is traveling at a rate of 66 feet per second (45 miles per hour) when the brakes are applied. The position function for the car is

$$s = -8.25t^2 + 66t,$$

where s is measured in feet and t is measured in seconds. Complete the table showing the position, velocity, and acceleration for the given values of t. What can you conclude?

t	0	1	2	3	4	5
$s(t)$						
$v(t)$						
$a(t)$						

In Exercises 43 and 44, use a graphing utility to graph f, f', and f'' on the same viewing rectangle. What is the relationship between the degree of f and the degrees of its successive derivatives? In general, what is the relationship between the degree of a polynomial function and the degrees of its successive derivatives?

43. $f(x) = x^2 - 6x + 6$ **44.** $f(x) = 3x^3 - 9x$

In Exercises 45 and 46, the graphs of f, f', and f'' are shown together. Which is which? Explain your reasoning.

45.   **46.**

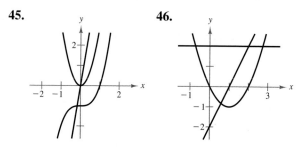

True or False In Exercises 47–53, determine whether the statement is true or false. Explain your reasoning.

47. If $y = f(x)g(x)$, then $y' = f'(x)g'(x)$.

48. If $y = (x + 1)(x + 2)(x + 3)(x + 4)$, then $y^{(5)} = 0$.

49. If $f'(c)$ and $g'(c)$ are zero and $h(x) = f(x)g(x)$, then $h'(c) = 0$.

50. If $f(x)$ is an nth-degree polynomial, then $f^{(n+1)}(x) = 0$.

51. The second derivative represents the rate of change of the first derivative.

52. If the velocity of an object is constant, then its acceleration is zero.

53. If $f'(x) = g'(x)$, then $f(x) = g(x)$.

54. *Projectile Motion* An object is thrown upward from the top of a 64-foot building with an initial velocity of 48 feet per second.

(a) Write the position function of the object.

(b) Write the velocity and acceleration functions.

(c) When will the object hit the ground?

(d) When is the velocity of the object zero?

(e) How high does the object go?

Implicit Differentiation

Implicit and Explicit Functions ∎ *Implicit Differentiation* ∎
Extended Application: Demand Function

Implicit and Explicit Functions

So far in the text, functions involving two variables have generally been expressed in the **explicit form** $y = f(x)$. That is, one of the two variables has been explicitly given in terms of the other. For example,

$$y = 3x - 5, \quad s = -16t^2 + 20t, \quad \text{and} \quad u = 3w - w^2$$

are each written in explicit form, and we say that y, s, and u are functions of x, t, and w, explicitly. Many functions, however, are not given explicitly and are only implied by a given equation, as illustrated in Example 1.

EXAMPLE 1 *Finding a Derivative Explicitly*

Find dy/dx for the equation $xy = 1$.

Solution
In this equation, y is **implicitly** defined as a function of x. One way to find dy/dx is to first solve the equation for y, then differentiate as usual.

$xy = 1$	*Implicit form*
$y = \dfrac{1}{x}$	*Solve for y.*
$\quad = x^{-1}$	*Rewrite.*
$\dfrac{dy}{dx} = -x^{-2}$	*Differentiate with respect to x.*
$\quad = -\dfrac{1}{x^2}$	*Simplify.*

The procedure shown in Example 1 works well whenever you can easily write the given function explicitly. You cannot, however, use this procedure when you are unable to solve for y as a function of x. For instance, how would you find dy/dx in the equation

$$x^2 - 2y^3 + 4y = 2,$$

where it is very difficult to express y as a function of x explicitly? To do this, you can use a procedure called **implicit differentiation.**

Implicit Differentiation

To understand how to find dy/dx implicitly, you must realize that the differentiation is taking place *with respect to x*. This means that when you differentiate terms involving x alone, you can differentiate as usual. *But* when you differentiate terms involving y, you must apply the Chain Rule because you are assuming that y is defined implicitly as a function of x. Study the next example carefully. Note in particular how the Chain Rule is used to introduce the dy/dx factors in parts b and d.

EXAMPLE 2 *Applying the Chain Rule*

Differentiate the following with respect to x.

a. $3x^2$ **b.** $2y^3$ **c.** $x + 3y$ **d.** xy^2

Solution

a. The only variable in this expression is x. Thus, to differentiate with respect to x, you can use the Simple Power Rule and the Constant Multiple Rule to obtain

$$\frac{d}{dx}[3x^2] = 6x.$$

b. This case is different. The variable in the expression is y, and yet you are asked to differentiate with respect to x. To do this, assume that y is a differentiable function of x and use the Chain Rule.

$$\frac{d}{dx}[2y^3] = \overbrace{2}^{c}\ \overbrace{(3)}^{n}\ \overbrace{y^2}^{u^{n-1}}\ \overbrace{\frac{dy}{dx}}^{u'} \qquad \textit{Chain Rule}$$

$$= 6y^2\frac{dy}{dx}$$

where $\overbrace{}^{cu^n}$.

c. This expression involves both x and y. By the Sum Rule and the Constant Multiple Rule, you can write

$$\frac{d}{dx}[x + 3y] = 1 + 3\frac{dy}{dx}.$$

d. By the Product Rule and the Chain Rule, you can write

$$\frac{d}{dx}[xy^2] = x\frac{d}{dy}[y^2] + y^2\frac{d}{dx}[x] \qquad \textit{Product Rule}$$

$$= x\left(2y\frac{dy}{dx}\right) + y^2(1) \qquad \textit{Chain Rule}$$

$$= 2xy\frac{dy}{dx} + y^2.$$

Implicit Differentiation

Consider an equation involving x and y in which y is a differentiable function of x. You can use the following steps to find dy/dx.

1. Differentiate both sides of the equation *with respect to x.*
2. Write the result so that all terms involving dy/dx are on the left side of the equation and all other terms are on the right side of the equation.
3. Factor dy/dx out of the terms on the left side of the equation.
4. Solve for dy/dx by dividing both sides of the equation by the left-hand factor that does not contain dy/dx.

In Example 3, note that implicit differentiation can produce an expression for dy/dx that contains both x and y.

EXAMPLE 3 *Finding the Slope of a Graph Implicitly*

Find the slope of the tangent line to the ellipse given by $x^2 + 4y^2 = 4$ at the point $\left(\sqrt{2}, -1/\sqrt{2}\right)$, as shown in Figure 2.32.

Solution

$$x^2 + 4y^2 = 4 \qquad \text{\textit{Original equation}}$$

$$\frac{d}{dx}[x^2 + 4y^2] = \frac{d}{dx}[4] \qquad \text{\textit{Differentiate with respect to x.}}$$

$$2x + 8y\left(\frac{dy}{dx}\right) = 0 \qquad \text{\textit{Implicit differentiation}}$$

$$8y\left(\frac{dy}{dx}\right) = -2x \qquad \text{\textit{Subtract 2x from both sides.}}$$

$$\frac{dy}{dx} = \frac{-2x}{8y} \qquad \text{\textit{Divide both sides by 8y.}}$$

$$\frac{dy}{dx} = -\frac{x}{4y} \qquad \text{\textit{Simplify.}}$$

To find the slope at the given point, substitute $x = \sqrt{2}$ and $y = -1/\sqrt{2}$ into the derivative.

$$\frac{dy}{dx} = -\frac{x}{4y} \qquad \text{\textit{Derivative}}$$

$$= -\frac{\sqrt{2}}{4\left(-1/\sqrt{2}\right)} \qquad \text{\textit{Substitute for x and y.}}$$

$$= \frac{1}{2} \qquad \text{\textit{Simplify.}}$$

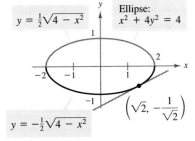

$y = \frac{1}{2}\sqrt{4 - x^2}$

Ellipse:
$x^2 + 4y^2 = 4$

$\left(\sqrt{2}, -\dfrac{1}{\sqrt{2}}\right)$

$y = -\frac{1}{2}\sqrt{4 - x^2}$

FIGURE 2.32 Slope of tangent line is $\frac{1}{2}$

Note To see the benefit of implicit differentiation, try reworking Example 3 using the explicit function

$$y = -\frac{1}{2}\sqrt{4 - x^2}.$$

The graph of this function is the lower half of the ellipse.

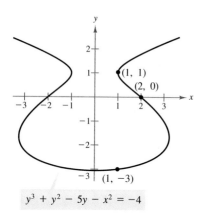

$$y^3 + y^2 - 5y - x^2 = -4$$

FIGURE 2.33

EXAMPLE 4 *Using Implicit Differentiation*

Find dy/dx for the equation $y^3 + y^2 - 5y - x^2 = -4$.

Solution

$$y^3 + y^2 - 5y - x^2 = -4 \qquad \text{\textit{Original equation}}$$

$$\frac{d}{dx}[y^3 + y^2 - 5y - x^2] = \frac{d}{dx}[-4] \qquad \begin{array}{l}\textit{Differentiate}\\ \textit{both sides.}\end{array}$$

$$3y^2\frac{dy}{dx} + 2y\frac{dy}{dx} - 5\frac{dy}{dx} - 2x = 0 \qquad \begin{array}{l}\textit{Implicit}\\ \textit{differentiation}\end{array}$$

$$3y^2\frac{dy}{dx} + 2y\frac{dy}{dx} - 5\frac{dy}{dx} = 2x \qquad \begin{array}{l}\textit{Collect}\\ dy/dx \textit{ terms.}\end{array}$$

$$\frac{dy}{dx}(3y^2 + 2y - 5) = 2x \qquad \textit{Factor.}$$

$$\frac{dy}{dx} = \frac{2x}{3y^2 + 2y - 5}$$

The graph of the original equation is shown in Figure 2.33. What is the slope of the graph at the points $(1, -3)$, $(2, 0)$, and $(1, 1)$?   ■

EXAMPLE 5 *Finding the Slope of a Graph Implicitly*

Find the slope of the graph of $3(x^2 + y^2)^2 = 100xy$ at the point $(3, 1)$.

Solution

$$3(x^2 + y^2)^2 = 100xy$$

$$\frac{d}{dx}[3(x^2 + y^2)^2] = \frac{d}{dx}[100xy]$$

$$3(2)(x^2 + y^2)\left(2x + 2y\frac{dy}{dx}\right) = 100\left[x\frac{dy}{dx} + y(1)\right]$$

$$12y(x^2 + y^2)\frac{dy}{dx} - 100x\frac{dy}{dx} = 100y - 12x(x^2 + y^2)$$

$$[12y(x^2 + y^2) - 100x]\frac{dy}{dx} = 100y - 12x(x^2 + y^2)$$

$$\frac{dy}{dx} = \frac{100y - 12x(x^2 + y^2)}{12y(x^2 + y^2) - 100x}$$

$$\frac{dy}{dx} = \frac{25y - 3x(x^2 + y^2)}{-25x + 3y(x^2 + y^2)}$$

At the point $(3, 1)$, the slope of the graph is

$$\frac{dy}{dx} = \frac{25(1) - 3(3)(3^2 + 1^2)}{-25(3) + 3(1)(3^2 + 1^2)} = \frac{25 - 90}{-75 + 30} = \frac{-65}{-45} = \frac{13}{9},$$

as shown in Figure 2.34. The graph is called a **lemniscate**.   ■

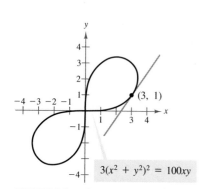

$$3(x^2 + y^2)^2 = 100xy$$

FIGURE 2.34   Lemniscate

Extended Application: Demand Function

Real
Life

EXAMPLE 6 *Using a Demand Function*

The demand function for a product is modeled by

$$p = \frac{3}{0.000001x^3 + 0.01x + 1},$$

where p is measured in dollars and x is measured in thousands of units, as shown in Figure 2.35. Find the rate of change of the demand x with respect to the price p when $x = 100$.

Solution

To simplify the differentiation, begin by rewriting the function. Then, differentiate *with respect to p.*

$$p = \frac{3}{0.000001x^3 + 0.01x + 1}$$

$$0.000001x^3 + 0.01x + 1 = \frac{3}{p}$$

$$0.000003x^2\frac{dx}{dp} + 0.01\frac{dx}{dp} = -\frac{3}{p^2}$$

$$(0.000003x^2 + 0.01)\frac{dx}{dp} = -\frac{3}{p^2}$$

$$\frac{dx}{dp} = -\frac{3}{p^2(0.000003x^2 + 0.01)}$$

When $x = 100$, the price is

$$p = \frac{3}{0.000001(100)^3 + 0.01(100) + 1} = \$1.$$

Thus, when $x = 100$ and $p = 1$, the rate of change of the demand with respect to the price is

$$\frac{dx}{dp} = -\frac{3}{(1)^2[0.000003(100)^2 + 0.01]} = -75.$$

This means that when $x = 100$, the demand is dropping at the rate of 75 thousand units for each dollar increase in price. ■

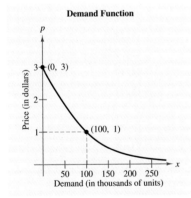

Demand Function

(0, 3)

(100, 1)

Price (in dollars)

50 100 150 200 250
Demand (in thousands of units)

FIGURE 2.35

Discussion Problem Comparing Derivatives

In Example 6, the derivative dx/dp does not represent the slope of the graph of the demand function. Because the demand function is given by $p = f(x)$, the slope of the graph is given by dp/dx. Find dp/dx and show that the two derivatives are related by $dx/dp = 1/(dp/dx)$.

Warm Up

The following warm-up exercises involve skills that were covered in earlier sections. You will use these skills in the exercise set for this section.

In Exercises 1–6, solve the equation for *y*.

1. $x - \dfrac{y}{x} = 2$

2. $\dfrac{4}{x-3} = \dfrac{1}{y}$

3. $xy - x + 6y = 6$

4. $12 + 3y = 4x^2 + x^2y$

5. $x^2 + y^2 = 5$

6. $x = \pm\sqrt{6 - y^2}$

In Exercises 7–10, evaluate the expression at the indicated point.

7. $\dfrac{3x^2 - 4}{3y^2}$, $(2, 1)$

8. $\dfrac{x^2 - 2}{1 - y}$, $(0, -3)$

9. $\dfrac{5x}{3y^2 - 12y + 5}$, $(-1, 2)$

10. $\dfrac{1}{y^2 - 2xy + x^2}$, $(4, 3)$

E X E R C I S E S 2.7

means that technology can help you solve or check the exercise(s).

In Exercises 1–6, find *dy/dx*.

1. $5xy = 1$

2. $\frac{1}{2}x^2 - y = 6x$

3. $y^2 = 1 - x^2$, $0 \le x \le 1$

4. $4x^2y - \dfrac{3}{y} = 0$

5. $\dfrac{2 - x}{y - 3} = 5$

6. $\dfrac{xy - y^2}{y - x} = 1$

In Exercises 7-22, find *dy/dx* by implicit differentiation and evaluate the derivative at the indicated point.

Equation	*Point*
7. $x^2 + y^2 = 49$	$(0, 7)$
8. $x^2 - y^2 = 16$	$(4, 0)$
9. $y + xy = 4$	$(-5, -1)$
10. $4xy + x^2 = 5$	$(1, 1)$
11. $x^2 - y^3 = 3$	$(2, 1)$
12. $x^3 + y^3 = 8$	$(0, 2)$
13. $x^3 - xy + y^2 = 4$	$(0, -2)$
14. $x^2y + y^2x = -2$	$(2, -1)$
15. $x^3y^3 - y = x$	$(0, 0)$

	Equation	*Point*
16.	$x^3 + y^3 = 2xy$	$(1, 1)$
17.	$x^{1/2} + y^{1/2} = 9$	$(16, 25)$
18.	$\sqrt{xy} = x - 2y$	$(4, 1)$
19.	$x^{2/3} + y^{2/3} = 5$	$(8, 1)$
20.	$(x + y)^3 = x^3 + y^3$	$(-1, 1)$
21.	$x^3 - 3x^2y + 3xy^2 = 26$	$(2, 3)$
22.	$y^2(x^2 + 9) = x^2 - 9$	$(3, 0)$

In Exercises 23–26, find the slope of the graph at the indicated point.

23. $3x^2 - 2y + 5 = 0$

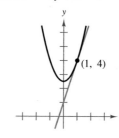

24. $x^2 + y^2 = 4$

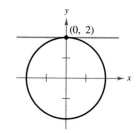

25. $4x^2 + 9y^2 = 36$

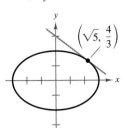

26. $x^2 - y^3 = 0$

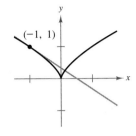

In Exercises 27–30, find dy/dx implicitly and explicitly (the explicit functions are shown on the graph) and show that the results are equivalent. Use the graph to estimate the slope of the tangent line at the labeled point. Then verify your result analytically by evaluating dy/dx at the point.

27. $x^2 + y^2 = 25$

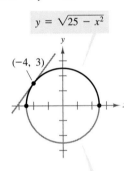

28. $x - y^2 - 1 = 0$

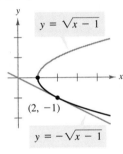

29. $9x^2 + 16y^2 = 144$

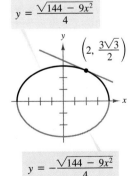

30. $4y^2 - x^2 = 7$

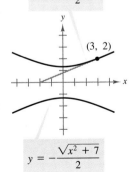

 In Exercises 31–34, find equations of the tangent lines to the graph at the given points. Graph the equation and the tangent lines on the same viewing rectangle.

	Equation	*Points*
31.	$x^2 + y^2 = 169$	$(5, 12)$ and $(-12, 5)$
32.	$x^2 + y^2 = 9$	$(0, 3)$ and $\left(2, \sqrt{5}\right)$
33.	$y^2 = 5x^3$	$\left(1, \sqrt{5}\right)$ and $\left(1, -\sqrt{5}\right)$
34.	$y^2 = \dfrac{x^3}{4 - x}$	$(2, 2)$ and $(2, -2)$

Demand In Exercises 35–38, find the rate of change of x with respect to p.

35. $p = 0.006x^4 + 0.02x^2 + 10, \quad x \geq 0$

36. $p = \dfrac{725}{0.02x}$

37. $p = \dfrac{200 - x}{2x}, \quad 0 < x \leq 200$

38. $p = \sqrt{\dfrac{500 - x}{2x}}, \quad 0 < x \leq 500$

39. *Production* Let x represent the units of labor and y the capital invested in a manufacturing process. When 135,540 units are produced, the relationship between labor and capital can be modeled by

$$100x^{0.75}y^{0.25} = 135,540.$$

(a) Find the rate of change of y with respect to x when $x = 1500$ and $y = 1000$.

(b) The model used in the problem is called the *Cobb-Douglas production function*. Graph the model and describe the relationship between labor and capital.

 40. *AIDS Epidemic* The number of cases y of AIDS reported in the United States from 1984 to 1991 can be modeled by

$$y = \dfrac{115.01t^2 + 25.84t + 5.92}{0.01t^2 - 0.18t + 1}, \quad 4 \leq t \leq 11,$$

where $t = 4$ represents 1984.

(a) Graph the model and describe the results.

(b) Use the graph to determine the year during which the number of reported cases increased most rapidly.

(c) Complete the table to confirm your estimate.

t	4	5	6	7	8	9	10	11
y								
y'								

2.8 Related Rates

Related Variables ▪ *Solving Related-Rate Problems*

Related Variables

In this section you will study problems involving variables that are changing with respect to time. If two (or more) such variables are related to each other, then their rates of change with respect to time are also related.

For instance, suppose that x and y are related by the equation $y = 2x$. If both variables are changing with respect to time, then their rates of change will also be related.

$$x \text{ and } y \text{ are related.}$$

$$y = 2x \quad \longrightarrow \quad \frac{dy}{dt} = 2\frac{dx}{dt}$$

The rates of change of x and y are related.

In this simple example, you can see that because y always has twice the value of x, it follows that the rate of change of y with respect to time is always twice the rate of change of x with respect to time.

EXAMPLE 1 *Examining Two Rates That Are Related*

The variables x and y are differentiable functions of t and are related by the equation $y = x^2 + 3$. When $x = 1$, $dx/dt = 2$. Find dy/dt when $x = 1$.

Solution
Use the Chain Rule to differentiate both sides of the equation *with respect to t*.

$$y = x^2 + 3 \qquad\qquad \textit{Original equation}$$

$$\frac{d}{dt}[y] = \frac{d}{dt}[x^2 + 3] \qquad\qquad \textit{Differentiate with respect to t.}$$

$$\frac{dy}{dt} = 2x\frac{dx}{dt} \qquad\qquad \textit{Chain Rule}$$

When $x = 1$ and $dx/dt = 2$, you have

$$\frac{dy}{dt} = 2x\frac{dx}{dt} = 2(1)(2) = 4.$$

Solving Related-Rate Problems

In Example 1, you are *given* the mathematical model.

> *Given equation:* $y = x^2 + 3$
>
> $\qquad$ *Given rate:* $\dfrac{dx}{dt} = 2$ when $x = 1$
>
> $\qquad\qquad$ *Find:* $\dfrac{dy}{dt}$ when $x = 1$

In the next example, you are asked to *create* a similar mathematical model.

Real Life

EXAMPLE 2 *Changing Area*

A pebble is dropped into a calm pool of water, causing ripples in the form of concentric circles, as shown in Figure 2.36. The radius r of the outer ripple is increasing at a constant rate of 1 foot per second. When the radius is 4 feet, at what rate is the total area A of the disturbed water changing?

Solution
The variables r and A are related by the equation for the area of a circle, $A = \pi r^2$. To solve this problem, use the fact that the rate of change of the radius is given by dr/dt.

> $\qquad$ *Equation:* $A = \pi r^2$
>
> $\qquad$ *Given rate:* $\dfrac{dr}{dt} = 1$ when $r = 4$
>
> $\qquad\qquad$ *Find:* $\dfrac{dA}{dt}$ when $r = 4$

Using this model, you can proceed as in Example 1.

$$A = \pi r^2 \qquad\qquad\qquad \textit{Equation}$$

$$\frac{d}{dt}[A] = \frac{d}{dt}[\pi r^2] \qquad\qquad \textit{Differentiate with respect to t.}$$

$$\frac{dA}{dt} = 2\pi r \frac{dr}{dt} \qquad\qquad \textit{Chain Rule}$$

When $r = 4$ and $dr/dt = 1$, you have

$$\frac{dA}{dt} = 2\pi r \frac{dr}{dt} = 2\pi(4)(1) = 8\pi \ \text{ft}^2/\text{sec}.$$

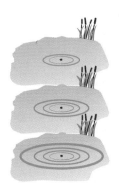

FIGURE 2.36

Note In Example 2, note that the radius changes at a *constant* rate ($dr/dt = 1$ for all t), but the area changes at a *nonconstant* rate.

When $r = 1$ ft	When $r = 2$ ft	When $r = 3$ ft	When $r = 4$ ft
$\dfrac{dA}{dt} = 2\pi$ ft^2/sec	$\dfrac{dA}{dt} = 4\pi$ ft^2/sec	$\dfrac{dA}{dt} = 6\pi$ ft^2/sec	$\dfrac{dA}{dt} = 8\pi$ ft^2/sec

The solution shown in Example 2 illustrates the steps for solving a related-rate problem.

Guidelines for Solving a Related-Rate Problem

1. Assign symbols to all *given* quantities and all quantities *to be determined*. Make a sketch and label the quantities if feasible.
2. Write an equation that relates all variables whose rates of change are either given or to be determined.
3. Use the Chain Rule to differentiate both sides of the equation *with respect to time*.
4. Substitute into the resulting equation all known values of the variables and their rates of change. Then solve for the required rate of change.

Note Be sure you notice the order of Steps 3 and 4 in the guidelines. You do not substitute the known values for the variables until after you have differentiated.

In Step 2 of the guidelines, note that you must write an equation that relates the given variables. To help you with this step we have included, in the back of the book, reference tables that summarize many common formulas. For instance, the volume of a sphere of radius r is given by the formula

$$V = \frac{4}{3}\pi r^3,$$

as listed in the Appendix in Reference Table 15.

The following table shows the mathematical models for some common rates of change which can be used in the first step of the solution of a related-rate problem.

Verbal Statement	Mathematical Model
The velocity of a car after traveling 1 hour is 50 miles per hour.	$x =$ distance traveled $\frac{dx}{dt} = 50$ when $t = 1$
Water is being pumped into a swimming pool at the rate of 10 cubic feet per minute.	$V =$ volume of water in pool $\frac{dV}{dt} = 10$ ft³/min
A population of bacteria is increasing at the rate of 2000 per hour.	$x =$ number in population $\frac{dx}{dt} = 2000$ bacteria per hour
Revenue is increasing at the rate of $4000 per month.	$R =$ revenue $\frac{dR}{dt} = 4000$ dollars per month

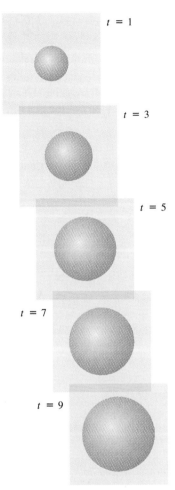

$t = 1$

$t = 3$

$t = 5$

$t = 7$

$t = 9$

FIGURE 2.37 Expanding Balloon

EXAMPLE 3 *Inflating a Balloon*

Air is being pumped into a spherical balloon at the rate of 4.5 cubic inches per minute, as indicated in Figure 2.37. Find the rate of change of the radius when the radius is 2 inches.

Solution

Let V represent the volume of the balloon and let r represent the radius. Because the volume is increasing at the rate of 4.5 cubic inches per minute, you know that $dV/dt = 4.5$. An equation that relates V and r is $V = \frac{4}{3}\pi r^3$. Thus, the problem can be represented by the following model.

$$\text{Equation: } V = \frac{4}{3}\pi r^3$$

$$\text{Given rate: } \frac{dV}{dt} = 4.5$$

$$\text{Find: } \frac{dr}{dt} \text{ when } r = 2$$

By differentiating the equation, you obtain

$$V = \frac{4}{3}\pi r^3 \qquad \textit{Equation}$$

$$\frac{d}{dt}[V] = \frac{d}{dt}\left[\frac{4}{3}\pi r^3\right] \qquad \textit{Differentiate with respect to t.}$$

$$\frac{dV}{dt} = \frac{4}{3}\pi(3r^2)\frac{dr}{dt} \qquad \textit{Chain Rule}$$

$$\frac{1}{4\pi r^2}\frac{dV}{dt} = \frac{dr}{dt}. \qquad \textit{Solve for dr/dt.}$$

When $r = 2$ and $dV/dt = 4.5$, the rate of change of the radius is

$$\frac{dr}{dt} = \frac{1}{4\pi r^2}\frac{dV}{dt} = \frac{1}{4\pi(2^2)}(4.5) \approx 0.09 \text{ in./min.}$$

In Example 3, note that the volume is increasing at a *constant* rate but the radius is increasing at a *variable* rate. In this particular example, the radius is increasing more and more slowly as t increases. This is illustrated in Figure 2.37 and Table 2.7.

TABLE 2.7

t	1	3	5	7	9	11
$V = 4.5t$	4.5	13.5	22.5	31.5	40.5	49.5
$r = \sqrt[3]{\dfrac{3V}{4\pi}}$	1.02	1.48	1.75	1.96	2.13	2.28
$\dfrac{dr}{dt}$	0.34	0.16	0.12	0.09	0.08	0.07

EXAMPLE 4 *Tracking an Airplane by Radar*

An airplane is flying at an altitude of 6 miles on a flight path that will take it directly over a radar tracking station. Let s represent the distance (in miles) between the radar station and the plane. The distance s is decreasing at a rate of 400 miles per hour when s is 10 miles. What is the velocity of the plane?

Solution

Begin by drawing a figure and labeling the distances, as shown in Figure 2.38. The Pythagorean Theorem can be used to write an equation that relates x and s.

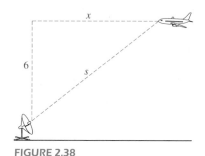

FIGURE 2.38

$$\text{Equation: } x^2 + 6^2 = s^2$$

$$\text{Given rate: } \frac{ds}{dt} = -400 \text{ when } s = 10$$

$$\text{Find: } \frac{dx}{dt} \text{ when } s = 10$$

By differentiating the equation, you obtain

$x^2 + 6^2 = s^2$	*Equation*
$\dfrac{d}{dt}[x^2 + 6^2] = \dfrac{d}{dt}[s^2]$	*Differentiate with respect to t.*
$2x\dfrac{dx}{dt} = 2s\dfrac{ds}{dt}$	*Chain Rule*
$\dfrac{dx}{dt} = \dfrac{s}{x}\left(\dfrac{ds}{dt}\right).$	*Solve for dx/dt.*

To find dx/dt, you must first find x when $s = 10$.

$x^2 + 6^2 = s^2$	*Equation*
$x^2 = s^2 - 36$	*Subtract 36 from both sides.*
$x = \sqrt{s^2 - 36}$	*Take square root of both sides.*
$x = \sqrt{10^2 - 36}$	*Substitute 10 for s.*
$x = \sqrt{64}$	
$x = 8$	*Solve for x.*

Using $s = 10$, $ds/dt = -400$, and $x = 8$, you can find dx/dt as follows.

$$\frac{dx}{dt} = \frac{s}{x}\left(\frac{ds}{dt}\right) = \frac{10}{8}(-400) = -500 \text{ mi/hr}$$

Because the velocity of the airplane is -500 miles per hour, it follows that its *speed* is 500 miles per hour. ▬

Note In Example 4, note that the rate of change of the distance x is negative because x is decreasing.

Real Life

EXAMPLE 5 *Increasing Production*

A company is increasing the production of a product at the rate of 200 units per week. The weekly demand function is modeled by

$$p = 100 - 0.001x,$$

where p is the price per unit and x is the number of units produced in a week. Find the rate of change of the revenue with respect to time when the weekly production is 2000 units.

Solution

Equation: $R = xp = x(100 - 0.001x) = 100x - 0.001x^2$

Given rate: $\dfrac{dx}{dt} = 200$

Find: $\dfrac{dR}{dt}$ when $x = 2000$

By differentiating the equation, you obtain

$$R = 100x - 0.001x^2 \qquad \text{\textit{Equation}}$$

$$\frac{d}{dt}[R] = \frac{d}{dt}[100x - 0.001x^2] \qquad \text{\textit{Differentiate with respect to t.}}$$

$$\frac{dR}{dt} = (100 - 0.002x)\frac{dx}{dt}. \qquad \text{\textit{Chain Rule}}$$

Using $x = 2000$ and $dx/dt = 200$, you have

$$\frac{dR}{dt} = (100 - 0.002x)\frac{dx}{dt}$$

$$= [100 - 0.002(2000)](200)$$

$$= \$19,200 \text{ per week.}$$

Discussion Problem   Extending the Example

In Example 5, the company is increasing its revenue for the product by decreasing the price. The number of units produced and the price per unit during the first 10 weeks are shown in Table 2.8.

TABLE 2.8

t	1	2	3	4	5	6	7	8	9	10
x	200	400	600	800	1000	1200	1400	1600	1800	2000
p	\$99.8	\$99.6	\$99.4	\$99.2	\$99.0	\$98.8	\$98.6	\$98.4	\$98.2	\$98.0

Is the revenue increasing during this 10-week period? For how many weeks can the company continue to drop the price before the revenue begins to decline?

Warm Up

The following warm-up exercises involve skills that were covered in earlier sections. You will use these skills in the exercise set for this section.

In Exercises 1–6, write a formula for the given quantity.

1. Area of a circle

2. Volume of a sphere

3. Surface area of a cube

4. Volume of a cube

5. Volume of a cone

6. Area of a triangle

In Exercises 7–10, find *dy/dx* by implicit differentiation.

7. $x^2 + y^2 = 9$

8. $3xy - x^2 = 6$

9. $x^2 + 2y + xy = 12$

10. $x + xy^2 - y^2 = xy$

EXERCISES 2.8

In Exercises 1–4, find the indicated values of *dy/dt* and *dx/dt*.

Equation		Find	Given
1. $y = x^2 - \sqrt{x}$	(a)	$\dfrac{dy}{dt}$	$x = 4, \dfrac{dx}{dt} = 8$
	(b)	$\dfrac{dx}{dt}$	$x = 16, \dfrac{dy}{dt} = 12$
2. $y = x^2 - 3x$	(a)	$\dfrac{dy}{dt}$	$x = 3, \dfrac{dx}{dt} = 2$
	(b)	$\dfrac{dx}{dt}$	$x = 1, \dfrac{dy}{dt} = 5$
3. $xy = 4$	(a)	$\dfrac{dy}{dt}$	$x = 8, \dfrac{dx}{dt} = 10$
	(b)	$\dfrac{dx}{dt}$	$x = 1, \dfrac{dy}{dt} = -6$
4. $x^2 + y^2 = 25$	(a)	$\dfrac{dy}{dt}$	$x = 3, y = 4, \dfrac{dx}{dt} = 8$
	(b)	$\dfrac{dx}{dt}$	$x = 4, y = 3, \dfrac{dy}{dt} = -2$

5. *Area* The radius *r* of a circle is increasing at a rate of 2 inches per minute. Find the rate of change of the area when (a) *r* = 6 inches and (b) *r* = 24 inches.

6. *Volume* The radius *r* of a sphere is increasing at a rate of 2 inches per minute. Find the rate of change of the volume when (a) *r* = 6 inches and (b) *r* = 24 inches.

7. *Area* Let *A* be the area of a circle of radius *r* that is changing with respect to time. If *dr/dt* is constant, is *dA/dt* constant? Explain your reasoning.

8. *Volume* Let *V* be the volume of a sphere of radius *r* that is changing with respect to time. If *dr/dt* is constant, is *dV/dt* constant? Explain your reasoning.

9. *Inflating Balloon* A spherical balloon is inflated with gas at the rate of 20 cubic feet per minute. How fast is the radius of the balloon changing at the instant the radius is (a) 1 foot and (b) 2 feet?

10. *Volume of Cone* The radius *r* of a right circular cone is increasing at a rate of 2 inches per minute. The height *h* of the cone is related to the radius by *h* = 3*r*. Find the rate of change of the volume when (a) *r* = 6 inches and (b) *r* = 24 inches.

11. *Falling Sand* Sand is falling off a conveyer and is forming a conical pile at the rate of 20 cubic feet per minute. The diameter of the base of the cone is approximately three times the height of the cone. At what rate is the height of the pile changing when the pile is 10 feet high?

12. *Water Tank* A conical tank (with vertex down) is 10 feet across the top and 12 feet deep. Water is flowing into the tank at a rate of 10 cubic feet per minute. Find the rate of change of the water depth when the depth is 8 feet.

13. *Expanding Cube* All edges of a cube are expanding at a rate of 3 centimeters per second. How fast is the volume changing when each edge is (a) 1 centimeter and (b) 10 centimeters?

14. *Expanding Cube* All edges of a cube are expanding at a rate of 3 centimeters per second. How fast is the surface area changing when each edge is (a) 1 centimeter and (b) 10 centimeters?

15. *Moving Point* A point is moving along the graph of $y = x^2$ so that dx/dt is 2 centimeters per minute. Find dy/dt for the following.
   (a) $x = -3$ (b) $x = 0$ (c) $x = 1$ (d) $x = 3$

16. *Moving Point* A point is moving along the graph of $y = 1/(1 + x^2)$ so that dx/dt is 2 centimeters per minute. Find dy/dt for the following.
   (a) $x = -2$ (b) $x = 2$ (c) $x = 0$ (d) $x = 10$

17. *Sliding Ladder* A 25-foot ladder is leaning against a house (see figure). The base of the ladder is pulled away from the house at a rate of 2 feet per second. How fast is the top of the ladder moving down the wall when the base is (a) 7 feet, (b) 15 feet, and (c) 24 feet from the house?

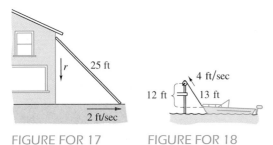

FIGURE FOR 17   FIGURE FOR 18

18. *Boat Speed* A boat is pulled by a winch on a dock, and the winch is 12 feet above the deck of the boat (see figure). The winch pulls the rope at a rate of 4 feet per second. Find the speed of the boat when 13 feet of rope is out. What happens to the speed of the boat as it gets closer and closer to the dock?

19. *Air Traffic Control* An air traffic controller spots two airplanes at the same altitude converging to a point as they fly at right angles to each other. One airplane is 150 miles from the point and has a speed of 450 miles per hour. The other is 200 miles from the point and has a speed of 600 miles per hour.
   (a) At what rate is the distance between the planes changing?
   (b) How much time does the controller have to get one of the airplanes on a different flight path?

20. *Airplane Speed* An airplane flying at an altitude of 6 miles passes directly over a radar antenna (see figure). When the airplane is 10 miles away ($s = 10$), the radar detects that the distance s is changing at a rate of 240 miles per hour. What is the speed of the airplane?

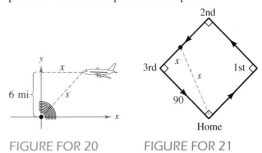

FIGURE FOR 20   FIGURE FOR 21

21. *Baseball* A (square) baseball diamond has sides that are 90 feet long (see figure). A player 26 feet from third base is running at a speed of 30 feet per second. At what rate is the player's distance from home plate changing?

22. *Environment* An accident at an oil drilling platform is causing a circular oil slick. Engineers determine that the slick is 0.08 feet thick, and when the radius is 750 feet, the slick is increasing at the rate of 0.5 feet per minute. At what rate (in cubic feet per minute) is oil flowing from the site of the accident?

23. *Profit* A company is increasing the production of a product at the rate of 25 units per week. The demand and cost functions for the product are
$$p = 50 - 0.01x \quad \text{and} \quad C = 4000 + 40x - 0.02x^2.$$
Find the rate of change of the profit with respect to time when the weekly sales are $x = 800$ units.

24. *Sales* The profit for a product is increasing at a rate of $6384 per week. The demand and cost functions for the product are
$$p = 6000 - 0.4x^2 \quad \text{and} \quad C = 2400x + 5200.$$
Find the rate of change of sales with respect to time when the weekly sales are $x = 44$ units.

25. *Drug Costs* The annual cost (in millions of dollars) for a government agency to seize p% of an illegal drug is
$$C = \frac{528p}{100 - p}, \quad 0 \le p < 100.$$
The agency's goal is to increase p by 5% per year. Find the rate of change of the cost when (a) $p = 30$% and (b) $p = 60$%.

Chapter Summary and Study Tips

After studying this chapter, you should have acquired the following skills. The exercise numbers are keyed to the Review Exercises that begin on page 166. Answers to odd-numbered Review Exercises are given in the back of the text. Worked-out solutions are listed in the *Student Study Guide*.

■ Approximate the slope of the tangent line to a graph at a point. (*Section 2.1*) Review Exercises 1–4

■ Interpret the slope of a graph in a real-life setting. (*Section 2.1*) Review Exercises 5, 6

■ Use the limit definition to find the slope of a graph at a point. (*Section 2.1*) Review Exercises 7–10

■ Use the limit definition to find the derivative of a function. (*Section 2.1*) Review Exercises 11–14

$$f'(x) = \lim_{\Delta x \to 0} \frac{f(x + \Delta x) - f(x)}{\Delta x}$$

■ Use the derivative to find the slope of a graph at a point. (*Section 2.1*) Review Exercises 15–18

■ Use the derivative to find an equation of a tangent line to a graph at a point. (*Section 2.1*) Review Exercises 19, 20

$$y - y_1 = m(x - x_1)$$

■ Use the graph of a function to recognize points at which the function is not differentiable. (*Section 2.1*) Review Exercises 21–24

■ Use the Constant Rule for differentiation. (*Section 2.2*) Review Exercises 25, 26

$$\frac{d}{dx}[c] = 0$$

■ Use the Simple Power Rule for differentiation. (*Section 2.2*) Review Exercises 27–30

$$\frac{d}{dx}[x^n] = nx^{n-1}$$

■ Use the Constant Multiple Rule for differentiation. (*Section 2.2*) Review Exercises 31–34

$$\frac{d}{dx}[cf(x)] = cf'(x)$$

■ Use the Sum and Difference Rules for differentiation. (*Section 2.2*) Review Exercises 35–38

$$\frac{d}{dx}[f(x) \pm g(x)] = f'(x) \pm g'(x)$$

■ Find the average rate of change of a function over an interval and the instantaneous rate of change at a point. (*Section 2.3*) Review Exercises 39, 40

$$\text{Average rate of change} = \frac{f(b) - f(a)}{b - a}$$

$$\text{Instantaneous rate of change} = \lim_{\Delta x \to 0} \frac{f(x + \Delta x) - f(x)}{\Delta x}$$

■ Find the average and instantaneous rates of change of a quantity in a real-life problem. (*Section 2.3*) Review Exercises 41, 42

*Several student study aids are available with this text. The *Student Solutions Guide* includes detailed solutions to all odd-numbered exercises, as well as practice chapter tests with answers. The *Graphics Calculator Guide* offers instructions on the use of a variety of graphing calculators and computer graphing software. The *Brief Calculus TUTOR* includes additional examples for selected exercises in the text.

- Find the velocity of an object that is moving in a straight line. (*Section 2.3*)
- Create mathematical models for the revenue, cost, and profit for a product. (*Section 2.3*)

$$P = R - C, \qquad R = xp$$

- Find the marginal revenue, marginal cost, and marginal profit for a product. (*Section 2.3*)
- Use the Product Rule for differentiation. (*Section 2.4*)

$$\frac{d}{dx}[f(x)g(x)] = f(x)g'(x) + g(x)f'(x)$$

- Use the Quotient Rule for differentiation. (*Section 2.4*)

$$\frac{d}{dx}\left[\frac{f(x)}{g(x)}\right] = \frac{g(x)f'(x) - f(x)g'(x)}{[g(x)]^2}$$

- Use the General Power Rule for differentiation. (*Section 2.5*)

$$\frac{d}{dx}[u^n] = nu^{n-1}u'$$

- Use differentiation rules efficiently to find the derivative of any algebraic function. Then simplify the result. (*Section 2.5*)
- Use derivatives to answer questions about real-life situations. (*Sections 2.1–2.5*)
- Find higher-order derivatives. (*Section 2.6*)
- Find and use the position function to determine the velocity and acceleration of a moving object. (*Section 2.6*)
- Find derivatives implicitly. (*Section 2.7*)
- Solve related rate problems. (*Section 2.8*)

- **Simplify Your Derivatives**   Often our students ask if they have to simplify their derivatives. Our answer is "Yes, if you expect to use them!" In the next chapter, you will see that almost all applications of derivatives require that the derivatives be written in simplified form. It is not difficult to see the advantage of a derivative in simplified form. Consider, for instance, the derivative of $f(x) = x/\sqrt{x^2 + 1}$. The "raw form" produced by the Quotient and Chain Rules

$$f'(x) = \frac{(x^2 + 1)^{1/2}(1) - (x)\left(\frac{1}{2}\right)(x^2 + 1)^{-1/2}(2x)}{(\sqrt{x^2 + 1})^2}$$

is obviously much more difficult to use than the simplified form

$$f'(x) = \frac{1}{(x^2 + 1)^{3/2}}.$$

- **List Units of Measure in Applied Problems**   When using derivatives in real-life applications, be sure to list the units of measure for each variable. For instance, if R is measured in dollars and t is measured in years, then the derivative dR/dt is measured in dollars per year.

Review Exercises

 means that technology can help you solve or check the exercise(s).

In Exercises 1–4, approximate the slope of the tangent line to the graph at (x, y).

1.

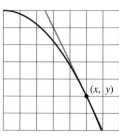

2.

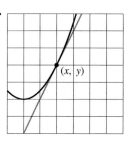

3.

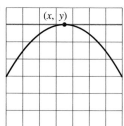

4.

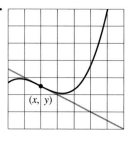

5. *Revenue* The graph below approximates the annual revenue (in millions of dollars) of Bethlehem Steel Corporation for the years 1985–1992, with $t = 0$ corresponding to 1980. Estimate the slope of the graph when $t = 7$ and $t = 10$. Interpret each slope in the context of the problem. *(Source: Bethlehem Steel Corporation)*

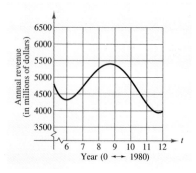

6. *Cellular Telephones* The graph below approximates the number of subscribers (in thousands) of cellular telephones for the years 1984–1991, with $t = 0$ corresponding to 1980. Estimate the slope of the graph when $t = 6$ and $t = 10$. Interpret each slope in the context of the problem. *(Source: Cellular Telecommunications Industry Association)*

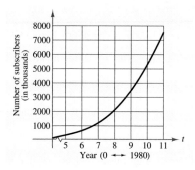

In Exercises 7–10, use the limit definition to find the slope of the tangent line to the graph of f at the indicated point.

7. $f(x) = -3x - 5$; $(-2, 1)$

8. $f(x) = x^2 + 10$; $(2, 14)$

9. $f(x) = \sqrt{x + 9}$; $(-5, 2)$

10. $f(x) = \dfrac{x + 1}{x}$; $(1, 2)$

In Exercises 11–14, use the limit definition to find the derivative of the function.

11. $f(x) = 7x + 3$

12. $f(x) = x^2 - 7x - 8$

13. $f(x) = \dfrac{1}{x - 5}$

14. $f(x) = \sqrt{x - 5}$

In Exercises 15–18, find the slope of the graph of f at the indicated point.

15. $f(x) = 8 - 5x$; $(3, -7)$

16. $f(x) = -\frac{1}{2}x^2 + 2x$; $(2, 2)$

17. $f(x) = \sqrt{x} + 2$; $(9, 5)$ **18.** $f(x) = \dfrac{5}{x}$; $(1, 5)$

In Exercises 19 and 20, use the derivative to find an equation of the tangent line to the graph of f at the indicated point.

19. $f(x) = \dfrac{x^2 + 3}{x}$; (1, 4)

20. $f(x) = -x^2 - 4x - 4$; (−4, −4)

In Exercises 21–24, determine the x-value at which the function is not differentiable.

21. $y = \dfrac{x+1}{x-1}$

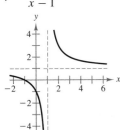

22. $y = -|x| + 3$

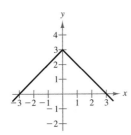

23. $y = \begin{cases} -x - 2, & x \le 0 \\ x^3 + 2, & x > 0 \end{cases}$

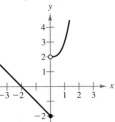

24. $y = (x+1)^{2/3}$

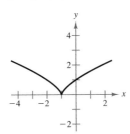

In Exercises 25–38, find the derivative of the function.

25. $f(x) = \sqrt{5}$

26. $f(x) = -10^2$

27. $y = x^5$

28. $g(x) = -\dfrac{1}{x^3}$

29. $y = \sqrt{x}$

30. $h(t) = \dfrac{1}{\sqrt{t}}$

31. $f(x) = 3x^4$

32. $f(x) = \dfrac{4}{x^4}$

33. $g(t) = \dfrac{2}{3t^2}$

34. $h(x) = \dfrac{2}{(3x)^2}$

35. $y = 11x^4 - 5x^2 + 1$

36. $y = x^3 - 5 + \dfrac{3}{x^3}$

37. $f(x) = \sqrt{x} - \dfrac{1}{\sqrt{x}}$

38. $f(x) = 2x^{-3} + 4 - \sqrt{x}$

In Exercises 39 and 40, find the average rate of change of the function over the indicated interval. Then compare the average rate of change with the instantaneous rates of change at the endpoints of the interval.

39. $f(x) = x^2 + 3x - 4$; [0, 1]

40. $f(x) = x^3 + x$; [−2, 2]

41. *Revenue* The annual revenue R (in millions of dollars) of Bethlehem Steel Corporation for the years 1984–1992 can be modeled by

$$R = 18.45t^4 - 650.94t^3 + 8292.83t^2 - 45{,}165.42t + 93{,}432.28,$$

where $t = 0$ corresponds to 1980. *(Source: Bethlehem Steel Corporation)*

(a) Find the average rate of change for the interval from 1986 to 1990.

(b) Find the instantaneous rate of change of the model in 1986 and 1990.

(c) Interpret the results of parts (a) and (b) in the context of the problem.

42. *Cellular Telephones* The number of subscribers S (in thousands) for cellular telephones for the years 1984–1991 can be modeled by

$$S = -2.3t^4 + 82.23t^3 - 865.34t^2 + 3890.15t + 6301.7,$$

where $t = 0$ corresponds to 1980. *(Source: Cellular Telecommunications Industry Association)*

(a) Find the average rate of change for the interval from 1986 to 1990.

(b) Find the instantaneous rate of change of the model in 1986 and 1990.

(c) Interpret the results of parts (a) and (b) in the context of the problem.

43. *Velocity* A rock is dropped from a tower on the Brooklyn Bridge, 276 feet above the East River. Let t represent the time in seconds.

   (a) Write a model for the position function (assume that air resistance is negligible).

   (b) Find the average velocity during the first 2 seconds.

   (c) Find the instantaneous velocity when $t = 2$ and $t = 3$.

   (d) How long will it take for the rock to hit the water?

   (e) When it hits the water, what is the rock's speed?

44. *Velocity* The straight-line distance s (in feet) traveled by an accelerating bicyclist can be modeled by

$$s = 2t^{3/2}, \qquad 0 \le t \le 8,$$

where t is the time (in seconds). Complete the table showing the velocity of the bicyclist at 2-second intervals.

Time, t	0	2	4	6	8
Velocity					

45. *Revenue, Cost, and Profit* The fixed cost of operating a small flower shop is $2500 per month. The average cost of a floral arrangement is $15 and the average price is $27.50. Write the monthly revenue, cost, and profit functions for the floral shop in terms of x, the number of arrangements sold.

46. *Profit* The weekly demand and cost functions for a product are

$$p = 1.89 - 0.0083x \quad \text{and} \quad C = 21 + 0.65x.$$

Write the profit function for this product.

Marginal Cost In Exercises 47 and 48, find the marginal cost function.

47. $C = 2500 + 320x$

48. $C = 475 + 5.25x^{2/3}$

Marginal Revenue In Exercises 49 and 50, find the marginal revenue function.

49. $R = \dfrac{35x}{\sqrt{x-2}}, \quad x \ge 6$

50. $R = x\left(5 + \dfrac{10}{\sqrt{x}}\right)$

Marginal Profit In Exercises 51 and 52, find the marginal profit function.

51. $P = -0.0002x^3 + 6x^2 - x - 2000$

52. $P = -\frac{1}{15}x^3 + 4000x^2 - 120x - 144{,}000$

In Exercises 53–76, find the derivative of the function. Simplify your result.

53. $f(x) = x^3(5 - 3x^2)$

54. $y = (3x^2+7)(x^2-2x)$

55. $y = (4x-3)(x^3-2x^2)$

56. $s = \left(4 - \dfrac{1}{t^2}\right)(t^2-3t)$

57. $f(x) = \dfrac{6x-5}{x^2+1}$

58. $f(x) = \dfrac{x^2+x-1}{x^2-1}$

59. $g(t) = \dfrac{(1/t)-2t}{t^3-1}$

60. $h(x) = \dfrac{4x^2+\sqrt{x}}{3x^2-2}$

61. $f(x) = (5x^2+2)^3$

62. $f(x) = \sqrt[3]{x^2-1}$

63. $h(x) = \dfrac{2}{\sqrt{x+1}}$

64. $g(x) = \sqrt{x^6-12x^3+9}$

65. $g(x) = x\sqrt{x^2+1}$

66. $g(t) = \dfrac{t}{(1-t)^3}$

67. $f(t) = (t+1)\sqrt{t^2+1}$

68. $y = 3x(x+2)^3$

69. $f(x) = -2(1-4x^2)^2$

70. $f(x) = \left(x^2+\dfrac{1}{x}\right)^5$

71. $h(x) = [x^2(2x+3)]^3$

72. $f(x) = [(x-2)(x+4)]^2$

73. $f(x) = x^2(x-1)^5$

74. $f(s) = s^3(s^2-1)^{5/2}$

75. $h(t) = \dfrac{\sqrt{3t+1}}{(1-3t)^2}$

76. $g(x) = \dfrac{(3x+1)^2}{(x^2+1)^2}$

 77. *Refrigeration* The temperature T (in degrees Fahrenheit) of food placed in a freezer can be modeled by

$$T = \dfrac{1300}{t^2+2t+25},$$

where t is the time (in hours).

   (a) Find the rate of change of T when $t = 1$, $t = 3$, $t = 5$, and $t = 10$.

   (b) Graph the model and describe the rate at which the temperature is changing.

78. *Forestry* According to the *Doyle Log Rule*, the volume V (in board feet) of a log of length L (feet) and diameter D (inches) at the small end is

$$V = \left(\dfrac{D-4}{4}\right)^2 L.$$

Find the rate at which the volume is changing for a 12-foot-long log whose smallest diameter is (a) 8 inches, (b) 16 inches, (c) 24 inches, and (d) 36 inches.

In Exercises 79–86, find the indicated derivative.

79. Given $f(x) = 3x^2 + 7x + 1$, find $f''(x)$.

80. Given $f'(x) = 5x^4 - 6x^2 + 2x$, find $f'''(x)$.

81. Given $f'''(x) = -\dfrac{6}{x^4}$, find $f^{(5)}(x)$.

82. Given $f(x) = \sqrt{x}$, find $f^{(4)}(x)$.

83. Given $f'(x) = 7x^{5/2}$, find $f''(x)$.

84. Given $f(x) = x^2 + \dfrac{3}{x}$, find $f''(x)$.

85. Given $f''(x) = 6\sqrt[3]{x}$, find $f'''(x)$.

86. Given $f'''(x) = 20x^4 - \dfrac{2}{x^3}$, find $f^{(5)}(x)$.

87. *Diver* A person dives from a 30-foot platform with an initial velocity of 5 feet per second (upward).

   (a) Find the position function of the diver.

   (b) How long will it take for the diver to hit the water?

   (c) What is the diver's velocity at impact?

   (d) What is the diver's acceleration at impact?

88. *Velocity and Acceleration* The position function of a particle is given by

$$s = \frac{1}{t^2 + 2t + 1},$$

where s is the height (in feet) and t is the time (in seconds). Find the velocity and acceleration functions.

In Exercises 89–96, use implicit differentiation to find dy/dx.

89. $x^2 + 3xy + y^3 = 10$

90. $x^2 + 9y^2 - 4x + 3y - 7 = 0$

91. $y^2 - x^2 = 49$

92. $y^2 + x^2 - 6y - 2x - 5 = 0$

93. $y^2 = (x - y)(x^2 + y)$ **94.** $y^2 - 3 = 2\sqrt{xy}$

95. $2\sqrt[3]{x} + 3\sqrt{y} = 10$ **96.** $y^3 - 2x^2y + 3xy^2 = -1$

97. *Water Level* A swimming pool is 40 feet long, 20 feet wide, 4 feet deep at the shallow end, and 9 feet deep at the deep end (see figure). Water is being pumped into the pool at the rate of 10 cubic feet per minute. How fast is the water level rising when there is 4 feet of water in the deep end?

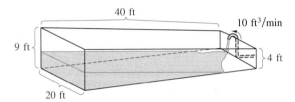

98. *Profit* The demand and cost functions for a product can be modeled by

$$p = 211 - 0.002x$$

and

$$C = 30x + 1{,}500{,}000,$$

where x is the number of units produced.

   (a) Write the profit function for this product.

   (b) Find the marginal profit when 80,000 units are produced.

   (c) Graph the profit function and use the graph to determine the price you would charge for the product. Explain your reasoning.

Sample Post-Graduation Exam Questions

The following questions were taken from certified public accountant (CPA) exams, graduate management admission tests (GMAT), graduate records exams (GRE), actuarial exams, or college-level academic skills tests (CLAST). The answers to the questions are given in the back of the book.

1. What is the length of the line segment that connects A to B (see graph)?
(a) 2   (b) 4   (c) $2\sqrt{2}$ (d) 6   (e) $\sqrt{3}$

For Questions 2–4, refer to the following table.

Participation in National Elections (millions of persons)

Characteristic	1984 Persons of voting age	1984 Percent voted	1988 Persons of voting age	1988 Percent voted	1992 Persons of voting age	1992 Percent voted
Total	170.0	59.9	178.1	57.4	185.7	61.3
Male	80.3	59.0	84.5	56.4	88.6	60.2
Female	89.6	60.8	93.6	58.3	97.1	62.3
Age 18 to 20	11.2	36.7	10.7	33.2	9.7	38.5
21 to 24	16.7	43.5	14.8	38.3	14.6	45.7
25 to 34	40.3	54.5	42.7	48.0	41.6	53.2
35 to 44	30.7	63.5	35.2	61.3	39.7	63.6
45 to 64	44.3	69.8	45.9	67.9	49.1	70.0
65 years and over	26.7	67.7	28.8	68.8	30.8	70.1

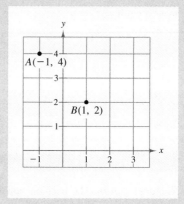

FIGURE FOR 1

2. Which of the following groups had the highest percent of voters in 1992?
(a) male   (b) age 35 to 44   (c) age 25 to 34
(d) age 18 to 20   (e) female

3. In 1984, what percent (to the nearest percent) of persons of voting age were female?
(a) 47   (b) 53   (c) 55   (d) 57   (e) 52

4. In 1988, how many males of voting age voted?
(a) 47,658,000   (b) 48,503,000   (c) 102,229,400
(d) 47,377,000   (e) 54,568,800

For Questions 5 and 6, refer to the following example.

Suppose the position s at time t of an accelerating object is given by $s(t) = t(t^2 - 2)^2$.

5. The acceleration function is
(a) $5t^4 - 12t^2 + 4$ (b) $5t^4 - 8t$ (c) $20t^3 - 24t$
(d) $20t^3 - 8$ (e) $t^5 - 4t^2 + 4$

6. The velocity of the object at time $t = 3$ is
(a) 301   (b) 381   (c) 147   (d) 532   (e) 468

$$\frac{d^2y}{dx^2} < 0$$

3

Applications of the Derivative

Calculus, like most other branches of mathematics, was created as a language for modeling real life. All of the early work in calculus dealt with real-life problems such as finding the maximum distance between the earth and the sun. The applied nature of calculus was evident in one of the first comprehensive texts on the subject, published in 1748 by the Italian mathematician Maria Gaetana Agnesi (1718–1799).

Today, calculus, combined with technology, is still an important tool for solving optimization problems, such as finding the minimum cost, the maximum profit, or the minimum distance.

$$f'(x) = 0$$

Agnesi

The Bettmann Archive

3.1 Increasing and Decreasing Functions

Increasing and Decreasing Functions ■ *Critical Numbers and Their Use* ■ *Extended Application: Profit, Revenue, and Cost*

Increasing and Decreasing Functions

A function is **increasing** if its graph moves up as x moves to the right and **decreasing** if its graph moves down as x moves to the right. The following definition states this more formally.

Definition of Increasing and Decreasing Functions

A function f is **increasing** on an interval if for any x_1 and x_2 in the interval

$$x_2 > x_1 \quad \text{implies} \quad f(x_2) > f(x_1).$$

A function f is **decreasing** on an interval if for any x_1 and x_2 in the interval

$$x_2 > x_1 \quad \text{implies} \quad f(x_2) < f(x_1).$$

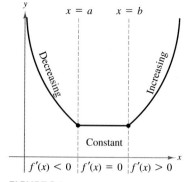

FIGURE 3.1

The function in Figure 3.1 is decreasing on the interval $(-\infty, a)$, constant on the interval (a, b), and increasing on the interval (b, ∞). Actually, from the definition of increasing and decreasing functions, the function shown in Figure 3.1 is decreasing on the interval $(-\infty, a]$ and increasing on the interval $[b, \infty)$. In this text, however, we restrict the discussion to finding *open* intervals on which a function is increasing or decreasing.

The derivative of a function can be used to determine whether the function is increasing or decreasing on an interval.

Test for Increasing or Decreasing Functions

Let f be differentiable on the interval (a, b).

1. If $f'(x) > 0$ for all x in (a, b), then f is increasing on (a, b).
2. If $f'(x) < 0$ for all x in (a, b), then f is decreasing on (a, b).
3. If $f'(x) = 0$ for all x in (a, b), then f is constant on (a, b).

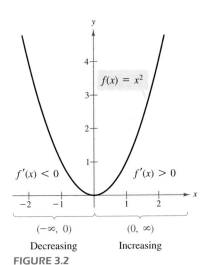

$f(x) = x^2$

$f'(x) < 0$ $f'(x) > 0$

$\underbrace{\qquad\qquad}_{(-\infty,\ 0)}$ $\underbrace{\qquad\qquad}_{(0,\ \infty)}$

Decreasing   Increasing

FIGURE 3.2

Discovery

Use a graphing utility to graph $f(x) = 2 - x^2$ and $f'(x) = -2x$ on the same coordinate plane. On what interval is f increasing? On what interval is f' positive? Describe how the first derivative can be used to determine where a function is increasing and decreasing. Repeat this analysis for $g(x) = x^3 - x$ and $g'(x) = 3x^2 - 1$.

EXAMPLE 1 *Testing for Increasing and Decreasing Functions*

Show that the function

$$f(x) = x^2$$

is decreasing on the open interval $(-\infty, 0)$ and increasing on the open interval $(0, \infty)$.

Solution

The derivative of f is

$$f'(x) = 2x.$$

On the open interval $(-\infty, 0)$, the fact that x is negative implies that $f'(x) = 2x$ is also negative. Hence, by the test for a decreasing function, you can conclude that f is *decreasing* on this interval. Similarly, on the open interval $(0, \infty)$, the fact that x is positive implies that $f'(x) = 2x$ is also positive. Hence, it follows that f is *increasing* on this interval, as shown in Figure 3.2.   ▪

Real Life

EXAMPLE 2 *Modeling Poultry Consumption*

From 1970 through 1990, the consumption C of poultry (in boneless pounds per person per year) can be modeled by

$$C = 33.5 + 0.074t^2, \qquad 0 \le t \le 20,$$

where $t = 0$ represents 1970 (see Figure 3.3). Show that the consumption of poultry was increasing from 1970 through 1990. *(Source: U.S. Department of Agriculture)*

Solution

The derivative of this model is $dC/dt = 0.148t$. As long as t is positive, the derivative is positive. Therefore, the function is increasing, which implies that the consumption of poultry was increasing from 1970 through 1990.

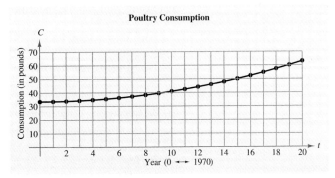

Poultry Consumption

FIGURE 3.3

Critical Numbers and Their Use

In Example 1, you were given two intervals—one on which the function is decreasing and one on which it is increasing. Suppose you had been asked to determine these intervals. To do this, you could have used the fact that for a continuous function, $f'(x)$ can change signs only at x-values where $f'(x) = 0$ or at x-values where $f'(x)$ is undefined, as shown in Figure 3.4. These two types of numbers are called the **critical numbers** of f.

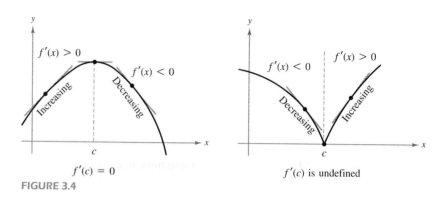

FIGURE 3.4

Definition of a Critical Number

Note This definition requires that a critical number be in the domain of the function.

If f is defined at c, then c is a critical number of f if $f'(c) = 0$ or if f' is undefined at c.

To determine the intervals on which a continuous function is increasing or decreasing, you can use the following guidelines.

Guidelines for Applying Increasing/Decreasing Test

1. Find the derivative of f.

2. Locate the critical numbers of f and use these numbers to determine test intervals. That is, find all x for which $f'(x) = 0$ or $f'(x)$ is undefined.

3. Test the sign of $f'(x)$ at an arbitrary number in each of the test intervals.

4. Use the test for increasing or decreasing functions to decide whether f is increasing or decreasing on each interval.

EXAMPLE 3 *Finding Increasing and Decreasing Intervals*

Find the open intervals on which the function

$$f(x) = x^3 - \tfrac{3}{2}x^2$$

is increasing or decreasing.

Solution

Begin by finding the derivative of f. Then set the derivative equal to zero and solve for the critical numbers.

$f'(x) = 3x^2 - 3x$	*Differentiate original function.*
$3x^2 - 3x = 0$	*Set derivative equal to 0.*
$3(x)(x - 1) = 0$	*Factor.*
$x = 0, \ x = 1$	*Critical numbers*

Because there are no x-values for which f' is undefined, it follows that $x = 0$ and $x = 1$ are the *only* critical numbers. Thus, the intervals that need to be tested are $(-\infty, 0)$, $(0, 1)$, and $(1, \infty)$. Table 3.1 summarizes the testing of these three intervals.

TABLE 3.1

Interval	$-\infty < x < 0$	$0 < x < 1$	$1 < x < \infty$
Test value	$x = -1$	$x = \tfrac{1}{2}$	$x = 2$
Sign of $f'(x)$	$f'(-1) = 6 > 0$	$f'\left(\tfrac{1}{2}\right) = -\tfrac{3}{4} < 0$	$f'(2) = 6 > 0$
Conclusion	Increasing	Decreasing	Increasing

The graph of f is shown in Figure 3.5. Note that the test values in the intervals were chosen for convenience—other x-values could have been used. ▬

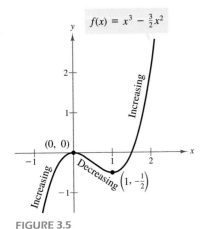

$f(x) = x^3 - \tfrac{3}{2}x^2$

(0, 0)

Increasing

Decreasing

Increasing

$\left(1, -\tfrac{1}{2}\right)$

FIGURE 3.5

📷 TECHNOLOGY

You can use the trace feature of a graphing utility to confirm the result of Example 3. Begin by graphing the function, as shown at the right. Then activate the trace feature and move the cursor from left to right. In intervals in which the function is increasing, note that the y-values increase as the x-values increase, whereas in intervals in which the function is decreasing, the y-values decrease as the x-values increase.

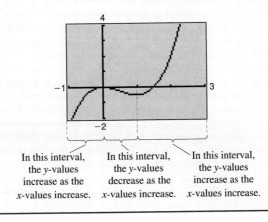

In this interval, the y-values increase as the x-values increase.

In this interval, the y-values decrease as the x-values increase.

In this interval, the y-values increase as the x-values increase.

Not only is the function in Example 3 continuous on the entire real line, it is also differentiable there. For such functions, the only critical numbers are those for which $f'(x) = 0$. The next example considers a continuous function that has *both* types of critical numbers—those for which $f'(x) = 0$ and those for which f' is undefined.

EXAMPLE 4 *Finding Increasing and Decreasing Intervals*

Find the open intervals on which the function

$$f(x) = (x^2 - 4)^{2/3}$$

is increasing or decreasing.

Solution

Begin by finding the derivative of the function.

$$f'(x) = \frac{2}{3}(x^2 - 4)^{-1/3}(2x) \qquad\qquad \text{\textit{Differentiate.}}$$

$$= \frac{4x}{3(x^2 - 4)^{1/3}} \qquad\qquad \text{\textit{Simplify.}}$$

From this, you can see that the derivative is zero when $x = 0$ and the derivative is undefined when $x = \pm2$. Thus, the critical numbers are

$$x = -2, \quad x = 0, \quad \text{and} \quad x = 2. \qquad \text{\textit{Critical numbers}}$$

This implies that the test intervals are

$$(-\infty, -2), \quad (-2, 0), \quad (0, 2), \quad \text{and} \quad (2, \infty). \quad \text{\textit{Test intervals}}$$

Table 3.2 summarizes the testing of these four intervals, and the graph of the function is shown in Figure 3.6.

TABLE 3.2

Interval	$-\infty < x < -2$	$-2 < x < 0$	$0 < x < 2$	$2 < x < \infty$
Test value	$x = -3$	$x = -1$	$x = 1$	$x = 3$
Sign of $f'(x)$	$f'(-3) < 0$	$f'(-1) > 0$	$f'(1) < 0$	$f'(3) > 0$
Conclusion	Decreasing	Increasing	Decreasing	Increasing

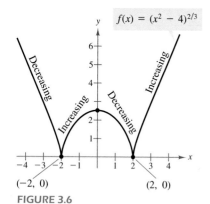

$f(x) = (x^2 - 4)^{2/3}$

$(-2, 0)$ $(2, 0)$

FIGURE 3.6

Note In Table 3.2, it is not necessary to *evaluate* $f'(x)$ at the test values—you need only to determine its sign. For example, you can determine the sign of $f'(-3)$ as follows.

$$f'(-3) = \frac{4(-3)}{3(9 - 4)^{1/3}} = \frac{\text{negative}}{\text{positive}} = \text{negative}$$

The functions in Examples 1–4 are continuous on the entire real line. If there are isolated x-values at which a function is not continuous, then these x-values should be used along with the critical numbers to determine the test intervals. For example, the function

$$f(x) = \frac{x^4 + 1}{x^2}$$

is not continuous when $x = 0$. Because the derivative of f,

$$f'(x) = \frac{2(x^4 - 1)}{x^3},$$

is zero when $x = \pm 1$, you should use the following numbers to determine the test intervals.

$x = -1, \ x = 1$ *Critical numbers*

$x = 0$ *Discontinuity*

After testing $f'(x)$, you can determine that the function is decreasing on the intervals $(-\infty, -1)$ and $(0, 1)$, and increasing on the intervals $(-1, 0)$ and $(1, \infty)$, as shown in Figure 3.7.

The converse of the test for increasing and decreasing functions is *not* true. For instance, it is possible for a function to be increasing on an interval even though its derivative is not positive at every point in the interval.

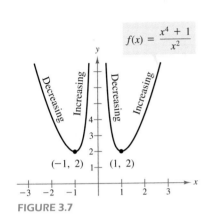

$$f(x) = \frac{x^4 + 1}{x^2}$$

$(-1, 2)$ $(1, 2)$

FIGURE 3.7

EXAMPLE 5 *Testing an Increasing Function*

Show that $f(x) = x^3 - 3x^2 + 3x$ is increasing on the entire real line.

Solution
From the derivative of f,

$$f'(x) = 3x^2 - 6x + 3 = 3(x - 1)^2,$$

you can see that the only critical number is $x = 1$. Thus, the test intervals are $(-\infty, 1)$ and $(1, \infty)$. Table 3.3 summarizes the testing of these two intervals. From Figure 3.8, you can see that f is increasing on the entire real line—even though $f'(1) = 0$. To convince yourself of this, look back at the definition of an increasing function.

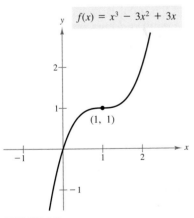

$f(x) = x^3 - 3x^2 + 3x$

$(1, 1)$

FIGURE 3.8

TABLE 3.3

Interval	$-\infty < x < 1$	$1 < x < \infty$
Test value	$x = 0$	$x = 2$
Sign of $f'(x)$	$f'(0) = 3(-1)^2 > 0$	$f'(2) = 3(2)^2 > 0$
Conclusion	Increasing	Increasing

Extended Application: Profit, Revenue, and Cost

EXAMPLE 6 *Profit Analysis*

Real Life

A national toy distributor determines the following cost and revenue models for one of its games.

$$C = 2.4x - 0.0002x^2, \qquad 0 \le x \le 6000$$

$$R = 7.2x - 0.001x^2, \qquad 0 \le x \le 6000$$

Determine the interval on which the profit function is increasing.

Solution

The profit for producing x units is

$$P = R - C$$
$$= (7.2x - 0.001x^2) - (2.4x - 0.0002x^2)$$
$$= 4.8x - 0.0008x^2.$$

To find the interval on which the profit is increasing, set the marginal profit P' equal to zero and solve for x.

$$P' = 4.8 - 0.0016x \qquad \text{\textit{Differentiate profit function.}}$$
$$4.8 - 0.0016x = 0 \qquad \text{\textit{Set } } P' \text{ \textit{equal to 0.}}$$
$$-0.0016x = -4.8 \qquad \text{\textit{Subtract 4.8 from both sides.}}$$
$$x = \frac{-4.8}{-0.0016} \qquad \text{\textit{Divide both sides by } } -0.0016.$$
$$x = 3000 \text{ units} \qquad \text{\textit{Simplify.}}$$

In the interval $(0, 3000)$, P' is positive and the profit is *increasing*. In the interval $(3000, 6000)$, P' is negative and the profit is *decreasing*. The graphs of the cost, revenue, and profit functions are shown in Figure 3.9. ■

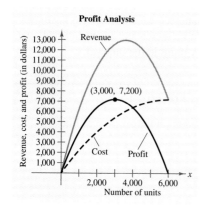

Profit Analysis

FIGURE 3.9

Discussion Problem   **Comparing Cost, Revenue, and Profit**

Use the models given in Example 6 to answer the following questions.

1. What is the demand function for the product described in the example?
2. What price would you set to obtain a maximum profit?
3. What price would you set to obtain a maximum revenue?
4. Why doesn't the maximum revenue occur at the same x-value as the maximum profit?

Warm Up

The following warm-up exercises involve skills that were covered in earlier sections. You will use these skills in the exercise set for this section.

In Exercises 1–4, solve the equation.

1. $x^2 = 8x$ **2.** $15x = \dfrac{5}{8}x^2$ **3.** $\dfrac{x^2 - 25}{x^3} = 0$ **4.** $\dfrac{2x}{\sqrt{1 - x^2}} = 0$

In Exercises 5–8, find the domain of the expression.

5. $\dfrac{x + 3}{x - 3}$ **6.** $\dfrac{2}{\sqrt{1 - x}}$ **7.** $\dfrac{2x + 1}{x^2 - 3x - 10}$ **8.** $\dfrac{3x}{\sqrt{9 - 3x^2}}$

In Exercises 9 and 10, evaluate the expression when $x = -2$, 0, and 2.

9. $\dfrac{4}{3}x^3 - 6x^2 + \dfrac{1}{3}$ **10.** $\dfrac{2x - 3}{x^2 - 3x - 1}$

E X E R C I S E S   3.1   means that technology can help you solve or check the exercise(s).

In Exercises 1–4, evaluate the derivative of the function at the indicated points on the graph.

1. $f(x) = \dfrac{x^2}{x^2 + 4}$

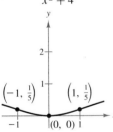

2. $f(x) = x + \dfrac{32}{x^2}$

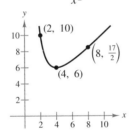

3. $f(x) = (x + 2)^{2/3}$

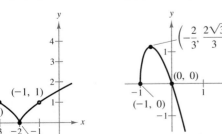

4. $f(x) = -3x\sqrt{x + 1}$

In Exercises 5–8, identify the open intervals on which the function is increasing or decreasing.

5. $f(x) = -(x + 1)^2$

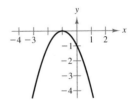

6. $f(x) = \dfrac{x^3}{4} - 3x$

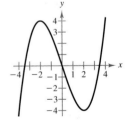

7. $f(x) = x^4 - 2x^2$

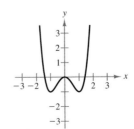

8. $f(x) = x^2/(x + 1)$

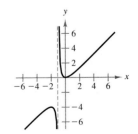

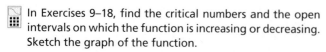

In Exercises 9–18, find the critical numbers and the open intervals on which the function is increasing or decreasing. Sketch the graph of the function.

9. $f(x) = 2x - 3$

10. $f(x) = 5 - 3x$

11. $g(x) = -(x - 1)^2$

12. $g(x) = (x + 2)^2$

13. $y = x^2 - 5x$

14. $y = -(x^2 - 2x)$

15. $y = x^3 - 6x^2$

16. $y = (x - 2)^3$

17. $f(x) = -(x + 1)^3$

18. $f(x) = \sqrt{4 - x^2}$

In Exercises 19–30, find the critical numbers and the open intervals on which the function is increasing or decreasing.

19. $f(x) = -2x^2 + 4x + 3$

20. $f(x) = x^2 + 8x + 10$

21. $y = 3x^3 + 12x^2 + 15x$

22. $y = (x - 1)^2(x + 2)$

23. $h(x) = x^{2/3}$

24. $h(x) = \sqrt[3]{x - 1}$

25. $f(x) = x^4 - 2x^3$

26. $f(x) = x\sqrt{x + 1}$

27. $f(x) = 2x\sqrt{3 - x}$

28. $f(x) = 2(x+2)\sqrt{1 - x}$

29. $f(x) = \dfrac{x}{x^2 + 4}$

30. $f(x) = \dfrac{x^2}{x^2 + 4}$

In Exercises 31–36, find the critical numbers and the open intervals on which the function is increasing or decreasing. (*Hint:* Check for discontinuities.)

31. $f(x) = x + \dfrac{1}{x}$

32. $f(x) = \dfrac{x}{x + 1}$

33. $f(x) = \dfrac{2x}{16 - x^2}$

34. $y = \begin{cases} 2x + 1, & x \le -1 \\ x^2 - 2, & x > -1 \end{cases}$

35. $y = \begin{cases} 4 - x^2, & x \le 0 \\ -2x, & x > 0 \end{cases}$

36. $y = \begin{cases} -x^3 + 1, & x \le 0 \\ -x^2 + 2x, & x > 0 \end{cases}$

37. *Ordering and Transportation Cost* The ordering and transportation cost C (in hundreds of dollars) for an automobile dealership is

$$C = 10\left(\dfrac{1}{x} + \dfrac{x}{x + 3}\right), \qquad 1 \le x,$$

where x is the number of automobiles ordered.

(a) Find the intervals on which C is increasing or decreasing.

(b) Use a graphing utility to graph the cost function.

(c) Use the trace feature to determine the order sizes for which the cost is $900. Assuming that the revenue function is increasing for $x \ge 0$, which order size would you use? Explain your reasoning.

38. *Drug Concentration* A drug is administered to a patient. A model giving the drug concentration C (in milligrams per milliliter) in the patient's bloodstream over a 2-hour period is

$$C = 0.29483t + 0.04253t^2 - 0.00035t^3, \qquad 0 \le t \le 120,$$

where t is the time (in minutes).

(a) Find the intervals on which C is increasing or decreasing.

(b) Use a graphing utility to graph the function.

(c) Explain how to use the test for increasing and decreasing functions to determine that the concentration will never reach 120 milligrams.

Position Function In Exercises 39 and 40, the position function gives the height s (in feet) of a ball, where the time t is measured in seconds. Find the time interval in which the ball is rising and the interval in which it is falling.

39. $s = 96t - 16t^2, \qquad 0 \le t \le 6$

40. $s = -16t^2 + 64t, \qquad 0 \le t \le 4$

41. *Bankers* The number y (in thousands) of people employed in banking and financing can be modeled by

$$y = -1.27t^3 + 6.63t^{5/2} + 1706.88, \qquad 0 \le t \le 21,$$

where $t = 0$ corresponds to 1970. (*Source: U.S. Bureau of Labor Statistics*)

(a) Use a graphing utility to graph the model. Then graphically estimate the years during which the model is increasing and the years during which it is decreasing.

(b) Use the test for increasing and decreasing functions to verify the result of part (a).

42. *Profit* The profit P made by a cinema for selling x bags of popcorn can be modeled by

$$P = 2.36x - \dfrac{x^2}{25{,}000} - 3500, \qquad 0 \le x \le 50{,}000.$$

(a) Find the intervals on which P is increasing or decreasing.

(b) If you owned the cinema, what price would you charge to obtain a maximum profit for popcorn? Explain.

In Exercises 43 and 44, use a graphing utility to (a) sketch the graphs of f and f' on the same viewing rectangle, (b) find the critical numbers of f, and (c) find the intervals on which f' is positive and on which it is negative. Describe the behavior of f relative to the sign of f'.

Function	*Interval*
43. $f(x) = 2x\sqrt{9 - x^2}$	$[-3, 3]$
44. $f(x) = 10\left(5 - \sqrt{x^2 - 3x + 16}\right)$	$[0, 5]$

3.2 Extrema and the First-Derivative Test

Relative Extrema ■ *The First-Derivative Test* ■ *Absolute Extrema* ■ *Applications of Extrema*

Relative Extrema

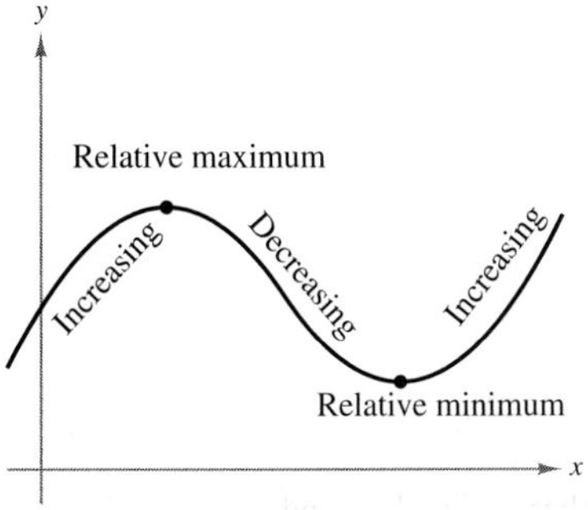

In the preceding section, you used the derivative to determine the intervals on which a function is increasing or decreasing. In this section, you will examine the points at which a function changes from increasing to decreasing, or vice versa. At such a point, the function has a **relative extremum.** (The plural of extremum is *extrema.*) The **relative extrema** of a function include the **relative minima** and **relative maxima** of the function. For instance, the function shown in Figure 3.10 has two relative extrema—the left point is a relative maximum and the right point is a relative minimum.

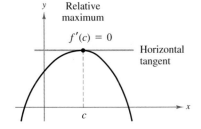

FIGURE 3.10

Definition of Relative Extrema

Let f be a function defined at c.

1. $f(c)$ is a **relative maximum** of f if there exists an interval (a, b) containing c such that $f(x) \leq f(c)$ for all x in (a, b).
2. $f(c)$ is a **relative minimum** of f if there exists an interval (a, b) containing c such that $f(x) \geq f(c)$ for all x in (a, b).

If $f(c)$ is a relative extremum of f, then the relative extremum is said to *occur* at $x = c$.

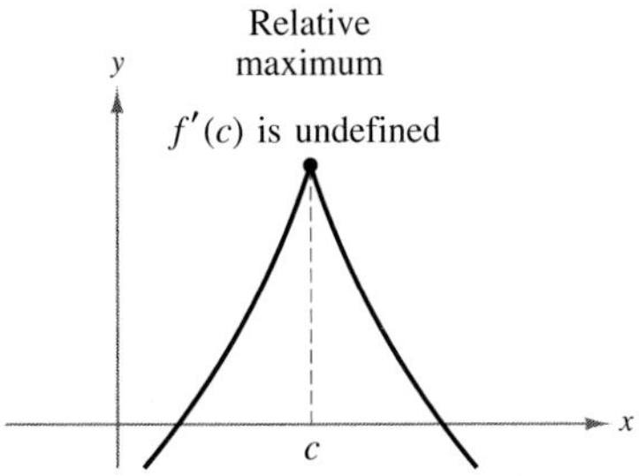

For a continuous function, the relative extrema must occur at critical numbers of the function, as shown in Figure 3.11.

Occurrence of Relative Extrema

If f has a relative minimum or relative maximum when $x = c$, then c is a critical number of f. That is, either

$$f'(c) = 0 \quad \text{or} \quad f'(c) \text{ is undefined.}$$

FIGURE 3.11

181

The First-Derivative Test

The result on the previous page implies that in your search for relative extrema of a continuous function, you need only test the critical numbers of the function. Once you have determined that c is a critical number of a function f, the **First-Derivative Test** for relative extrema enables you to classify $f(c)$ as a relative minimum, a relative maximum, or neither.

First-Derivative Test for Relative Extrema

Let f be continuous on the interval (a, b) in which c is the only critical number. If f is differentiable on the interval (except possibly at c), then $f(c)$ can be classified as a relative minimum, a relative maximum, or neither, as follows.

1. In the interval (a, b), if $f'(x)$ is negative to the left of $x = c$ and positive to the right of $x = c$, then $f(c)$ is a relative minimum.
2. In the interval (a, b), if $f'(x)$ is positive to the left of $x = c$ and negative to the right of $x = c$, then $f(c)$ is a relative maximum.
3. In the interval (a, b), if $f'(x)$ has the same sign to the left and right of $x = c$, then $f(c)$ is not a relative extremum of f.

A graphical interpretation of the First-Derivative Test is shown below in Figure 3.12.

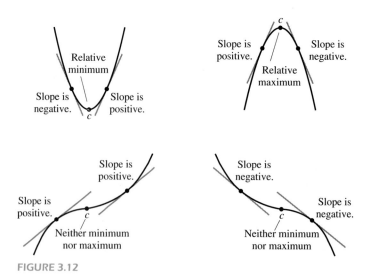

FIGURE 3.12

EXAMPLE 1 *Finding Relative Extrema*

Find all relative extrema of the function

$$f(x) = 2x^3 - 3x^2 - 36x + 14.$$

Solution

Begin by finding the critical numbers of f.

$$f'(x) = 6x^2 - 6x - 36 \qquad \textit{Find derivative of } f.$$

$$6x^2 - 6x - 36 = 0 \qquad\qquad \textit{Set derivative equal to } 0.$$

$$6(x^2 - x - 6) = 0 \qquad\qquad \textit{Factor out common factor.}$$

$$6(x - 3)(x + 2) = 0 \qquad\qquad \textit{Factor.}$$

$$x = -2, \ 3 \qquad\qquad \textit{Critical numbers}$$

Because $f'(x)$ is defined for all x, the only critical numbers of f are $x = -2$ and $x = 3$. Using these numbers, you can form the three test intervals $(-\infty, -2)$, $(-2, 3)$, and $(3, \infty)$. The testing of the three intervals is shown in Table 3.4.

TABLE 3.4

Interval	$-\infty < x < -2$	$-2 < x < 3$	$3 < x < \infty$
Test value	$x = -3$	$x = 0$	$x = 4$
Sign of $f'(x)$	$f'(-3) = 36 > 0$	$f'(0) = -36 < 0$	$f'(4) = 36 > 0$
Conclusion	Increasing	Decreasing	Increasing

Using the First-Derivative Test, you can conclude that the critical number -2 yields a relative maximum [$f'(x)$ changes sign from positive to negative], and the critical number 3 yields a relative minimum [$f'(x)$ changes sign from negative to positive].

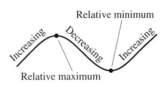

The graph of f is shown in Figure 3.13. To find the y-coordinates of the relative extrema, substitute the x-coordinates into the function. For instance, the relative maximum is $f(-2) = 58$ and the relative minimum is $f(3) = -67$.   ▬

Note In Section 2.2, Example 8, you examined the graph of the function $f(x) = x^3 - 4x + 2$ and discovered that it does *not* have a relative minimum at the point $(1, -1)$. Try using the First-Derivative Test to find the point at which the graph *does* have a relative minimum.

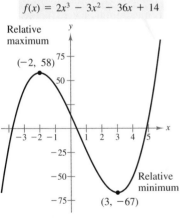

$f(x) = 2x^3 - 3x^2 - 36x + 14$

Relative maximum

$(-2, 58)$

Relative minimum

$(3, -67)$

FIGURE 3.13

In Example 1, both critical numbers yielded relative extrema. In the next example, only one of the two critical numbers yields a relative extremum.

EXAMPLE 2 *Finding Relative Extrema*

Find all relative extrema of the function

$$f(x) = x^4 - x^3.$$

Solution

From the derivative of the function,

$$f'(x) = 4x^3 - 3x^2 = x^2(4x - 3),$$

you can see that the function has only two critical numbers: $x = 0$ and $x = \frac{3}{4}$. These numbers produce the test intervals $(-\infty, 0)$, $\left(0, \frac{3}{4}\right)$, and $\left(\frac{3}{4}, \infty\right)$, which are tested in Table 3.5.

TABLE 3.5

Interval	$-\infty < x < 0$	$0 < x < \frac{3}{4}$	$\frac{3}{4} < x < \infty$
Test value	$x = -1$	$x = \frac{1}{2}$	$x = 1$
Sign of $f'(x)$	$f'(-1) = -7 < 0$	$f'\left(\frac{1}{2}\right) = -\frac{1}{4} < 0$	$f'(1) = 1 > 0$
Conclusion	Decreasing	Decreasing	Increasing

By the First-Derivative Test, it follows that f has a relative minimum when $x = \frac{3}{4}$, as shown in Figure 3.14. Note that the critical number $x = 0$ does not yield a relative extremum. ▬

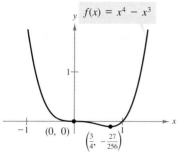

FIGURE 3.14

$f(x) = x^4 - x^3$

$(0, 0)$

$\left(\frac{3}{4}, -\frac{27}{256}\right)$

Relative minimum

EXAMPLE 3 *Finding Relative Extrema*

Find all relative extrema of the function

$$f(x) = 2x - 3x^{2/3}.$$

Solution

From the derivative of the function,

$$f'(x) = 2 - \frac{2}{x^{1/3}} = \frac{2(x^{1/3} - 1)}{x^{1/3}},$$

you can see that $f'(1) = 0$ and f' is undefined at $x = 0$. Therefore, the function has two critical numbers: $x = 1$ and $x = 0$. These numbers produce the test intervals

$$(-\infty, 0), \quad (0, 1), \quad \text{and} \quad (1, \infty).$$

By testing these intervals, you can conclude that f has a relative maximum at $(0, 0)$ and a relative minimum at $(1, -1)$, as shown in Figure 3.15. ▬

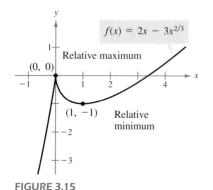

FIGURE 3.15

$f(x) = 2x - 3x^{2/3}$

Relative maximum

$(0, 0)$

$(1, -1)$ Relative minimum

▦ TECHNOLOGY

Finding Relative Extrema

There are several ways to use technology to find relative extrema of a function. One way is to use a graphing utility to graph the function, and then use the zoom and trace features to find the relative minimum and relative maximum points. For instance, consider the graph of $f(x) = 3.1x^3 - 7.3x^2 + 1.2x + 2.5$, as shown at the left. From the graph, you can see that the function has one relative maximum and one relative minimum. You can approximate the coordinates of these points by zooming in and using the trace feature, as shown below.

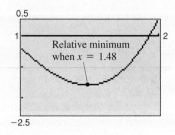

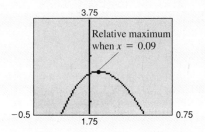

A second way to use technology to find relative extrema is to perform the First-Derivative Test with a symbolic differentiation utility. For instance, you can use *Derive* to differentiate the function, set the derivative equal to zero, and solve the resulting equation. After obtaining the critical numbers, 0.087015 and 1.482878, you can graph the function and observe that the first yields a relative maximum and the second yields a relative minimum.

1: $3.1x^3 - 7.3x^2 + 1.2x + 2.5$	*Author.*
2: $\dfrac{d}{dx}(3.1x^3 - 7.3x^2 + 1.2x + 2.5)$	*Differentiate.*
3: $\dfrac{93x^2}{10} - \dfrac{73x}{5} + \dfrac{6}{5}$	*Simplify.*
4: $x = \dfrac{73}{93} - \dfrac{\sqrt{4213}}{93}$	*Solve(1st solution).*
5: $x = \dfrac{\sqrt{4213}}{93} + \dfrac{73}{93}$	*Solve(2nd solution).*
6: $x = 0.0870148$	*Approximate(1st solution).*
7: $x = 1.48287$	*Approximate(2nd solution).*

Absolute Extrema

The terms *relative minimum* and *relative maximum* describe the *local* behavior of a function. To describe the *global* behavior of the function on an entire interval, you can use the terms **absolute maximum** and **absolute minimum.**

Definition of Absolute Extrema

Let f be defined on an interval I containing c.

1. $f(c)$ is an **absolute minimum of f** on I if $f(c) \leq f(x)$ for every x in I.
2. $f(c)$ is an **absolute maximum of f** on I if $f(c) \geq f(x)$ for every x in I.

The absolute minimum and absolute maximum values of a function on an interval are sometimes simply called the **minimum** and **maximum** of f on I.

Be sure that you understand the distinction between relative extrema and absolute extrema. For instance, in Figure 3.16, the function has a relative minimum that also happens to be an absolute minimum on the interval $[a, b]$. The relative maximum of f, however, is not the absolute maximum on the interval $[a, b]$. The next theorem points out that if a continuous function has a closed interval as its domain, then it *must* have both an absolute minimum and an absolute maximum on the interval. From Figure 3.16, note that these can occur at an endpoint of the interval.

Extreme Value Theorem

If f is continuous on $[a, b]$, then f takes on both a minimum value and a maximum value on $[a, b]$.

Although a continuous function has just one minimum and one maximum value on a closed interval, either of these values can occur for more than one x-value. For instance, on the interval $[-3, 3]$, the function

$$f(x) = 9 - x^2$$

has a minimum value of zero when $x = -3$ *and* when $x = 3$, as shown in Figure 3.17.

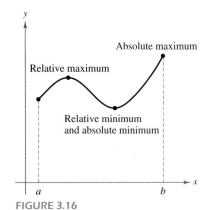

FIGURE 3.16

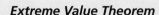

FIGURE 3.17

When looking for the extreme values of a function on a *closed* interval, remember that you must consider the values of the function at the endpoints as well as at the critical numbers of the function. You can use the following guidelines to find extrema on a closed interval.

▦ **TECHNOLOGY**

A graphing utility can help you locate the extrema of a function on a closed interval. For instance, try using a graphing utility to confirm the results of Example 4. (Set the viewing rectangle to $-1 \leq x \leq 6$ and $-8 \leq y \leq 4$.)

Guidelines for Finding Extrema on a Closed Interval

To find the extrema of a continuous function f on a closed interval $[a, b]$, use the following steps.

1. Evaluate f at each of its critical numbers in (a, b).
2. Evaluate f at each endpoint, a and b.
3. The least of these values is the minimum, and the greatest is the maximum.

EXAMPLE 4 *Finding Extrema on a Closed Interval*

Find the minimum and maximum values of

$$f(x) = x^2 - 6x + 2$$

on the interval $[0, 5]$.

Solution

Begin by finding the critical numbers of the function.

$$f'(x) = 2x - 6 \qquad \text{\textit{Find derivative of } f.}$$
$$2x - 6 = 0 \qquad \text{\textit{Set derivative equal to } 0.}$$
$$2x = 6 \qquad \text{\textit{Add 6 to both sides.}}$$
$$x = 3 \qquad \text{\textit{Solve for } x.}$$

From this, you can see that the only critical number of f is $x = 3$. Because this number lies in the interval under question, you should test the values of $f(x)$ at this number *and* at the endpoints of the interval, as shown in Table 3.6.

TABLE 3.6

x-value	Endpoint: $x = 0$	Critical number: $x = 3$	Endpoint: $x = 5$
$f(x)$	$f(0) = 2$	$f(3) = -7$	$f(5) = -3$
Conclusion	Maximum	Minimum	

From the table, you can see that the minimum of f on the interval $[0, 5]$ is $f(3) = -7$. Moreover, the maximum of f on the interval $[0, 5]$ is $f(0) = 2$. This is confirmed by the graph of f, as shown in Figure 3.18. ▬

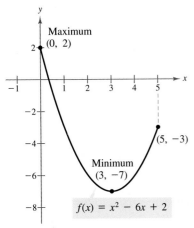

FIGURE 3.18

Applications of Extrema

Finding the minimum and maximum values of a function is one of the most common applications of calculus.

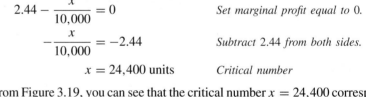

EXAMPLE 5   *Finding the Maximum Profit*

In Examples 7 and 8 in Section 2.3, we discussed a fast-food restaurant whose profit function for hamburgers is

$$P = 2.44x - \frac{x^2}{20,000} - 5000, \qquad 0 \le x \le 50,000.$$

Find the production level that produces a maximum profit.

Solution
Begin by setting the marginal profit equal to zero and solving for x.

$$P' = 2.44 - \frac{x}{10,000} \qquad \textit{Find marginal profit.}$$

$$2.44 - \frac{x}{10,000} = 0 \qquad \textit{Set marginal profit equal to 0.}$$

$$-\frac{x}{10,000} = -2.44 \qquad \textit{Subtract 2.44 from both sides.}$$

$$x = 24,400 \text{ units} \qquad \textit{Critical number}$$

From Figure 3.19, you can see that the critical number $x = 24,400$ corresponds to the production level that produces a maximum profit. To find the maximum profit, substitute $x = 24,400$ into the profit function.

$$P = 2.44x - \frac{x^2}{20,000} - 5000$$

$$= 2.44(24,400) - \frac{(24,400)^2}{20,000} - 5000$$

$$= \$24,768$$

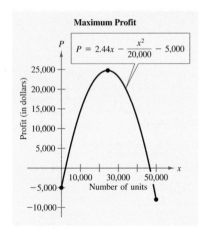

Maximum Profit

$$P = 2.44x - \frac{x^2}{20,000} - 5,000$$

FIGURE 3.19

Discussion Problem   **Setting the Price of a Product**

In Example 5, we discovered that a production level of 24,400 units corresponds to a maximum profit. Remember that this model assumes that the quantity demanded and the price per unit are related by a demand function. Thus, the only way to sell more hamburgers is to lower the price, and consequently lower the profit. What is the demand function for Example 5? What is the price per unit that produces a maximum profit?

Warm Up

The following warm-up exercises involve skills that were covered in earlier sections. You will use these skills in the exercise set for this section.

In Exercises 1–4, solve the equation $f'(x) = 0$.

1. $f(x) = 4x^4 - 2x^2 + 1$

2. $f(x) = \frac{1}{3}x^3 - \frac{3}{2}x^2 - 10x$

3. $f(x) = 5x^{4/5} - 4x$

4. $f(x) = \frac{1}{2}x^2 - 3x^{5/3}$

In Exercises 5–8, use $g(x) = -x^5 - 2x^4 + 4x^3 + 2x - 1$ to determine the sign of the derivative.

5. $g'(-4)$

6. $g'(0)$

7. $g'(1)$

8. $g'(3)$

In Exercises 9 and 10, decide whether the function is increasing or decreasing on the given interval.

9. $f(x) = 2x^2 - 11x - 6$, $(3, 6)$

10. $f(x) = x^3 + 2x^2 - 4x - 8$, $(-2, 0)$

E X E R C I S E S   3.2   means that technology can help you solve or check the exercise(s).

In Exercises 1–4, use a table similar to that in Example 1 to find all relative extrema of the function.

1. $f(x) = -2x^2 + 4x + 3$

2. $f(x) = x^2 + 8x + 10$

3. $f(x) = x^2 - 6x$

4. $f(x) = (x-1)^2(x+2)$

In Exercises 5–20, find all relative extrema of the function.

5. $g(x) = 6x^3 - 15x^2 + 12x$

6. $g(x) = \frac{1}{5}(x^5 - 5x)$

7. $h(x) = -(x+4)^3$

8. $h(x) = (x-3)^3$

9. $f(x) = x^3 - 6x^2 + 15$

10. $f(x) = x^4 - 1$

11. $f(x) = x^4 - 2x^3$

12. $f(x) = x^4 - 32x + 4$

13. $f(x) = x^{1/5} + 2$

14. $f(t) = (t - 1)^{1/3}$

15. $g(t) = t^{2/3}$

16. $h(t) = (t - 1)^{2/3}$

17. $f(x) = x + \dfrac{1}{x}$

18. $f(x) = \dfrac{x}{x+1}$

19. $h(x) = \dfrac{4}{x^2 + 1}$

20. $g(x) = 4\left(1 + \dfrac{1}{x} + \dfrac{1}{x^2}\right)$

In Exercises 21–30, find the absolute extrema of the function on the closed interval.

Function	Interval
21. $f(x) = 2(3 - x)$	$[-1, 2]$
22. $f(x) = \frac{1}{3}(2x + 5)$	$[0, 5]$
23. $f(x) = 5 - 2x^2$	$[0, 3]$
24. $f(x) = x^2 + 2x - 4$	$[-1, 1]$
25. $f(x) = x^3 - 3x^2$	$[-1, 3]$
26. $f(x) = x^3 - 12x$	$[0, 4]$
27. $f(x) = 3x^{2/3} - 2x$	$[-1, 2]$
28. $g(t) = \dfrac{t^2}{t^2 + 3}$	$[-1, 1]$
29. $h(s) = \dfrac{1}{3 - s}$	$[0, 2]$
30. $h(t) = \dfrac{t}{t - 2}$	$[3, 5]$

In Exercises 31–34, use a graphing utility to graphically find the absolute extrema of the function on the closed interval.

Function	Interval
31. $f(x) = 0.4x^3 - 1.8x^2 + x - 3$	$[0, 5]$
32. $f(x) = \frac{4}{3}x\sqrt{3 - x}$	$[0, 3]$
33. $f(x) = 3.2x^5 + 5x^3 - 3.5x$	$[0, 1]$
34. $f(x) = 4\sqrt{x} - 2x + 1$	$[0, 6]$

In Exercises 35–38, find the absolute extrema of the function on the interval $[0, \infty)$.

35. $f(x) = \dfrac{4x}{x^2 + 1}$ **36.** $f(x) = \dfrac{8}{x + 1}$

37. $f(x) = 3 - \dfrac{2}{x^2 - 4x + 5}$

38. $f(x) = \dfrac{4}{(x - 1)^2 + 1}$

In Exercises 39 and 40, find the maximum value of $|f''(x)|$ on the closed interval. (You will use this skill in Section 6.5 to estimate the error in the Trapezoidal Rule.)

Function	Interval
39. $f(x) = x^3(3x^2 - 10)$	$[0, 1]$
40. $f(x) = \dfrac{1}{x^2 + 1}$	$[0, 3]$

In Exercises 41 and 42, find the maximum value of $|f^{(4)}(x)|$ on the closed interval. (You will use this skill in Section 6.5 to estimate the error in Simpson's Rule.)

Function	Interval
41. $f(x) = 15x^4 - \left(\dfrac{2x - 1}{2}\right)^6$	$[0, 1]$
42. $f(x) = \dfrac{1}{x^2}$	$[1, 2]$

43. *Cost* A retailer has determined the cost C for ordering and storing x units of a product to be

$$C = 3x + \frac{20,000}{x}, \qquad 0 < x \le 200.$$

The delivery truck can bring at most 200 units per order. Find the order size that will minimize the cost.

44. *Profit* The quantity demanded x for a product is inversely proportional to the cube of the price p for $p > 1$. When the price is $10 per unit, the quantity demanded is eight units. The initial cost is $100 and the cost per unit is $4. What price will yield a maximum profit?

45. *Profit* When soft drinks sold for $0.80 per can at football games, approximately 6000 cans were sold. When the price was raised to $1.00 per can, the quantity demanded dropped to 5600. The initial cost is $5000 and the cost per unit is $0.40. Assuming that the demand function is linear, what price will yield a maximum profit?

46. *Trachea Contraction* Coughing forces the trachea (windpipe) to contract, which in turn affects the velocity of the air through the trachea. The velocity of the air during coughing can be modeled by

$$v = k(R - r)r^2, \qquad 0 \le r < R,$$

where k is a constant, R is the normal radius of the trachea, and r is the radius during coughing. What radius r will produce the maximum air velocity?

47. *Demographics* From 1940 to 1991, the number r of males to 100 females in the United States can be modeled by

$$r = 0.000045t^3 - 0.2295t + 100.84,$$

where $t = 0$ corresponds to 1940. *(Source: U.S. Bureau of the Census)* Determine the year in which the number r was a minimum. In that year, were there more females or more males in the population? Explain.

— *Business Capsule* —

AMERICAN DEMOGRAPHICS.

MAY 1993   A PUBLICATION OF DOW JONES & COMPANY INC.

A Guide to Interactivity

BUILDING BRANDS   PAGE 26

Courtesy of *American Demographics*

American Demographics magazine is a monthly publication that analyzes population attitudes and characteristics in the United States.

3.3 Concavity and the Second-Derivative Test

Concavity ■ *Points of Inflection* ■ *The Second-Derivative Test* ■
Extended Application: Diminishing Returns

Concavity

You already know that locating the intervals in which a function f increases or decreases is helpful in determining its graph. In this section, you will see that locating the intervals in which f' increases or decreases, can determine where the graph of f is curving upward or curving downward. This notion of curving upward or downward is defined formally as the **concavity** of the graph of the function.

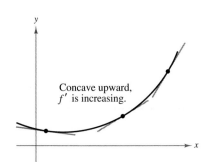

Concave upward, f' is increasing.

Definition of Concavity

Let f be differentiable on an open interval I. The graph of f is

1. **concave upward** on I if f' is increasing on the interval.
2. **concave downward** on I if f' is decreasing on the interval.

From Figure 3.20, you can observe the following graphical interpretation of concavity.

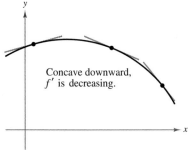

Concave downward, f' is decreasing.

1. A curve that is concave upward lies *above* its tangent line.
2. A curve that is concave downward lies *below* its tangent line.

This visual test for concavity is useful when the graph of a function is given. To determine concavity without seeing a graph, you need an analytic test. It turns out that you can use the second derivative to determine these intervals in much the same way that you used the first derivative to determine the intervals in which f is increasing or decreasing.

FIGURE 3.20

Test for Concavity

Let f be a function whose second derivative exists on an open interval I.

1. If $f''(x) > 0$ for all x in I, then f is concave upward on I.
2. If $f''(x) < 0$ for all x in I, then f is concave downward on I.

Discovery

Use a graphing utility to graph the function $f(x) = x^3 - x$ and its second derivative $f''(x) = 6x$ on the same coordinate plane. On what interval is f concave upward? On what interval is f'' positive? Describe how the second derivative can be used to determine where a function is concave upward and concave downward. Repeat this analysis for the functions $g(x) = x^4 - 6x^2$ and $g''(x) = 12x^2 - 12$.

For a *continuous* function f, you can find the intervals on which the graph of f is concave upward and concave downward as follows. [For a function that is not continuous, the test intervals should be formed using points of discontinuity, along with the points at which $f''(x)$ is zero or undefined.]

Guidelines for Applying Concavity Test

1. Locate the x-values at which $f''(x) = 0$ or $f''(x)$ is undefined.
2. Use these x-values to determine the test intervals.
3. Test the sign of $f''(x)$ in each test interval.

EXAMPLE 1 *Applying the Test for Concavity*

a. The graph of the function

$$f(x) = x^2 \qquad\qquad Function$$

is concave upward on the entire real line because its second derivative

$$f''(x) = 2 \qquad\qquad Second\ derivative$$

is positive for all x. (See Figure 3.21.)

b. The graph of the function

$$f(x) = \sqrt{x} \qquad\qquad Function$$

is concave downward for $x > 0$, because its second derivative

$$f''(x) = -\frac{1}{4}x^{-3/2} \qquad\qquad Second\ derivative$$

is negative for all $x > 0$. (See Figure 3.22.)

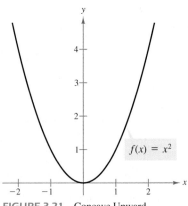

FIGURE 3.21 Concave Upward

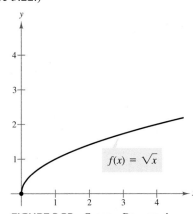

FIGURE 3.22 Concave Downward

EXAMPLE 2 *Determining Concavity*

Determine the intervals in which the graph of the function is concave upward or concave downward.

$$f(x) = \frac{6}{x^2 + 3}$$

Solution

Begin by finding the second derivative of f.

$$f(x) = 6(x^2 + 3)^{-1} \qquad\qquad \text{\textit{Original function}}$$

$$f'(x) = (-6)(2x)(x^2 + 3)^{-2} \qquad\qquad \text{\textit{Chain Rule}}$$

$$= \frac{-12x}{(x^2 + 3)^2} \qquad\qquad \text{\textit{Simplify.}}$$

$$f''(x) = \frac{(x^2 + 3)^2(-12) - (-12x)(2)(2x)(x^2 + 3)}{(x^2 + 3)^4} \qquad \text{\textit{Quotient Rule}}$$

$$= \frac{-12(x^2 + 3) + (48x^2)}{(x^2 + 3)^3} \qquad\qquad \text{\textit{Simplify.}}$$

$$= \frac{36(x^2 - 1)}{(x^2 + 3)^3} \qquad\qquad \text{\textit{Simplify.}}$$

From this, you can see that $f''(x)$ is defined for all real numbers and $f''(x) = 0$ when $x = \pm 1$. Thus, you can test the concavity of f by testing the intervals $(-\infty, -1)$, $(-1, 1)$, and $(1, \infty)$, as shown in Table 3.7. The graph of f is shown in Figure 3.23.

TABLE 3.7

Note In Example 2, f' is increasing on the interval $(1, \infty)$ even though f is decreasing there. Be sure you see that the increasing or decreasing of f' does not necessarily correspond to the increasing or decreasing of f.

Interval	$-\infty < x < -1$	$-1 < x < 1$	$1 < x < \infty$
Test value	$x = -2$	$x = 0$	$x = 2$
Sign of $f''(x)$	$f''(-2) > 0$	$f''(0) < 0$	$f''(2) > 0$
Conclusion	Concave upward	Concave downward	Concave upward

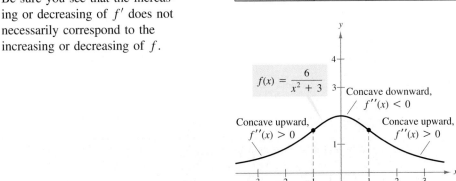

FIGURE 3.23

Discovery

Use a graphing utility to graph $f(x) = x^3 - 6x^2 + 12x - 6$ and $f''(x) = 6x - 12$ on the same coordinate plane. At what x-value does $f''(x) = 0$? At what x-value does the point of inflection occur? Repeat this analysis for $g(x) = x^4 - 5x^2 + 7$ and $g''(x) = 12x^2 - 10$. Make a general statement about the relationship of the point of inflection of a function and the second derivative of a function.

Points of Inflection

If the tangent line to a graph exists at a point at which the concavity changes, then the point is a **point of inflection.** Three examples of inflection points are shown in Figure 3.24. (Note that the third graph has a vertical tangent line at its point of inflection.)

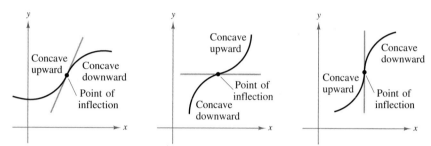

FIGURE 3.24   The graph *crosses* its tangent line at a point of inflection.

Definition of Point of Inflection

If the graph of a continuous function possesses a tangent line at a point where its concavity changes from upward to downward (or vice versa), then the point is a **point of inflection.**

Note   As shown in Figure 3.24, a graph crosses its tangent line at a point of inflection.

Because a point of inflection occurs where the concavity of a graph changes, it must be true that at such points the sign of f'' changes. Thus, to locate possible points of inflection, you need only determine the values of x for which $f''(x) = 0$ or for which $f''(x)$ does not exist. This parallels the procedure for locating the relative extrema of f by determining the critical numbers of f.

Property of Points of Inflection

If $(c, f(c))$ is a point of inflection of the graph of f, then either

$$f''(c) = 0 \quad \text{or} \quad f''(c) \text{ does not exist.}$$

$f(x) = x^4 + x^3 - 3x^2 + 1$

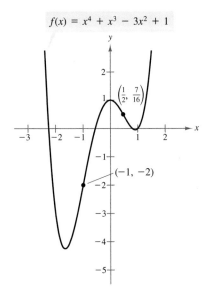

FIGURE 3.25   Two Points of Inflection

EXAMPLE 3 *Finding Points of Inflection*

Discuss the concavity of the graph of f and find its points of inflection.

$$f(x) = x^4 + x^3 - 3x^2 + 1$$

Solution

Begin by finding the second derivative of f.

$$f'(x) = 4x^3 + 3x^2 - 6x \qquad \text{\textit{Find first derivative.}}$$
$$f''(x) = 12x^2 + 6x - 6 \qquad \text{\textit{Find second derivative.}}$$
$$= 6(2x - 1)(x + 1) \qquad \text{\textit{Factor.}}$$

From this, you can see that the possible points of inflection occur at $x = \frac{1}{2}$ and $x = -1$. After testing the intervals

$$(-\infty, -1), \quad \left(-1, \frac{1}{2}\right), \quad \text{and} \quad \left(\frac{1}{2}, \infty\right),$$

you can determine that the graph is concave upward in $(-\infty, -1)$, concave downward in $\left(-1, \frac{1}{2}\right)$, and concave upward in $\left(\frac{1}{2}, \infty\right)$. Because the concavity changes at $x = -1$ and $x = \frac{1}{2}$, you can conclude that the graph has points of inflection at these x-values, as shown in Figure 3.25. ▪

It is possible for the second derivative to be zero at a point that is *not* a point of inflection. For example, compare the graphs of $f(x) = x^3$ and $g(x) = x^4$, as shown in Figure 3.26. Both second derivatives are zero when $x = 0$, but only the graph of f has a point of inflection at $x = 0$. This shows that before concluding that a point of inflection exists at a value of x for which $f''(x) = 0$, you must test to be certain that the concavity actually changes at the point.

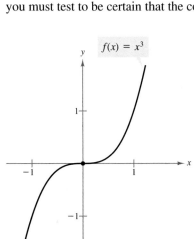

$f''(0) = 0$, and $(0, 0)$ is a point of inflection.

$g''(0) = 0$, but $(0, 0)$ is not a point of inflection.

FIGURE 3.26

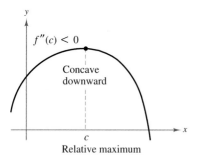

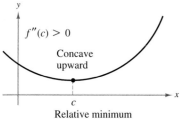

FIGURE 3.27

The Second-Derivative Test

The second derivative can be used to perform a simple test for relative minima and relative maxima. If f is a function such that $f'(c) = 0$ and the graph of f is concave upward at $x = c$, then $f(c)$ is a relative minimum of f. Similarly, if f is a function such that $f'(c) = 0$, and the graph of f is concave downward at $x = c$, then $f(c)$ is a relative maximum of f, as shown in Figure 3.27.

Second-Derivative Test

Let $f'(c) = 0$, and let f'' exist on an open interval containing c.

1. If $f''(c) > 0$, then $f(c)$ is a relative minimum.
2. If $f''(c) < 0$, then $f(c)$ is a relative maximum.
3. If $f''(c) = 0$, then the test fails. In such cases, you can use the First-Derivative Test to determine whether $f(c)$ is a relative minimum or relative maximum.

EXAMPLE 4 *Using the Second-Derivative Test*

Find the relative extrema of
$$f(x) = -3x^5 + 5x^3.$$

Solution
Begin by finding the first derivative of f.
$$f'(x) = -15x^4 + 15x^2 = 15x^2(1 - x^2)$$

From this derivative, you can see that $x = 0$, $x = -1$, and $x = 1$ are the only critical numbers of f. Using the second derivative,
$$f''(x) = -60x^3 + 30x,$$

you can apply the Second-Derivative Test, as follows.

Point	Sign of $f''(x)$	Conclusion
$(-1, -2)$	$f''(-1) = 30 > 0$	Relative minimum
$(0, 0)$	$f''(0) = 0$	Test fails.
$(1, 2)$	$f''(1) = -30 < 0$	Relative maximum

Because the test fails at $(0, 0)$, you can apply the First-Derivative Test to conclude that the point $(0, 0)$ is neither a relative minimum nor a relative maximum—a test for concavity would show that this point is a point of inflection. The graph of f is shown in Figure 3.28.

FIGURE 3.28

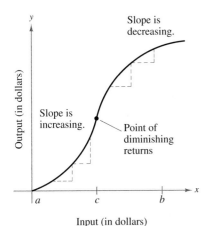

FIGURE 3.29

Extended Application: Diminishing Returns

In economics, the notion of concavity is related to the concept of **diminishing returns.** Consider a function

Output ⌐   ⌐ Input
↓   ↓
$$y = f(x),$$

where x measures the input (in dollars) and y measures the output (in dollars). In Figure 3.29, notice that the graph of this input-output function is concave upward in the interval (a, c) and is concave downward in the interval (c, b). In the interval (a, c), each additional dollar of input returns more than the previous input dollar. By contrast, in the interval (c, b), each additional dollar of input returns less than the previous input dollar. The point $(c, f(c))$ is called the **point of diminishing returns.** An increased investment beyond this point is usually considered a poor use of capital.

Real Life

EXAMPLE 5 *Exploring Diminishing Returns*

By increasing its advertising cost x (in thousands of dollars) for a product, a company discovers that it can increase the sales y (in thousands of dollars) according to the model

$$y = \frac{1}{10,000}(300x^2 - x^3), \qquad 0 \le x \le 200.$$

Find the point of diminishing returns for this product.

Solution

Begin by finding the first and second derivatives.

$$y' = \frac{1}{10,000}(600x - 3x^2) \qquad \text{*First derivative*}$$

$$y'' = \frac{1}{10,000}(600 - 6x) \qquad \text{*Second derivative*}$$

The second derivative is zero only when $x = 100$. By testing the intervals $(0, 100)$ and $(100, 200)$, you can conclude that the graph has a point of diminishing returns when $x = 100$, as shown in Figure 3.30. ■

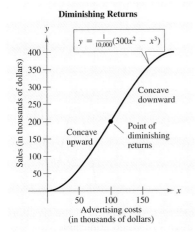

FIGURE 3.30

Discussion Problem   **Diminishing Returns**

Use the function in Example 5 to verify that additional dollars of input return more than the previous input dollars in the interval $(0, 100)$, and less than the previous input dollars in the interval $(100, 200)$.

Warm Up   The following warm-up exercises involve skills that were covered in earlier sections. You will use these skills in the exercise set for this section.

In Exercises 1–6, find the second derivative of the function.

1. $f(x) = 4x^4 - 9x^3 + 5x - 1$

2. $g(s) = (s^2 - 1)(s^2 - 3s + 2)$

3. $g(x) = (x^2 + 1)^4$

4. $f(x) = (x - 3)^{4/3}$

5. $h(x) = \dfrac{5}{(2x)^3}$

6. $f(x) = \dfrac{2x - 1}{3x + 2}$

In Exercises 7–10, find the critical numbers of the function.

7. $f(x) = 5x^3 - 5x + 11$

8. $f(x) = x^4 - 4x^3 - 10$

9. $g(t) = \dfrac{16 + t^2}{t}$

10. $h(x) = \dfrac{x^4 - 50x^2}{8}$

E X E R C I S E S   3.3   means that technology can help you solve or check the exercise(s).

In Exercises 1–6, find the intervals on which the graph is concave upward and those on which it is concave downward.

1. $y = x^2 - x - 2$

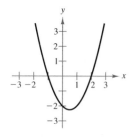

2. $y = -x^3 + 3x^2 - 2$

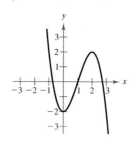

3. $f(x) = \dfrac{x^2 - 1}{2x + 1}$

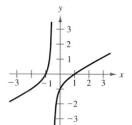

4. $f(x) = \dfrac{x^2 + 4}{4 - x^2}$

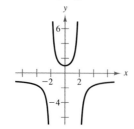

5. $f(x) = \dfrac{24}{x^2 + 12}$

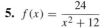

6. $y = x^5 + 5x^4 - 40x^2$

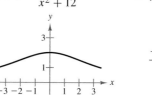

 In Exercises 7–18, find all relative extrema of the function. Use the Second-Derivative Test when applicable.

7. $f(x) = 6x - x^2$

8. $f(x) = x^2 + 3x - 8$

9. $f(x) = (x - 5)^2$

10. $f(x) = -(x - 5)^2$

11. $f(x) = x^3 - 5x^2 + 7x$

12. $f(x) = 5 + 3x^2 - x^3$

13. $f(x) = x^4 - 4x^3 + 2$

14. $f(x) = 3x^3 + 5x^2 - 2$

15. $f(x) = x^{2/3} - 3$

16. $f(x) = \sqrt{x^2 + 1}$

17. $f(x) = x + \dfrac{4}{x}$

18. $f(x) = \dfrac{x}{x - 1}$

In Exercises 19–22, state the sign of $f'(x)$ and $f''(x)$ in the interval $(0, 2)$.

19.
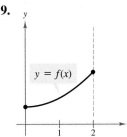

20.
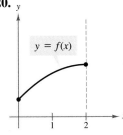

21.
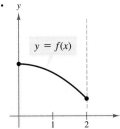

22.
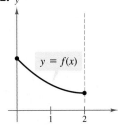

In Exercises 23–30, find the points of inflection of the graph.

23. $f(x) = x^3 - 9x^2 + 24x - 18$

24. $f(x) = x(6 - x)^2$

25. $f(x) = (x - 1)^3(x - 5)$

26. $f(x) = x^4 - 18x^2 + 5$

27. $g(x) = 2x^4 - 8x^3 + 12x^2 + 12x$

28. $f(x) = -4x^3 - 8x^2 + 32$

29. $h(x) = (x - 2)^3(x - 1)$

30. $f(t) = (1 - t)(t - 4)(t^2 - 4)$

In Exercises 31–42, sketch the graph of the function and identify all relative extrema and points of inflection.

31. $f(x) = x^3 - 12x$

32. $f(x) = x^3 + 1$

33. $f(x) = x^3 - 6x^2 + 12x$

34. $f(x) = x^3 - \frac{3}{2}x^2 - 6x$

35. $f(x) = \frac{1}{4}x^4 - 2x^2$

36. $f(x) = 2x^4 - 8x + 3$

37. $g(x) = (x - 2)(x + 1)^2$

38. $g(x) = (x - 6)(x + 2)^3$

39. $g(x) = x\sqrt{x + 3}$

40. $g(x) = \frac{x^3}{16}(x + 4)$

41. $f(x) = \frac{4}{1 + x^2}$

42. $f(x) = x - 4\sqrt{x + 1}$

In Exercises 43 and 44, sketch a graph of a function f having the given characteristics.

	Function	First Derivative	Second Derivative
43.	$f(2) = 0$ $f(4) = 0$	$f'(x) < 0, \ x < 3$ $f'(3) = 0$ $f'(x) > 0, \ x > 3$	$f''(x) > 0$
44.	$f(2) = 0$ $f(4) = 0$	$f'(x) > 0, \ x < 3$ $f'(3)$ is undefined $f'(x) < 0, \ x > 3$	$f''(x) > 0, \ x \neq 3$

In Exercises 45 and 46, use the graph to sketch the graph of f'. Find the intervals on which (a) $f'(x)$ is positive, (b) $f'(x)$ is negative, (c) f' is increasing, and (d) f' is decreasing. For each of these intervals, describe the corresponding behavior of f.

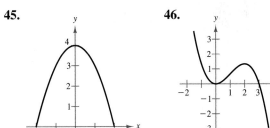

45.

46.

Point of Diminishing Returns In Exercises 47–50, identify the point of diminishing returns for the input-output function. For each function, R is the revenue and x is the amount spent on advertising.

47. $R = \dfrac{1}{50,000}(600x^2 - x^3), \quad 0 \leq x \leq 400$

48. $R = -\frac{4}{27}(x^3 - 33x^2 + 120x - 845), \quad 0 \leq x \leq 20$

49. $R = -\frac{4}{9}(x^3 - 9x^2 - 27), \quad 0 \leq x \leq 5$

50. $R = \dfrac{200x^2}{100 + x^2}, \quad 0 \leq x \leq 50$

Average Cost In Exercises 51–54, you are given the total cost of producing x units. Find the production level that minimizes the average cost per unit.

51. $C = 0.5x^2 + 15x + 5000$

52. $C = 0.16x^2 + 10x + 256$

53. $C = 0.002x^3 + 20x + 500$

54. $C = 0.001x^3 + 0.10x + 750$

Productivity In Exercises 55 and 56, consider a college student who works from 7 P.M. to 11 P.M. assembling mechanical components. The number N of components assembled after t hours is given by the function. At what time is the student assembling components at the greatest rate?

55. $N = -0.12t^3 + 0.54t^2 + 8.22t$, $0 \le t \le 4$

56. $N = \dfrac{20t^2}{4 + t^2}$, $0 \le t \le 4$

Sales Growth In Exercises 57 and 58, find the time t in years when the annual sales x of a new product are increasing at the greatest rate.

57. $x = \dfrac{10,000t^2}{9 + t^2}$ **58.** $x = \dfrac{500,000t^2}{36 + t^2}$

59. *Sales* A company introduces a new product line. The annual sales of the product increase over time, as shown in the graph. Visually locate the point of inflection of the graph. Describe its significance.

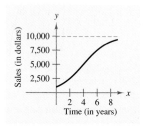

60. Show that the point of inflection of the graph of $f(x) = x(x - 6)^2$ lies midway between the relative extrema of f.

In Exercises 61–64, use a graphing utility to graph f, f', and f'' on the same viewing rectangle. Graphically locate the relative extrema and points of inflection of the graph.

Function	*Interval*
61. $f(x) = \dfrac{2x}{x^2 - x + 1}$	$[-2, 2]$
62. $f(x) = \dfrac{4x^2}{x^2 + 2}$	$[-3, 3]$
63. $f(x) = \frac{1}{2}x^3 - x^2 + 3x - 5$	$[0, 3]$
64. $f(x) = -\frac{1}{20}x^5 - \frac{1}{12}x^2 - \frac{1}{3}x + 1$	$[-2, 2]$

65. *Millionaires* The number N (in thousands) of millionaires having a net worth of x dollars (in millions) in 1986 can be modeled by

$$N = -\frac{4211.02}{x^2} + \frac{5897.24}{x^{3/2}} - 135.91.$$

Use a graphing utility to graph the function. *(Source: Internal Revenue Service)*

66. *Research Project* Use your school's library or some other reference source to research the financial history of a business person. Gather the data on the person's net worth over a period of time, and use a graphing utility to graph a scatter plot of the data. Fit a model to the data. Does the model appear to be concave up or down? Does it appear to be increasing or decreasing? Discuss the implications of your answers.

— *Business Capsule* —

©1993 Kathleen King

Bill Gates, president of Microsoft Corporation, was the wealthiest person in America in 1992. He wrote his first computer program at age 13, and scored a perfect 800 in math on his SAT test. In 1992, his net worth was $6.3 billion.

Optimization Problems

Solving Optimization Problems

Solving Optimization Problems

One of the most common applications of calculus is the determination of optimum (minimum or maximum) values. Before outlining a general method for solving optimization problems, we present an example.

Real Life

EXAMPLE 1 *Finding the Maximum Volume*

A manufacturer wants to design an open box that has a square base and a surface area of 108 square inches, as shown in Figure 3.31. What dimensions will produce a box with a maximum volume?

Solution

Because the base of the box is square, the volume is

$$V = x^2h. \qquad \textit{Primary equation}$$

This equation is called the **primary equation** because it gives a formula for the quantity to be optimized. The surface area of the box is

$$S = (\text{area of base}) + (\text{area of four sides})$$

$$108 = x^2 + 4xh. \qquad \textit{Secondary equation}$$

Because V is to be optimized, it helps to express V as a function of just one variable. To do this, solve the secondary equation for h in terms of x to obtain $h = (108 - x^2)/4x$ and substitute into the primary equation.

$$V = x^2h = x^2 \left(\frac{108 - x^2}{4x} \right) = 27x - \frac{1}{4}x^3 \qquad \textit{Function of one variable}$$

Before finding which x-value yields a maximum value of V, you need to determine the **feasible domain** of the function. That is, what values of x make sense in the problem? Because x must be nonnegative and the area of the base $(A = x^2)$ is at most 108, you can conclude that the feasible domain is

$$0 \le x \le \sqrt{108}. \qquad \textit{Feasible domain}$$

Using the techniques described in the first three sections of this chapter, you can determine that (in the interval $0 \le x \le \sqrt{108}$) this function has an absolute maximum when $x = 6$ inches and $h = 3$ inches.

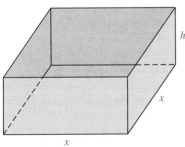

FIGURE 3.31 Open Box with Square Base
$S = x^2 + 4xh = 108$

In studying Example 1, be sure that you understand the basic question that it asks. Some students have trouble with optimization problems because they are too eager to start solving the problem by using a pat formula. For instance, in Example 1, you should realize that there are infinitely many open boxes having 108 square inches of surface area. You might begin to solve this problem by asking yourself which basic shape would seem to yield a maximum volume. Should the box be tall, cubical, or squat? You might even try calculating a few volumes, as shown in Figure 3.32, to see if you can get a good feeling for what the optimum dimensions should be.

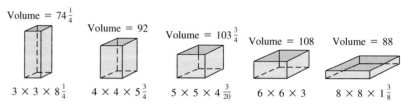

Volume = $74\frac{1}{4}$ Volume = 92   Volume = $103\frac{3}{4}$ Volume = 108   Volume = 88

$3 \times 3 \times 8\frac{1}{4}$ $4 \times 4 \times 5\frac{3}{4}$ $5 \times 5 \times 4\frac{3}{20}$ $6 \times 6 \times 3$ $8 \times 8 \times 1\frac{3}{8}$

FIGURE 3.32   Which size box has the maximum volume?

Remember that you are not ready to begin solving an optimization problem until you have clearly identified what the problem is. Once you are sure you understand what is being asked, you are ready to begin considering a method for solving the problem.

There are several steps in the solution of Example 1. The first step is to sketch a diagram and assign symbols to all *known* quantities and to all quantities *to be determined*. The second step is to write a primary equation for the quantity to be optimized. Then, a secondary equation is used to rewrite the primary equation as a function of one variable. Finally, calculus is used to determine the optimum value. These steps are summarized below.

Note   When performing Step 5, remember that to determine the maximum or minimum value of a continuous function f on a closed interval, you need to compare the values of f at its critical numbers with the values of f at the endpoints of the interval. The largest of these values is the desired maximum and the smallest is the desired minimum.

Guidelines for Solving Optimization Problems

1. Assign symbols to all given quantities and to all quantities to be determined. When feasible, sketch a diagram.
2. Write a **primary equation** for the quantity to be maximized or minimized. (A summary of several common formulas is given in Appendix B.)
3. Reduce the primary equation to one having a single independent variable: this may involve the use of a **secondary equation** that relates the independent variables of the primary equation.
4. Determine the **feasible domain** of the primary equation. That is, determine the values for which the stated problem makes sense.
5. Use calculus to find the desired maximum or minimum value.

EXAMPLE 2 *Finding a Maximum Sum*

The product of two positive numbers is 288. Minimize the sum of the second number and twice the first number.

Solution

1. Let x be the first number, y the second, and S the sum to be minimized.

2. Because you want to minimize S, the primary equation is

$$S = 2x + y. \qquad\qquad \textit{Primary equation}$$

3. Because the product of the two numbers is 288, you can write the following secondary equation.

$$xy = 288 \qquad\qquad \textit{Secondary equation}$$

$$y = \frac{288}{x}$$

Using this result, you can rewrite the primary equation as a function of one variable.

$$S = 2x + \frac{288}{x} \qquad\qquad \textit{Function of one variable}$$

4. Because the numbers are nonnegative, the feasible domain is

$$0 < x. \qquad\qquad \textit{Feasible domain}$$

5. To find the maximum value of S, begin by finding its critical numbers.

$$\frac{dS}{dx} = 2 - \frac{288}{x^2} \qquad\qquad \textit{Find derivative of S.}$$

$$0 = 2 - \frac{288}{x^2} \qquad\qquad \textit{Set derivative equal to 0.}$$

$$x^2 = 144 \qquad\qquad \textit{Simplify.}$$

$$x = \pm 12 \qquad\qquad \textit{Critical numbers}$$

Choosing the positive x-value, you can use the First-Derivative Test to conclude that S is decreasing in the interval $(0, 12)$ and increasing in the interval $(12, \infty)$, as shown in Table 3.8. Therefore, $x = 12$ yields a minimum and the two numbers are

$$x = 12 \quad \text{and} \quad y = \frac{288}{12} = 24.$$

TABLE 3.8

Interval	$0 < x < 12$	$12 < x < \infty$
Test value	$x = 1$	$x = 13$
Sign of $\dfrac{dS}{dx}$	$\dfrac{dS}{dx} < 0$	$\dfrac{dS}{dx} > 0$
Conclusion	S is decreasing	S is increasing

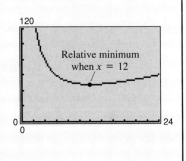

▦ TECHNOLOGY

After you have written the primary equation as a function of a single variable, you can estimate the optimum value by graphing the function. For instance, the graph of

$$S = 2x + \frac{288}{x},$$

shown below, indicates that the minimum value of S occurs when x is about 12.

Relative minimum when $x = 12$

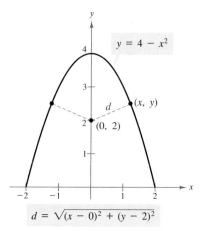

$y = 4 - x^2$

(x, y)

$(0, 2)$

$d = \sqrt{(x - 0)^2 + (y - 2)^2}$

FIGURE 3.33

EXAMPLE 3 *Finding a Minimum Distance*

Find the points on the graph of $y = 4 - x^2$ that are closest to $(0, 2)$.

Solution

1. Figure 3.33 indicates that there are two points at a minimum distance from the point $(0, 2)$.
2. You are asked to minimize the distance d. Hence, you can use the Distance Formula to obtain a primary equation.

$$d = \sqrt{(x - 0)^2 + (y - 2)^2}. \qquad \textit{Primary equation}$$

3. Using the secondary equation $y = 4 - x^2$, you can rewrite the primary equation as a function of a single variable.

$$d = \sqrt{x^2 + (4 - x^2 - 2)^2} \qquad \textit{Substitute } 4 - x^2 \textit{ for } y.$$
$$= \sqrt{x^4 - 3x^2 + 4} \qquad \textit{Simplify.}$$

Because d is smallest when the expression under the radical is smallest, you simplify the problem by finding the minimum value of

$$f(x) = x^4 - 3x^2 + 4.$$

4. The domain of f is the entire real line.
5. To find the minimum value of $f(x)$, first find the critical numbers of f.

$$f'(x) = 4x^3 - 6x \qquad \textit{Find derivative of } f.$$
$$0 = 4x^3 - 6x \qquad \textit{Set derivative equal to } 0.$$
$$0 = 2x(2x^2 - 3) \qquad \textit{Factor.}$$
$$x = 0, \ \sqrt{\tfrac{3}{2}}, \ -\sqrt{\tfrac{3}{2}} \qquad \textit{Critical numbers}$$

By the First-Derivative Test, you can conclude that $x = 0$ yields a relative maximum, whereas both $\sqrt{3/2}$ and $-\sqrt{3/2}$ yield a minimum. Hence, on the graph of $y = 4 - x^2$, the points that are closest to the point $(0, 2)$ are

$$\left(\sqrt{\tfrac{3}{2}}, \tfrac{5}{2}\right) \quad \text{and} \quad \left(-\sqrt{\tfrac{3}{2}}, \tfrac{5}{2}\right).$$

Note To confirm the result in Example 3, try computing the distance between several points on the graph of $y = 4 - x^2$ and the point $(0, 2)$. For instance, the distance between $(1, 3)$ and $(0, 2)$ is

$$d = \sqrt{(0 - 1)^2 + (2 - 3)^2} = \sqrt{2} \approx 1.414.$$

Note that this is greater than the distance between $\left(\sqrt{3/2}, 5/2\right)$ and $(0, 2)$, which is

$$d = \sqrt{\left(0 - \sqrt{\tfrac{3}{2}}\right)^2 + \left(2 - \tfrac{5}{2}\right)^2} = \sqrt{\tfrac{7}{4}} \approx 1.323.$$

Real
Life

EXAMPLE 4 *Finding a Minimum Area*

A rectangular page is to contain 24 square inches of print. The margins at the top and bottom of the page are $1\frac{1}{2}$ inches wide. The margins on each side are 1 inch. What should the dimensions of the page be so that the least amount of paper is used?

Solution

1. A diagram of the page is shown in Figure 3.34.

2. Letting A be the area to be minimized, the primary equation is

$$A = (x + 3)(y + 2).$$ *Primary equation*

3. The printed area inside the margins is given by

$$24 = xy.$$ *Secondary equation*

Solving this equation for y produces

$$y = \frac{24}{x}.$$

By substituting this into the primary equation, you obtain

$$A = (x + 3)\left(\frac{24}{x} + 2\right)$$ *Function of one variable*

$$= (x + 3)\left(\frac{24 + 2x}{x}\right)$$

$$= \frac{2x^2 + 30x + 72}{x}$$

$$= 30 + 2x + \frac{72}{x}.$$ *Simplify.*

4. Because x must be positive, the feasible domain is $x > 0$.

5. To find the minimum area, begin by finding the critical numbers of A.

$$\frac{dA}{dx} = 2 - \frac{72}{x^2}$$ *Find derivative of A.*

$$0 = 2 - \frac{72}{x^2}$$ *Set derivative equal to 0.*

$$x^2 = 36$$

$$x = \pm 6$$ *Critical numbers*

Because $x = -6$ is not in the feasible domain, you need only consider the critical number $x = 6$. Using the First-Derivative Test, it follows that A is a minimum when $x = 6$. Thus, the dimensions of the page should be

$$x + 3 = 6 + 3 = 9 \text{ inches}$$

by

$$y + 2 = \frac{24}{6} + 2 = 6 \text{ inches.}$$

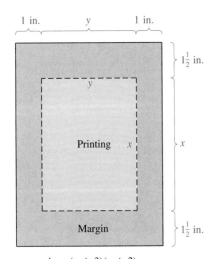

1 in. y 1 in.

$1\frac{1}{2}$ in.

y

Printing x x

Margin $1\frac{1}{2}$ in.

$A = (x + 3)(y + 2)$

FIGURE 3.34

As applications go, the four examples described in this section are fairly simple, and yet the resulting primary equations are quite complicated. Real-life applications often involve equations that are at least as complex as these four, and you should expect this. Remember that one of the main goals of this course is to enable you to use the power of calculus to analyze equations that at first glance seem formidable.

Also remember that once you have found the primary equation, you can use the graph of the equation as an aid to solving the optimization problem. For instance, the graphs of the primary equations in Examples 1 through 4 are shown in Figure 3.35.

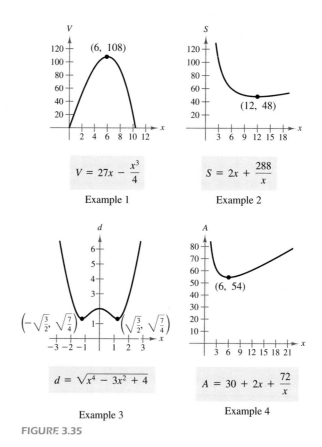

$$V = 27x - \frac{x^3}{4}$$

Example 1

$$S = 2x + \frac{288}{x}$$

Example 2

$$d = \sqrt{x^4 - 3x^2 + 4}$$

Example 3

$$A = 30 + 2x + \frac{72}{x}$$

Example 4

FIGURE 3.35

Discussion Problem   **Using Technology**

Use a graphing utility to reproduce the graphs shown above. Then, use the zoom and trace features to approximate the optimum values.

Warm Up

The following warm-up exercises involve skills that were covered in earlier sections. You will use these skills in the exercise set for this section.

In Exercises 1–4, write a formula for the written statement.

1. The sum of one number and half a second number is 12.

2. The product of one number and twice another is 24.

3. The area of a rectangle is 24 square units.

4. The distance between two points is 10 units.

In Exercises 5–10, find the critical numbers of the function.

5. $y = x^2 + 6x - 9$

6. $y = 2x^3 - x^2 - 4x$

7. $y = 5x + \dfrac{125}{x}$

8. $y = 3x + \dfrac{96}{x^2}$

9. $y = \dfrac{x^2 + 1}{x}$

10. $y = \dfrac{x}{x^2 + 9}$

E X E R C I S E S   3.4 means that technology can help you solve or check the exercise(s).

In Exercises 1–6, find two positive numbers satisfying the given requirements.

1. The sum is 110 and the product is a maximum.

2. The sum is S and the product is a maximum.

3. The sum of the first and twice the second is 36 and the product is a maximum.

4. The sum of the first and twice the second is 100 and the product is a maximum.

5. The product is 192 and the sum is a minimum.

6. The product is 192 and the sum of the first plus three times the second is a minimum.

7. What positive number x minimizes the sum of x and its reciprocal?

8. The difference of two numbers is 50. Find the two numbers so that their product is a minimum.

9. Verify that each rectangular solid has a surface area of 150 square inches (see figure). Find the volume of each.

10. Verify that each right circular cylinder has a surface area of 24π (see figure). Find the volume of each.

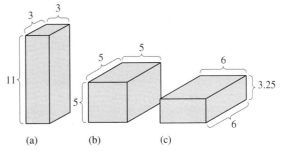

(a)　　(b)　　(c)

FIGURE FOR 9

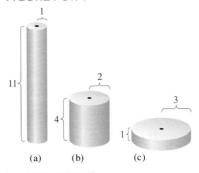

(a)　　(b)　　(c)

FIGURE FOR 10

In Exercises 11 and 12, find the length and width of the specified rectangle.

11. Maximum area; Perimeter: 100 feet

12. Minimum perimeter; Area: 64 square feet

13. *Area* A rancher has 200 feet of fencing to enclose two adjacent rectangular corrals (see figure). What dimensions should be used so that the enclosed area will be a maximum?

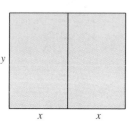

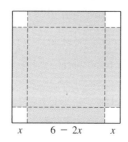

FIGURE FOR 13   FIGURE FOR 15

14. *Fence Length* A dairy farmer plans to enclose a rectangular pasture adjacent to a river. To provide enough grass for the herd, the pasture must contain 180,000 square meters. No fencing is required along the river. What dimensions use the least amount of fencing?

15. *Volume* An open box is to be made from a 6-inch by 6-inch square piece of material by cutting equal squares from each corner and turning up the sides (see figure). Find the volume of the largest box that can be made in this manner.

16. *Volume* An open box is to be made from a 2-foot by 3-foot rectangular piece of material by cutting equal squares from each corner and turning up the sides. Find the volume of the largest box that can be made in this manner.

17. *Surface Area* A net enclosure for golf practice is open at one end (see figure). The volume of the enclosure is $83\frac{1}{3}$ cubic meters. Find the dimensions that require the least amount of netting.

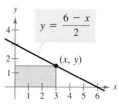

FIGURE FOR 17   FIGURE FOR 21

18. *Volume* For the golf practice enclosure in Exercise 17, assume that 192 square feet of netting is available. What dimensions will maximize the volume?

19. *Area* An indoor physical fitness room consists of a rectangular region with a semicircle on each end. The perimeter of the room is to be a 200-meter running track. Find the dimensions that will make the area of the rectangular region as large as possible.

20. *Area* A page is to contain 30 square inches of print. The margins at the top and bottom of the page are each 2 inches wide. The margins on each side are only 1 inch wide. What dimensions will minimize the amount of paper used?

21. *Area* A rectangle is bounded by the x- and y-axes and the graph of $y = (6-x)/2$ (see figure). What length and width should the rectangle have so that its area is a maximum?

22. *Area* A right triangle is formed in the first quadrant by the x- and y-axes and a line through the point $(2, 3)$. Find the vertices of the triangle so that its area is a minimum.

23. *Area* A rectangle is bounded by the x-axis and the semicircle $y = \sqrt{25 - x^2}$ (see figure). What length and width should the rectangle have so that its area is a maximum?

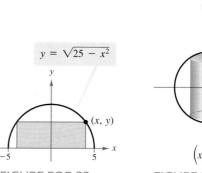

FIGURE FOR 23   FIGURE FOR 26

24. *Area* Find the dimensions of the largest rectangle that can be inscribed in a semicircle of radius r. (See Exercise 23.)

25. *Volume* You are designing a soft drink container that has the shape of a right circular cylinder. The container is to hold 12 fluid ounces (1 fluid ounce is approximately 1.80469 cubic inches). Find the dimensions that will use a minimum amount of construction material.

26. *Volume* Find the volume of the largest right circular cylinder that can be inscribed in a sphere of radius r (see figure).

27. *Closest Point* Find the point on the graph of $y = x^2 + 1$ that is closest to the point $(0, 4)$.

28. *Closest Point* Find the point on the graph of $y = x^2$ that is closest to the point $\left(2, \frac{1}{2}\right)$.

29. *Volume* A rectangular package to be sent by a postal service can have a maximum combined length and girth of 108 inches. Find the dimensions of the package with maximum volume. As shown in the figure, assume the package's dimensions are x by x by y.

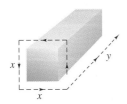

FIGURE FOR 29

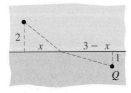

FIGURE FOR 33

30. *Surface Area* A container is formed by adjoining two hemispheres to each end of a right circular cylinder. The total volume of the container is 12 cubic inches. Find the radius of the cylinder that produces the minimum surface area.

31. *Area* The combined perimeter of a circle and a square is 16. Find the dimensions of the circle and square that produce a minimum total area.

32. *Area* The combined perimeter of an equilateral triangle and a square is 10. Find the dimensions of the triangle and square that produce a minimum total area.

33. *Time* You are in a boat 2 miles from the nearest point on the coast. You are to go to point Q, 3 miles down the coast and 1 mile inland (see figure). You can row at a rate of 2 miles per hour and can walk at a rate of 4 miles per hour. Toward which point on the coast should you row in order to reach point Q in the least time?

34. *Time* The conditions are the same as in Exercise 33, except that you can row at a rate of 4 miles per hour. How does this change the solution?

35. *Strength of a Beam* A wooden beam has a rectangular cross section of height h and width w (see figure). The strength S of the beam is directly proportional to its width and the square of its height. What are the dimensions of the strongest beam that can be cut from a round log of diameter 24 inches? (*Hint:* $S = kh^2w$, where k is the proportionality constant.)

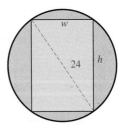

FIGURE FOR 35

36. *Surface Area* A right circular cylinder of radius r and height h has a volume of 25 cubic inches. The surface area of the cylinder is given by

$$S = 2\pi r \left(r + \frac{25}{\pi r^2} \right).$$

Use a graphing utility to graph S and S' and find the value of r that yields the minimum surface area.

37. *Area* Use a graphing utility to graph the primary equation and its first derivative to find the dimensions of the rectangle of maximum area that can be inscribed in a semicircle of radius 10.

38. *Area* Four feet of wire is to be used to form a square and a circle.

    (a) Express the sum of the areas of the square and the circle as a function $A(x)$ of the side of the square x.

    (b) What is the domain of $A(x)$?

    (c) Use a graphing utility to graph A on its domain.

    (d) How much wire should be used for the square and how much for the circle in order to enclose the least amount of total area?

    (e) How much wire should be used for the square and how much for the circle in order to enclose the most amount of total area?

3.5 Business and Economics Applications

Optimization in Business and Economics ■ *Price Elasticity of Demand* ■ *Business Terms and Formulas*

Optimization in Business and Economics

The problems in this section are primarily optimization problems. Hence, the five-step procedure used in Section 3.4 is an appropriate strategy to follow.

EXAMPLE 1 *Finding the Maximum Revenue*

Real Life

A company has determined that its total revenue (in dollars) for a product can be modeled by

$$R = -x^3 + 450x^2 + 52,500x,$$

where x is the number of units produced (and sold). What production level will yield a maximum revenue?

Solution

1. A sketch of the revenue function is given in Figure 3.36.
2. The primary equation is the given revenue function.

$$R = -x^3 + 450x^2 + 52,500x \qquad \textit{Primary equation}$$

3. Because R is already given as a function of one variable, you do not need a secondary equation.
4. The feasible domain of the primary equation is

$$0 \le x \le 546. \qquad \textit{Feasible domain}$$

This is determined by finding the x-intercepts of the revenue function, as shown in Figure 3.36.

5. To maximize the revenue, find the critical numbers.

$$\frac{dR}{dx} = -3x^2 + 900x + 52,500 = 0 \qquad \textit{Set derivative equal to 0.}$$

$$-3(x - 350)(x + 50) = 0 \qquad \textit{Factor.}$$

$$x = 350, \ -50 \qquad \textit{Critical numbers}$$

The only critical number in the feasible domain is $x = 350$. From the graph of the function, you can see that a production level of 350 units corresponds to a maximum revenue. ▬

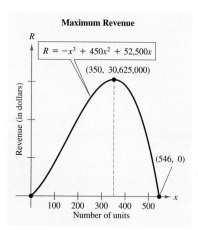

FIGURE 3.36 Maximum revenue occurs when $dR/dx = 0$.

To study the effects of production levels on cost, economists use the **average cost function** $\overline{C}$, which is defined as

$$\overline{C} = \frac{C}{x},$$ *Average cost function*

where $C = f(x)$ is the total cost function and x is the number of units produced.

EXAMPLE 2 *Finding the Minimum Average Cost*   *Real Life*

A company estimates that the cost (in dollars) of producing x units of a product can be modeled by

$$C = 800 + 0.04x + 0.0002x^2.$$

Find the production level that minimizes the average cost per unit.

Solution

1. C represents the total cost, x represents the number of units produced, and $\overline{C}$ represents the average cost per unit.

2. The primary equation is

$$\overline{C} = \frac{C}{x}.$$ *Primary equation*

3. Substituting the given equation for C produces

$$\overline{C} = \frac{800 + 0.04x + 0.0002x^2}{x}$$ *Substitute for C.*

$$= \frac{800}{x} + 0.04 + 0.0002x.$$ *Function of one variable*

4. The feasible domain for this function is

$$0 < x.$$ *Feasible domain*

5. You can find the critical numbers as follows.

$$\frac{d\overline{C}}{dx} = -\frac{800}{x^2} + 0.0002 = 0$$ *Set derivative equal to 0.*

$$0.0002 = \frac{800}{x^2}$$

$$x^2 = \frac{800}{0.0002}$$

$$x^2 = 4,000,000$$

$$x = \pm 2000$$ *Critical numbers*

By choosing the positive value of x and sketching the graph of $\overline{C}$, as shown in Figure 3.37, you can see that a production level of $x = 2000$ minimizes the average cost per unit. ■

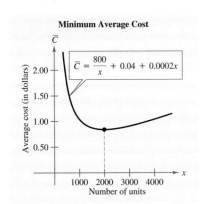

Minimum Average Cost

$\overline{C} = \frac{800}{x} + 0.04 + 0.0002x$

FIGURE 3.37 Minimum average cost when $d\overline{C}/dx = 0$.

Note To see that $x = 2000$ corresponds to a minimum average cost in Example 2, try evaluating $\overline{C}$ for several values of x. For instance, when $x = 400$, the average cost per unit is $\overline{C} = \$2.12$, but when $x = 2000$, the average cost per unit is $\overline{C} = \$0.84$.

Real
Life

EXAMPLE 3 *Finding the Maximum Revenue*

A business sells 2000 units per month at a price of $10 each. It can sell 250 more items per month for each $0.25 reduction in price. What price per unit will maximize the monthly revenue?

Solution

1. Let x represent the number of units sold in a month, let p represent the price per unit, and let R represent the monthly revenue.
2. Because the revenue is to be maximized, the primary equation is

$$R = xp. \qquad \textit{Primary equation}$$

3. A price of $p = \$10$ corresponds to $x = 2000$ and a price of $p = \$9.75$ corresponds to $x = 2250$. Using this information, you can use the point-slope form to create the demand equation.

$$p - 10 = \frac{10 - 9.75}{2000 - 2250}(x - 2000) \qquad \textit{Point-slope form}$$

$$p - 10 = -0.001(x - 2000) \qquad \textit{Simplify.}$$

$$p = -0.001x + 12 \qquad \textit{Secondary equation}$$

Substituting this value into the revenue equation produces

$$R = x(-0.001x + 12) \qquad \textit{Substitute for p.}$$

$$= -0.001x^2 + 12x. \qquad \textit{Function of one variable}$$

4. The feasible domain of the revenue function is

$$0 \le x \le 12{,}000. \qquad \textit{Feasible domain}$$

5. To maximize the revenue, find the critical numbers.

$$\frac{dR}{dx} = 12 - 0.002x = 0 \qquad \textit{Set derivative equal to 0.}$$

$$-0.002x = -12$$

$$x = 6000 \qquad \textit{Critical number}$$

From the graph of R in Figure 3.38, you can see that this production level yields a maximum revenue. The price that corresponds to this production level is

$$p = 12 - 0.001(x) \qquad \textit{Demand function}$$

$$= 12 - 0.001(6000) \qquad \textit{Substitute 6000 for x.}$$

$$= \$6. \qquad \textit{Price per unit} \qquad \blacksquare$$

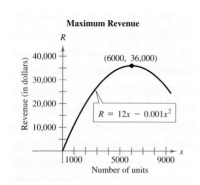

Maximum Revenue

(6000, 36,000)

$R = 12x - 0.001x^2$

Revenue (in dollars)

40,000
30,000
20,000
10,000

1000   5000   9000

Number of units

FIGURE 3.38

Note In Example 3, the revenue function was written as a function of x. It could also have been written as a function of p. That is, $R = 1000(12p - p^2)$. By finding the critical numbers of this function, you can determine that the maximum revenue occurs when $p = 6$.

Real Life

EXAMPLE 4 *Finding the Maximum Profit*

The marketing department for a business has determined that the demand for a product can be modeled by

$$p = \frac{50}{\sqrt{x}}.$$

The cost of producing x units is given by $C = 0.5x + 500$. What price will yield a maximum profit?

Solution

1. Let R represent the revenue, P the profit, p the price per unit, x the number of units, and C the total cost of producing x units.
2. Because you are maximizing the profit, the primary equation is

$$P = R - C. \qquad \textit{Primary equation}$$

3. Because the revenue is $R = xp$, you can write the profit function as

$$
\begin{aligned}
P &= R - C \\
&= xp - (0.5x + 500) &\textit{Substitute for C.} \\
&= x\left(\frac{50}{\sqrt{x}}\right) - 0.5x - 500 &\textit{Substitute for p.} \\
&= 50\sqrt{x} - 0.5x - 500. &\textit{Function of one variable}
\end{aligned}
$$

4. The feasible domain of the function is $0 < x \le 7872$. (When x is greater than 7872, the profit is negative.)
5. To maximize the profit, find the critical numbers.

$$
\begin{aligned}
\frac{dP}{dx} &= \frac{25}{\sqrt{x}} - 0.5 = 0 &\textit{Set derivative equal to 0.} \\
\sqrt{x} &= 50 \\
x &= 2500 &\textit{Critical number}
\end{aligned}
$$

From the graph of the profit function shown in Figure 3.39, you can see that $x = 2500$ corresponds to a maximum profit. The price that corresponds to $x = 2500$ is

$$p = \frac{50}{\sqrt{x}} = \frac{50}{\sqrt{2500}} = \frac{50}{50} = \$1.00. \qquad \textit{Price per unit}$$

Note To find the maximum profit in Example 4, we differentiated the equation $P = R - C$ and set dP/dx equal to zero. From the equation

$$\frac{dP}{dx} = \frac{dR}{dx} - \frac{dC}{dx} = 0,$$

it follows that the maximum profit occurs when the marginal revenue is equal to the marginal cost, as shown in Figure 3.40.

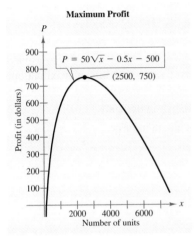

Maximum Profit

$P = 50\sqrt{x} - 0.5x - 500$

(2500, 750)

Profit (in dollars)

Number of units

FIGURE 3.39

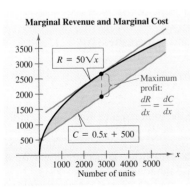

Marginal Revenue and Marginal Cost

$R = 50\sqrt{x}$

Maximum profit:
$\dfrac{dR}{dx} = \dfrac{dC}{dx}$

$C = 0.5x + 500$

Number of units

FIGURE 3.40

Price Elasticity of Demand

One way economists measure the responsiveness of consumers to a change in the price of a product is with **price elasticity of demand.** For example, a drop in the price of vegetables might result in a much greater demand for vegetables—such a demand is called **elastic.** On the other hand, items such as milk and water are relatively unresponsive to changes in price—the demand for such items is called **inelastic**.

More formally, the elasticity of demand is the percent change of a quantity demanded x, divided by the percent change in its price p. You can develop a formula for price elasticity of demand using the approximation

$$\frac{\Delta p}{\Delta x} \approx \frac{dp}{dx},$$

which is based on the definition of the derivative. Using this approximation, you can write

$$\text{Price elasticity of demand} = \frac{\text{rate of change in demand}}{\text{rate of change in price}}$$

$$= \frac{\Delta x / x}{\Delta p / p}$$

$$= \frac{p/x}{\Delta p / \Delta x}$$

$$\approx \frac{p/x}{dp/dx}.$$

Note The following list shows some estimates of elasticities of demand for common products.

(Source: Principles of Economics *by James Kearl)*

Item	Absolute Value of Elasticity
Cottonseed oil	6.92
Tomatoes	4.60
Restaurant meals	1.63
Automobiles	1.35
Cable TV	1.20
Beer	1.13
Housing	1.00
Movies	0.87
Clothing	0.60
Cigarettes	0.51
Coffee	0.25
Gasoline	0.15
Newspapers	0.10
Mail	0.05

Which of these items are elastic? Which are inelastic?

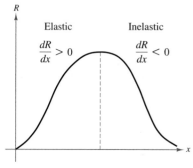

FIGURE 3.41   Revenue Curve

Definition of Price Elasticity of Demand

If $p = f(x)$ is a differentiable function, then the **price elasticity of demand** is given by

$$\eta = \frac{p/x}{dp/dx},$$

where η is the lowercase Greek letter eta. For a given price, the demand is **elastic** if $|\eta| > 1$, the demand is **inelastic** if $|\eta| < 1$, and the demand has **unit elasticity** if $|\eta| = 1$.

Price elasticity of demand is related to the total revenue function, as indicated in Figure 3.41 and the following list.

1. If the demand is *elastic*, then a decrease in price is accompanied by a sufficient increase in unit sales to increase the total revenue.
2. If the demand is *inelastic*, then a decrease in price is not accompanied by a sufficient increase in unit sales to increase the total revenue.

Real Life

EXAMPLE 5 *Comparing Elasticity and Revenue*

The demand function for a product is modeled by

$$p = \sqrt{450 - x}, \qquad 0 \le x \le 450.$$

a. Find the intervals on which the demand is elastic, inelastic, and of unit elasticity.

b. Use the result of part a to describe the behavior of the revenue function.

Solution

a. The price elasticity of demand is given by

$$\eta = \frac{p/x}{dp/dx}$$

$$= \frac{\left(\dfrac{\sqrt{450 - x}}{x}\right)}{\left(\dfrac{-1}{2\sqrt{450 - x}}\right)}$$

$$= \frac{-900 + 2x}{x}$$

$$= -\frac{900}{x} + 2.$$

The demand is of unit elasticity when $|\eta| = 1$. In the interval $[0, 450]$, the only solution of the equation

$$|\eta| = \left|-\frac{900}{x} + 2\right| = 1 \qquad\qquad \textit{Unit elasticity}$$

is $x = 300$. Thus, the demand is of unit elasticity when $x = 300$. For x-values in the interval $(0, 300)$,

$$|\eta| = \left|-\frac{900}{x} + 2\right| > 1, \qquad 0 < x < 300, \qquad \textit{Elastic}$$

which implies that the demand is elastic when $0 < x < 300$. For x-values in the interval $(300, 450)$,

$$|\eta| = \left|-\frac{900}{x} + 2\right| < 1, \qquad 300 < x < 450, \qquad \textit{Inelastic}$$

which implies that the demand is inelastic when $300 < x < 450$.

b. From part a, you can conclude that the revenue function R is increasing on the open interval $(0, 300)$, is decreasing on the open interval $(300, 450)$, and has a maximum when $x = 300$, as indicated in Figure 3.42. ■

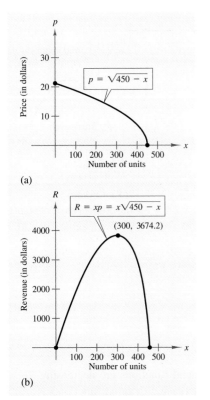

$p = \sqrt{450 - x}$

(a)

$R = xp = x\sqrt{450 - x}$

$(300, 3674.2)$

(b)

FIGURE 3.42

Note   In our discussion of price elasticity of demand, the price is assumed to decrease as the quantity demanded increases. Thus, the demand function $p = f(x)$ is decreasing and dp/dx is negative.

Business Terms and Formulas

We conclude with a summary of the basic business terms and formulas used in this section.

Summary of Business Terms and Formulas

x = number of units produced (or sold)

p = price per unit

R = total revenue from selling x units = xp

C = total cost of producing x units

P = total profit from selling x units = $R - C$

$\overline{C}$ = average cost per unit = $\dfrac{C}{x}$

η = price elasticity of demand = $\dfrac{p/x}{dp/dx}$

dR/dx = marginal revenue

dC/dx = marginal cost

dP/dx = marginal profit

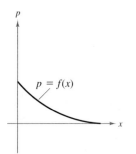

Demand function

Quantity demanded increases as price decreases.

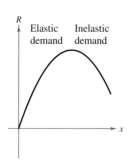

Revenue function

The low prices required to sell more units eventually result in a decreasing revenue.

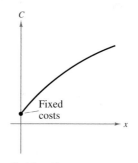

Cost function

As x increases, the average cost per unit decreases.

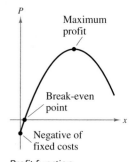

Profit function

The break-even point occurs when R = C.

FIGURE 3.43

Discussion Problem *Demand Function*

Throughout this text, we assume that demand functions are decreasing. Can you think of a product that has an increasing demand function? That is, can you think of a product that becomes more in demand as its price increases?

Warm Up

The following warm-up exercises involve skills that were covered in earlier sections. You will use these skills in the exercise set for this section.

In Exercises 1–4, evaluate the expression with $x = 150$.

1. $\left| -\dfrac{300}{x} + 3 \right|$

2. $\left| -\dfrac{600}{5x} + 2 \right|$

3. $\left| \dfrac{(20x^{-1/2})/x}{-10x^{-3/2}} \right|$

4. $\left| \dfrac{(4000/x^2)/x}{-8000x^{-3}} \right|$

In Exercises 5–10, find the marginal revenue, cost, or profit.

5. $C = 650 + 1.2x + 0.003x^2$

6. $P = 0.01x^2 + 11x$

7. $R = 14x - \dfrac{x^2}{2000}$

8. $R = 3.4x - \dfrac{x^2}{1500}$

9. $P = -0.7x^2 + 7x - 50$

10. $C = 1700 + 4.2x + 0.001x^3$

E X E R C I S E S   3.5   means that technology can help you solve or check the exercise(s).

In Exercises 1–4, find the number of units x that produces a maximum revenue R.

1. $R = 800x - 0.2x^2$

2. $R = 48x^2 - 0.02x^3$

3. $R = 400x - x^2$

4. $R = 30x^{2/3} - 2x$

In Exercises 5–8, find the number of units x that produces the minimum average cost per unit $\overline{C}$.

5. $C = 1.25x^2 + 25x + 8000$

6. $C = 0.001x^3 - 5x + 250$

7. $C = 3000x - x^2\sqrt{300 - x}$

8. $C = (2x^3 - x^2 + 5000x)/(x^2 + 2500)$

In Exercises 9–12, find the price per unit p that produces the maximum profit P.

Cost Function	Demand Function
9. $C = 100 + 30x$	$p = 90 - x$
10. $C = 2400x + 5200$	$p = 6000 - 0.4x^2$
11. $C = 8000 + 50x + 0.03x^2$	$p = 70 - 0.001x$
12. $C = 35x + 500$	$p = 50 - 0.1\sqrt{x}$

Average Cost In Exercises 13 and 14, use the cost function to find the production level for which the average cost is a minimum. For this production level, show that the marginal cost and average cost are equal.

13. $C = 2x^2 + 5x + 18$ **14.** $C = x^3 - 6x^2 + 13x$

15. *Profit* A commodity has a demand function modeled by $p = 100 - 0.5x^2$ and a total cost function modeled by $C = 40x + 37.5$.

(a) What price yields a maximum profit?

(b) When the profit is maximized, what is the average cost per unit?

16. *Profit* How would the answer to Exercise 15 change if the marginal cost rose from \$40 per unit to \$50 per unit? In other words, rework Exercise 15 using the cost function $C = 50x + 37.5$.

Profit In Exercises 17 and 18, find the amount s of advertising that maximizes the profit P. (s and P are measured in thousands of dollars.) Find the point of diminishing returns.

17. $P = -2s^3 + 35s^2 - 100s + 200$

18. $P = -0.1s^3 + 6s^2 + 400$

19. *Profit* The cost per unit in the production of a type of radio is $60. The manufacturer charges $90 per unit for orders of 100 or less. To encourage large orders, however, the manufacturer reduces the charge by $0.10 per radio for each order in excess of 100 units. For instance, an order of 101 radios would be $89.90 per radio, an order of 102 radios would be $89.80 per radio, and so on. Find the largest order the manufacturer should allow to obtain a maximum profit.

20. *Profit* A real estate office handles 50 apartment units. When the rent is $540 per month, all units are occupied. For each $30 increase in rent, however, an average of one unit becomes vacant. Each occupied unit requires an average of $36 per month for service and repairs. What rent should be charged to obtain a maximum profit?

21. *Revenue* When a wholesaler sold a certain product at $40 per unit, sales were 300 units each week. After a price increase of $5, however, the average number of units sold dropped to 275 each week. Assuming that the demand function is linear, what price per unit will yield a maximum total revenue?

22. *Profit* Assume that the amount of money deposited in a bank is proportional to the square of the interest rate the bank pays on the money. Furthermore, the bank can reinvest the money at 12% simple interest. Find the interest rate the bank should pay to maximize its profit.

23. *Cost* A power station is on one side of a river that is 0.5 mile wide, and a factory is 6 miles downstream on the other side of the river (see figure). It costs $6 per foot to run power lines overland and $8 per foot to run them underwater. Find the most economical path for the power line from the power station to the factory.

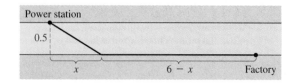

24. *Cost* An offshore oil well is 1 mile off the coast. The refinery is 2 miles down the coast. Laying pipe in the ocean is twice as expensive as laying it on land. Find the most economical path for the pipe from the well to the refinery.

25. *Cost* A small business uses a minivan to make deliveries. The cost per hour for fuel for the van is $C = v^2/600$, where v is the speed of the minivan (in miles per hour). The driver of the minivan is paid $10 per hour. Find the speed that minimizes the cost of a 110-mile trip. (Assume there are no costs other than fuel and wages.)

26. *Cost* Repeat Exercise 25 for a fuel cost per hour of

$$C = \frac{v^2 + 360}{720}$$

and a wage of $8 per hour.

Elasticity In Exercises 27–32, find the price elasticity of demand for the demand function at the indicated x-value. Is the demand elastic, inelastic, or of unit elasticity at the indicated x-value?

Demand Function	Quantity Demanded
27. $p = 400 - 3x$	$x = 20$
28. $p = 5 - 0.03x$	$x = 100$
29. $p = 300 - 0.2x^2$	$x = 30$
30. $p = \dfrac{500}{x+2}$	$x = 23$
31. $p = \dfrac{100}{x^2} + 2$	$x = 10$
32. $p = 100 - \sqrt{0.2x}$	$x = 125$

33. *Elasticity* The demand function for a product is

$$x = 20 - 2p^2.$$

(a) Consider a price of $2. If the price increases by 5%, what is the corresponding percent change in the quantity demanded?

(b) Average elasticity of demand is defined to be the percent change in quantity divided by the percent change in price. Use the percent of part (a) to find the average elasticity over the interval [2, 2.1].

(c) Find the elasticity for a price of $2 and compare the result with that of part (b).

(d) Find an expression for the total revenue and find the values of x and p that maximize the total revenue.

34. *Elasticity* The demand function for a product is

$$p^3 + x^3 = 9.$$

(a) Find the price elasticity of demand when $x = 2$.

(b) Find the values of x and p that maximize the total revenue.

(c) For the value of x found in part (b), show that the price elasticity of demand has unit elasticity.

35. *Elasticity* The demand function for a product is

$$p = (16 - x)^{1/2}, \qquad 0 < x < 16.$$

(a) Find the price elasticity of demand when $x = 9$.

(b) Find the values of x and p that maximize the total revenue.

(c) For the value of x found in part (b), show that the price elasticity of demand has unit elasticity.

36. *Revenue* The demand for a car wash is

$$x = 600 - 50p,$$

where the current price is $5.00. Can revenue be increased by lowering the price and thus attracting more customers? Use price elasticity of demand to determine your answer.

37. *Revenue* Repeat Exercise 36 with a demand function of

$$x = 800 - 40p.$$

38. A demand function is modeled by $x = a/p^m$, where a is a constant and $m > 1$. Show that $\eta = -m$. In other words, show that a 1% increase in price results in an $m\%$ decrease in the quantity demanded.

39. *Railroad Revenues* The annual revenue (in millions of dollars) for Union Pacific for the years 1985–1994 can be modeled by

$$R = 4.7t^4 - 193.5t^3 + 2941.7t^2 - 19,294.7t + 52,012,$$

where $t = 0$ corresponds to 1980. *(Source: Union Pacific Corporation)*

(a) During which year, between 1985 and 1994, was Union Pacific's revenue the least?

(b) During which year was the revenue the greatest?

(c) Find the revenue for each year in parts (a) and (b).

(d) Use a graphing utility to graph the revenue function. Then use the zoom and trace features to confirm the results in parts (a), (b), and (c).

40. *Banana Sales* The annual net profit (in millions of dollars) for Chiquita Brands for the years 1983–1991 can be modeled by

$$P = 0.33t^4 - 9.35t^3 + 94.88t^2 - 399.98t + 616.62,$$

where $t = 0$ corresponds to 1980. *(Source: Chiquita Brands International, Inc.)*

(a) During which year, between 1983 and 1991, was Chiquita's profit the least?

(b) During which year was the profit the greatest?

(c) Find the profit for each year in parts (a) and (b).

(d) Use a graphing utility to graph the profit function. Then use the zoom and trace features to confirm the results in parts (a), (b), and (c).

41. Match each graph with the function it best represents—a demand function, a revenue function, a cost function, or a profit function. Explain your reasoning. [The graphs are labeled (a)–(d).]

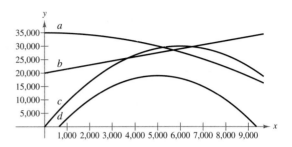

42. *Research Project* Choose an innovative product like the one describe below. Use your school's library or some other reference source to research the history of the product. Collect data about the revenue that product has generated, and find a mathematical model for the data. Summarize your findings.

— *Business Capsule* —

Courtesy of the University of Michigan Economics Club, Flint Michigan

The economics club at the University of Michigan at Flint marketed "Economist Greats" trading cards, featuring statistics on 29 famous economists, such as Milton Friedman. Profits from the cards were used to finance trips to the Chicago Federal Reserve Bank. Pictured are: (left to right, kneeling) Thomas Lowell, vice president; Michael Kidder, president; Bryan Klawuhir, secretary; Bonnie Kincaid, treasurer; and (left to right standing) Vikac Kapur; Professor Dale Matcheck, club advisor; and Michael Wickerham.

Case Study: Consumer Response to Price Changes

PROFILE

COMPANY
Central and South West Services, Inc.
■
LOCATION
Dallas, TX
■
TYPE OF BUSINESS
A holding company that owns four electric utilities
■

Economists and business managers are often interested in how consumers respond to a change in the price of a product or service. Knowledge of the precise magnitude of the response or *price elasticity of demand* for its product is often useful in helping a firm formulate its pricing strategy.

Central and South West (C & SW), Inc. is a public utility holding company in Dallas, Texas that currently owns four electric utilities. Among its many customers are a large number of commercial and industrial purchasers of electricity. Analysts at Central and South West Services, a subsidiary of C & SW, are sometimes called upon to construct models of the demand for electricity by commercial and industrial users. Such models, in turn, can be used to generate price elasticities of demand for electrical usage.

The quantity of electricity demanded is likely to be a function of many variables in addition to the price. To estimate commercial and industrial electricity demand in the state of Texas, C & SW Services constructed a model including the price of electricity and several other "explanatory" variables. In order to hold constant all of the other variables so that the focus is solely on the effects of *price* changes on electricity demand, the values of the explanatory variables for the second quarter of 1993 were plugged into the model to yield the following, more compact electricity demand equation:

$$Y = 429937.843 - 7.4363X.$$

where:

Y = Total quarterly sales of electricity to commercial and industrial users (kilowatt hours, KWH)

X = Price of electricity ($/KWH), obtained from typical bills of commercial and industrial electricity users

This equation describes how the quantity of electricity demanded by commercial and industrial users (Y) varies with the price (X), *given the economic conditions that prevailed in the second quarter of 1993.* Moreover, taking the derivative of electricity demand with respect to price yields

$$dY/dX = -7.4363.$$

Applying the definition of price elasticity of demand (η) yields the elasticity coefficient:

$$\eta = (dY/dX) \cdot (X/Y)$$
$$= -7.4363 \cdot (X/Y)$$

This equation can be used to compute the price elasticity of demand at each price (X) and the corresponding quantity of electricity demanded (Y). The table below shows these calculations over a sample range of electricity prices. Note that the elasticity coefficient is negative, indicating an inverse relation between price and quantity of electricity demanded. Also, note that $|\eta| < 1$, indicating that electricity demand is *inelastic* over this range. For example, at a price of 3645.73, the quantity of electricity demanded falls by about 0.07 percent for a 1 percent increase in price.

Price ($/KWH) (X)	Commercial and Industrial Electricity Sales (KWH) (Y)	Price Elasticity of Demand (η)
3645.73	402,827.101	−0.067
3745.73	402,083.471	−0.069
3845.73	401,339.841	−0.071
3945.73	400,596.211	−0.073
4045.73	399,852.581	−0.075
4145.73	399,108.951	−0.077
4245.73	398,365.321	−0.079
4345.73	397,621.691	−0.081
4445.73	396,878.061	−0.083
4545.73	396,134.431	−0.085
4645.73	395,390.801	−0.087
4745.73	394,647.171	−0.089
4845.73	393,903.541	−0.091
4945.73	393,159.911	−0.094
5045.73	392,416.281	−0.096
5145.73	391,672.651	−0.098
5245.73	390,929.021	−0.100
5345.73	390,185.391	−0.102
5445.73	389,441.761	−0.104
5545.73	388,698.131	−0.106

Courtesy of CSW, Inc.

Dr. Enrique Mejorada is a Senior Strategic Planning Consultant for Central and South West Services, Inc. ■

What Would You Do?

1. Plot the data on the quantity of electricity sales at various prices to show that the commercial and industrial demand curve for electricity is *linear.* Why does the price elasticity of demand change along a linear demand curve? What implications might this have on the pricing decisions of an electric utility?

2. Why do you think the demand for electricity by commercial and industrial users is *inelastic* with respect to price?

3. Given that the demand for electricity is inelastic, what would you expect to happen to an electric utility's total revenue as it sold more electricity by lowering its price? Show this using some of the price and sales figures presented in the table.

3.6 Asymptotes

Vertical Asymptotes and Infinite Limits ■ Horizontal Asymptotes
and Limits at Infinity ■ Applications of Asymptotes

Vertical Asymptotes and Infinite Limits

In the first three sections of this chapter, you studied ways in which you can use calculus to help analyze the graph of a function. In this section, you will study another valuable aid to curve sketching—the determination of vertical and horizontal asymptotes.

Recall from Section 1.5, Example 10, that the function

$$f(x) = \frac{3}{x - 2}$$

is unbounded as x approaches 2 (see Figure 3.44). This type of behavior is described by saying that the line $x = 2$ is a **vertical asymptote** of the graph of f. The type of limit in which $f(x)$ approaches infinity (or negative infinity) as x approaches c from the left or from the right is an **infinite limit.** The infinite limits for the function $f(x) = 3/(x - 2)$ can be written as

$$\lim_{x \to 2^-} \frac{3}{x - 2} = -\infty \quad \text{and} \quad \lim_{x \to 2^+} \frac{3}{x - 2} = \infty.$$

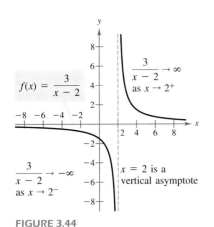

$f(x) = \dfrac{3}{x - 2}$

$\dfrac{3}{x - 2} \to \infty$ as $x \to 2^+$

$\dfrac{3}{x - 2} \to -\infty$ as $x \to 2^-$

$x = 2$ is a vertical asymptote

FIGURE 3.44

Definition of Vertical Asymptote

If $f(x)$ approaches infinity (or negative infinity) as x approaches c from the right or from the left, then the line $x = c$ is a **vertical asymptote** of the graph of f.

 ## TECHNOLOGY

When you use a graphing utility to graph a function that has a vertical asymptote, the utility may try to connect separate branches of the graph. For instance, the figure at the right shows the graph of $f(x) = 3/(x - 2)$ on a graphing calculator.

This line is not part of the graph of the function.

The graph of the function has two branches.

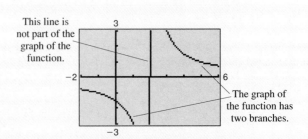

One of the most common instances of a vertical asymptote is the graph of a *rational function*—that is, a function of the form $f(x) = p(x)/q(x)$, where $p(x)$ and $q(x)$ are polynomials. If c is a real number such that $q(c) = 0$ and $p(c) \neq 0$, then the graph of f has a vertical asymptote at $x = c$. Example 1 shows four typical cases.

EXAMPLE 1 *Finding Infinite Limits*

Limit from the Left   *Limit from the Right*

a. $\displaystyle \lim_{x \to 1^-} \frac{1}{x-1} = -\infty$ $\displaystyle \lim_{x \to 1^+} \frac{1}{x-1} = \infty$ (See Figure 3.45a.)

b. $\displaystyle \lim_{x \to 1^-} \frac{-1}{x-1} = \infty$ $\displaystyle \lim_{x \to 1^+} \frac{-1}{x-1} = -\infty$ (See Figure 3.45b.)

c. $\displaystyle \lim_{x \to 1^-} \frac{-1}{(x-1)^2} = -\infty$ $\displaystyle \lim_{x \to 1^+} \frac{-1}{(x-1)^2} = -\infty$ (See Figure 3.45c.)

d. $\displaystyle \lim_{x \to 1^-} \frac{1}{(x-1)^2} = \infty$ $\displaystyle \lim_{x \to 1^+} \frac{1}{(x-1)^2} = \infty$ (See Figure 3.45d.)

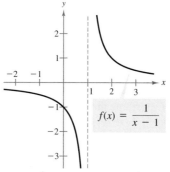

$$\lim_{x \to 1^-} \frac{1}{x-1} = -\infty \qquad \lim_{x \to 1^+} \frac{1}{x-1} = \infty$$

(a)

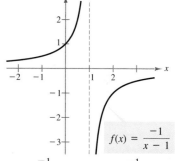

$$\lim_{x \to 1^-} \frac{-1}{x-1} = \infty \qquad \lim_{x \to 1^+} \frac{-1}{x-1} = -\infty$$

(b)

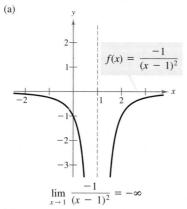

$$\lim_{x \to 1} \frac{-1}{(x-1)^2} = -\infty$$

(c)

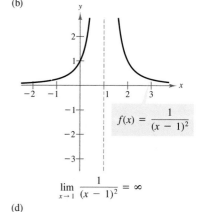

$$\lim_{x \to 1} \frac{1}{(x-1)^2} = \infty$$

(d)

FIGURE 3.45

Each of the graphs in Example 1 has only one vertical asymptote. As shown in the next example, the graph of a rational function can have more than one vertical asymptote.

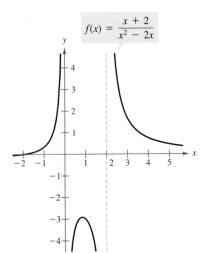

$$f(x) = \frac{x+2}{x^2 - 2x}$$

FIGURE 3.46   Vertical Asymptotes at $x = 0$ and $x = 2$

EXAMPLE 2 *Finding Vertical Asymptotes*

Determine the vertical asymptotes of the graph of

$$f(x) = \frac{x+2}{x^2 - 2x}.$$

Solution

The possible vertical asymptotes correspond to the x-values for which the denominator is zero.

$x^2 - 2x = 0$	*Set denominator equal to 0.*
$x(x - 2) = 0$	*Factor.*
$x = 0,\ 2$	*Zeros of denominator*

Because the numerator of $f(x)$ is not zero at either of these x-values, you can conclude that the graph of f has two vertical asymptotes—one at $x = 0$ and one at $x = 2$, as shown in Figure 3.46. ▪

EXAMPLE 3 *Finding a Vertical Asymptote*

Determine the vertical asymptotes of the graph of

$$f(x) = \frac{x^2 + 2x - 8}{x^2 - 4}.$$

Solution

First factor the numerator and denominator. Then cancel common factors.

$f(x) = \dfrac{x^2 + 2x - 8}{x^2 - 4}$	*Original function*
$= \dfrac{(x+4)(x-2)}{(x+2)(x-2)}$	*Factor numerator and denominator.*
$= \dfrac{(x+4)\cancel{(x-2)}}{(x+2)\cancel{(x-2)}}$	*Cancel common factors.*
$= \dfrac{x+4}{x+2},\quad x \neq 2$	*Simplify.*

For all values of x other than $x = 2$, the graph of this simplified function is the same as the graph of f. Thus, you can conclude that the graph of f has only one vertical asymptote. This occurs at $x = -2$, as shown in Figure 3.47. ▪

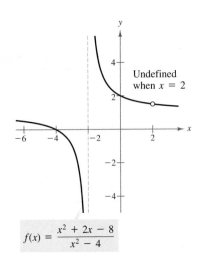

Undefined when $x = 2$

$$f(x) = \frac{x^2 + 2x - 8}{x^2 - 4}$$

FIGURE 3.47   Vertical Asymptote at $x = -2$

From Example 3, you know that the graph of

$$f(x) = \frac{x^2 + 2x - 8}{x^2 - 4}$$

has a vertical asymptote at $x = -2$. This implies that the limit of $f(x)$ as $x \to -2$ from the right (or from the left) is either ∞ or $-\infty$. But without looking at the graph, how can you determine that the limit from the left is *negative* infinity and the limit from the right is *positive* infinity? That is, why is

$$\lim_{x \to 2^-} \frac{x^2 + 2x - 8}{x^2 - 4} = -\infty \qquad \textit{Limit from the left}$$

and

$$\lim_{x \to 2^+} \frac{x^2 + 2x - 8}{x^2 - 4} = \infty? \qquad \textit{Limit from the right}$$

It is cumbersome to determine these limits analytically, and you may find the graphical method shown in Example 4 to be more efficient.

EXAMPLE 4 *Determining Infinite Limits*

Find the limits.

$$\lim_{x \to 1^-} \frac{x^2 - 3x}{x - 1} \quad \text{and} \quad \lim_{x \to 1^+} \frac{x^2 - 3x}{x - 1}$$

Solution

Begin by considering the function

$$f(x) = \frac{x^2 - 3x}{x - 1}.$$

Because the denominator is zero when $x = 1$ and the numerator is not zero when $x = 1$, it follows that the graph of the function has a vertical asymptote at $x = 1$. This implies that each of the given limits is either ∞ or $-\infty$. To determine which, use a graphing utility to graph the function, as shown in Figure 3.48. From the graph, you can see that the limit from the left is positive infinity and the limit from the right is negative infinity. That is,

$$\lim_{x \to 1^-} \frac{x^2 - 3x}{x - 1} = \infty \qquad \textit{Limit from the left}$$

and

$$\lim_{x \to 1^+} \frac{x^2 - 3x}{x - 1} = -\infty. \qquad \textit{Limit from the right}$$

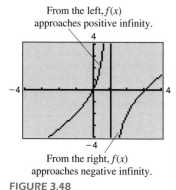

From the left, $f(x)$ approaches positive infinity.

From the right, $f(x)$ approaches negative infinity.

FIGURE 3.48

Note In Example 4, try evaluating $f(x)$ at x-values that are just barely to the left of 1. You will find that you can make the values of $f(x)$ arbitrarily large by choosing x sufficiently close to 1. For instance, $f(0.99999) = 199,999$.

Horizontal Asymptotes and Limits at Infinity

Another type of limit, called a **limit at infinity,** specifies a finite value approached by a function as x increases (or decreases) without bound.

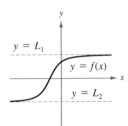

$y = L_1$

$y = f(x)$

$y = L_2$

Definition of Horizontal Asymptote

If f is a function and L_1 and L_2 are real numbers, the statements

$$\lim_{x \to \infty} f(x) = L_1 \quad \text{and} \quad \lim_{x \to -\infty} f(x) = L_2$$

denote **limits at infinity.** The lines $y = L_1$ and $y = L_2$ are **horizontal asymptotes** of the graph of f.

Figure 3.49 shows two ways in which the graph of a function can approach one or more horizontal asymptotes. Note that it is possible for the graph of a function to cross its horizontal asymptote.

Limits at infinity share many of the properties of limits discussed in Section 1.5. When finding horizontal asymptotes, you can use the property that

$$\lim_{x \to \infty} \frac{1}{x^r} = 0, \quad r > 0 \qquad \text{and} \qquad \lim_{x \to -\infty} \frac{1}{x^r} = 0, \quad r > 0.$$

(The second limit assumes that x^r is defined when $x < 0$.)

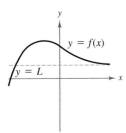

$y = f(x)$

$y = L$

FIGURE 3.49

EXAMPLE 5 *Finding Limits at Infinity*

Find the limit: $\displaystyle \lim_{x \to \infty} \left(5 - \frac{2}{x^2} \right)$.

Solution

$$\lim_{x \to \infty} \left(5 - \frac{2}{x^2} \right) = \lim_{x \to \infty} 5 - \lim_{x \to \infty} \frac{2}{x^2}$$

$$= \lim_{x \to \infty} 5 - 2 \left(\lim_{x \to \infty} \frac{1}{x^2} \right)$$

$$= 5 - 2(0)$$

$$= 5$$

You can verify this limit by sketching the graph of

$$f(x) = 5 - \frac{2}{x^2},$$

as shown in Figure 3.50. Note that the graph has $y = 5$ as a horizontal asymptote to the right. By evaluating the limit of $f(x)$ as $x \to -\infty$, you can show that this line is also a horizontal asymptote to the left.

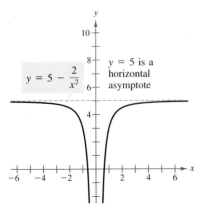

$y = 5 - \dfrac{2}{x^2}$

$y = 5$ is a horizontal asymptote

FIGURE 3.50

There is an easy way to determine whether the graph of a *rational* function has a horizontal asymptote. This shortcut is based on a comparison of the degrees of the numerator and denominator of the rational function.

⊞ **TECHNOLOGY**

Some functions have two horizontal asymptotes: one to the right and one to the left. For instance, try sketching the graph of

$$f(x) = \frac{x}{\sqrt{x^2 + 1}}.$$

What horizontal asymptotes does the function appear to have?

Horizontal Asymptotes of Rational Functions

Let $f(x) = p(x)/q(x)$ be a rational function.

1. If the degree of the numerator is less than the degree of the denominator, then $y = 0$ is a horizontal asymptote of the graph of f (to the left and to the right).
2. If the degree of the numerator is equal to the degree of the denominator, then $y = a/b$ is a horizontal asymptote of the graph of f (to the left and to the right), where a and b are the leading coefficients of $p(x)$ and $q(x)$, respectively.
3. If the degree of the numerator is greater than the degree of the denominator, then the graph of f has no horizontal asymptote.

EXAMPLE 6 *Finding Horizontal Asymptotes*

Find the horizontal asymptotes of the graphs of the functions.

a. $y = \dfrac{-2x + 3}{3x^2 + 1}$ **b.** $y = \dfrac{-2x^2 + 3}{3x^2 + 1}$ **c.** $y = \dfrac{-2x^3 + 3}{3x^2 + 1}$

Solution

a. Because the degree of the numerator is less than the degree of the denominator horizontal asymptote. (See Figure 3.51a.)
b. Because the degree of the numerator is equal to the degree of the denominator, the line $y = -\frac{2}{3}$ is a horizontal asymptote. (See Figure 3.51b.)
c. Because the degree of the numerator is greater than the degree of the denominator, the graph has no horizontal asymptote. (See Figure 3.51c.)

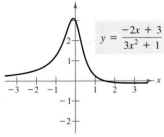

(a) $y = 0$ is a horizontal asymptote.

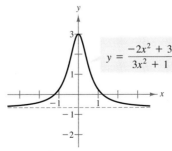

(b) $y = -\frac{2}{3}$ is a horizontal asymptote.

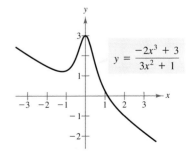

(c) No horizontal asymptote.

FIGURE 3.51

Applications of Asymptotes

There are many examples of asymptotic behavior in real life. For instance, Example 7 describes the asymptotic behavior of an average cost function.

Real Life

EXAMPLE 7 *Modeling Average Cost*

A small business invests $5000 in a new product. In addition to this initial investment, the product will cost $0.50 per unit to produce. Find the average cost per unit if 1000 units are produced, if 10,000 units are produced, and if 100,000 units are produced. What is the limit of the average cost as the number of units produced increases?

Solution

From the given information, you can model the total cost C (in dollars) by

$$C = 0.5x + 5000, \qquad \text{\textit{Total cost function}}$$

where x is the number of units produced. This implies that the average cost function is

$$\overline{C} = \frac{C}{x} = 0.5 + \frac{5000}{x}. \qquad \text{\textit{Average cost function}}$$

If only 1000 units are produced, then the average cost per unit is

$$\overline{C} = 0.5 + \frac{5000}{1000} = \$5.50. \qquad \text{\textit{Average cost for 1000 units}}$$

If 10,000 units are produced, then the average cost per unit is

$$\overline{C} = 0.5 + \frac{5000}{10,000} = \$1.00. \qquad \text{\textit{Average cost for 10,000 units}}$$

If 100,000 units are produced, then the average cost per unit is

$$\overline{C} = 0.5 + \frac{5000}{100,000} = \$0.55. \qquad \text{\textit{Average cost for 100,000 units}}$$

As x approaches infinity, the limiting average cost per unit is

$$\lim_{x \to \infty} \left(0.5 + \frac{5000}{x} \right) = \$0.50.$$

As shown in Figure 3.52, this example points out one of the major problems of small businesses. That is, it is difficult to have competitively low prices when the production level is low. ■

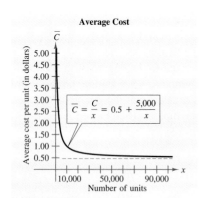

Average Cost

$$\overline{C} = \frac{C}{x} = 0.5 + \frac{5,000}{x}$$

FIGURE 3.52 As $x \longrightarrow \infty$, the average cost per unit approaches $0.50.

Note In Example 7, suppose that the small business had made an initial investment of $50,000. How would this change the answers to the questions? Would it change the average cost of producing x units? Would it change the limiting average cost per unit?

Real
Life

EXAMPLE 8 *Modeling Smokestack Emission*

During the 1980s, industries in the United States spent billions of dollars to reduce air pollution. For instance, the estimated cost of air pollution control in 1989 was 1.8 billion dollars. *(Source: U.S. Bureau of the Census, Current Industrial Reports)*

A manufacturing plant has determined that the cost C (in dollars) of removing $p\%$ of the smokestack pollutants of its main smokestack is modeled by

$$C = \frac{80{,}000p}{100 - p}.$$

What is the vertical asymptote of this function? What does the vertical asymptote mean to the plant owners?

Solution

The graph of the cost function is shown in Figure 3.53. From the graph, you can see that $p = 100$ is the vertical asymptote. This means that as the plant attempts to remove higher and higher percents of the pollutants, the cost increases dramatically. For instance, the cost of removing 85% of the pollutants is

$$C = \frac{80{,}000(85)}{100 - 85} \approx \$453{,}333 \qquad \textit{Cost for 85\% removal}$$

but the cost of removing 90% is

$$C = \frac{80{,}000(90)}{100 - 90} \approx \$720{,}000. \qquad \textit{Cost for 90\% removal}$$

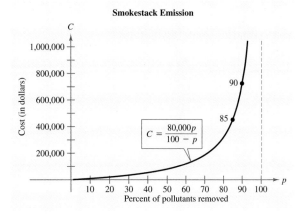

Smokestack Emission

FIGURE 3.53

Discussion Problem   Indirect Costs

In Example 8, the given cost model considers only the direct cost for the manufacturing plant. Describe the possible indirect costs incurred by society as the plant attempts to remove more and more of the pollutants from its smokestack emission.

Warm Up

The following warm-up exercises involve skills that were covered in earlier sections. You will use these skills in the exercise set for this section.

In Exercises 1–6, find the limit.

1. $\lim\limits_{x \to 2} (x + 1)$

2. $\lim\limits_{x \to -1} (3x + 4)$

3. $\lim\limits_{x \to -3} \dfrac{2x^2 + x - 15}{x + 3}$

4. $\lim\limits_{x \to 2} \dfrac{3x^2 - 8x + 4}{x - 2}$

5. $\lim\limits_{x \to 0^+} \sqrt{x}$

6. $\lim\limits_{x \to 1^+} (x + \sqrt{x - 1})$

In Exercises 7–10, find the average cost and the marginal cost.

7. $C = 150 + 3x$

8. $C = 1900 + 1.7x + 0.002x^2$

9. $C = 0.005x^2 + 0.5x + 1375$

10. $C = 760 + 0.05x$

EXERCISES 3.6   means that technology can help you solve or check the exercise(s).

In Exercises 1–8, find the vertical and horizontal asymptotes.

1. $f(x) = \dfrac{x^2 + 1}{x^2}$

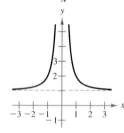

2. $f(x) = \dfrac{4}{(x - 2)^3}$

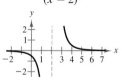

5. $f(x) = \dfrac{x^3}{x^2 - 1}$

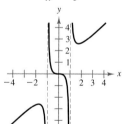

6. $f(x) = \dfrac{-4x}{x^2 + 4}$

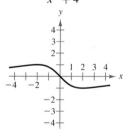

3. $f(x) = \dfrac{x^2 - 2}{x^2 - x - 2}$

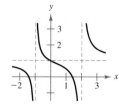

4. $f(x) = \dfrac{2 + x}{1 - x}$

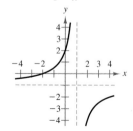

7. $f(x) = \dfrac{x^2 - 1}{2x^2 - 8}$

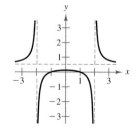

8. $f(x) = \dfrac{x^2 + 1}{x^3 - 8}$

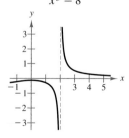

In Exercises 9–14, match the function with its graph. Use horizontal asymptotes as an aid. [The graphs are labeled (a)–(f).]

9. $f(x) = \dfrac{3x^2}{x^2 + 2}$

10. $f(x) = \dfrac{2x}{\sqrt{x^2 + 2}}$

11. $f(x) = \dfrac{x}{x^2 + 2}$

12. $f(x) = 2 + \dfrac{x^2}{x^4 + 1}$

13. $f(x) = 5 - \dfrac{1}{x^2 + 1}$

14. $f(x) = \dfrac{2x^2 - 3x + 5}{x^2 + 1}$

(a)
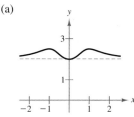

(b)
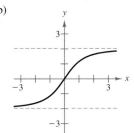

(c)
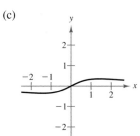

(d)
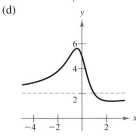

(e)
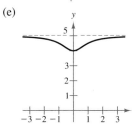

(f)
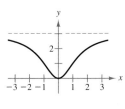

In Exercises 15–22, find the limit.

15. $\displaystyle\lim_{x \to -2^-} \dfrac{1}{(x + 2)^2}$

16. $\displaystyle\lim_{x \to -2^-} \dfrac{1}{x + 2}$

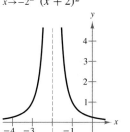

17. $\displaystyle\lim_{x \to 3^+} \dfrac{x - 4}{x - 3}$

18. $\displaystyle\lim_{x \to 1^+} \dfrac{2 + x}{1 - x}$

19. $\displaystyle\lim_{x \to 4^-} \dfrac{x^2}{x^2 - 16}$

20. $\displaystyle\lim_{x \to 4} \dfrac{x^2}{x^2 + 16}$

21. $\displaystyle\lim_{x \to 0^-} \left(1 + \dfrac{1}{x}\right)$

22. $\displaystyle\lim_{x \to 0^-} \left(x^2 - \dfrac{1}{x}\right)$

In Exercises 23–32, find the limit.

23. $\displaystyle\lim_{x \to \infty} \dfrac{2x - 1}{3x + 2}$

24. $\displaystyle\lim_{x \to \infty} \dfrac{5x^3 + 1}{10x^3 - 3x^2 + 7}$

25. $\displaystyle\lim_{x \to \infty} \dfrac{3x}{4x^2 - 1}$

26. $\displaystyle\lim_{x \to \infty} \dfrac{2x^{10} - 1}{10x^{11} - 3}$

27. $\displaystyle\lim_{x \to -\infty} \dfrac{5x^2}{x + 3}$

28. $\displaystyle\lim_{x \to \infty} \dfrac{x^3 - 2x^2 + 3x + 1}{x^2 - 3x + 2}$

29. $\displaystyle\lim_{x \to \infty} (2x - x^{-2})$

30. $\displaystyle\lim_{x \to \infty} (x + 3)^{-2}$

31. $\displaystyle\lim_{x \to -\infty} \left(\dfrac{2x}{x - 1} + \dfrac{3x}{x + 1}\right)$

32. $\displaystyle\lim_{x \to \infty} \left(\dfrac{2x^2}{x - 1} + \dfrac{3x}{x + 1}\right)$

In Exercises 33 and 34, complete the table. Then use the result to estimate the limit of $f(x)$ as x approaches infinity.

33. $f(x) = \dfrac{x + 1}{x\sqrt{x}}$

x	10^0	10^1	10^2	10^3	10^4	10^5	10^6
$f(x)$							

34. $f(x) = x^2 - x\sqrt{x(x - 1)}$

x	10^0	10^1	10^2	10^3	10^4	10^5	10^6
$f(x)$							

In Exercises 35 and 36, complete the table and use the result to estimate the limit of $f(x)$ as x approaches infinity *and* as x approaches negative infinity.

35. $f(x) = \dfrac{2x}{\sqrt{x^2 + 4}}$

x	-10^6	-10^4	-10^2	10^0	10^2	10^4	10^6
$f(x)$							

36. $f(x) = x - \sqrt{x(x - 1)}$

x	-10^6	-10^4	-10^2	10^0	10^2	10^4	10^6
$f(x)$							

In Exercises 37–54, sketch the graph of the equation. Use intercepts, extrema, and asymptotes as sketching aids.

37. $y = \dfrac{2+x}{1-x}$

38. $y = \dfrac{x-3}{x-2}$

39. $f(x) = \dfrac{x^2}{x^2+9}$

40. $f(x) = \dfrac{x^2}{x^2-9}$

41. $g(x) = \dfrac{x^2}{x^2-16}$

42. $g(x) = \dfrac{x}{x^2-4}$

43. $xy^2 = 4$

44. $x^2 y = 4$

45. $y = \dfrac{2x}{1-x}$

46. $y = \dfrac{2x}{1-x^2}$

47. $y = 3(1 - x^{-2})$

48. $y = 1 + x^{-1}$

49. $f(x) = \dfrac{1}{x^2-x-2}$

50. $f(x) = \dfrac{x-2}{x^2-4x+3}$

51. $g(x) = \dfrac{x^2-x-2}{x-2}$

52. $g(x) = \dfrac{x^2-9}{x+3}$

53. $y = \dfrac{2x^2-6}{(x-1)^2}$

54. $y = \dfrac{x}{(x+1)^2}$

55. *Cost* The cost C (in dollars) of producing x units of a product is $C = 1.35x + 4570$.

(a) Find $\overline{C}$ when $x = 100$ and when $x = 1000$.

(b) What is the limit of $\overline{C}$ as x approaches infinity?

56. *Average Cost* A business has a cost (in dollars) of $C = 0.5x + 500$ for producing x units. Find the limit of the average cost per unit as x approaches infinity.

57. *Cost* The cost C (in millions of dollars) for the federal government to seize $p\%$ of a type of illegal drug as it enters the country is modeled by

$$C = \frac{528p}{100-p}, \qquad 0 \le p < 100.$$

(a) Find the cost of seizing 25%.

(b) Find the cost of seizing 50%.

(c) Find the cost of seizing 75%.

(d) Find the limit of C as $p \to 100^-$.

58. *Cost* The cost C (in dollars) of removing $p\%$ of the air pollutants in the stack emission of a utility company that burns coal is modeled by

$$C = \frac{80{,}000p}{100-p}, \qquad 0 \le p < 100.$$

(a) Find the cost of removing 15%.

(b) Find the cost of removing 50%.

(c) Find the cost of removing 90%.

(d) Find the limit of C as $p \to 100^-$.

59. *Learning Curve* Psychologists have developed mathematical models to predict performance P (the percent of correct responses) as a function of n, the number of times a task is performed. One such model is

$$P = \frac{b + \theta a(n-1)}{1 + \theta(n-1)},$$

where a, b, and θ are constants that depend on the actual learning situation. Find the limit of P as n approaches infinity.

 60. *Learning Curve* Consider the learning curve given by

$$P = \frac{0.5 + 0.9(n-1)}{1 + 0.9(n-1)}, \qquad 0 < n.$$

(a) Complete the table for the model.

n	1	2	3	4	5	6	7	8	9	10
P										

(b) Find the limit as n approaches infinity.

(c) Graph this learning curve, and interpret the graph in the context of the problem.

61. *Wildlife Management* The state game commission introduces 30 elk into a new state park. The population N of the herd is modeled by

$$N = \frac{10(3 + 4t)}{1 + 0.1t},$$

where t is the time in years.

(a) Find the size of the herd after 5, 10, and 25 years.

(b) According to this model, what is the limiting size of the herd as time progresses?

62. *Average Profit* The cost and revenue functions for a product are

$$C = 34.5x + 15{,}000 \quad \text{and} \quad R = 69.9x.$$

(a) Find the average profit function

$$\overline{P} = \frac{R - C}{x}.$$

(b) Find the average profit when x is 1000, 10,000, and 100,000.

(c) What is the limit of the average profit function as x approaches infinity?

 63. Use a graphing utility to graph

$$f(x) = \frac{3x}{\sqrt{4x^2 + 1}}$$

and locate any horizontal asymptotes.

Curve Sketching: A Summary

Summary of Curve Sketching Techniques ■ *Summary of
Simple Polynomial Graphs*

Summary of Curve Sketching Techniques

It would be difficult to overstate the importance of using graphs in mathematics. Descartes's introduction of analytic geometry contributed significantly to the rapid advances of calculus that began during the mid-seventeenth century.

So far, you have studied several concepts that are useful in analyzing the graph of a function.

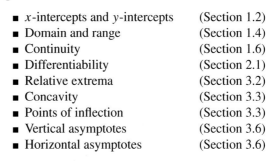

- x-intercepts and y-intercepts   (Section 1.2)
- Domain and range   (Section 1.4)
- Continuity   (Section 1.6)
- Differentiability   (Section 2.1)
- Relative extrema   (Section 3.2)
- Concavity   (Section 3.3)
- Points of inflection   (Section 3.3)
- Vertical asymptotes   (Section 3.6)
- Horizontal asymptotes   (Section 3.6)

When you are sketching the graph of a function, either by hand or with a graphing utility, remember that you cannot normally show the *entire* graph. The decision as to which part of the graph to show is crucial. For instance, which of the viewing rectangles in Figure 3.54 better represents the graph of

$$f(x) = x^3 - 25x^2 + 74x - 20?$$

FIGURE 3.54

The lower viewing rectangle gives a more complete view of the graph, but the context of the problem might indicate that the upper view is better. Here are some guidelines for analyzing the graph of a function.

Guidelines for Analyzing the Graph of a Function

1. Determine the domain and range of the function. If the function models a real-life situation, consider the context.
2. Determine the intercepts and asymptotes of the graph.
3. Locate the x-values where $f'(x)$ and $f''(x)$ are zero or undefined. Use the results to determine the relative extrema and points of inflection.

EXAMPLE 1 *Analyzing a Graph*

Analyze the graph of

$$f(x) = x^3 + 3x^2 - 9x + 5. \qquad \textit{Original function}$$

Solution

Begin by finding the intercepts of the graph. This function factors as

$$f(x) = (x-1)^2(x+5). \qquad \textit{Factored form}$$

Thus, the x-intercepts occur when $x = 1$ and $x = -5$. The derivative is

$$f'(x) = 3x^2 + 6x - 9 \qquad \textit{First derivative}$$
$$= 3(x-1)(x+3). \qquad \textit{Factored form}$$

Thus, the critical numbers of f are $x = 1$ and $x = -3$. The second derivative of f is

$$f''(x) = 6x + 6 \qquad \textit{Second derivative}$$
$$= 6(x+1), \qquad \textit{Factored form}$$

which implies that the second derivative is zero when $x = -1$. Testing the values of $f'(x)$ and $f''(x)$, as shown in Table 3.9, you can see that f has one relative minimum, one relative maximum, and one point of inflection. The graph of f is shown in Figure 3.55.

TABLE 3.9

	$f(x)$	$f'(x)$	$f''(x)$	Shape of graph
x in $(-\infty, -3)$		$+$	$-$	Increasing, concave downward
$x = -3$	32	0	$-$	Relative maximum
x in $(-3, -1)$		$-$	$-$	Decreasing, concave downward
$x = -1$	16	$-$	0	Point of inflection
x in $(-1, 1)$		$-$	$+$	Decreasing, concave upward
$x = 1$	0	0	$+$	Relative minimum
x in $(1, \infty)$		$+$	$+$	Increasing, concave upward

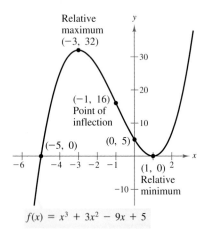

FIGURE 3.55

Note In Example 1, you are able to find the zeros of f, f', and f'' algebraically (by factoring). When this is not feasible, you can use a graphing utility to find the zeros. For instance, the function

$$g(x) = x^3 + 3x^2 - 9x + 6$$

is similar to the function in the example, but it doesn't factor with integer coefficients. Using a graphing utility, you can determine that the function has only one x-intercept, $x \approx -5.0275$.

EXAMPLE 2 *Analyzing a Graph*

Analyze the graph of

$$f(x) = x^4 - 12x^3 + 48x^2 - 64x. \qquad \textit{Original function}$$

Solution

Begin by finding the intercepts of the graph. This function factors as

$$f(x) = x(x^3 - 12x^2 + 48x - 64)$$
$$= x(x - 4)^3. \qquad \textit{Factored form}$$

Thus, the x-intercepts occur when $x = 0$ and $x = 4$. The derivative is

$$f'(x) = 4x^3 - 36x^2 + 96x - 64 \qquad \textit{First derivative}$$
$$= 4(x - 1)(x - 4)^2. \qquad \textit{Factored form}$$

Thus, the critical numbers of f are $x = 1$ and $x = 4$. The second derivative of f is

$$f''(x) = 12x^2 - 72x + 96 \qquad \textit{Second derivative}$$
$$= 12(x - 4)(x - 2), \qquad \textit{Factored form}$$

which implies that the second derivative is zero when $x = 2$ and $x = 4$. Testing the values of $f'(x)$ and $f''(x)$, as shown in Table 3.10, you can see that f has a minimum and two points of inflection. The graph is shown in Figure 3.56.

TABLE 3.10

	$f(x)$	$f'(x)$	$f''(x)$	Shape of graph
x in $(-\infty, 1)$		$-$	$+$	Decreasing, concave upward
$x = 1$	-27	0	$+$	Minimum
x in $(1, 2)$		$+$	$+$	Increasing, concave upward
$x = 2$	-16	$+$	0	Point of inflection
x in $(2, 4)$		$+$	$-$	Increasing, concave downward
$x = 4$	0	0	0	Point of inflection
x in $(4, \infty)$		$+$	$+$	Increasing, concave upward

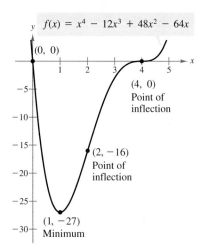

$f(x) = x^4 - 12x^3 + 48x^2 - 64x$

(0, 0)

(4, 0)
Point of inflection

(2, −16)
Point of inflection

(1, −27)
Minimum

FIGURE 3.56

Note A polynomial function of degree n can have at most $n - 1$ relative extrema and at most $n - 2$ points of inflection. For instance, the third-degree polynomial in Example 1 has two relative extrema and one point of inflection. Similarly, the fourth-degree polynomial function in Example 2 has one relative extremum and two points of inflection. Is it possible for a third-degree function to have no relative extrema? Is it possible for a fourth-degree function to have no relative extrema?

Discovery

Show that the function in Example 3 can be rewritten as

$$f(x) = \frac{x^2 - 2x + 4}{x - 2}$$

$$= x + \frac{4}{x - 2}.$$

Use a graphing utility to graph f together with the line $y = x$. How do the two graphs compare as you zoom out? Describe what is meant by a "slant asymptote." Find the slant asymptote of the function $g(x) = \dfrac{x^2 - x - 1}{x - 1}$.

EXAMPLE 3 *Analyzing a Graph*

Analyze the graph of

$$f(x) = \frac{x^2 - 2x + 4}{x - 2}. \qquad \textit{Original function}$$

Solution

The y-intercept occurs at $(0, -2)$. Using the Quadratic Formula on the numerator, you can see that there are no x-intercepts. Because the denominator is zero when $x = 2$ (and the numerator is not zero when $x = 2$), it follows that $x = 2$ is a vertical asymptote of the graph. There are no horizontal asymptotes because the degree of the numerator is greater than the degree of the denominator. The derivative is

$$f'(x) = \frac{(x - 2)(2x - 2) - (x^2 - 2x + 4)}{(x - 2)^2} \qquad \textit{First derivative}$$

$$= \frac{x(x - 4)}{(x - 2)^2}. \qquad \textit{Factored form}$$

Thus, the critical numbers of f are $x = 0$ and $x = 4$. The second derivative is

$$f''(x) = \frac{(x - 2)^2(2x - 4) - (x^2 - 4x)(2)(x - 2)}{(x - 2)^4} \qquad \textit{Second derivative}$$

$$= \frac{(x - 2)(2x^2 - 8x + 8 - 2x^2 + 8x)}{(x - 2)^4}$$

$$= \frac{8}{(x - 2)^3}. \qquad \textit{Factored form}$$

Because the second derivative has no zeros and because $x = 2$ is not in the domain of the function, you can conclude that the graph has no points of inflection. Testing the values of $f'(x)$ and $f''(x)$, as shown in Table 3.11, you can see that f has one relative minimum and one relative maximum. The graph of f is shown in Figure 3.57.

TABLE 3.11

	$f(x)$	$f'(x)$	$f''(x)$	Shape of graph
x in $(-\infty, 0)$		$+$	$-$	Increasing, concave downward
$x = 0$	-2	0	$-$	Relative maximum
x in $(0, 2)$		$-$	$-$	Decreasing, concave downward
$x = 2$	Undef.	Undef.	Undef.	Vertical asymptote
x in $(2, 4)$		$-$	$+$	Decreasing, concave upward
$x = 4$	6	0	$+$	Relative minimum
x in $(4, \infty)$		$+$	$+$	Increasing, concave upward

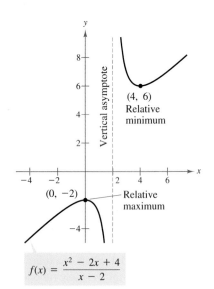

$$f(x) = \frac{x^2 - 2x + 4}{x - 2}$$

FIGURE 3.57

EXAMPLE 4 *Analyzing a Graph*

Analyze the graph of

$$f(x) = \frac{2(x^2 - 9)}{x^2 - 4}.$$ *Original function*

Solution

Begin by writing the function in factored form.

$$f(x) = \frac{2(x - 3)(x + 3)}{(x - 2)(x + 2)}$$ *Factored form*

The y-intercept is $\left(0, \frac{9}{2}\right)$, and the x-intercepts are $(-3, 0)$ and $(3, 0)$. There are vertical asymptotes at $x = \pm 2$ and a horizontal asymptote at $y = 2$. The first derivative is

$$f'(x) = \frac{2[(x^2 - 4)(2x) - (x^2 - 9)(2x)]}{(x^2 - 4)^2}$$ *First derivative*

$$= \frac{20x}{(x^2 - 4)^2}.$$ *Factored form*

Thus, the critical number of f is $x = 0$. The second derivative of f is

$$f''(x) = \frac{(x^2 - 4)^2(20) - (20x)(2)(2x)(x^2 - 4)}{(x^2 - 4)^4}$$ *Second derivative*

$$= \frac{20(x^2 - 4)(x^2 - 4 - 4x^2)}{(x^2 - 4)^4}$$

$$= -\frac{20(3x^2 + 4)}{(x^2 - 4)^3}.$$ *Factored form*

Because the second derivative has no zeros and $x = \pm 2$ is not in the domain of the function, you can conclude that the graph has no points of inflection. Testing the values of $f'(x)$ and $f''(x)$, as shown in Table 3.12, you can see that f has one relative minimum. The graph of f is shown in Figure 3.58.

TABLE 3.12

	$f(x)$	$f'(x)$	$f''(x)$	Shape of graph
x in $(-\infty, -2)$		$-$	$-$	Decreasing, concave downward
$x = -2$	Undef.	Undef.	Undef.	Vertical asymptote
x in $(-2, 0)$		$-$	$+$	Decreasing, concave upward
$x = 0$	$\frac{9}{2}$	0	$+$	Relative minimum
x in $(0, 2)$		$+$	$+$	Increasing, concave upward
$x = 2$	Undef.	Undef.	Undef.	Vertical asymptote
x in $(2, \infty)$		$+$	$-$	Increasing, concave downward

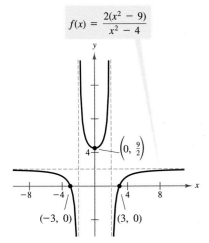

$$f(x) = \frac{2(x^2 - 9)}{x^2 - 4}$$

FIGURE 3.58

EXAMPLE 5 *Analyzing a Graph*

Analyze the graph of

$$f(x) = 2x^{5/3} - 5x^{4/3}.$$ *Original function*

Solution

Begin by writing the function in factored form.

$$f(x) = x^{4/3}(2x^{1/3} - 5)$$ *Factored form*

One of the intercepts is $(0, 0)$. A second x-intercept occurs when $2x^{1/3} = 5$.

$$2x^{1/3} = 5$$

$$x^{1/3} = \tfrac{5}{2}$$

$$x = \left(\tfrac{5}{2}\right)^3$$

$$x = \tfrac{125}{8}$$

The first derivative is

$$f'(x) = \tfrac{10}{3}x^{2/3} - \tfrac{20}{3}x^{1/3}$$ *First derivative*

$$= \tfrac{10}{3}x^{1/3}(x^{1/3} - 2).$$ *Factored form*

Thus, the critical numbers of f are $x = 0$ and $x = 8$. The second derivative is

$$f''(x) = \tfrac{20}{9}x^{-1/3} - \tfrac{20}{9}x^{-2/3}$$ *Second derivative*

$$= \tfrac{20}{9}x^{-2/3}(x^{1/3} - 1)$$

$$= \frac{20(x^{1/3} - 1)}{9x^{2/3}}.$$ *Factored form*

Thus, possible points of inflection occur when $x = 1$ and when $x = 0$. Testing the values of $f'(x)$ and $f''(x)$, as shown in Table 3.13, you can see that f has one relative maximum, one relative minimum, and one point of inflection. The graph of f is shown in Figure 3.59.

TABLE 3.13

	$f(x)$	$f'(x)$	$f''(x)$	Shape of graph
x in $(-\infty, 0)$		$+$	$-$	Increasing, concave downward
$x = 0$	0	0	Undef.	Relative maximum
x in $(0, 1)$		$-$	$-$	Decreasing, concave downward
$x = 1$	-3	$-$	0	Point of inflection
x in $(1, 8)$		$-$	$+$	Decreasing, concave upward
$x = 8$	-16	0	$+$	Relative minimum
x in $(8, \infty)$		$+$	$+$	Increasing, concave upward

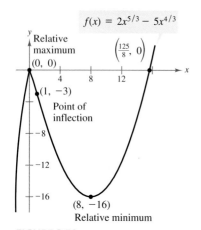

$f(x) = 2x^{5/3} - 5x^{4/3}$

Relative maximum $(0, 0)$

$\left(\tfrac{125}{8}, 0\right)$

$(1, -3)$

Point of inflection

$(8, -16)$

Relative minimum

FIGURE 3.59

Summary of Simple Polynomial Graphs

A summary of the graphs of polynomial functions of degrees 0, 1, 2, and 3 is shown in Figure 3.60. Because of their simplicity, lower-degree polynomial functions are commonly used as mathematical models.

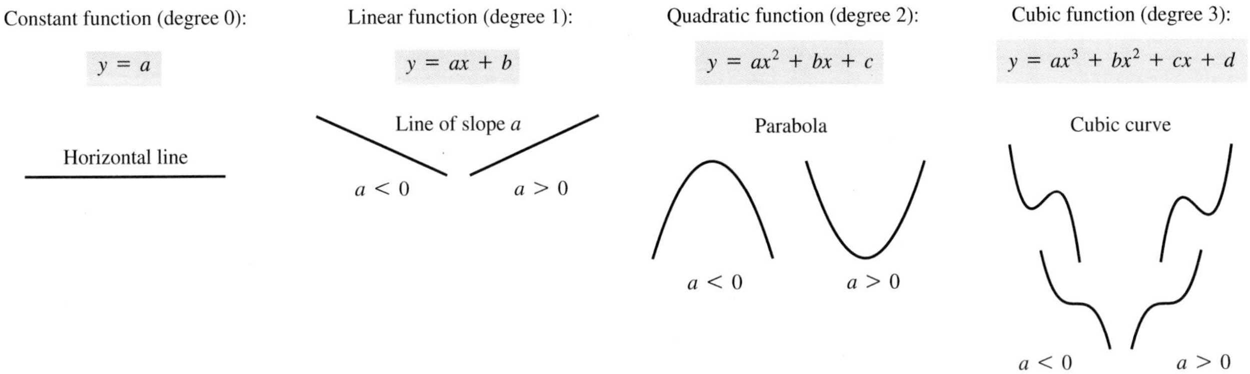

Constant function (degree 0):

$$y = a$$

Horizontal line

Linear function (degree 1):

$$y = ax + b$$

Line of slope a

$a < 0$ $a > 0$

Quadratic function (degree 2):

$$y = ax^2 + bx + c$$

Parabola

$a < 0$ $a > 0$

Cubic function (degree 3):

$$y = ax^3 + bx^2 + cx + d$$

Cubic curve

$a < 0$ $a > 0$

FIGURE 3.60

Note   The graph of any cubic polynomial has one point of inflection. The slope of the graph at the point of inflection may be zero or nonzero.

Discussion Problem   Graphs of Fourth-Degree Polynomial Functions

In the summary shown above, the graphs of cubic functions have been classified as four basic types. How many basic types of graphs are possible with fourth-degree polynomial functions? Make a rough sketch of each type. Then use a graphing utility to classify each of the following. (In each case, use a viewing rectangle that shows all the basic characteristics of the graph.)

a. $y = x^4$

b. $y = -x^4 + 5x^2$

c. $y = x^4 - x^3 + x$

d. $y = -x^4 - 4x^3 - 3x^2 + x$

e. $y = -x^4 + 2x^2$

f. $y = -x^4 + x^3$

g. $y = x^4 - 5x$

h. $y = x^4 - 8x^3 + 8x^2$

Warm Up

The following warm-up exercises involve skills that were covered in earlier sections. You will use these skills in the exercise set for this section.

In Exercises 1–4, find the vertical and horizontal asymptotes of the graph.

1. $f(x) = \dfrac{1}{x^2}$ **2.** $f(x) = \dfrac{8}{(x-2)^2}$

3. $f(x) = \dfrac{40x}{x+3}$ **4.** $f(x) = \dfrac{x^2-3}{x^2-4x+3}$

In Exercises 5–10, determine the open intervals on which the function is increasing or decreasing.

5. $f(x) = x^2 + 4x + 2$ **6.** $f(x) = -x^2 - 8x + 1$

7. $f(x) = x^3 - 3x + 1$ **8.** $f(x) = \dfrac{-x^3 + x^2 - 1}{x^2}$

9. $f(x) = \dfrac{x-2}{x-1}$ **10.** $f(x) = -x^3 - 4x^2 + 3x + 2$

E X E R C I S E S   3.7

 means that technology can help you solve or check the exercise(s).

In Exercises 1–30, sketch the graph of the function. Choose a scale that allows all relative extrema and points of inflection to be identified on the graph.

1. $y = -x^2 - 2x + 3$ **2.** $y = 2x^2 - 4x + 1$

3. $y = x^3 - 4x^2 + 6$ **4.** $y = -\frac{1}{3}(x^3 - 3x + 2)$

5. $y = 2 - x - x^3$ **6.** $y = x^3 + 3x^2 + 3x + 2$

7. $y = 3x^3 - 9x + 1$ **8.** $y = \frac{1}{3}(x-1)^2 + 2$

9. $y = 3x^4 + 4x^3$ **10.** $y = 3x^4 - 6x^2$

11. $y = (x+1)(x-2)(x-5)$

12. $y = -x^3 + 3x^2 + 9x - 2$

13. $y = x^4 - 8x^3 + 18x^2 - 16x + 5$

14. $y = x^4 - 4x^3 + 16x - 16$

15. $y = x^4 - 4x^3 + 16x$ **16.** $y = x^5 + 1$

17. $y = x^5 - 5x$ **18.** $y = (x-1)^5$

19. $y = |2x - 3|$ **20.** $y = |x^2 - 6x + 5|$

21. $y = \dfrac{x^2+2}{x^2+1}$ **22.** $y = \dfrac{x}{x^2+1}$

23. $y = 3x^{2/3} - 2x$ **24.** $y = 3x^{2/3} - x^2$

25. $y = 1 - x^{2/3}$ **26.** $y = (1-x)^{2/3}$

27. $y = x^{1/3} + 1$ **28.** $y = x^{-1/3}$

29. $y = x^{5/3} - 5x^{2/3}$ **30.** $y = x^{4/3}$

In Exercises 31–40, sketch the graph of the function. Label the intercepts, relative extrema, points of inflection, and asymptotes. Then state the domain of the function.

31. $y = \dfrac{1}{x-2} - 3$ **32.** $y = \dfrac{x^2+1}{x^2-2}$

33. $y = \dfrac{2x}{x^2-1}$ **34.** $y = \dfrac{x^2-6x+12}{x-4}$

35. $y = x\sqrt{4-x}$ **36.** $y = x\sqrt{4-x^2}$

37. $y = \dfrac{x-3}{x}$ **38.** $y = x + \dfrac{32}{x^2}$

39. $y = \dfrac{x^3}{x^3-1}$ **40.** $y = \dfrac{x^4}{x^4-1}$

In Exercises 41–44, find values of *a*, *b*, *c*, and *d* so that the graph of

$$f(x) = ax^3 + bx^2 + cx + d$$

will resemble the given graph. Then use a graphing utility to confirm your result. (There are many correct answers.)

41.

42.

43.

44.

In Exercises 45–48, use the graph of *f'* or *f''* to sketch the graph of *f*. (There are many correct answers.)

45.

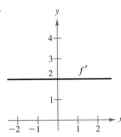

46.

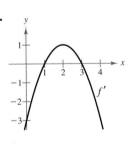

47.

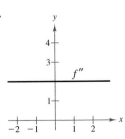

48.

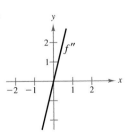

In Exercises 49 and 50, sketch a graph of a function *f* having the given characteristics. (There are many correct answers.)

Function	*First Derivative*
49. $f(-2) = 0$	$f'(x) > 0, \ -\infty < x < -1$
$f(0) = 0$	$f'(-1) = 0$
$f(2) = 0$	$f'(x) < 0, \ -1 < x < 0$
	$f'(0) = 0$
	$f'(x) > 0, \ 0 < x < 1$
	$f'(1) = 0$
	$f'(x) < 0, \ 1 < x < \infty$

Function	*Second Derivative*
50. $f(4) = 2$	$f''(x) > 0, \ x \neq 4$

51. *Cost* An employee of a delivery company earns \$9 per hour driving a delivery van in an area where gasoline costs \$1.20 per gallon. When the van is driven at a constant speed *s* (in miles per hour with $40 \leq s \leq 65$), the van gets $500/s$ miles per gallon.

(a) Find the cost *C* as a function of *s* for a 100-mile trip on an interstate highway.

(b) Sketch the graph of the function found in part (a) and determine the most economical speed.

52. *Cost* Repeat Exercise 51 with a gasoline cost of \$1.50 per gallon.

53. *Consumer Preference* Use a graphing utility to graph the consumer indifference curve

$$5 = \frac{xy}{x+y}, \qquad x > 5, \ y > 5,$$

where *x* and *y* are measures of preference of two products.

54. *Profit* The management of a company is considering three possible models to predict the company's profit from 1995 through 2000. Model I gives the expected annual profit if the current trends continue. Models II and III give the expected annual profit for various combinations of increased labor and energy costs. In each model, *P* is the profit (in billions of dollars) and $t = 0$ corresponds to 1995.

Model I: $P = 0.03t^2 - 0.01t + 3.39$
Model II: $P = 0.08t + 3.36$
Model III: $P = -0.07t^2 + 0.05t + 3.38$

(a) Use a graphing utility to graph all three models on the same viewing rectangle.

(b) For which models are profits increasing during the interval from 1995 through 2000?

(c) Which model is the most optimistic? Which is the most pessimistic?

Differentials

When the derivative was defined in Section 2.1 as the limit of the ratio $\Delta y/\Delta x$, it seemed natural to retain the quotient symbolism for the limit itself. Thus, the derivative of y with respect to x was denoted by

$$\frac{dy}{dx} = \lim_{\Delta x \to 0} \frac{\Delta y}{\Delta x}$$

even though we did not interpret dy/dx as the quotient of two separate quantities. In this section, you will see that the quantities dy and dx can be assigned meanings in such a way that their quotient, when $dx \neq 0$, is equal to the derivative of y with respect to x.

Definition of Differentials

Let $y = f(x)$ represent a differentiable function. The **differential of x** (denoted by dx) is any nonzero real number. The **differential of y** (denoted by dy) is

$$dy = f'(x)\,dx.$$

Note In this definition, dx can have any nonzero value. In most applications, however, dx is chosen to be small and this choice is denoted by $dx = \Delta x$.

One use of differentials is in approximating the change in $f(x)$ that corresponds to a change in x, as shown in Figure 3.61. This change is denoted by

$$\Delta y = f(x + \Delta x) - f(x). \qquad \textit{Change in } y$$

In Figure 3.61, notice that as Δx gets smaller and smaller, the values of dy and Δy get closer and closer. That is, when Δx is small,

$$dy \approx \Delta y.$$

This approximation is the basis for most applications of differentials.

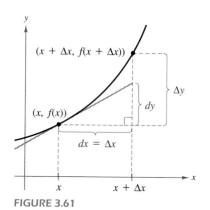

FIGURE 3.61

242

EXAMPLE 1 *Interpreting Differentials Graphically*

Consider the function

$$f(x) = x^2.$$ *Original function*

Find the value of dy when $x = 1$ and $dx = 0.01$. Compare this with the value of Δy when $x = 1$ and $\Delta x = 0.01$. Interpret the results graphically.

Solution

Begin by finding the derivative of f.

$$f'(x) = 2x$$ *Derivative of f*

When $x = 1$ and $dx = 0.01$, the value of the differential dy is

$$
\begin{aligned}
dy &= f'(x)\,dx && \text{\textit{Differential of y}} \\
&= f'(1)(0.01) && \text{\textit{Substitute 1 for x and 0.01 for dx.}} \\
&= 2(1)(0.01) && \text{\textit{Use } } f'(x) = 2x. \\
&= 0.02. && \text{\textit{Simplify.}}
\end{aligned}
$$

When $x = 1$ and $\Delta x = 0.01$, the value of Δy is

$$
\begin{aligned}
\Delta y &= f(x + \Delta x) - f(x) && \text{\textit{Change in y}} \\
&= f(1.01) - f(1) && \text{\textit{Substitute 1 for x and 0.01 for } } \Delta x. \\
&= (1.01)^2 - (1)^2 \\
&= 0.0201. && \text{\textit{Simplify.}}
\end{aligned}
$$

Note that $dy \approx \Delta y$, as shown in Figure 3.62.   ▬

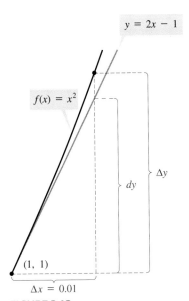

$y = 2x - 1$

$f(x) = x^2$

Δy

dy

$(1, 1)$

$\Delta x = 0.01$

FIGURE 3.62

The validity of the approximation

$$dy \approx \Delta y, \qquad dx \neq 0$$

stems from the definition of the derivative. That is, the existence of the limit

$$f'(x) = \lim_{\Delta x \to 0} \frac{f(x + \Delta x) - f(x)}{\Delta x}$$

implies that when Δx is close to zero, then $f'(x)$ is close to the difference quotient. Thus, you can write

$$
\begin{aligned}
\frac{f(x + \Delta x) - f(x)}{\Delta x} &\approx f'(x) \\
f(x + \Delta x) - f(x) &\approx f'(x)\,\Delta x \\
\Delta y &\approx f'(x)\,\Delta x.
\end{aligned}
$$

Substituting dx for Δx and dy for $f'(x)\,dx$ produces

$$\Delta y \approx dy.$$

Marginal Analysis

Differentials are used in economics to approximate changes in revenue, cost, and profit. Suppose that $R = f(x)$ is the total revenue for selling x units of a product. When the number of units increases by 1, the change in x is $\Delta x = 1$, and the change in R is

$$\Delta R = f(x + \Delta x) - f(x) \approx dR = \frac{dR}{dx}\,dx.$$

In other words, you can use the differential dR to approximate the change in the revenue that accompanies the sale of one additional unit. Similarly, the differentials dC and dP can be used to approximate the change in cost or profit that accompanies the sale (or production) of one additional unit.

Real Life

EXAMPLE 2 *Using Marginal Analysis*

The demand function for a product is modeled by

$$p = \sqrt{400 - x}.$$

Use differentials to approximate the change in revenue as sales increase from 256 units to 257 units. Compare this with the actual change in revenue.

Solution

Begin by finding the marginal revenue, dR/dx.

$$R = xp \qquad\qquad\qquad\qquad\qquad \textit{Formula for revenue}$$

$$= x\sqrt{400 - x} \qquad\qquad\qquad\qquad \textit{Use } p = \sqrt{400 - x}.$$

$$\frac{dR}{dx} = x\left(\frac{1}{2}\right)(400 - x)^{-1/2}(-1) + (400 - x)^{1/2}(1) \quad \textit{Product Rule}$$

$$= \frac{800 - 3x}{2\sqrt{400 - x}} \qquad\qquad\qquad\qquad \textit{Simplify.}$$

When $x = 256$ and $dx = \Delta x = 1$, you can approximate the change in the revenue to be

$$dR = \frac{dR}{dx}\,dx$$

$$= \frac{800 - 3(256)}{2\sqrt{400 - 256}}(1)$$

$$\approx \$1.33.$$

When x increases from 256 to 257, the actual change in revenue is

$$\Delta R = 257\sqrt{400 - 257} - 256\sqrt{400 - 256}$$

$$= 3073.27 - 3072.00$$

$$\approx \$1.27.$$

Real Life

EXAMPLE 3 *Using Marginal Analysis*

The profit derived from selling x units of an item is modeled by

$$P = 0.0002x^3 + 10x.$$

Use the differential dP to approximate the change in profit when the production level changes from 50 to 51 units. Compare this with the actual gain in profit obtained by increasing the production level from 50 to 51 units.

Solution

The marginal profit is

$$\frac{dP}{dx} = 0.0006x^2 + 10.$$

When $x = 50$ and $dx = 1$, the differential dP is

$$\begin{aligned} dP &= \frac{dP}{dx}\, dx \\ &= [0.0006(50)^2 + 10](1) \\ &= \$11.50. \end{aligned}$$

When x changes from 50 to 51 units, the actual change in profit is

$$\begin{aligned} \Delta P &= [(0.0002)(51)^3 + 10(51)] - [(0.0002)(50)^3 + 10(50)] \\ &\approx 536.53 - 525.00 \\ &= \$11.53. \end{aligned}$$

These values are shown graphically in Figure 3.63.

Note Example 3 uses differentials to solve the same problem that was solved in Example 5 in Section 2.3. Look back at that solution. Which approach do you prefer?

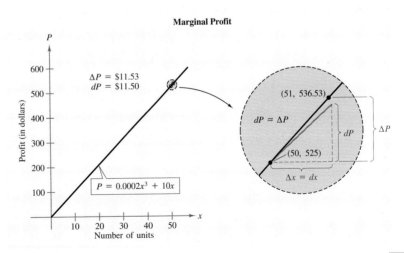

Marginal Profit

$\Delta P = \$11.53$
$dP = \$11.50$

$P = 0.0002x^3 + 10x$

Profit (in dollars)

Number of units

$(51, 536.53)$

$dP \approx \Delta P$

dP

ΔP

$(50, 525)$

$\Delta x = dx$

FIGURE 3.63

Formulas for Differentials

You can use the definition of differentials to rewrite each differentiation rule in **differential form.** For example, if u and v are differentiable functions of x, then $du = u'\,dx$ and $dv = v'\,dx$, which implies that you can write the Product Rule in the following differential form.

$$d[uv] = \frac{d}{dx}[uv]\,dx \qquad\qquad \text{Differential of } uv$$

$$= [uv' + vu']\,dx \qquad\qquad \text{Product Rule}$$

$$= uv'\,dx + vu'\,dx$$

$$= u\,dv + v\,du \qquad\qquad \text{Differential form of Product Rule}$$

The following summary gives the differential forms of the differentiation rules presented so far in the text.

Differential Forms of Differentiation Rules

Constant Multiple Rule:	$d[cu] = c\,du$
Sum or Difference Rule:	$d[u \pm v] = du \pm dv$
Product Rule:	$d[uv] = u\,dv + v\,du$
Quotient Rule:	$d\left[\dfrac{u}{v}\right] = \dfrac{v\,du - u\,dv}{v^2}$
Constant Rule:	$d[c] = 0$
Power Rule:	$d[x^n] = nx^{n-1}\,dx$
	$d[x] = 1$

The next example compares the derivatives and differentials of several simple functions.

EXAMPLE 4 *Finding Differentials*

	Function	*Derivative*	*Differential*
a.	$y = x^2$	$\dfrac{dy}{dx} = 2x$	$dy = 2x\,dx$
b.	$y = \dfrac{3x + 2}{5}$	$\dfrac{dy}{dx} = \dfrac{3}{5}$	$dy = \dfrac{3}{5}\,dx$
c.	$y = 2x^2 - 3x$	$\dfrac{dy}{dx} = 4x - 3$	$dy = (4x - 3)\,dx$
d.	$y = \dfrac{1}{x}$	$\dfrac{dy}{dx} = -\dfrac{1}{x^2}$	$dy = -\dfrac{1}{x^2}\,dx$

Error Propagation

A common use of differentials is the estimation of errors that are propagated by physical measuring devices. This is illustrated in Example 5.

Real
Life

EXAMPLE 5 *Estimating Measurement Errors*

The radius of a ball bearing is measured to be 0.7 inch, as shown in Figure 3.64. This implies that the volume of the ball bearing is $\frac{4}{3}\pi(0.7)^3 \approx 1.4368$ cubic inches. You are told that the measurement of the radius is correct to within 0.01 inch. How far off could the calculation of the volume be?

Solution

Because the value of r can be off by 0.01 inch, it follows that

$$-0.01 \leq \Delta r \leq 0.01. \qquad \textit{Possible error in measuring}$$

Using $\Delta r = dr$, you can estimate the possible error in the volume.

$$V = \tfrac{4}{3}\pi r^3 \qquad\qquad \textit{Formula for volume}$$

$$dV = \frac{dV}{dr}\,dr \qquad\qquad \textit{Formula for differential of V}$$

$$= 4\pi r^2\,dr$$

$$= 4\pi(0.7)^2(\pm 0.01) \qquad \textit{Substitute for r and dr.}$$

$$\approx \pm 0.0616 \text{ cubic inch} \qquad \textit{Possible error}$$

Thus, the volume of the ball bearing could range between $(1.4368 - 0.0616) = 1.3752$ cubic inches and $(1.4368 + 0.0616) = 1.4984$ cubic inches. ▬

In Example 5, the **relative error** in the volume is defined to be the ratio of dV to V. This ratio is

$$\frac{dV}{V} \approx \frac{\pm 0.0616}{1.4368} \approx \pm 0.0429.$$

This corresponds to a **percentage error** of 4.29%.

FIGURE 3.64

0.7 in.

Discussion Problem   Finding Propagated Errors

a. In Example 5, if the radius of the ball bearing were measured to be 1.5 inches, correct to within 0.01 inch, would the percentage error of the volume be greater or smaller than in the example?

b. In Example 5, if the radius of the ball bearing were measured to be 0.7 inches, correct to within 0.02 inch, would the percentage error of the volume be greater or smaller than in the example?

Warm Up

The following warm-up exercises involve skills that were covered in earlier sections. You will use these skills in the exercise set for this section.

In Exercises 1–6, find the derivative.

1. $C = 44 + 0.09x^2$

2. $R = x(1.25 + 0.02\sqrt{x})$

3. $P = -0.03x^{1/3} + 1.4x - 2250$

4. $A = \frac{1}{4}\sqrt{3}x^2$

5. $C = 2\pi r$

6. $S = 4\pi r^2$

In Exercises 7–10, write a formula for the quantity.

7. Area A of a circle of radius r

8. Area A of a square of side x

9. Volume V of a cube of edge x

10. Volume V of a sphere of radius r

E X E R C I S E S   3.8 means that technology can help you solve or check the exercise(s).

In Exercises 1–10, find the differential dy.

1. $y = 3x^2 - 4$

2. $y = 2x^{3/2}$

3. $y = (4x - 1)^3$

4. $y = (1 - 2x^2)^{-2}$

5. $y = \sqrt{x^2 + 1}$

6. $y = \sqrt[3]{6x^2}$

7. $y = x\sqrt{1 - x^2}$

8. $y = -4x^{-1}$

9. $y = \dfrac{x - 3}{1 - 3x}$

10. $y = \dfrac{x}{x^2 + 1}$

In Exercises 11–14, let $x = 1$ and $\Delta x = 0.01$. Find Δy.

11. $f(x) = 5x^2 - 1$

12. $f(x) = \sqrt{3x}$

13. $f(x) = \dfrac{4}{\sqrt[3]{x}}$

14. $f(x) = \dfrac{x}{x^2 + 1}$

In Exercises 15–18, compare the values of dy and Δy.

15. $y = x^3$ $x = 1$ $\Delta x = dx = 0.1$

16. $y = 1 - 2x^2$ $x = 0$ $\Delta x = dx = -0.1$

17. $y = x^4 + 1$ $x = -1$ $\Delta x = dx = 0.01$

18. $y = 2x + 1$ $x = 2$ $\Delta x = dx = 0.01$

In Exercises 19–22, let $x = 2$ and complete the table for the function.

$dx = \Delta x$	dy	Δy	$\Delta y - dy$	$dy/\Delta y$
1.000				
0.500				
0.100				
0.010				
0.001				

19. $y = x^2$

20. $y = \dfrac{1}{x^2}$

21. $y = x^5$

22. $y = \sqrt{x}$

23. *Demand* The demand function for a product is modeled by

$$p = 75 - 0.25x.$$

(a) If x changes from 7 to 8, what is the corresponding change in p? Compare the values of Δp and dp.

(b) Repeat part (a) when x changes from 70 to 71 units.

24. *Wildlife Management* A state game commission introduces 50 deer into newly acquired state game lands. The population N of the herd can be modeled by

$$N = \frac{10(5 + 3t)}{1 + 0.04t},$$

where t is the time in years. Use differentials to approximate the change in the herd size from $t = 5$ to $t = 6$.

Marginal Analysis In Exercises 25–28, use differentials to approximate the change in cost, revenue, or profit corresponding to an increase in sales of *one unit*. For instance, in Exercise 25, approximate the change in cost as *x* increases from 12 units to 13 units.

Function	*x-Value*
25. $C = 0.05x^2 + 4x + 10$	$x = 12$
26. $R = 30x - 0.15x^2$	$x = 75$
27. $P = -0.5x^3 + 2500x - 6000$	$x = 50$
28. $P = -x^2 + 60x - 100$	$x = 25$

29. *Marginal Analysis* A retailer has determined that the monthly sales x of a watch is 150 units when the price is $50, but decreases to 120 units when the price is $60. Assume that the demand is a linear function of the price. Find the revenue R as a function of x and approximate the change in revenue for a one-unit increase in sales when $x = 141$. Make a sketch showing dR and ΔR.

30. *Marginal Analysis* A manufacturer determines that the demand x for a product is inversely proportional to the square of the price p. When the price is $10, the demand is 2500. Find the revenue R as a function of x and approximate the change in revenue for a one-unit increase in sales when $x = 3000$. Make a sketch showing dR and ΔR.

31. *Marginal Analysis* The demand x for a radio is 30,000 units per week when the price is $25 and 40,000 units when the price is $20. The initial investment is $275,000 and the cost per unit is $17. Assume that the demand is a linear function of the price. Find the profit P as a function of x and approximate the change in profit for a one-unit increase in sales when $x = 28,000$. Make a sketch showing dP and ΔP.

32. *Marginal Analysis* The variable cost for the production of a calculator is $14.25 and the initial investment is $110,000. Find the total cost C as a function of x, the number of units produced. Then use differentials to approximate the change in the cost for a one-unit increase in production when $x = 50,000$. Make a sketch showing dC and ΔC. Explain why $dC = \Delta C$ in this problem.

33. The area A of a square of side x is $A = x^2$.
   (a) Compute dA and ΔA in terms of x and Δx.
   (b) In the figure, identify the region whose area is dA.
   (c) In the figure, identify the region whose area is $\Delta A - dA$.

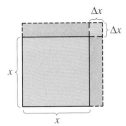

34. *Area* The side of a square is measured to be 12 inches, with a possible error of $\frac{1}{64}$ inch. Use differentials to approximate the possible error and the relative error in computing the area of the square.

35. *Area* The radius of a circle is measured to be 10 inches, with a possible error of $\frac{1}{8}$ inch. Use differentials to approximate the possible error and the relative error in computing the area of the circle.

36. *Volume and Surface Area* The edge of a cube is measured to be 12 inches, with a possible error of 0.03 inch. Use differentials to approximate the possible error and the relative error in computing (a) the volume of the cube, and (b) the surface area of the cube.

37. *Volume* The radius of a sphere is measured to be 6 inches, with a possible error of 0.02 inch. Use differentials to approximate the possible error and the relative error in computing the volume of the sphere.

38. *Drug Concentration* The concentration C (in milligrams per milliliter) in a patient's bloodstream t hours after injection into muscle tissue is modeled by

$$C = \frac{3t}{27 + t^3}.$$

Use differentials to approximate the change in the concentration when t changes from $t = 1$ to $t = 1.5$.

39. *Profit* The profit P for a company selling x units of a product is modeled by

$$P = R - C = (500x - x^2) - (0.5x^2 - 77x + 3000).$$

Approximate the change in the profit as the production changes from $x = 115$ to $x = 120$.

Chapter Summary and Study Tips

After studying this chapter, you should have acquired the following skills. The exercise numbers are keyed to the Review Exercises that begin on page 252. Answers to odd-numbered Review Exercises are given in the back of the text.

- Find the critical numbers of a function. (*Section 3.1*)   Review Exercises 1–4
  c is a critical number of f if $f'(c) = 0$ or $f'(c)$ is undefined.

- Find the open intervals on which a function is increasing or decreasing. (*Section 3.1*)   Review Exercises 5–8
  Increasing if $f'(x) > 0$
  Decreasing if $f'(x) < 0$

- Find intervals on which a real-life model is increasing or decreasing, and interpret the results in context. (*Section 3.1*)   Review Exercises 9, 10

- Use the First-Derivative Test to find the relative extrema of a function. (*Section 3.2*)   Review Exercises 11–20

- Find the absolute extrema of a continuous function on a closed interval. (*Section 3.2*)   Review Exercises 21–28

- Find minimum and maximum values of a real-life model and interpret the results in context. (*Section 3.2*)   Review Exercises 29, 30

- Find the open intervals on which the graph of a function is concave upward or concave downward. (*Section 3.3*)   Review Exercises 31–34
  Concave upward if $f''(x) > 0$
  Concave downward if $f''(x) < 0$

- Find the points of inflection of the graph of a function. (*Section 3.3*)   Review Exercises 35–38

- Use the Second-Derivative Test to find the relative extrema of a function. (*Section 3.3*)   Review Exercises 39–42

- Find the point of diminishing returns of an input-output model. (*Section 3.3*)   Review Exercises 43, 44

- Solve real-life optimization problems. (*Section 3.4*)   Review Exercises 45–48

- Solve business and economics optimization problems. (*Section 3.5*)   Review Exercises 49–54

- Find the price elasticity of demand for a demand function. (*Section 3.5*)   Review Exercises 55, 56

- Find the vertical and horizontal asymptotes of a rational function and sketch its graph. (*Section 3.6*)   Review Exercises 57–62

- Find infinite limits and limits at infinity. (*Section 3.6*)   Review Exercises 63–68

- Use asymptotes to answer questions about real life. (*Section 3.6*)   Review Exercises 69, 70

- Analyze the graph of a function. (*Section 3.7*)   Review Exercises 71–80

- Find the differential of a function. (*Section 3.8*)   Review Exercises 81–84

- Use differentials to approximate changes in a function. (*Section 3.8*)   Review Exercises 85–88

- Use differentials to approximate changes in real-life models. (*Section 3.8*)   Review Exercises 89, 90

*Several student study aids are available with this text. The *Student Solutions Guide* includes detailed solutions to all odd-numbered exercises, as well as practice chapter tests with answers. The *Graphics Calculator Guide* offers instructions on the use of a variety of graphing calculators and computer graphing software. The *Brief Calculus TUTOR* includes additional examples for selected exercises in the text.

■ Solve Problems Graphically, Analytically, and Numerically   When analyzing the graph of a function, use a variety of problem-solving strategies. For instance, if you were asked to analyze the graph of $f(x) = x^3 - 4x^2 + 5x - 4$, you could begin *graphically*. That is, you could use a graphing utility to find a viewing rectangle that appears to show the important characteristics of the graph. From the graph shown below, the function appears to have one relative minimum, one relative maximum, and one point of inflection.

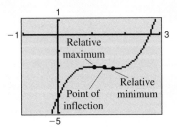

Next, you could use calculus to *analyze* the graph. Because the derivative of f is $f'(x) = 3x^2 - 8x + 5 = (3x - 5)(x - 1)$, the critical numbers of f are $x = \frac{5}{3}$ and $x = 1$. By the First-Derivative Test, you can conclude that $x = \frac{5}{3}$ yields a relative minimum and $x = 1$ yields a relative maximum. Because $f''(x) = 6x - 8$, you can conclude that $x = \frac{4}{3}$ yields a point of inflection. Finally, you could analyze the graph *numerically*. For instance, you could construct a table of values and observe that f is increasing in the interval $(-\infty, 1)$, decreasing in the interval $(1, \frac{5}{3})$, and increasing in the interval $(\frac{5}{3}, \infty)$.

■ Problem-Solving Strategies   If you get stuck when trying to solve an optimization problem, consider the following strategies.

1. *Draw a Diagram.* If feasible, draw a diagram that represents the problem. Label all known values and unknown values on the diagram.
2. *Solve a Simpler Problem.* Simplify the problem, or write several simple examples of the problem. For instance, if you are asked to find the dimensions that will produce a maximum area, try calculating the area of several examples.
3. *Rewrite the Problem in Your Own Words.* Rewriting a problem can help you understand it better.
4. *Guess and Check.* Try guessing the answer, then check your guess in the statement of the original problem. By refining your guesses, you may be able to think of a general strategy for solving the problem.

Review Exercises

means that technology can help you solve or check the exercise(s).

In Exercises 1–4, find the critical numbers of the function.

1. $f(x) = -x^2 + 2x + 4$

2. $g(x) = (x-1)^2(x-3)$

3. $h(x) = \sqrt{x}(x-3)$

4. $f(x) = (x+1)^3$

 In Exercises 5–8, determine the open intervals on which the function is increasing or decreasing. Solve the problem analytically and graphically.

5. $f(x) = x^2 + x - 2$

6. $g(x) = -x^2 + 7x - 12$

7. $h(x) = \dfrac{x^2 - 3x - 4}{x - 3}$

8. $f(x) = -x^3 + 6x^2 - 2$

9. *Temperature* The daily maximum temperature T (in degrees Fahrenheit) for New York City can be modeled by

$$T = 0.036t^4 - 0.909t^3 + 5.874t^2 - 2.599t + 37.789,$$

where $0 \le t \le 12$ and $t = 0$ corresponds to January 1. *(Source: National Oceanic and Atmospheric Administration)*

(a) Find the intervals in which the model is increasing.
(b) Find the intervals in which the model is decreasing.
(c) Interpret the results of parts (a) and (b).
(d) Sketch a graph of the model.

10. *Newspaper Publishers* The number N of morning newspaper publishers in the United States from 1940 through 1990 can be modeled by

$$N = 0.0063t^3 - 0.0281t^{5/2} - 14.968t^{1/2} + 375.018,$$

where $0 \le t \le 50$ and $t = 0$ corresponds to 1940. *(Source: American Newspaper Publishers Association)*

(a) Find the intervals in which the model is increasing.
(b) Find the intervals in which the model is decreasing.
(c) Interpret the results of parts (a) and (b).
(d) Sketch a graph of the model.

 In Exercises 11–20, use the First-Derivative Test to find the relative extrema of the function. Then use a graphing utility to confirm your result.

11. $f(x) = 4x^3 - 6x^2 - 2$

12. $f(x) = \frac{1}{4}x^4 - 8x$

13. $g(x) = x^2 - 16x + 12$

14. $h(x) = 4 + 10x - x^2$

15. $h(x) = 2x^2 - x^4$

16. $s(x) = x^4 - 8x^2 + 3$

17. $f(x) = \dfrac{6}{x^2 + 1}$

18. $f(x) = \dfrac{2}{x^2 - 1}$

19. $h(x) = \dfrac{x^2}{x - 2}$

20. $g(x) = x - 6\sqrt{x}$, $x > 0$

 In Exercises 21–28, find the absolute extrema of the function on the indicated interval. Then use a graphing utility to confirm your result.

21. $f(x) = x^2 + 5x + 6$; $[-3, 0]$

22. $f(x) = x^4 - 2x^3$; $[0, 2]$

23. $f(x) = x^3 - 12x + 1$; $[-4, 4]$

24. $f(x) = \dfrac{8}{x} + x$; $[1, 4]$

25. $f(x) = 3x^4 - 6x^2 + 2$; $[0, 2]$

26. $f(x) = -x^4 + x^2 + 2$; $[0, 2]$

27. $f(x) = \dfrac{2x}{x^2 + 1}$; $[-1, 2]$

28. $f(x) = 4\sqrt{x} - x^2$; $[0, 3]$

29. *Diamond Prices* During the 1980s, the average price p (in dollars) on the Antwerp Index of a 1-carat diamond can be modeled by

$$p = -0.7t^3 + 16.25t^2 - 106t + 388,$$

where $2 \le t \le 10$ and $t = 2$ corresponds to 1982. *(Source: Diamond High Council)*

(a) Estimate the maximum average price between 1982 and 1990. When did this price occur?
(b) Estimate the minimum average price between 1982 and 1990. When did this price occur?

30. *Waste Oxidation* When organic waste is dumped into a pond, the decomposition of the waste consumes oxygen. A model for the oxygen level O (where 1 is the normal level) of a pond as waste material oxidizes is

$$O = \dfrac{t^2 - t + 1}{t^2 + 1}, \qquad 0 \le t,$$

where t is the time in weeks.

(a) When is the oxygen level lowest? What is this level?
(b) When is the oxygen level highest? What is this level?
(c) Describe the oxygen level as t increases.

In Exercises 31–34, determine the open intervals on which the graph of the function is concave upward or concave downward. Then use a graphing utility to confirm your result.

31. $f(x) = (x - 2)^3$

32. $h(x) = x^5 - 10x^2$

33. $g(x) = \frac{1}{4}(-x^4 + 8x^2 - 12)$

34. $h(x) = x^3 - 6x$

In Exercises 35–38, find the points of inflection of the graph of the function.

35. $f(x) = \frac{1}{2}x^4 - 4x^3$

36. $f(x) = (x + 2)^2(x - 4)$

37. $f(x) = x^3(x - 3)^2$

38. $f(x) = \frac{1}{4}x^4 - 2x^2 - x$

In Exercises 39–42, use the Second-Derivative Test to find the relative extrema of the function.

39. $f(x) = x^5 - 5x^3$ **40.** $f(x) = x(x^2 - 3x - 9)$

41. $f(x) = (x - 1)^3(x + 4)^2$ **42.** $f(x) = \dfrac{(x - 2)^2}{4 - x}$

In Exercises 43 and 44, identify the point of diminishing returns for the input-output function. For each function, R is the revenue (in thousands of dollars) and x is the amount spent on advertising (in thousands of dollars).

43. $R = \frac{1}{1500}(150x^2 - x^3)$, $0 \le x \le 100$

44. $R = -\frac{2}{3}(x^3 - 12x^2 - 6)$, $0 \le x \le 8$

45. *Minimum Sum* Find two positive numbers whose product is 169 and whose sum is a minimum. Solve the problem analytically *and* graphically.

46. *Length* The wall of a building is to be braced by a beam that must pass over a 5-foot fence that is parallel to the building and 4 feet from the building. Find the length of the shortest beam that can be used.

47. *Charitable Contributions* The percent P of income that Americans give to charities can be modeled by

$$P = 0.0014x^2 - 0.1529x + 5.855, \quad 5 \le x \le 100,$$

where x is the annual income in thousands of dollars. (*Source: Independent Sector*)

(a) What income level corresponds to the least percent of charitable contributions?

(b) What income level corresponds to the greatest percent of charitable contributions?

48. *Construction Costs* A fence is to be built to enclose a rectangular region of 4800 square feet. The fencing material along three sides costs $3 per foot. The fencing material along the fourth side cost $4 per foot.

(a) Find the most economical dimensions of the region.

(b) How would the result of part (a) change if the fencing material costs for all sides increased by $1 per foot?

49. *Profit* The demand and cost functions for a product are

$$p = 36 - 4x \quad \text{and} \quad C = 2x^2 + 6.$$

(a) What level of production will produce a maximum profit?

(b) What level of production will produce a minimum average cost per unit?

50. *Profit* The demand and cost functions for a product are

$$p = 75 - x \quad \text{and} \quad C = 0.5x^2 + 62x + 125.$$

(a) What level of production will produce a maximum profit?

(b) What level of production will produce a minimum average cost per unit?

51. *Revenue* For groups of 20 or more, a theater determines the ticket price p according to the formula

$$p = 15 - 0.1(n - 20), \quad 20 \le n \le N,$$

where n is the number in the group. What should the value of N be? Explain your reasoning.

52. *Cost* The cost of fuel to run a locomotive is proportional to the $\frac{3}{2}$ power of the speed. At a speed of 25 miles per hour, the cost of fuel is $50 per hour. Other costs amount to $100 per hour. Find the speed that will minimize the cost per mile.

53. *Inventory Cost* The cost C of inventory depends on ordering and storage costs,

$$C = \left(\frac{Q}{x}\right)s + \left(\frac{x}{2}\right)r,$$

where Q is the number of units sold per year, r is the cost of storing one unit for 1 year, s is the cost of placing an order, and x is the number of units in the order. Determine the order size that will minimize the cost when $Q = 10,000$, $s = 4.5$, and $r = 5.76$.

54. *Profit* The demand and cost functions for a product are

$$p = 600 - 3x \quad \text{and} \quad C = 0.3x^2 + 6x + 600,$$

where p is the price per unit, x is the number of units, and C is the total cost. The profit for producing x units is

$$P = xp - C - xt,$$

where t is the excise tax per unit. Find the maximum profit for excise taxes of $t = \$5$, $t = \$10$, and $t = \$20$.

In Exercises 55 and 56, find the intervals on which the demand is elastic, inelastic, and of unit elasticity.

55. $p = \dfrac{36}{\sqrt[3]{x}}, \qquad x > 0$

56. $p = \sqrt{1900 - x}, \qquad 0 \le x \le 1900$

 In Exercises 57–62, find the vertical and horizontal asymptotes of the graph. Then sketch the graph.

57. $h(x) = \dfrac{2x + 3}{x - 4}$ **58.** $g(x) = \dfrac{5x^2}{x^2 + 2}$

59. $f(x) = \dfrac{1 - 3x}{x}$ **60.** $h(x) = \dfrac{3x}{\sqrt{x^2 + 2}}$

61. $f(x) = \dfrac{3}{x^2 - 5x + 4}$ **62.** $h(x) = \dfrac{2x^2 + 3x - 5}{x - 1}$

In Exercises 63–68, find the limit, if it exists.

63. $\displaystyle \lim_{x \to 0^+} \left(x - \frac{1}{x^3} \right)$

64. $\displaystyle \lim_{x \to -1^+} \frac{x^2 - 2x + 1}{x + 1}$

65. $\displaystyle \lim_{x \to \infty} \frac{5x^2 + 3}{2x^2 - x + 1}$

66. $\displaystyle \lim_{x \to \infty} \frac{3x^2 - 2x + 3}{x + 1}$

67. $\displaystyle \lim_{x \to -\infty} \frac{3x^2}{x + 2}$

68. $\displaystyle \lim_{x \to -\infty} \left(\frac{x}{x - 2} + \frac{2x}{x + 2} \right)$

 69. *Ultraviolet Radiation* For a person with sensitive skin, the amount T (in hours) of exposure to the sun can be modeled by

$$T = \frac{0.37s + 23.8}{s}, \qquad 0 < s \le 120,$$

where s is the Sunsor Scale reading. *(Source: Sunsor, Inc.)*

(a) Use a graphing utility to graph the model. Compare your result with the graph below.
(b) Describe the value of T as s increases.

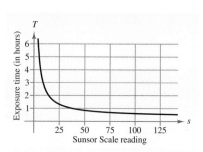

70. *Average Cost and Profit* The cost and revenue functions for a product are

$$C = 10{,}000 + 48.9x \quad \text{and} \quad R = 68.5x.$$

(a) Find the average cost function.
(b) What is the limit of the average cost as x approaches infinity?
(c) Find the average profit when x is 1 million, 2 million, and 10 million.
(d) What is the limit of the average profit as x increases without bound?

In Exercises 71–80, sketch the graph of the function. Label the intercepts, relative extrema, points of inflection, and asymptotes. State the domain of the function.

71. $f(x) = 4x - x^2$ **72.** $f(x) = 4x^3 - x^4$

73. $f(x) = x\sqrt{16 - x^2}$ **74.** $f(x) = x + \dfrac{4}{x^2}$

75. $f(x) = \dfrac{x + 1}{x - 1}$ **76.** $f(x) = x^2 + \dfrac{2}{x}$

77. $f(x) = x^3(x + 1)$ **78.** $f(x) = \dfrac{2x}{1 + x^2}$

79. $f(x) = x^{4/5}$ **80.** $f(x) = x^3 + \dfrac{243}{x}$

In Exercises 81–84, find the differential dy.

81. $y = 6x^2 - 5$ **82.** $y = (3x^2 - 2)^3$

83. $y = -\dfrac{5}{\sqrt[3]{x}}$ **84.** $y = \dfrac{2 - x}{x + 5}$

In Exercises 85–88, use differentials to approximate the change in cost, revenue, or profit corresponding to an increase in sales of one unit.

85. $C = 40x^2 + 1225$, $x = 10$

86. $C = 1.5\sqrt[3]{x} + 500$, $x = 125$

87. $R = 6.25x + 0.4x^{3/2}$, $x = 225$

88. $P = 0.003x^2 + 0.019x - 1200$, $x = 750$

89. *Surface Area and Volume* The diameter of a sphere is measured to be 18 inches with a possible error of 0.05 inch. Use differentials to approximate the possible error in the surface area and the volume of the sphere.

90. *Demand* A company finds that the demand for its product is modeled by $p = 85 - 0.125x$. If x changes from 7 to 8, what is the corresponding change in p? Compare the values of Δp and dp.

Sample Post-Graduation Exam Questions

The following questions were taken from certified public accountant (CPA) exams, graduate management admission tests (GMAT), graduate records exams (GRE), actuarial exams, or college-level academic skills tests (CLAST). The answers to the questions are given in the back of the book.

For Questions 1–3, use the data given in the graph. *(Source: National Sporting Goods Association)*

1. The percent increase in revenue between 1985 and 1987 is about
(a) 40%   (b) 95%   (c) 200%   (d) 125%

2. Which of the following statements about the sale of walking shoes can be inferred from the graph?
  I. The greatest increase in revenue occurs between 1988 and 1989.
  II. In 1991, the revenue was greater than $1.3 billion.
  III. Between 1985 and 1988, the average increase in revenue per year was about $122 million.

(a) I and II   (b) I and III   (c) II and III   (d) I, II, and III

3. Let $f(x)$ represent the revenue function. In 1990, $f'(x) =$
(a) 1   (b) 1509   (c) 0   (d) undefined

4. In 1985, a company issued 75,000 shares of stock. Each share of stock was worth $85.75. Five years later, each share of the stock was worth $72.21. How much less were the shares worth in 1990 than in 1985?
(a) $1,115,500   (b) $1,051,500   (c) $1,155,000
(d) $1,015,500

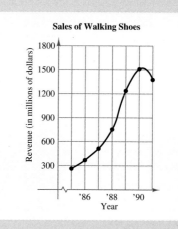

FIGURE FOR 1–4

5. Which of these figure most resembles the graph of $f(x) = \dfrac{x}{x^2 - 4}$?

(a) 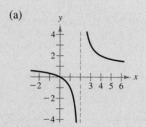   (b) 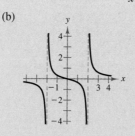   (c) 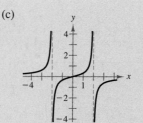   (d)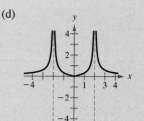

6. On April 1, 1994, Starn Corp. issued 300 of its $1000 face value bonds at $101 plus accrued interest. The bonds were dated November 1, 1993, and bear interest at an annual rate of 8% payable semiannually on November 1 and May 1. What amount did Starn receive from the bond issuance?
(a) $313,000   (b) $327,000   (c) $303,000   (d) $300,000

$$\frac{d}{dx}[\ln x] = \frac{1}{x}$$

4 Exponential and Logarithmic Functions

During the past 300 years, as calculus was developing into its modern-day form, three categories of functions have proved most useful for modeling real life. These are algebraic functions (polynomial functions, rational functions, and radical functions), exponential and logarithmic functions, and trigonometric functions.

Algebraic functions are the oldest. In fact, records of polynomial functions date back to Babylonian times. Nonalgebraic functions, such as exponential and logarithmic functions, are much newer. Many mathematicians, including the Swiss mathematician Leonhard Euler (1707–1783) and the Indian mathematician Srinivasa Ramanujan (1887–1920), have contributed to the representation of nonalgebraic functions.

$$e \approx 2.71828$$

Ramanujan

THE GRANGER COLLECTION, New York

Exponential Functions

Exponential Functions

You are already familiar with the behavior of algebraic functions such as

$$f(x) = x^2, \quad g(x) = \sqrt{x} = x^{1/2}, \quad \text{and} \quad h(x) = \frac{1}{x} = x^{-1},$$

which involve a variable raised to a constant power. By interchanging roles and raising a constant to a variable power, you obtain another important class of functions called **exponential functions.** Some simple examples are

$$f(x) = 2^x, \quad g(x) = \left(\frac{1}{10}\right)^x = \frac{1}{10^x}, \quad \text{and} \quad h(x) = 3^{2x} = 9^x.$$

In general, you can use any positive base $a \neq 1$ as the base of an exponential function.

Definition of Exponential Function

If $a > 0$ and $a \neq 1$, then the **exponential function** with base a is given by

$$f(x) = a^x.$$

For working with exponential functions, the following properties of exponents are useful.

Properties of Exponents

Let a and b be positive numbers.

1. $a^0 = 1$ **2.** $a^x a^y = a^{x+y}$ **3.** $\dfrac{a^x}{a^y} = a^{x-y}$

4. $(a^x)^y = a^{xy}$ **5.** $(ab)^x = a^x b^x$ **6.** $\left(\dfrac{a}{b}\right)^x = \dfrac{a^x}{b^x}$

7. $a^{-x} = \dfrac{1}{a^x}$

EXAMPLE 1 *Applying Properties of Exponents*

a. $(2^2)(2^3) = 2^{2+3} = 2^5 = 32$ *Property 2*

b. $(2^2)(2^{-3}) = 2^{2-3} = 2^{-1} = \frac{1}{2}$ *Properties 2 and 7*

c. $(3^2)^3 = 3^{2(3)} = 3^6 = 729$ *Property 4*

d. $\left(\frac{1}{3}\right)^{-2} = \frac{1}{(1/3)^2} = \left(\frac{1}{1/3}\right)^2 = 3^2 = 9$ *Properties 6 and 7*

e. $\dfrac{3^2}{3^3} = 3^{2-3} = 3^{-1} = \dfrac{1}{3}$ *Properties 3 and 7*

f. $(2^{1/2})(3^{1/2}) = [(2)(3)]^{1/2} = 6^{1/2} = \sqrt{6}$ *Property 5* ▬

Although Example 1 demonstrates the properties of exponents with integer and rational exponents, it is important to realize that the properties hold for *all* real exponents. With a calculator, you can obtain approximations of a^x for any base a and any real exponent x. Here are some examples.

$$2^{-0.6} \approx 0.660, \qquad \pi^{0.75} \approx 2.360, \qquad (1.56)^{\sqrt{2}} \approx 1.876$$

Real Life

EXAMPLE 2 *Dating Organic Material*

In living organic material, the ratio of radioactive carbon isotopes to the total number of carbon atoms is about 1 to 10^{12}. When organic material dies, its radioactive carbon isotopes begin to decay, with a half-life of about 5700 years. This means that after 5700 years, the ratio of isotopes to atoms will have decreased to one-half the original ratio, after a second 5700 years the ratio will have decreased to one-fourth of the original, and so on. Figure 4.1 shows this decreasing ratio. The formula for the ratio R of carbon isotopes to carbon atoms is

$$R = \left(\frac{1}{10^{12}}\right)\left(\frac{1}{2}\right)^{t/5700},$$

where t is the time (in years). Find the value of R for the following times.

a. 10,000 years   **b.** 20,000 years   **c.** 25,000 years

Solution

a. $R = \left(\dfrac{1}{10^{12}}\right)\left(\dfrac{1}{2}\right)^{10,000/5700} \approx 2.964 \times 10^{-13}$ *Ratio for 10,000 years*

b. $R = \left(\dfrac{1}{10^{12}}\right)\left(\dfrac{1}{2}\right)^{20,000/5700} \approx 8.785 \times 10^{-14}$ *Ratio for 20,000 years*

c. $R = \left(\dfrac{1}{10^{12}}\right)\left(\dfrac{1}{2}\right)^{25,000/5700} \approx 4.783 \times 10^{-14}$ *Ratio for 25,000 years* ▬

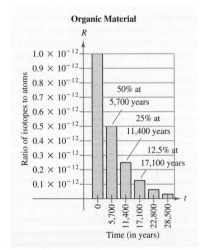

Organic Material

50% at 5,700 years

25% at 11,400 years

12.5% at 17,100 years

Time (in years)

FIGURE 4.1

EXAMPLE 3 *Graphing Exponential Functions*

Sketch the graphs of the following exponential functions.

a. $f(x) = 2^x$ **b.** $g(x) = \left(\frac{1}{2}\right)^x = 2^{-x}$ **c.** $h(x) = 3^x$

Solution

To sketch these functions by hand, you can begin by constructing a table of values, as shown in Table 4.1.

TABLE 4.1

x	-3	-2	-1	0	1	2	3	4
$f(x) = 2^x$	$\frac{1}{8}$	$\frac{1}{4}$	$\frac{1}{2}$	1	2	4	8	16
$g(x) = 2^{-x}$	8	4	2	1	$\frac{1}{2}$	$\frac{1}{4}$	$\frac{1}{8}$	$\frac{1}{16}$
$h(x) = 3^x$	$\frac{1}{27}$	$\frac{1}{9}$	$\frac{1}{3}$	1	3	9	27	81

The graphs of the three functions are shown in Figure 4.2. Note that the graphs of $f(x) = 2^x$ and $h(x) = 3^x$ are increasing, whereas the graph of $g(x) = 2^{-x}$ is decreasing.

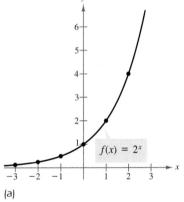

(a)

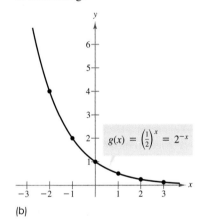

(b)

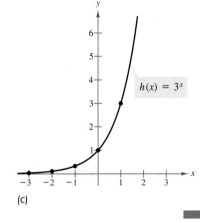

(c)

FIGURE 4.2

TECHNOLOGY

Try graphing the functions $f(x) = 2^x$ and $h(x) = 3^x$ on the same viewing rectangle, as shown at the right. From the display, you can see that the graph of h is increasing more rapidly than the graph of f.

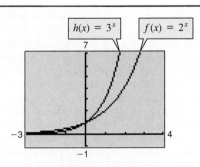

The forms of the graphs in Figure 4.2 are typical of the graphs of the exponential functions $y = a^{-x}$ and $y = a^x$, where $a > 1$. The basic characteristics of such graphs are summarized in Figure 4.3.

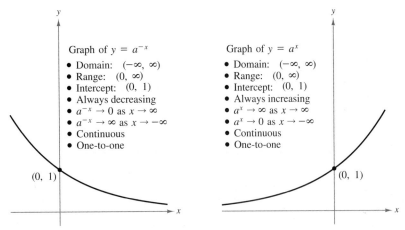

Graph of $y = a^{-x}$
- Domain: $(-\infty, \infty)$
- Range: $(0, \infty)$
- Intercept: $(0, 1)$
- Always decreasing
- $a^{-x} \to 0$ as $x \to \infty$
- $a^{-x} \to \infty$ as $x \to -\infty$
- Continuous
- One-to-one

Graph of $y = a^x$
- Domain: $(-\infty, \infty)$
- Range: $(0, \infty)$
- Intercept: $(0, 1)$
- Always increasing
- $a^x \to \infty$ as $x \to \infty$
- $a^x \to 0$ as $x \to -\infty$
- Continuous
- One-to-one

$(0, 1)$ $(0, 1)$

FIGURE 4.3   Characteristics of the Exponential Functions a^{-x} and $a^x (a > 1)$

EXAMPLE 4 *Graphing an Exponential Function*

Sketch the graph of

$$f(x) = 3^{-x} - 1.$$

Solution
Begin by creating a table of values, as shown in Table 4.2.

TABLE 4.2

x	-2	-1	0	1	2
$f(x)$	$3^2 - 1 = 8$	$3^1 - 1 = 2$	$3^0 - 1 = 0$	$3^{-1} - 1 = -\frac{2}{3}$	$3^{-2} - 1 = -\frac{8}{9}$

From the limit

$$\lim_{x \to \infty} (3^{-x} - 1) = \lim_{x \to \infty} 3^{-x} - \lim_{x \to \infty} 1$$

$$= \lim_{x \to \infty} \frac{1}{3^x} - \lim_{x \to \infty} 1$$

$$= 0 - 1$$

$$= -1,$$

you can see that $y = -1$ is a horizontal asymptote of the graph. The graph is shown in Figure 4.4. ■

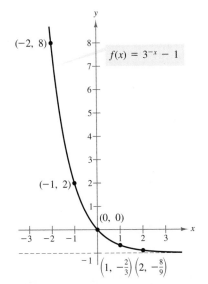

$(-2, 8)$

$f(x) = 3^{-x} - 1$

$(-1, 2)$

$(0, 0)$

$\left(1, -\frac{2}{3}\right) \left(2, -\frac{8}{9}\right)$

FIGURE 4.4

Natural Exponential Functions

At the beginning of this section, exponential functions were introduced using an unspecified base a. In calculus, the most convenient (or natural) choice for a base is the irrational number e, whose decimal approximation is

$$e \approx 2.71828182846.$$

Although this choice of base may seem unusual, its convenience will become apparent as the rules for differentiating exponential functions are developed in Section 4.2. In that development, you will encounter the limit used in the following definition of e.

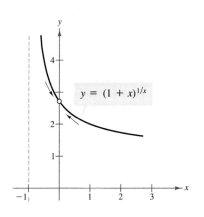

$$\lim_{x \to 0}(1 + x)^{1/x} \approx 2.71828$$

FIGURE 4.5

Limit Definition of e

The irrational number e is defined to be the limit of $(1 + x)^{1/x}$ as $x \to 0$. That is,

$$\lim_{x \to 0}(1 + x)^{1/x} = e.$$

The graph of $y = (1 + x)^{1/x}$ is shown in Figure 4.5. Try reproducing this graph with a graphing utility. Then, use the zoom and trace features to find values of y near $x = 0$. You will find that the y-values get closer and closer to the number $e \approx 2.71828$.

EXAMPLE 5 *Graphing the Natural Exponential Function*

Sketch the graph of

$$f(x) = e^x.$$

Solution

Begin by evaluating the function for several values of x, as shown in Table 4.3.

TABLE 4.3

x	-2	-1	0	1	2
$f(x)$	$e^{-2} \approx 0.135$	$e^{-1} \approx 0.368$	$e^0 = 1$	$e^1 \approx 2.718$	$e^2 \approx 7.389$

The graph of $f(x) = e^x$ is shown in Figure 4.6. Note that e^x is positive for all values of x. Moreover, the graph has the x-axis as a horizontal asymptote to the left. That is,

$$\lim_{x \to -\infty} e^x = 0.$$

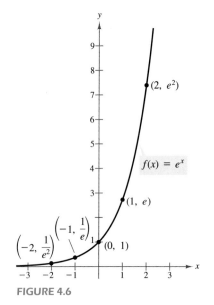

FIGURE 4.6

Exponential functions are often used to model the growth of a quantity or a population. When the quantity's growth is *not* restricted, an exponential model is often used. When the quantity's growth *is* restricted, the best model is often a **logistics growth function** of the form

$$f(t) = \frac{a}{1 + be^{-kt}}.$$

Graphs of both types of population growth models are shown in Figure 4.7.

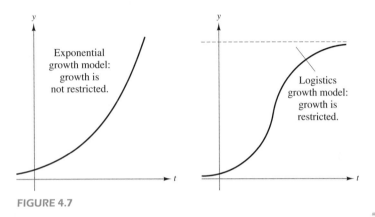

Exponential growth model: growth is not restricted.

Logistics growth model: growth is restricted.

FIGURE 4.7

FPG International/© Telegraph Colour Library

When a culture is grown in a dish, the size of the dish and the available food limit the culture's growth.

Real Life

EXAMPLE 6 *Modeling a Population*

A bacterial culture is growing according to the *logistics growth model*

$$y = \frac{1.25}{1 + 0.25e^{-0.4t}}, \qquad 0 \le t,$$

where y is the culture weight (in grams) and t is the time (in hours). Find the weight of the culture after 0 hours, 1 hour, and 10 hours. What is the limit of the model as t increases without bound?

Solution

$$y = \frac{1.25}{1 + 0.25e^{-0.4(0)}} = 1 \text{ gram} \qquad \textit{Weight when } t = 0$$

$$y = \frac{1.25}{1 + 0.25e^{-0.4(1)}} \approx 1.071 \text{ grams} \qquad \textit{Weight when } t = 1$$

$$y = \frac{1.25}{1 + 0.25e^{-0.4(10)}} \approx 1.244 \text{ grams} \qquad \textit{Weight when } t = 10$$

As t approaches infinity, the limit of y is

$$\lim_{t \to \infty} \frac{1.25}{1 + 0.25e^{-0.4t}} = \lim_{t \to \infty} \frac{1.25}{1 + (0.25/e^{0.4t})} = \frac{1.25}{1 + 0} = 1.25.$$

Thus, as t increases without bound, the weight of the culture approaches 1.25 grams. The graph of the model is shown in Figure 4.8. ■

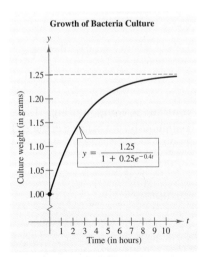

Growth of Bacteria Culture

$$y = \frac{1.25}{1 + 0.25e^{-0.4t}}$$

Culture weight (in grams)

Time (in hours)

FIGURE 4.8

Extended Application: Compound Interest

If P dollars is deposited in an account at an annual interest rate of r (in decimal form), what is the balance after 1 year? The answer depends on the number of times the interest is compounded, according to the formula

$$A = P\left(1 + \frac{r}{n}\right)^n,$$

where n is the number of compoundings per year. The balances for a deposit of $1000 at 8%, at various compounding periods, are shown in Table 4.4.

TABLE 4.4

Number of times compounded per year, n	Balance (in dollars), A
Annually, $n = 1$	$A = P\left(1 + \frac{0.08}{1}\right)^1 = \$1080.00$
Semiannually, $n = 2$	$A = P\left(1 + \frac{0.08}{2}\right)^2 = \$1081.60$
Quarterly, $n = 4$	$A = P\left(1 + \frac{0.08}{4}\right)^4 = \$1082.43$
Monthly, $n = 12$	$A = P\left(1 + \frac{0.08}{12}\right)^{12} = \$1083.00$
Daily, $n = 365$	$A = P\left(1 + \frac{0.08}{365}\right)^{365} = \$1083.28$

It may surprise you to discover that as n increases, the balance A approaches a limit, as indicated in the following development. In this development, let $x = r/n$. Then $x \to 0$ as $n \to \infty$, and you have

$$A = \lim_{n \to \infty} P\left(1 + \frac{r}{n}\right)^n$$

$$= P \lim_{n \to \infty} \left[\left(1 + \frac{r}{n}\right)^{n/r}\right]^r$$

$$= P\left[\lim_{x \to 0} (1 + x)^{1/x}\right]^r \qquad \text{\textit{Substitute x for r/n.}}$$

$$= Pe^r.$$

This limit is the balance after one year of **continuous compounding.** Thus, for a deposit of $1000 at 8%, compounded continuously, the balance at the end of the year would be $A = 1000e^{0.08} \approx \$1083.29$.

Summary of Compound Interest Formulas

Let P be the amount deposited, t be the number of years, A be the balance, and r be the annual interest rate (in decimal form).

1. Compounded n times per year: $A = P\left[1 + (r/n)\right]^{nt}$
2. Compounded continuously: $A = Pe^{rt}$

The average interest rates paid by banks for savings accounts have varied greatly during the past 30 years. At times, savings accounts have earned as much as 12% annual interest and at times they have earned as little as 3%. The next example shows how the annual interest rate can affect the balance of an account.

Real Life

EXAMPLE 7 *Finding Account Balances*

You are creating a trust fund for your newborn nephew. You deposit $25,000 in an account, with instructions that the account will be turned over to your nephew on his 25th birthday. Compare the balances in the account for the following.

a. 6%, compounded continuously   **b.** 6%, compounded quarterly

c. 10%, compounded continuously   **d.** 10%, compounded quarterly

Solution

a. $25,000e^{0.06(25)} = \$112,042.23$ *6%, compounded continuously*

b. $25,000\left(1 + \dfrac{0.06}{4}\right)^{4(25)} = \$110,801.14$ *6%, compounded quarterly*

c. $25,000e^{0.1(25)} = \$304,562.35$ *10%, compounded continuously*

d. $25,000\left(1 + \dfrac{0.1}{4}\right)^{4(25)} = \$295,342.91$ *10%, compounded quarterly*

The growth of the account for parts a and c is shown in Figure 4.9. Notice the dramatic difference between the balances at 6% and at 10%. ▬

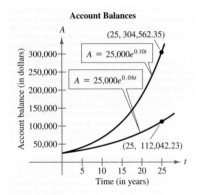

Account Balances

$A = 25,000e^{0.10t}$

$A = 25,000e^{0.06t}$

(25, 304,562.35)

(25, 112,042.23)

Account balance (in dollars)

Time (in years)

FIGURE 4.9

Discussion Problem   Compound Interest

You want to invest $5000 in a certificate of deposit for 10 years. You are given the following options. Which would you choose?

a. You can buy a 10-year certificate of deposit that earns 7%, compounded continuously. You are guaranteed a 7% rate for the entire 10 years, but you cannot withdraw the money early without paying a substantial penalty.

b. You can buy a 5-year certificate of deposit that earns 6%, compounded continuously. After 5 years, you can reinvest your money at whatever the current interest rate is at that time.

c. You can buy a 2-year certificate of deposit that earns 5%, compounded continuously. After 2 years, you can reinvest your money at whatever the current interest rate is at that time.

Warm Up The following warm-up exercises involve skills that were covered in earlier sections. You will use these skills in the exercise set for this section.

In Exercises 1 and 2, discuss the continuity of the function.

1. $f(x) = \dfrac{3x^2 + 2x + 1}{x^2 + 1}$

2. $g(x) = \dfrac{x^2 - 9x + 20}{x - 4}$

In Exercises 3–10, find the limit.

3. $\displaystyle\lim_{x \to \infty} \frac{25}{1 + 4x}$

4. $\displaystyle\lim_{x \to \infty} \frac{16x}{3 + x^2}$

5. $\displaystyle\lim_{x \to \infty} \frac{8x^3 + 2}{2x^3 + x}$

6. $\displaystyle\lim_{x \to \infty} \frac{x}{2x}$

7. $\displaystyle\lim_{x \to \infty} \frac{3}{2 + (1/x)}$

8. $\displaystyle\lim_{x \to \infty} \frac{6}{1 + x^{-2}}$

9. $\displaystyle\lim_{x \to \infty} 2^{-x}$

10. $\displaystyle\lim_{x \to \infty} \frac{7}{1 + 5x}$

E X E R C I S E S   4.1   means that technology can help you solve or check the exercise(s).

In Exercises 1 and 2, evaluate each expression.

1. (a) $5(5^3)$ (b) $27^{2/3}$
(c) $64^{3/4}$ (d) $81^{1/2}$
(e) $25^{3/2}$ (f) $32^{2/5}$

2. (a) $\left(\frac{1}{5}\right)^3$ (b) $\left(\frac{1}{8}\right)^{1/3}$
(c) $64^{2/3}$ (d) $\left(\frac{5}{8}\right)^2$
(e) $100^{3/2}$ (f) $4^{5/2}$

In Exercises 3–8, use the properties of exponents to simplify the expression.

3. (a) $(5^2)(5^3)$ (b) $(5^2)(5^{-3})$
(c) $(5^2)^2$ (d) 5^{-3}

4. (a) $\dfrac{5^3}{5^6}$ (b) $\left(\dfrac{1}{5}\right)^{-2}$
(c) $(8^{1/2})(2^{1/2})$ (d) $(32^{3/2})\left(\frac{1}{2}\right)^{3/2}$

5. (a) $\dfrac{5^3}{25^2}$ (b) $(9^{2/3})(3)(3^{2/3})$
(c) $[(25^{1/2})(5^2)]^{1/3}$ (d) $(8^2)(4^3)$

6. (a) $(4^3)(4^2)$ (b) $\left(\frac{1}{4}\right)^2 (4^2)$
(c) $(4^6)^{1/2}$ (d) $[(8^{-1})(8^{2/3})]^3$

7. (a) $(e^3)(e^4)$ (b) $(e^3)^4$
(c) $(e^3)^{-2}$ (d) e^0

8. (a) $\left(\dfrac{1}{e}\right)^{-2}$ (b) $\left(\dfrac{e^5}{e^2}\right)^{-1}$
(c) $\dfrac{e^5}{e^3}$ (d) $\dfrac{1}{e^{-3}}$

In Exercises 9–22, solve the equation for x.

9. $3^x = 81$ **10.** $5^{x+1} = 125$

11. $\left(\frac{1}{3}\right)^{x-1} = 27$ **12.** $\left(\frac{1}{5}\right)^{2x} = 625$

13. $4^3 = (x + 2)^3$ **14.** $4^2 = (x + 2)^2$

15. $x^{3/4} = 8$ **16.** $(x + 3)^{4/3} = 16$

17. $e^{-3x} = e$ **18.** $e^x = 1$

19. $e^{\sqrt{x}} = e^3$ **20.** $e^{-1/x} = \sqrt{e}$

21. $x^{2/3} = \sqrt[3]{e^2}$ **22.** $\dfrac{x^2}{2} = e^2$

In Exercises 23–28, match the function with its graph. [The graphs are labeled (a)–(f).]

23. $f(x) = 3^x$

24. $f(x) = 3^{-x/2}$

25. $f(x) = -3^x$

26. $f(x) = 3^{x-2}$

27. $f(x) = 3^{-x} - 1$

28. $f(x) = 3^x + 2$

(a)
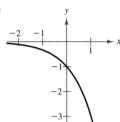

(b)
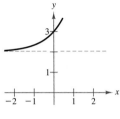

(c)
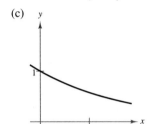

(d)
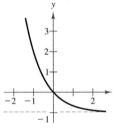

(e)
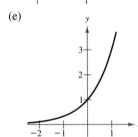

(f)
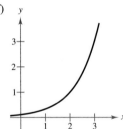

In Exercises 29–44, sketch the graph of the function.

29. $f(x) = 6^x$

30. $f(x) = 4^x$

31. $f(x) = \left(\frac{1}{5}\right)^x = 5^{-x}$

32. $f(x) = \left(\frac{1}{4}\right)^x = 4^{-x}$

33. $y = 3^{-x^2}$

34. $y = 2^{-x^2}$

35. $y = 3^{-|x|}$

36. $y = 3^{|x|}$

37. $s(t) = \frac{1}{4}(3^{-t})$

38. $s(t) = 2^{-t} + 3$

39. $h(x) = e^{x-2}$

40. $f(x) = e^{2x}$

41. $N(t) = 500e^{-0.2t}$

42. $A(t) = 500e^{0.15t}$

43. $g(x) = \frac{2}{1 + e^{x^2}}$

44. $g(x) = \frac{10}{1 + e^{-x}}$

Compound Interest In Exercises 45–48, complete the table to determine the balance A for P dollars invested at rate r for t years, compounded n times per year.

n	1	2	4	12	365	Continuous compounding
A						

45. $P = \$1000, r = 3\%, t = 10$ years

46. $P = \$2500, r = 5\%, t = 20$ years

47. $P = \$1000, r = 3\%, t = 40$ years

48. $P = \$2500, r = 5\%, t = 40$ years

Compound Interest In Exercises 49–52, complete the table to determine the amount of money P that should be invested at rate r to produce a final balance of $\$100,000$ in t years.

t	1	10	20	30	40	50
P						

49. $r = 4\%$, compounded continuously

50. $r = 3\%$, compounded continuously

51. $r = 5\%$, compounded monthly

52. $r = 6\%$, compounded daily

53. *Demand* The demand function for a product is modeled by

$$p = 5000\left(1 - \frac{4}{4 + e^{-0.002x}}\right),$$

as shown in the figure. Find the price of the product if the quantity demanded is (a) $x = 100$ units and (b) $x = 500$ units. What is the limit of the price as x increases without bound?

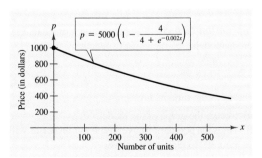

54. *Demand* The demand function for a product is modeled by $p = 500 - 0.5e^{0.004x}$, as shown in the figure. Find the price of the product if the quantity demanded is (a) $x = 1000$ units and (b) $x = 1500$ units.

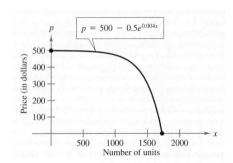

$p = 500 - 0.5e^{0.004x}$

Price (in dollars)

Number of units

55. *Probability* The average time between incoming calls at a switchboard is 3 minutes. If a call has just come in, the probability that the next call will come within the next t minutes is $P(t) = 1 - e^{-t/3}$. Find the probability of the following.

(a) A call comes in within $\frac{1}{2}$ minute.
(b) A call comes in within 2 minutes.
(c) A call comes in within 5 minutes.

56. *Fuel Efficiency* An automobile gets 28 miles per gallon at speeds of up to 50 miles per hour. At speeds of 50 miles per hour or greater, the number of miles per gallon drops at the rate of 12% for each 10 miles per hour. If s is the speed (in miles per hour) and y is the miles per gallon, then

$$y = 28e^{0.6-0.012s}, \qquad s \geq 50.$$

Use this information to complete the table. What can you conclude?

Speed (s)	50	55	60	65	70
Miles per gallon (y)					

57. *Population Growth* The population y of a bacterial culture is modeled by the logistics function

$$y = \frac{925}{1 + e^{-0.3t}},$$

where t is the time in days.

(a) Sketch the graph of the model.
(b) Does the population have a limit as t increases without bound? Explain your answer.

58. *Forestry* The yield V (in millions of cubic feet per acre) of a forest at age t years is modeled by

$$V = 6.7e^{-48/t}.$$

(a) Find the yield after 20 years and after 50 years.
(b) Sketch the graph of the model.
(c) Does the yield have a limit as t increases without bound? Explain your answer.

59. *Learning Theory* In a learning theory project, the proportion P of correct responses after n trials can be modeled by

$$P = \frac{0.83}{1 + e^{-0.2n}}.$$

(a) Use a graphing utility to estimate the proportion of correct responses after ten trials. Check your result analytically.
(b) Use a graphing utility to estimate the number of trials required to have a proportion of correct responses of 0.75.
(c) Does the proportion of correct responses have a limit as n increases without bound? Explain your answer.

60. *Learning Theory* In a typing class, the average number N of words per minute typed after t weeks of lessons can be modeled by

$$N = \frac{95}{1 + 8.5e^{-0.12t}}.$$

(a) Use a graphing utility to estimate the average number of words per minute typed after 10 weeks. Check your result analytically.
(b) Use a graphing utility to estimate the number of weeks required to produce an average of 70 words per minute.
(c) Does the number of words per minute have a limit as t increases without bound? Explain your answer.

In Exercises 61–64, use a graphing utility to graph the function. Determine whether the function has any horizontal asymptotes and discuss the continuity of the function.

61. $f(x) = \dfrac{e^x + e^{-x}}{2}$

62. $f(x) = \dfrac{e^x - e^{-x}}{2}$

63. $f(x) = \dfrac{2}{1 + e^{1/x}}$

64. $f(x) = \dfrac{2}{1 + 2e^{-0.2x}}$

Derivatives of Exponential Functions

Derivatives of Exponential Functions ■ *Applications* ■
The Normal Probability Density Function

Derivatives of Exponential Functions

In Section 4.1, we stated that the most convenient base for exponential functions is the irrational number e. The convenience of this base stems primarily from the fact that the function $f(x) = e^x$ *is its own derivative.* You will see that this is not true of other exponential functions of the form $y = a^x$ where $a \neq e$. To verify that $f(x) = e^x$ is its own derivative, notice that the limit

$$\lim_{\Delta x \to 0} (1 + \Delta x)^{1/\Delta x} = e$$

implies that for small values of Δx, $e \approx (1 + \Delta x)^{1/\Delta x}$, or $e^{\Delta x} \approx 1 + \Delta x$. This approximation is used in the following derivation.

$$f'(x) = \lim_{\Delta x \to 0} \frac{f(x + \Delta x) - f(x)}{\Delta x} \qquad \text{\textit{Definition of derivative}}$$

$$= \lim_{\Delta x \to 0} \frac{e^{x + \Delta x} - e^x}{\Delta x} \qquad \text{\textit{Use } } f(x) = e^x.$$

$$= \lim_{\Delta x \to 0} \frac{e^x(e^{\Delta x} - 1)}{\Delta x} \qquad \text{\textit{Factor numerator.}}$$

$$= \lim_{\Delta x \to 0} \frac{e^x[(1 + \Delta x) - 1]}{\Delta x} \qquad \text{\textit{Substitute } } 1 + \Delta x \text{ \textit{for } } e^{\Delta x}.$$

$$= \lim_{\Delta x \to 0} \frac{e^x(\Delta x)}{\Delta x} \qquad \text{\textit{Cancel common factor.}}$$

$$= \lim_{\Delta x \to 0} e^x \qquad \text{\textit{Simplify.}}$$

$$= e^x \qquad \text{\textit{Evaluate limit.}}$$

By applying the Chain Rule, you can obtain the derivative of e^u with respect to x. Both formulas are summarized as follows.

Derivatives of Exponential Functions

Let u be a differentiable function of x.

1. $\dfrac{d}{dx}[e^x] = e^x$ **2.** $\dfrac{d}{dx}[e^u] = e^u u'$

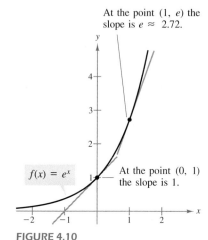

At the point $(1, e)$ the slope is $e \approx 2.72$.

$f(x) = e^x$

At the point $(0, 1)$ the slope is 1.

FIGURE 4.10

EXAMPLE 1 *Interpreting Derivatives Graphically*

Find the slope of the tangent line of

$$f(x) = e^x \hspace{4em} \textit{Original function}$$

at the points $(0, 1)$ and $(1, e)$. What conclusion can you make?

Solution

Because the derivative of f is

$$f'(x) = e^x, \hspace{4em} \textit{Derivative}$$

it follows that the slope of the tangent line to the graph of f is

$$f'(0) = e^0 = 1 \hspace{4em} \textit{Slope at point} (0, 1)$$

at the point $(0, 1)$, and the slope of the tangent line to the graph of f is

$$f'(1) = e^1 = e \hspace{4em} \textit{Slope at point} (1, e)$$

at the point $(1, e)$, as shown in Figure 4.10. From this pattern, you can see that the slope of the tangent line to the graph of $f(x) = e^x$ at any point (x, e^x) is equal to the y-coordinate of the point. ▪

EXAMPLE 2 *Differentiating Exponential Functions*

Differentiate the following functions.

a. $f(x) = e^{2x}$ \hspace{4em} **b.** $f(x) = e^{-3x^2}$

c. $f(x) = 6e^{x^3}$ \hspace{4em} **d.** $f(x) = e^{-x}$

Solution

a. Let $u = 2x$. Then $u' = 2$, and you can apply the Chain Rule.

$$f'(x) = e^u u' = e^{2x}(2) = 2e^{2x}$$

b. Let $u = -3x^2$. Then $u' = -6x$, and you can apply the Chain Rule.

$$f'(x) = e^u u' = e^{-3x^2}(-6x) = -6xe^{-3x^2}$$

c. Let $u = x^3$. Then $u' = 3x^2$, and you can apply the Chain Rule.

$$f'(x) = 6e^u u' = 6e^{x^3}(3x^2) = 18x^2 e^{x^3}$$

d. Let $u = -x$. Then $u' = -1$, and you can apply the Chain Rule.

$$f'(x) = e^u u' = e^{-x}(-1) = -e^{-x}$$ ▪

Note In Example 2, notice that when you differentiate an exponential function, the exponent does not change. For instance, the derivative of $y = e^{3x}$ is $y' = 3e^{3x}$. In both the function and its derivative, the exponent is $3x$.

The differentiation rules that you studied in Chapter 2 can be used with exponential functions, as illustrated in Example 3.

EXAMPLE 3 *Differentiating Exponential Functions*

Differentiate the following functions.

a. $f(x) = xe^x$

b. $f(x) = \dfrac{e^x - e^{-x}}{2}$

c. $f(x) = \dfrac{e^x}{x}$

d. $f(x) = xe^x - e^x$

Solution

a. $f(x) = xe^x$ *Original function*

$f'(x) = xe^x + e^x(1)$ *Product Rule*

$= xe^x + e^x$ *Simplify.*

b. $f(x) = \dfrac{e^x - e^{-x}}{2}$ *Original function*

$= \tfrac{1}{2}(e^x - e^{-x})$ *Rewrite.*

$f'(x) = \tfrac{1}{2}(e^x + e^{-x})$ *Constant Multiple Rule*

c. $f(x) = \dfrac{e^x}{x}$ *Original function*

$f'(x) = \dfrac{xe^x - e^x(1)}{x^2}$ *Quotient Rule*

$= \dfrac{e^x(x - 1)}{x^2}$ *Simplify.*

d. $f(x) = xe^x - e^x$ *Original function*

$f'(x) = [xe^x + e^x(1)] - e^x$ *Product and Difference Rules*

$= xe^x + e^x - e^x$

$= xe^x$ *Simplify.* ▬

▦ TECHNOLOGY

If you have access to a symbolic differentiation utility such as *Derive*, *Maple*, or *Mathematica*, try using it to find the derivatives of the functions in Example 3. For example, the display at the right shows how *Derive* can be used to find the derivative of $f(x) = xe^x - e^x$.

1:	x EXP (x) - EXP (x)	*Author.*
2:	$\dfrac{d}{dx}$ (x EXP (x) - EXP (x))	*Differentiate.*
3:	x ê^x	*Simplify.*

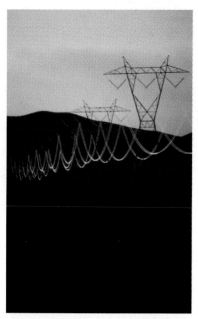

FPG International/© Campbell & Boulanger

Utility wires strung between poles have the shape of a catenary.

Applications

In Chapter 3, you learned how to use derivatives to analyze the graph of a function. The next example applies these techniques to a function composed of exponential functions. In the example, notice that $e^a = e^b$ implies that $a = b$.

Real Life

EXAMPLE 4 *Analyzing a Catenary*

When a telephone wire is hung between two poles, the wire forms a ∪-shaped curve called a **catenary.** For instance, the function

$$y = 30(e^{x/60} + e^{-x/60}), \qquad -30 \le x \le 30$$

models the shape of a telephone wire that is strung between two poles that are 60 feet apart (x and y are measured in feet). Show that the lowest point on the wire is midway between the two poles. How much does the wire sag between the two poles?

Solution

The derivative of the function is

$$y' = 30\left[e^{x/60}\left(\tfrac{1}{60}\right) + e^{-x/60}\left(-\tfrac{1}{60}\right)\right] = \tfrac{1}{2}(e^{x/60} - e^{-x/60}).$$

To find the critical numbers, set the derivative equal to zero.

$$\tfrac{1}{2}(e^{x/60} - e^{-x/60}) = 0 \qquad \textit{Set derivative equal to 0.}$$

$$e^{x/60} - e^{-x/60} = 0 \qquad \textit{Multiply both sides by 2.}$$

$$e^{x/60} = e^{-x/60} \qquad \textit{Add } e^{-x/60} \textit{ to both sides.}$$

$$\frac{x}{60} = -\frac{x}{60} \qquad \textit{If } e^a = e^b, \textit{ then } a = b.$$

$$x = -x \qquad \textit{Multiply both sides by 60.}$$

$$2x = 0 \qquad \textit{Add } x \textit{ to both sides.}$$

$$x = 0 \qquad \textit{Divide both sides by 2.}$$

Using the First-Derivative Test, you can determine that the critical number $x = 0$ yields a relative minimum of the function. From the graph in Figure 4.11, you can see that this relative minimum is actually a minimum on the interval $[-30, 30]$. To find how much the wire sags between the two poles, you can compare its height at each pole with the height at the midpoint.

$$y = 30(e^{-30/60} + e^{-(-30)/60}) \approx 67.7 \text{ feet} \qquad \textit{Height at left pole}$$

$$y = 30(e^{0/60} + e^{-(0)/60}) = 60 \text{ feet} \qquad \textit{Height at midpoint}$$

$$y = 30(e^{30/60} + e^{-(30)/60}) \approx 67.7 \text{ feet} \qquad \textit{Height at right pole}$$

From this, you can see that the wire sags about 7.7 feet.

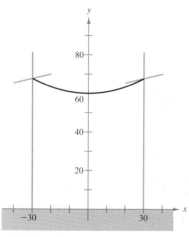

FIGURE 4.11

EXAMPLE 5 *Finding a Maximum Revenue*

The demand function for a product is modeled by

$$p = 56e^{-0.000012x}, \qquad \text{\textit{Demand function}}$$

where p is the price per unit (in dollars) and x is the number of units. What price will yield a maximum revenue?

Solution

The revenue function is

$$R = xp = 56xe^{-0.000012x}. \qquad \text{\textit{Revenue function}}$$

To find the maximum revenue *analytically*, you would set the marginal revenue, dR/dx, equal to zero and solve for x. In this problem, it is easier to use a *graphical* approach. After experimenting to find a reasonable viewing rectangle, you can obtain a graph of R that is similar to that shown in Figure 4.12. Using the zoom and trace features, you can conclude that the maximum revenue occurs when x is about 83,300 units. To find the price that corresponds to this production level, substitute $x \approx 83,300$ into the demand function.

$$p \approx 56e^{-0.000012(83,300)}$$

$$\approx \$20.61.$$

Thus, a price of about $20.61 will yield a maximum revenue.

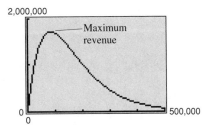

FIGURE 4.12 Use the zoom and trace features to approximate the x-value that corresponds to the maximum revenue.

Note   Try solving the problem in Example 5 analytically. When you do this, you obtain

$$\frac{dR}{dx} = 56xe^{-0.000012x}(-0.000012) + e^{-0.000012x}(56) = 0.$$

Explain how you would solve this equation. What is the solution?

The Normal Probability Density Function

If you take a course in statistics or quantitative business analysis, you will spend quite a bit of time studying the characteristics and use of the **normal probability density function** given by

$$f(x) = \frac{1}{\sigma\sqrt{2\pi}}e^{-(x-\mu)^2/2\sigma^2},$$

where σ is the lowercase Greek letter sigma, and μ is the lowercase Greek letter mu. In this formula, σ represents the standard deviation of the probability distribution, and μ represents the mean of the probability distribution.

EXAMPLE 6   *Exploring a Probability Density Function*

Show that the graph of the normal probability density function

$$f(x) = \frac{1}{\sqrt{2\pi}}e^{-x^2/2} \qquad\qquad \textit{Original function}$$

has points of inflection at $x = \pm 1$.

Solution

Begin by finding the second derivative of the function.

$$f'(x) = \frac{1}{\sqrt{2\pi}}(-x)e^{-x^2/2} \qquad\qquad \textit{First derivative}$$

$$f''(x) = \frac{1}{\sqrt{2\pi}}[(-x)(-x)e^{-x^2/2} + (-1)e^{-x^2/2}] \qquad \textit{Second derivative}$$

$$= \frac{1}{\sqrt{2\pi}}(e^{-x^2/2})(x^2 - 1) \qquad\qquad \textit{Simplify.}$$

By setting the second derivative equal to zero, you can determine that $x = \pm 1$. By testing the concavity of the graph, you can then conclude that these x-values yield points of inflection, as shown in Figure 4.13.　■

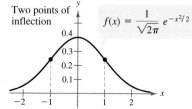

Two points of inflection

$$f(x) = \frac{1}{\sqrt{2\pi}}\, e^{-x^2/2}$$

FIGURE 4.13   The graph of the normal probability density function is bell-shaped.

Discussion Problem   The Normal Probability Density Function

Use Example 6 as a model to show that the graph of the normal probability density function with $\mu = 0$,

$$f(x) = \frac{1}{\sigma\sqrt{2\pi}}e^{-x^2/2\sigma^2},$$

has points of inflection at $x = \pm\sigma$. What is the maximum of the function?

Warm Up

The following warm-up exercises involve skills that were covered in earlier sections. You will use these skills in the exercise set for this section.

In Exercises 1–4, factor the expression.

1. $x^2 e^x - \frac{1}{2} e^x$

2. $(xe^{-x})^{-1} + e^x$

3. $xe^x - e^{2x}$

4. $e^x - xe^{-x}$

In Exercises 5–8, find the derivative of the function.

5. $f(x) = \dfrac{3}{7x^2}$

6. $g(x) = 3x^2 - \dfrac{x}{6}$

7. $f(x) = (4x - 3)(x^2 + 9)$

8. $f(t) = \dfrac{t - 2}{\sqrt{t}}$

In Exercises 9 and 10, find the relative extrema of the function.

9. $f(x) = \frac{1}{8}x^3 - 2x$

10. $f(x) = x^4 - 2x^2 + 5$

E X E R C I S E S 4.2 means that technology can help you solve or check the exercise(s).

In Exercises 1–4, find the slope of the tangent line to the exponential function at the point (0, 1).

1. $y = e^{3x}$

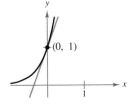

2. $y = e^{2x}$

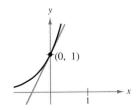

3. $y = e^{-x}$

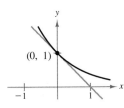

4. $y = e^{-2x}$

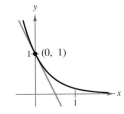

In Exercises 5–20, find the derivative of the function.

5. $y = e^{4x}$

6. $y = e^{1-x}$

7. $y = e^{-2x+x^2}$

8. $y = e^{-x^2}$

9. $f(x) = e^{1/x}$

10. $f(x) = e^{-1/x^2}$

11. $g(x) = e^{\sqrt{x}}$

12. $g(x) = e^{x^3}$

13. $f(x) = (x^2 + 1)e^{4x}$

14. $y = x^2 e^{-x}$

15. $y = (e^{-x} + e^x)^3$

16. $y = (10e^{-x})^2$

17. $f(x) = \dfrac{2}{e^x + e^{-x}}$

18. $f(x) = \dfrac{e^x + e^{-x}}{2}$

19. $y = xe^x - 4e^{-x}$

20. $y = x^2 e^x - 2xe^x + 2e^x$

In Exercises 21 and 22, find dy/dx implicitly.

21. $xe^x + 2ye^x = 0$

22. $e^{xy} + x^2 - y^2 = 10$

In Exercises 23–26, find the second derivative.

23. $f(x) = 2e^{3x} + 3e^{-2x}$

24. $f(x) = (1 + 2x)e^{4x}$

25. $f(x) = 5e^{-x} - 2e^{-5x}$

26. $f(x) = (3 + 2x)e^{-3x}$

In Exercises 27–30, graph and analyze the function. Include extrema, points of inflection, and asymptotes in your analysis.

27. $f(x) = \dfrac{1}{2 - e^{-x}}$ **28.** $f(x) = \dfrac{e^x - e^{-x}}{2}$

29. $f(x) = x^2 e^{-x}$ **30.** $f(x) = xe^{-x}$

 In Exercises 31 and 32, use a graphing utility to graph the function. Determine any asymptotes of the graph.

31. $f(x) = \dfrac{8}{1 + e^{-0.5x}}$ **32.** $g(x) = \dfrac{8}{1 + e^{-0.5/x}}$

 Depreciation In Exercises 33 and 34, the value V (in dollars) of an item is a function of the time t (in years).

(a) Graph the function over the interval [0, 10].
(b) Find the rate of change of V when $t = 1$.
(c) Find the rate of change of V when $t = 5$.
(d) Compare this depreciation model with a linear model. What are the advantages of each?

33. $V = 15,000e^{-0.6286t}$ **34.** $V = 500,000e^{-0.2231t}$

 35. *Forest Defoliation* To estimate the defoliation p (in percent of foliage) caused by gypsy moths during a year, a forester counts the number x (in thousands) of egg masses on $\frac{1}{40}$ of an acre the preceding fall. The defoliation is modeled by

$$p = \frac{300}{3 + 17e^{-1.57x}}.$$

(Source: National Forest Service)

(a) Use a graphing utility to graph the model.
(b) Estimate the percent of defoliation if 2000 egg masses are counted.
(c) Estimate the number of egg masses for which the amount of defoliation is increasing most rapidly.

36. *Learning Theory* The average typing speed N (in words per minute) after t weeks of lessons is modeled by

$$N = \frac{95}{1 + 8.5e^{-0.12t}}.$$

Find the rate at which the typing speed is changing when (a) $t = 5$ weeks, (b) $t = 10$ weeks, and (c) $t = 30$ weeks.

37. *Compound Interest* The balance A (in dollars) in a savings account is given by $A = 5000e^{0.08t}$, where t is measured in years. Find the rate at which the balance is changing when (a) $t = 1$ year, (b) $t = 10$ years, and (c) $t = 50$ years.

38. *Ebbinghaus Model* The *Ebbinghaus Model* for human memory is

$$p = (100 - a)e^{-bt} + a,$$

where p is the percent retained after t weeks. (The constants a and b vary from one person to another.) If $a = 20$ and $b = 0.5$, at what rate is information being retained after 1 week? After 3 weeks?

39. *Orchard Yield* The yield V (in pounds per acre) for an orchard at age t (in years) is modeled by

$$V = 7955.6e^{-0.0458/t}.$$

At what rate is the yield changing when $t = 5$ years? When $t = 10$ years? When $t = 25$ years?

 40. *Self-Employed Workers* From 1980 to 1991, the number y (in thousands) of self-employed workers in the United States can be modeled by

$$y = 6787.05 + 1851.06e^{0.06t},$$

where $t = 0$ corresponds to 1980. *(Source: U.S. Bureau of Labor Statistics)*

(a) Use a graphing utility to graph the model.
(b) Use the graph to estimate the rate of change in the number of self-employed people in 1980 and in 1990.
(c) Confirm the result of part (b) analytically.

 41. Use a graphing utility to graph the normal probability density function with $\mu = 0$ and $\sigma = 2$, 3, and 4 on the same viewing rectangle. What effect does the standard deviation σ have on the function? Explain.

 42. Use a graphing utility to graph the normal probability density function with $\sigma = 1$ and $\mu = -2$, 1, and 3 on the same viewing rectangle. What effect does the mean μ have on the function? Explain.

Logarithmic Functions

The Natural Logarithmic Function ▪ *Properties of Logarithmic Functions* ▪
Solving Exponential and Logarithmic Equations ▪ *Applications*

The Natural Logarithmic Function

From your previous algebra courses, you should be somewhat familiar with logarithms. For instance, the **common logarithm** $\log_{10} x$ is defined as

$$\log_{10} x = b \quad \text{if and only if} \quad 10^b = x.$$

The base of common logarithms is 10. In calculus, the most useful base for logarithms is the number e.

Definition of the Natural Logarithmic Function

The **natural logarithmic function**, denoted by $\ln x$, is defined as

$$\ln x = b \quad \text{if and only if} \quad e^b = x.$$

$\ln x$ is read as "el-en of x" or as "the natural log of x."

This definition implies that the natural logarithmic function and the natural exponential function are inverses of each other. Thus, every logarithmic equation can be written in an equivalent exponential form and every exponential equation can be written in logarithmic form. Here are some examples.

Logarithmic form $\quad \ln 1 = 0 \quad \ln e = 1 \quad \ln \dfrac{1}{e} = -1 \quad \ln 2 \approx 0.693$

Exponential form $\quad e^0 = 1 \qquad e^1 = e \qquad e^{-1} = \dfrac{1}{e} \qquad e^{0.693} \approx 2$

Because the functions $f(x) = e^x$ and $g(x) = \ln x$ are inverses of each other, their graphs are reflections of each other in the line $y = x$. This reflective property is illustrated in Figure 4.14. The figure also contains a summary of several properties of the graph of the natural logarithmic function.

Notice that the domain of the natural logarithmic function is the set of *positive real numbers*—be sure you see that $\ln x$ is not defined for zero or for negative numbers. You can test this on your calculator. If you try evaluating $\ln(-1)$ or $\ln 0$, your calculator should indicate that the value is not a real number.

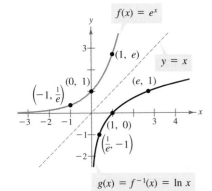

$g(x) = \ln x$

- Domain: $(0, \infty)$
- Range: $(-\infty, \infty)$
- Intercept: $(1, 0)$
- Always increasing
- $\ln x \to \infty$ as $x \to \infty$
- $\ln x \to -\infty$ as $x \to 0^+$
- Continuous
- One-to-one

FIGURE 4.14

EXAMPLE 1 *Graphing Logarithmic Functions*

Sketch the graphs of the following functions.

a. $f(x) = \ln(x + 1)$ **b.** $f(x) = 2 \ln(x - 2)$

Solution

a. Because the natural logarithmic function is defined only for positive values, the domain of the function is $0 < x + 1$, or

$$-1 < x. \qquad \textit{Domain}$$

To sketch the graph, begin by constructing a table of values, as shown in Table 4.5. Then, plot the points in the table and connect them with a smooth curve, as shown in Figure 4.15(a).

TABLE 4.5

x	−0.5	0	0.5	1	1.5	2
$\ln(x + 1)$	−0.693	0	0.405	0.693	0.916	1.099

b. The domain of this function is $0 < x - 2$, or

$$2 < x. \qquad \textit{Domain}$$

Table 4.6 shows a table of values for the function, and its graph is shown in Figure 4.15(b).

TABLE 4.6

x	2.5	3	3.5	4	4.5	5
$2 \ln(x - 2)$	−1.386	0	0.811	1.386	1.833	2.197

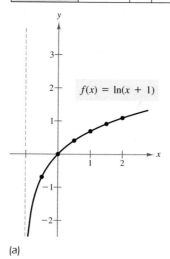
(a)

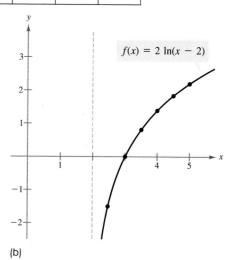
(b)

FIGURE 4.15

Properties of Logarithmic Functions

Recall from Section 1.4 that inverse functions possess the property that

$$f(f^{-1}(x)) = x \quad \text{and} \quad f^{-1}(f(x)) = x.$$

The properties listed below follow from the fact that the natural logarithmic function and natural exponential function are inverses of each other.

Inverse Properties of Logarithms and Exponents

1. $\ln e^x = x$ **2.** $e^{\ln x} = x$

EXAMPLE 2 *Applying Inverse Properties*

Simplify the following expressions.

a. $\ln e^{\sqrt{2}}$ **b.** $e^{\ln 3x}$

Solution

a. Because $\ln e^x = x$, it follows that

$$\ln e^{\sqrt{2}} = \sqrt{2}.$$

b. Because $e^{\ln x} = x$, it follows that

$$e^{\ln 3x} = 3x.$$

Most of the properties of exponential functions can be rewritten in terms of logarithmic functions. For instance, the property

$$e^x e^y = e^{x+y}$$

states that you can multiply two exponential expressions by adding their exponents. In terms of logarithms, this property becomes

$$\ln xy = \ln x + \ln y.$$

This property and two other properties of logarithms are summarized below.

Properties of Logarithms

1. $\ln xy = \ln x + \ln y$ **2.** $\ln \dfrac{x}{y} = \ln x - \ln y$

3. $\ln x^n = n \ln x$

Rewriting a logarithm of a single quantity as the sum, difference, or multiple of logarithms is called *expanding* the logarithmic expression. The reverse procedure is called *condensing* a logarithmic expression.

TECHNOLOGY

Try using a graphing utility to verify the results of Example 3b. That is, try graphing the functions

$$y = \ln \sqrt{x^2 + 1}$$

and

$$y = \frac{1}{2} \ln(x^2 + 1).$$

Because these two functions are equivalent, their graphs should coincide.

EXAMPLE 3 *Expanding Logarithmic Expressions*

Use the properties of logarithms to write each expression as a sum, difference, or multiple of logarithms. (Assume $x > 0$ and $y > 0$.)

a. $\ln \dfrac{10}{9}$ **b.** $\ln \sqrt{x^2 + 1}$ **c.** $\ln \dfrac{xy}{5}$ **d.** $\ln \dfrac{x^2}{6y^3}$

Solution

a. $\ln \dfrac{10}{9} = \ln 10 - \ln 9$ *Property 2*

b. $\ln \sqrt{x^2 + 1} = \ln(x^2 + 1)^{1/2}$ *Write with rational exponents.*

$\qquad\qquad\quad = \frac{1}{2} \ln(x^2 + 1)$ *Property 3*

c. $\ln \dfrac{xy}{5} = \ln(xy) - \ln 5$ *Property 2*

$\qquad\quad = \ln x + \ln y - \ln 5$ *Property 1*

d. $\ln \dfrac{x^2}{6y^3} = \ln x^2 - \ln 6y^3$ *Property 2*

$\qquad\qquad = \ln x^2 - (\ln 6 + \ln y^3)$ *Property 1*

$\qquad\qquad = \ln x^2 - \ln 6 - \ln y^3$ *Simplify.*

$\qquad\qquad = 2 \ln x - \ln 6 - 3 \ln y$ *Property 3*

EXAMPLE 4 *Condensing Logarithmic Expressions*

Use the properties of logarithms to rewrite each expression as the logarithm of a single quantity. (Assume $x > 0$ and $y > 0$.)

a. $\ln x + 2 \ln y$ **b.** $\ln(x + 1) + \ln(x + 2) - 3 \ln x$

Solution

a. $\ln x + 2 \ln y = \ln x + \ln y^2$ *Property 3*

$\qquad\qquad\quad = \ln xy^2$ *Property 1*

b. $\ln(x + 1) + \ln(x + 2) - 3 \ln x = \ln(x + 1)(x + 2) - 3 \ln x$

$\qquad\qquad\qquad\qquad\qquad = \ln(x^2 + 3x + 2) - \ln x^3$

$\qquad\qquad\qquad\qquad\qquad = \ln \dfrac{x^2 + 3x + 2}{x^3}$

Solving Exponential and Logarithmic Equations

The inverse properties of logarithms and exponents can be used to solve exponential and logarithmic equations, as illustrated in the next two examples.

EXAMPLE 5 *Solving Exponential Equations*

Solve the following equations.

a. $e^x = 5$ **b.** $10 + 3e^{0.1t} = 14$

Solution

a.
$$e^x = 5 \qquad \text{\textit{Original equation}}$$
$$\ln e^x = \ln 5 \qquad \text{\textit{Take log of both sides.}}$$
$$x = \ln 5 \qquad \text{\textit{Inverse property: } } \ln e^x = x$$

b.
$$10 + 3e^{0.1t} = 14 \qquad \text{\textit{Original equation}}$$
$$3e^{0.1t} = 4 \qquad \text{\textit{Subtract 10 from both sides.}}$$
$$e^{0.1t} = \tfrac{4}{3} \qquad \text{\textit{Divide both sides by 3.}}$$
$$\ln e^{0.1t} = \ln \tfrac{4}{3} \qquad \text{\textit{Take log of both sides.}}$$
$$0.1t = \ln \tfrac{4}{3} \qquad \text{\textit{Inverse property: } } \ln e^{0.1t} = 0.1t$$
$$t = 10 \ln \tfrac{4}{3} \qquad \text{\textit{Multiply both sides by 10.}} \quad ■$$

Note In the examples on this page, note that the key step to solving an exponential equation is to "take the log of both sides," and the key step to solving a logarithmic equation is to "exponentiate both sides."

EXAMPLE 6 *Solving Logarithmic Equations*

Solve the following equations.

a. $\ln x = 5$ **b.** $3 + 2 \ln x^2 = 7$

Solution

a.
$$\ln x = 5 \qquad \text{\textit{Original equation}}$$
$$e^{\ln x} = e^5 \qquad \text{\textit{Exponentiate both sides.}}$$
$$x = e^5 \qquad \text{\textit{Inverse property: } } e^{\ln x} = x$$

b.
$$3 + 2 \ln x^2 = 7 \qquad \text{\textit{Original equation}}$$
$$2 \ln x^2 = 4 \qquad \text{\textit{Subtract 3 from both sides.}}$$
$$\ln x^2 = 2 \qquad \text{\textit{Divide both sides by 2.}}$$
$$e^{\ln x^2} = e^2 \qquad \text{\textit{Exponentiate both sides.}}$$
$$x^2 = e^2 \qquad \text{\textit{Inverse property: } } e^{\ln x^2} = x^2$$
$$x = \pm e \qquad \text{\textit{Solve for } } x. \quad ■$$

Applications

Real Life

EXAMPLE 7 *Finding Doubling Time*

You deposit P dollars in an account whose annual interest rate is r, compounded continuously. How long will it take for your balance to double?

Solution

The balance in the account after t years is

$$A = Pe^{rt}.$$

Thus, the balance will have doubled when $Pe^{rt} = 2P$. To find the "doubling time," solve this equation for t.

$Pe^{rt} = 2P$	*Balance in account has doubled.*
$e^{rt} = 2$	*Divide both sides by P.*
$\ln e^{rt} = \ln 2$	*Take log of both sides.*
$rt = \ln 2$	*Inverse property:* $\ln e^{rt} = rt$
$t = \dfrac{1}{r} \ln 2$	*Divide both sides by r.*

From this result, you can see that the time it takes for the balance to double is inversely proportional to the interest rate r. Table 4.7 shows the doubling times for several interest rates. Notice that the doubling time decreases as the rate increases. The relationship between doubling time and the interest rate is shown graphically in Figure 4.16.

TABLE 4.7

r	3%	4%	5%	6%	7%	8%	9%	10%	11%	12%
t	23.1	17.3	13.9	11.6	9.9	8.7	7.7	6.9	6.3	5.8

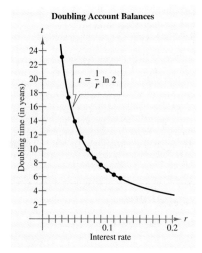

Doubling Account Balances

$t = \dfrac{1}{r} \ln 2$

Doubling time (in years)

Interest rate

FIGURE 4.16

Discussion Problem   Tripling Time

You deposit P dollars in an account whose annual interest rate is r, compounded continuously. Use Example 7 as a model to determine how long it will take for your balance to triple. Complete the following table and use a graph to illustrate your answer.

Rate, r	3%	4%	5%	6%	7%	8%	9%	10%	11%	12%
Tripling time, t										

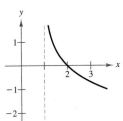

The following warm-up exercises involve skills that were covered in earlier sections. You will use these skills in the exercise set for this section.

In Exercises 1–4, use the properties of exponents to simplify the expression.

1. $(4^2)(4^{-3})$

2. $(2^3)^2$

3. e^0

4. $(3e)^4$

In Exercises 5–8, solve for *x*.

5. $0 < x + 4$

6. $0 < x^2 + 1$

7. $0 < \sqrt{x^2 - 1}$

8. $0 < x - 5$

In Exercises 9 and 10, find the balance in the account after 10 years.

9. $P = \$1900$, $r = 6\%$, compounded continuously

10. $P = \$2500$, $r = 3\%$, compounded continuously

E X E R C I S E S   4.3   means that technology can help you solve or check the exercise(s).

In Exercises 1–8, write the logarithmic equation as an exponential equation, or vice versa.

1. $\ln 2 = 0.6931\ldots$

2. $\ln 8.4 = 2.1282\ldots$

3. $\ln 0.2 = -1.6094\ldots$

4. $\ln 0.056 = -2.8824\ldots$

5. $e^0 = 1$

6. $e^2 = 7.3891\ldots$

7. $e^{-3} = 0.0498\ldots$

8. $e^{0.25} = 1.2840\ldots$

 In Exercises 9–12, match the function with its graph. [The graphs are labeled (a)–(d).]

9. $f(x) = 2 + \ln x$

10. $f(x) = -\ln x$

11. $f(x) = \ln(x + 2)$

12. $f(x) = -\ln(x - 1)$

(a)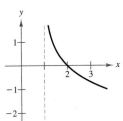

(b)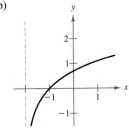

(c)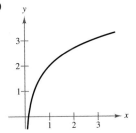

(d)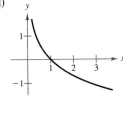

In Exercises 13–18, sketch the graph of the function.

13. $y = \ln(x - 1)$

14. $y = \ln|x|$

15. $y = \ln 2x$

16. $y = 5 + \ln x$

17. $y = 3 \ln x$

18. $y = \frac{1}{4} \ln x$

In Exercises 19–22, show, analytically *and* graphically, that the functions are inverses of each other.

19. $f(x) = e^{2x}$
$g(x) = \ln \sqrt{x}$

20. $f(x) = e^x - 1$
$g(x) = \ln(x + 1)$

21. $f(x) = e^{2x-1}$
$g(x) = \frac{1}{2} + \ln \sqrt{x}$

22. $f(x) = e^{x/3}$
$g(x) = \ln x^3$

In Exercises 23–28, apply the inverse properties of logarithmic and exponential functions to simplify the expression.

23. $\ln e^{x^2}$

24. $\ln e^{2x-1}$

25. $e^{\ln(5x+2)}$

26. $-1 + \ln e^{2x}$

27. $e^{\ln \sqrt{x}}$

28. $-8 + e^{\ln x^3}$

In Exercises 29 and 30, use the properties of logarithms and the fact that $\ln 2 \approx 0.6931$ and $\ln 3 \approx 1.0986$ to approximate the logarithm. Then use a calculator to confirm your approximation.

29. (a) $\ln 6$ (b) $\ln \frac{3}{2}$

    (c) $\ln 81$ (d) $\ln \sqrt{3}$

30. (a) $\ln 0.25$ (b) $\ln 24$

    (c) $\ln \sqrt[3]{12}$ (d) $\ln \frac{1}{72}$

In Exercises 31–40, use the properties of logarithms to write the expression as a sum, difference, or multiple of logarithms.

31. $\ln \frac{2}{3}$

32. $\ln \frac{1}{5}$

33. $\ln xyz$

34. $\ln \frac{xy}{z}$

35. $\ln \sqrt{x^2 + 1}$

36. $\ln \sqrt{2^3}$

37. $\ln \frac{2x}{\sqrt{x^2 - 1}}$

38. $\ln \frac{1}{e}$

39. $\ln \frac{3x(x+1)}{(2x+1)^2}$

40. $\ln z(z-1)^2$

In Exercises 41–50, write the expression as the logarithm of a single quantity.

41. $\ln(x - 2) - \ln(x + 2)$

42. $\ln(2x + 1) + \ln(2x - 1)$

43. $3 \ln x + 2 \ln y - 4 \ln z$

44. $\frac{1}{3}[2 \ln(x + 3) + \ln x - \ln(x^2 - 1)]$

45. $3[\ln x + \ln(x + 3) - \ln(x + 4)]$

46. $2 \ln 3 - \frac{1}{2} \ln(x^2 + 1)$

47. $\frac{3}{2}[\ln x(x^2 + 1) - \ln(x + 1)]$

48. $2 \ln x + \frac{1}{2} \ln(x + 1)$

49. $2[\ln x - \ln(x + 1)] - 3[\ln x - \ln(x - 1)]$

50. $\frac{1}{2} \ln(x - 2) + \frac{3}{2} \ln(x + 2)$

In Exercises 51–66, solve for x or t.

51. $e^{\ln x} = 4$

52. $e^{\ln x^2} - 9 = 0$

53. $\ln x = 0$

54. $2 \ln x = 4$

55. $e^{x+1} = 4$

56. $e^{-0.5x} = 0.075$

57. $300e^{-0.2t} = 700$

58. $e^{-0.0174t} = 0.5$

59. $5^{2x} = 15$

60. $2^{1-x} = 6$

61. $500(1.07)^t = 1000$

62. $400(1.06)^t = 1300$

63. $1000\left(1 + \dfrac{0.07}{12}\right)^{12t} = 3000$

64. $1000\left(1 + \dfrac{0.06}{12}\right)^{12t} = 3000$

65. $\dfrac{36}{1 + e^{-t}} = 20$

66. $\dfrac{45}{1 + e^{-2t}} = 24$

67. *Compound Interest* A deposit of $1000 is made into an account that earns interest at an annual rate of 5%. How long will it take for the balance to double if the interest is compounded (a) annually, (b) monthly, (c) daily, and (d) continuously?

68. *Compound Interest* A deposit of $1000 is made into an account that earns interest at an annual rate of 4.5%. How long will it take for the balance to triple if the interest is compounded (a) annually, (b) monthly, (c) daily, and (d) continuously?

69. *Compound Interest* Complete the table showing the time t necessary for P dollars to triple if the interest is compounded continuously at the rate of r.

r	2%	4%	6%	8%	10%	12%	14%
t							

70. *Demand* The demand function for a product is

$$p = 250 - 0.8e^{0.005x},$$

where p is the price per unit and x is the number of units sold. Find the number of units sold for a price of (a) $p = \$200$ and (b) $p = \$125$.

71. *Population Growth* The population P of Pensacola, Florida from 1970 through 1990 can be modeled by

$$P = 243{,}000e^{0.01737t},$$

where $t = 0$ corresponds to 1970. According to this model, in what year will Pensacola have a population of 400,000? *(Source: U.S. Bureau of Census)*

72. *Population Growth* The population P of Middletown, Connecticut from 1970 through 1990 can be modeled by

$$P = 74{,}000e^{0.00979t},$$

where $t = 0$ corresponds to 1970. According to this model, in what year will Middletown have a population of 100,000? *(Source: U.S. Bureau of Census)*

Carbon Dating In Exercises 73–76, you are given the ratio of carbon atoms in a fossil. Use the following information to estimate the age of the fossil. In living organic material, the ratio of radioactive carbon isotopes to the total number of carbon atoms is about 1 to 10^{12}. (See Example 2 in Section 4.1.) When organic material dies, its radioactive carbon isotopes begin to decay, with a half-life of about 5700 years. Thus, the ratio R of carbon isotopes to carbon-14 atoms is

$$R = 10^{-12} \left(\tfrac{1}{2}\right)^{t/5700},$$

where t is the time (in years) and $t = 0$ represents the time when the organic material died.

73. $R = 0.32 \times 10^{-12}$ **74.** $R = 0.27 \times 10^{-12}$

75. $R = 0.22 \times 10^{-12}$ **76.** $R = 0.13 \times 10^{-12}$

© Kenneth Garrett/Courtesy University of Innsbruck, Austria

In 1991, forensic expert Rainer Henn uncovered one of the most important discoveries of modern archaeology— the body of a man that had been frozen in the Alps 5300 years ago. The age of the "iceman" was estimated by carbon dating.

77. *Human Memory Model* Students in a mathematics class were given an exam and then retested monthly with equivalent exams. The average score S (on a 100-point scale) for the class can be modeled by

$$S = 80 - 14 \ln(t + 1), \qquad 0 \le t \le 12,$$

where t is the time in months.

(a) What was the average score on the original exam?

(b) What was the average score after 4 months?

(c) After how many months was the average score 46?

78. *Research Project* Use a graphing utility to sketch the graph of

$$y = 10 \ln \left(\frac{10 + \sqrt{100 - x^2}}{10} \right) - \sqrt{100 - x^2}$$

over the interval $(0, 10]$. This graph is called a *tractrix* or *pursuit curve*. Use your school's library or some other reference source to find information about a tractrix. Explain how such a curve can arise in a real-life setting.

79. Demonstrate that

$$\frac{\ln x}{\ln y} \neq \ln \frac{x}{y} = \ln x - \ln y$$

by completing the table.

x	y	$\dfrac{\ln x}{\ln y}$	$\ln \dfrac{x}{y}$	$\ln x - \ln y$
1	2			
3	4			
10	5			
4	0.5			

80. Complete the table using $f(x) = \dfrac{\ln x}{x}$.

x	1	5	10	10^2	10^4	10^6
$f(x)$						

(a) Use the table to estimate the limit: $\displaystyle \lim_{x \to \infty} f(x)$.

(b) Use a graphing utility to estimate the relative extrema of f.

In Exercises 81 and 82, use a graphing utility to verify that the functions are equivalent for $x > 0$.

81. $f(x) = \ln \dfrac{x^2}{4}$

    $g(x) = 2 \ln x - \ln 4$

82. $f(x) = \ln \sqrt{x(x^2 + 1)}$

    $g(x) = \tfrac{1}{2}[\ln x + \ln(x^2 + 1)]$

Derivatives of Logarithmic Functions

Derivatives of Logarithmic Functions ■ *Applications* ■
Other Bases

Derivatives of Logarithmic Functions

Implicit differentiation can be used to develop the derivative of the natural logarithmic function.

$$y = \ln x \qquad\qquad \text{\textit{Natural logarithmic function}}$$
$$e^y = x \qquad\qquad \text{\textit{Write in exponential form.}}$$
$$\frac{d}{dx}[e^y] = \frac{d}{dx}[x] \qquad\qquad \text{\textit{Differentiate with respect to } x.}$$
$$e^y \frac{dy}{dx} = 1 \qquad\qquad \text{\textit{Chain Rule}}$$
$$\frac{dy}{dx} = \frac{1}{e^y} \qquad\qquad \text{\textit{Divide both sides by } } e^y.$$
$$\frac{dy}{dx} = \frac{1}{x} \qquad\qquad \text{\textit{Substitute } x \text{ for } e^y.}$$

This result and its Chain Rule version are summarized below.

Derivative of the Natural Logarithmic Function

Let u be a differentiable function of x.

1. $\dfrac{d}{dx}[\ln x] = \dfrac{1}{x}$ $\qquad\qquad$ **2.** $\dfrac{d}{dx}[\ln u] = \dfrac{1}{u}u'$

EXAMPLE 1 *Differentiating a Logarithmic Function*

Find the derivative of

$$f(x) = \ln 2x.$$

Solution
Let $u = 2x$. Then $u' = 2$, and you can apply the Chain Rule as follows.

$$f'(x) = \frac{1}{u}u' = \frac{1}{2x}(2) = \frac{1}{x}$$

EXAMPLE 2 *Differentiating Logarithmic Functions*

Find the derivatives of the functions.

a. $f(x) = \ln(2x^2 + 4)$ **b.** $f(x) = x \ln x$

Solution

a. Let $u = 2x^2 + 4$. Then $u' = 4x$, and you can apply the Chain Rule.

$$f'(x) = \frac{1}{u} u' \qquad\qquad \textit{Chain Rule}$$

$$= \frac{1}{2x^2 + 4}(4x)$$

$$= \frac{2x}{x^2 + 2} \qquad\qquad \textit{Simplify.}$$

b. Using the Product Rule, you can find the derivative.

$$f'(x) = x\frac{d}{dx}[\ln x] + (\ln x)\frac{d}{dx}[x] \qquad \textit{Product Rule}$$

$$= x\left(\frac{1}{x}\right) + (\ln x)(1) \qquad\qquad \textit{Derivative of } \ln x$$

$$= 1 + \ln x \qquad\qquad\qquad \textit{Simplify.} \qquad\qquad ▬$$

When differentiating logarithmic functions, it is often helpful to use the properties of logarithms to rewrite the function *before* differentiating, as shown in the next example.

EXAMPLE 3 *Rewriting Before Differentiating*

Find the derivative of $f(x) = \ln\sqrt{x+1}$.

Solution

$$f(x) = \ln\sqrt{x+1} \qquad\qquad\qquad \textit{Original function}$$

$$= \ln(x+1)^{1/2} \qquad\qquad\qquad \textit{Rewrite with rational exponent.}$$

$$= \frac{1}{2}\ln(x+1) \qquad\qquad\qquad \textit{Property of logarithms}$$

$$f'(x) = \frac{1}{2}\left(\frac{1}{x+1}\right) \qquad\qquad \textit{Differentiate.}$$

$$= \frac{1}{2(x+1)} \qquad\qquad\qquad \textit{Simplify.} \qquad\qquad ▬$$

Note   To see the advantage of rewriting before differentiating, try using the Chain Rule to differentiate $f(x) = \ln\sqrt{x+1}$ and compare your work with that shown in Example 3.

Discovery

What is the domain of the function $f(x) = \ln\sqrt{x+1}$ in Example 3? What is the domain of the function $f'(x) = 1/[2(x+1)]$? In general, you must be careful to understand the domains of functions involving logarithms. For example, are the domains of the functions $y_1 = \ln x^2$ and $y_2 = 2\ln x$ the same? Try graphing them on your graphing utility.

The next example is an even more dramatic illustration of the benefit of rewriting a function before differentiating.

EXAMPLE 4 *Rewriting Before Differentiating*

Find the derivative of

$$f(x) = \ln \frac{x(x^2 + 1)^2}{\sqrt{2x^3 + 1}}.$$

Solution

$$f(x) = \ln \frac{x(x^2 + 1)^2}{\sqrt{2x^3 + 1}} \qquad\qquad \textit{Original function}$$

$$= \ln x(x^2 + 1)^2 - \ln(2x^3 + 1)^{1/2} \qquad \textit{Logarithmic properties}$$

$$= \ln x + ln(x^2 + 1)^2 - \ln(2x^3 + 1)^{1/2} \qquad \textit{Logarithmic properties}$$

$$= \ln x + 2\ln(x^2 + 1) - \frac{1}{2}\ln(2x^3 + 1) \qquad \textit{Logarithmic properties}$$

$$f'(x) = \frac{1}{x} + 2\left(\frac{2x}{x^2 + 1}\right) - \frac{1}{2}\left(\frac{6x^2}{2x^3 + 1}\right) \qquad \textit{Differentiate.}$$

$$= \frac{1}{x} + \frac{4x}{x^2 + 1} - \frac{3x^2}{2x^3 + 1} \qquad \textit{Simplify.}$$

▬

Note Finding the derivative of the function in Example 4 without first rewriting would be a formidable task.

$$f'(x) = \frac{1}{x(x^2 + 1)^2/\sqrt{2x^3 + 1}} \frac{d}{dx}\left[\frac{x(x^2 + 1)^2}{\sqrt{2x^3 + 1}}\right]$$

You might try showing that this yields the same result obtained in Example 4, but be careful—the algebra is messy.

▦ TECHNOLOGY

If you use a symbolic differentiation utility to differentiate logarithmic functions, you should be aware that it will not generally list the derivative in the form given in Example 4. For instance, the result listed by *Derive* is shown at the right. To show that the two forms are equivalent, you could rewrite the answer given in Example 4 as a single fraction.

1: $\quad \text{LN}\left[\dfrac{x(x^2 + 1)^2}{\sqrt{(2x^3 - 1)}}\right]$ $\qquad$ *Author.*

2: $\quad \dfrac{d}{dx}\text{LN}\left[\dfrac{x(x^2 + 1)^2}{\sqrt{(2x^3 - 1)}}\right]$ $\qquad$ *Differentiate.*

3: $\quad \dfrac{7x^5 - x^3 - 5x^2 - 1}{x(2x^3 - 1)(x^2 + 1)}$ $\qquad$ *Simplify.*

Applications

EXAMPLE 5 *Analyzing a Graph*

Analyze the graph of the function $f(x) = x^2 - \ln x$.

Solution

From Figure 4.17, it appears that the function has a minimum just to the left of $x = 1$. To find the minimum analytically, find the critical numbers by setting the derivative of f equal to zero and solving for x.

$$f(x) = x^2 - \ln x \qquad \text{\textit{Original function}}$$

$$f'(x) = 2x - \frac{1}{x} \qquad \text{\textit{Differentiate.}}$$

$$2x - \frac{1}{x} = 0 \qquad \text{\textit{Set derivative equal to 0.}}$$

$$2x = \frac{1}{x} \qquad \text{\textit{Add $1/x$ to both sides.}}$$

$$2x^2 = 1 \qquad \text{\textit{Multiply both sides by x.}}$$

$$x^2 = \tfrac{1}{2} \qquad \text{\textit{Divide both sides by 2.}}$$

$$x = \pm\sqrt{\tfrac{1}{2}} \qquad \text{\textit{Take square root of both sides.}}$$

Of these two possible critical numbers, only the positive one lies in the domain of f. By applying the First-Derivative Test, you can confirm that the function has a relative minimum when $x = 1/\sqrt{2} \approx 0.707$.

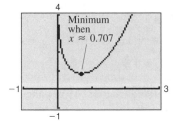

FIGURE 4.17

EXAMPLE 6 *Finding a Rate of Change*

Real Life

A group of 200 college students were tested every 6 months over a 4-year period. The group was composed of students who took French during the fall semester of their freshman year and did not take subsequent French courses. The average test score p (in percent) is modeled by

$$p = 91.6 - 15.6\ln(t + 1), \qquad 0 \le t \le 48,$$

where t is the time in months, as shown in Figure 4.18. At what rate was the average score changing after 1 year?

Solution

The rate of change is

$$\frac{dp}{dt} = -\frac{15.6}{t + 1}.$$

When $t = 12$, $dp/dt = -1.2$, which means that the average score was decreasing at the rate of 1.2% per month.

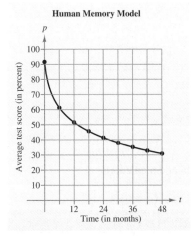

FIGURE 4.18

Other Bases

This chapter began with a definition of a general exponential function

$$f(x) = a^x,$$

where a is a positive number such that $a \neq 1$. The corresponding **logarithm to the base a** is defined by

$$\log_a x = b \quad \text{if and only if} \quad a^b = x.$$

As with the natural logarithmic function, the domain of the logarithmic function to the base a is the set of positive numbers.

EXAMPLE 7 *Evaluating Logarithms*

a. $\log_2 8 = 3$ This is true because $2^3 = 8$.

b. $\log_{10} 100 = 2$ This is true because $10^2 = 100$.

c. $\log_{10} \frac{1}{10} = -1$ This is true because $10^{-1} = \frac{1}{10}$.

d. $\log_3 81 = 4$ This is true because $3^4 = 81$.

e. $\log_{10} 4 \approx 0.602$ This is true because $10^{0.602} \approx 4$.   ▬

Logarithms to the base 10 are called **common logarithms.** Most calculators have only two logarithm keys—a natural logarithm key denoted by $\boxed{\text{LN}}$ and a common logarithm key denoted by $\boxed{\text{LOG}}$. Logarithms to other bases can be evaluated with the following change of base formula.

$$\log_a x = \frac{\ln x}{\ln a} \qquad\qquad \textit{Change of base formula}$$

EXAMPLE 8 *Evaluating Logarithms*

Evaluate the following logarithms.

a. $\log_2 3$ **b.** $\log_3 6$ **c.** $\log_2(-1)$

Solution

In each case, use the change of base formula and a calculator.

a. $\log_2 3 = \dfrac{\ln 3}{\ln 2} \approx 1.585$

b. $\log_3 6 = \dfrac{\ln 6}{\ln 3} \approx 1.631$

c. $\log_2(-1)$ is not defined.   ▬

To find derivatives of exponential or logarithmic functions to bases other than e, you can either convert to base e using the formulas

$$a^x = e^{(\ln a)x} \quad \text{and} \quad \log_a x = \left(\frac{1}{\ln a}\right)\ln x$$

or use the following differentiation rules.

Other Bases and Differentiation

Let u be a differentiable function of x.

1. $\dfrac{d}{dx}[a^x] = (\ln a)a^x$ **2.** $\dfrac{d}{dx}[a^u] = (\ln a)a^u u'$

3. $\dfrac{d}{dx}[\log_a x] = \left(\dfrac{1}{\ln a}\right)\dfrac{1}{x}$ **4.** $\dfrac{d}{dx}[\log_a u] = \left(\dfrac{1}{\ln a}\right)\dfrac{u'}{u}$

Real Life

EXAMPLE 9 *Finding a Rate of Change*

Radioactive carbon isotopes have a half-life of 5700 years. If 1 gram of the isotopes is present in an object now, the amount A (in grams) that will be present after t years is

$$A = \left(\frac{1}{2}\right)^{t/5700}.$$

At what rate is the amount changing when $t = 10,000$ years?

Solution
The derivative of A with respect to t is

$$\frac{dA}{dt} = \left(\ln \frac{1}{2}\right)\left(\frac{1}{2}\right)^{t/5700}\left(\frac{1}{5700}\right).$$

When $t = 10,000$, the value of the derivative is

$$\frac{dA}{dt} = \left(\ln \frac{1}{2}\right)\left(\frac{1}{2}\right)^{10,000/5700}\left(\frac{1}{5700}\right) \approx -0.000036,$$

which implies that the amount of isotopes in the object is decreasing at the rate of 0.000036 gram per year.

Discussion Problem   *Deriving Differentiation Rules*

Explain how to use the change of base formulas listed at the top of this page to derive the formulas for the derivatives of $f(x) = a^x$ and $f(x) = \log_a x$.

Warm Up

The following warm-up exercises involve skills that were covered in earlier sections. You will use these skills in the exercise set for this section.

In Exercises 1–6, expand the logarithmic expression.

1. $\ln(x + 1)^2$

2. $\ln x(x + 1)$

3. $\ln \dfrac{x}{x + 1}$

4. $\ln \left(\dfrac{x}{x - 3} \right)^3$

5. $\ln \dfrac{4x(x - 7)}{x^2}$

6. $\ln \dfrac{x^3(x + 1)}{\sqrt{x - 2}}$

In Exercises 7 and 8, find *dy/dx* implicitly.

7. $y^x + xy = 7$

8. $x^2 y - xy^2 = 3x$

In Exercises 9 and 10, find the second derivative of *f*.

9. $f(x) = x^2(x + 1) - 3x^3$

10. $f(x) = -\dfrac{1}{x^2}$

E X E R C I S E S   4.4 means that technology can help you solve or check the exercise(s).

In Exercises 1–6, find the slope of the tangent line to the graph of the function at the point $(1, 0)$.

1. $y = \ln x^3$

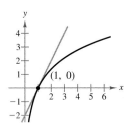

2. $y = \ln x^{5/2}$

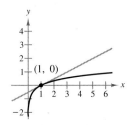

5. $y = \ln x^{3/2}$

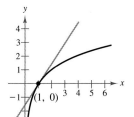

6. $y = \ln x$

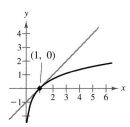

In Exercises 7–34, find the derivative of the function.

7. $y = \ln x^2$

8. $y = \ln(x^2 + 3)$

9. $f(x) = \ln 2x$

10. $f(x) = \ln(1 - x^2)$

11. $y = \ln \sqrt{x^4 - 4x}$

12. $y = \ln(1 - x)^{3/2}$

13. $y = \frac{1}{2}(\ln x)^6$

14. $y = (\ln x^2)^2$

15. $y = x \ln x$

16. $f(x) = x^2 \ln x$

17. $y = \ln(x\sqrt{x^2 - 1})$

18. $y = \ln[x^2(x + 1)^3]$

19. $y = \ln \dfrac{x}{x^2 + 1}$

20. $y = \ln \dfrac{x}{x + 1}$

3. $y = \ln x^2$

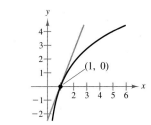

4. $y = \ln x^{1/2}$

21. $y = \ln \dfrac{x+1}{x-1}$

22. $y = \ln \dfrac{x-1}{x+1}$

23. $y = \ln \sqrt{\dfrac{x+1}{x-1}}$

24. $y = \ln \sqrt[3]{\dfrac{x-1}{x+1}}$

25. $y = \dfrac{\ln x}{x^2}$

26. $y = \dfrac{\ln x}{x}$

27. $y = \ln \dfrac{\sqrt{4+x^2}}{x}$

28. $y = \ln(x + \sqrt{4+x^2})$

29. $y = \ln \sqrt{2x^2 - 3}$

30. $y = \dfrac{-\sqrt{x^2+1}}{x} + \ln(x + \sqrt{x^2+1})$

31. $g(x) = e^{-x} \ln x$

32. $g(x) = \ln \dfrac{e^x + e^{-x}}{2}$

33. $f(x) = \ln e^{x^2}$

34. $f(x) = \ln \dfrac{1 + e^x}{1 - e^x}$

In Exercises 35–38, write the expression with base e.

35. 2^x

36. 3^x

37. $\log_4 x$

38. $\log_3 x$

In Exercises 39–44, evaluate the logarithm.

39. $\log_2 4$

40. $\log_5 12$

41. $\log_3 \frac{1}{2}$

42. $\log_7 343$

43. $\log_{10} 31$

44. $\log_9 3$

In Exercises 45–56, find the derivative of the function.

45. $y = 3^x$

46. $y = \left(\frac{1}{4}\right)^x$

47. $f(x) = \log_2 x$

48. $g(x) = \log_5 x$

49. $h(x) = 4^{2x-3}$

50. $y = 6^{5x}$

51. $y = \log_3(3x + 7)$

52. $y = \log_{10}(x^2 + 6x)$

53. $f(x) = 10^{x^2}$

54. $g(x) = 25^{2x^2}$

55. $y = x2^x$

56. $y = 3x3^x$

In Exercises 57–60, find dy/dx implicitly.

57. $x^2 - 3 \ln y + y^2 = 10$

58. $\ln xy + 5x = 30$

59. $4x^3 + \ln y^2 + 2y = 2x$

60. $4xy + \ln(x^2 y) = 7$

In Exercises 61–64, find the second derivative of the function.

61. $f(x) = x \ln \sqrt{x} + 2x$

62. $f(x) = 3 + 2 \ln x$

63. $f(x) = 5^x$

64. $f(x) = \log_{10} x$

In Exercises 65–70, find the slope of the graph at the indicated point. Then write an equation of the tangent line at the point.

65. $f(x) = 1 + 2x \ln x$

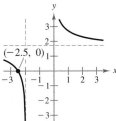

66. $f(x) = 2 \ln x^3$

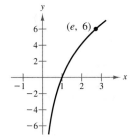

67. $f(x) = \ln \dfrac{5(x+2)}{x}$

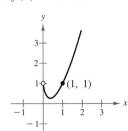

68. $f(x) = \ln(x\sqrt{x+3})$

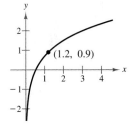

69. $f(x) = \log_2 x$

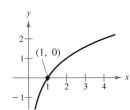

70. $f(x) = x^2 \log_3 x^2$

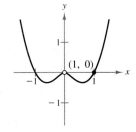

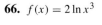 In Exercises 71–76, graph and analyze the function. Include any relative extrema and points of inflection in your analysis.

71. $y = x - \ln x$

72. $y = \frac{1}{2}x^2 - \ln x$

73. $y = \dfrac{\ln x}{x}$

74. $y = x \ln x$

75. $y = x^2 \ln x$

76. $y = (\ln x)^2$

In Exercises 77–80, find dx/dp for the demand function. Interpret this rate of change when the price is $10.

77. $x = \ln \dfrac{1000}{p}$

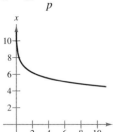

78. $x = 1000 - p \ln p$

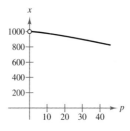

79. $x = \dfrac{500}{\ln(p^2 + 1)}$

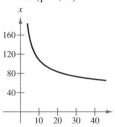

80. $x = 300 - 50 \ln(\ln p)$

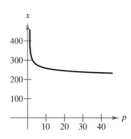

81. Solve the demand function in Exercise 77 for p. Use the result to find dp/dx. Then find the value of dp/dx when $p = \$10$. What is the relationship between this derivative and dx/dp?

82. Solve the demand function in Exercise 79 for p. Use the result to find dp/dx. Then find the value of dp/dx when $p = \$10$. What is the relationship between this derivative and dx/dp?

 83. *Minimum Average Cost* The cost of producing x units of a product is

$$C = 500 + 300x - 300 \ln x, \qquad x \geq 1.$$

Use a graphing utility to find the minimum average cost. Then confirm your result analytically.

84. *Chemistry* Radioactive Einsteinium isotopes have a half-life of 276 days. If 1 gram of the isotopes is present in an object now, the amount A (in grams) present after t days is

$$A = \left(\frac{1}{2}\right)^{t/276}.$$

At what rate is the amount changing when $t = 500$ days?

85. *Nonprofit Organizations* The number T (in thousands) of income tax returns filed by nonprofit organizations from 1984 to 1992 can be modeled by

$$T = -\frac{682.83 \ln t}{t} + 0.025t^3 - 15.19t + 317.92,$$

where $t = 4$ corresponds to 1984. *(Source: Internal Revenue Service)*

(a) Use a graphing utility to graph T over the interval [4, 12].

(b) Estimate the number of returns filed by nonprofit organizations in 1991.

(c) At what rate was the number of filed returns changing in 1991?

86. *Research Project* Use your school's library or some other reference source to research information about a nonprofit organization, such as that mentioned below. Collect data about the organization (revenue or membership over a 20-year period, for example) and find a mathematical model to represent the data.

Courtesy of Edward Santalone

$\mathcal{B}$*usiness* $\mathcal{C}$*apsule*

In 1993, James A. Cunningham was the senior vice president for finance and administration of the National Audubon Society, in New York City. Founded in 1905, the National Audubon Society is one of the oldest and largest national conservation organizations in the world. In 1992, the society received revenue of more than $43 million.

Exponential Growth and Decay

Exponential Growth and Decay ■ *Applications*

Exponential Growth and Decay

In this section you will learn to create models of *exponential growth and decay*. Real-life situations that involve exponential growth and decay deal with a substance or population whose *rate of change at any time t is proportional to the amount of the substance present at that time.* For example, the rate of decomposition of a radioactive substance is proportional to the amount of radioactive substance at a given instant. In its simplest form, this relationship is described by the following equation.

Rate of change of y ⟶ is ⟶ proportional to y.

$$\frac{dy}{dt} = ky$$

In this equation, k is a constant and y is a function of t. The solution of this equation is given below.

Note In the model $y = Ce^{kt}$, C is called the "initial value" because when $t = 0$,

$$y = Ce^{k(0)} = C(1) = C.$$

Law of Exponential Growth and Decay

If y is a positive quantity whose rate of change with respect to time is proportional to the quantity present at any time t, then y is of the form

$$y = Ce^{kt},$$

where C is the **initial value** and k is the **constant of proportionality.** **Exponential growth** is indicated by $k > 0$ and **exponential decay** by $k < 0$.

Discovery

Use a graphing utility to graph $y = Ce^{2t}$ for $C = 1, 2,$ and 5. How does the value of C affect the shape of the graph? Now graph $y = 2e^{kt}$ for $k = -2, -1, 0, 1,$ and 2. How does the value of k affect the shape of the graph? Which function grows faster, $y = e^x$ or $y = x^{10}$?

Proof Because the rate of change of y is proportional to y, you can write

$$\frac{dy}{dt} = ky.$$

You can see that $y = Ce^{kt}$ is a solution of this equation by differentiating to obtain $dy/dt = kCe^{kt}$ and substituting

$$\frac{dy}{dt} = kCe^{kt} = k(Ce^{kt}) = ky.$$

■

Applications

Radioactive decay is measured in terms of **half-life,** the number of years required for half of the atoms in a sample of radioactive material to decay. The half-lives of some common radioactive isotopes are as follows.

Uranium (U^{238})	4,510,000,000 years
Plutonium (Pu^{239})	24,360 years
Carbon (C^{14})	5,730 years
Radium (Ra^{226})	1,620 years
Einsteinium (Es^{254})	270 days
Nobelium (No^{257})	23 seconds

Real Life

EXAMPLE 1 *Modeling Radioactive Decay*

Much of the cost of nuclear energy is the cost of disposing of radioactive waste. Because of the long half-life of the waste, it must be stored in containers that will remain undisturbed for thousands of years.

A sample contains 1 gram of radium. How much radium will remain after 1000 years?

Solution

Let y represent the mass (in grams) of the radium in the sample. Because the rate of decay is proportional to y, you can apply the Law of Exponential Decay to conclude that y is of the form $y = Ce^{kt}$, where t is measured in years. From the given information, you know that $y = 1$ when $t = 0$. Substituting these values into the model produces

$$1 = Ce^{k(0)}, \qquad \textit{Substitute 1 for y and 0 for t.}$$

which implies that $C = 1$. Because radium has a half-life of 1620 years, you know that $y = \frac{1}{2}$ when $t = 1620$. Substituting these values into the model allows to you solve for k.

$$y = e^{kt} \qquad \textit{Exponential decay model}$$

$$\tfrac{1}{2} = e^{k(1620)} \qquad \textit{Substitute } \tfrac{1}{2} \textit{ for y and 1620 for t.}$$

$$\ln \tfrac{1}{2} = 1620k \qquad \textit{Take log of both sides.}$$

$$\tfrac{1}{1620} \ln \tfrac{1}{2} = k \qquad \textit{Divide both sides by 1620.}$$

Thus, $k \approx -0.0004279$, and the exponential decay model is

$$y = e^{-0.0004279t}.$$

To find the amount of radium remaining in the sample after 1000 years, substitute $t = 1000$ into the model. This produces

$$y = e^{-0.0004279(1000)} \approx 0.652 \text{ grams.}$$

The graph of the model is shown in Figure 4.19.

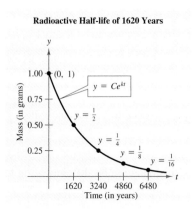

Radioactive Half-life of 1620 Years

FIGURE 4.19

The steps used in Example 1 are summarized below.

Guidelines for Modeling Exponential Growth and Decay

1. Use the given information to write *two* sets of conditions involving y and t.
2. Substitute the given conditions into the model $y = Ce^{kt}$ and use the results to solve for the constants C and k. (If one of the conditions involves $t = 0$, substitute that value first to solve for C.)
3. Use the model $y = Ce^{kt}$ to answer the question.

Real Life

EXAMPLE 2 *Modeling Population Growth*

In a research experiment, a population of fruit flies is increasing according to the law of exponential growth. After 2 days, there are 100 flies, and after 4 days, there are 300 flies. How many flies will there be after 5 days?

Solution

Let y be the number of flies at time t. From the given information, you know that $y = 100$ when $t = 2$ and $y = 300$ when $t = 4$. Substituting this information into the model $y = Ce^{kt}$ produces

$$100 = Ce^{2k} \quad \text{and} \quad 300 = Ce^{4k}.$$

To solve for k, solve for C in the first equation and substitute into the second equation.

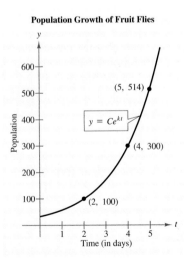

Population Growth of Fruit Flies

$300 = Ce^{4k}$	*Second equation*
$300 = \left(\dfrac{100}{e^{2k}}\right)e^{4k}$	*Substitute $100/e^{2k}$ for C.*
$\dfrac{300}{100} = e^{2k}$	*Divide both sides by 100.*
$\ln 3 = 2k$	*Take log of both sides.*
$\dfrac{1}{2}\ln 3 = k$	*Solve for k.*

Using $k = \frac{1}{2}\ln 3 \approx 0.5493$, you can determine that $C \approx 100/e^{2(0.5493)} \approx 33$. Thus, the exponential growth model is

$$y = 33e^{0.5493t},$$

as shown in Figure 4.20. This implies that, after 5 days, the population is

$$y = 33e^{0.5493(5)} \approx 514 \text{ flies.}$$

FIGURE 4.20

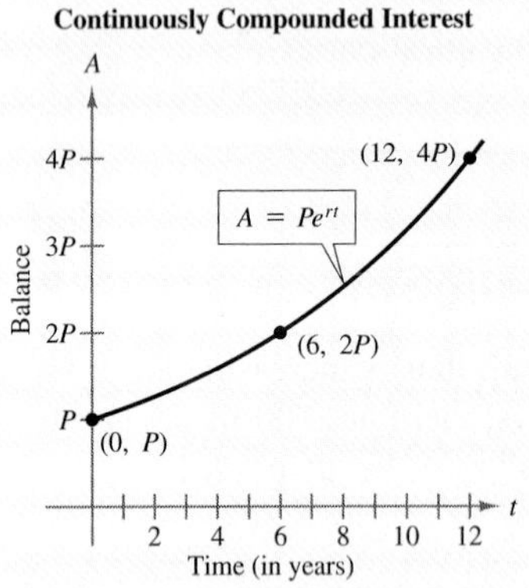

Real Life

EXAMPLE 3 *Modeling Compound Interest*

Money is deposited in an account for which the interest is compounded continuously. The balance in the account doubles in 6 years. What is the annual interest rate?

Solution

The balance A in an account with continuously compounded interest is given by the exponential growth model

$$A = Pe^{rt},$$ *Exponential growth model*

where P is the original deposit, r is the annual interest rate (in decimal form), and t is the time (in years). From the given information, you know that $A = 2P$ when $t = 6$, as shown in Figure 4.21. Use this information to solve for r.

$A = Pe^{rt}$	*Exponential growth model*
$2P = Pe^{r(6)}$	*Substitute $2P$ for A and 6 for t.*
$2 = e^{6r}$	*Divide both sides by P.*
$\ln 2 = 6r$	*Take log of both sides.*
$\dfrac{1}{6}\ln 2 = r$	*Divide both sides by 6.*

Thus, the annual interest rate is $r = \frac{1}{6}\ln 2 \approx 0.1155$ or about 11.55%.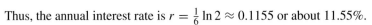

Each of the examples in this section uses the exponential growth model in which the base is e. Exponential growth, however, can be modeled with *any* base. That is, the model

$$y = Ca^{bt}$$

also represents exponential growth. (To see this, note that the model can be written in the form $y = Ce^{(\ln a)bt}$.) In some real-life settings, bases other than e are more convenient. For instance, in Example 1, knowing that the half-life of radium is 1620 years, you can immediately write the exponential decay model as

$$y = \left(\frac{1}{2}\right)^{t/1620}.$$

Using this model, the amount of radium left in the sample after 1000 years is

$$y = \left(\frac{1}{2}\right)^{1000/1620} \approx 0.652 \text{ grams,}$$

which is the same answer obtained in Example 1.

Note Can you see why you can immediately write the model $y = \left(\frac{1}{2}\right)^{t/1620}$ for the radioactive decay described in Example 1? Notice that when $t = 1620$, the value of y is $\frac{1}{2}$, when $t = 3240$, the value of y is $\frac{1}{4}$, and so on.

Continuously Compounded Interest

$A = Pe^{rt}$

Points shown on graph: $(0, P)$, $(6, 2P)$, $(12, 4P)$

Balance (vertical axis): P, $2P$, $3P$, $4P$
Time (in years) (horizontal axis): 2, 4, 6, 8, 10, 12

FIGURE 4.21

TECHNOLOGY

Fitting an Exponential Model to Data

Most graphing utilities have programs that allow you to find the *least-squares regression exponential model* for data. Depending on the type of graphing utility, the program may fit the data to a model of the form

$$y = ab^x, \qquad \textit{Exponential model with base b}$$

or it may fit the data to a model of the form

$$y = ae^{bx}. \qquad \textit{Exponential model with base e}$$

To see how to use such a program, consider the following.

The total amount y (in billions of dollars) of life insurance in the United States from 1950 through 1990 is listed in Table 4.8. *(Source: American Council of Life Insurance)*

TABLE 4.8

Year	1950	1955	1960	1965	1970	1975	1980	1985	1990
Amount	234	372	586	901	1402	2140	3541	6053	9393

To fit an exponential model to this data, let $x = 0$ represent 1950. Then enter the following coordinates into the statistical data bank of the graphing utility.

$$(0, 234), (5, 372), (10, 586), (15, 901), (20, 1402)$$

$$(25, 2140), (30, 3541), (35, 6053), (40, 9393)$$

After running the exponential regression program with a graphing utility that uses the model $y = ab^x$, the display should read $a \approx 229.8$ and $b \approx 1.0964$. (The correlation of $r \approx 0.9996$ tells you that the fit is very good.) Thus, a model for the data is

$$y = 229.8(1.0964)^x. \qquad \textit{Exponential model with base b}$$

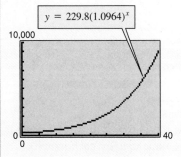

$y = 229.8(1.0964)^x$

The graph of the model is shown at the left. Notice that one way to interpret the model is that the total amount of life insurance in the United States increased by about 9.64% each year from 1950 through 1990.

If you use a graphing utility that uses the model $y = ae^{bx}$, then the display should read $a \approx 229.8$ and $b \approx 0.0920$. The corresponding model is

$$y = 229.8e^{0.0920x}. \qquad \textit{Exponential model with base e}$$

You can use either model to predict the total amount of life insurance in future years. For instance, in 1995 $(x = 45)$, the amount is predicted to be

$$y = 229.8(1.0964)^{45} \approx \$14{,}450 \text{ billion.}$$

Real
Life

EXAMPLE 4 *Modeling Sales*

Four months after discontinuing advertising on national television, a manufacturer notices that sales have dropped from 100,000 units per month to 80,000 units. If the sales follow an exponential pattern of decline, what will they be after another 4 months?

Solution

Let y represent the number of units, let t represent the time (in months), and consider the exponential growth model

$$y = Ce^{kt}. \qquad \text{\textit{Exponential growth model}}$$

From the given information, you know that $y = 100{,}000$ when $t = 0$. Using this information, you have

$$100{,}000 = Ce^0,$$

which implies that $C = 100{,}000$. To solve for k, use the fact that $y = 80{,}000$ when $t = 4$.

$y = 100{,}000e^{kt}$	*Exponential decay model*
$80{,}000 = 100{,}000e^{k(4)}$	*Substitute* 80,000 *for y and* 4 *for t.*
$0.8 = e^{4k}$	*Divide both sides by* 100,000.
$\ln 0.8 = 4k$	*Take log of both sides.*
$\dfrac{1}{4} \ln 0.8 = k$	*Divide both sides by* 4.

Thus, $k = \frac{1}{4} \ln 0.8 \approx -0.0558$, which means that the model is

$$y = 100{,}000e^{-0.0558t}.$$

After 4 more months ($t = 8$), you can expect sales to drop to

$$y = 100{,}000e^{-0.0558(8)} \approx 64{,}000 \text{ units,}$$

as shown in Figure 4.22.

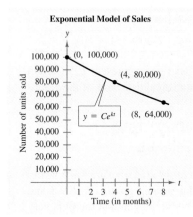

Exponential Model of Sales

(0, 100,000)

(4, 80,000)

$y = Ce^{kt}$ (8, 64,000)

Number of units sold

Time (in months)

FIGURE 4.22

Discussion Problem   ## Comparing Exponential and Linear Models

In Example 4, it was assumed that the sales were following an exponential pattern of decline. This implies that the sales were dropping by the same *percent* each month. Suppose instead that the sales were following a linear pattern of decline (dropping by the same *amount* each month). How would this affect the answer to the question given in Example 4? Which of the two models do you think might be more realistic? Explain your reasoning.

Warm Up

The following warm-up exercises involve skills that were covered in earlier sections. You will use these skills in the exercise set for this section.

In Exercises 1–4, solve the equation for *k*.

1. $12 = 24e^{4k}$

2. $10 = 3e^{5k}$

3. $25 = 16e^{3k}$

4. $22 = 32e^{20k}$

In Exercises 5–8, find the derivative of the function.

5. $y = 32e^{0.23t}$

6. $y = 24e^{-1.4t}$

7. $y = 18e^{0.072t}$

8. $y = 25e^{-0.001t}$

In Exercises 9 and 10, simplify the expression.

9. $e^{\ln 4}$

10. $4e^{\ln 3}$

E X E R C I S E S   4.5   means that technology can help you solve or check the exercise(s).

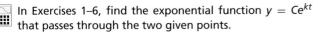

In Exercises 1–6, find the exponential function $y = Ce^{kt}$ that passes through the two given points.

1. $y = Ce^{kt}$

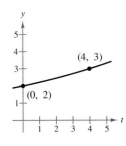

2. $y = Ce^{kt}$

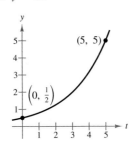

3. $y = Ce^{kt}$

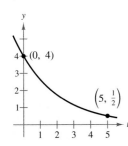

4. $y = Ce^{kt}$

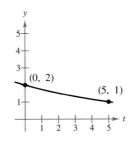

5. $y = Ce^{kt}$

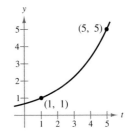

6. $y = Ce^{kt}$

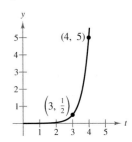

In Exercises 7–10, use the given information to write an equation for *y*. Confirm your result analytically by showing that the function satisfies the equation $dy/dt = Cy$. Does the function represent exponential growth or decay?

7. $\dfrac{dy}{dt} = 2y, \quad y = 10$ when $t = 0$

8. $\dfrac{dy}{dt} = -\dfrac{2}{3}y, \quad y = 20$ when $t = 0$

9. $\dfrac{dy}{dt} = -4y, \quad y = 30$ when $t = 0$

10. $\dfrac{dy}{dt} = 5.2y, \quad y = 18$ when $t = 0$

Radioactive Decay In Exercises 11–16, complete the table for the given radioactive isotope.

	Isotope	Half-Life (in Years)	Initial Quantity	Amount After 1,000 Years	Amount After 10,000 Years
11.	Ra^{226}	1,620	10 grams		
12.	Ra^{226}	1,620		1.5 grams	
13.	C^{14}	5,730			2 grams
14.	C^{14}	5,730	3 grams		
15.	Pu^{239}	24,360		2.1 grams	
16.	Pu^{239}	24,360			0.4 grams

17. *Radioactive Decay* What percent of a present amount of radioactive radium (Ra^{226}) will remain after 900 years?

18. *Radioactive Decay* Find the half-life of a radioactive material if after 1 year 99.57% of the initial amount remains.

19. *Carbon Dating* C^{14} dating assumes that the carbon dioxide on the earth today has the same radioactive content as it did centuries ago. If this is true, then the amount of C^{14} absorbed by a tree that grew several centuries ago should be the same as the amount of C^{14} absorbed by a similar tree today. A piece of ancient charcoal contains only 15% as much of the radioactive carbon as a piece of modern charcoal. How long ago was the tree burned to make the ancient charcoal? (The half-life of C^{14} is 5730 years.)

20. *Carbon Dating* Repeat Exercise 19 for a piece of charcoal that contains 30% as much radioactive carbon as a modern piece.

21. *Population Growth* The number of a certain type of bacteria increases continuously at a rate proportional to the number present. There are 150 present at a given time and 450 present 5 hours later.

   (a) How many will there be 10 hours after the initial time?

   (b) How long will it take for the population to double?

   (c) Does the answer to part (b) depend on the starting time? Explain.

22. *School Expenditures* In 1980, the total amount spent on schools in the United States was $270.9 billion. By 1988, the total amount had risen to $343.1 billion. Assume the total amount spent can be modeled by exponential growth. *(Source: U.S. National Center for Education Statistics)*

   (a) Estimate the total amount spent in 1985.

   (b) How many years would it take for the amount to double?

   (c) By what percent is the amount increasing each year?

Compound Interest In Exercises 23–28, complete the table for an account in which interest is compounded continuously.

	Initial Investment	Annual Rate	Time to Double	Amount After 10 Years	Amount After 25 Years
23.	$1,000	12%			
24.	$20,000	$10\frac{1}{2}$%			
25.	$750		$7\frac{3}{4}$ years		
26.	$10,000		5 years		
27.	$500			$1,292.85	
28.	$2,000				$6,008.33

29. *Effective Yield* The effective yield is the annual rate i that will produce the same interest per year as the nominal rate r compounded n times per year.

   (a) For a rate r that is compounded n times per year, show that the effective yield is

$$i = \left(1 + \frac{r}{n}\right)^n - 1.$$

   (b) Find the effective yield for a nominal rate of 6%, compounded monthly.

30. *Effective Yield* The effective yield is the annual rate i that will produce the same interest per year as the nominal rate r.

   (a) For a rate r that is compounded continuously, show that the effective yield is

$$i = e^r - 1.$$

   (b) Find the effective yield for a nominal rate of 6%, compounded continuously.

Effective Yield In Exercises 31 and 32, use the results of Exercises 29 and 30 to complete the table showing the effective yield for a nominal rate of r.

Number of compoundings per year	4	12	365	Continuous
Effective yield				

31. $r = 5\%$ **32.** $r = 7\frac{1}{2}\%$

33. *Rule of 70* Verify that the time necessary for an investment to double its value is approximately $70/r$, where r is the annual interest rate. (This formula requires that r be entered as a percent, not as a decimal.)

34. *Rule of 70* Use the Rule of 70 given in Exercise 33 to approximate the time necessary for an investment to double in value if (a) $r = 10\%$ and (b) $r = 7\%$.

35. *Revenue* The revenue for La-Z-Boy Chair Company was $254.9 million in 1983 and $684.1 million in 1992. *(Source: La-Z-Boy Chair Company)*

(a) Use an exponential growth model to predict the 1996 revenue.
(b) Use a linear model to predict the 1996 revenue.

36. *Net Worth* The net worth for American Greetings Corporation was $365.5 million in 1983 and $952.5 million in 1992. *(Source: American Greetings Corporation)*

(a) Use an exponential growth model to predict the 1996 net worth.
(b) Use a linear model to predict the 1996 net worth.

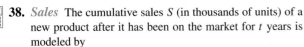 **37.** *Sales* The cumulative sales S (in thousands of units) of a new product after it has been on the market for t years is modeled by

$$S = Ce^{k/t}.$$

During the first year, 5000 units were sold. The saturation point for the market is 30,000 units. That is, the limit of S as $t \to \infty$ is 30,000.

(a) Solve for C and k in the model.
(b) How many units will be sold after 5 years?
(c) Sketch the graph of the sales function.

38. *Sales* The cumulative sales S (in thousands of units) of a new product after it has been on the market for t years is modeled by

$$S = 30(1 - 3^{kt}).$$

During the first year, 5000 units were sold.

(a) Solve for k in the model.
(b) What is the saturation point for this product?
(c) How many units will be sold after 5 years?
(d) Sketch the graph of the sales function.

39. *Learning Curve* The management of a factory finds that the maximum number of units a worker can produce in a day is 30. The learning curve for the number of units N produced per day after a new employee has worked t days is modeled by

$$N = 30(1 - e^{kt}).$$

After 20 days on the job, a worker is producing 19 units in a day. How many days should pass before this worker is producing 25 units per day?

40. *Learning Curve* The management in Exercise 39 requires that a new employee be producing at least 20 units per day after 30 days on the job.

(a) Find a learning curve model that describes this minimum requirement.
(b) Find the number of days before a minimal achiever is producing 25 units per day.

41. *Revenue* A small business assumes that the demand function for one of its products can be modeled by

$$p = Ce^{kx}.$$

When $p = \$45$, $x = 1000$ units, and when $p = \$40$, $x = 1200$ units.

(a) Solve for C and k.
(b) Find the values of x and p that will maximize the revenue for this product.

42. *Revenue* Repeat Exercise 41 given that when $p = \$5$, $x = 300$ units, and when $p = \$4$, $x = 400$ units.

43. *Forestry* The value V (in dollars) of a tract of timber can be modeled by

$$V = 100{,}000e^{0.75\sqrt{t}},$$

where $t = 0$ corresponds to 1990. If money earns interest at a rate of 4%, compounded continuously, then the present value A of the timber at any time t is

$$A = Ve^{-0.04t}.$$

Find the year in which the timber should be harvested to maximize the present value.

44. *Forestry* Repeat Exercise 43 using the model

$$V = 100{,}000e^{0.6\sqrt{t}}.$$

45. *Earthquake Intensity* On the Richter Scale, the magnitude R of an earthquake of intensity I is given by

$$R = \frac{\ln I - \ln I_0}{\ln 10},$$

where I_0 is the minimum intensity used for comparison. Assume $I_0 = 1$.

(a) Find the intensity of the 1906 San Francisco earthquake in which $R = 8.3$.
(b) Find the factor by which the intensity is increased when the value of R is doubled.
(c) Find dR/dI.

Case Study: Investing for Retirement

PROFILE

COMPANY
Towers Perrin
■

YEARS IN BUSINESS
60
■

LOCATION
Valhalla, NY
■

TYPE OF BUSINESS
*Management,
compensation, benefit
and actuarial consulting
for businesses*
■

EMPLOYEES
Approx. 5000
■

SALES
*Approx. $700 million
annually*
■

Many workers today are faced with decisions about how to invest for retirement, as companies offer them more choices and discretion in managing their pension funds. Because workers are often unfamiliar with these kinds of decisions, they frequently opt for safety—investing in assets such as U.S. Government bonds that virtually guarantee that none of their money will be lost.

However, safety has its cost. Investments typically offer a tradeoff between safety and rate of return. Companies that are safer—those that have consistently paid dividends on their stock and have always paid the interest and principal on their bonds—can typically pay a lower rate of return and still acquire all the funds they need. Companies that are riskier— perhaps because it is not clear that they will have profits to share with stockholders, or even enough income to pay their creditors on time—typically have to pay a higher rate of return to get the investment capital that they need.

In choosing their retirement investments, workers opting for safety without taking account of the cost, may be unknowingly choosing a much lower standard of living in retirement than they would otherwise enjoy. Owing to the effects of compounding, small differences in the rate of return on different investments can make major differences in the value of a retirement savings fund over the worklife of an employee.

Towers Perrin, a management consulting firm with a branch office in Valhalla, New York, offers such a retirement savings plan to its employees. The board of directors was concerned that employees had chosen to invest approximately 50 percent of their profit-sharing funds in fixed-income, low-return investments. It was suspected that some employees made their choices without a full understanding of the investment alternatives and risk-return relationships associated with them. As a result, Towers Perrin set up a task force to develop options and explain them to the employees. Five different investment portfolios, embodying different combinations of risk and return, were developed (see table and graph).

	Portfolios				
No. of Years	Mix A 6.6% 0.066	Mix B 7.8% 0.078	Mix C 9.0% 0.09	Mix D 10.2% 0.102	Mix E 11.0% 0.11
1	1,066	1,078	1,090	1,102	1,110
2	1,136	1,162	1,188	1,214	1,232
3	1,211	1,253	1,295	1,338	1,368
4	1,291	1,350	1,412	1,475	1,518
5	1,377	1,456	1,539	1,625	1,685
6	1,467	1,569	1,677	1,791	1,870
7	1,564	1,692	1,828	1,974	2,076
8	1,667	1,824	1,993	2,175	2,305
9	1,778	1,966	2,172	2,397	2,558
10	1,895	2,119	2,367	2,641	2,839
15	2,608	3,085	3,642	4,293	4,785
20	3,590	4,491	5,604	6,976	8,062
25	4,942	6,538	8,623	11,338	13,585
30	6,803	9,518	13,268	18,427	22,892

The most conservative investment portfolio, called "Mix A," has 80% of its funds invested in conservative securities, such as bonds and fixed- income investments. This portfolio has a very small chance of losing money in any given year, but it is expected to return only 6.6% per year on average in the future. At the other end of the spectrum is "Mix E," with a high proportion of small-company and international stocks, which are much riskier in the short run. In any given year, this portfolio is much more likely to drop in value, but it is also expected to average 11.0% annually over the long term. Mixes B, C, and D represent intermediate positions between these extremes.

You might think that the Mix A portfolio would provide about 60% (6.6% / 11.0%) of the income of Mix E over the years. But, owing to the effects of compounding, the difference is significantly greater than this. The accompanying table and graph show the value of $1000 invested in each of the five portfolios, after various amounts of time. This $1000 will grow to $6803 after 30 years at 6.6%, but it will grow to $22,892 after the same number of years at 11.0%—more than three times as much!

After they were educated about the tradeoff between risk and return and the impact of compounding, only 5% of active Towers Perrin employees chose to invest 100% of their money in fixed-income investments. This contrasts with 24% before the education program.

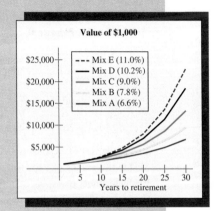

What Would You Do?

1. Which one of the five portfolios would you choose if you were a Towers Perrin employee, and why? Suppose you could also choose to divide your retirement savings among any combination of the five portfolios. Compare the following options. Which would you prefer? List your assumptions. Explain your reasoning and calculations.
(a) 20% of your retirement savings invested in Mix A, 20% in B, 20% in C, 20% in D, and 20% in E.
(b) 0% in A, 30% in B, 40% in C, 20% in D, and 10% in E
(c) 10% in A, 20% in B, 20% in C, 25% in D, and 25% in E

2. What are your own feelings about risk? Are you willing to take a greater risk of seeing your savings plummet in value in some years, in return for a higher rate of growth on average? Or would your blood pressure rise with every drop in the stock market?

3. Would your risk/return decision be different if you were two years away from retirement? . . . if you were a 25-year-old just starting with the company? Why?

4. Many people put off saving for retirement until a higher interest rate comes along or until they feel financially able to put aside a larger sum of money. Compare the following options using the portfolio chart. Which is better and why? Make a generalization on retirement savings strategy based on your findings.
(a) Putting aside $2000 (lump sum) at 6.6% 30 years before retirement.
(b) Putting aside $5000 (lump sum) at 10.2% 10 years before retirement.
(c) Putting aside $3000 (lump sum) at 7.8% 20 years before retirement.

Chapter Summary and Study Tips

After studying this chapter, you should have acquired the following skills. The exercise numbers are keyed to the Review Exercises that begin on page 308. Answers to odd-numbered Review Exercises are given in the back of the text.

■ Use the properties of exponents to evaluate and simplify exponential expressions. (*Section 4.1*)   Review Exercises 1–12

$$a^0 = 1, \quad a^x a^y = a^{x+y}, \quad \frac{a^x}{a^y} = a^{x-y}, \quad (a^x)^y = a^{xy}$$

$$(ab)^x = a^x b^x, \quad \left(\frac{a}{b}\right)^x = \frac{a^x}{b^x}, \quad a^{-x} = \frac{1}{a^x}$$

■ Use properties of exponents to answer questions about real life. (*Section 4.1*)   Review Exercises 13, 14

■ Sketch the graphs of exponential functions. (*Section 4.1*)   Review Exercises 15–18

■ Evaluate limits of exponential functions in real life. (*Section 4.1*)   Review Exercises 19, 20

■ Evaluate and graph functions involving the natural exponential function. (*Section 4.1*)   Review Exercises 21–24

$$\lim_{x \to 0} (1 + x)^{1/x} = e \approx 2.71828$$

■ Sketch the graphs of logistics growth functions. (*Section 4.1*)   Review Exercises 25, 26

■ Solve compound interest problems. (*Section 4.1*)   Review Exercises 27–30

$$A = P\left(1 + \frac{r}{n}\right)^{nt}, \quad A = Pe^{rt}$$

■ Answer questions involving the natural exponential function as a real-life model. (*Section 4.1*)   Review Exercises 31, 32

■ Find the derivatives of natural exponential functions. (*Section 4.2*)   Review Exercises 33–40

$$\frac{d}{dx}[e^x] = e^x, \quad \frac{d}{dx}[e^u] = e^u u'$$

■ Use calculus to analyze the graphs of functions that involve the natural exponential function. (*Section 4.2*)   Review Exercises 41–44

■ Use the definition of the natural logarithmic function to write exponential equations in logarithmic form, and vice versa. (*Section 4.3*)   Review Exercises 45–48

$$\ln x = b \quad \text{if and only if} \quad e^b = x.$$

■ Sketch the graphs of natural logarithmic functions. (*Section 4.3*)   Review Exercises 49–52

■ Use properties of logarithms to expand and condense logarithmic expressions. (*Section 4.3*)   Review Exercises 53–56

$$\ln xy = \ln x + \ln y, \quad \ln \frac{x}{y} = \ln x - \ln y, \quad \ln x^n = n \ln x$$

■ Use inverse properties of exponential and logarithmic functions to solve exponential and logarithmic equations. (*Section 4.3*)   Review Exercises 57–64

$$\ln e^x = x, \quad e^{\ln x} = x$$

*Several student study aids are available with this text. The *Student Solutions Guide* includes detailed solutions to all odd-numbered exercises, as well as practice chapter tests with answers. The *Graphics Calculator Guide* offers instructions on the use of a variety of graphing calculators and computer graphing software. The *Brief Calculus TUTOR* includes additional examples for selected exercises in the text.

■ Use properties of natural logarithms to answer questions about real life. (*Section 4.3*)

■ Find the derivatives of natural logarithmic functions. (*Section 4.4*)

$$\frac{d}{dx}[\ln x] = \frac{1}{x}, \qquad \frac{d}{dx}[\ln u] = \frac{1}{u}u'$$

■ Use calculus to analyze the graphs of functions that involve the natural logarithmic function. (*Section 4.4*)

■ Use the definition of logarithms to evaluate logarithmic expressions involving other bases. (*Section 4.4*)

$$\log_a x = b \quad \text{if and only if} \quad a^b = x$$

■ Use the change of base formula to evaluate logarithmic expressions involving other bases. (*Section 4.4*)

$$\log_a x = \frac{\ln x}{\ln a}$$

■ Find the derivatives of exponential and logarithmic functions involving other bases. (*Section 4.4*)

$$\frac{d}{dx}[a^x] = (\ln a)a^x, \qquad \frac{d}{dx}[a^u] = (\ln a)a^u u'$$

$$\frac{d}{dx}[\log_a x] = \left(\frac{1}{\ln a}\right)\frac{1}{x}, \qquad \frac{d}{dx}[\log_a u] = \left(\frac{1}{\ln a}\right)\frac{u'}{u}$$

■ Use calculus to answer questions about real-life rates of change. (*Section 4.4*)

■ Use exponential growth and decay to model real-life situations. (*Section 4.5*)

■ Classifying Differentiation Rules   Differentiation rules fall into two basic classes: (1) general rules that apply to all differentiable functions and (2) specific rules that apply to special types of functions. At this point in the course, you have studied six general rules: the Constant Rule, the Constant Multiple Rule, the Sum Rule, the Difference Rule, the Product Rule, and the Quotient Rule. Although these rules were introduced in the context of algebraic functions, remember that they can also be used with exponential and logarithmic functions. You have also studied three specific rules: the Power Rule, the Exponential Rule, and the Log Rule. Each of these rules comes in two forms: the "simple" version, such as $D_x[e^x] = e^x$, and the Chain Rule version, such as $D_x[e^u] = e^u u'$.

■ To Memorize or Not to Memorize?   When studying mathematics, you need to memorize some formulas and rules. Much of this will come from practice—the formulas that you use most often will be committed to memory. Some formulas, however, are used only infrequently. With these, it is helpful to be able to *derive* the formula from a *known* formula. For instance, knowing the Log Rule for differentiation and change of base formula, $\log_a x = (\ln x)/(\ln a)$, allows you to derive the formula for the derivative of a logarithmic function to base a.

Review Exercises

 means that technology can help you solve or check the exercise(s).

In Exercises 1–4, evaluate the expression.

1. $4(4^4)$

2. $16^{3/2}$

3. $\left(\frac{1}{5}\right)^4$

4. $\left(\frac{27}{8}\right)^{-1/3}$

In Exercises 5–8, use the properties of exponents to simplify the expression.

5. $\left(\frac{25}{4}\right)^0$

6. $(9^{1/3})(3^{1/3})$

7. $\frac{6^3}{36^2}$

8. $\frac{1}{4}\left(\frac{1}{2}\right)^{-3}$

In Exercises 9–12, solve the equation for x.

9. $5^x = 625$

10. $x^{3/4} = 8$

11. $e^{-1/2} = e^{x-1}$

12. $\frac{x^3}{3} = e^3$

13. *New York Stock Exchange* The total number y (in millions) of shares of stocks listed on the New York Stock Exchange between 1940 and 1990 can be modeled by

$$y = 29.619(1.0927)^t, \quad 40 \le t \le 90,$$

where $t = 40$ corresponds to 1940. Use this model to estimate the total number of shares listed in 1950, 1970, and 1990. (*Source: New York Stock Exchange*)

14. *Recycling* Between 1960 and 1988, the percent p of materials in municipal waste that was recycled can be modeled by

$$p = 4.92 + 1.29(1.097)^t, \quad 0 \le t \le 28,$$

where $t = 0$ corresponds to 1960. Use the model to estimate the percent of recycled materials in 1960, 1970, and 1988. (*Source: Franklin Associates, Ltd.*)

 In Exercises 15–18, sketch the graph of the function.

15. $f(x) = 9^{x/2}$

16. $f(t) = \left(\frac{1}{6}\right)^t$

17. $f(x) = \left(\frac{1}{2}\right)^{2x} + 4$

18. $g(x) = \frac{1}{4}(2)^{-3x}$

19. *Demand* The demand function for a product is

$$p = 12{,}500 - \frac{10{,}000}{2 + e^{-0.001x}},$$

where p is the price per unit and x is the number of units produced (see figure). What is the limit of the price as x increases without bound? Explain what this means in the context of the problem.

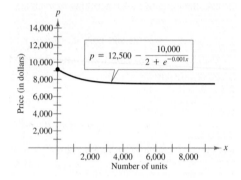

20. *Endangered Species* Biologists consider a species of a plant or animal to be endangered if it is expected to become extinct in less than 20 years. The population y of a certain species is modeled by

$$y = 1096e^{-0.39t}$$

(see figure). Is this species endangered? Explain your reasoning.

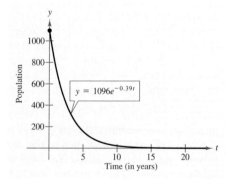

In Exercises 21–24, evaluate the function at the indicated value. Then sketch its graph.

21. $f(x) = 2e^{x-1}$, $x = 2$

22. $f(t) = e^{4t} - 1$, $t = 0$

23. $g(t) = 12e^{-0.2t}$, $t = 17$

24. $g(x) = \dfrac{24}{1 + e^{-0.3x}}$, $x = 300$

25. *Stocking a Lake with Fish* A lake is stocked with 500 fish and the fish population P begins to increase according to the logistics model

$$P = \frac{10,000}{1 + 19e^{-t/5}}, \qquad 0 \le t,$$

where t is measured in months.

(a) Use a graphing utility to graph the function.

(b) Estimate the number of fish in the lake after 4 months.

(c) Does the population have a limit as t increases without bound? Explain your reasoning.

(d) After how many months is the population increasing most rapidly? Explain your reasoning.

26. *Spread of a Virus* On a college campus of 5000 students, the spread of a flu virus through the student body is modeled by

$$P = \frac{5000}{1 + 4999e^{-0.8t}}, \qquad 0 \le t,$$

where P is the total number of infected people and t is the time, measured in days.

(a) Use a graphing utility to graph the function.

(b) How many students will be infected after 5 days?

(c) According to this model, will all the students on campus become infected with the flu? Explain your reasoning.

In Exercises 27 and 28, complete the table to determine the balance A when P dollars is invested at an annual rate of r for t years, compounded n times per year.

n	1	2	4	12	365	Continuous compounding
A						

27. $P = \$1000$, $r = 4\%$, $t = 5$ years

28. $P = \$5000$, $r = 6\%$, $t = 20$ years

In Exercises 29 and 30, $2000 is deposited in an account. Decide which account, (a) or (b), will have the greater balance after 10 years.

29. (a) 5%, compounded continuously

   (b) 6%, compounded quarterly

30. (a) $6\frac{1}{2}\%$, compounded monthly

   (b) $6\frac{1}{4}\%$, compounded continuously

31. *Age at First Marriage* The average age A of an American man at his first marriage from 1970 to 1990 can be modeled by

$$A = 22.38 + \frac{1}{0.1513 + 9.6155e^{-0.3708t}}, \qquad 0 \le t \le 20,$$

where $t = 0$ corresponds to 1970. Use this model to estimate the average age of an American man at his first marriage in 1970, 1980, and 1990. *(Source: U.S. Center for Health Statistics)*

32. *Profit* The profit P (in millions of dollars) for Blockbuster Video from 1986 to 1993 can be modeled by

$$P = 1.96e^{0.3696t} - 17.24,$$

where $t = 6$ corresponds to 1986. Use this model to estimate the profit for Blockbuster Video in 1986, 1990, and 1993. *(Source: Blockbuster Entertainment Corporation)*

In Exercises 33–40, find the derivative of the function.

33. $y = 4e^{x^2}$

34. $y = x^2 e^x$

35. $y = \dfrac{x}{e^{2x}}$

36. $y = \dfrac{x - 1}{e^x + 2x^2}$

37. $y = \sqrt{4e^{4x}}$

38. $y = \sqrt[3]{xe^{3x}}$

39. $y = \dfrac{5}{1 + e^{2x}}$

40. $y = \dfrac{e^x}{1 - xe^x}$

In Exercises 41–44, graph and analyze the function. Include any relative extrema, points of inflection, and asymptotes in your analysis.

41. $f(x) = 4e^{-x}$

42. $f(x) = x^3 e^x$

43. $f(x) = \dfrac{e^x}{x^2}$

44. $f(x) = \dfrac{1}{xe^x}$

In Exercises 45 and 46, write the logarithmic equation as an exponential equation.

45. $\ln 12 \approx 2.4849$

46. $\ln 0.6 \approx -0.5108$

In Exercises 47 and 48, write the exponential equation as a logarithmic equation.

47. $e^{1.5} \approx 4.4817$ **48.** $e^{-4} \approx 0.0183$

 In Exercises 49–52, sketch the graph of the function.

49. $y = \ln(4 - x)$ **50.** $y = 5 + \ln x$

51. $y = \ln \dfrac{x}{3}$ **52.** $y = -\dfrac{5}{6} \ln x$

In Exercises 53–56, use the properties of logarithms to write the expression as a sum, difference, or multiple of logarithms.

53. $\ln \sqrt{x^2(x-1)}$ **54.** $\ln \dfrac{x^2}{(x+1)^3}$

55. $\ln \left(\dfrac{1-x}{3x} \right)^3$ **56.** $\ln \dfrac{e^4}{5x}$

In Exercises 57–64, solve the equation for x.

57. $e^{\ln x} = 3$ **58.** $\ln x = 3e^{-1}$

59. $\ln 2x - \ln(3x - 1) = 0$ **60.** $e^{2x-1} - 6 = 0$

61. $\ln x + \ln(x - 3) = 0$ **62.** $e^{-1.386x} = 0.25$

63. $9^{6x} - 27 = 0$ **64.** $100(1.21)^x = 110$

 65. *Home Mortgage* The monthly payment M for a home mortgage of P dollars for t years at an annual interest rate of $r\%$ is given by

$$M = P \left(\frac{\dfrac{r}{12}}{1 - \left(\dfrac{1}{(r/12) + 1} \right)^{12t}} \right).$$

(a) Use a graphing utility to graph the model when $P = \$100,000$ and $r = 8\%$.

(b) You are given a choice of a 20-year term or a 30-year term. Which would you choose? Explain.

 66. *Hourly Wages* The average hourly wage w in the United States from 1970 to 1991 can be modeled by

$$w = 50.92 - 47.7e^{-0.0077t},$$

where $t = 0$ corresponds to 1970. *(Source: U.S. Bureau of Labor Statistics)*

(a) Use a graphing utility to graph the model.

(b) Use the model to determine the year in which the average hourly wage was $10.

(c) For how many years past 1991 do you think this equation might be a good model for the average hourly wage? Explain your reasoning.

In Exercises 67–74, find the derivative of the function.

67. $f(x) = \ln 3x^2$ **68.** $y = \ln \dfrac{x(x-1)}{x-2}$

69. $y = x\sqrt{\ln x}$ **70.** $f(x) = \ln \sqrt{4x}$

71. $y = \dfrac{\ln x}{x^3}$ **72.** $y = \ln(x^2 - 2)^{2/3}$

73. $f(x) = \ln e^{-x^2}$ **74.** $y = \ln \dfrac{e^x}{1 + e^x}$

In Exercises 75–78, graph and analyze the function. Include any relative extrema and points of inflection in your analysis.

75. $y = \ln(x + 3)$ **76.** $y = \dfrac{8 \ln x}{x^2}$

77. $y = \ln \dfrac{10}{x + 2}$ **78.** $y = \ln \dfrac{x^2}{9 - x^2}$

In Exercises 79–82, evaluate the logarithm.

79. $\log_7 49$ **80.** $\log_2 32$

81. $\log_{10} 1$ **82.** $\log_4 \frac{1}{64}$

In Exercises 83–86, use the change of base formula to evaluate the logarithm. Round the result to four decimal places.

83. $\log_5 10$ **84.** $\log_4 12$

85. $\log_{16} 64$ **86.** $\log_{10} 125$

In Exercises 87–90, find the derivative of the function.

87. $y = \log_3(2x - 1)$ **88.** $y = \log_{10} \dfrac{3}{x}$

89. $y = \log_2 \dfrac{1}{x^2}$ **90.** $y = \log_{16}(x^2 - 3x)$

 91. *Depreciation* After t years, the value V of a car purchased for $20,000 is

$$V = 20,000(0.75)^t.$$

(a) Sketch a graph of the function and determine the value of the car 2 years after it was purchased.

(b) Find the rate of change of V with respect to t when $t = 1$ and when $t = 4$.

(c) After how many years will the car be worth $5000?

92. *Inflation* If the annual rate of inflation averages 5% over the next 10 years, then the approximate cost of goods or services C during any year in that decade is

$$C = P(1.05)^t,$$

where t is the time in years and P is the present cost.

(a) The price of an oil change is presently $24.95. Estimate the price of an oil change 10 years from now.

(b) Find the rate of change of C with respect to t when $t = 1$.

93. *Drug Decomposition* A medical solution contains 500 milligrams of a drug per milliliter when the solution is prepared. After 40 days, it contains only 300 milligrams per milliliter. Assuming that the rate of decomposition is proportional to the concentration present, find an equation giving the concentration A after t days.

94. *Population Growth* A population is growing continuously at the rate of $2\frac{1}{2}\%$ per year. Find the time necessary for the population to (a) double in size and (b) triple in size.

95. *Half-Life* A sample of radioactive waste is taken from a nuclear plant. The sample contains 50 grams of strontium-90 at time $t = 0$ years and 42.031 grams after 7 years. What is the half-life of strontium-90?

96. *Half-Life* The half-life of cobalt-60 is 5.2 years. Find the time it would take for a sample of 0.5 grams of cobalt-60 to decay to 0.1 grams.

97. *Profit* The profit for Wendy's was $3.5 million in 1987 and $64.7 million in 1992 (see figure). Use an exponential growth model to predict the 1996 profit. *(Source: Wendy's International, Inc.)*

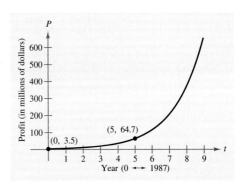

FIGURE FOR 97

98. *Net Worth* The net worth of Toys'R'Us was $459.7 million in 1983 and $2889.0 million in 1992 (see figure). Use an exponential growth model to predict the 1996 net worth. *(Source: Toys'R'Us)*

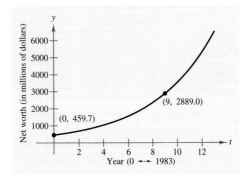

Sample Post-Graduation Exam Questions

The following questions were taken from certified public accountant (CPA) exams, graduate management admission tests (GMAT), graduate records exams (GRE), actuarial exams, or college-level academic skills tests (CLAST). The answers to the questions are given in the back of the book.

1. The rate of decay of a radioactive substance is proportional to the amount of the substance present. Three years ago there were 6 grams of substance. Now there are 5 grams. How many grams will there be 3 years from now?
  (a) 4   (b) $\frac{25}{6}$ (c) $\frac{125}{36}$ (d) $\frac{75}{36}$

2. In a certain town, 45% of the people have brown hair, 30% have brown eyes, and 15% have both brown hair and brown eyes. What percent of the people in the town have neither brown hair nor brown eyes?
  (a) 25%   (b) 35%   (c) 40%   (d) 50%

3. You have $900 and deposit it in a savings account that is compounded continuously at 4.76%. After 16 years the amount in the account will be
  (a) $1927.53   (b) $1077.81   (c) $943.88   (d) $2827.53

4. A bookstore orders 75 books. Each book costs the bookstore $29 and is sold for $42. The bookstore must pay a $4 service charge for each unsold book returned. If the bookstore returns seven books, how much profit will the bookstore make?
  (a) $975   (b) $947   (c) $653   (d) $681

For Questions 5–7, use the data given in the graph.

5. In how many of the years was spending larger than in the preceding year?
  (a) 2   (b) 4   (c) 1   (d) 3

6. In which year was the profit the greatest?
  (a) 1988   (b) 1991   (c) 1987   (d) 1989

7. In 1990, profits decreased by x percent from 1989 with x equal to
  (a) 60%   (b) 140%   (c) 340%   (d) 40%

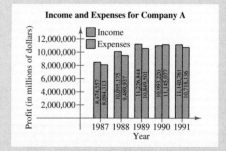

FIGURE FOR 5–7

$$\int_a^b f'(x)\, dx = f(b) - f(a)$$

5 Integration and Its Applications

Calculus involves two basic operations: differentiation and integration (or antidifferentiation). Because these operations are inverses of each other, some applications of integration are straightforward. For instance, integrating an acceleration function produces a velocity function. Other applications are surprisingly unrelated to derivatives and rates of change. For instance, integration can be used to find the area of a region, the average value of a function, and the volume of a solid.

Although integration was used by the early developers of calculus (such as Newton), its rigorous development is usually associated with the German mathematician Georg Friedrich Riemann (1826–1866).

$$\int f'(x)\, dx = f(x) + C$$

Riemann

The Bettmann Archive

Antiderivatives and Indefinite Integrals

Antiderivatives ■ *Notation for Antiderivatives and Indefinite Integrals* ■
Finding Antiderivatives ■ *Particular Solutions* ■ *Applications*

Antiderivatives

Up to this point in the text, you have been concerned primarily with this problem: given a function, find its derivative. Many important applications of calculus involve the inverse problem: given the derivative of a function, find the function. For example, suppose you are given

$$f'(x) = 2, \quad g'(x) = 3x^2, \quad \text{and} \quad s'(t) = 4t.$$

Your goal is to determine the functions f, g, and s. By making educated guesses, you might come up with the following.

$$f(x) = 2x \quad \text{because} \quad \frac{d}{dx}[2x] = 2$$

$$g(x) = x^3 \quad \text{because} \quad \frac{d}{dx}[x^3] = 3x^2$$

$$s(t) = 2t^2 \quad \text{because} \quad \frac{d}{dt}[2t^2] = 4t$$

This operation of determining the original function from its derivative is the inverse operation of differentiation. It is called **antidifferentiation.**

Definition of Antiderivative

A function F is an **antiderivative** of a function f if for every x in the domain of f, it follows that $F'(x) = f(x)$.

Note In this text, we use the phrase "$F(x)$ is an antiderivative of $f(x)$" synonymously with "F is an antiderivative of f."

If $F(x)$ is an antiderivative of $f(x)$, then $F(x) + C$, where C is any constant, is also an antiderivative of $f(x)$. For example,

$$F(x) = x^3, \quad G(x) = x^3 - 5, \quad \text{and} \quad H(x) = x^3 + 0.3$$

are antiderivatives of $3x^2$ because the derivative of each is $3x^2$. As it turns out, *all* antiderivatives of $3x^2$ are of the form $x^3 + C$. Thus, the process of antidifferentiation does not determine a single function, but rather a *family* of functions, each differing from the others by a constant.

Notation for Antiderivatives and Indefinite Integrals

The antidifferentiation process is also called **integration** and is denoted by the symbol

$$\int,$$ *Integral sign*

called an **integral sign.** The symbol

$$\int f(x)\,dx$$ *Indefinite integral*

is the **indefinite integral** of $f(x)$, and it denotes the family of antiderivatives of $f(x)$. That is, if $F'(x) = f(x)$ for all x, then you can write

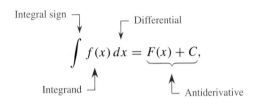

where $f(x)$ is the **integrand** and C is the **constant of integration.** The differential dx in the indefinite integral identifies the variable of integration. That is, the symbol $\int f(x)\,dx$ denotes the "antiderivative of f *with respect to* x" just as the symbol dy/dx denotes the "derivative of y *with respect to x.*"

Integral Notation of Antiderivatives

The notation

$$\int f(x)\,dx = F(x) + C,$$

where C is an arbitrary constant, means that F is an antiderivative of f. That is, $F'(x) = f(x)$ for all x in the domain of f.

Discovery

Verify that $F_1(x) = x^2 - 2x$, $F_2(x) = x^2 - 2x - 1$, and $F_3(x) = (x - 1)^2$ are antiderivatives of $f(x) = 2x - 2$. Use a graphing utility to graph F_1, F_2, and F_3 on the same coordinate plane. How are their graphs related? What can you say about the graph of any other antiderivative of f?

EXAMPLE 1 *Notation for Antiderivatives*

Using integral notation, you can write the three antiderivatives given at the beginning of this section as follows.

a. $\displaystyle\int 2\,dx = 2x + C$ **b.** $\displaystyle\int 3x^2\,dx = x^3 + C$ **c.** $\displaystyle\int 4t\,dt = 2t^2 + C$

Finding Antiderivatives

The inverse relationship between the operations of integration and differentiation can be shown symbolically, as follows.

$$\frac{d}{dx}\left[\int f(x)\,dx\right] = f(x)$$ *Differentiation is the inverse of integration.*

$$\int f'(x)\,dx = f(x) + C$$ *Integration is the inverse of differentiation.*

This inverse relationship between integration and differentiation allows you to obtain integration formulas directly from differentiation formulas. The following summary lists the integration formulas that correspond to some of the differentiation formulas you have studied.

Basic Integration Rules

1. $\int k\,dx = kx + C$, k is a constant   *Constant Rule*

2. $\int kf(x)\,dx = k\int f(x)\,dx$ *Constant Multiple Rule*

3. $\int [f(x) + g(x)]\,dx = \int f(x)\,dx + \int g(x)\,dx$ *Sum Rule*

4. $\int [f(x) - g(x)]\,dx = \int f(x)\,dx - \int g(x)\,dx$ *Difference Rule*

5. $\int x^n\,dx = \frac{x^{n+1}}{n+1} + C$, $n \neq -1$ *Simple Power Rule*

Note You will study the General Power Rule for integration in Section 5.2 and the Exponential and Log Rules in Section 5.3.

Be sure you see that the Simple Power Rule has the restriction that n cannot be -1. Thus, you *cannot* use the Simple Power Rule to evaluate the integral

$$\int \frac{1}{x}\,dx.$$

To evaluate this integral, you need the Log Rule (Section 5.3).

Note In Example 2b, the integral $\int 1\,dx$ is usually shortened to the form $\int dx$.

EXAMPLE 2 *Finding Indefinite Integrals*

a. $\int 2\,dx = 2x + C$ **b.** $\int 1\,dx = x + C$ **c.** $\int -5\,dt = -5t + C$

EXAMPLE 3 *Finding an Indefinite Integral*

Find the indefinite integral $\int 3x \, dx$.

Solution

$$
\begin{aligned}
\int 3x \, dx &= 3 \int x \, dx && \textit{Constant Multiple Rule} \\[4pt]
&= 3 \int x^1 \, dx && \textit{Rewrite } x \textit{ as } x^1. \\[4pt]
&= 3 \left(\frac{x^2}{2} \right) + C && \textit{Power Rule with } n = 1 \\[4pt]
&= \frac{3}{2} x^2 + C && \textit{Simplify.}
\end{aligned}
$$

In finding indefinite integrals, a strict application of the basic integration rules tends to produce cumbersome constants of integration. For instance, in Example 3, you could have written

$$
\int 3x \, dx = 3 \int x \, dx = 3 \left(\frac{x^2}{2} + C \right) = \frac{3}{2} x^2 + 3C.
$$

However, because C represents *any* constant, it is unnecessary to write $3C$ as the constant of integration. You can simply write $\frac{3}{2} x^2 + C$.

In Example 3, note that the general pattern of integration is similar to that of differentiation.

Given:	Rewrite:	Integrate:	Simplify:
$\int 3x \, dx$	$3 \int x^1 \, dx$	$3 \left(\dfrac{x^2}{2} \right) + C$	$\dfrac{3}{2} x^2 + C$

⊞ TECHNOLOGY

If you have access to a symbolic integration program such as *Derive*, *Maple*, or *Mathematica*, try using it to find antiderivatives. For instance, the following steps show how *Derive* can be used to find the antiderivative in Example 3. Note that the program does not list a constant of integration.

1: **3x**   *Author.*

2: $\int$ **3x dx**   *Integrate.*

3: $\dfrac{\mathbf{3x^2}}{\mathbf{2}}$ *Simplify.*

EXAMPLE 4 *Rewriting Before Integrating*

	Given Integral	Rewrite	Integrate	Simplify
a.	$\int \dfrac{1}{x^3} \, dx$	$\int x^{-3} \, dx$	$\dfrac{x^{-2}}{-2} + C$	$-\dfrac{1}{2x^2} + C$
b.	$\int \sqrt{x} \, dx$	$\int x^{1/2} \, dx$	$\dfrac{x^{3/2}}{3/2} + C$	$\dfrac{2}{3} x^{3/2} + C$

Note   Remember that you can check your answer to an antidifferentiation problem by differentiating. For instance, in Example 4b, you can check that $\frac{2}{3} x^{3/2}$ is the correct antiderivative by differentiating to obtain

$$
\frac{d}{dx} \left[\frac{2}{3} x^{3/2} \right] = \left(\frac{2}{3} \right) \left(\frac{3}{2} \right) x^{1/2} = \sqrt{x}.
$$

With the five basic integration rules, you can integrate *any* polynomial function, as demonstrated in the next example.

EXAMPLE 5 *Integrating Polynomial Functions*

Find the following indefinite integrals.

a. $\int (x + 2)\,dx$ **b.** $\int (3x^4 - 5x^2 + x)\,dx$

Solution

a. Use the Sum Rule to integrate each part separately.

$$\int (x + 2)\,dx = \int x\,dx + \int 2\,dx$$

$$= \frac{x^2}{2} + 2x + C$$

b. Try to identify each basic integration rule used to evaluate this integral.

$$\int (3x^4 - 5x^2 + x)\,dx = 3\left(\frac{x^5}{5}\right) - 5\left(\frac{x^3}{3}\right) + \frac{x^2}{2} + C$$

$$= \frac{3}{5}x^5 - \frac{5}{3}x^3 + \frac{1}{2}x^2 + C$$

EXAMPLE 6 *Rewriting Before Integrating*

Find the indefinite integral

$$\int \frac{x + 1}{\sqrt{x}}\,dx.$$

Solution

Begin by rewriting the quotient in the integrand as a sum. Then rewrite each term using rational exponents.

Note When integrating quotients, don't make the mistake of integrating the numerator and denominator separately. For instance, in Example 6, be sure you see that

$$\int \frac{x + 1}{\sqrt{x}}\,dx \neq \frac{\int (x + 1)\,dx}{\int \sqrt{x}\,dx}.$$

$$\int \frac{x + 1}{\sqrt{x}}\,dx = \int \left(\frac{x}{\sqrt{x}} + \frac{1}{\sqrt{x}}\right)\,dx \qquad \text{\textit{Rewrite as a sum.}}$$

$$= \int \left(x^{1/2} + x^{-1/2}\right)\,dx \qquad \text{\textit{Use rational exponents.}}$$

$$= \frac{x^{3/2}}{3/2} + \frac{x^{1/2}}{1/2} + C \qquad \text{\textit{Apply Power Rule.}}$$

$$= \frac{2}{3}x^{3/2} + 2x^{1/2} + C \qquad \text{\textit{Simplify.}}$$

If you have access to a symbolic integration program, try using it for this example. When we tried it, the program listed the antiderivative as $2\sqrt{x}(x + 3)/3$, which is equivalent to the result listed above.

Particular Solutions

You have already seen that the equation $y = \int f(x)\,dx$ has many solutions, each differing from the others by a constant. This means that the graphs of any two antiderivatives of f are vertical translations of each other. For instance, Figure 5.1 shows the graphs of several antiderivatives of the form

$$y = F(x) = \int (3x^2 - 1)\,dx = x^3 - x + C.$$

Each of these antiderivatives is a solution of $dy/dx = 3x^2 - 1$.

In many applications of integration, you are given enough information to determine a **particular solution.** To do this, you need only know the value of $F(x)$ for one value of x. (This information is called an **initial condition.**) For example, in Figure 5.1, there is only one curve that passes through the point $(2, 4)$. To find this curve, use the following information.

$$F(x) = x^3 - x + C \qquad \text{\textit{General solution}}$$

$$F(2) = 4 \qquad \text{\textit{Initial condition}}$$

By using the initial condition in the general solution, you can determine that $F(2) = 2^3 - 2 + C = 4$, which implies that $C = -2$. Thus, the particular solution is

$$F(x) = x^3 - x - 2. \qquad \text{\textit{Particular solution}}$$

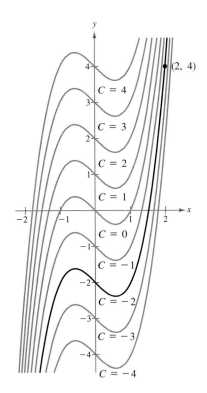

$$F(x) = x^3 - x + C$$

FIGURE 5.1

EXAMPLE 7 Finding a Particular Solution

Find the general solution of $F'(x) = 2x - 2$, and find the particular solution that satisfies the initial condition $F(1) = 2$.

Solution
Begin by integrating to find the general solution.

$$F(x) = \int (2x - 2)\,dx \qquad \text{\textit{Integrate } } F'(x) \text{ \textit{to obtain } } F(x).$$

$$= x^2 - 2x + C \qquad \text{\textit{General solution}}$$

Using the initial condition $F(1) = 2$, you can write

$$F(1) = 1^2 - 2(1) + C = 2,$$

which implies that $C = 3$. Thus, the particular solution is

$$F(x) = x^2 - 2x + 3. \qquad \text{\textit{Particular solution}}$$

This solution is shown graphically in Figure 5.2. Note that each of the gray curves represents a solution of the equation $F'(x) = 2x - 2$. The black curve, however, is the only solution that passes through the point $(1, 2)$, which means that $F(x) = x^2 - 2x + 3$ is the only solution that satisfies the initial condition.

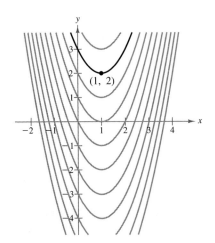

FIGURE 5.2

Applications

In Chapter 2, you used the general position function for a falling object, $s(t) = -16t^2 + v_0 t + s_0$, where $s(t)$ is the height (in feet) and t is the time (in seconds). In the next example, integration is used to *derive* this function.

Real Life

EXAMPLE 8 *Deriving a Position Function*

A ball is thrown upward with an initial velocity of 64 feet per second from an initial height of 80 feet, as shown in Figure 5.3. Derive the position function giving the height s (in feet) as a function of the time t (in seconds). When does the ball hit the ground?

Solution

Let $t = 0$ represent the initial time. Then the two given conditions can be written as follows.

$s(0) = 80$	*Initial height is 80 feet.*
$s'(0) = 64$	*Initial velocity is 64 feet per second.*

Because the acceleration due to gravity is -32 feet per second per second, you have the following.

$s''(t) = -32$	*Acceleration due to gravity*
$s'(t) = \displaystyle\int -32\,dt$	*Integrate $s''(t)$ to obtain $s'(t)$.*
$\quad = -32t + C_1$	*Velocity function*

Using the initial velocity, you can conclude that $C_1 = 64$.

$s'(t) = -32t + 64$	*Velocity function*
$s(t) = \displaystyle\int (-32t + 64)\,dt$	*Integrate $s'(t)$ to obtain $s(t)$.*
$\quad = -16t^2 + 64t + C_2$	*Position function*

Using the initial height, it follows that $C_2 = 80$. Thus, the position function is

$$s(t) = -16t^2 + 64t + 80. \qquad \textit{Position function}$$

To find the time when the ball hits the ground, set the position function equal to zero and solve for t.

$-16t^2 + 64t + 80 = 0$	*Set $s(t)$ equal to zero.*
$-16(t + 1)(t - 5) = 0$	*Factor.*
$t = -1,\ t = 5$	*Solve for t.*

Because the time must be positive, you can conclude that the ball hits the ground 5 seconds after it is thrown.

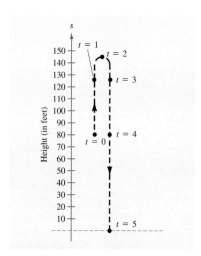

FIGURE 5.3

Real
Life

EXAMPLE 9 *Finding a Cost Function*

The marginal cost for producing x units of a product is modeled by

$$\frac{dC}{dx} = 32 - 0.04x. \qquad \text{\textit{Marginal cost}}$$

It costs \$50 to produce one unit. Find the total cost of producing 200 units.

Solution

To find the cost function, integrate the marginal cost function.

$$C = \int (32 - 0.04x)\,dx \qquad \text{\textit{Integrate } } dC/dx \text{ \textit{to obtain } } C.$$

$$= 32x - 0.04\left(\frac{x^2}{2}\right) + K$$

$$= 32x - 0.02x^2 + K \qquad \text{\textit{Cost function}}$$

To solve for K, use the initial condition that $C = 50$ when $x = 1$.

$$50 = 32(1) - 0.02(1)^2 + K \qquad \text{\textit{Substitute 50 for } } C \text{ \textit{and 1 for } } x.$$

$$18.02 = K \qquad \text{\textit{Solve for } } K.$$

Thus, the total cost function is

$$C = 32x - 0.02x^2 + 18.02, \qquad \text{\textit{Cost function}}$$

which implies that the cost of producing 200 units is

$$C = 32(200) - 0.02(200)^2 + 18.02$$

$$= \$5618.02.$$

Discussion Problem **Investigating Marginal Cost**

In Example 9, you were given a marginal cost function of

$$\frac{dC}{dx} = 32 - 0.04x.$$

This means that the cost of making each additional unit decreases by about \$0.04. You can confirm this by finding the cost of making different amounts of the product.

x	1	2	3	4	5	6
C	\$50.00	\$81.94	\$113.84	\$145.70	\$177.52	\$209.30

From this, you can see that the first unit cost \$50, the second cost \$31.94, the third cost \$31.90, the fourth cost \$31.86, the fifth cost \$31.82, and so on. At what production level would this costing scheme cease to make sense? Explain your reasoning.

Warm Up

The following warm-up exercises involve skills that were covered in earlier sections. You will use these skills in the exercise set for this section.

In Exercises 1–6, rewrite the expression using rational exponents.

1. $\dfrac{\sqrt{x}}{x}$

2. $\sqrt[3]{2x}\,(2x)$

3. $\sqrt{5x^3} + \sqrt{x^5}$

4. $\dfrac{1}{\sqrt{x}} + \dfrac{1}{\sqrt[3]{x^2}}$

5. $\dfrac{(x+1)^3}{\sqrt{x+1}}$

6. $\dfrac{\sqrt{x}}{\sqrt[3]{x}}$

In Exercises 7–10, let $(x, y) = (2, 2)$, and solve the equation for C.

7. $y = x^2 + 5x + C$

8. $y = 3x^3 - 6x + C$

9. $y = -16x^2 + 26x + C$

10. $y = -\frac{1}{4}x^4 - 2x^2 + C$

E X E R C I S E S   5.1   means that technology can help you solve or check the exercise(s).

In Exercises 1–8, verify the statement by showing that the derivative of the right side is equal to the integrand of the left side.

1. $\displaystyle\int \left(-\frac{9}{x^4}\right) dx = \frac{3}{x^3} + C$

2. $\displaystyle\int \frac{4}{\sqrt{x}}\, dx = 8\sqrt{x} + C$

3. $\displaystyle\int \left(4x^3 - \frac{1}{x^2}\right) dx = x^4 + \frac{1}{x} + C$

4. $\displaystyle\int \left(1 - \frac{1}{\sqrt[3]{x^2}}\right) dx = x - 3\sqrt[3]{x} + C$

5. $\displaystyle\int 2x^3 \sqrt{x}\, dx = \frac{4}{9}x^{9/2} + C$

6. $\displaystyle\int (x-2)(x+2)\, dx = \frac{x^3}{3} - 4x + C$

7. $\displaystyle\int \frac{x^2 - 1}{x^{3/2}}\, dx = \frac{2(x^2+3)}{3\sqrt{x}} + C$

8. $\displaystyle\int \frac{2x-1}{x^{4/3}}\, dx = \frac{3(x+1)}{\sqrt[3]{x}} + C$

In Exercises 9–18, find the indefinite integral and check your result by differentiation.

9. $\displaystyle\int 6\, dx$

10. $\displaystyle\int -4\, dx$

11. $\displaystyle\int 3t^2\, dt$

12. $\displaystyle\int t^4\, dt$

13. $\displaystyle\int 5x^{-3}\, dx$

14. $\displaystyle\int 4y^{-3}\, dy$

15. $\displaystyle\int du$

16. $\displaystyle\int e\, dt$

17. $\displaystyle\int x^{3/2}\, dx$

18. $\displaystyle\int v^{-1/2}\, dv$

In Exercises 19–24, complete a table using Example 4 as a model.

19. $\displaystyle\int \sqrt[3]{x}\, dx$

20. $\displaystyle\int \frac{1}{x^2}\, dx$

21. $\displaystyle\int \frac{1}{x\sqrt{x}}\, dx$

22. $\displaystyle\int x(x^2 + 3)\, dx$

23. $\displaystyle\int \frac{1}{2x^3}\, dx$

24. $\displaystyle\int \frac{1}{(2x)^3}\, dx$

In Exercises 25 and 26, find two functions that have the given derivative, and sketch the graph of each. (There is more than one correct answer.)

25.

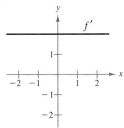

26.

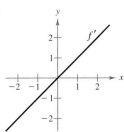

In Exercises 27–44, find the indefinite integral and check your result by differentiation.

27. $\displaystyle\int (x^3 + 2)\,dx$

28. $\displaystyle\int (x^2 - 2x + 3)\,dx$

29. $\displaystyle\int (2x^{4/3} + 3x - 1)\,dx$

30. $\displaystyle\int \left(\sqrt{x} + \dfrac{1}{2\sqrt{x}}\right)dx$

31. $\displaystyle\int \sqrt[3]{x^2}\,dx$

32. $\displaystyle\int (\sqrt[4]{x^3} + 1)\,dx$

33. $\displaystyle\int \dfrac{1}{x^3}\,dx$

34. $\displaystyle\int \dfrac{1}{x^4}\,dx$

35. $\displaystyle\int \dfrac{1}{4x^2}\,dx$

36. $\displaystyle\int (2x + x^{-1/2})\,dx$

37. $\displaystyle\int \dfrac{t^2 + 2}{t^2}\,dt$

38. $\displaystyle\int \dfrac{x^2 + 1}{x^2}\,dx$

39. $\displaystyle\int u(3u^2 + 1)\,du$

40. $\displaystyle\int \sqrt{x}(x + 1)\,dx$

41. $\displaystyle\int (x - 1)(6x - 5)\,dx$

42. $\displaystyle\int (2t^2 - 1)^2\,dt$

43. $\displaystyle\int y^2\sqrt{y}\,dy$

44. $\displaystyle\int (1 + 3t)t^2\,dt$

In Exercises 45–50, find the particular solution $y = f(x)$ that satisfies the differential equation and initial condition.

45. $f'(x) = 3\sqrt{x} + 3, \quad f(1) = 4$

46. $f'(x) = \frac{1}{5}x - 2, \quad f(10) = -10$

47. $f'(x) = 6x(x - 1), \quad f(1) = -1$

48. $f'(x) = (2x - 3)(2x + 3), \quad f(3) = 0$

49. $f'(x) = \dfrac{2 - x}{x^3}, \quad x > 0, \quad f(2) = \dfrac{3}{4}$

50. $f'(x) = -\dfrac{5}{(x - 2)^2}, \quad x \neq 2, \quad f(7) = 1$

In Exercises 51 and 52, you are shown a family of graphs, each of which is a solution of the given differential equation. Find the equation of the particular solution that passes through the indicated point.

51. $\dfrac{dy}{dx} = -5x - 2$

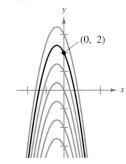

(0, 2)

52. $\dfrac{dy}{dx} = 2(x - 1)$

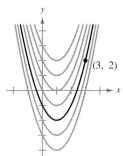

(3, 2)

In Exercises 53 and 54, find the equation of the function f whose graph passes through the point.

Derivative	*Point*
53. $f'(x) = 6\sqrt{x} - 10$	$(4, 2)$
54. $f'(x) = \dfrac{6}{x^2}$	$(2, 5)$

In Exercises 55–58, find a function f that satisfies the given conditions.

55. $f''(x) = 2, \quad f'(2) = 5, \quad f(2) = 10$

56. $f''(x) = x^2, \quad f'(0) = 6, \quad f(0) = 3$

57. $f''(x) = x^{-2/3}, \quad f'(8) = 6, \quad f(0) = 0$

58. $f''(x) = x^{-3/2}, \quad f'(1) = 2, \quad f(9) = -4$

Cost In Exercises 59–62, find the cost function for the marginal cost and fixed cost.

Marginal Cost	*Fixed Cost* ($x = 0$)
59. $\dfrac{dC}{dx} = 85$	$\$5500$
60. $\dfrac{dC}{dx} = \dfrac{1}{50}x + 10$	$\$1000$
61. $\dfrac{dC}{dx} = \dfrac{1}{20\sqrt{x}} + 4$	$\$750$
62. $\dfrac{dC}{dx} = \dfrac{\sqrt[4]{x}}{10} + 10$	$\$2300$

Demand Function In Exercises 63 and 64, find the revenue and demand functions for the given marginal revenue. (Use the fact that $R = 0$ when $x = 0$.)

63. $\dfrac{dR}{dx} = 225 - 3x$

64. $\dfrac{dR}{dx} = 100 - 6x - 2x^2$

Profit In Exercises 65 and 66, find the profit function for the given marginal profit and initial condition.

Marginal Profit	Initial Condition

65. $\dfrac{dP}{dx} = -18x + 1650$ $P(15) = \$22{,}725$

66. $\dfrac{dP}{dx} = -40x + 250$ $P(5) = \$650$

Vertical Motion In Exercises 67–70, use $a(t) = -32$ feet per second^2 as the acceleration due to gravity. (Neglect air resistance.)

67. A ball is thrown vertically upward with an initial velocity of 60 feet per second. How high will the ball go?

68. *Grand Canyon* The Grand Canyon is 6000 feet deep at the deepest part. A rock is dropped from this height. Express the height of the rock as a function of the time t (in seconds). How long will it take the rock to hit the canyon floor?

69. With what initial velocity must an object be thrown upward from the ground to reach the height of the Washington Monument (550 feet)?

70. A balloon, rising vertically with a velocity of 16 feet per second, releases a sandbag at an instant when the balloon is 64 feet above the ground.

    (a) How many seconds after its release will the bag strike the ground?

    (b) With what velocity will the bag strike the ground?

71. *Cost* A company produces a product for which the marginal cost of producing x units is

$$\frac{dC}{dx} = 2x - 12$$

and the fixed costs are $125.

    (a) Find the total cost function and the average cost function.

    (b) Find the total cost of producing 50 units.

    (c) In part (b), how much of the total cost is fixed? How much is variable? Give examples of fixed costs associated with the manufacturing of a product. Give examples of variable costs.

72. *Population Growth* The growth rate of a city's population is modeled by

$$\frac{dP}{dt} = 500t^{1.06},$$

where t is the time in years. The city's population is now 50,000. What will the population be in 10 years?

73. *Natural Gas Consumption* The consumption S (in quadrillion Btu's) of natural gas in the United States increased steadily between 1986 and 1992. The rate of increase can be modeled by

$$\frac{dS}{dt} = -0.175t^2 + 0.4t + 0.81, \qquad 0 \le t \le 6,$$

where $t = 0$ represents 1986. In 1986, the consumption was 16.7 quadrillion Btu's. Find a model for the consumption from 1986 through 1992 and determine the consumption in 1992. *(Source: U.S. Energy Information Administration)*

74. *Research Project* Use your school's library or some other reference source to research a company that markets a natural resource. Find data on the revenue of the company and on the consumption of the resource. Then find a model for each. Is the company's revenue related to the consumption of the resource? Explain.

The General Power Rule ■ *Substitution* ■
Extended Application: Propensity to Consume

The General Power Rule

In Section 5.1, you used the Simple Power Rule

$$\int x^n \, dx = \frac{x^{n+1}}{n+1} + C, \qquad n \neq -1$$

to find antiderivatives of functions expressed as powers of x alone. In this section, you will study a technique for finding antiderivatives of more complicated functions.

To begin, consider how you might find the antiderivative of $2x(x^2 + 1)^3$. Because you are hunting for a function whose derivative is $2x(x^2 + 1)^3$, you might discover the following.

$$\frac{d}{dx}[(x^2 + 1)^4] = 4(x^2 + 1)^3(2x) \qquad \textit{Use Chain Rule.}$$

$$\frac{d}{dx}\left[\frac{(x^2 + 1)^4}{4}\right] = (x^2 + 1)^3(2x) \qquad \textit{Divide both sides by 4.}$$

$$\frac{(x^2 + 1)^4}{4} + C = \int 2x(x^2 + 1)^3 \, dx \qquad \textit{Write in integral form.}$$

The key to this solution is the presence of the factor $2x$ in the integrand. In other words, this solution works because $2x$ is precisely the derivative of $(x^2 + 1)$. Letting $u = x^2 + 1$, you can write

$$\int \overbrace{2x}(x^2 + 1)^3 \, dx = \int u^3 \frac{du}{dx} \, dx = \frac{u^4}{4} + C.$$

This is an example of the **General Power Rule** for integration.

General Power Rule for Integration

If u is a differentiable function of x, then

$$\int u^n \frac{du}{dx} \, dx = \frac{u^{n+1}}{n+1} + C, \qquad n \neq -1.$$

When using the General Power Rule, you must first identify a factor u of the integrand that is raised to a power. Then, you must show that its derivative du/dx is also a factor of the integrand. This is demonstrated in Example 1.

EXAMPLE 1 *Applying the General Power Rule*

Find the following indefinite integrals.

a. $\displaystyle \int 3(3x-1)^4\,dx$ **b.** $\displaystyle \int (2x+1)(x^2+x)\,dx$

c. $\displaystyle \int 3x^2\sqrt{x^3-2}\,dx$ **d.** $\displaystyle \int \frac{-4x}{(1-2x^2)^2}\,dx$

Solution

a. $\displaystyle \int 3(3x-1)^4\,dx = \int \overset{u^n}{\overbrace{(3x-1)^4}}\,\overset{\frac{du}{dx}}{\overbrace{(3)}}\,dx$ *Let $u = 3x - 1$.*

$\displaystyle \qquad\qquad\qquad\quad = \frac{(3x-1)^5}{5} + C$ *General Power Rule*

Note Example 1b illustrates a case of the General Power Rule that is sometimes overlooked— when the power is $n = 1$. In this case, the rule takes the form

$$\int u\frac{du}{dx}\,dx = \frac{u^2}{2} + C.$$

b. $\displaystyle \int (2x+1)(x^2+x)\,dx = \int \overset{u^n}{\overbrace{(x^2+x)}}\,\overset{\frac{du}{dx}}{\overbrace{(2x+1)}}\,dx$ *Let $u = x^2 + x$.*

$\displaystyle \qquad\qquad\qquad\qquad\quad = \frac{(x^2+x)^2}{2} + C$ *General Power Rule*

c. $\displaystyle \int 3x^2\sqrt{x^3-2}\,dx = \int \overset{u^n}{\overbrace{(x^3-2)^{1/2}}}\,\overset{\frac{du}{dx}}{\overbrace{(3x^2)}}\,dx$ *Let $u = x^3 - 2$.*

$\displaystyle \qquad\qquad\qquad\quad = \frac{(x^3-2)^{3/2}}{3/2} + C$ *General Power Rule*

$\displaystyle \qquad\qquad\qquad\quad = \frac{2}{3}(x^3-2)^{3/2} + C$ *Simplify.*

d. $\displaystyle \int \frac{-4x}{(1-2x^2)^2}\,dx = \int \overset{u^n}{\overbrace{(1-2x^2)^{-2}}}\,\overset{\frac{du}{dx}}{\overbrace{(-4x)}}\,dx$ *Let $u = 1 - 2x^2$.*

$\displaystyle \qquad\qquad\qquad\qquad = \frac{(1-2x^2)^{-1}}{-1} + C$ *General Power Rule*

$\displaystyle \qquad\qquad\qquad\qquad = -\frac{1}{1-2x^2} + C$ *Simplify.*

Many times, part of the derivative du/dx is missing from the integrand, and in *some* cases you can make the necessary adjustments to apply the General Power Rule.

EXAMPLE 2 *Multiplying and Dividing by a Constant*

Find the indefinite integral $\displaystyle\int x(3 - 4x^2)^2\,dx$.

Solution

Let $u = 3 - 4x^2$. To apply the General Power Rule, you need to create $du/dx = -8x$ as a factor of the integrand. You can accomplish this by multiplying and dividing by the constant -8.

Note Try using the Chain Rule to check the result of Example 2. After differentiating $-\frac{1}{24}(3 - 4x^2)^3$ and simplifying, you should obtain the original integrand.

$$\int x(3 - 4x^2)^2\,dx = \int \left(-\frac{1}{8}\right)\overbrace{(3 - 4x^2)^2}^{u^n}\overbrace{(-8x)}^{\frac{du}{dx}}\,dx \qquad \text{\textit{Multiply and divide by} }-8.$$

$$= -\frac{1}{8}\int (3 - 4x^2)^2(-8x)\,dx \qquad \text{\textit{Factor} }-\tfrac{1}{8}\text{ \textit{out of integrand.}}$$

$$= \left(-\frac{1}{8}\right)\frac{(3 - 4x^2)^3}{3} + C \qquad \text{\textit{General Power Rule}}$$

$$= -\frac{(3 - 4x^2)^3}{24} + C \qquad \text{\textit{Simplify.}}$$

EXAMPLE 3 *A Failure of the General Power Rule*

Find the indefinite integral $\displaystyle\int -8(3 - 4x^2)^2\,dx$.

Solution

Let $u = 3 - 4x^2$. As in Example 2, to apply the General Power Rule you must create $du/dx = -8x$ as a factor of the integrand. In Example 2, you could do this by multiplying and dividing by a constant, and then factoring that constant out of the integrand. This strategy doesn't work with variables. That is,

Note In Example 3, be sure you see that you cannot factor variable quantities outside the integral sign. After all, if this were permissible, then you could move the entire integrand outside the integral sign and eliminate the need for all integration rules except the rule $\int dx = x + C$.

$$\int -8(3 - 4x^2)^2\,dx \neq \frac{1}{x}\int (3 - 4x^2)^2(-8x)\,dx.$$

To find this indefinite integral, you can expand the integrand and use the Simple Power Rule.

$$\int -8(3 - 4x^2)^2\,dx = \int (-72 + 192x^2 - 128x^4)\,dx$$

$$= -72x + 64x^3 - \frac{128}{5}x^5 + C$$

When an integrand contains an extra constant factor that is not needed as part of u', you can simply move the factor outside the integral sign, as illustrated in the next example.

EXAMPLE 4 *Applying the General Power Rule*

Find the indefinite integral $\displaystyle\int 7x^2\sqrt{x^3+1}\,dx$.

Solution
Let $u = x^3 + 1$. Then you need to create $u' = 3x^2$ by multiplying and dividing by 3. The constant factor $\frac{7}{3}$ is not needed as part of u', and can be moved outside the integral sign.

$$\int 7x^2\sqrt{x^3+1}\,dx = \int 7x^2(x^3+1)^{1/2}\,dx \qquad \text{\textit{Write with rational exponents.}}$$

$$= \int \frac{7}{3}(x^3+1)^{1/2}(3x^2)\,dx \qquad \text{\textit{Multiply and divide by 3.}}$$

$$= \frac{7}{3}\int (x^3+1)^{1/2}(3x^2)\,dx \qquad \text{\textit{Factor $\frac{7}{3}$ outside integral.}}$$

$$= \frac{7}{3}\frac{(x^3+1)^{3/2}}{3/2} + C \qquad \text{\textit{General Power Rule}}$$

$$= \frac{14}{9}(x^3+1)^{3/2} \qquad \text{\textit{Simplify.}}$$

TECHNOLOGY

If you use a symbolic integration program, such as *Derive*, *Maple*, or *Mathematica*, to find indefinite integrals, you should be in for some surprises. This is true because integration is not nearly as straightforward as differentiation. By trying different integrands, you should be able to find several that the program cannot solve: in such situations, it may list a new indefinite integral. You should also be able to find several that have horrendous antiderivatives, some with functions that you may not recognize. Two examples, using *Derive*, are shown at the right.

1: $\sqrt{(x^3 + 1)}$ *Author.*

2: $\int \sqrt{(x^3 + 1)}\ dx$ *Integrate.*

3: $\dfrac{3\int \dfrac{1}{\sqrt{(x^3 + 1)}}\ dx}{5} + \dfrac{2x\sqrt{(x^3 + 1)}}{5}$ *Simplify.*

1: $\sqrt{(x^2 + 1)}$ *Author.*

2: $\int \sqrt{(x^2 + 1)}\ dx$ *Integrate.*

3: $\dfrac{LN(\sqrt{(x^2 + 1)} + x)}{2} + \dfrac{x\sqrt{(x^2 + 1)}}{2}$ *Simplify.*

Substitution

The integration technique used in Examples 1, 2, and 4 depends on your ability to recognize or create an integrand of the form $u^n u'$. With more complicated integrands, it is difficult to recognize the steps needed to fit the integrand to a basic integration formula. When this occurs, an alternative procedure called **substitution** or **change of variables** can be helpful. With this procedure, you completely rewrite the integral in terms of u and du. That is, if $u = f(x)$, then $du = f'(x)\,dx$, and the General Power Rule takes the form

$$\int u^n \frac{du}{dx}\,dx = \int u^n\,du.$$ *General Power Rule*

EXAMPLE 5 *Integrating by Substitution*

Find the indefinite integral $\displaystyle\int \sqrt{1 - 3x}\,dx$.

Solution

Begin by letting $u = 1 - 3x$. Then, $u' = -3$ and $du = -3\,dx$. This implies that $dx = -\frac{1}{3}\,du$, and you can find the indefinite integral as follows.

$$\int \sqrt{1 - 3x}\,dx = \int (1 - 3x)^{1/2}\,dx \qquad \text{Write with rational exponent.}$$

$$= \int u^{1/2} \left(-\frac{1}{3}\,du\right) \qquad \text{Substitute } u \text{ and } du.$$

$$= -\frac{1}{3}\int u^{1/2}\,du \qquad \text{Factor } -\tfrac{1}{3} \text{ out of integrand.}$$

$$= -\frac{1}{3}\frac{u^{3/2}}{3/2} + C \qquad \text{Apply Power Rule.}$$

$$= -\frac{2}{9}u^{3/2} + C \qquad \text{Simplify.}$$

$$= -\frac{2}{9}(1 - 3x)^{3/2} + C \qquad \text{Substitute } 1 - 3x \text{ for } u. \quad \blacksquare$$

To become efficient at integration, you should learn to use *both* techniques discussed in this section. For simpler integrals, you should use pattern recognition and create du/dx by multiplying and dividing by an appropriate constant. For more complicated integrals, you should use a formal change of variables, as illustrated in Example 5. (You will learn more about this technique in Chapter 6.) For the integrals in this section's exercise set, try working several of the problems twice—once with pattern recognition and once using formal substitution.

Extended Application: Propensity to Consume

In 1990, the U. S. poverty level for a family of four was about $13,400. Families at or below the poverty level tend to consume 100% of their income—that is, they use all their income to purchase necessities such as food, clothing, and shelter. As income level increases, the average consumption tends to drop below 100%. For instance, a family earning $15,000 may be able to save $318 and thus consume only $14,682 (97.9%) of their income. As the income increases, the ratio of consumption to savings tends to decrease. The rate of change of consumption with respect to income is called the **marginal propensity to consume.** *(Source: U.S. Bureau of Census)*

Real Life

EXAMPLE 6 *Analyzing Consumption*

For a family of four in 1990, the marginal propensity to consume income x can be modeled by

$$\frac{dQ}{dx} = \frac{0.97}{(x - 13,399)^{0.03}}, \qquad x \geq 13,400,$$

where Q represents the income consumed. Use the model to estimate the amount consumed by a family of four whose 1990 income was $26,000.

Solution

Begin by integrating dQ/dx to find a model for the consumption Q. Use the initial condition that $Q = 13,400$ when $x = 13,400$.

$$\frac{dQ}{dx} = \frac{0.97}{(x - 13,399)^{0.03}} \qquad \textit{Marginal propensity to consume}$$

$$Q = \int \frac{0.97}{(x - 13,399)^{0.03}}\, dx \qquad \textit{Integrate to obtain Q.}$$

$$= \int 0.97(x - 13,399)^{-0.03}\, dx \quad \textit{Rewrite.}$$

$$= (x - 13,399)^{0.97} + C \qquad \textit{General Power Rule}$$

$$= (x - 13,399)^{0.97} + 13,399 \quad \textit{Use initial condition to find C.}$$

Using this model, you can estimate that a family of four with an income of $x = 26,000$ consumed about $22,892. The graph of Q is shown in Figure 5.4.

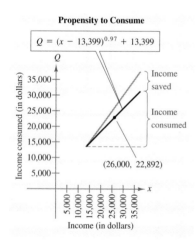

Propensity to Consume

$Q = (x - 13,399)^{0.97} + 13,399$

(26,000, 22,892)

Income consumed (in dollars)

Income (in dollars)

Income saved

Income consumed

FIGURE 5.4

Discussion Problem   **Extending the Example**

According to the model in Example 6, at what income level would a family of four consume $30,000?

Warm Up

The following warm-up exercises involve skills that were covered in earlier sections. You will use these skills in the exercise set for this section.

In Exercises 1–6, find the indefinite integral.

1. $\displaystyle\int (2x^3 + 1)\,dx$

2. $\displaystyle\int (x^{1/2} + 3x - 4)\,dx$

3. $\displaystyle\int \frac{1}{x^2}\,dx$

4. $\displaystyle\int \frac{1}{3t^3}\,dt$

5. $\displaystyle\int (1 + 2t)t^{3/2}\,dt$

6. $\displaystyle\int \sqrt{x}(2x - 1)\,dx$

In Exercises 7–10, simplify the expression.

7. $\left(-\dfrac{5}{4}\right)\dfrac{(x - 2)^4}{4}$

8. $\left(\dfrac{1}{6}\right)\dfrac{(x - 1)^{-2}}{-2}$

9. $(6)\dfrac{(x^2 + 3)^{2/3}}{2/3}$

10. $\left(\dfrac{5}{2}\right)\dfrac{(1 - x^3)^{-1/2}}{-1/2}$

EXERCISES 5.2

 means that technology can help you solve or check the exercise(s).

In Exercises 1–6, identify u and du/dx for the integral $\int u^n(du/dx)\,dx$.

1. $\displaystyle\int (5x^2 + 1)^2(10x)\,dx$

2. $\displaystyle\int (3 - 4x^2)^3(-8x)\,dx$

3. $\displaystyle\int \sqrt{1 - x^2}(-2x)\,dx$

4. $\displaystyle\int 3x^2\sqrt{x^3 + 1}\,dx$

5. $\displaystyle\int \left(4 + \frac{1}{x^2}\right)\left(\frac{-2}{x^3}\right)dx$

6. $\displaystyle\int \frac{1}{(1 + 2x)^2}(2)\,dx$

In Exercises 7–32, find the indefinite integral and check the result by differentiation.

7. $\displaystyle\int (1 + 2x)^4(2)\,dx$

8. $\displaystyle\int (x^2 - 1)^3(2x)\,dx$

9. $\displaystyle\int \sqrt{5x^2 - 4}(10x)\,dx$

10. $\displaystyle\int \sqrt{3 - x^3}(3x^2)\,dx$

11. $\displaystyle\int (x - 1)^4\,dx$

12. $\displaystyle\int (x - 3)^{5/2}\,dx$

13. $\displaystyle\int x(x^2 - 1)^7\,dx$

14. $\displaystyle\int x(1 - 2x^2)^3\,dx$

15. $\displaystyle\int \frac{x^2}{(1 + x^3)^2}\,dx$

16. $\displaystyle\int \frac{x^2}{(x^3 - 1)^2}\,dx$

17. $\displaystyle\int \frac{x + 1}{(x^2 + 2x - 3)^2}\,dx$

18. $\displaystyle\int \frac{6x}{(1 + x^2)^3}\,dx$

19. $\displaystyle\int \frac{x - 2}{\sqrt{x^2 - 4x + 3}}\,dx$

20. $\displaystyle\int \frac{4x + 6}{(x^2 + 3x + 7)^3}\,dx$

21. $\displaystyle\int 5x\sqrt[3]{1 - x^2}\,dx$

22. $\displaystyle\int u^3\sqrt{u^4 + 2}\,du$

23. $\displaystyle\int \frac{4x}{\sqrt{1 + x^2}}\,dx$

24. $\displaystyle\int \frac{x^2}{\sqrt{1 - x^3}}\,dx$

25. $\displaystyle\int \frac{-3}{\sqrt{2x + 3}}\,dx$

26. $\displaystyle\int \frac{t + 2t^2}{\sqrt{t}}\,dt$

27. $\displaystyle\int \frac{x^3}{\sqrt{1 - x^4}}\,dx$

28. $\displaystyle\int \left(1 + \frac{1}{t}\right)^3\left(\frac{1}{t^2}\right)dt$

29. $\displaystyle\int \frac{1}{\sqrt{2x}}\,dx$

30. $\displaystyle\int \frac{1}{3x^2}\,dx$

31. $\displaystyle\int (x^3 + 3x)(x^2 + 1)\,dx$

32. $\displaystyle\int (3 - 2x - 4x^2)(1 + 4x)\,dx$

In Exercises 33–40, use formal substitution (as illustrated in Example 5) to find the indefinite integral.

33. $\int x(6x^2 - 1)^3 \, dx$

34. $\int x^2(1 - x^3)^2 \, dx$

35. $\int x^2(2 - 3x^3)^{3/2} \, dx$

36. $\int t\sqrt{t^2 + 1} \, dt$

37. $\int \frac{x}{\sqrt{x^2 + 25}} \, dx$

38. $\int \frac{3}{\sqrt{2x + 1}} \, dx$

39. $\int \frac{x^2 + 1}{\sqrt{x^3 + 3x + 4}} \, dx$

40. $\int \sqrt{x}(4 - x^{3/2})^2 \, dx$

In Exercises 41–44, perform the integration in two ways: once using the Simple Power Rule and once using the General Power Rule. Explain the difference in the results.

41. $\int (2x - 1)^2 \, dx$

42. $\int (3 - 2x)^2 \, dx$

43. $\int x(x^2 - 1)^2 \, dx$

44. $\int x(2x^2 + 1)^2 \, dx$

 45. Find the equation of the function f whose graph passes through the point $(0, \frac{4}{3})$ and whose derivative is

$$f'(x) = x\sqrt{1 - x^2}.$$

 46. Find the equation of the function f whose graph passes through the point $(0, \frac{7}{3})$ and whose derivative is

$$f'(x) = x\sqrt{1 - x^2}.$$

 47. *Cost* The marginal cost of a product is modeled by

$$\frac{dC}{dx} = \frac{4}{\sqrt{x + 1}}.$$

When $x = 15$, $C = 50$.

(a) Find the cost function.

(b) Graph dC/dx and C on the same set of axes.

48. *Cost* The marginal cost of a product is modeled by

$$\frac{dC}{dx} = \frac{12}{\sqrt[3]{12x + 1}}.$$

When $x = 13$, $C = 100$.

(a) Find the cost function.

(b) Graph dC/dx and C on the same set of axes.

Supply Function In Exercises 49 and 50, find the supply function $x = f(p)$ that satisfies the given conditions.

49. $\frac{dx}{dp} = p\sqrt{p^2 - 25}$, $\quad x = 600$ when $p = \$13$

50. $\frac{dx}{dp} = \frac{10}{\sqrt{p - 3}}$, $\quad x = 100$ when $p = \$3$

Demand Function In Exercises 51 and 52, find the demand function $x = f(p)$ that satisfies the given conditions.

51. $\frac{dx}{dp} = -\frac{6000p}{(p^2 - 16)^{3/2}}$, $\quad x = 5000$ when $p = \$5$

52. $\frac{dx}{dp} = -\frac{400}{(0.02p - 1)^3}$, $\quad x = 10,000$ when $p = \$100$

53. *Depreciation* The rate of depreciation dV/dt of a machine is inversely proportional to the square of $t + 1$, where V is the value of the machine t years after it is purchased. The initial value of the machine is $500,000. By the end of the first year, the value of the machine decreases $100,000. Estimate its value after 4 years.

54. *Cash Flow* The rate of disbursement dQ/dt of a \$2 million federal grant is proportional to the square of $100 - t$, where t is the time (in days, $0 \leq t \leq 100$) and Q is the amount that remains to be disbursed. Find the amount that remains to be disbursed after 50 days. Assume that the entire grant will be disbursed after 100 days.

 In Exercises 55 and 56, (a) use the marginal propensity to consume, dQ/dx, to write Q as a function of x, where x is the income and Q is the income consumed. Assume that 100% of the income is consumed for families that have an annual income of \$20,000 or less. (b) Use the result of part (a) to complete the table showing the income consumed and the income saved, $x - Q$, for various incomes. (c) Graphically represent the income consumed and saved.

x	20,000	50,000	100,000	150,000
Q				
$x - Q$				

55. $\frac{dQ}{dx} = \frac{0.95}{(x - 19,999)^{0.05}}$, $\quad 20,000 \leq x$

56. $\frac{dQ}{dx} = \frac{0.93}{(x - 19,999)^{0.07}}$, $\quad 20,000 \leq x$

 In Exercises 57 and 58, use a symbolic integration utility to find the indefinite integral. Verify the result by differentiating.

57. $\int \frac{1}{\sqrt{x} + \sqrt{x + 1}} \, dx$

58. $\int \frac{x}{\sqrt{3x + 2}} \, dx$

5.3 Exponential and Logarithmic Integrals

Using the Exponential Rule ■ *Using the Log Rule*

Using the Exponential Rule

Each of the differentiation rules for exponential functions has its corresponding integration rule.

Integrals of Exponential Functions

Let u be a differentiable function of x.

$$\int e^x \, dx = e^x + C \qquad \text{Simple Exponential Rule}$$

$$\int e^u \frac{du}{dx} \, dx = \int e^u \, du = e^u + C \qquad \text{General Exponential Rule}$$

EXAMPLE 1 *Integrating Exponential Functions*

a. $\displaystyle\int 2e^x \, dx = 2 \int e^x \, dx$ ⟶ *Constant Multiple Rule*

$\qquad\qquad = 2e^x + C$ ⟶ *Simple Exponential Rule*

b. $\displaystyle\int 2e^{2x} \, dx = \int e^{2x}(2) \, dx$ ⟶ *Let $u = 2x$, then $u' = 2$.*

$\qquad\qquad = \int e^u u' \, dx$

$\qquad\qquad = e^{2x} + C$ ⟶ *General Exponential Rule*

c. $\displaystyle\int (e^x + x) \, dx = \int e^x \, dx + \int x \, dx$ ⟶ *Sum Rule*

$\qquad\qquad = e^x + \dfrac{x^2}{2} + C$ ⟶ *Simple Exponential and Power Rules*

You can check each of these results by differentiating. ■

EXAMPLE 2 *Integrating an Exponential Function*

Find the indefinite integral $\int e^{3x+1}\, dx$.

Solution

Let $u = 3x + 1$, then $du/dx = 3$. You can introduce the missing factor of 3 in the integrand by multiplying and dividing by 3.

$$\int e^{3x+1}\, dx = \frac{1}{3}\int e^{3x+1}(3)\, dx \qquad \textit{Multiply and divide by 3.}$$

$$= \frac{1}{3}\int e^{u}\frac{du}{dx}\, dx \qquad \textit{Substitute u and du/dx.}$$

$$= \frac{1}{3}e^{u} + C \qquad \textit{General Exponential Rule}$$

$$= \frac{1}{3}e^{3x+1} + C \qquad \textit{Substitute for u.}$$

⊞ **TECHNOLOGY**

If you use a symbolic integration utility to find antiderivatives of exponential or logarithmic functions, you can easily obtain results that are beyond the scope of this course. For instance, the following antiderivative of e^{x^2} involves the imaginary unit i and the probability function called "ERF." In this course, you are not expected to interpret or use such results. You can simply state that the function cannot be integrated using elementary functions.

1: EXP (x^2)

2: $\int$ EXP (x^2) dx

3: $-\dfrac{i\sqrt{\pi}\ \text{ERF}\ (ix)}{2}$

EXAMPLE 3 *Integrating an Exponential Function*

Find the indefinite integral $\int 5xe^{-x^2}\, dx$.

Solution

Let $u = -x^2$, then $du/dx = -2x$. You can create the factor of $(-2x)$ in the integrand by multiplying and dividing by -2.

$$\int 5xe^{-x^2}\, dx = \int \left(-\frac{5}{2}\right)e^{-x^2}(-2x)\, dx \qquad \textit{Multiply and divide by }-2.$$

$$= -\frac{5}{2}\int e^{-x^2}(-2x)\, dx \qquad \begin{array}{l}\textit{Factor the constant }-\frac{5}{2}\\ \textit{out of the integrand.}\end{array}$$

$$= -\frac{5}{2}\int e^{u}\frac{du}{dx}\, dx \qquad \textit{Substitute u and du/dx.}$$

$$= -\frac{5}{2}e^{u} + C \qquad \textit{General Exponential Rule}$$

$$= -\frac{5}{2}e^{-x^2} + C \qquad \textit{Substitute for u.}$$

Note   Remember that you cannot introduce a missing *variable* in the integrand. For instance, you cannot find $\int e^{x^2}\, dx$ by multiplying and dividing by $2x$ and then factoring $1/(2x)$ out of the integral. That is,

$$\int e^{x^2}\, dx \neq \frac{1}{2x}\int e^{x^2}(2x)\, dx.$$

Using the Log Rule

When the Power Rules for integration were introduced in Sections 5.1 and 5.2, you saw that they work for powers other than $n = -1$.

$$\int x^n \, dx = \frac{x^{n+1}}{n+1} + C, \qquad n \neq -1 \qquad \textit{Simple Power Rule}$$

$$\int u^n \frac{du}{dx} \, dx = \frac{u^{n+1}}{n+1} + C, \qquad n \neq -1 \qquad \textit{General Power Rule}$$

The Log Rules for integration allow you to integrate functions of the form $\int x^{-1} \, dx$ and $\int u^{-1} \, du$.

Note Notice the absolute values in the Log Rules. For those special cases in which u or x cannot be negative, you can omit the absolute value. For instance, in Example 4b, it is not necessary to write the antiderivative as $\ln |x^2| + C$ because x^2 cannot be negative.

Integrals of Logarithmic Functions

Let u be a differentiable function of x.

$$\int \frac{1}{x} \, dx = \ln |x| + C \qquad\qquad \textit{Simple Logarithmic Rule}$$

$$\int \frac{u'}{u} \, dx = \int \frac{1}{u} \, du = \ln |u| + C \qquad \textit{General Logarithmic Rule}$$

You can verify each of these rules by differentiating. For instance, to verify that $d/dx[\ln |x|] = 1/x$, notice that

$$\frac{d}{dx}[\ln x] = \frac{1}{x} \quad \text{and} \quad \frac{d}{dx}[\ln(-x)] = \frac{-1}{-x} = \frac{1}{x}.$$

EXAMPLE 4 *Integrating Logarithmic Functions*

a. $\displaystyle \int \frac{4}{x} \, dx = 4 \int \frac{1}{x} \, dx$ *Constant Multiple Rule*

$\qquad\qquad = 4 \ln |x| + C$ *Simple Logarithmic Rule*

b. $\displaystyle \int \frac{2x}{x^2} \, dx = \int \frac{u'}{u} \, dx$ *Let $u = x^2$, then $u' = 2x$.*

$\qquad\qquad = \ln |u| + C$ *General Logarithmic Rule*

$\qquad\qquad = \ln x^2 + C$ *Substitute for u.*

c. $\displaystyle \int \frac{3}{3x+1} \, dx = \int \frac{u'}{u} \, dx$ *Let $u = 3x + 1$, then $u' = 3$.*

$\qquad\qquad = \ln |u| + C$ *General Logarithmic Rule*

$\qquad\qquad = \ln |3x + 1| + C$ *Substitute for u.*

EXAMPLE 5 *Using the Log Rule*

Find the indefinite integral $\int \dfrac{1}{2x - 1} \, dx$.

Solution

Let $u = 2x - 1$, then $du/dx = 2$. You can create the necessary factor of 2 in the integrand by multiplying and dividing by 2.

$$\int \frac{1}{2x - 1} \, dx = \frac{1}{2} \int \frac{2}{2x - 1} \, dx \qquad \text{\textit{Multiply and divide by 2.}}$$

$$= \frac{1}{2} \int \frac{u'}{u} \, dx \qquad \text{\textit{Substitute u and u'.}}$$

$$= \frac{1}{2} \ln |u| + C \qquad \text{\textit{General Log Rule}}$$

$$= \frac{1}{2} \ln |2x - 1| + C \qquad \text{\textit{Substitute for u.}}$$

EXAMPLE 6 *Using the Log Rule*

Find the indefinite integral $\int \dfrac{6x}{x^2 + 1} \, dx$.

Solution

Let $u = x^2 + 1$, then $du/dx = 2x$. You can create the necessary factor of $2x$ in the integrand by factoring a 3 out of the integrand.

$$\int \frac{6x}{x^2 + 1} \, dx = 3 \int \frac{2x}{x^2 + 1} \, dx \qquad \text{\textit{Factor 3 out of integrand.}}$$

$$= 3 \int \frac{u'}{u} \, dx \qquad \text{\textit{Substitute u and u'.}}$$

$$= 3 \ln |u| + C \qquad \text{\textit{General Log Rule}}$$

$$= 3 \ln(x^2 + 1) + C \qquad \text{\textit{Substitute for u.}}$$

Integrals to which the Log Rule can be applied are often given in disguised form. For instance, if a rational function has a numerator of degree greater than or equal to that of the denominator, you should use long division to rewrite the integrand. Here is an example.

$$\int \frac{x^2 + 6x + 1}{x^2 + 1} \, dx = \int \left(1 + \frac{6x}{x^2 + 1} \right) dx$$

$$= x + 3 \ln(x^2 + 1) + C.$$

The next example summarizes some additional situations in which it is helpful to rewrite the integrand in order to recognize the antiderivative.

EXAMPLE 7 *Rewriting Before Integrating*

Find the following indefinite integrals.

a. $\int \dfrac{3x^2 + 2x - 1}{x^2}\, dx$ **b.** $\int \dfrac{1}{1 + e^{-x}}\, dx$ **c.** $\int \dfrac{x^2 + x + 1}{x - 1}\, dx$

Solution

a. Begin by rewriting the integrand as the sum of three fractions.

$$\int \frac{3x^2 + 2x - 1}{x^2}\, dx = \int \left(\frac{3x^2}{x^2} + \frac{2x}{x^2} - \frac{1}{x^2} \right) dx$$

$$= \int \left(3 + \frac{2}{x} - \frac{1}{x^2} \right) dx$$

$$= 3x + 2 \ln|x| + \frac{1}{x} + C$$

b. Begin by rewriting the integrand by multiplying and dividing by e^x.

$$\int \frac{1}{1 + e^{-x}}\, dx = \int \left(\frac{e^x}{e^x} \right) \frac{1}{1 + e^{-x}}\, dx$$

$$= \int \frac{e^x}{e^x + 1}\, dx$$

$$= \ln(e^x + 1) + C$$

c. Begin by dividing the numerator by the denominator.

$$\int \frac{x^2 + x + 1}{x - 1}\, dx = \int \left(x + 2 + \frac{3}{x - 1} \right) dx$$

$$= \frac{x^2}{2} + 2x + 3 \ln|x - 1| + C$$

Discussion Problem   *Using the General Log Rule*

One of the most common applications of the Log Rule is to find indefinite integrals of the form

$$\int \frac{a}{bx + c}\, dx.$$

Describe a quick way to find this indefinite integral. Then apply your technique to the following.

a. $\int \dfrac{1}{2x - 5}\, dx$ **b.** $\int \dfrac{4}{3x + 2}\, dx$ **c.** $\int \dfrac{7}{8x - 3}\, dx$

Warm Up

The following warm-up exercises involve skills that were covered in earlier sections. You will use these skills in the exercise set for this section.

In Exercises 1 and 2, find the domain of the function.

1. $y = \ln(2x - 5)$

2. $y = \ln(x^2 - 5x + 6)$

In Exercises 3–6, use long division to rewrite the quotient.

3. $\dfrac{x^2 + 4x + 2}{x + 2}$

4. $\dfrac{x^2 - 6x + 9}{x - 4}$

5. $\dfrac{x^3 + 4x^2 - 30x - 4}{x^2 - 4x}$

6. $\dfrac{x^4 - x^3 + x^2 + 15x + 2}{x^2 + 5}$

In Exercises 7–10, evaluate the integral.

7. $\displaystyle\int \left(x^3 + \frac{1}{x^2} \right) dx$

8. $\displaystyle\int \frac{x^2 + 2x}{x}\, dx$

9. $\displaystyle\int \frac{x^3 + 4}{x^2}\, dx$

10. $\displaystyle\int \frac{x + 3}{x^3}\, dx$

EXERCISES 5.3

means that technology can help you solve or check the exercise(s).

In Exercises 1–16, use the Exponential Rule to find the indefinite integral.

1. $\displaystyle\int 2e^{2x}\, dx$

2. $\displaystyle\int -3e^{-3x}\, dx$

3. $\displaystyle\int e^{4x}\, dx$

4. $\displaystyle\int e^{-0.25x}\, dx$

5. $\displaystyle\int 9xe^{-x^2}\, dx$

6. $\displaystyle\int 3xe^{0.5x^2}\, dx$

7. $\displaystyle\int 5x^2 e^{x^3}\, dx$

8. $\displaystyle\int (2x + 1)e^{x^2 + x}\, dx$

9. $\displaystyle\int (x^2 + 2x)e^{x^3 + 3x^2 - 1}\, dx$

10. $\displaystyle\int 3(x - 4)e^{x^2 - 8x}\, dx$

11. $\displaystyle\int 5e^{2-x}\, dx$

12. $\displaystyle\int 3e^{-(x+1)/2}\, dx$

13. $\displaystyle\int \frac{1}{x^2} e^{2/x}\, dx$

14. $\displaystyle\int \frac{1}{x^3} e^{1/4x^2}\, dx$

15. $\displaystyle\int \frac{1}{\sqrt{x}} e^{\sqrt{x}}\, dx$

16. $\displaystyle\int (e^x - e^{-x})^2\, dx$

In Exercises 17–32, use the Log Rule to find the indefinite integral.

17. $\displaystyle\int \frac{1}{x + 1}\, dx$

18. $\displaystyle\int \frac{1}{x - 5}\, dx$

19. $\displaystyle\int \frac{1}{3 - 2x}\, dx$

20. $\displaystyle\int \frac{1}{6x - 5}\, dx$

21. $\displaystyle\int \frac{x}{x^2 + 1}\, dx$

22. $\displaystyle\int \frac{x^2}{3 - x^3}\, dx$

23. $\displaystyle\int \frac{x^2}{x^3 + 1}\, dx$

24. $\displaystyle\int \frac{x}{x^2 + 4}\, dx$

25. $\displaystyle\int \frac{x + 3}{x^2 + 6x + 7}\, dx$

26. $\displaystyle\int \frac{x^2 + 2x + 3}{x^3 + 3x^2 + 9x + 1}\, dx$

27. $\displaystyle\int \frac{1}{x \ln x}\, dx$

28. $\displaystyle\int \frac{e^x}{1 + e^x}\, dx$

29. $\displaystyle\int \frac{e^{-x}}{1 + e^{-x}}\, dx$

30. $\displaystyle\int \frac{3e^x}{2 + e^x}\, dx$

31. $\displaystyle\int \frac{4e^{2x}}{5 - e^{2x}}\, dx$

32. $\displaystyle\int \frac{-e^{3x}}{2 - e^{3x}}\, dx$

In Exercises 33–50, use any basic integration formula or formulas to find the indefinite integral.

33. $\displaystyle\int \frac{e^{2x} + 2e^x + 1}{e^x}\,dx$ **34.** $\displaystyle\int (6x + e^x)\sqrt{3x^2 + e^x}\,dx$

35. $\displaystyle\int e^x\sqrt{1 - e^x}\,dx$ **36.** $\displaystyle\int \frac{2(e^x - e^{-x})}{(e^x + e^{-x})^2}\,dx$

37. $\displaystyle\int \frac{1}{(x - 1)^2}\,dx$ **38.** $\displaystyle\int \frac{1}{\sqrt{x + 1}}\,dx$

39. $\displaystyle\int \frac{x^2 - 4}{x}\,dx$ **40.** $\displaystyle\int \frac{x + 5}{x}\,dx$

41. $\displaystyle\int \frac{x^3 - 8x}{2x^2}\,dx$ **42.** $\displaystyle\int \frac{x - 1}{4x}\,dx$

43. $\displaystyle\int \frac{2}{1 + e^{-x}}\,dx$ **44.** $\displaystyle\int \frac{3}{1 + e^{-3x}}\,dx$

45. $\displaystyle\int \frac{x^2 + 2x + 5}{x - 1}\,dx$ **46.** $\displaystyle\int \frac{x - 3}{x + 3}\,dx$

47. $\displaystyle\int \frac{x^3 + 4x^2 + 3x}{x^2 - x}\,dx$ **48.** $\displaystyle\int \frac{x^4 - x^3 + x^2 + 2}{x + 5}\,dx$

49. $\displaystyle\int \frac{1 + e^{-x}}{1 + xe^{-x}}\,dx$ **50.** $\displaystyle\int \frac{5}{e^{-5x} + 7}\,dx$

51. *Bacteria Growth* A population of bacteria is growing at the rate of

$$\frac{dP}{dt} = \frac{3000}{1 + 0.25t},$$

where t is the time in days. When $t = 0$, the population is 1000.

(a) Write an equation that models the population P in terms of the time t.

(b) What is the population after 3 days?

(c) After how many days will the population be 12,000?

52. *Population Decline* Because of an insufficient oxygen supply, the trout population in a lake is dying. The population's rate of change can be modeled by

$$\frac{dP}{dt} = -125e^{-t/20},$$

where t is the time in days. When $t = 0$, the population is 2500.

(a) Write an equation that models the population P in terms of the time t.

(b) What is the population after 15 days?

(c) According to this model, how long will it take for the entire trout population to die?

53. *Demand* The marginal price for the demand of a product can be modeled by

$$\frac{dp}{dx} = 0.1e^{-x/500},$$

where x is the quantity demanded. When the demand is 600 units, the price is \$30.

(a) Find the demand function, $p = f(x)$.

(b) Use a graphing utility to graph the demand function. Does the price increase or decrease as the demand increases?

(c) Use the zoom and trace features of the graphing utility to find the quantity demanded when the price is \$22.

54. *Revenue* The marginal revenue for the sale of a product can be modeled by

$$\frac{dR}{dx} = 50 - 0.02x + \frac{100}{x + 1},$$

where x is the quantity demanded.

(a) Find the revenue function.

(b) Use a graphing utility to graph the revenue function.

(c) Find the revenue when 1500 units are sold.

(d) Use the zoom and trace features of the graphing utility to find the number of units sold when the revenue is \$60,230.

55. *ATM Transactions* From 1986 through 1992, the number of automatic teller machine (ATM) transactions T (in millions) in the United States changed at the rate of

$$\frac{dT}{dt} = 23.23t^{3/2} - 7.89t^2 + 44.71e^{-t},$$

where $t = 0$ corresponds to 1986. In 1992, there were 600 million transactions. *(Source: Bank Network News)*

(a) Write a model that gives the total number of ATM transactions per year.

(b) Use the model to find the number of ATM transactions in 1987.

56. *Revenue* The rate of change in revenue for Lotus Development Corporation from 1983 through 1993 can be modeled by

$$\frac{dR}{dt} = \frac{2358.96}{t} - \frac{12{,}756.78}{t^2} + 15{,}456.42e^{-t},$$

where R is the revenue (in millions) and $t = 3$ corresponds to 1983. In 1989, the revenue for Lotus was \$556 million. *(Source: Lotus Development Corporation.)*

(a) Find a model for Lotus's revenue from 1983 through 1993.

(b) Find Lotus's revenue in 1991.

Area and the Fundamental Theorem of Calculus

Area and Definite Integrals ■ *The Fundamental Theorem of Calculus* ■
Marginal Analysis ■ *Average Value* ■ *Even and Odd Functions*

Area and Definite Integrals

From your study of geometry, you know that area is a number that suggests the size of a bounded region. For simple regions, such as rectangles, triangles, and circles, area can be found using a geometric formula.

In this section, you will learn how to use calculus to find the area of nonstandard regions, such as the region R shown in Figure 5.5.

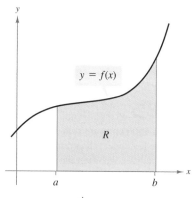

FIGURE 5.5 $\int_a^b f(x)\,dx = $ Area

Definition of a Definite Integral

Let f be nonnegative and continuous on the closed interval $[a, b]$. The area of the region bounded by the graph of f, the x-axis, and the lines $x = a$ and $x = b$ is denoted by

$$\text{Area} = \int_a^b f(x)\,dx.$$

The expression $\int_a^b f(x)\,dx$ is called the **definite integral** from a to b, where a is the **lower limit of integration** and b is the **upper limit of integration.**

EXAMPLE 1 *Evaluating a Definite Integral*

Evaluate the definite integral $\int_0^2 2x\,dx$.

Solution
This definite integral represents the area of the region bounded by the graph of $f(x) = 2x$, the x-axis, and the line $x = 2$, as shown in Figure 5.6. The region is triangular, with a height of four units and a base of two units.

$$\int_0^2 2x\,dx = \frac{1}{2}(\text{base})(\text{height}) \qquad \textit{Formula for area of triangle}$$

$$= \frac{1}{2}(2)(4)$$

$$= 4 \qquad\qquad\qquad \textit{Simplify.}$$

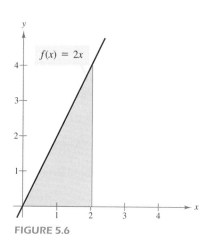

FIGURE 5.6

The Fundamental Theorem of Calculus

Consider the function $A(x)$, which denotes the area of the region shown in Figure 5.7. To discover the relationship between $A(x)$ and $f(x)$, let x increase by an amount Δx. This increases the area by ΔA. Let $f(m)$ and $f(M)$ denote the minimum and maximum values of f on the interval $[x, x + \Delta x]$.

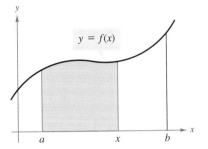

FIGURE 5.7 $A(x) =$ Area from a to x

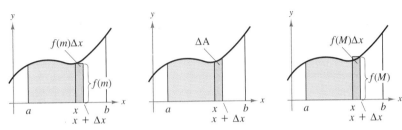

FIGURE 5.8

As indicated in Figure 5.8, you can write the following inequality.

$$f(m)\,\Delta x \leq \quad \Delta A \quad \leq f(M)\,\Delta x \qquad \textit{See Figure 5.8.}$$

$$f(m) \leq \quad \frac{\Delta A}{\Delta x} \quad \leq f(M) \qquad \textit{Divide each term by } \Delta x.$$

$$\lim_{\Delta x \to 0} f(m) \leq \lim_{\Delta x \to 0} \frac{\Delta A}{\Delta x} \leq \lim_{\Delta x \to 0} f(M) \qquad \textit{Take limit of each term.}$$

$$f(x) \leq \quad A'(x) \quad \leq f(x) \qquad \textit{Definition of derivative of } A(x)$$

Thus, $f(x) = A'(x)$, and $A(x) = F(x) + C$, where $F'(x) = f(x)$. Because $A(a) = 0$, it follows that $C = -F(a)$. Thus, $A(x) = F(x) - F(a)$, which implies that

$$A(b) = \int_a^b f(x)\,dx = F(b) - F(a).$$

This equation tells you that *if you can find an antiderivative for* f, then you can use the antiderivative to evaluate the definite integral $\int_a^b f(x)\,dx$. This result is called the **Fundamental Theorem of Calculus.**

The Fundamental Theorem of Calculus

If f is nonnegative and continuous on the closed interval $[a, b]$, then

$$\int_a^b f(x)\,dx = F(b) - F(a),$$

where F is any function such that $F'(x) = f(x)$ for all x in $[a, b]$.

Three comments about the Fundamental Theorem of Calculus are in order.

1. The Fundamental Theorem of Calculus describes a way of *evaluating* a definite integral, not a procedure for finding antiderivatives.
2. In applying the Fundamental Theorem, it is helpful to use the notation

$$\int_a^b f(x)\,dx = F(x) \Bigg]_a^b = F(b) - F(a).$$

3. The constant of integration C can be dropped because

$$\int_a^b f(x)\,dx = \left[F(x) + C \right]_a^b$$
$$= [F(b) + C] - [F(a) + C]$$
$$= F(b) - F(a) + C - C$$
$$= F(b) - F(a).$$

In the development of the Fundamental Theorem of Calculus, f was assumed to be nonnegative on the closed interval $[a, b]$. As such, the definite integral was defined as an area. Now, with the Fundamental Theorem, the definition can be extended to include functions that are negative on all or part of the closed interval $[a, b]$. Specifically, if f is *any* function that is continuous on a closed interval $[a, b]$, then the **definite integral** of $f(x)$ from a to b is defined to be

$$\int_a^b f(x)\,dx = F(b) - F(a),$$

where F is an antiderivative of f. Remember that definite integrals do not necessarily represent areas and can be negative, zero, or positive.

Note Be sure you see the distinction between indefinite and definite integrals. The *indefinite integral*

$$\int f(x)\,dx$$

denotes a family of *functions*, each of which is an antiderivative of f, whereas the *definite integral*

$$\int_a^b f(x)\,dx$$

is a *number*.

Properties of Definite Integrals

Let f and g be continuous on the closed interval $[a, b]$.

1. $\displaystyle\int_a^b kf(x)\,dx = k \int_a^b f(x)\,dx, \quad k$ is a constant.

2. $\displaystyle\int_a^b [f(x) \pm g(x)]\,dx = \int_a^b f(x)\,dx \pm \int_a^b g(x)\,dx$

3. $\displaystyle\int_a^b f(x)\,dx = \int_a^c f(x)\,dx + \int_c^b f(x)\,dx, \quad a < c < b$

4. $\displaystyle\int_a^a f(x)\,dx = 0$

5. $\displaystyle\int_a^b f(x)\,dx = -\int_b^a f(x)\,dx$

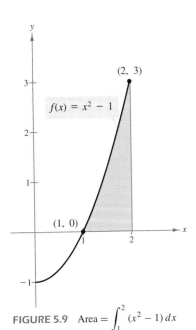

FIGURE 5.9   Area $= \int_{1}^{2} (x^2 - 1)\,dx$

EXAMPLE 2   *Finding Area by the Fundamental Theorem*

Find the area of the region bounded by the x-axis and the graph of

$$f(x) = x^2 - 1, \qquad 1 \le x \le 2.$$

Solution
Note that $f(x) \ge 0$ on the interval $1 \le x \le 2$, as shown in Figure 5.9. Therefore, you can represent the area of the region by a definite integral. To find the area, use the Fundamental Theorem of Calculus.

$$\text{Area} = \int_{1}^{2} (x^2 - 1)\,dx \qquad \textit{Definition of definite integral}$$

$$= \left[\frac{x^3}{3} - x \right]_{1}^{2} \qquad \textit{Find antiderivative.}$$

$$= \left(\frac{(2)^3}{3} - 2 \right) - \left(\frac{(1)^3}{3} - 1 \right) \qquad \textit{Apply Fundamental Theorem.}$$

$$= \frac{4}{3} \qquad \textit{Simplify.}$$

Thus, the area of the region is $\frac{4}{3}$ square units.

Note   It is easy to make errors in signs when evaluating definite integrals. To avoid such errors, enclose the values of the antiderivative at the upper and lower limits of integration in separate sets of parentheses, as shown above.

EXAMPLE 3   *Evaluating a Definite Integral*

Evaluate the definite integral

$$\int_{0}^{1} (4t + 1)^2\,dt,$$

and sketch the region whose area is represented by the integral.

Solution

$$\int_{0}^{1} (4t + 1)^2\,dt = \frac{1}{4} \int_{0}^{1} (4t + 1)^2 (4)\,dt \qquad \textit{Multiply and divide by 4.}$$

$$= \frac{1}{4} \left[\frac{(4t + 1)^3}{3} \right]_{0}^{1} \qquad \textit{Find antiderivative.}$$

$$= \frac{1}{4} \left[\left(\frac{5^3}{3} \right) - \left(\frac{1}{3} \right) \right] \qquad \textit{Apply Fundamental Theorem.}$$

$$= \frac{31}{3} \qquad \textit{Simplify.}$$

The region is shown in Figure 5.10.

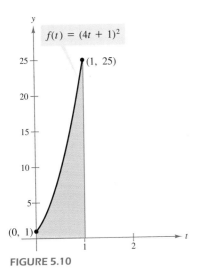

FIGURE 5.10

EXAMPLE 4 *Evaluating Definite Integrals*

a. $\displaystyle\int_0^3 e^{2x}\,dx = \left.\frac{1}{2}e^{2x}\right]_0^3 = \frac{1}{2}(e^6 - e^0) \approx 201.21$

b. $\displaystyle\int_1^2 \frac{1}{x}\,dx = \left.\ln x\right]_1^2 = \ln 2 - \ln 1 = \ln 2 \approx 0.69$

c. $\displaystyle\int_1^4 -3\sqrt{x}\,dx = -3\int_1^4 x^{1/2}\,dx$ *Write with rational exponent.*

$\qquad\qquad = -3\left[\dfrac{x^{3/2}}{3/2}\right]_1^4$ *Find antiderivative.*

$\qquad\qquad = \left.-2x^{3/2}\right]_1^4$

$\qquad\qquad = -2(4^{3/2} - 1^{3/2})$ *Apply Fundamental Theorem.*

$\qquad\qquad = -2(8 - 1)$

$\qquad\qquad = -14$ *Simplify.*

Note In Example 4c, note that the value of a definite integral can be negative.

EXAMPLE 5 *Interpreting Absolute Value*

Evaluate the definite integral $\displaystyle\int_0^2 |2x - 1|\,dx$.

Solution

The region represented by the definite integral is shown in Figure 5.11. From the definition of absolute value, you can write

$$|2x - 1| = \begin{cases} -(2x - 1), & x < \frac{1}{2} \\ 2x - 1, & x \ge \frac{1}{2}. \end{cases}$$

Using Property 3 of definite integrals, you can rewrite the integral as two definite integrals.

$$\int_0^2 |2x - 1|\,dx = \int_0^{1/2} -(2x - 1)\,dx + \int_{1/2}^2 (2x - 1)\,dx$$

$$= \left[-x^2 + x\right]_0^{1/2} + \left[x^2 - x\right]_{1/2}^2$$

$$= \left(-\frac{1}{4} + \frac{1}{2}\right) - (0 + 0) + (4 - 2) - \left(\frac{1}{4} - \frac{1}{2}\right)$$

$$= \frac{5}{2}$$

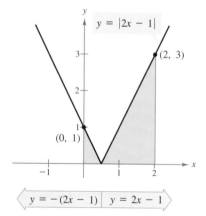

FIGURE 5.11

Marginal Analysis

You have already studied *marginal analysis* in the context of derivatives and differentials (Sections 2.3 and 3.8). There, you were given a cost, revenue, or profit function, and you used the derivative to approximate the additional cost, revenue, or profit obtained by selling one additional unit. In this section, you will examine the reverse process. That is, you will be given the marginal cost, marginal revenue, or marginal profit and will be asked to use a definite integral to find the exact increase or decrease in cost, revenue, or profit obtained by selling one or several additional units.

For instance, suppose you wanted to find the additional revenue obtained by increasing sales from x_1 to x_2 units. If you knew the revenue function $R(x)$, you could simply subtract $R(x_1)$ from $R(x_2)$. If you didn't know the revenue function, but did know the marginal revenue function, you could still find the additional revenue by using a definite integral, as follows.

$$\int_{x_1}^{x_2} \frac{dR}{dx}\, dx = R(x_2) - R(x_1)$$

Real Life

EXAMPLE 6 *Analyzing a Profit Function*

TECHNOLOGY

Symbolic integration programs, such as *Derive*, *Maple*, and *Mathematica*, can be used to evaluate definite integrals as well as indefinite integrals. If you have access to such a program, try using it to evaluate several of the definite integrals in this section.

The marginal profit for a product is modeled by

$$\frac{dP}{dx} = -0.0005x + 12.2.$$

a. Find the change in profit when sales increase from 100 to 101 units.
b. Find the change in profit when sales increase from 100 to 110 units.

Solution
a. The change in profit obtained by increasing sales from 100 to 101 units is

$$\int_{100}^{101} \frac{dP}{dx}\, dx = \int_{100}^{101} (-0.0005x + 12.2)\, dx$$

$$= \left[-0.00025x^2 + 12.2x \right]_{100}^{101}$$

$$\approx \$12.15.$$

b. The change in profit obtained by increasing sales from 100 to 110 units is

$$\int_{100}^{110} \frac{dP}{dx}\, dx = \int_{100}^{110} (-0.0005x + 12.2)\, dx$$

$$= \left[-0.00025x^2 + 12.2x \right]_{100}^{110}$$

$$\approx \$121.48.$$

Average Value

The *average value* of a function on a closed interval is defined as follows.

Definition of the Average Value of a Function

If f is continuous on $[a, b]$, then the **average value** of f on $[a, b]$ is

$$\text{Average value of } f \text{ on } [a, b] = \frac{1}{b - a} \int_a^b f(x)\, dx.$$

Real Life

EXAMPLE 7 *Finding the Average Cost*

The cost per unit c of producing a product over a 2-year period is modeled by

$$c = 0.005t^2 + 0.01t + 13.15, \qquad 0 \le t \le 24,$$

where t is the time in months. Approximate the average cost per unit over the 2-year period.

Solution

The average cost can be found by integrating c over the interval $[0, 24]$.

$$\text{Average cost per unit} = \frac{1}{24} \int_0^{24} (0.005t^2 + 0.01t + 13.15)\, dt$$

$$= \frac{1}{24} \left[\frac{0.005t^3}{3} + \frac{0.01t^2}{2} + 13.15t \right]_0^{24}$$

$$= \frac{1}{24}(341.52)$$

$$= \$14.23$$

To check the reasonableness of the average value found in Example 7, assume that one unit is produced each month, beginning with $t = 0$ and ending with $t = 24$. When $t = 0$, the cost is

$$c = 0.005(0)^2 + 0.01(0) + 13.15 = \$13.15.$$

Similarly, when $t = 1$, the cost is

$$c = 0.005(1)^2 + 0.01(1) + 13.15 \approx \$13.17.$$

Each month, the cost increases, and the average of the 25 costs is

$$\frac{13.15 + 13.17 + 13.19 + 13.23 + \cdots + 16.27}{25} \approx \$14.25.$$

Even and Odd Functions

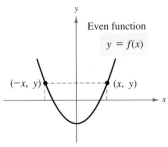

Even function
$y = f(x)$

$(-x, y)$ (x, y)

(a) y-Axis symmetry

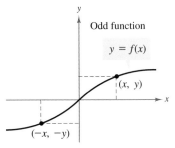

Odd function
$y = f(x)$

(x, y)

$(-x, -y)$

(b) Origin symmetry
FIGURE 5.12

Several common functions have graphs that are symmetric to the y-axis or the origin, as shown in Figure 5.12. If the graph of f is symmetric to the y-axis, as in Figure 5.12(a), then

$$f(-x) = f(x), \qquad\qquad \textit{Even function}$$

and f is called an **even** function. If the graph of f is symmetric to the origin, as in Figure 5.12(b), then

$$f(-x) = -f(x), \qquad\qquad \textit{Odd function}$$

and f is called an **odd** function.

Integration of Even and Odd Functions

1. If f is an *even* function, then $\displaystyle\int_{-a}^{a} f(x)\,dx = 2\int_{0}^{a} f(x)\,dx.$

2. If f is an *odd* function, then $\displaystyle\int_{-a}^{a} f(x)\,dx = 0.$

EXAMPLE 8 *Integrating Even and Odd Functions*

a. Because $f(x) = x^2$ is even,

$$\int_{-2}^{2} x^2\,dx = 2\int_{0}^{2} x^2\,dx = 2\left[\frac{x^3}{3}\right]_{0}^{2} = 2\left(\frac{8}{3} - 0\right) = \frac{16}{3}.$$

b. Because $f(x) = x^3$ is odd,

$$\int_{-2}^{2} x^3\,dx = 0.$$

Discussion Problem   Using Geometry to Evaluate Definite Integrals

When using the Fundamental Theorem of Calculus to evaluate $\int_{a}^{b} f(x)\,dx$, remember that you must first be able to find an antiderivative of $f(x)$. If you are unable to find an antiderivative, you cannot use the Fundamental Theorem. In some cases, you can still evaluate the definite integral. For instance, explain how you can use geometry to evaluate

$$\int_{-2}^{2} \sqrt{4 - x^2}\,dx.$$

$\mathcal{Warm\ Up}$

The following warm-up exercises involve skills that were covered in earlier sections. You will use these skills in the exercise set for this section.

In Exercises 1–4, find the indefinite integral.

1. $\int (3x + 7)\, dx$

2. $\int (x^{3/2} + 2\sqrt{x})\, dx$

3. $\int \dfrac{1}{5x}\, dx$

4. $\int e^{-6x}\, dx$

In Exercises 5 and 6, evaluate the expression when $a = 5$ and $b = 3$.

5. $\left(\dfrac{a}{5} - a\right) - \left(\dfrac{b}{5} - b\right)$

6. $\left(6a - \dfrac{a^3}{3}\right) - \left(6b - \dfrac{b^3}{3}\right)$

In Exercises 7–10, integrate the marginal function.

7. $\dfrac{dC}{dx} = 0.02x^{3/2} + 29{,}500$

8. $\dfrac{dR}{dx} = 9000 + 2x$

9. $\dfrac{dP}{dx} = 25{,}000 - 0.01x$

10. $\dfrac{dC}{dx} = 0.03x^2 + 4600$

E X E R C I S E S   5.4   means that technology can help you solve or check the exercise(s).

In Exercises 1–4, sketch the region whose area is represented by the definite integral. Then use a geometric formula to evaluate the integral.

1. $\int_0^2 3\, dx$

2. $\int_0^3 2x\, dx$

3. $\int_0^5 (x + 1)\, dx$

4. $\int_{-3}^3 \sqrt{9 - x^2}\, dx$

In Exercises 5–10, find the area of the region.

5. $y = x - x^2$

6. $y = \dfrac{1}{x^2}$

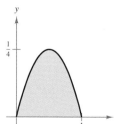

7. $y = 1 - x^4$

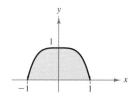

8. $y = 3e^{-x/2}$

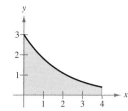

9. $y = \sqrt[3]{2x}$

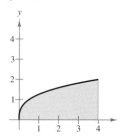

10. $y = \dfrac{x^2 + 4}{x}$

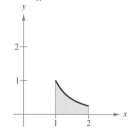
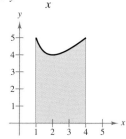

In Exercises 11–38, evaluate the definite integral.

11. $\displaystyle\int_0^1 2x\,dx$

12. $\displaystyle\int_2^7 3\,dv$

13. $\displaystyle\int_{-1}^0 (2x+1)\,dx$

14. $\displaystyle\int_2^5 (-3x+4)\,dx$

15. $\displaystyle\int_{-1}^1 (2t-1)^2\,dt$

16. $\displaystyle\int_0^3 (3x^2+x-2)\,dx$

17. $\displaystyle\int_0^1 (2t-1)^2\,dt$

18. $\displaystyle\int_2^2 (x-3)^4\,dx$

19. $\displaystyle\int_{-1}^1 (\sqrt[3]{t}-2)\,dt$

20. $\displaystyle\int_1^4 \sqrt{\frac{2}{x}}\,dx$

21. $\displaystyle\int_1^4 \frac{2u-1}{\sqrt{u}}\,du$

22. $\displaystyle\int_0^1 \frac{x-\sqrt{x}}{3}\,dx$

23. $\displaystyle\int_{-1}^0 (t^{1/3}-t^{2/3})\,dt$

24. $\displaystyle\int_0^4 (x^{1/2}+x^{1/4})\,dx$

25. $\displaystyle\int_0^4 \frac{1}{\sqrt{2x+1}}\,dx$

26. $\displaystyle\int_0^2 \frac{x}{\sqrt{1+2x^2}}\,dx$

27. $\displaystyle\int_0^1 e^{-2x}\,dx$

28. $\displaystyle\int_1^2 e^{1-x}\,dx$

29. $\displaystyle\int_1^3 \frac{e^{3/x}}{x^2}\,dx$

30. $\displaystyle\int_{-1}^1 (e^x-e^{-x})\,dx$

31. $\displaystyle\int_{-1}^1 |4x|\,dx$

32. $\displaystyle\int_0^3 |2x-3|\,dx$

33. $\displaystyle\int_0^4 (2-|x-2|)\,dx$

34. $\displaystyle\int_{-4}^4 (4-|x|)\,dx$

35. $\displaystyle\int_{-1}^2 \frac{x}{x^2-9}\,dx$

36. $\displaystyle\int_2^3 \frac{x+1}{x^2+2x-3}\,dx$

37. $\displaystyle\int_0^3 \frac{2e^x}{2+e^x}\,dx$

38. $\displaystyle\int_1^2 \frac{(2+\ln x)^3}{x}\,dx$

In Exercises 39–44, evaluate the definite integral. Then sketch the region whose area is represented by the integral.

39. $\displaystyle\int_1^3 (4x-3)\,dx$

40. $\displaystyle\int_0^2 (x+4)\,dx$

41. $\displaystyle\int_0^1 (x-x^3)\,dx$

42. $\displaystyle\int_0^1 \sqrt{x}(1-x)\,dx$

43. $\displaystyle\int_2^4 \frac{3x^2}{x^3-1}\,dx$

44. $\displaystyle\int_0^{\ln 6} \frac{e^x}{2}\,dx$

45. Find the area of the region bounded by the graphs of $y=3x^2+1$, $y=0$, $x=0$, and $x=2$.

46. Find the area of the region bounded by the graphs of $y=1+\sqrt{x}$, $y=0$, $x=0$, and $x=4$.

47. Find the area of the region bounded by the graphs of $y=(x+5)/x$, $y=0$, $x=1$, and $x=5$.

48. Find the area of the region bounded by the graphs of $y=3e^x$, $y=0$, $x=-2$, and $x=1$.

In Exercises 49 and 50, use the values $\int_0^5 f(x)\,dx=8$ and $\int_0^5 g(x)\,dx=3$ to evaluate the definite integral.

49. (a) $\displaystyle\int_0^5 [f(x)+g(x)]\,dx$ (b) $\displaystyle\int_0^5 [f(x)-g(x)]\,dx$

(c) $\displaystyle\int_0^5 -4f(x)\,dx$ (d) $\displaystyle\int_0^5 [f(x)-3g(x)]\,dx$

50. (a) $\displaystyle\int_0^5 2g(x)\,dx$ (b) $\displaystyle\int_5^0 f(x)\,dx$

(c) $\displaystyle\int_5^5 f(x)\,dx$ (d) $\displaystyle\int_0^5 [f(x)-f(x)]\,dx$

In Exercises 51–56, sketch the graph of the function over the interval. Find the average value of the function over the interval. Then find all x-values in the interval for which the function is equal to its average value.

Function	*Interval*
51. $f(x)=6-x^2$	$[-2,2]$
52. $f(x)=5e^{0.2(x-10)}$	$[0,10]$
53. $f(x)=x\sqrt{4-x^2}$	$[0,2]$
54. $f(x)=\dfrac{5x}{x^2+1}$	$[0,7]$
55. $f(x)=x-2\sqrt{x}$	$[0,4]$
56. $f(x)=\dfrac{1}{(x-3)^2}$	$[0,2]$

In Exercises 57–60, state whether the function is even, odd, or neither.

57. $f(x)=3x^4$

58. $g(x)=x^3-2x$

59. $g(t)=2t^5-3t^2$

60. $f(t)=5t^4+1$

61. Use the value $\int_0^2 x^2\,dx=\frac{8}{3}$ to evaluate the definite integral. Explain your reasoning.

(a) $\displaystyle\int_{-2}^0 x^2\,dx$ (b) $\displaystyle\int_{-2}^2 x^2\,dx$ (c) $\displaystyle\int_0^2 -x^2\,dx$

62. Use the value $\int_0^2 x^3\,dx=4$ to evaluate the definite integral. Explain your reasoning.

(a) $\displaystyle\int_{-2}^0 x^3\,dx$ (b) $\displaystyle\int_{-2}^2 x^3\,dx$ (c) $\displaystyle\int_0^2 3x^3\,dx$

Marginal Analysis In Exercises 63–68, find the change in cost C, revenue R, or profit P, for the given marginal. In each case, assume the number of units x increases by 3 from the specified value of x.

Marginal	Number of Units, x
63. $\dfrac{dC}{dx} = 2.25$	$x = 100$
64. $\dfrac{dC}{dx} = \dfrac{20{,}000}{x^2}$	$x = 10$
65. $\dfrac{dR}{dx} = 48 - 3x$	$x = 12$
66. $\dfrac{dR}{dx} = 75\left(20 + \dfrac{900}{x}\right)$	$x = 500$
67. $\dfrac{dP}{dx} = \dfrac{400 - x}{150}$	$x = 200$
68. $\dfrac{dP}{dx} = 12.5(40 - 3\sqrt{x})$	$x = 125$

Capital Accumulation In Exercises 69–72, you are given the rate of investment dI/dt. Find the capital accumulation over a 5-year period by evaluating the definite integral

$$\text{Capital accumulation} = \int_0^5 \frac{dI}{dt}\, dt,$$

where t is the time in years.

69. $\dfrac{dI}{dt} = 500$ **70.** $\dfrac{dI}{dt} = 500\sqrt{t + 1}$

71. $\dfrac{dI}{dt} = 100t$ **72.** $\dfrac{dI}{dt} = \dfrac{12{,}000t}{(t^2 + 2)^2}$

73. *Cost* The total cost of purchasing and maintaining a piece of equipment for x years can be modeled by

$$C = 5000\left(25 + 3\int_0^x t^{1/4}\, dt\right).$$

Find the total cost after (a) 1 year, (b) 5 years, and (c) 10 years.

74. *Depreciation* A company purchases a new machine for which the rate of depreciation can be modeled by

$$\frac{dV}{dt} = 10{,}000(t - 6), \qquad 0 \le t \le 5,$$

where V is the value of the machine after t years. Set up and evaluate the definite integral that yields the total loss of value of the machine over the first 3 years.

75. *Compound Interest* A deposit of \$2250 is made in a savings account at an annual rate of 12%, compounded continuously. Find the average balance in the account during the first 5 years.

76. *Blood Flow* The velocity v of blood at a distance r from the center of an artery of radius R can be modeled by

$$v = k(R^2 - r^2),$$

where k is a constant. Find the average velocity along a radius of the artery. (Use 0 and R as the limits of integration.)

77. *Computer Industry* The rate of change in revenue for the computer and data processing industry in the United States from 1985 through 1990 can be modeled by

$$\frac{dR}{dt} = 6.972\sqrt{t} - 0.40t^2, \qquad 0 \le t \le 5,$$

where R is the revenue (in billions of dollars) and $t = 0$ represents 1985. In 1985, the revenue was \$45.2 billion. *(Source: U.S. Bureau of Census)*

(a) Write a model for the revenue as a function of t.

(b) What was the average revenue for 1985 through 1990?

78. *Death Rate* In the United States, the annual death rate R (in deaths per 1000 people x years old) can be modeled by

$$R = 0.036x^2 - 2.8x + 58.14, \qquad 40 \le x \le 60$$

(see figure). Find the average death rate for people between (a) 40 and 50 years of age, and (b) 50 and 60 years of age.

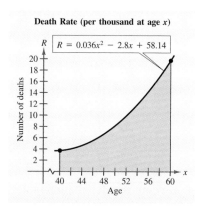

Death Rate (per thousand at age x)

$R = 0.036x^2 - 2.8x + 58.14$

In Exercises 79–82, use a symbolic integration utility to evaluate the definite integral.

79. $\displaystyle\int_3^6 \frac{x}{3\sqrt{x^2 - 8}}\, dx$ **80.** $\displaystyle\int_{1/2}^1 (x + 1)\sqrt{1 - x}\, dx$

81. $\displaystyle\int_2^5 \left(\frac{1}{x^2} - \frac{1}{x^3}\right) dx$ **82.** $\displaystyle\int_0^1 x^3(x^3 + 1)^3\, dx$

5.5

The Area of a Region Bounded by Two Graphs

Area of a Region Bounded by Two Graphs ■
Consumer and Producer Surplus ■ *Applications*

Area of a Region Bounded by Two Graphs

With a few modifications, you can extend the use of definite integrals from finding the area of a region *under a graph* to finding the area of a region *bounded by two graphs*. To see how this is done, consider the region bounded by the graphs of f, g, $x = a$, and $x = b$, as shown in Figure 5.13. If the graphs of both f and g lie above the x-axis, then you can interpret the area of the region between the graphs as the area of the region under the graph of g subtracted from the area of the region under the graph of f, as shown in Figure 5.13.

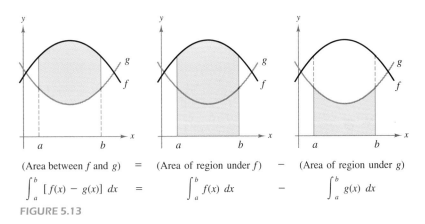

(Area between f and g)	=	(Area of region under f)	−	(Area of region under g)

$$\int_a^b [f(x) - g(x)]\, dx \quad = \quad \int_a^b f(x)\, dx \quad - \quad \int_a^b g(x)\, dx$$

FIGURE 5.13

Although Figure 5.13 depicts the graphs of f and g lying above the x-axis, this is not necessary, and the same integrand $[f(x) - g(x)]$ can be used as long as both functions are continuous and $g(x) \le f(x)$ on the interval $[a, b]$.

Area of a Region Bounded by Two Graphs

If f and g are continuous on $[a, b]$ and $g(x) \le f(x)$ for all x in the interval, then the area of the region bounded by the graphs of f, g, $x = a$, and $x = b$ is given by

$$A = \int_a^b [f(x) - g(x)]\, dx.$$

351

EXAMPLE 1 *Finding the Area Bounded by Two Graphs*

Find the area of the region bounded by the graphs of $y = x^2 + 2$ and $y = x$, for $0 \leq x \leq 1$.

Solution

Begin by sketching the graphs of both functions, as shown in Figure 5.14. From the figure, you can see that $x \leq x^2 + 2$ for all x in $[0, 1]$. Thus, you can let $f(x) = x^2 + 2$ and $g(x) = x$. Then compute the area as follows.

$$
\begin{aligned}
\text{Area} &= \int_a^b [f(x) - g(x)]\,dx && \textit{Area between } f \textit{ and } g \\
&= \int_0^1 [(x^2 + 2) - (x)]\,dx && \textit{Substitute for } f \textit{ and } g. \\
&= \int_0^1 (x^2 - x + 2)\,dx \\
&= \left[\frac{x^3}{3} - \frac{x^2}{2} + 2x \right]_0^1 && \textit{Find antiderivative.} \\
&= \frac{11}{6} \text{ square units} && \textit{Apply Fundamental Theorem.}
\end{aligned}
$$

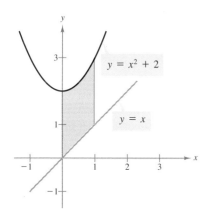

FIGURE 5.14

EXAMPLE 2 *Finding Area Between Intersecting Graphs*

Find the area of the region bounded by the graphs of $y = 2 - x^2$ and $y = x$.

Solution

In this problem, the values of a and b are not given and you must compute them by finding the points of intersection of the two graphs. To do this, equate the two functions and solve for x. When you do this, you will obtain $x = -2$ and $x = 1$. From Figure 5.15, you can see that the graph of $f(x) = 2 - x^2$ lies above the graph of $g(x) = x$ for all x in the interval $[-2, 1]$.

$$
\begin{aligned}
\text{Area} &= \int_a^b [f(x) - g(x)]\,dx && \textit{Area between } f \textit{ and } g \\
&= \int_{-2}^1 [(2 - x^2) - (x)]\,dx && \textit{Substitute for } f \textit{ and } g. \\
&= \int_{-2}^1 (-x^2 - x + 2)\,dx \\
&= \left[-\frac{x^3}{3} - \frac{x^2}{2} + 2x \right]_{-2}^1 && \textit{Find antiderivative.} \\
&= \frac{9}{2} \text{ square units} && \textit{Apply Fundamental Theorem.}
\end{aligned}
$$

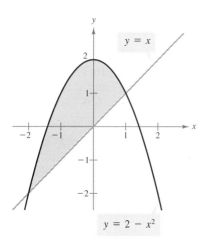

FIGURE 5.15

EXAMPLE 3 *Finding Area Below the x-Axis*

Find the area of the region bounded by the graph of $y = x^2 - 3x - 4$ and the x-axis.

Solution

Begin by finding the intercepts of the graph. To do this, set the function equal to zero and solve for x.

$$x^2 - 3x - 4 = 0 \qquad \text{\textit{Set function equal to 0.}}$$
$$(x - 4)(x + 1) = 0 \qquad \text{\textit{Factor.}}$$
$$x = 4, \ -1 \qquad \text{\textit{Solve for x.}}$$

From Figure 5.16, you can see that $x^2 - 3x - 4 \le 0$ for all x in the interval $[-1, 4]$. Therefore, you can let $f(x) = 0$ and $g(x) = x^2 - 3x - 4$, and compute the area as follows.

$$\text{Area} = \int_a^b [f(x) - g(x)]\,dx \qquad \text{\textit{Area between f and g}}$$

$$= \int_{-1}^4 [(0) - (x^2 - 3x - 4)]\,dx \qquad \text{\textit{Substitute for f and g.}}$$

$$= \int_{-1}^4 (-x^2 + 3x + 4)\,dx$$

$$= \left[-\frac{x^3}{3} + \frac{3x^2}{2} + 4x \right]_{-1}^4 \qquad \text{\textit{Find antiderivative.}}$$

$$= \frac{125}{6} \qquad \text{\textit{Apply Fundamental Theorem.}}$$

Therefore, the area of the region is $\frac{125}{6}$ square units. ■

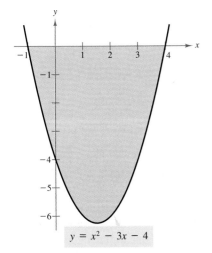

$y = x^2 - 3x - 4$

FIGURE 5.16

▦ TECHNOLOGY

Most graphing utilities can display regions that are bounded by two graphs. For instance, to graph the region in Example 3 on a *TI-81* or *TI-82*, set the viewing rectangle to $-1 \le x \le 4$ and $-7 \le y \le 1$. Then enter

Shade($X^2 - 3X - 4$, 0)

and press ENTER. You should obtain the graph shown at the right.

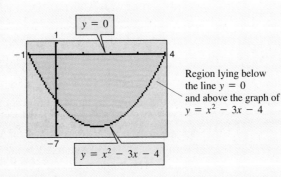

$y = 0$

Region lying below the line $y = 0$ and above the graph of $y = x^2 - 3x - 4$

$y = x^2 - 3x - 4$

Sometimes two graphs intersect at more than two points. To determine the area of the region bounded by two such graphs, you must find *all* points of intersection and check to see which graph is above the other in each interval determined by the points.

EXAMPLE 4 *Using Multiple Points of Intersection*

Find the area of the region bounded by the graphs of

$$f(x) = 3x^3 - x^2 - 10x \quad \text{and} \quad g(x) = -x^2 + 2x.$$

Solution

To find the points of intersection of the two graphs, set the functions equal to each other and solve for x.

$f(x) = g(x)$	*Set $f(x)$ equal to $g(x)$.*
$3x^3 - x^2 - 10x = -x^2 + 2x$	*Substitute for $f(x)$ and $g(x)$.*
$3x^3 - 12x = 0$	*Write in standard form.*
$3x(x^2 - 4) = 0$	
$3x(x - 2)(x + 2) = 0$	*Factor.*
$x = 0, \ 2, \ -2$	*Solve for x.*

These three points of intersection determine two intervals of integration: $[-2, 0]$ and $[0, 2]$. In Figure 5.17, you can see that $g(x) \leq f(x)$ in the interval $[-2, 0]$, and that $f(x) \leq g(x)$ in the interval $[0, 2]$. Therefore, you must use two integrals to determine the area of the region bounded by the graphs of f and g—one for the interval $[-2, 0]$ and one for the interval $[0, 2]$.

$$\text{Area} = \int_{-2}^{0} [f(x) - g(x)]\, dx + \int_{0}^{2} [g(x) - f(x)]\, dx$$

$$= \int_{-2}^{0} (3x^3 - 12x)\, dx + \int_{0}^{2} (-3x^3 + 12x)\, dx$$

$$= \left[\frac{3x^4}{4} - 6x^2 \right]_{-2}^{0} + \left[-\frac{3x^4}{4} + 6x^2 \right]_{0}^{2}$$

$$= (0 - 0) - (12 - 24) + (-12 + 24) - (0 + 0)$$

$$= 24$$

Thus, the region has an area of 24 square units. ▬

Note It is easy to make an error when calculating areas such as that in Example 4. To give yourself some idea about the reasonableness of your solution, you could make a careful sketch of the region on graph paper and then use the grid on the graph paper to approximate the area. Try doing this with the graph shown in Figure 5.17. Is your approximation close to 24 square units?

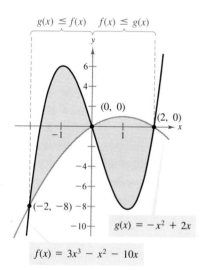

$g(x) \leq f(x)$ $f(x) \leq g(x)$

$(0, 0)$

$(2, 0)$

$(-2, -8)$

$g(x) = -x^2 + 2x$

$f(x) = 3x^3 - x^2 - 10x$

FIGURE 5.17

Consumer and Producer Surplus

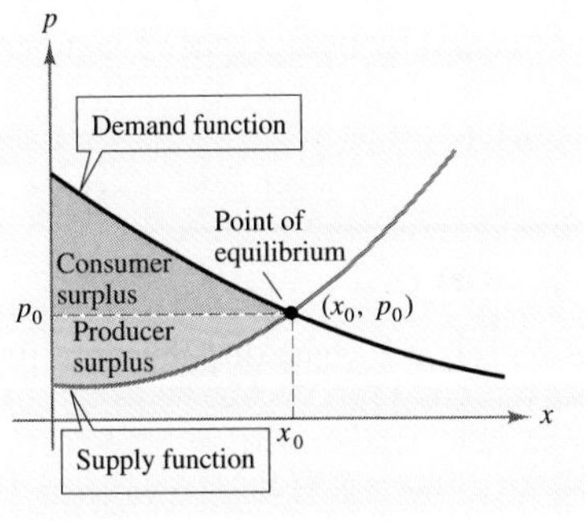

You already know that a demand function relates the price of a product to the consumer demand. A **supply function** relates the price of a product to producers' willingness to supply the product. Whereas a typical demand function is decreasing, a typical supply function is increasing. That is, as the price increases, more and more producers become willing to supply the product. The point (x_0, p_0) at which a demand function $p = D(x)$ and a supply function $p = S(x)$ intersect is called the **point of equilibrium.**

Economists call the area of the region bounded by the graph of the demand function, the horizontal line $p = p_0$, and the vertical line $x = 0$ the **consumer surplus.** Similarly, the area of the region bounded by the graph of the supply function, the horizontal line $p = p_0$, and the vertical line $x = 0$ is called the **producer surplus,** as shown in Figure 5.18.

FIGURE 5.18

Real Life

EXAMPLE 5 *Finding Surplus*

The demand and supply functions for a product are modeled by

Demand: $p = -0.36x + 9$ and   *Supply*: $p = 0.14x + 2,$

where x is the number of units (in millions). Find the consumer and producer surplus for this product.

Solution

By equating the demand and supply functions, you can determine that the point of equilibrium occurs when $x = 14$ (million) and the price is $3.96 per unit. The consumer surplus and producer surplus are shown in Figure 5.19.

$$
\begin{aligned}
\text{Consumer surplus} &= \int_0^{14} (\text{demand function} - \text{price})\, dx \\
&= \int_0^{14} [(-0.36x + 9) - 3.96]\, dx \\
&= \left[-0.18x^2 + 5.04x \right]_0^{14} \\
&= 35.28
\end{aligned}
$$

$$
\begin{aligned}
\text{Producer surplus} &= \int_0^{14} (\text{price} - \text{supply function})\, dx \\
&= \int_0^{14} [3.96 - (0.14x + 2)]\, dx \\
&= \left[-0.07x^2 + 1.96x \right]_0^{14} \\
&= 13.72
\end{aligned}
$$

Supply and Demand

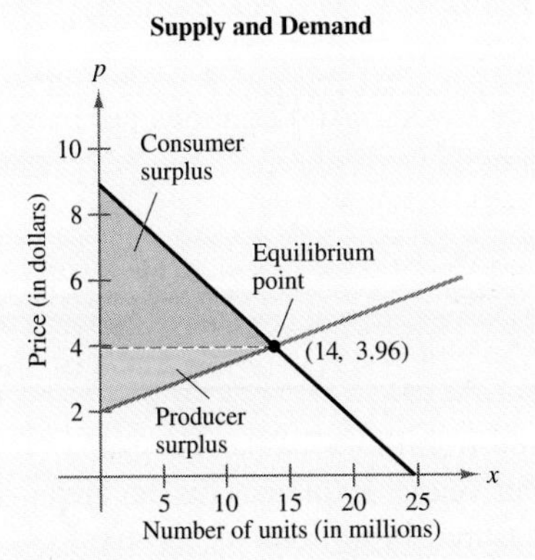

FIGURE 5.19

Tony Stone Images/Martin Rogers

In 1992, the United States imported more than 9 million barrels of oil per day. This accounted for about half of the daily U.S. oil consumption.

Applications

In addition to consumer and producer surplus, there are many other types of applications involving the area of a region bounded by two graphs. Example 6 illustrates one of these applications.

Real Life

EXAMPLE 6 *Finding Fuel Consumption*

The total U. S. consumption (in billions of barrels per year) of petroleum fuel for transportation from 1950 to 1979 can be modeled by

$$f(t) = 0.000433t^2 + 0.0962t + 2.76, \qquad -20 \le t \le 9,$$

where $t = 0$ represents January 1, 1970. When crude oil prices increased dramatically in the late 1970s, fuel consumption changed and began following the model

$$g(t) = -0.00831t^2 + 0.152t + 2.81, \qquad 9 \le t \le 16,$$

as shown in Figure 5.20. Find the total amount of fuel saved from 1979 to 1986 as a result of fuel being consumed at the post-1979 rate rather than at the pre-1979 rate. *(Source: U.S. Department of Energy)*

Solution

Because the graph of the pre-1979 model f lies above the graph of the post-1979 model g on the interval [9, 16], the amount of fuel saved can be approximated by the following integral.

$$\text{Fuel saved} = \int_9^{16} [f(t) - g(t)]\, dt$$

$$= \int_9^{16} (0.008743t^2 - 0.0558t - 0.05)\, dt$$

$$= \left[\frac{0.008743t^3}{3} - \frac{0.0558t^2}{2} - 0.05t \right]_9^{16}$$

$$= 4.58 \text{ billion barrels}$$

Thus, about 4.58 billion barrels of fuel were saved. (At 42 gallons per barrel, that is a savings of almost 200 billion gallons!) ■

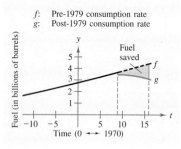

U.S. Fuel Consumption

f: Pre-1979 consumption rate
g: Post-1979 consumption rate

Fuel (in billions of barrels)

Time (0 ↔ 1970)

FIGURE 5.20

Discussion Problem **Finding Units for Area**

In Example 6, the vertical axis is measured in billions of barrels per year and the horizontal axis is measured in years. Explain why the area of the region shown in Figure 5.20 is measured in billions of barrels.

Warm Up

The following warm-up exercises involve skills that were covered in earlier sections. You will use these skills in the exercise set for this section.

In Exercises 1–4, simplify the expression.

1. $(-x^2 + 4x + 3) - (x + 1)$

2. $(-2x^2 + 3x + 9) - (-x + 5)$

3. $(-x^3 + 3x^2 - 1) - (x^2 - 4x + 4)$

4. $(3x + 1) - (-x^3 + 9x + 2)$

In Exercises 5–10, find the points of intersection of the graphs.

5. $f(x) = x^2 - 4x + 4$, $g(x) = 4$

6. $f(x) = -3x^2$, $g(x) = 6 - 9x$

7. $f(x) = x^2$, $g(x) = -x + 6$

8. $f(x) = \frac{1}{2}x^3$, $g(x) = 2x$

9. $f(x) = x^2 - 3x$, $g(x) = 3x - 5$

10. $f(x) = e^x$, $g(x) = e$

EXERCISES 5.5   means that technology can help you solve or check the exercise(s).

In Exercises 1–8, find the area of the region.

1. $f(x) = x^2 - 6x$
$g(x) = 0$

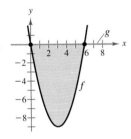

2. $f(x) = x^2 + 2x + 1$
$g(x) = 2x + 5$

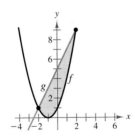

5. $f(x) = 3(x^3 - x)$
$g(x) = 0$

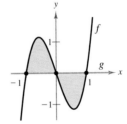

6. $f(x) = (x - 1)^3$
$g(x) = x - 1$

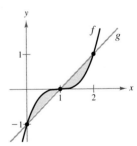

3. $f(x) = x^2 - 4x + 3$
$g(x) = -x^2 + 2x + 3$

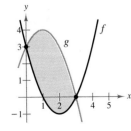

4. $f(x) = x^2$
$g(x) = x^3$

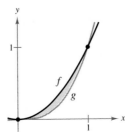

7. $f(x) = e^x - 1$
$g(x) = 0$

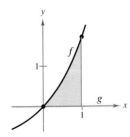

8. $f(x) = -x + 3$
$g(x) = 2x^{-1}$

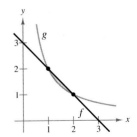

In Exercises 9 and 10, sketch the region whose area is represented by the definite integral.

9. $\int_0^4 [(x+1) - \frac{1}{2}x]\,dx$

10. $\int_{-1}^1 [(1-x^2) - (x^2-1)]\,dx$

In Exercises 11–34, sketch the region bounded by the graphs of the functions and find the area of the region.

11. $f(x) = x^2 - 4x,\ g(x) = 0$

12. $f(x) = 3 - 2x - x^2,\ g(x) = 0$

13. $f(x) = x^2 + 2x + 1,\ g(x) = x + 1$

14. $f(x) = -x^2 + 4x + 2,\ g(x) = x + 2$

15. $y = 2x,\ y = 4 - 2x,\ y = 0$

16. $y = \dfrac{1}{x^2},\ y = 0,\ x = 1,\ x = 5$

17. $f(x) = x^2 - x,\ g(x) = 2(x+2)$

18. $f(x) = x(x^2 - 3x + 3),\ g(x) = x^2$

19. $y = x^3 - 2x + 1,\ y = -2x,\ x = 1$

20. $f(x) = \sqrt[3]{x},\ g(x) = x$

21. $f(x) = \sqrt{3x} + 1,\ g(x) = x + 1$

22. $f(x) = x^2 + 5x - 6,\ g(x) = 6x - 6$

23. $y = x^2 - 4x + 3,\ y = 3 + 4x - x^2$

24. $y = x^4 - 2x^2,\ y = 2x^2$

25. $f(y) = y^2,\ g(y) = y + 2$

26. $f(y) = y(2 - y),\ g(y) = -y$

27. $x = y^2 + 2,\ x = 0,\ y = -1,\ y = 2$

28. $f(y) = \sqrt{y},\ y = 9,\ x = 0$

29. $f(x) = e^{0.5x},\ g(x) = -\dfrac{1}{x},\ x = 1,\ x = 2$

30. $f(x) = \dfrac{1}{x},\ g(x) = -e^x,\ x = \dfrac{1}{2},\ x = 1$

31. $y = \dfrac{4}{x},\ y = x,\ x = 1,\ x = 4$

32. $y = \dfrac{8}{x},\ y = x^2,\ x = 1,\ x = 4$

33. $y = xe^{-x^2},\ y = 0,\ x = 0,\ x = 1$

34. $y = \dfrac{e^{1/x}}{x^2},\ y = 0,\ x = 1,\ x = 3$

In Exercises 35 and 36, use integration to find the area of the triangular region having the given vertices.

35. $(0, 0),\ (4, 0),\ (4, 4)$

36. $(0, 0),\ (4, 0),\ (6, 4)$

Consumer and Producer Surplus In Exercises 37–46, find the consumer and producer surplus.

Demand Function	Supply Function
37. $p_1(x) = 50 - 0.5x$	$p_2(x) = 0.125x$
38. $p_1(x) = 60 - x$	$p_2(x) = 10 + \frac{7}{3}x$
39. $p_1(x) = 300 - x$	$p_2(x) = 100 + x$
40. $p_1(x) = -0.05x + 100$	$p_2(x) = 0.1x + 25$
41. $p_1(x) = 200 - 0.02x^2$	$p_2(x) = 100 + x$
42. $p_1(x) = 1000 - 0.4x^2$	$p_2(x) = 42x$
43. $p_1(x) = -0.125x + 50$	$p_2(x) = 10 + 0.025x$
44. $p_1(x) = 10 - 0.00001x^2$	$p_2(x) = 5 + 0.005x$
45. $p_1(x) = \dfrac{10{,}000}{\sqrt{x + 100}}$	$p_2(x) = 100\sqrt{0.05x + 10}$
46. $p_1(x) = \sqrt{25 - 0.1x}$	$p_2(x) = \sqrt{9 + 0.1x} - 2$

47. *Essay* Describe the characteristics of typical demand and supply functions.

48. *Essay* Suppose that the demand and supply functions for a product did not intersect. What could you conclude?

Revenue In Exercises 49 and 50, two models, R_1 and R_2, are given for revenue (in billions of dollars) for a large corporation. Both models are estimates of revenues for 1996–2000, with $t = 6$ representing 1996. Which model is projecting the greater revenue? How much more total revenue does that model project over the 4-year period?

49. $R_1 = 7.21 + 0.58t,\ R_2 = 7.21 + 0.45t$

50. $R_1 = 7.21 + 0.26t + 0.02t^2,\ R_2 = 7.21 + 0.1t + 0.01t^2$

51. *Fuel Cost* The projected fuel cost C (in millions of dollars) for an airline company from 1995 through 2005 is

$$C_1 = 568.5 + 7.15t,$$

where $t = 5$ represents 1995. If the company purchases more efficient airplane engines, fuel costs are expected to decrease and follow the model

$$C_2 = 525.6 + 6.43t.$$

How much can the company save on the more efficient engines? Explain your reasoning.

52. *Decreasing Profits* The profit (in millions of dollars) for Federal Express from 1985 through 1988 can be modeled by

$$P = 461.43 - 0.076t^3 - 1879.35t^{-1}, \qquad 5 \le t \le 8,$$

where $t = 5$ represents 1985. From 1989 through 1992, the company's profit can be modeled by

$$P = 603.45 - 59.34t + 0.122t^3 - 0.00023e^t, \ 8 < t \le 12.$$

How much more or less profit would Federal Express have made if the first model had continued to be valid from 1989 through 1992? *(Source: Federal Express)*

53. *Consumer and Producer Surplus* Factory orders for an air conditioner are about 6000 units per week when the price is $331 and about 8000 units per week when the price is $303. The supply function is given by $p = 0.0275x$. Find the consumer and producer surplus. (Assume the demand function is linear.)

54. *Consumer and Producer Surplus* Repeat Exercise 53 with a demand of about 6000 units per week when the price is $325 and about 8000 units per week when the price is $300. Find the consumer and producer surplus. (Assume the demand function is linear.)

55. *Profit* The revenue from a manufacturing process (in millions of dollars) is projected to follow the model $R = 100$ for 10 years. Over the same period of time, the cost (in millions of dollars) is projected to follow the model $C = 60 + 0.2t^2$, where t is the time (in years). Approximate the profit over the 10-year period.

56. *Profit* Repeat Exercise 55 for revenue and cost models given by $R = 100 + 0.08t$ and $C = 60 + 0.2t^2$.

57. *Lorenz Curve* Economists use *Lorenz curves* to illustrate the distribution of income in a country. Letting x represent the percent of families in a country and y represent the percent of total income, the model $y = x$ would represent a country in which each family has the same income. The Lorenz curve, $y = f(x)$, represents the actual income distribution. The area between these two models, for $0 \le x \le 100$, indicates the "income inequality" of a country. In 1990, the Lorenz curve for the United States could be modeled by

$$y = 0.344x + 0.0658x^{3/2}, \qquad 0 \le x \le 100,$$

where x is measured from the poorest to the wealthiest families. Find the income inequality for the United States in 1990. *(Source: U.S. Bureau of Census)*

58. *Income Distribution* Use the Lorenz curve given in Exercise 57 to complete the table, which lists the percent of total income earned by each quintile in the United States in 1990.

Quintile	Lowest	2nd	3rd	4th	Highest
Percent					

59. *Research Project* Use your school's library or some other reference source to research a small company similar to that described below. Describe the costs, market conditions, and competition that affect the company's success.

Pacific Drifters, Inc.

In July of 1987, accountant Doug Stavoe invested $25,000 to start Pacific Drifters, Inc. in Newport Beach, California. Pacific Drifters does sky-writing and banner-pulling for business and personal messages including birthdays, marriage proposals, and anniversaries. Between 1987 and 1993, Stavoe's sales increased by about 20% per year.

Case Study: Airline Passenger Demand

PROFILE

COMPANY
Delta Air Lines, Inc.
■

YEARS IN BUSINESS
65
■

LOCATION
Atlanta, GA
■

TYPE OF BUSINESS
*Commercial
passenger airline*
■

EMPLOYEES
Approx. 72,000
■

REVENUE
*Approx. $12 billion
annually*
■

With over 2500 domestic flight legs daily, one major task faced by Delta Air Lines is deciding, in light of cost and other constraints, how to assign its fleet of over 500 aircraft along its daily routes so as to match the number of available seats to the expected number of passengers on any given flight. With the help of mathematical programming algorithms, experts at Delta have developed a solution to the fleet assignment problem that is expected to save Delta millions of dollars over the next few years.

Although the overall process of cost minimization is very complex to analyze because it has so many facets, an important element is the attempt to minimize costs due to "spill"— i.e., the number of passengers that cannot be accommodated because of insufficient aircraft capacity resulting from incorrect fleet assignments. While some of these spilled passengers may be "recaptured" on other Delta flights, those passengers who are lost to competing airlines represent opportunity costs in the form of forgone revenues for Delta. Of course, assigning too *large* an aircraft for a particular flight could create the opposite problem—losses due to empty seats and higher operating costs. An important objective is therefore to estimate the expected size of the spill for any given size of aircraft.

Making the assumption that passenger demand (x) is normally distributed (a common airline-industry assumption) with a given mean and standard deviation, spill is represented diagrammatically as the truncation of the passenger demand function at the point of an aircraft's capacity. In this case, standard deviation is a measure of the amount of variation in passenger demand. In the figure on page 361, where passenger demand is assumed to have a mean of 120 passengers and a standard deviation of 25 passengers, the probability that there will be spill for a Boeing 727 (which has a capacity of 148) is given by the shaded area under the normally distributed demand function to the right of 148 passengers. We can use calculus to determine the size of this area. Converting the given normally distributed demand function $f(x)$ into a *standard* normal function $f(z)$ with a mean of zero and a standard deviation of 1, we have:

$$P(148 < x) = P(1.12 < z) = \int_{1.12}^{\infty} \frac{1}{\sqrt{2\pi}} e^{-z^2/2}\, dz = 0.1314.$$

Thus, with the given passenger demand distribution, there is a 13 percent chance that there will be more passengers than can be accommodated by a Boeing 727. The definite integral above cannot actually be evaluated by finding an antiderivative, so the probability must be obtained using a statistical table such as the one found in Reference Table 24 of Appendix B. This probability can then be used to estimate the expected *number* of spilled passengers for a Boeing 727. If the distribution of passenger demand is known, this procedure can yield the expected spill for any given size of aircraft.

Multiplying the expected number of spilled passengers by the percent that is expected to be lost to competitors (i.e., not "recaptured" on other Delta flights) gives the estimated total number of lost spilled passengers. Finally, when the number of lost spilled passengers is multiplied by the average revenue per spilled passenger, the result is an estimate of spill cost to Delta.

Source: The Operations Research Society of America and The Institute of Management Sciences, 290 Westminster Street, Providence, RI 02903.

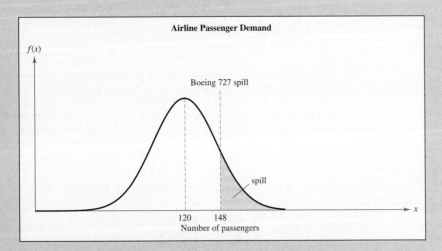

What Would You Do?

1. With a given passenger demand function for a particular scheduled flight, explain what happens to the expected spill as larger-sized aircraft are assigned to that flight.

2. Explain what would happen to the expected size of spill for a given size of aircraft if the *variation* in passenger demand were to increase. Describe factors that might affect the variations in passenger demand.

3. For a hypothetical aircraft, construct an estimate of spill cost. Invent and justify your own figures required to use the process given in the last paragraph. How could you improve this situation?

The Definite Integral as the Limit of a Sum

The Midpoint Rule ■ *The Definite Integral as the Limit of a Sum*

The Midpoint Rule

In Section 5.4, you learned that you cannot use the Fundamental Theorem of Calculus to evaluate a definite integral unless you can find an antiderivative of the integrand. In cases when this cannot be done, you can approximate the value of the integral using an approximation technique. One such technique is called the **Midpoint Rule.** (Two other techniques are discussed in Section 6.5.)

EXAMPLE 1 *Approximating the Area of a Plane Region*

Use the five rectangles in Figure 5.21 to approximate the area of the region bounded by the graph of $f(x) = -x^2 + 5$, the x-axis, and the lines $x = 0$ and $x = 2$.

Solution

You can find the heights of the five rectangles by evaluating f at the midpoint of each of the following intervals.

$$\left[0, \frac{2}{5}\right], \quad \left[\frac{2}{5}, \frac{4}{5}\right], \quad \left[\frac{4}{5}, \frac{6}{5}\right], \quad \left[\frac{6}{5}, \frac{8}{5}\right], \quad \left[\frac{8}{5}, \frac{10}{5}\right]$$

Evaluate $f(x)$ at the midpoints of these intervals.

The width of each rectangle is $\frac{2}{5}$. Hence, the sum of the five areas is

$$\text{Area} \approx \frac{2}{5} f\left(\frac{1}{5}\right) + \frac{2}{5} f\left(\frac{3}{5}\right) + \frac{2}{5} f\left(\frac{5}{5}\right) + \frac{2}{5} f\left(\frac{7}{5}\right) + \frac{2}{5} f\left(\frac{9}{5}\right)$$

$$= \frac{2}{5} \left[f\left(\frac{1}{5}\right) + f\left(\frac{3}{5}\right) + f\left(\frac{5}{5}\right) + f\left(\frac{7}{5}\right) + f\left(\frac{9}{5}\right) \right]$$

$$= \frac{2}{5} \left(\frac{124}{25} + \frac{116}{25} + \frac{100}{25} + \frac{76}{25} + \frac{44}{25} \right)$$

$$= \frac{920}{125}$$

$$= 7.36.$$

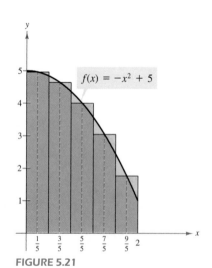

$f(x) = -x^2 + 5$

FIGURE 5.21

For the region in Example 1, you can find the exact area with a definite integral. That is,

$$\text{Area} = \int_0^2 (-x^2 + 5)\,dx$$

$$= \frac{22}{3}$$

$$\approx 7.33.$$

The approximation procedure used in Example 1 is the **Midpoint Rule.** You can use the Midpoint Rule to approximate *any* definite integral—not just those representing area. The basic steps are summarized below.

Guidelines for Using the Midpoint Rule

To approximate the definite integral $\int_a^b f(x)\,dx$ with the Midpoint Rule, use the following steps.

1. Divide the interval $[a, b]$ into n subintervals, each of width
$$\Delta x = \frac{b-a}{n}.$$

2. Find the midpoint of each subinterval.
$$\text{Midpoints} = \{x_1, x_2, x_3, \ldots, x_n\}$$

3. Evaluate f at each midpoint and form the following sum.
$$\int_a^b f(x)\,dx \approx \frac{b-a}{n}[f(x_1) + f(x_2) + f(x_3) + \cdots + f(x_n)]$$

Note   In Example 1, the Midpoint Rule is used to approximate an integral whose exact value can be found with the Fundamental Theorem of Calculus. This was done to illustrate the accuracy of the rule. In practice, of course, you would use the Midpoint Rule to approximate the values of definite integrals for which you cannot find an antiderivative. Examples 2 and 3 illustrate such integrals.

An important characteristic of the Midpoint Rule is that the approximation tends to improve as n increases. Table 5.1 shows the approximations for the area of the region described in Example 1 for various values of n. For example, for $n = 10$, the Midpoint Rule yields

$$\int_0^2 (-x^2 + 5)\,dx = \frac{2}{10}\left[f\left(\frac{1}{10}\right) + f\left(\frac{3}{10}\right) + \cdots + f\left(\frac{19}{10}\right)\right]$$

$$\approx 7.3400.$$

TABLE 5.1

n	5	10	15	20	25	30
Approximation	7.3600	7.3400	7.3363	7.3350	7.3344	7.3341

Note that as n increases, the approximation gets closer and closer to the exact value of the integral, which was found to be $\frac{22}{3} \approx 7.3333$.

Programming the Midpoint Rule

The easiest way to use the Midpoint Rule to approximate the definite integral $\int_a^b f(x)\,dx$ is to program it into a computer or programmable calculator. For instance, the following program will evaluate the Midpoint Rule on a *TI-81* or *TI-82* calculator. See Appendix D for other models.

```
Prgm1:MIDPOINT
:Disp ''LOWER LIMIT''        Prompt for value of a.
:Input A                     Input value of a.
:Disp ''UPPER LIMIT''        Prompt for value of b.
:Input B                     Input value of b.
:Disp ''N DIVISIONS''        Prompt for value of n.
:Input N                     Input value of n.
:0→S                         Initialize sum of areas.
:(B−A)/N→W                   Calculate width of subinterval.
:1→J                         Initialize counter.
:Lbl 1                       Begin loop.
:A+(J−1)W→L                  Calculate left endpoint.
:A+JW→R                      Calculate right endpoint.
:(L+R)/2→X                   Calculate midpoint of subinterval.
:S+WY₁ →S                    Add area to sum.
:IS>(J,N)                    Test counter.
:Goto 1                      End loop.
:Disp ''APPROXIMATION''
:Disp S                      Display approximation.
```

Before executing the program, enter the function as Y_1. When the program is executed, you will be prompted to enter the lower and upper limits of integration, and the number of subintervals you want to use.

With most integrals, you can determine the accuracy of the approximation by using increasingly larger values of n. For example, if you use the program to approximate the value of $\int_0^2 \sqrt{x^3 + 1}\,dx$, you will obtain the following approximations.

n	10	20	30	40	50	60	70
Approximation	3.238	3.240	3.241	3.241	3.241	3.241	3.241

Thus, it seems clear that the approximation of 3.241 is accurate to three decimal places.

EXAMPLE 2 *Using the Midpoint Rule*

Use the Midpoint Rule with $n = 5$ to approximate

$$\int_0^1 \frac{1}{x^2 + 1} \, dx.$$

Solution

With $n = 5$, the interval $[0, 1]$ is divided into five subintervals.

$$\left[0, \frac{1}{5}\right], \quad \left[\frac{1}{5}, \frac{2}{5}\right], \quad \left[\frac{2}{5}, \frac{3}{5}\right], \quad \left[\frac{3}{5}, \frac{4}{5}\right], \quad \left[\frac{4}{5}, 1\right]$$

The midpoints of these intervals are $\frac{1}{10}$, $\frac{3}{10}$, $\frac{5}{10}$, $\frac{7}{10}$, and $\frac{9}{10}$. Because each subinterval has a width of $\Delta x = (1 - 0)/5 = \frac{1}{5}$, you can approximate the value of the definite integral as follows.

$$\int_0^1 \frac{1}{x^2 + 1} \, dx = \frac{1}{5}\left(\frac{1}{1.01} + \frac{1}{1.09} + \frac{1}{1.25} + \frac{1}{1.49} + \frac{1}{1.81}\right)$$

$$\approx 0.786$$

The region whose area is represented by the definite integral is shown in Figure 5.22. The actual area of this region is $\pi/4 \approx 0.785$. Thus, the approximation is off by only 0.001.

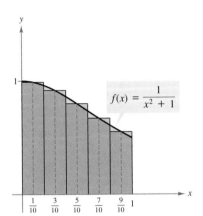

$f(x) = \dfrac{1}{x^2 + 1}$

FIGURE 5.22

EXAMPLE 3 *Using the Midpoint Rule*

Use the Midpoint Rule with $n = 10$ to approximate $\displaystyle\int_1^3 \sqrt{x^2 + 1} \, dx.$

Solution

Begin by dividing the interval $[1, 3]$ into ten subintervals. The midpoints of these intervals are

$$\frac{11}{10}, \quad \frac{13}{10}, \quad \frac{3}{2}, \quad \frac{17}{10}, \quad \frac{19}{10}, \quad \frac{21}{10}, \quad \frac{23}{10}, \quad \frac{5}{2}, \quad \frac{27}{10}, \quad \text{and} \quad \frac{29}{10}.$$

Because each subinterval has a width of $\Delta x = (3 - 1)/10 = \frac{1}{5}$, you can approximate the value of the definite integral as follows.

$$\int_1^3 \sqrt{x^2 + 1} \, dx = \frac{1}{5}\left[\sqrt{(1.1)^2 + 1} + \sqrt{(1.3)^2 + 1} + \cdots + \sqrt{(2.9)^2 + 1}\right]$$

$$\approx 4.504$$

The region whose area is represented by the definite integral is shown in Figure 5.23. Using techniques that are not within the scope of this course, it can be shown that the actual area is

$$\frac{1}{2}\left[3\sqrt{10} + \ln\left(3 + \sqrt{10}\right) - \sqrt{2} - \ln\left(1 + \sqrt{2}\right)\right] \approx 4.505.$$

Thus, the approximation is off by only 0.001.

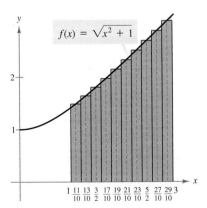

$f(x) = \sqrt{x^2 + 1}$

FIGURE 5.23

The Definite Integral as the Limit of a Sum

Consider the closed interval $[a, b]$, divided into n subintervals whose midpoints are x_i and whose widths are $\Delta x = (b - a)/n$. In this section you have seen that the midpoint approximation

$$\int_a^b f(x)\,dx \approx f(x_1)\Delta x + f(x_2)\Delta x + f(x_3)\Delta x + \cdots + f(x_n)\Delta x$$

$$= [f(x_1) + f(x_2) + f(x_3) + \cdots + f(x_n)]\Delta x$$

becomes better and better as n increases. In fact, the limit of this sum as n approaches infinity is exactly equal to the definite integral. That is,

$$\int_a^b f(x)\,dx = \lim_{n \to \infty} [f(x_1) + f(x_2) + f(x_3) + \cdots + f(x_n)]\Delta x.$$

It can be shown that this limit is valid as long as x_i is *any* point in the ith interval.

EXAMPLE 4 *Approximating a Definite Integral*

Use a computer or programmable calculator to approximate the definite integral

$$\int_0^1 e^{-x^2}\,dx.$$

Solution

Using the program on page 364, with $n = 10, 20, 30, 40$, and 50, it appears that the value of the integral is approximately 0.7468. If you have access to a computer or calculator with a built-in program to approximate definite integrals, try using it to approximate this integral. When we used *Derive* to approximate the integral, we obtained 0.746824. ■

Discussion Problem A Failure of the Midpoint Rule

Suppose you use the Midpoint Rule to approximate the definite integral

$$\int_0^1 \frac{1}{x^2}\,dx.$$

The table shows the approximations that are obtained with $n = 10, 20, 30, 40, 50$, and 60.

n	10	20	30	40	50	60
Approximation	48.3	97.7	147.0	196.4	245.7	295.1

Why are the approximations getting larger and larger? Why isn't the Midpoint Rule working?

Warm Up

The following warm-up exercises involve skills that were covered in earlier sections. You will use these skills in the exercise set for this section.

In Exercises 1–6, find the midpoint of the interval.

1. $[0, \frac{1}{3}]$

2. $[\frac{1}{10}, \frac{2}{10}]$

3. $[\frac{3}{20}, \frac{4}{20}]$

4. $[1, \frac{7}{6}]$

5. $[2, \frac{31}{15}]$

6. $[\frac{26}{9}, 3]$

In Exercises 7–10, find the limit.

7. $\lim\limits_{x \to \infty} \dfrac{2x^2 + 4x - 1}{3x^2 - 2x}$

8. $\lim\limits_{x \to \infty} \dfrac{4x + 5}{7x - 5}$

9. $\lim\limits_{x \to \infty} \dfrac{x - 7}{x^2 + 1}$

10. $\lim\limits_{x \to \infty} \dfrac{5x^3 + 1}{x^3 + x^2 + 4}$

E X E R C I S E S 5.6   means that technology can help you solve or check the exercise(s).

In Exercises 1–4, use the Midpoint Rule with $n = 4$ to approximate the area of the region. Compare your result with the exact area obtained with a definite integral.

1. $f(x) = -2x + 3,\ [0, 1]$ **2.** $f(x) = \sqrt{x} + 1,\ [0, 2]$

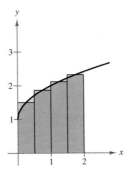

3. $f(x) = \sqrt{x},\ [0, 1]$ **4.** $f(x) = 1 - x^2,\ [-1, 1]$

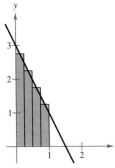

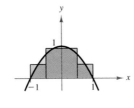

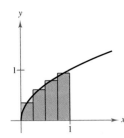

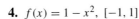

In Exercises 5–14, use the Midpoint Rule with $n = 4$ to approximate the area of the region bounded by the graph of f and the x-axis over the interval. Compare your result with the exact area. Sketch the region.

Function	*Interval*
5. $f(x) = x^2 + 2$	$[-1, 1]$
6. $f(x) = 3x - 4$	$[2, 5]$
7. $f(x) = 2x^2$	$[1, 3]$
8. $f(x) = 2x - x^3$	$[0, 1]$
9. $f(x) = x^3 - 1$	$[1, 2]$
10. $f(x) = x^2 - x^3$	$[0, 1]$
11. $f(x) = x^2 - x^3$	$[-1, 0]$
12. $f(x) = 2x^2 - x + 1$	$[0, 2]$
13. $f(x) = x(1 - x)^2$	$[0, 1]$
14. $f(x) = x^2(3 - x)$	$[0, 3]$

In Exercises 15 and 16, use a program similar to that on page 364 to approximate the area of the region. How large must n be to obtain an approximation that is correct to within 0.01?

15. $\displaystyle\int_0^4 (2x^2 + 3)\, dx$ **16.** $\displaystyle\int_0^4 (2x^3 + 3)\, dx$

In Exercises 17 and 18, use the Midpoint Rule with $n = 4$ to approximate the area of the region. Compare your result with the exact area obtained with a definite integral.

17. $f(y) = \frac{1}{4}y$, $[2, 4]$

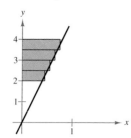

18. $f(y) = 4y - y^2$, $[0, 4]$

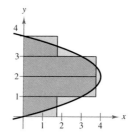

In Exercises 19–22, use the Midpoint Rule to complete the table.

Number of subintervals	2	4	6	8	10
Approximation	?	?	?	?	?

19. $\displaystyle\int_0^4 \sqrt{16 - x^2}\, dx$

20. $\displaystyle\int_{-1}^1 (3-x)\sqrt{1 - x^2}\, dx$

21. $\displaystyle\int_0^2 \sqrt{1 + x^3}\, dx$

22. $\displaystyle\int_0^2 x\sqrt{1 + x^3}\, dx$

Trapezoidal Rule In Exercises 23 and 24, use the Trapezoidal Rule with $n = 8$ to approximate the definite integral. Compare the result with the exact value and the approximation obtained with $n = 8$ and the Midpoint Rule. Which approximation technique appears to be better? Let f be continuous on $[a, b]$ and let b be the number of equal subintervals (see figure). Then the Trapezoidal Rule for approximating $\int_a^b f(x)\, dx$ is

$$\frac{b - a}{2n}[f(x_0) + 2f(x_1) + \cdots + 2f(x_{n-1}) + f(x_n)].$$

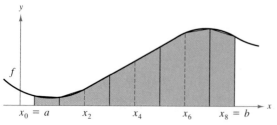

FIGURE FOR 23 AND 24

23. $\displaystyle\int_0^2 x^3\, dx$

24. $\displaystyle\int_1^3 \frac{1}{x^2}\, dx$

In Exercises 25–28, use the Trapezoidal Rule to approximate the definite integral with $n = 4$.

25. $\displaystyle\int_0^2 \frac{1}{x + 1}\, dx$

26. $\displaystyle\int_0^4 \sqrt{1 + x^2}\, dx$

27. $\displaystyle\int_{-1}^1 \frac{1}{x^2 + 1}\, dx$

28. $\displaystyle\int_1^5 \frac{\sqrt{x - 1}}{x}\, dx$

In Exercises 29 and 30, use a computer or programmable calculator to complete the table.

n	Midpoint rule	Trapezoidal rule
4		
8		
12		
16		
20		

29. $\displaystyle\int_0^4 \sqrt{2 + 3x^2}\, dx$

30. $\displaystyle\int_0^2 \frac{5}{x^3 + 1}\, dx$

In Exercises 31 and 32, use the Trapezoidal Rule with $n = 10$ to approximate the area of the region bounded by the graphs of the equations.

31. $y = \sqrt{\dfrac{x^3}{4 - x}}$, $y = 0$, $x = 3$

32. $y = x\sqrt{\dfrac{4 - x}{4 + x}}$, $y = 0$, $x = 4$

33. *Velocity and Acceleration* The tables lists the velocity v (in feet per second) of an accelerating car over a 20-second interval. Use the Trapezoidal Rule to approximate the distance in feet that the car travels during the 20 seconds. (The distance is given by $s = \int_0^{20} v\, dt$.)

Time, t	0	5	10	15	20
Velocity, v	0.0	29.3	51.3	66.0	73.3

Volumes of Solids of Revolution

The Disc Method ■ *The Washer Method* ■ *Applications*

The Disc Method

Plane region

Axis of revolution

FIGURE 5.24

As shown in Figure 5.24, a **solid of revolution** is formed by revolving a plane region about a line. The line is called the **axis of revolution.**

To develop a formula for finding the volume of a solid of revolution, consider a continuous function f that is nonnegative on the interval $[a, b]$. Suppose that the area of the region is approximated by n rectangles, each of width Δx, as shown in Figure 5.25. By revolving the rectangles about the x-axis, you obtain n circular discs, each with a volume of $\pi [f(x_i)]^2 \Delta x$. The volume of the solid formed by revolving the region about the x-axis is approximately equal to the sum of the volumes of the n discs. Moreover, by taking the limit as n approaches infinity, you can see that the exact volume is given by a definite integral. This result is called the **Disc Method.**

The Disc Method

The volume of the solid formed by revolving the region bounded by the graph of f and the x-axis ($a \le x \le b$) about the x-axis is

$$\text{Volume} = \pi \int_a^b [f(x)]^2 \, dx.$$

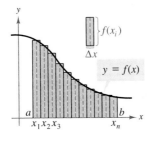

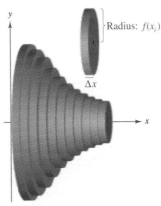

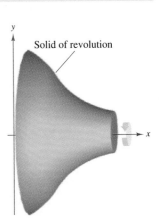

Approximation by n rectangles

Approximation by n discs

FIGURE 5.25

369

EXAMPLE 1 *Finding the Volume of a Solid of Revolution*

Find the volume of the solid formed by revolving the region bounded by the graph of $f(x) = -x^2 + x$ and the x-axis about the x-axis.

Solution

Begin by sketching the region bounded by the graph of f and the x-axis. As shown in Figure 5.26(a), sketch a representative rectangle whose height is $f(x)$ and whose width is Δx. From this rectangle, you can see that the radius of the solid is

$$\text{Radius} = f(x) = -x^2 + x.$$

Using the Disc Method, you can find the volume of the solid of revolution.

$$\begin{aligned}
\text{Volume} &= \pi \int_0^1 [f(x)]^2 \, dx & &\textit{Disc Method}\\[2mm]
&= \pi \int_0^1 (-x^2 + x)^2 \, dx & &\textit{Substitute for } f(x).\\[2mm]
&= \pi \int_0^1 (x^4 - 2x^3 + x^2) \, dx & &\textit{Expand integrand.}\\[2mm]
&= \pi \left[\frac{x^5}{5} - \frac{x^4}{2} + \frac{x^3}{3} \right]_0^1 & &\textit{Find antiderivative.}\\[2mm]
&= \frac{\pi}{30} & &\textit{Apply Fundamental Theorem.}\\[2mm]
&\approx 0.105 & &\textit{Round to three decimal places.}
\end{aligned}$$

Thus, the volume of the solid is about 0.105 cubic units.

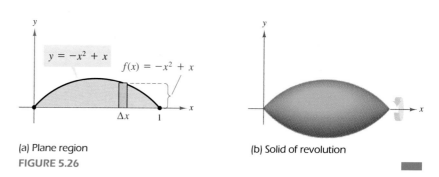

(a) Plane region   (b) Solid of revolution
FIGURE 5.26

Note In Example 1, the entire problem was solved *without* referring to the three-dimensional sketch given in Figure 5.26(b). In general, to set up the integral for calculating the volume of a solid of revolution, a sketch of the plane region is more useful than a sketch of the solid, because the radius is more readily visualized in the plane region.

The Washer Method

You can extend the Disc Method to find the volume of a solid of revolution with a *hole*. Consider a region that is bounded by the graphs of f and g, as shown in Figure 5.27(a). If the region is revolved about the x-axis, then the volume of the resulting solid can be found by applying the Disc Method to f and g and subtracting the results.

$$\text{Volume} = \pi \int_a^b [f(x)]^2 \, dx - \pi \int_a^b [g(x)]^2 \, dx$$

Writing this as a single integral produces the **Washer Method.**

The Washer Method

Let f and g be continuous and nonnegative on the closed interval $[a, b]$, as shown in Figure 5.27(a). If $g(x) \le f(x)$ for all x in the interval, then the volume of the solid formed by revolving the region bounded by the graphs of f and g $(a \le x \le b)$ about the x-axis is

$$\text{Volume} = \pi \int_a^b ([f(x)]^2 - [g(x)]^2) \, dx.$$

$f(x)$ is the **outer radius** and $g(x)$ is the **inner radius.**

In Figure 5.27(b), note that the solid of revolution has a hole. Moreover, the radius of the hole is $g(x)$, the inner radius.

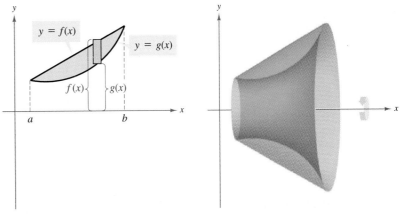

(a) Plane region

(b) Solid of revolution with hole

FIGURE 5.27

EXAMPLE 2 *Using the Washer Method*

Find the volume of the solid formed by revolving the region bounded by the graphs of $f(x) = \sqrt{25 - x^2}$ and $g(x) = 3$ about the x-axis (see Figure 5.28).

Solution

First find the points of intersection of f and g by setting $f(x)$ equal to $g(x)$ and solving for x.

$$f(x) = g(x) \qquad \text{Set } f(x) \text{ equal to } g(x).$$
$$\sqrt{25 - x^2} = 3 \qquad \text{Substitute for } f(x) \text{ and } g(x).$$
$$25 - x^2 = 9 \qquad \text{Square both sides.}$$
$$16 = x^2$$
$$\pm 4 = x \qquad \text{Solve for } x.$$

Using $f(x)$ as the outer radius and $g(x)$ as the inner radius, you can find the volume of the solid as follows.

$$\text{Volume} = \pi \int_{-4}^{4} ([f(x)]^2 - [g(x)]^2)\, dx \qquad \text{Washer Method}$$

$$= \pi \int_{-4}^{4} \left[\left(\sqrt{25 - x^2} \right)^2 - (3)^2 \right] dx \quad \text{Substitute for } f(x) \text{ and } g(x).$$

$$= \pi \int_{-4}^{4} (16 - x^2)\, dx \qquad \text{Simplify.}$$

$$= \pi \left[16x - \frac{x^3}{3} \right]_{-4}^{4} \qquad \text{Find antiderivative.}$$

$$= \frac{256\pi}{3} \qquad \text{Apply Fundamental Theorem.}$$

$$\approx 268.08 \qquad \text{Round to two decimal places.}$$

Thus, the volume of the solid is about 268.08 cubic inches.

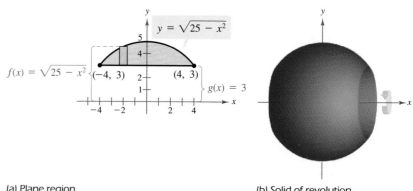

(a) Plane region

(b) Solid of revolution

FIGURE 5.28

Applications

American football, in its modern form, is a twentieth-century invention. In the 1800s a rough, soccer-like game was played with a "round football." In 1905, at the request of President Theodore Roosevelt, the Intercollegiate Athletic Association (which became the NCAA in 1910) was formed. With the introduction of the forward pass in 1906, the shape of the ball was altered to make it easier to grip.

Real Life

EXAMPLE 3 *Finding a Football's Volume*

A regulation-size football can be modeled as a solid of revolution formed by revolving the graph of

$$f(x) = -0.0944x^2 + 3.4, \qquad -5.5 \le x \le 5.5$$

about the x-axis, as shown in Figure 5.29. Use this model to find the volume of a football. (In the model, x and y are measured in inches.)

Solution

To find the volume of the solid of revolution, use the Disc Method.

$$\text{Volume} = \pi \int_{-5.5}^{5.5} [f(x)]^2 \, dx \qquad \textit{Disc Method}$$

$$= \pi \int_{-5.5}^{5.5} (-0.0944x^2 + 3.4)^2 \, dx \quad \textit{Substitute for } f(x).$$

$$\approx 232 \text{ cubic inches} \qquad \textit{Volume}$$

Thus, the volume of the football is about 232 cubic inches.

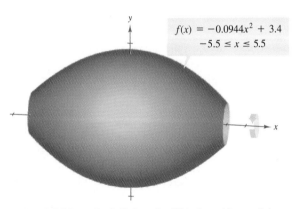

$$f(x) = -0.0944x^2 + 3.4$$
$$-5.5 \le x \le 5.5$$

FIGURE 5.29 A football-shaped solid is formed by revolving a parabolic segment about the x-axis.

Discussion Problem Testing the Reasonableness of an Answer

A football is about 11 inches long and has a diameter of about 7 inches. In Example 3, the volume of a football was approximated to be 232 cubic inches. Explain how you can determine whether this answer is reasonable.

Warm Up

The following warm-up exercises involve skills that were covered in earlier sections. You will use these skills in the exercise set for this section.

In Exercises 1–6, solve for x.

1. $x^2 = 2x$

2. $-x^2 + 4x = x^2$

3. $x = -x^3 + 5x$

4. $x^2 + 1 = x + 3$

5. $-x + 4 = \sqrt{4x - x^2}$

6. $\sqrt{x - 1} = \frac{1}{2}(x - 1)$

In Exercises 7–10, evaluate the integral.

7. $\displaystyle\int_0^2 2e^{2x}\, dx$

8. $\displaystyle\int_{-1}^3 \frac{2x + 1}{x^2 + x + 2}\, dx$

9. $\displaystyle\int_0^2 x\sqrt{x^2 + 1}\, dx$

10. $\displaystyle\int_1^5 \frac{(\ln x)^2}{x}\, dx$

EXERCISES 5.7   means that technology can help you solve or check the exercise(s).

In Exercises 1–16, find the volume of the solid formed by revolving the region bounded by the graph(s) of the equation(s) about the x-axis.

1. $y = \sqrt{4 - x^2}$

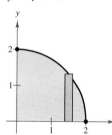

2. $y = x^2$

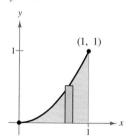

3. $y = \sqrt{x}$

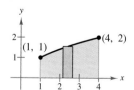

4. $y = \sqrt{4 - x^2}$

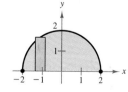

5. $y = 4 - x^2$, $y = 0$

6. $y = x$, $y = 0$, $x = 4$

7. $y = 1 - \frac{1}{4}x^2$, $y = 0$

8. $y = x^2 + 1$, $y = 5$

9. $y = -x + 1$, $y = 0$, $x = 0$

10. $y = x$, $y = e^{x-1}$, $x = 0$

11. $y = \dfrac{1}{x} - \dfrac{1}{2}$, $y = -\dfrac{1}{2}x + 1$

12. $y = \sqrt{x}$, $y = 0$, $x = 4$

13. $y = 2x^2$, $y = 0$, $x = 2$

14. $y = \dfrac{1}{x}$, $y = 0$, $x = 1$, $x = 3$

15. $y = e^x$, $y = 0$, $x = 0$, $x = 1$

16. $y = x^2$, $y = 4x - x^2$

In Exercises 17–24, find the volume of the solid formed by revolving the region bounded by the graph(s) of the equation(s) about the y-axis.

17. $y = x^2$, $y = 4$, $0 \le x \le 2$

18. $y = \sqrt{16 - x^2}$, $y = 0$, $0 \le x \le 4$

19. $x = 1 - \frac{1}{2}y$, $x = 0$, $y = 0$

20. $x = y(y - 1)$, $x = 0$

21. $y = x^{2/3}$

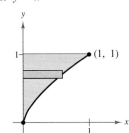

22. $x = -y^2 + 4y$

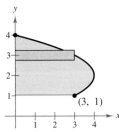

23. $y = \sqrt{4 - x}, \; y = 0, \; x = 0$

24. $y = 4, \; y = 0, \; x = 2, \; x = 0$

25. The line segment from $(0, 0)$ to $(6, 3)$ is revolved about the x-axis to form a cone. What is the volume of the cone?

26. Use the Disc Method to verify that the volume of a right circular cone is $\frac{1}{3}\pi r^2 h$, where r is the radius of the base and h is the height.

27. Use the Disc Method to verify that the volume of a sphere of radius r is $\frac{4}{3}\pi r^3$.

28. The right half of the ellipse $9x^2 + 25y^2 = 225$ is revolved about the y-axis to form an oblate spheroid (shaped like an M&M candy). Find the volume of the spheroid.

29. The upper half of the ellipse $9x^2 + 16y^2 = 144$ is revolved about the x-axis to form a prolate spheroid (shaped like a football). Find the volume of the spheroid.

30. *Fuel Tank* A tank on the wing of a jet is modeled by revolving the region bounded by the graph of $y = \frac{1}{8}x^2\sqrt{2 - x}$ and the x-axis about the x-axis, where x and y are measured in meters (see figure). Find the volume of the tank.

Jet Wing Tank

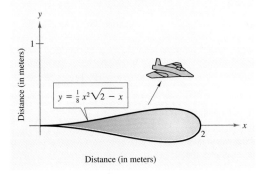

Distance (in meters)

31. *Fish Population* A pond is to be stocked with a species of fish. The food supply in 500 cubic feet of pond water can adequately support one fish. The pond is nearly circular, is 20 feet deep at its center, and has a radius of 200 feet. The bottom of the pond can be modeled by

$$y = 20[(0.005x)^2 - 1].$$

(a) How much water is in the pond?

(b) Estimate the maximum number of fish the pond can support.

32. *Wedding Ring* A jewelry manufacturer plans to make gold wedding rings from a gold alloy that costs $50,000 per cubic inch. A cross section of the ring is bounded by the graphs of

$$y = \frac{7}{16} - 16x^2 \quad \text{and} \quad y = \frac{3}{8},$$

as shown in the figure.

(a) Find the volume (in cubic inches) of a ring.

(b) Estimate the cost of material for each ring.

Model of a Gold Ring

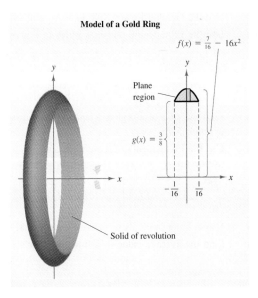

In Exercises 33 and 34, use the Midpoint Rule with $n = 4$ to approximate the volume of the solid generated by revolving the region bounded by the graphs of the equations about the x-axis.

33. $y = \sqrt[3]{x + 1}, \; y = 0, \; x = 0, \; x = 7$

34. $y = \dfrac{10}{x^2 + 1}, \; y = 0, \; x = 0, \; x = 3$

Chapter Summary and Study Tips

After studying this chapter, you should have acquired the following skills. The exercise numbers are keyed to the Review Exercises that begin on page 378. Answers to odd-numbered Review Exercises are given in the back of the text.

■ Use basic integration formulas to find antiderivatives. (*Section 5.1*) Review Exercises 1–8

$$\int k\,dx = kx + C \qquad\qquad \int [f(x) - g(x)]\,dx = \int f(x)\,dx - \int g(x)\,dx$$

$$\int kf(x)\,dx = k\int f(x)\,dx \qquad\qquad \int x^n\,dx = \frac{x^{n+1}}{n+1} + C, \qquad n \neq -1$$

$$\int [f(x) + g(x)]\,dx = \int f(x)\,dx + \int g(x)\,dx$$

■ Use initial conditions to find particular solutions of indefinite integrals. (*Section 5.1*) Review Exercises 9–12
■ Use antiderivatives to solve real-life problems. (*Section 5.1*) Review Exercises 13, 14
■ Use the General Power Rule to find antiderivatives. (*Section 5.2*) Review Exercises 15–22

$$\int u^n \frac{du}{dx}\,dx = \frac{u^{n+1}}{n+1} + C, \qquad n \neq -1$$

■ Use the General Power Rule to solve real-life problems. (*Section 5.2*) Review Exercises 23, 24
■ Use the Exponential and Log Rules to find antiderivatives. (*Section 5.3*) Review Exercises 25–32

$$\int e^x\,dx = e^x + C \qquad\qquad \int \frac{1}{x}\,dx = \ln|x| + C$$

$$\int e^u \frac{du}{dx}\,dx = \int e^u\,du = e^u + C \qquad \int \frac{u'}{u}\,dx = \int \frac{1}{u}\,du = \ln|u| + C$$

■ Find the area of a region bounded by the graph of a function and the *x*-axis. (*Section 5.4*) Review Exercises 33–38
■ Use the Fundamental Theorem of Calculus to evaluate a definite integral. (*Section 5.4*) Review Exercises 39–50

$$\int_a^b f(x)\,dx = F(x)\Big]_a^b = F(b) - F(a), \qquad \text{where } F'(x) = f(x)$$

■ Use definite integrals to solve marginal analysis problems. (*Section 5.4*) Review Exercises 51, 52
■ Find the average value of a function over a closed interval. (*Section 5.4*) Review Exercises 53–56

$$\text{Average value} = \frac{1}{b-a}\int_a^b f(x)\,dx$$

■ Use average values to solve real-life problems. (*Section 5.4*) Review Exercises 57, 58

*Several student study aids are available with this text. The *Student Solutions Guide* includes detailed solutions to all odd-numbered exercises, as well as practice chapter tests with answers. The *Graphics Calculator Guide* offers instructions on the use of a variety of graphing calculators and computer graphing software. The *Brief Calculus TUTOR* includes additional examples for selected exercises in the text.

■ Use properties of even and odd functions to help evaluate definite integrals. *(Section 5.4)*
Review Exercises 59–62

Even function: $f(-x) = f(x)$ Odd function: $f(-x) = -f(x)$

■ Find the area of a region bounded by two graphs. *(Section 5.5)*
Review Exercises 63–70

$$A = \int_a^b [f(x) - g(x)]\, dx$$

■ Find consumer and producer surplus. *(Section 5.5)*
Review Exercises 71, 72

■ Use the area of a region bounded by two graphs to solve real-life problems. *(Section 5.5)*
Review Exercises 73, 74

■ Use the Midpoint Rule to approximate the value of a definite integral. *(Section 5.6)*
Review Exercises 75–78

$$\int_a^b f(x)\, dx \approx \frac{b-a}{n}[f(x_1) + f(x_2) + f(x_3) + \cdots + f(x_n)]$$

■ Use the Disc Method to find the volume of a solid of revolution. *(Section 5.7)*
Review Exercises 79–82

$$\text{Volume} = \pi \int_a^b [f(x)]^2\, dx$$

■ Use the Washer Method to find the volume of a solid of revolution with a hole. *(Section 5.7)*
Review Exercises 83–86

$$\text{Volume} = \pi \int_a^b ([f(x)]^2 - [g(x)]^2)\, dx$$

■ Use solids of revolution to solve real-life problems. *(Section 5.7)*
Review Exercises 87, 88

■ Indefinite and Definite Integrals   When evaluating integrals, remember that an indefinite integral is a *family of antiderivatives*, each differing by a constant C, whereas a definite integral is a number.

■ Check Antiderivatives by Differentiating   When finding an antiderivative, remember that you can check your result by differentiating. For example, you can check that the antiderivative

$$\int (3x^3 - 4x)\, dx = \frac{3}{4}x^4 - 2x^2 + C$$

is correct by differentiating to obtain

$$\frac{d}{dx}\left[\frac{3}{4}x^4 - 2x^2 + C\right] = 3x^3 - 4x.$$

Because the derivative is equal to the original integrand, you know that the antiderivative is correct.

■ Grouping Symbols and the Fundamental Theorem   When using the Fundamental Theorem of Calculus to evaluate a definite integral, you can avoid sign errors by using grouping symbols. Here is an example.

$$\int_1^3 (x^3 - 9x)\, dx = \left[\frac{x^4}{4} - \frac{9x^2}{2}\right]_1^3 = \left[\frac{3^4}{4} - \frac{9(3^2)}{2}\right] - \left[\frac{1^4}{4} - \frac{9(1^2)}{2}\right] = \frac{81}{4} - \frac{81}{2} - \frac{1}{4} + \frac{9}{2} = -16$$

Review Exercises

 means that technology can help you solve or check the exercise(s).

In Exercises 1–8, find the indefinite integral.

1. $\int 16\,dx$

2. $\int \frac{3}{5}x\,dx$

3. $\int (2x^2 + 5x)\,dx$

4. $\int (5 - 6x^2)\,dx$

5. $\int \frac{2}{3\sqrt[3]{x}}\,dx$

6. $\int 6x^2\sqrt{x}\,dx$

7. $\int (\sqrt[3]{x^4} + 3x)\,dx$

8. $\int \left(\frac{4}{\sqrt{x}} + \sqrt{x}\right)\,dx$

In Exercises 9–12, find the particular solution, $y = f(x)$, that satisfies the given conditions.

9. $f'(x) = 3x + 1$, $f(2) = 6$

10. $f'(x) = x^{-1/3} - 1$, $f(8) = 4$

11. $f''(x) = 2x^2$, $f'(3) = 10$, $f(3) = 6$

12. $f''(x) = \frac{6}{\sqrt{x}} + 3$, $f'(1) = 12$, $f(4) = 56$

13. *Vertical Motion* An object is projected upward from the ground with an initial velocity of 80 feet per second.

(a) How long does it take the object to rise to its maximum height?

(b) What is the maximum height?

(c) When is the velocity of the object half of its initial velocity?

(d) What is the height of the object when its velocity is one-half the initial velocity?

14. *Forecasting* The weekly revenue for a new product has been increasing. The rate of change of the revenue can be modeled by

$$\frac{dR}{dt} = 0.675t^{3/2}, \qquad 0 \le t \le 225,$$

where t is the time (in weeks). When $t = 0$, $R = 0$.

(a) Find a model for the revenue function.

(b) When will the weekly revenue be $27,000?

In Exercises 15–22, find the indefinite integral.

15. $\int (1 + 5x)^2\,dx$

16. $\int (x - 6)^{4/3}\,dx$

17. $\int \frac{1}{\sqrt{5x - 1}}\,dx$

18. $\int \frac{4x}{\sqrt{1 - 3x^2}}\,dx$

19. $\int x(1 - 4x^2)\,dx$

20. $\int \frac{x^2}{(x^3 - 4)^2}\,dx$

21. $\int (x^4 - 2x)(2x^3 - 1)\,dx$

22. $\int \frac{\sqrt{x}}{(1 - x^{3/2})^3}\,dx$

23. *Production* The output P (in board feet) of a small sawmill changes according to the model

$$\frac{dP}{dt} = 2t(0.001t^2 + 0.5)^{1/4}, \qquad 0 \le t \le 40,$$

where t is measured in hours. Find the number of board feet produced in (a) 6 hours and (b) 12 hours.

24. *Cost* The marginal cost for a catering service to cater to x people can be modeled by

$$\frac{dC}{dx} = \frac{5x}{\sqrt{x^2 + 1000}}.$$

When $x = 225$, the cost is $1136.06. Find the cost of catering for (a) 500 people and (b) 1000 people.

In Exercises 25–32, find the indefinite integral.

25. $\int 3e^{-3x}\,dx$

26. $\int (2t - 1)e^{t^2 - t}\,dt$

27. $\int (x - 1)e^{x^2 - 2x}\,dx$

28. $\int \frac{4}{6x - 1}\,dx$

29. $\int \frac{x^2}{1 - x^3}\,dx$

30. $\int \frac{x - 4}{x^2 - 8x}\,dx$

31. $\int \frac{(\sqrt{x} + 1)^2}{\sqrt{x}}\,dx$

32. $\int \frac{e^{5x}}{5 + e^{5x}}\,dx$

In Exercises 33–38, find the area of the region.

33. $f(x) = 4 - 2x$

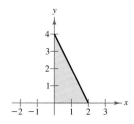

34. $f(x) = 4 - x^2$

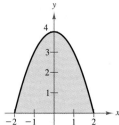

35. $f(y) = (y - 2)^2$

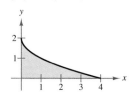

36. $f(x) = \sqrt{9 - x^2}$

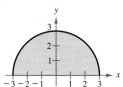

37. $f(x) = \dfrac{2}{x+1}$

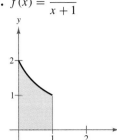

38. $f(x) = 2xe^{x^2-4}$

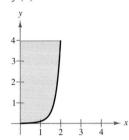

In Exercises 39–50, use the Fundamental Theorem of Calculus to evaluate the definite integral.

39. $\displaystyle\int_0^4 (2+x)\,dx$

40. $\displaystyle\int_{-1}^1 (t^2+2)\,dt$

41. $\displaystyle\int_{-1}^1 (4t^3 - 2t)\,dt$

42. $\displaystyle\int_1^4 2x\sqrt{x}\,dx$

43. $\displaystyle\int_0^3 \dfrac{1}{\sqrt{1+x}}\,dx$

44. $\displaystyle\int_3^6 \dfrac{x}{3\sqrt{x^2-8}}\,dx$

45. $\displaystyle\int_1^2 \left(\dfrac{1}{x^2} - \dfrac{1}{x^3}\right)dx$

46. $\displaystyle\int_0^1 x^2(x^3+1)^3\,dx$

47. $\displaystyle\int_1^3 \dfrac{(3+\ln x)}{x}\,dx$

48. $\displaystyle\int_0^{\ln 5} e^{x/5}\,dx$

49. $\displaystyle\int_{-1}^1 3xe^{x^2-1}\,dx$

50. $\displaystyle\int_1^3 \dfrac{1}{x(\ln x + 2)^2}\,dx$

 51. *Cost* The marginal cost for a typical additional client at a law firm can be modeled by

$$\dfrac{dC}{dx} = 675 + 0.5x,$$

where x is the number of clients. How does the cost C change when x increases from 50 to 51 clients?

 52. *Profit* The marginal profit obtained by selling x dollars of automobile insurance can be modeled by

$$\dfrac{dP}{dx} = 0.4\left(1 - \dfrac{5000}{x}\right), \qquad x \ge 5000.$$

Find the change in the profit when x increases from \$75,000 to \$100,000.

In Exercises 53–56, find the average value of the function on the closed interval. Then find the values of x in the interval where the function assumes its average value.

53. $f(x) = \dfrac{4}{\sqrt{x-1}}$, $[5, 10]$
54. $f(x) = \dfrac{20\ln x}{x}$, $[2, 10]$

55. $f(x) = e^{5-x}$, $[2, 5]$
56. $f(x) = x^3$, $[0, 2]$

57. *Checking Account* An interest-bearing checking account yields 4% interest compounded continuously. If you deposit \$500 into such an account, and never write checks, what will the average value of the account be over a period of 2 years? Explain your reasoning.

58. *Fuel Cost* Suppose that the price p of gasoline can be modeled by

$$p = 1.00 + 0.1t + 0.02t^2,$$

where $t = 0$ represents January 1, 1990. Find the cost of driving an automobile that is driven 15,000 miles per year and gets 33 miles per gallon from 1990 through 1994.

In Exercises 59–62, explain how the given value can be used to evaluate the second integral.

59. $\displaystyle\int_0^2 6x^5\,dx = 64,\quad \int_{-2}^2 6x^5\,dx$

60. $\displaystyle\int_0^3 (x^4+x^2)\,dx = 57.6,\quad \int_{-3}^3 (x^4+x^2)\,dx$

61. $\displaystyle\int_1^2 \dfrac{4}{x^2}\,dx = 2,\quad \int_{-2}^{-1} \dfrac{4}{x^2}\,dx$

62. $\displaystyle\int_0^1 (x^3-x)\,dx = -\dfrac{1}{4},\quad \int_{-1}^0 (x^3-x)\,dx$

In Exercises 63–70, sketch the region bounded by the graphs of the equations. Then find the area of the region.

63. $y = \dfrac{1}{x^2}$, $y = 4$, $x = 3$

64. $y = x$, $y = 2 - x^2$

65. $y = 1 - \dfrac{1}{2}x$, $y = x - 2$, $y = 1$

66. $y = \dfrac{4}{\sqrt{x+1}}$, $y = 0$, $x = 0$, $x = 8$

67. $y = \sqrt{x}(x - 1)$, $y = 0$

68. $y = x$, $y = x^5$

69. $y = (x-3)^2$, $y = 8 - (x-3)^2$

70. $y = 4 - x$, $y = x^2 - 5x + 8$, $x = 0$

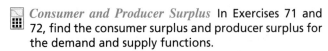

Consumer and Producer Surplus In Exercises 71 and 72, find the consumer surplus and producer surplus for the demand and supply functions.

71. Supply function: $p_2(x) = 500 - x$

Demand function: $p_1(x) = 1.25x + 162.5$

72. Supply function: $p_2(x) = \sqrt{100{,}000 - 0.15x^2}$

Demand function: $p_1(x) = \sqrt{0.01x^2 + 36{,}000}$

73. *Ice Cream Consumption* The per capita consumption y (in pounds) of ice cream in the United States from 1981 through 1987 can be modeled by

$$y = 18.94 - \frac{5.175}{t} + \frac{3.632}{t^{3/2}}, \qquad 1 \le t \le 7,$$

where $t = 1$ represents December 31, 1981. The consumption from 1988 through 1990 can be modeled by

$$y = 0.019t^3 - 0.294t^2 + 26.465, \qquad 7 < t \le 10.$$

If the consumption had continued to follow the first model from 1988 through 1990, how much more or less ice cream would have been consumed? *(Source: U.S. Department of Agriculture)*

74. *Credit Card Sales* The annual sales s (in millions of dollars) for VISA, Mastercard, and American Express from 1983 through 1992 can be modeled by

$s = 15.969t - 6.318$ *VISA*

$s = 8.581t + 6.965$ *Mastercard*

$s = 6.214t + 10.345$ *American Express*

where $3 \le t \le 12$ represents the 10-year period from 1983 through 1992. *(Source: Credit Card News)*

(a) From 1983 through 1992, how much more were VISA's sales than Mastercard's sales?

(b) From 1983 through 1992, how much more were VISA's sales than American Express's sales?

In Exercises 75–78, use the Midpoint Rule with $n = 4$ to approximate the definite integral. Then use a programmable calculator or computer with $n = 20$. Compare the two approximations.

75. $\displaystyle\int_0^2 (x^2 + 1)^2 \, dx$

76. $\displaystyle\int_{-1}^1 \sqrt{1 - x^2} \, dx$

77. $\displaystyle\int_0^1 \frac{1}{x^2 + 1} \, dx$

78. $\displaystyle\int_{-1}^1 e^{3-x^2} \, dx$

In Exercises 79–82, use the Disc Method to find the volume of the solid of revolution formed by revolving the region about the x-axis.

79. $y = \dfrac{1}{\sqrt{x}}$

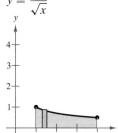

80. $y = \sqrt{16 - x}$

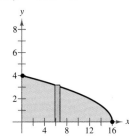

81. $y = e^{1-x}$

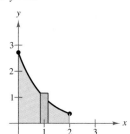

82. $y = \dfrac{1}{x}$

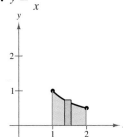

In Exercises 83–86, use the Washer Method to find the volume of the solid of revolution formed by revolving the region about the x-axis.

83. $y = 2x + 1,\ y = 1$

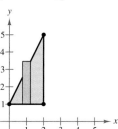

84. $y = \sqrt{x},\ y = 2$

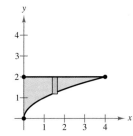

85. $y = x^2,\ y = x^3$

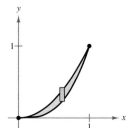

86. $y = 4 - \tfrac{1}{2}x^2,\ y = 2$

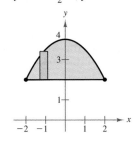

87. *Manufacturing* To create a part for an engine, a manufacturer drills a hole through the center of a metal sphere whose radius is 1 inch. The hole has a radius of 0.25 inch. What is the volume of the resulting ring?

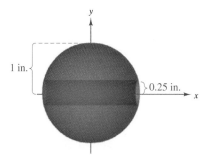

88. *Design* To create a computer design for a funnel, an engineer revolves the region bounded by the lines $y = 3 - x$, $y = 0$, and $x = 0$ about the y-axis, where x and y are measured in feet. Find the volume of the funnel.

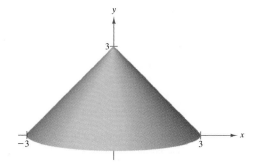

Sample Post-Graduation Exam Questions

The following questions were taken from certified public accountant (CPA) exams, graduate management admission tests (GMAT), graduate records exams (GRE), actuarial exams, or college-level academic skills tests (CLAST). The answers to the questions are given in the back of the book.

For Questions 1–3, use the data given in the table.

Number of Students	Number of Correct Answers
12	45 to 50
11	40 to 44
14	35 to 39
10	30 to 34
7	0 to 29

1. To pass the exam, a student must correctly answer 75% of the questions. What is the maximum number of students that could have passed the exam?
(a) 23   (b) 47   (c) 30   (d) 37

2. What percent of the class answered 40 or more questions correctly?
(a) 69%   (b) 43%   (c) 22%   (d) 31%

3. The number of students who answered 35 to 39 questions correctly is y times the number who answered 29 or fewer correctly, where y is
(a) $\frac{1}{2}$ (b) 1   (c) 2   (d) $\frac{7}{5}$

4. For which of the following statements is $x = -2$ a solution?
 I. $x^2 + 4x + 4 \leq 0$
 II. $|3x + 6| = 0$
 III. $x^2 + 7x + 10 > 0$

(a) I, II, and III   (b) I and II   (c) II and III   (d) I and III

5. The area of the shaded region in the graph is A square units, with A equal to
(a) $\frac{50}{3}$ (b) $\frac{25}{3}$ (c) 50   (d) $\frac{4}{3}$

For Questions 6–8, use the following data.
56, 58, 54, 54, 59, 56, 55, 57, 56, 62

6. The mean of the data is
(a) 55.5   (b) 56.7   (c) 56.5   (d) 56

7. The median of the data is
(a) 55.5   (b) 56.7   (c) 56.5   (d) 56

8. The mode of the data is
(a) 55.5   (b) 56.7   (c) 56.5   (d) 56

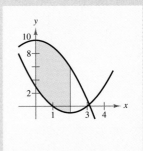

FIGURE FOR 5

$$\int_{-\infty}^{\infty} \frac{1}{\sigma\sqrt{2\pi}}\, e^{-x^2/2\sigma^2}\, dx = 1$$

6

Techniques of Integration

All algebraic, exponential, and logarithmic functions have derivatives that are algebraic, exponential, or logarithmic. This is not true with antiderivatives. For instance, an algebraic function can have an antiderivative that is an inverse trigonometric function. What is even more diabolical is that the antiderivatives of some algebraic, exponential, and logarithmic functions are not elementary functions.

Mathematicians, such as the Russian Sonya Kovalevsky (1850–1891), have devised ingenious techniques for *finding* antiderivatives of those functions that possess elementary antiderivatives, and for *approximating* definite integrals of functions that do not possess elementary antiderivatives.

$$\int u\, dv = uv - \int v\, du$$

Kovalevsky

The Bettmann Archive

383

6.1 Integration by Substitution

Review of Basic Integration Formulas ■ *Integration by Substitution* ■ *Substitution and Definite Integrals* ■ *Applications*

Review of Basic Integration Formulas

Each of the basic integration rules you studied in Chapter 5 was derived from a corresponding differentiation rule. It may surprise you to learn that, although you now have all the necessary tools for *differentiating* algebraic, exponential, and logarithmic functions, your set of tools for *integrating* these functions is by no means complete! The primary objective of this chapter is to develop several techniques that greatly expand the set of integrals to which the basic integration formulas can be applied.

Basic Integration Formulas

1. Constant Rule:
$$\int k \, dx = kx + C$$

2. Simple Power Rule ($n \neq -1$):
$$\int x^n \, dx = \frac{x^{n+1}}{n+1} + C$$

3. General Power Rule ($n \neq -1$):
$$\int u^n u' \, dx = \int u^n \, du = \frac{u^{n+1}}{n+1} + C$$

4. Simple Exponential Rule:
$$\int e^x \, dx = e^x + C$$

5. General Exponential Rule:
$$\int e^u u' \, dx = \int e^u \, du = e^u + C$$

6. Simple Log Rule:
$$\int \frac{1}{x} \, dx = \ln |x| + C$$

7. General Log Rule:
$$\int \frac{u'}{u} \, dx = \int \frac{1}{u} \, du = \ln |u| + C$$

You don't have to work many integration problems before you realize that integration is not nearly as straightforward as differentiation. A major part of any integration problem is determining which basic integration formula (or formulas) to use to solve the problem. This requires remembering the basic formulas, familiarity with various procedures for rewriting integrands in the basic forms, and lots of practice.

Integration by Substitution

There are several techniques for rewriting an integral so that it fits one or more of the basic formulas. One of the most powerful techniques is **integration by substitution.** With this technique, you choose part of the integrand to be u and then rewrite the entire integral in terms of u.

EXAMPLE 1 *Integration by Substitution*

Use the substitution $u = x + 1$ to find the indefinite integral

$$\int \frac{x}{(x + 1)^2} \, dx.$$

Solution

From the substitution $u = x + 1$,

$$x = u - 1, \quad u' = 1, \quad \text{and} \quad dx = du.$$

Replacing *all* instances of x and dx with the appropriate u-variable forms produces the following.

$$\int \frac{x}{(x + 1)^2} \, dx = \int \frac{u - 1}{u^2} \, du \qquad \text{Substitute for } x \text{ and } dx.$$

$$= \int \left(\frac{u}{u^2} - \frac{1}{u^2} \right) du \qquad \text{Write as separate fractions.}$$

$$= \int \left(\frac{1}{u} - \frac{1}{u^2} \right) du \qquad \text{Simplify.}$$

$$= \ln |u| + \frac{1}{u} + C \qquad \text{Find antiderivative.}$$

$$= \ln |x + 1| + \frac{1}{x + 1} + C \qquad \text{Substitute for } u.$$

The basic steps for integration by substitution are outlined in the following guidelines.

Guidelines for Integration by Substitution

1. Let u be a function of x (usually part of the integrand).
2. Solve for x and dx in terms of u and du.
3. Convert the entire integral to u-variable form and try to fit it to one or more of the basic integration formulas. If none fits, try a different substitution.
4. After integrating, rewrite the antiderivative as a function of x.

EXAMPLE 2 *Integration by Substitution*

Find the indefinite integral

$$\int x\sqrt{x^2 - 1}\,dx.$$

Solution
Consider the substitution $u = x^2 - 1$, which produces $du = 2x\,dx$. To create $2x\,dx$ as part of the integral, multiply and divide by 2.

$$\int x\sqrt{x^2 - 1}\,dx = \frac{1}{2}\int \overbrace{(x^2 - 1)^{1/2}}^{u^{1/2}}\overbrace{2x\,dx}^{du} \qquad \textit{Multiply and divide by 2.}$$

$$= \frac{1}{2}\int u^{1/2}\,du \qquad\qquad \textit{Substitute for x and dx.}$$

$$= \frac{1}{2}\frac{u^{3/2}}{3/2} + C \qquad\qquad \textit{Power Rule}$$

$$= \frac{1}{3}u^{3/2} + C \qquad\qquad \textit{Simplify.}$$

$$= \frac{1}{3}(x^2 - 1)^{3/2} + C \qquad\qquad \textit{Substitute for u.}$$

You can check this result by differentiating. ▬

EXAMPLE 3 *Integration by Substitution*

Find the indefinite integral

$$\int \frac{e^{3x}}{1 + e^{3x}}\,dx.$$

Solution
Consider the substitution $u = 1 + e^{3x}$, which produces $du = 3e^{3x}\,dx$. To create $3e^{3x}\,dx$ as part of the integral, multiply and divide by 3.

$$\int \frac{e^{3x}}{1 + e^{3x}}\,dx = \frac{1}{3}\int \overbrace{\frac{1}{1 + e^{3x}}}^{1/u}\overbrace{3e^{3x}\,dx}^{du} \qquad \textit{Multiply and divide by 3.}$$

$$= \frac{1}{3}\int \frac{1}{u}\,du \qquad\qquad \textit{Substitute for x and dx.}$$

$$= \frac{1}{3}\ln|u| + C \qquad\qquad \textit{Log Rule}$$

$$= \frac{1}{3}\ln(1 + e^{3x}) + C \qquad\qquad \textit{Substitute for u.}$$

Note that the absolute value is not necessary in the final answer because the quantity $(1 + e^{3x})$ is positive for all values of x. ▬

EXAMPLE 4 *Integration by Substitution*

Find the indefinite integral

$$\int x\sqrt{x-1}\,dx.$$

Solution

Consider the substitution $u = x - 1$, which produces $du = dx$ and $x = u + 1$.

$$\int x\sqrt{x-1}\,dx = \int (u+1)(u^{1/2})\,du \qquad \text{Substitute for } x \text{ and } dx.$$

$$= \int (u^{3/2} + u^{1/2})\,du$$

$$= \frac{u^{5/2}}{5/2} + \frac{u^{3/2}}{3/2} + C \qquad \text{Power Rule}$$

$$= \frac{2}{5}(x-1)^{5/2} + \frac{2}{3}(x-1)^{3/2} + C \qquad \text{Substitute for } u.$$

This form of the antiderivative can be further simplified.

$$\frac{2}{5}(x-1)^{5/2} + \frac{2}{3}(x-1)^{3/2} + C = \frac{6}{15}(x-1)^{5/2} + \frac{10}{15}(x-1)^{3/2} + C$$

$$= \frac{2}{15}(x-1)^{3/2}[3(x-1)+5] + C$$

$$= \frac{2}{15}(x-1)^{3/2}(3x+2) + C$$

You can check this answer by differentiating.   ▬

Example 4 demonstrates one of the characteristics of integration by substitution. That is, the form of the antiderivative as it exists immediately after resubstitution into x-variable form can often be simplified. Thus, when working the exercises in this section, don't assume that your answer is incorrect just because it doesn't look exactly like the answer given in the back of the text. You may be able to reconcile the two answers by algebraic simplification.

▦ TECHNOLOGY

If you have access to a symbolic integration utility such as *Derive*, *Maple*, or *Mathematica*, try using it to solve several of the exercises in this section. For instance, the display at the right shows how *Derive* can be used to find the indefinite integral in Example 4.

1: $x\sqrt{(x-1)}$ — *Author.*

2: $\int x\sqrt{(x-1)}\,dx$ — *Integrate.*

3: $\dfrac{2(x-1)^{3/2}(3x+2)}{15}$ — *Simplify.*

Substitution and Definite Integrals

The fourth step outlined in the guidelines for integration by substitution suggests that you convert back to the variable x. To evaluate *definite* integrals, however, it is often more convenient to determine the limits of integration for the variable u. This is often easier than converting back to the variable x and evaluating the antiderivative at the original limits.

EXAMPLE 5   *Using Substitution with a Definite Integral*

Evaluate the definite integral $\displaystyle\int_1^5 \frac{x}{\sqrt{2x-1}}\, dx$.

Solution
Use the substitution $u = \sqrt{2x-1}$, which implies that $u^2 = 2x - 1$, $x = \frac{1}{2}(u^2 + 1)$, and $dx = u\, du$. Before substituting, determine the new upper and lower limits of integration.

> *Lower limit:* When $x = 1$, $u = \sqrt{2(1) - 1} = 1$.
> *Upper limit:* When $x = 5$, $u = \sqrt{2(5) - 1} = 3$.

Now, substitute and integrate, as follows.

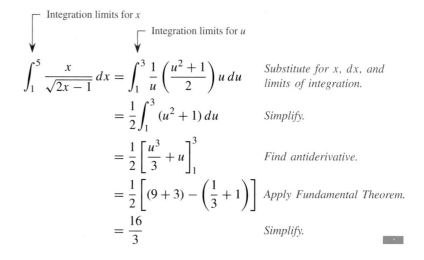

$$\underbrace{\int_1^5}\ \frac{x}{\sqrt{2x-1}}\, dx = \underbrace{\int_1^3}\ \frac{1}{u}\left(\frac{u^2+1}{2}\right) u\, du \qquad \text{Substitute for } x, dx, \text{ and limits of integration.}$$

$$= \frac{1}{2}\int_1^3 (u^2 + 1)\, du \qquad \text{Simplify.}$$

$$= \frac{1}{2}\left[\frac{u^3}{3} + u\right]_1^3 \qquad \text{Find antiderivative.}$$

$$= \frac{1}{2}\left[(9 + 3) - \left(\frac{1}{3} + 1\right)\right] \qquad \text{Apply Fundamental Theorem.}$$

$$= \frac{16}{3} \qquad \text{Simplify.}$$

Note   In Example 5, you can interpret the equation

$$\int_1^5 \frac{x}{\sqrt{2x-1}}\, dx = \int_1^3 \frac{1}{u}\left(\frac{u^2+1}{2}\right) u\, du$$

graphically to mean that the two different regions shown in Figures 6.1 and 6.2 have the same area.

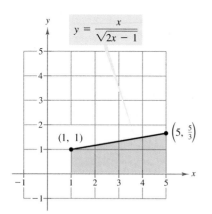

FIGURE 6.1   Region Before Substitution

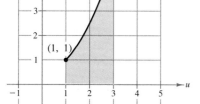

FIGURE 6.2   Region After Substitution

Applications

Psychologists use definite integrals to represent the probability that an event will occur. For instance, a probability of 0.5 means that the event will occur about 50% of the time.

Integration can be used to find the probability that an event will occur. In such an application, the real-life situation is modeled by a *probability density function* f, and the probability that x will lie between a and b is represented by

$$P(a \le x \le b) = \int_a^b f(x)\,dx.$$

The probability $P(a \le x \le b)$ must be a number between 0 and 1.

Real Life

EXAMPLE 6 *Finding a Probability*

A psychologist finds that the probability that a participant in a memory experiment will recall between a and b percent (in decimal form) of the material is

$$P(a \le x \le b) = \int_a^b \frac{28}{9}x\sqrt[3]{1-x}\,dx, \qquad 0 \le a \le b \le 1.$$

Find the probability that a randomly chosen participant will recall between 0% and 87.5% of the material.

Solution
Let $u = \sqrt[3]{1-x}$. Then $u^3 = 1 - x$, $x = 1 - u^3$, and $dx = -3u^2\,du$.

Lower limit: When $x = 0$, $u = \sqrt[3]{1-0} = 1$.
Upper limit: When $x = 0.875$, $u = \sqrt[3]{1-0.875} = 0.5$.

To find the probability, substitute and integrate, as follows.

$$\int_0^{0.875} \frac{28}{9}x\sqrt[3]{1-x}\,dx = \int_1^{1/2}\left[\frac{28}{9}(1-u^3)(u)(-3u^2)\right]du$$

$$= 3\left(\frac{28}{9}\right)\int_1^{1/2}(u^6 - u^3)\,du$$

$$= \frac{28}{3}\left[\frac{u^7}{7} - \frac{u^4}{4}\right]_1^{1/2}$$

$$\approx 0.865$$

Thus, the probability is about 86.5%, as indicated in Figure 6.3.

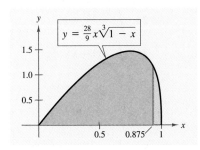

FIGURE 6.3

Discussion Problem   ### Extending the Example

In Example 6, explain how you could find a value of b such that $P(0 \le x \le b) = 0.5$.

Warm Up

The following warm-up exercises involve skills that were covered in earlier sections. You will use these skills in the exercise set for this section.

In Exercises 1–6, evaluate the indefinite integral.

1. $\displaystyle\int 5\,dx$

2. $\displaystyle\int x^{3/2}\,dx$

3. $\displaystyle\int 2x(x^2+1)^3\,dx$

4. $\displaystyle\int 6e^{6x}\,dx$

5. $\displaystyle\int \frac{2}{2x+1}\,dx$

6. $\displaystyle\int 2xe^{-x^2}\,dx$

In Exercises 7–10, simplify the expression.

7. $2x(x-1)^2 + x(x-1)$

8. $6x(x+4)^3 - 3x^2(x+4)^2$

9. $3(x+7)^{1/2} - 2x(x+7)^{-1/2}$

10. $(x+5)^{1/3} - 5(x+5)^{-2/3}$

EXERCISES 6.1

 means that technology can help you solve or check the exercise(s).

In Exercises 1–40, evaluate the indefinite integral.

1. $\displaystyle\int (x-2)^4\,dx$

2. $\displaystyle\int (x+5)^{3/2}\,dx$

3. $\displaystyle\int \frac{2}{(t-9)^2}\,dt$

4. $\displaystyle\int \frac{2t-1}{t^2-t+2}\,dt$

5. $\displaystyle\int \sqrt{1+x}\,dx$

6. $\displaystyle\int (3+x)^5\,dx$

7. $\displaystyle\int \frac{12x+2}{3x^2+x}\,dx$

8. $\displaystyle\int \frac{6x^2+2}{x^3+x}\,dx$

9. $\displaystyle\int \frac{1}{(5x+1)^3}\,dx$

10. $\displaystyle\int \frac{1}{(3x+1)^2}\,dx$

11. $\displaystyle\int \frac{1}{\sqrt{x+1}}\,dx$

12. $\displaystyle\int \frac{1}{\sqrt{5x+1}}\,dx$

13. $\displaystyle\int \frac{e^{3x}}{1-e^{3x}}\,dx$

14. $\displaystyle\int 2xe^{3x^2}\,dx$

15. $\displaystyle\int \frac{x^2}{x-1}\,dx$

16. $\displaystyle\int \frac{2x}{x-4}\,dx$

17. $\displaystyle\int x\sqrt{x^2+4}\,dx$

18. $\displaystyle\int \frac{t}{\sqrt{1-t^2}}\,dt$

19. $\displaystyle\int e^{5x}\,dx$

20. $\displaystyle\int \frac{e^x}{1+e^x}\,dx$

21. $\displaystyle\int \frac{x}{(x+1)^4}\,dx$

22. $\displaystyle\int \frac{x^2}{(x+1)^3}\,dx$

23. $\displaystyle\int \frac{x}{(3x-1)^2}\,dx$

24. $\displaystyle\int \frac{5x}{(x-4)^3}\,dx$

25. $\displaystyle\int x(1-x)^4\,dx$

26. $\displaystyle\int x^2(1-x)^3\,dx$

27. $\displaystyle\int \frac{x-1}{x^2-2x}\,dx$

28. $\displaystyle\int \frac{x}{(x^2-1)^2}\,dx$

29. $\displaystyle\int x\sqrt{x-3}\,dx$

30. $\displaystyle\int x\sqrt{2x+1}\,dx$

31. $\displaystyle\int x^2\sqrt{1-x}\,dx$

32. $\displaystyle\int x^3\sqrt{x+2}\,dx$

33. $\displaystyle\int \frac{x^2-1}{\sqrt{2x-1}}\,dx$

34. $\displaystyle\int \frac{2x-1}{\sqrt{x+3}}\,dx$

35. $\displaystyle\int t\sqrt[3]{t+1}\,dt$

36. $\displaystyle\int x\sqrt[3]{x-2}\,dx$

37. $\displaystyle\int \frac{1}{\sqrt{t}-1}\,dt$

38. $\displaystyle\int \frac{1}{\sqrt{x}+1}\,dx$

39. $\displaystyle\int \frac{2\sqrt{t}+1}{t}\,dt$

40. $\displaystyle\int \frac{6x+\sqrt{2x}}{x}\,dx$

In Exercises 41–52, evaluate the definite integral.

41. $\displaystyle\int_0^4 \sqrt{2x+1}\,dx$

42. $\displaystyle\int_2^4 \sqrt{4x+1}\,dx$

43. $\displaystyle\int_0^1 3xe^{x^2}\,dx$

44. $\displaystyle\int_0^2 e^{-2x}\,dx$

45. $\displaystyle\int_0^4 \frac{x}{(x+4)^2}\,dx$

46. $\displaystyle\int_0^1 x(x+5)^4\,dx$

47. $\displaystyle\int_0^{0.5} x(1-x)^3\,dx$

48. $\displaystyle\int_0^{0.5} x^2(1-x)^3\,dx$

49. $\displaystyle\int_3^7 x\sqrt{x-3}\,dx$

50. $\displaystyle\int_0^4 \frac{x}{\sqrt{2x+1}}\,dx$

51. $\displaystyle\int_0^7 x\sqrt[3]{x+1}\,dx$

52. $\displaystyle\int_1^2 (x-1)\sqrt{2-x}\,dx$

In Exercises 53–56, find the area of the region bounded by the graphs of the equations.

53. $y = -x\sqrt{x+2}$, $y = 0$

54. $y = x\sqrt[3]{1-x}$, $y = 0$

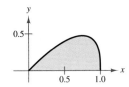

55. $y^2 = x^2(1-x^2)$

(*Hint:* Find the area of the region bounded by $y = x\sqrt{1-x^2}$ and $y = 0$. Then multiply by 4.)

56. $y = 1/\left(1+\sqrt{x}\right)$,
$y = 0$, $x = 0$, $x = 4$

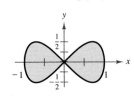

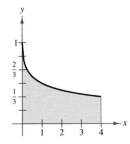

In Exercises 57 and 58, find the volume of the solid generated by revolving the region bounded by the graphs of the equations about the *x*-axis.

57. $y = x\sqrt{1-x^2}$

58. $y = \sqrt{x}(1-x)^2$, $y = 0$

In Exercises 59 and 60, find the average amount by which the function *f* exceeds the function *g* on the interval.

59. $f(x) = \dfrac{1}{x+1}$, $g(x) = \dfrac{x}{(x+1)^2}$, $[0, 1]$

60. $f(x) = x\sqrt{4x+1}$, $g(x) = 2\sqrt{x^3}$, $[0, 2]$

61. *Probability* The probability of recall in an experiment is modeled by

$$P(a \le x \le b) = \int_a^b \frac{15}{4}x\sqrt{1-x}\,dx,$$

where *x* is the percent of recall (see figure).

(a) What is the probability of recalling between 40% and 80%?

(b) What is the median percent recall? That is, for what value of *b* is $P(0 \le x \le b) = 0.5$?

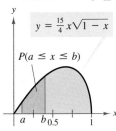

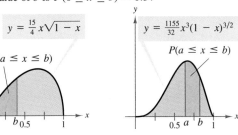

FIGURE FOR 61   FIGURE FOR 62

62. *Probability* The probability of finding between *a* and *b* percent of iron in ore samples is modeled by

$$P(a \le x \le b) = \int_a^b \frac{1155}{32}x^3(1-x)^{3/2}\,dx$$

(see figure). Find the probability that a sample will contain between (a) 0% and 25%, and (b) 50% and 100% of iron.

63. *Revenue* A company sells a seasonal product that has a daily revenue modeled by

$$R = 0.06t^2(365-t)^{1/2} + 1250, \qquad 0 \le t \le 365.$$

Find the average daily revenue over the period of 1 year.

64. *Revenue* Describe a product whose seasonal sales pattern resembles the model in Exercise 63. Explain your reasoning.

In Exercises 65 and 66, use the Midpoint Rule with $n = 10$ to approximate the area of the region bounded by the graph(s) of the equation(s).

65. $y = \sqrt[3]{x}\sqrt{4-x}$, $y = 0$

66. $y^2 = x^2(1-x^2)$

Integration by Parts and Present Value

Integration by Parts

In this section, you will study an integration technique called **integration by parts.** This technique is particularly useful for integrands involving the products of algebraic and exponential or logarithmic functions, such as $\int x^2 e^x \, dx$ and $\int x \ln x \, dx$. Integration by parts is based on the Product Rule.

$$\frac{d}{dx}[uv] = uv' + vu' \qquad \text{Product Rule}$$

$$uv = \int uv' \, dx + \int vu' \, dx \qquad \text{Integrate both sides.}$$

$$uv = \int u \, dv + \int v \, du \qquad \text{Write in differential form.}$$

$$\int u \, dv = uv - \int v \, du \qquad \text{Rewrite.}$$

Note When using integration by parts, note that you can first choose dv or first choose u. After you do, however, the choice of the other factor is determined—it must be the remaining portion of the integrand. Also note that dv *must* contain the differential dx of the original integral.

Integration by Parts

Let u and v be differentiable functions of x.

$$\int u \, dv = uv - \int v \, du$$

Note that the formula for integration by parts expresses the original integral in terms of another integral. Depending on the choices for u and dv, it may be easier to evaluate the second integral than the original one.

Guidelines for Integration by Parts

1. Let dv be the most complicated portion of the integrand that fits a basic integration formula. Let u be the remaining factor.
2. Let u be the portion of the integrand whose derivative is a simpler function than u itself. Let dv be the remaining factor.

EXAMPLE 1 *Integration by Parts*

Find the indefinite integral $\int xe^x \, dx$.

Solution

To apply integration by parts, you must rewrite the original integral in the form $\int u \, dv$. That is, you must break $xe^x \, dx$ into two factors—one "part" representing u and the other "part" representing dv. There are several ways to do this.

$$\int \underbrace{(x)}_{u}\underbrace{(e^x \, dx)}_{dv} \qquad \int \underbrace{(e^x)}_{u}\underbrace{(x \, dx)}_{dv} \qquad \int \underbrace{(1)}_{u}\underbrace{(xe^x \, dx)}_{dv} \qquad \int \underbrace{(xe^x)}_{u}\underbrace{(dx)}_{dv}$$

Following the guidelines, you should choose the first option because $dv = e^x \, dx$ is the most complicated portion of the integrand that fits a basic integration formula *and* because the derivative of $u = x$ is simpler than x.

$$dv = e^x \, dx \quad \longrightarrow \quad v = \int dv = \int e^x \, dx = e^x$$

$$u = x \quad \longrightarrow \quad du = dx$$

With these substitutions, you can apply the integration by parts formula as follows.

$$\int xe^x \, dx = xe^x - \int e^x \, dx \qquad\qquad \scriptstyle \int u \, dv = uv - \int v \, du$$

$$= xe^x - e^x + C \qquad\qquad \scriptstyle Integrate \int e^x \, dx.$$ ■

Note In Example 1, notice that you do not need to include a constant of integration when solving $v = \int e^x \, dx = e^x$. To see why this is true, try replacing e^x by $e^x + C_1$ in the solution.

$$\int xe^x \, dx = x(e^x + C_1) - \int (e^x + C_1) \, dx$$

After integrating, you can see that the terms involving C_1 cancel.

TECHNOLOGY

If you have access to a symbolic integration utility such as *Derive*, *Maple*, or *Mathematica*, try using it to solve several of the exercises in this section. For instance, the display at the right shows how *Derive* can be used to find the indefinite integral in Example 1. Note that the form of the integral is slightly different from that obtained in Example 1.

```
1: xê^x                              Author.

2: ∫xê^x dx                          Integrate.

3: ê^x(x - 1)                        Simplify.
```

Note To remember the integration by parts formula, you might like to use the following "Z" pattern. The top row represents the original integral, the diagonal row represents uv, and the bottom row represents the new integral.

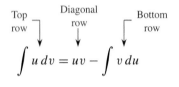

$$\int u\, dv = uv - \int v\, du$$

EXAMPLE 2 *Integration by Parts*

Find the indefinite integral $\int x^2 \ln x\, dx$.

Solution
For this integral, x^2 is more easily integrated than $\ln x$. Furthermore, the derivative of $\ln x$ is simpler than $\ln x$. Therefore, you should choose $dv = x^2\, dx$.

$$dv = x^2\, dx \quad \longrightarrow \quad v = \int dv = \int x^2\, dx = \frac{x^3}{3}$$

$$u = \ln x \quad \longrightarrow \quad du = \frac{1}{x}\, dx$$

Using these substitutions, apply the integration by parts formula as follows.

$$\int x^2 \ln x\, dx = \frac{x^3}{3} \ln x - \int \left(\frac{x^3}{3}\right)\left(\frac{1}{x}\right) dx \qquad \int u\, dv = uv - \int v\, du$$

$$= \frac{x^3}{3} \ln x - \frac{1}{3}\int x^2\, dx \qquad\qquad \textit{Simplify.}$$

$$= \frac{x^3}{3} \ln x - \frac{x^3}{9} + C \qquad\qquad \textit{Integrate.}$$

EXAMPLE 3 *Integrating by Parts with a Single Factor*

Find the indefinite integral $\int \ln x\, dx$.

Solution
This integral is unusual because it has only one factor. In such cases, you should choose $dv = dx$ and choose u to be the single factor.

$$dv = dx \quad \longrightarrow \quad v = \int dv = \int dx = x$$

$$u = \ln x \quad \longrightarrow \quad du = \frac{1}{x}\, dx$$

Using these substitutions, apply the integration by parts formula as follows.

$$\int \ln x\, dx = x \ln x - \int \left(\frac{1}{x}\right)(x)\, dx \qquad \int u\, dv = uv - \int v\, du$$

$$= x \ln x - \int dx \qquad\qquad \textit{Simplify.}$$

$$= x \ln x - x + C \qquad\qquad \textit{Integrate.}$$

EXAMPLE 4 *Using Integration by Parts Repeatedly*

Find the indefinite integral $\int x^2 e^x \, dx$.

Solution

Using the guidelines, notice that the derivative of x^2 becomes simpler, whereas the derivative of e^x does not. Therefore, you should let $u = x^2$ and let $dv = e^x \, dx$.

$$dv = e^x \, dx \quad \longrightarrow \quad v = \int dv = \int e^x \, dx = e^x$$

$$u = x^2 \quad \longrightarrow \quad du = 2x \, dx$$

Using these substitutions, apply the integration by parts formula as follows.

$$\int x^2 e^x \, dx = x^2 e^x - \int 2x e^x \, dx \qquad \text{\textit{First application of integration by parts}}$$

To evaluate the new integral on the right, apply integration by parts a second time, using the following substitutions.

$$dv = e^x \, dx \quad \longrightarrow \quad v = \int dv = \int e^x \, dx = e^x$$

$$u = 2x \quad \longrightarrow \quad du = 2 \, dx$$

Note Remember that you can check an indefinite integral by differentiating. For instance, in Example 4, try differentiating the antiderivative

$$e^x (x^2 - 2x + 2) + C$$

to check that you obtain the original integrand, $x^2 e^x$.

Using these substitutions, apply the integration by parts formula as follows.

$$\int x^2 e^x \, dx = x^2 e^x - \int 2x e^x \, dx \qquad \text{\textit{First application of integration by parts}}$$

$$= x^2 e^x - \left(2x e^x - \int 2 e^x \, dx \right) \qquad \text{\textit{Second application of integration by parts}}$$

$$= x^2 e^x - 2x e^x + 2 e^x + C \qquad \text{\textit{Integrate.}}$$

$$= e^x (x^2 - 2x + 2) + C \qquad \text{\textit{Simplify.}}$$

You can confirm this result by differentiating. ▬

When making repeated applications of integration by parts, be careful not to interchange the substitutions in successive applications. For instance, in Example 4, the first substitutions were $dv = e^x \, dx$ and $u = x^2$. If in the second application you had switched to $dv = 2x \, dx$ and $u = e^x$, you would have reversed the previous integration and returned to the *original* integral.

$$\int x^2 e^x \, dx = x^2 e^x - \left(x^2 e^x - \int x^2 e^x \, dx \right)$$

$$= \int x^2 e^x \, dx$$

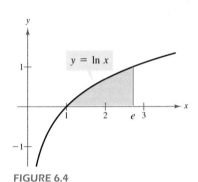

FIGURE 6.4

EXAMPLE 5 *Evaluating a Definite Integral*

Evaluate the definite integral

$$\int_1^e \ln x \, dx.$$

Solution

Integration by parts was used to find the antiderivative of $\ln x$ in Example 3. Using this result, you can evaluate the definite integral as follows.

$$\int_1^e \ln x \, dx = \left[x \ln x - x \right]_1^e \qquad \text{Use result of Example 3.}$$

$$= (e \ln e - e) - (1 \ln 1 - 1) \qquad \text{Apply Fundamental Theorem.}$$

$$= (e - e) - (0 - 1)$$

$$= 1 \qquad \text{Simplify.}$$

The area represented by this definite integral is shown in Figure 6.4. ▬

Before starting the exercises in this section, remember that it is not enough to know *how* to use the various integration techniques. You also must know *when* to use them. Integration is first and foremost a problem of recognition—recognizing which formula or technique to apply to obtain an antiderivative. Often, a slight alteration of an integrand will necessitate the use of a different integration technique. Here are some examples.

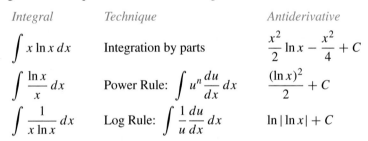

Integral	*Technique*	*Antiderivative*
$\displaystyle\int x \ln x \, dx$	Integration by parts	$\dfrac{x^2}{2} \ln x - \dfrac{x^2}{4} + C$
$\displaystyle\int \dfrac{\ln x}{x} \, dx$	Power Rule: $\displaystyle\int u^n \dfrac{du}{dx} \, dx$	$\dfrac{(\ln x)^2}{2} + C$
$\displaystyle\int \dfrac{1}{x \ln x} \, dx$	Log Rule: $\displaystyle\int \dfrac{1}{u} \dfrac{du}{dx} \, dx$	$\ln |\ln x| + C$

As you gain experience with integration by parts, your skill in determining u and dv will improve. The summary below gives suggestions for choosing u and dv.

Summary of Common Uses of Integration by Parts

1. $\displaystyle\int x^n e^{ax} \, dx$ Let $u = x^n$ and $dv = e^{ax} \, dx$. (Examples 1 and 4)

2. $\displaystyle\int x^n \ln x \, dx$ Let $u = \ln x$ and $dv = x^n \, dx$. (Examples 2 and 3)

Present Value

In some business applications, the future income for a business is known. For instance, the annual income might be given by $c(t)$. During a period of t_1 years, the total income would then be given by integrating $c(t)$ between $t = 0$ and $t = t_1$. To assign a **present value** to this income, you must use an exponential *damping factor* that takes into account future inflation. If the annual inflation rate is assumed to be r, then the damping factor is e^{-rt}.

Present Value

If $c(t)$ represents an income function over time and the annual rate of inflation is r, then the actual total income over t_1 years is

$$\text{Actual income over } t_1 \text{ years} = \int_0^{t_1} c(t)\,dt$$

and its **present value** is

$$\text{Present value} = \int_0^{t_1} c(t)e^{-rt}\,dt.$$

© Kent Sievers

In 1990, 32 states had state lotteries, and they totaled about $19 billion in revenue. Of this, 54% was paid out in prizes and 6% was used to cover the state lotteries' administration costs. These costs include the cost of making the lottery tickets. Chris and Chuck Harter, who own American Games, Inc., sell more than $15 million worth of lottery tickets and bingo cards to states and charitable organizations.

Real Life

EXAMPLE 6 *Finding Present Value*

You have just won a state lottery for $1,000,000. You will be paid $50,000 a year for 20 years. Assuming an annual inflation rate of 6%, what is the present value of this income?

Solution

The income function for your winnings is $c(t) = 50,000$. Thus, your actual income is

$$\text{Actual income} = \int_0^{20} 50,000\,dt = \left[\,50,000t\,\right]_0^{20} = \$1,000,000.$$

Because you do not receive this entire amount now, its present value is less than $1,000,000, and is

$$\text{Present value} = \int_0^{20} 50,000e^{-0.06t}\,dt = \left[\frac{50,000}{-0.06}e^{-0.06t}\right]_0^{20} \approx \$582,338.$$

This present value represents the amount that the state must deposit now to cover your payments over the next 20 years. This shows why state lotteries are so profitable—for the states!

Real Life

EXAMPLE 7 *Finding Present Value*

A company expects its income during the next 5 years to be given by

$$c(t) = 100,000t.$$

Assuming an annual inflation rate of 10%, what is the present value of this income?

Solution

The present value is

$$\text{Present value} = \int_0^5 100,000te^{-0.1t}\,dt = 100,000\int_0^5 te^{-0.1t}\,dt.$$

Using integration by parts, let $dv = e^{-0.1t}\,dt$ and obtain the following.

$$dv = e^{-0.1t}\,dt \quad \longrightarrow \quad v = \int dv = \int e^{-0.1t}\,dt = -10e^{-0.1t}$$

$$u = t \quad \longrightarrow \quad du = dt$$

This implies that

$$\int te^{-0.1t}\,dt = -10te^{-0.1t} + 10\int e^{-0.1t}\,dt$$

$$= -10te^{-01.t} - 100e^{-0.1t}$$

$$= -10e^{-0.1t}(t + 10).$$

Therefore, the present value is

$$\text{Present value} = 100,000\int_0^5 te^{-0.1t}\,dt$$

$$= 100,000\left[-10e^{-0.1t}(t + 10)\right]_0^5$$

$$\approx \$902,040.$$

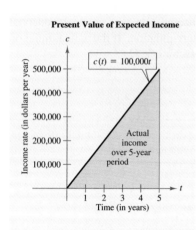

Present Value of Expected Income

$c(t) = 100,000t$

Actual income over 5-year period

Income rate (in dollars per year)

Time (in years)

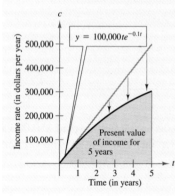

$y = 100,000te^{-0.1t}$

Present value of income for 5 years

Income rate (in dollars per year)

Time (in years)

FIGURE 6.5

Discussion Problem   *Present Value*

As shown in Figure 6.5, the actual income for the business in Example 7, over the next 5 years, is

$$\text{Actual income} = \int_0^5 100,000t\,dt = \$1,250,000.$$

If you were selling this company for its present value over a 5-year period, would it be to your advantage to argue that the inflation rate was lower than 10%? Explain your reasoning.

Warm Up

The following warm-up exercises involve skills that were covered in earlier sections. You will use these skills in the exercise set for this section.

In Exercises 1–6, find $f'(x)$.

1. $f(x) = 3x^4 + 2x - 1$

2. $f(x) = \frac{1}{4}x^2 + \frac{1}{8}x + 4$

3. $f(x) = \ln(x^4 + x^2)$

4. $f(x) = e^{x^3}$

5. $f(x) = x^2 e^x$

6. $f(x) = \dfrac{x+1}{x-3}$

In Exercises 7–10, find the area between the graphs of f and g.

7. $f(x) = -x^2 + 4,\ g(x) = x^2 - 4$

8. $f(x) = -x^2 + 2,\ g(x) = 1$

9. $f(x) = 4x,\ g(x) = x^2$

10. $f(x) = x^3 - 3x^2 + 2,\ g(x) = x - 1$

E X E R C I S E S 6.2

means that technology can help you solve or check the exercise(s).

In Exercises 1–6, use integration by parts to find the indefinite integral.

1. $\displaystyle\int xe^{3x}\,dx$

2. $\displaystyle\int xe^{-x}\,dx$

3. $\displaystyle\int x^2 e^{-x}\,dx$

4. $\displaystyle\int x^2 e^{2x}\,dx$

5. $\displaystyle\int \ln 2x\,dx$

6. $\displaystyle\int \ln x^2\,dx$

In Exercises 7–30, find the indefinite integral. (*Hint:* Integration by parts is not required for all the integrals.)

7. $\displaystyle\int e^{4x}\,dx$

8. $\displaystyle\int e^{-2x}\,dx$

9. $\displaystyle\int xe^{4x}\,dx$

10. $\displaystyle\int xe^{-2x}\,dx$

11. $\displaystyle\int xe^{x^2}\,dx$

12. $\displaystyle\int x^2 e^{x^3}\,dx$

13. $\displaystyle\int x^2 e^x\,dx$

14. $\displaystyle\int \frac{x}{e^x}\,dx$

15. $\displaystyle\int x^3 e^x\,dx$

16. $\displaystyle\int \frac{e^{1/t}}{t^2}\,dt$

17. $\displaystyle\int x^3 \ln x\,dx$

18. $\displaystyle\int x^2 \ln x\,dx$

19. $\displaystyle\int t \ln(t+1)\,dt$

20. $\displaystyle\int \frac{1}{x(\ln x)^3}\,dx$

21. $\displaystyle\int x(\ln x)^2\,dx$

22. $\displaystyle\int \ln 3x\,dx$

23. $\displaystyle\int \frac{(\ln x)^2}{x}\,dx$

24. $\displaystyle\int \frac{\ln x}{x^2}\,dx$

25. $\displaystyle\int x\sqrt{x-1}\,dx$

26. $\displaystyle\int x(x+1)^2\,dx$

27. $\displaystyle\int (x^2-1)e^x\,dx$

28. $\displaystyle\int \frac{x}{\sqrt{2+3x}}\,dx$

29. $\displaystyle\int \frac{xe^{2x}}{(2x+1)^2}\,dx$

30. $\displaystyle\int \frac{x^3 e^{x^2}}{(x^2+1)^2}\,dx$

In Exercises 31–34, evaluate the definite integral.

31. $\displaystyle\int_0^1 x^2 e^x\,dx$

32. $\displaystyle\int_0^2 \frac{x^2}{e^x}\,dx$

33. $\displaystyle\int_1^e x^5 \ln x\,dx$

34. $\displaystyle\int_0^1 \ln(1+2x)\,dx$

In Exercises 35–38, find the indefinite integral using the specified method.

35. $\int 2x\sqrt{2x-3}\,dx$

   (a) By parts, letting $dv = \sqrt{2x-3}\,dx$

   (b) By substitution, letting $u = \sqrt{2x-3}$

36. $\int x\sqrt{4+x}\,dx$

   (a) By parts, letting $dv = \sqrt{4+x}\,dx$

   (b) By substitution, letting $u = \sqrt{4+x}$

37. $\int \dfrac{x}{\sqrt{4+5x}}\,dx$

   (a) By parts, letting $dv = \dfrac{1}{\sqrt{4+5x}}\,dx$

   (b) By substitution, letting $u = \sqrt{4+5x}$

38. $\int x\sqrt{4-x}\,dx$

   (a) By parts, letting $dv = \sqrt{4-x}\,dx$

   (b) By substitution, letting $u = \sqrt{4-x}$

In Exercises 39 and 40, use integration by parts to verify the formula.

39. $\int x^n \ln x\,dx = \dfrac{x^{n+1}}{(n+1)^2}[-1 + (n+1)\ln x] + C,$

   $n \neq -1$

40. $\int x^n e^{ax}\,dx = \dfrac{x^n e^{ax}}{a} - \dfrac{n}{a}\int x^{n-1} e^{ax}\,dx$

In Exercises 41–44, use the results of Exercises 39 and 40 to find the indefinite integral.

41. $\int x^2 e^{5x}\,dx$

42. $\int xe^{-3x}\,dx$

43. $\int x^{-2} \ln x\,dx$

44. $\int x^{1/2} \ln x\,dx$

In Exercises 45 and 46, find the area of the region bounded by the graphs of the given equations.

45. $y = xe^{-x},\ y = 0,\ x = 4$

46. $y = \frac{1}{9}xe^{-x/3},\ y = 0,\ x = 0,\ x = 3$

47. Given the region bounded by the graphs of $y = 2\ln x$, $y = 0$, and $x = e$, find

   (a) the area of the region.

   (b) the volume of the solid generated by revolving the region about the x-axis.

48. Find the volume of the solid generated by revolving the region bounded by the graphs of $y = xe^x$, $y = 0$, $x = 0$, and $x = 1$ about the x-axis (see figure).

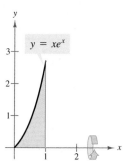

In Exercises 49–52, use a symbolic integrator to evaluate the integral.

49. $\int_0^2 t^3 e^{-4t}\,dt$

50. $\int_1^4 \ln x(x^2 + 4)\,dx$

51. $\int_0^5 x^4(25 - x^2)^{3/2}\,dx$

52. $\int_1^e x^9 \ln x\,dx$

53. *Product Demand* A manufacturing company forecasts that the demand x (in units) for its product over the next 10 years can be modeled by

$$x = 500(20 + te^{-0.1t}), \qquad 0 \le t \le 10,$$

where t is the time in years.

   (a) Is the company forecasting an increase or decrease in demand over the decade?

   (b) According to the model, what is the total demand over the next 10 years?

   (c) Find the average annual demand during the 10-year period.

54. *Capital Campaign* The board of trustees of a college is planning a 5-year capital gifts campaign to raise money for the college. The goal is to have an annual gift income I that is modeled by

$$I = 2000(375 + 68te^{-0.2t}), \qquad 0 \le t \le 5,$$

where t is the time in years.

   (a) Does the board of trustees expect the gift income to increase or decrease over the 5-year period?

   (b) Find the expected total gift income over the 5-year period.

   (c) Determine the average annual gift income over the 5-year period. Compare the result with the income given when $t = 3$.

55. *Learning Theory* A model for the ability M of a child to memorize, measured on a scale from 0 to 10, is

$$M = 1 + 1.6t \ln t, \qquad 0 < t \le 4,$$

where t is the child's age in years. Find the average value of this model between

(a) the child's first and second birthdays.
(b) the child's third and fourth birthdays.

56. *Revenue* A company sells a seasonal product. The revenue from the product can be modeled by

$$R = 410.5t^2 e^{-t/30} + 25,000, \qquad 0 \le t \le 365,$$

where t is the time in days.

(a) Find the average daily receipts during the first quarter, which is given by $0 \le t \le 91$.
(b) Find the average daily receipts during the fourth quarter, which is given by $274 \le t \le 365$.
(c) Find the total daily receipts during the year.

Present Value In Exercises 57–62, find the present value of the income c (measured in dollars) over t_1 years at the given annual inflation rate r.

57. $c = 5000, \ r = 5\%, \ t_1 = 4$ years

58. $c = 450, \ r = 4\%, \ t_1 = 10$ years

59. $c = 150,000 + 2500t, \ r = 4\%, \ t_1 = 10$ years

60. $c = 30,000 + 500t, \ r = 7\%, \ t_1 = 6$ years

61. $c = 1000 + 50e^{t/2}, \ r = 6\%, \ t_1 = 4$ years

62. $c = 5000 + 25te^{t/10}, \ r = 6\%, \ t_1 = 10$ years

63. *Present Value* A company expects its income c during the next 4 years to be modeled by

$$c = 150,000 + 75,000t.$$

(a) Find the actual income for the business over the 4 years.
(b) Assuming an annual inflation rate of 4%, what is the present value of this income?

64. *Contract Value* A professional athlete signs a 3-year contract in which the earnings can be modeled by

$$c = 300,000 + 125,000t.$$

(a) Find the actual value of the athlete's contract.
(b) Assuming an annual inflation rate of 5%, what is the present value of the contract?

 65. Use the Midpoint Rule with $n = 10$ to approximate

$$\int_1^4 \frac{4}{\sqrt{x} + \sqrt[3]{x}} \, dx.$$

 66. Use the Midpoint Rule with $n = 12$ to approximate the volume of the solid generated by revolving the region bounded by the graphs of

$$y = \frac{10}{\sqrt{x}e^x}, \ y = 0, \ x = 1, \text{ and } x = 4$$

about the x-axis.

Future Value In Exercises 67 and 68, find the future value of the income (in dollars) given by $f(t)$ over t_1 years at the annual interest rate of r. If f represents an investment function over a period of t_1 years that is continuously invested at an annual interest rate of r (compounded continuously), then the **future value** of the investment is given by

$$\text{Future value} = e^{rt_1} \int_0^{t_1} f(t)e^{-rt} \, dt.$$

67. $f(t) = 3000, \ r = 8\%, \ t_1 = 10$ years

68. $f(t) = 3000e^{0.05t}, \ r = 10\%, \ t_1 = 5$ years

69. *College Expenses* In 1993, the total cost to attend Notre Dame University for 1 year was estimated to be $19,937. If your grandparents had continuously invested in a college fund according to the model

$$f(t) = 250t$$

for 18 years, at an annual interest rate of 10%, would the fund have grown enough to allow you to cover 4 years of expenses at Notre Dame?

Case Study: Saving for a Child's College Education

PROFILE

COMPANY
*Provident Wealth
Management, Inc.*
▪
LOCATION
Erie, PA
▪
TYPE OF BUSINESS
*Financial planning
and investment
advisory services*
▪

We are often interested in determining the present value of a stream of income flows in the future. For example, an investor might wish to compute the present value of her accumulated savings over 20 years if she saves a fixed dollar amount every year. But Kay Johnson, a certified financial planner with Provident Wealth Management, Inc., explains that it is quite common for individuals seeking financial security to pose the question in reverse: In order to achieve a desired level of accumulated savings after, say, 20 years, how much needs to be saved every time period?

One typical scenario is that faced by parents who want to make sure that enough money will be saved for their newborn child's education. Suppose, for example, it is anticipated that $100,000 will be necessary for college expenses when the child turns 18. Generally, the following two options are available:

Option #1

The new parents could set aside one lump-sum amount today, investing it to earn the prevailing market interest rate so that, with the compounding of interest, the desired level of income would be achieved after 18 years. The concept of *present value* enables us to compute the lump-sum amount that needs to be invested today.

From Chapter 6, we know that if P is the amount deposited today and r is the annual interest rate (in decimal form), then the balance will grow to A after t years, if interest is compounded continuously, according to the following equation: $A = Pe^{rt}$. A is called the future value of P. But this equation can be rearranged to find the amount P that must be deposited today to achieve a specified level of A after t years: $P = Ae^{-rt}$. P is called the *present value* of A. So, for the new parents who want to accumulate $100,000 after 18 years, assuming an interest rate of 8 percent compounded continuously, the lump-sum deposit they must make today is: $P = (100,000)e^{-(0.08)(18)} = \$23,692.80$.

Option #2

Many young parents, however, are unable to set aside such a large lump-sum deposit all at once, despite the benefits from compound interest. For them, it might be more feasible to set aside a fixed dollar amount each year, over the course of 18 years. For these families, the question becomes: How much must be saved each year, if interest is compounded continuously, to accumulate $100,000 after 18 years? We know that the present value of a continuous income stream lasting for t_1 years at a constant rate of K dollars per year and an interest rate of r (in decimal form), compounded continuously, is:

$$\text{Present value} = \int_0^{t_1} Ke^{-rt}\, dt$$

$$= K \int_0^{t_1} e^{-rt}\, dt$$

$$= K \left[(-1/r)(e^{-rt}) \right]_0^{t_1}$$

$$= \left[(-K/r)(e^{-rt_1}) \right]_{t=0}^{t=t_1}$$

$$= \left[(-K/r)(e^{-rt_1} - 1) \right]$$

$$= (K/r)(1 - e^{-rt_1})$$

In option #1, we determined that it would be necessary to deposit $23,692.80 *today* to achieve the desired $100,000 18 years later. Using $23,692.80 as the present value in the equation just derived, we may solve for the constant deposit of K dollars per year that will lead to the desired present value of $23,692.80, and equivalently the desired future value of $100,000:

$$\$23,692.80 = (K/0.08)(1 - e^{-(0.08)(18)})$$

$$K = \$2483.94$$

Of course, if interest compounding and annual deposits are not continuous, but rather occur at discrete intervals, the amount of K will vary accordingly.

The following table summarizes and compares the two options:

Option	Total Accumulated Savings After 18 Years	Actual Savings	Interest Earned
#1	$100,000	$23,692.80	$76,307.20
#2	$100,000	$44,710.92 ($2483.94/year) · 18 years	$55,289.08

Edward D. Saloum

Donald L. Smith, Robin Smith, baby Alex, and Kay N. Johnson of the Provident Wealth Management Company ■

Assuming that option #2 is chosen, the actual dollar amount needed to be invested annually will still depend on both the interest rate and the desired level of total accumulated savings after 18 years. The table shows the required deposits per year for various levels of desired accumulated savings at different interest rates. Note that for any given interest rate, the amount of annual deposit will vary in direct proportion to the desired total accumulated savings.

Total Accumulated Savings After 18 Years	Annual Deposits Necessary at Various Interest Rates				
	$r = 6\%$	$r = 7\%$	$r = 8\%$	$r = 9\%$	$r = 10\%$
$100,000	3,085.35	2,771.82	2,483.94	2,220.53	1,980.34
$200,000	6,170.69	5,543.64	4,967.88	4,441.06	3,960.68
$300,000	9,256.03	8,315.46	7,451.82	6,661.59	5,941.02
$400,000	12,341.40	11,087.28	9,935.76	8,882.12	7,921.36

What Would You Do?

1. What effect would inflation have on the new parents' decisions concerning their saving for the education of their child? Recommend a strategy for overcoming inflation.

2. What would happen to the annual deposits under option #2 if the parents were able to deposit an increasing rather than a constant amount each year? Give examples.

3. Explore other circumstances to which the above analysis could be applicable.

Partial Fractions and Logistics Growth

Partial Fractions ■ *Logistics Growth Function*

Partial Fractions

In Sections 6.1 and 6.2, you studied integration by substitution and by parts. In this section you will study a third technique called **partial fractions.** This technique involves the decomposition of a rational function into the sum of two or more simple rational functions. For instance, suppose you know that

$$\frac{x+7}{x^2-x-6} = \frac{2}{x-3} - \frac{1}{x+2}.$$

Knowing the "partial fractions" on the right side would allow you to integrate the left side as follows.

$$\int \frac{x+7}{x^2-x-6}\,dx = \int \left(\frac{2}{x-3} - \frac{1}{x+2}\right)dx$$

$$= 2\int \frac{1}{x-3}\,dx - \int \frac{1}{x+2}\,dx$$

$$= 2\ln|x-3| - \ln|x+2| + C$$

To use this method, you must be able to factor the denominator of the original rational function *and* find the partial fraction decomposition of the function.

Note   A rational function $p(x)/q(x)$ is *proper* if the degree of the numerator is less than the degree of the denominator.

Partial Fractions

To find the partial fraction decomposition of the *proper* rational function $p(x)/q(x)$, factor $q(x)$ and write an equation that has the form

$$\frac{p(x)}{q(x)} = (\text{sum of partial fractions}).$$

For each *distinct* linear factor $(ax + b)$, the right side should include a term of the form

$$\frac{A}{ax+b}.$$

For each *repeated* linear factor $(ax+b)^n$, the right side should include n terms of the form

$$\frac{A_1}{ax+b} + \frac{A_2}{(ax+b)^2} + \cdots + \frac{A_n}{(ax+b)^n}.$$

EXAMPLE 1 *Finding a Partial Fraction Decomposition*

Write the partial fraction decomposition for

$$\frac{x+7}{x^2-x-6}.$$

Solution

Begin by factoring the denominator as $x^2 - x - 6 = (x - 3)(x + 2)$. Then, write the partial fraction decomposition as

$$\frac{x+7}{x^2-x-6} = \frac{A}{x-3} + \frac{B}{x+2}.$$

To solve this equation for A and B, multiply both sides of the equation by the least common denominator $(x - 3)(x + 2)$. This produces the following **basic equation.**

$$x + 7 = A(x + 2) + B(x - 3) \qquad \textit{Basic equation}$$

Because this equation is true for all x, you can substitute any convenient values of x into the equation. The x-values that are especially convenient are the ones that make a factor of the least common denominator zero: $x = -2$ and $x = 3$.

Substitute $x = -2$:

$$\begin{aligned} x + 7 &= A(x + 2) + B(x - 3) & \textit{Basic equation} \\ -2 + 7 &= A(-2 + 2) + B(-2 - 3) & \textit{Substitute } -2 \textit{ for } x. \\ 5 &= A(0) + B(-5) & \textit{Simplify.} \\ -1 &= B & \textit{Solve for } B. \end{aligned}$$

Substitute $x = 3$:

$$\begin{aligned} x + 7 &= A(x + 2) + B(x - 3) & \textit{Basic equation} \\ 3 + 7 &= A(3 + 2) + B(3 - 3) & \textit{Substitute 3 for } x. \\ 10 &= A(5) + B(0) & \textit{Simplify.} \\ 2 &= A & \textit{Solve for } A. \end{aligned}$$

Now that you have solved the basic equation for A and B, you can write the partial fraction decomposition as

$$\frac{x+7}{x^2-x-6} = \frac{2}{x-3} - \frac{1}{x+2},$$

as indicated at the beginning of this section. ▬

Note Be sure you see that the substitutions for x in Example 1 are chosen for their convenience in solving for A and B. The value $x = -2$ is chosen because it eliminates the term $A(x + 2)$, and the value $x = 3$ is chosen because it eliminates the term $B(x - 3)$.

EXAMPLE 2 *Integrating with Repeated Factors*

Find the indefinite integral $\displaystyle\int \frac{5x^2 + 20x + 6}{x^3 + 2x^2 + x}\, dx$.

Solution

Begin by factoring the denominator as $x(x + 1)^2$. Then, write the partial fraction decomposition as

$$\frac{5x^2 + 20x + 6}{x(x + 1)^2} = \frac{A}{x} + \frac{B}{x + 1} + \frac{C}{(x + 1)^2}.$$

To solve this equation for A, B, and C, multiply both sides of the equation by the least common denominator $x(x + 1)^2$.

$$5x^2 + 20x + 6 = A(x + 1)^2 + Bx(x + 1) + Cx \qquad \text{\textit{Basic equation}}$$

Now, solve for A and C by substituting $x = -1$ and $x = 0$ into the basic equation.

Substitute $x = -1$:

$$5(-1)^2 + 20(-1) + 6 = A(-1 + 1)^2 + B(-1)(-1 + 1) + C(-1)$$
$$-9 = A(0) + B(0) - C$$
$$9 = C \qquad \text{\textit{Solve for C.}}$$

Substitute $x = 0$:

$$5(0)^2 + 20(0) + 6 = A(0 + 1)^2 + B(0)(0 + 1) + C(0)$$
$$6 = A(1) + B(0) + C(0)$$
$$6 = A \qquad \text{\textit{Solve for A.}}$$

At this point, you have exhausted the convenient choices for x and have yet to solve for B. When this happens, you can use *any* other x-value along with the known values of A and C.

Substitute $x = 1$, $A = 6$, and $C = 9$:

$$5(1)^2 + 20(1) + 6 = (6)(1 + 1)^2 + B(1)(1 + 1) + (9)(1)$$
$$31 = 6(4) + B(2) + 9(1)$$
$$-1 = B \qquad \text{\textit{Solve for B.}}$$

Now that you have solved for A, B, and C, you can use the partial fraction decomposition to integrate.

$$\int \frac{5x^2 + 20x + 6}{x^3 + 2x^2 + x}\, dx = \int \left(\frac{6}{x} - \frac{1}{x + 1} + \frac{9}{(x + 1)^2} \right) dx$$
$$= 6 \ln |x| - \ln |x + 1| + 9 \frac{(x + 1)^{-1}}{-1} + C$$
$$= \ln \left| \frac{x^6}{x + 1} \right| - \frac{9}{x + 1} + C$$

▦ **TECHNOLOGY**

The use of partial fractions depends on the ability to factor the denominator. If this cannot be easily done, then partial fractions should not be used. For instance, consider the integral

$$\int \frac{5x^2 + 20x + 6}{x^3 + 2x^2 + x + 1}\, dx.$$

This integral is only slightly different from that in Example 2, yet it is immensely more difficult to solve. When we tried using a symbolic integration utility on this integral, it was unable to solve it. Of course, if the integral had been a definite integral (as is true in many applied problems), then you can use an approximation technique such as the Midpoint Rule.

You can use the partial fraction decomposition technique outlined in Examples 1 and 2 only with a *proper* rational function—that is, a rational function whose numerator is of lower degree than its denominator. If the numerator is of equal or greater degree, you must divide first. For instance, the rational function

$$\frac{x^3}{x^2 + 1}$$

is improper because the degree of the numerator is greater than the degree of the denominator. Before applying partial fractions to this function, you should divide the denominator into the numerator to obtain

$$\frac{x^3}{x^2 + 1} = x - \frac{x}{x^2 + 1}.$$

EXAMPLE 3 *Integrating an Improper Rational Function*

Find the indefinite integral $\displaystyle\int \frac{x^5 + x - 1}{x^4 - x^3}\, dx.$

Solution

This rational function is improper—its numerator has a degree greater than that of its denominator. Thus, you should begin by dividing the denominator into the numerator to obtain

$$\frac{x^5 + x - 1}{x^4 - x^3} = x + 1 + \frac{x^3 + x - 1}{x^4 - x^3}.$$

Now, applying partial fraction decomposition produces

$$\frac{x^3 + x - 1}{x^3(x - 1)} = \frac{A}{x} + \frac{B}{x^2} + \frac{C}{x^3} + \frac{D}{x - 1}.$$

Multiplying both sides by the least common denominator $x^3(x - 1)$ produces the basic equation.

$$x^3 + x - 1 = Ax^2(x - 1) + Bx(x - 1) + C(x - 1) + Dx^3 \quad \begin{array}{l}\textit{Basic}\\ \textit{equation}\end{array}$$

Using techniques similar to those in the first two examples, you can solve for A, B, C, and D to obtain

$$A = 0, \quad B = 0, \quad C = 1, \quad \text{and} \quad D = 1.$$

Thus, you can integrate as follows.

$$\int \frac{x^5 + x - 1}{x^4 - x^3}\, dx = \int \left(x + 1 + \frac{x^3 + x - 1}{x^4 - x^3} \right) dx$$

$$= \int \left(x + 1 + \frac{1}{x^3} + \frac{1}{x - 1} \right) dx$$

$$= \frac{x^2}{2} + x - \frac{1}{2x^2} + \ln|x - 1| + C$$

It often happens that you must use more than one integration technique to solve an integral. For instance, the next example uses both substitution and partial fractions.

EXAMPLE 4 *Using Substitution with Partial Fractions*

Find the indefinite integral $\int \dfrac{1}{e^x + 1}\, dx$.

Solution

Begin with the substitution $u = e^x$. Then, $du = e^x\, dx$. By multiplying and dividing by e^x *inside* the integral, you obtain the following.

$$\int \frac{1}{e^x + 1}\, dx = \int \frac{1}{e^x(e^x + 1)} e^x\, dx \qquad \textit{Multiply and divide by } e^x.$$

$$= \int \frac{1}{u(u + 1)}\, du \qquad \textit{Substitute } u \textit{ and } du.$$

To solve this integral, you can use a partial fraction decomposition to write

$$\frac{1}{u(u + 1)} = \frac{1}{u} - \frac{1}{u + 1}.$$

Thus, you can complete the integration as follows.

$$\int \frac{1}{e^x + 1}\, dx = \int \frac{1}{u(u + 1)}\, du \qquad \textit{Substitution}$$

$$= \int \left(\frac{1}{u} - \frac{1}{u + 1} \right) du \qquad \textit{Partial fractions}$$

$$= \ln|u| - \ln|u + 1| + C \qquad \textit{Find antiderivative.}$$

$$= \ln|e^x| - \ln|e^x + 1| + C \qquad \textit{Substitute for } u.$$

$$= x - \ln|e^x + 1| + C \qquad \textit{Simplify.}$$

Try checking this result by differentiating. ▬

When integrating rational functions, remember that some can be integrated *without* using partial fractions. Here are three examples.

1. $\displaystyle\int \frac{x}{x^2 - 1}\, dx = \frac{1}{2}\int \frac{2x}{x^2 - 1}\, dx = \frac{1}{2}\ln|x^2 - 1| + C$

2. $\displaystyle\int \frac{\cancel{x}}{\cancel{x}(x - 1)^2}\, dx = \int \frac{1}{(x - 1)^2}\, dx = -\frac{1}{x - 1} + C$

3. $\displaystyle\int \frac{x^2 + 2x}{x^3 + 3x^2 - 4}\, dx = \frac{1}{3}\int \frac{3x^2 + 6x}{x^3 + 3x^2 - 4}\, dx = \frac{1}{3}\ln|x^3 + 3x^2 - 4| + C$

In the second example, note that it is generally a good idea to reduce a rational function as a first step to integration.

Logistics Growth Function

In Section 4.5, you saw that exponential growth occurs in situations for which the rate of growth is proportional to the quantity present at any given time. That is, if y is the quantity at time t, then

$$\frac{dy}{dt} = ky \qquad\qquad \frac{dy}{dt} \text{ is proportional to } y.$$

$$y = Ce^{kt}. \qquad\qquad \textit{Exponential growth function}$$

Exponential growth is unlimited. As long as C and k are positive, the value of Ce^{kt} can be made arbitrarily large by choosing sufficiently large values of t.

In many real-life situations, however, the growth of a quantity is limited and cannot increase beyond a certain size L, as shown in Figure 6.6. The **logistics growth** model assumes that the rate of growth is proportional to both the quantity y and the difference between the quantity and the limit L. That is,

$$\frac{dy}{dt} = ky(L - y). \qquad\qquad \frac{dy}{dt} \text{ is proportional to } y \text{ and } (L - y).$$

The solution of this *differential equation* is given in Example 5.

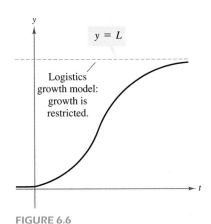

$y = L$

Logistics growth model: growth is restricted.

FIGURE 6.6

Note   The logistics growth model in Example 5 was simplified by assuming that the limit of the quantity y is 1. If the limit had been L, then the solution would have been

$$y = \frac{L}{1 + be^{-kt}}.$$

In the fourth step of the solution, notice that partial fractions were used to integrate the left side of the equation.

EXAMPLE 5 *Deriving the Logistics Growth Function*

Solve the equation $\dfrac{dy}{dt} = ky(1 - y)$. (Assume $y > 0$ and $1 - y > 0$.)

Solution

$$\frac{dy}{dt} = ky(1 - y) \qquad \textit{Differential equation}$$

$$\frac{1}{y(1 - y)}\, dy = k\, dt \qquad \textit{Write in differential form.}$$

$$\int \frac{1}{y(1 - y)}\, dy = \int k\, dt \qquad \textit{Integrate both sides.}$$

$$\int \left(\frac{1}{y} + \frac{1}{1 - y}\right) dy = \int k\, dt \qquad \textit{Rewrite using partial fractions.}$$

$$\ln y - \ln(1 - y) = kt + C_1 \qquad \textit{Find antiderivative.}$$

$$\ln \frac{y}{1 - y} = kt + C_1 \qquad \textit{Simplify.}$$

$$\frac{y}{1 - y} = Ce^{kt} \qquad \textit{Exponentiate and let } e^{C_1} = C.$$

Solving this equation for y produces

$$y = \frac{1}{1 + be^{-kt}}, \qquad \textit{Logistics growth function}$$

where $b = 1/C$.

Allstock/© Jim Brandenburg

In the United States, the National Park System has more than 330 protected areas for wildlife, and the National Wildlife Refuge System includes more than 400 refuges. The deer shown in this photo is a white-tailed buck.

EXAMPLE 6 *Modeling a Population*

Real Life

The state game commission releases 100 deer into a game refuge. During the first 5 years, the population increases to 432 deer. The commission believes that the population can be modeled by logistics growth with a limit of 2000 deer. Write the logistics growth model for this population. Then use the model to create a table showing the size of the deer population over the next 30 years.

Solution

Let y represent the number of deer in year t. Assuming a logistics growth model means that the rate of change in the population is proportional to both y and $(2000 - y)$. That is

$$\frac{dy}{dt} = ky(2000 - y).$$

The solution of this equation is

$$y = \frac{2000}{1 + be^{-kt}}.$$

Using the fact that $y = 100$ when $t = 0$, you can solve for b.

$$100 = \frac{2000}{1 + be^{-k(0)}} \quad \longrightarrow \quad b = 19$$

Then, using the fact that $y = 432$ when $t = 5$, you can solve for k.

$$432 = \frac{2000}{1 + 19e^{-k(5)}} \quad \longrightarrow \quad k \approx 0.33106$$

Thus, the logistics growth model for the population is

$$y = \frac{2000}{1 + 19e^{-0.33106t}}. \qquad \textit{Logistics growth model}$$

The population, in 5-year intervals, is shown in Table 6.1.

TABLE 6.1

Time, t	0	5	10	15	20	25	30
Population, y	100	432	1181	1766	1951	1990	1998

Discussion Problem Logistics Growth

Analyze the graph of the logistics growth function in Example 6. During which years is the *rate of growth* of the herd increasing? During which years is the *rate of growth* of the herd decreasing? How would these answers change if, instead of a limit of 2000 deer, the game refuge could contain a limit of 3000 deer?

Warm Up

The following warm-up exercises involve skills that were covered in earlier sections. You will use these skills in the exercise set for this section.

In Exercises 1–6, factor the expression.

1. $x^2 - x - 12$

2. $x^2 - 25$

3. $x^3 - x^2 - 2x$

4. $x^3 - 4x^2 + 4x$

5. $x^3 - 4x^2 + 5x - 2$

6. $x^3 - 5x^2 + 7x - 3$

In Exercises 7–10, rewrite the improper rational function as the sum of a proper rational function and a polynomial.

7. $\dfrac{x^2 - 2x + 1}{x - 2}$

8. $\dfrac{2x^2 - 4x + 1}{x - 1}$

9. $\dfrac{x^3 - 3x^2 + 2}{x - 2}$

10. $\dfrac{x^3 + 4x^2 + 5x + 2}{x^2 - 1}$

E X E R C I S E S   6.3

 means that technology can help you solve or check the exercise(s).

In Exercises 1–12, write the expression as a sum of partial fractions.

1. $\dfrac{2(x + 20)}{x^2 - 25}$

2. $\dfrac{10x + 3}{x^2 + x}$

3. $\dfrac{8x + 3}{x^2 - 3x}$

4. $\dfrac{3x + 11}{x^2 - 2x - 3}$

5. $\dfrac{4x - 13}{x^2 - 3x - 10}$

6. $\dfrac{7x + 5}{6(2x^2 + 3x + 1)}$

7. $\dfrac{3x^2 - 2x - 5}{x^3 + x^2}$

8. $\dfrac{3x^2 - x + 1}{x(x + 1)^2}$

9. $\dfrac{x + 1}{3(x - 2)^2}$

10. $\dfrac{3x - 4}{(x - 5)^2}$

11. $\dfrac{8x^2 + 15x + 9}{(x + 1)^3}$

12. $\dfrac{6x^2 - 5x}{x^3 + 6x^2 + 12x + 8}$

In Exercises 13–32, find the indefinite integral.

13. $\displaystyle\int \dfrac{1}{x^2 - 1}\, dx$

14. $\displaystyle\int \dfrac{9}{x^2 - 9}\, dx$

15. $\displaystyle\int \dfrac{-2}{x^2 - 16}\, dx$

16. $\displaystyle\int \dfrac{-4}{x^2 - 4}\, dx$

17. $\displaystyle\int \dfrac{1}{3x^2 - x}\, dx$

18. $\displaystyle\int \dfrac{3}{x^2 - 3x}\, dx$

19. $\displaystyle\int \dfrac{1}{2x^2 + x}\, dx$

20. $\displaystyle\int \dfrac{5}{x^2 + x - 6}\, dx$

21. $\displaystyle\int \dfrac{3}{x^2 + x - 2}\, dx$

22. $\displaystyle\int \dfrac{1}{4x^2 - 9}\, dx$

23. $\displaystyle\int \dfrac{5 - x}{2x^2 + x - 1}\, dx$

24. $\displaystyle\int \dfrac{x + 1}{x^2 + 4x + 3}\, dx$

25. $\displaystyle\int \dfrac{x^2 + 12x + 12}{x^3 - 4x}\, dx$

26. $\displaystyle\int \dfrac{3x^2 - 7x - 2}{x^3 - x}\, dx$

27. $\displaystyle\int \dfrac{x + 2}{x^2 - 4x}\, dx$

28. $\displaystyle\int \dfrac{4x^2 + 2x - 1}{x^3 + x^2}\, dx$

29. $\displaystyle\int \dfrac{4 - 3x}{(x - 1)^2}\, dx$

30. $\displaystyle\int \dfrac{x^4}{(x - 1)^3}\, dx$

31. $\displaystyle\int \dfrac{4x^2 - 1}{2x(x^2 + 2x + 1)}\, dx$

32. $\displaystyle\int \dfrac{3x}{x^2 - 6x + 9}\, dx$

In Exercises 33–40, evaluate the definite integral.

33. $\displaystyle\int_4^5 \dfrac{1}{9 - x^2}\, dx$

34. $\displaystyle\int_0^1 \dfrac{3}{2x^2 + 5x + 2}\, dx$

35. $\displaystyle\int_1^5 \dfrac{x - 1}{x^2(x + 1)}\, dx$

36. $\displaystyle\int_0^1 \dfrac{x^2 - x}{x^2 + x + 1}\, dx$

37. $\displaystyle\int_0^1 \frac{x^3}{x^2-2}\,dx$

38. $\displaystyle\int_0^1 \frac{x^3-1}{x^2-4}\,dx$

39. $\displaystyle\int_1^2 \frac{x^3-2x^2+1}{x^2-3x}\,dx$

40. $\displaystyle\int_2^4 \frac{x^4-4}{x^2-1}\,dx$

In Exercises 41–46, find the indefinite integral using the indicated substitution.

Integral	*Substitution*
41. $\displaystyle\int \frac{e^x}{(e^x-1)(e^x+4)}\,dx$	$u=e^x$
42. $\displaystyle\int \frac{e^x}{(e^{2x}-1)(e^x+1)}\,dx$	$u=e^x$
43. $\displaystyle\int \frac{1}{x\sqrt{4+x^2}}\,dx$	$u=\sqrt{4+x^2}$
44. $\displaystyle\int \frac{1}{x\sqrt{4+x}}\,dx$	$u=\sqrt{4+x}$
45. $\displaystyle\int \frac{1}{\sqrt{3x}\left(\sqrt{3x}+2\right)^2}\,dx$	$u=\sqrt{3x}$
46. $\displaystyle\int \frac{1}{x\left[1-(\ln x)^2\right]^2}\,dx$	$u=\ln x$

In Exercises 47–50, find the area of the shaded region.

47. $y=\dfrac{14}{16-x^2}$

48. $y=\dfrac{-4}{(x+2)(x-3)}$

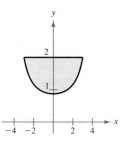

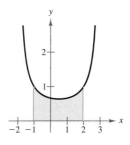

49. $y=\dfrac{x+1}{x^2-x}$

50. $y=\dfrac{x^2+2x-1}{x^2-4}$

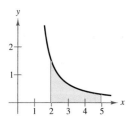

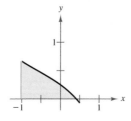

In Exercises 51–54, write the partial fraction decomposition for the rational expression. Check your result algebraically. Then assign a value to the constant a to check the result graphically.

51. $\dfrac{1}{a^2-x^2}$

52. $\dfrac{1}{x(x+a)}$

53. $\dfrac{1}{x(a-x)}$

54. $\dfrac{1}{(x+1)(a-x)}$

In Exercises 55–58, find the volume of the solid generated by revolving the region bounded by the graphs of the given equations about the x-axis.

55. $y=\dfrac{10}{x(x+10)}$, $y=0$, $x=1$, $x=5$

56. $y=\dfrac{-4}{(x+1)(x-4)}$, $y=0$, $x=0$, $x=3$

57. $y=\dfrac{2x}{x^3-4x}$, $x=1$, $x=-1$, $y=0$

58. $y=\dfrac{25x}{x^2+x-6}$, $x=-2$, $x=0$, $y=0$

59. *Population Growth* A conservation organization releases 100 animals of an endangered species into a game preserve. The organization believes that the preserve has a capacity of 1000 animals and that the growth of the herd will be logistic. That is, the size y of the herd will follow the equation

$$\int \frac{1}{y(1000-y)}\,dy = \int k\,dt,$$

where t is measured in years. Find this logistic curve. (To solve for the constant of integration C and the proportionality constant k, assume that $y=100$ when $t=0$ and $y=134$ when $t=2$.) Sketch the graph of your solution.

60. *Epidemic Model* A single infected individual enters a community of 500 individuals susceptible to the disease. The disease spreads at a rate proportional to the product of the total number infected and the number of susceptible individuals not yet infected. A model for the time it takes for the disease to spread to x individuals is

$$t=5010\int \frac{1}{(x+1)(500-x)}\,dx,$$

where t is the time in hours.

(a) Find the time it takes for 75% of the population to become infected (when $t=0$, $x=1$).

(b) Find the number of people who are infected after 100 hours.

61. *Marketing* After test-marketing a new menu item, a fast-food restaurant predicts that sales of the new item will grow according to the model

$$\frac{dS}{dt} = \frac{2t}{(t+4)^2},$$

where t is the time in weeks and S is the sales in thousands of dollars. Find the sales of the menu item at 10 weeks.

62. *Population Growth* One gram of a bacterial culture is present at time $t = 0$, and 10 grams is the upper limit of the culture's weight. The time required for the culture to grow to y grams is modeled by

$$kt = \int \frac{1}{y(10 - y)} dy,$$

where y is the weight of the culture in grams and t is the time in hours.

(a) Verify that the weight of the culture at time t is modeled by

$$y = \frac{10}{1 + 9e^{-10kt}}.$$

Use the fact that $y = 1$ when $t = 0$.

(b) Use the graph below to determine the constant k.

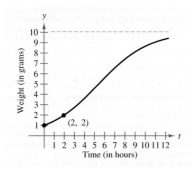

63. *GMAT* The number of people who took the Graduate Management Admission Test (GMAT) each year between 1982 and 1992 can be modeled by

$$N = \frac{242t^2 - 17{,}922t - 18{,}164}{4t^2 - 96t - 100},$$

where t is the time in years, with $t = 0$ corresponding to the year beginning September 1, 1982, and N is the number of test-takers in thousands. Find the total number of people who took the test between September 1, 1983 and August 31, 1990. *(Source: Graduate Management Admission Council)*

64. *Research Project* Use your school's library or some other reference source to research the opportunity cost of attending graduate school for 2 years to receive a Masters of Business Administration (MBA) degree rather than working for 2 years with a bachelor's degree. Write a short paper describing these costs.

Business Capsule

© Voldi Tanner from *Black Enterprise* Nov. 1993, Vol. 24, No. 3, page 34.

Right after receiving his undergraduate degree in history and economics from Stanford University in 1988, Greg Crossfield opened Field Copy and Printing, Inc. Crossfield launched the photocopy business by charging $300 on his VISA card for marketing. By the end of the first year in business, revenues reached $30,000, and in 1992 the company grossed $210,000.

6.4 Integration Tables and Completing the Square

Integration Tables ■ *Reduction Formulas* ■ *Completing the Square*

Integration Tables

So far in this chapter, you have studied three integration techniques to be used along with the basic integration formulas. Certainly these do not cover every possible method for finding an antiderivative, but they do cover most of the important ones.

In this section, you will expand the list of integration formulas to form a table of integrals. As you add new integration formulas to the basic list, two things occur. On one hand, it becomes increasingly difficult to memorize, or even become familiar with, the entire list of formulas. On the other hand, with a longer list you need fewer techniques for fitting an integral to one of the formulas on the list. The procedure of integrating by means of a long list of formulas is called **integration by tables.** (The tables in this section have only a partial listing of integration formulas. Much longer lists exist, some having several hundred formulas.)

Integration by tables should not be considered a trivial task. It requires considerable thought and insight, and it often requires substitution. Many people find a table of integrals to be a valuable supplement to the integration techniques discussed in the first three sections of this chapter. We encourage you to gain competence in the use of integration tables, as well as to continue to improve in the use of the various integration techniques. In doing so, you should find that a combination of techniques and tables is the most versatile approach to integration.

Each integration formula in the table on the following three pages can be developed using one or more of the techniques you have studied. You should try to verify several of the formulas. For instance, Formula 4

$$\int \frac{u}{(a + bu)^2} \, du = \frac{1}{b^2} \left(\frac{a}{a + bu} + \ln |a + bu| \right) + C \qquad \text{Formula 4}$$

can be verified using partial fractions, Formula 17

$$\int \frac{\sqrt{a + bu}}{u} \, du = 2\sqrt{a + bu} + a \int \frac{1}{u\sqrt{a + bu}} \, du \qquad \text{Formula 17}$$

can be verified using integration by parts, and Formula 37

$$\int \frac{1}{1 + e^u} \, du = u - \ln(1 + e^u) + C \qquad \text{Formula 37}$$

can be verified using substitution and partial fractions.

Note A symbolic integration utility, such as *Derive*, *Maple*, or *Mathematica*, consists, in part, of a database of integration tables. The primary difference between using a symbolic integrator and using a table of integrals is that with a symbolic integrator the computer searches through the database to find a fit. With a table of integrals, *you* must do the searching.

In the following table of integrals, the formulas have been grouped into eight different types according to the form of the integrand.

Forms involving u^n	Forms involving $a + bu$
Forms involving $\sqrt{a + bu}$	Forms involving $\sqrt{u^2 \pm a^2}$
Forms involving $u^2 - a^2$	Forms involving $\sqrt{a^2 - u^2}$
Forms involving e^u	Forms involving $\ln u$

Table of Integrals

Forms involving u^n

1. $\displaystyle \int u^n \, du = \frac{u^{n+1}}{n+1} + C, \quad n \neq -1$

2. $\displaystyle \int \frac{1}{u} \, du = \ln |u| + C$

Forms involving $a + bu$

3. $\displaystyle \int \frac{u}{a + bu} \, du = \frac{1}{b^2} (bu - a \ln |a + bu|) + C$

4. $\displaystyle \int \frac{u}{(a + bu)^2} \, du = \frac{1}{b^2} \left(\frac{a}{a + bu} + \ln |a + bu| \right) + C$

5. $\displaystyle \int \frac{u}{(a + bu)^n} \, du = \frac{1}{b^2} \left[\frac{-1}{(n-2)(a+bu)^{n-2}} + \frac{a}{(n-1)(a+bu)^{n-1}} \right] + C, \quad n \neq 1, 2$

6. $\displaystyle \int \frac{u^2}{a + bu} \, du = \frac{1}{b^3} \left[-\frac{bu}{2}(2a - bu) + a^2 \ln |a + bu| \right] + C$

7. $\displaystyle \int \frac{u^2}{(a + bu)^2} \, du = \frac{1}{b^3} \left(bu - \frac{a^2}{a + bu} - 2a \ln |a + bu| \right) + C$

8. $\displaystyle \int \frac{u^2}{(a + bu)^3} \, du = \frac{1}{b^3} \left[\frac{2a}{a + bu} - \frac{a^2}{2(a + bu)^2} + \ln |a + bu| \right] + C$

9. $\displaystyle \int \frac{u^2}{(a + bu)^n} \, du = \frac{1}{b^3} \left[\frac{-1}{(n-3)(a+bu)^{n-3}} + \frac{2a}{(n-2)(a+bu)^{n-2}} - \frac{a^2}{(n-1)(a+bu)^{n-1}} \right] + C, \quad n \neq 1, 2, 3$

10. $\displaystyle \int \frac{1}{u(a + bu)} \, du = \frac{1}{a} \ln \left| \frac{u}{a + bu} \right| + C$

11. $\displaystyle \int \frac{1}{u(a + bu)^2} \, du = \frac{1}{a} \left(\frac{1}{a + bu} + \frac{1}{a} \ln \left| \frac{u}{a + bu} \right| \right) + C$

12. $\displaystyle \int \frac{1}{u^2(a + bu)} \, du = -\frac{1}{a} \left(\frac{1}{u} + \frac{b}{a} \ln \left| \frac{u}{a + bu} \right| \right) + C$

13. $\displaystyle \int \frac{1}{u^2(a + bu)^2} \, du = -\frac{1}{a^2} \left[\frac{a + 2bu}{u(a + bu)} + \frac{2b}{a} \ln \left| \frac{u}{a + bu} \right| \right] + C$

Table of Integrals (continued)

Forms involving $\sqrt{a + bu}$

14. $\displaystyle \int u^n \sqrt{a + bu}\, du = \frac{2}{b(2n + 3)} \left[u^n (a + bu)^{3/2} - na \int u^{n-1} \sqrt{a + bu}\, du \right]$

15. $\displaystyle \int \frac{1}{u\sqrt{a + bu}}\, du = \frac{1}{\sqrt{a}} \ln \left| \frac{\sqrt{a + bu} - \sqrt{a}}{\sqrt{a + bu} + \sqrt{a}} \right| + C, \quad 0 < a$

16. $\displaystyle \int \frac{1}{u^n \sqrt{a + bu}}\, du = \frac{-1}{a(n - 1)} \left[\frac{\sqrt{a + bu}}{u^{n-1}} + \frac{(2n - 3)b}{2} \int \frac{1}{u^{n-1}\sqrt{a + bu}}\, du \right], \quad n \neq 1$

17. $\displaystyle \int \frac{\sqrt{a + bu}}{u}\, du = 2\sqrt{a + bu} + a \int \frac{1}{u\sqrt{a + bu}}\, du$

18. $\displaystyle \int \frac{\sqrt{a + bu}}{u^n}\, du = \frac{-1}{a(n - 1)} \left[\frac{(a + bu)^{3/2}}{u^{n-1}} + \frac{(2n - 5)b}{2} \int \frac{\sqrt{a + bu}}{u^{n-1}}\, du \right], \quad n \neq 1$

19. $\displaystyle \int \frac{u}{\sqrt{a + bu}}\, du = -\frac{2(2a - bu)}{3b^2} \sqrt{a + bu} + C$

20. $\displaystyle \int \frac{u^n}{\sqrt{a + bu}}\, du = \frac{2}{(2n + 1)b} \left(u^n \sqrt{a + bu} - na \int \frac{u^{n-1}}{\sqrt{a + bu}}\, du \right)$

Forms involving $\sqrt{u^2 \pm a^2}, 0 < a$

21. $\displaystyle \int \sqrt{u^2 \pm a^2}\, du = \frac{1}{2} \left(u\sqrt{u^2 \pm a^2} \pm a^2 \ln |u + \sqrt{u^2 \pm a^2}| \right) + C$

22. $\displaystyle \int u^2 \sqrt{u^2 \pm a^2}\, du = \frac{1}{8} \left[u(2u^2 \pm a^2)\sqrt{u^2 \pm a^2} - a^4 \ln |u + \sqrt{u^2 \pm a^2}| \right] + C$

23. $\displaystyle \int \frac{\sqrt{u^2 + a^2}}{u}\, du = \sqrt{u^2 + a^2} - a \ln \left| \frac{a + \sqrt{u^2 + a^2}}{u} \right| + C$

24. $\displaystyle \int \frac{\sqrt{u^2 \pm a^2}}{u^2}\, du = \frac{-\sqrt{u^2 \pm a^2}}{u} + \ln |u + \sqrt{u^2 \pm a^2}| + C$

25. $\displaystyle \int \frac{1}{\sqrt{u^2 \pm a^2}}\, du = \ln |u + \sqrt{u^2 \pm a^2}| + C$

26. $\displaystyle \int \frac{1}{u\sqrt{u^2 + a^2}}\, du = -\frac{1}{a} \ln \left| \frac{a + \sqrt{u^2 + a^2}}{u} \right| + C$

27. $\displaystyle \int \frac{u^2}{\sqrt{u^2 \pm a^2}}\, du = \frac{1}{2} \left(u\sqrt{u^2 \pm a^2} \mp a^2 \ln |u + \sqrt{u^2 \pm a^2}| \right) + C$

28. $\displaystyle \int \frac{1}{u^2 \sqrt{u^2 \pm a^2}}\, du = \mp \frac{\sqrt{u^2 \pm a^2}}{a^2 u} + C$

Table of Integrals (continued)

Forms involving $u^2 - a^2$, $0 < a$

29. $\displaystyle \int \frac{1}{u^2 - a^2} \, du = -\int \frac{1}{a^2 - u^2} \, du = \frac{1}{2a} \ln \left| \frac{u - a}{u + a} \right| + C$

30. $\displaystyle \int \frac{1}{(u^2 - a^2)^n} \, du = \frac{-1}{2a^2(n-1)} \left[\frac{u}{(u^2 - a^2)^{n-1}} + (2n - 3) \int \frac{1}{(u^2 - a^2)^{n-1}} \, du \right], \quad n \neq 1$

Forms involving $\sqrt{a^2 - u^2}, 0 < a$

31. $\displaystyle \int \frac{\sqrt{a^2 - u^2}}{u} \, du = \sqrt{a^2 - u^2} - a \ln \left| \frac{a + \sqrt{a^2 - u^2}}{u} \right| + C$

32. $\displaystyle \int \frac{1}{u\sqrt{a^2 - u^2}} \, du = -\frac{1}{a} \ln \left| \frac{a + \sqrt{a^2 - u^2}}{u} \right| + C$

33. $\displaystyle \int \frac{1}{u^2 \sqrt{a^2 - u^2}} \, du = \frac{-\sqrt{a^2 - u^2}}{a^2 u} + C$

Forms involving e^u

34. $\displaystyle \int e^u \, du = e^u + C$

35. $\displaystyle \int u e^u \, du = (u - 1)e^u + C$

36. $\displaystyle \int u^n e^u \, du = u^n e^u - n \int u^{n-1} e^u \, du$

37. $\displaystyle \int \frac{1}{1 + e^u} \, du = u - \ln(1 + e^u) + C$

38. $\displaystyle \int \frac{1}{1 + e^{nu}} \, du = u - \frac{1}{n} \ln(1 + e^{nu}) + C$

Forms involving $\ln u$

39. $\displaystyle \int \ln u \, du = u(-1 + \ln u) + C$

40. $\displaystyle \int u \ln u \, du = \frac{u^2}{4}(-1 + 2 \ln u) + C$

41. $\displaystyle \int u^n \ln u \, du = \frac{u^{n+1}}{(n+1)^2}[-1 + (n + 1) \ln u] + C, \quad n \neq -1$

42. $\displaystyle \int (\ln u)^2 \, du = u[2 - 2 \ln u + (\ln u)^2] + C$

43. $\displaystyle \int (\ln u)^n \, du = u(\ln u)^n - n \int (\ln u)^{n-1} \, du$

EXAMPLE 1 *Using Integration Tables*

Find the indefinite integral

$$\int \frac{x}{\sqrt{x-1}}\,dx.$$

Solution

Because the expression inside the radical is linear, you should consider forms involving $\sqrt{a+bu}$, as in Formula 19.

$$\int \frac{u}{\sqrt{a+bu}}\,du = -\frac{2(2a-bu)}{3b^2}\sqrt{a+bu} + C \qquad \textit{Formula 19}$$

Using this formula, let $a = -1$, $b = 1$, and $u = x$. Then $du = dx$, and you obtain the following.

$$\int \frac{x}{\sqrt{x-1}}\,dx = -\frac{2(-2-x)}{3}\sqrt{x-1} + C \qquad \textit{Substitute values of a, b, and u.}$$

$$= \frac{2}{3}(2+x)\sqrt{x-1} + C \qquad \textit{Simplify.}$$

■ **TECHNOLOGY**

Throughout this section, remember that a symbolic integration utility, such as *Derive*, *Mathematica*, or *Maple*, can be used instead of integration tables. If you have access to such a utility, try using it to find the indefinite integrals in Examples 1 and 2.

EXAMPLE 2 *Using Integration Tables*

Find the indefinite integral

$$\int x\sqrt{x^4 - 9}\,dx.$$

Solution

Because it is not clear which formula to use, you can begin by letting $u = x^2$ and $du = 2x\,dx$. With this substitution, you can write the integral as follows.

$$\int x\sqrt{x^4 - 9}\,dx = \frac{1}{2}\int \sqrt{(x^2)^2 - 9}(2x)\,dx \qquad \textit{Multiply and divide by 2.}$$

$$= \frac{1}{2}\int \sqrt{u^2 - 9}\,du \qquad \textit{Substitute u and du.}$$

Now, it appears that you can use Formula 21.

$$\int \sqrt{u^2 - a^2}\,du = \frac{1}{2}\left(u\sqrt{u^2 - a^2} - a^2 \ln |u + \sqrt{u^2 - a^2}|\right) + C.$$

Letting $a = 3$, you obtain

$$\int x\sqrt{x^4 - 9}\,dx = \frac{1}{2}\int \sqrt{u^2 - a^2}\,du$$

$$= \frac{1}{2}\left[\frac{1}{2}\left(u\sqrt{u^2 - a^2} - a^2 \ln |u + \sqrt{u^2 - a^2}|\right)\right] + C$$

$$= \frac{1}{4}\left(x^2\sqrt{x^4 - 9} - 9 \ln |x^2 + \sqrt{x^4 - 9}|\right) + C.$$

EXAMPLE 3 *Using Integration Tables*

Find the indefinite integral

$$\int \frac{1}{x\sqrt{x+1}}\,dx.$$

Solution
Considering forms involving $\sqrt{a+bu}$, where $a = 1$, $b = 1$, and $u = x$, you can use Formula 15.

$$\int \frac{1}{u\sqrt{a+bu}}\,du = \frac{1}{\sqrt{a}}\ln\left|\frac{\sqrt{a+bu}-\sqrt{a}}{\sqrt{a+bu}+\sqrt{a}}\right| + C, \qquad 0 < a.$$

Therefore,

$$\int \frac{1}{x\sqrt{x+1}}\,dx = \int \frac{1}{u\sqrt{a+bu}}\,du$$

$$= \frac{1}{\sqrt{a}}\ln\left|\frac{\sqrt{a+bu}-\sqrt{a}}{\sqrt{a+bu}+\sqrt{a}}\right| + C$$

$$= \ln\left|\frac{\sqrt{x+1}-1}{\sqrt{x+1}+1}\right| + C.$$

EXAMPLE 4 *Using Integration Tables*

Find the indefinite integral

$$\int \frac{x}{1+e^{-x^2}}\,dx.$$

Solution
Of the forms involving e^u, Formula 37

$$\int \frac{1}{1+e^u}\,du = u - \ln(1+e^u) + C$$

seems most appropriate. To use this formula, let $u = -x^2$ and $du = -2x\,dx$.

$$\int \frac{x}{1+e^{-x^2}}\,dx = -\frac{1}{2}\int \frac{1}{1+e^{-x^2}}(-2x)\,dx$$

$$= -\frac{1}{2}\int \frac{1}{1+e^u}\,du$$

$$= -\frac{1}{2}[u - \ln(1+e^u)] + C$$

$$= -\frac{1}{2}[-x^2 - \ln(1+e^{-x^2})] + C$$

$$= \frac{1}{2}[x^2 + \ln(1+e^{-x^2})] + C.$$

Reduction Formulas

Several of the formulas in the integration table have the form

$$\int f(x)\,dx = g(x) + \int h(x)\,dx,$$

where the right side contains an integral. Such integration formulas are called **reduction formulas** because they reduce the original integral to the sum of a function and a simpler integral.

EXAMPLE 5 *Using a Reduction Formula*

Find the indefinite integral $\int x^2 e^x\,dx$.

Solution
Using Formula 36,

$$\int u^n e^u\,du = u^n e^u - n\int u^{n-1} e^u\,du,$$

you can let $u = x$ and $n = 2$. Then $du = dx$, and you can write

$$\int x^2 e^x\,dx = x^2 e^x - 2\int x e^x\,dx.$$

Then, using Formula 35,

$$\int u e^u\,du = (u - 1)e^u + C,$$

you can write

$$\int x^2 e^x\,dx = x^2 e^x - 2\int x e^x\,dx$$
$$= x^2 e^x - 2(x - 1)e^x + C$$
$$= x^2 e^x - 2x e^x + 2e^x + C$$
$$= e^x(x^2 - 2x + 2) + C.$$

▬

TECHNOLOGY

You have now studied two ways to find the indefinite integral in Example 5. Example 5 uses an integration table, and Example 4 in Section 6.2 uses integration by parts. A third way would be to use a symbolic integrator. For instance, the screen at the right shows how to use *Derive* to solve the integral.

1: $\mathbf{x^2 \hat{e}^x}$ *Author.*

2: $\int \mathbf{x^2 \hat{e}^x dx}$ *Integrate.*

3: $\hat{e}^x(\mathbf{x^2 - 2x + 2})$ *Simplify.*

Completing the Square

Many integration formulas involve the sum or difference of two squares. You can extend the use of these formulas by an algebraic procedure called **completing the square.** This procedure is demonstrated in Example 6.

EXAMPLE 6 *Completing the Square*

Find the indefinite integral

$$\int \frac{1}{x^2 - 4x + 1} \, dx.$$

Solution

Begin by writing the denominator as the difference of two squares.

$$x^2 - 4x + 1 = (x^2 - 4x + 4) - 4 + 1 = (x - 2)^2 - 3.$$

Therefore, you can rewrite the original integral as

$$\int \frac{1}{x^2 - 4x + 1} \, dx = \int \frac{1}{(x - 2)^2 - 3} \, dx.$$

Considering $u = x - 2$ and $a = \sqrt{3}$, you can apply Formula 29

$$\int \frac{1}{u^2 - a^2} \, du = \frac{1}{2a} \ln \left| \frac{u - a}{u + a} \right| + C$$

to conclude that

$$\int \frac{1}{x^2 - 4x + 1} \, dx = \int \frac{1}{(x - 2)^2 - 3} \, dx$$

$$= \int \frac{1}{u^2 - a^2} \, du$$

$$= \frac{1}{2a} \ln \left| \frac{u - a}{u + a} \right| + C$$

$$= \frac{1}{2\sqrt{3}} \ln \left| \frac{x - 2 - \sqrt{3}}{x - 2 + \sqrt{3}} \right| + C.$$

Discussion Problem **Using Integration Tables**

Which integration formulas on pages 415–417 would you use to integrate the following? Explain your choice of u for each integral.

a. $\displaystyle\int \frac{e^x}{e^x + 1} \, dx$ b. $\displaystyle\int \frac{e^x}{e^{2x} + 1} \, dx$ c. $\displaystyle\int \frac{1}{e^{2x} + 1} \, dx$

Warm Up

The following warm-up exercises involve skills that were covered in earlier sections. You will use these skills in the exercise set for this section.

In Exercises 1–4, expand the expression.

1. $(x+4)^2$

2. $\left(x+\frac{1}{2}\right)^2$

3. $(x-1)^2$

4. $\left(x-\frac{1}{3}\right)^2$

In Exercises 5–8, write the partial fraction decomposition for the function.

5. $\dfrac{4}{x(x+2)}$

6. $\dfrac{3}{x(x-4)}$

7. $\dfrac{x+4}{x^2(x-2)}$

8. $\dfrac{3x^2+4x-8}{x(x-2)(x+1)}$

In Exercises 9 and 10, use integration by parts to find the indefinite integral.

9. $\displaystyle\int 2xe^x\,dx$

10. $\displaystyle\int 3x^2\ln x\,dx$

EXERCISES 6.4   means that technology can help you solve or check the exercise(s).

In Exercises 1–10, use the indicated formula from the table of integrals in this section to find the indefinite integral.

1. $\displaystyle\int \dfrac{x}{(2+3x)^2}\,dx$, Formula 4

2. $\displaystyle\int \dfrac{1}{x(2+3x)^2}\,dx$, Formula 11

3. $\displaystyle\int \dfrac{x}{\sqrt{2+3x}}\,dx$, Formula 19

4. $\displaystyle\int \dfrac{4}{x^2-9}\,dx$, Formula 29

5. $\displaystyle\int \dfrac{2x}{\sqrt{x^4-9}}\,dx$, Formula 25

6. $\displaystyle\int x^2\sqrt{x^2+9}\,dx$, Formula 22

7. $\displaystyle\int x^3e^{x^2}\,dx$, Formula 35

8. $\displaystyle\int \dfrac{x}{1+e^{x^2}}\,dx$, Formula 37

9. $\displaystyle\int x\ln(x^2+1)\,dx$, Formula 39

10. $\displaystyle\int x^3\ln x\,dx$, Formula 41

In Exercises 11–40, use the table of integrals in this section to find the indefinite integral.

11. $\displaystyle\int \dfrac{1}{x(1+x)}\,dx$

12. $\displaystyle\int \dfrac{1}{x(1+x)^2}\,dx$

13. $\displaystyle\int \dfrac{1}{x\sqrt{x^2+9}}\,dx$

14. $\displaystyle\int \dfrac{1}{\sqrt{x^2-1}}\,dx$

15. $\displaystyle\int \dfrac{1}{x\sqrt{4-x^2}}\,dx$

16. $\displaystyle\int \dfrac{\sqrt{x^2-9}}{x^2}\,dx$

17. $\displaystyle\int x\ln x\,dx$

18. $\displaystyle\int x^2(\ln x^3)^2\,dx$

19. $\displaystyle\int \dfrac{6x}{1+e^{3x^2}}\,dx$

20. $\displaystyle\int \dfrac{1}{1+e^x}\,dx$

21. $\displaystyle\int x\sqrt{x^4-4}\,dx$

22. $\displaystyle\int \dfrac{x}{x^4-9}\,dx$

23. $\displaystyle\int \frac{t^2}{(2+3t)^3}\,dt$

24. $\displaystyle\int \frac{\sqrt{3+4t}}{t}\,dt$

25. $\displaystyle\int \frac{s}{s^2\sqrt{3+s}}\,ds$

26. $\displaystyle\int \sqrt{3+x^2}\,dx$

27. $\displaystyle\int \frac{x^2}{(3+2x)^5}\,dx$

28. $\displaystyle\int \frac{1}{x^2\sqrt{x^2-4}}\,dx$

29. $\displaystyle\int \frac{1}{x^2\sqrt{1-x^2}}\,dx$

30. $\displaystyle\int \frac{2x}{(1-3x)^2}\,dx$

31. $\displaystyle\int x^2 \ln x\,dx$

32. $\displaystyle\int xe^{x^2}\,dx$

33. $\displaystyle\int \frac{x^2}{(3x-5)^2}\,dx$

34. $\displaystyle\int \frac{1}{2x^2(2x-1)^2}\,dx$

35. $\displaystyle\int x^2\sqrt{x^2+4}\,dx$

36. $\displaystyle\int \frac{1}{\sqrt{x}(1+2\sqrt{x})}\,dx$

37. $\displaystyle\int \frac{2}{1+e^{4x}}\,dx$

38. $\displaystyle\int \frac{e^x}{(1-e^{2x})}\,dx$

39. $\displaystyle\int \frac{\ln x}{x(4+3\ln x)}\,dx$

40. $\displaystyle\int (\ln x)^3\,dx$

In Exercises 41–44, evaluate the definite integral.

41. $\displaystyle\int_0^5 \frac{x}{\sqrt{5+2x}}\,dx$

42. $\displaystyle\int_0^5 \frac{x}{(4+x)^2}\,dx$

43. $\displaystyle\int_0^4 \frac{6}{1+e^{0.5x}}\,dx$

44. $\displaystyle\int_1^4 x\ln x\,dx$

In Exercises 45–48, find the indefinite integral (a) using integration tables and (b) using the specified method.

Integral	Method
45. $\displaystyle\int x^2 e^x\,dx$	Integration by parts
46. $\displaystyle\int x^4 \ln x\,dx$	Integration by parts
47. $\displaystyle\int \frac{1}{x^2(x+1)}\,dx$	Partial fractions
48. $\displaystyle\int \frac{1}{x^2-75}\,dx$	Partial fractions

In Exercises 49 and 50, complete the square to express each polynomial as the sum or difference of squares.

49. (a) x^2+6x (b) x^2-8x+9
(c) x^4+2x^2-5 (d) $3-2x-x^2$

50. (a) $2x^2+12x+14$ (b) $3x^2-12x-9$
(c) x^2-2x (d) $9+8x-x^2$

In Exercises 51–58, complete the square and then use integration tables to find the indefinite integral.

51. $\displaystyle\int \frac{1}{x^2+6x-8}\,dx$

52. $\displaystyle\int \frac{1}{(x^2+4x-5)}\,dx$

53. $\displaystyle\int \frac{1}{(x-1)\sqrt{x^2-2x+2}}\,dx$

54. $\displaystyle\int \sqrt{x^2-6x}\,dx$

55. $\displaystyle\int \frac{1}{2x^2-4x-6}\,dx$

56. $\displaystyle\int \frac{\sqrt{7-6x-x^2}}{x+3}\,dx$

57. $\displaystyle\int \frac{x}{\sqrt{x^4+2x^2+2}}\,dx$

58. $\displaystyle\int \frac{x\sqrt{x^4+4x^2+5}}{x^2+2}\,dx$

In Exercises 59 and 60, find the area of the region bounded by the graphs of the equations.

59. $y=\dfrac{x}{\sqrt{x+1}}$, $y=0$, $x=8$

60. $y=\dfrac{x}{1+e^{x^2}}$, $y=0$, $x=2$

Population Growth In Exercises 61 and 62, find the average value of the growth function over the interval, where N is the size of a population and t is the time in days.

61. $N=\dfrac{50}{1+e^{4.8-1.9t}}$, $[3,\ 4]$

62. $N=\dfrac{375}{1+e^{4.20-0.25t}}$, $[21,\ 28]$

63. *Revenue* The marginal revenue (in dollars per year) for a new product is modeled by
$$\frac{dR}{dt}=10{,}000\left[1-\frac{1}{(1+0.1t^2)^{1/2}}\right],$$
where t is the time in years. Estimate the total revenue from sales of the product over its first 2 years on the market.

64. *Consumer and Producer Surplus* Find the consumer surplus and the producer surplus for a product with the given demand and supply functions.
$$\text{Demand: } p=\frac{60}{\sqrt{x^2+81}}, \qquad \text{Supply: } p=\frac{x}{3}$$
(*Hint:* For a definition of consumer and producer surplus, see Section 5.5.)

6.5 Numerical Integration

Trapezoidal Rule ■ *Simpson's Rule* ■ *Error Analysis*

Trapezoidal Rule

FIGURE 6.7

FIGURE 6.8

In Section 5.6, you studied one technique for approximating the value of a *definite* integral—the Midpoint Rule. In this section, you will study two other approximation techniques: the **Trapezoidal Rule** and **Simpson's Rule.**

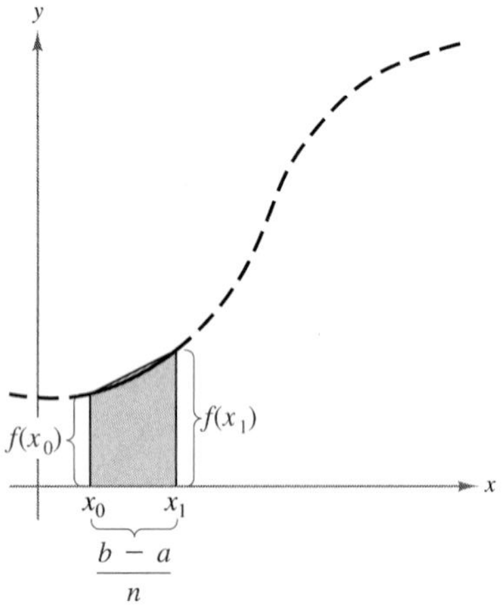

To develop the Trapezoidal Rule, consider a function f that is nonnegative and continuous on the closed interval $[a, b]$. To approximate the area represented by $\int_a^b f(x)\,dx$, partition the interval into n subintervals, each of width

$$\Delta x = \frac{b-a}{n}. \qquad \textit{Width of each subinterval}$$

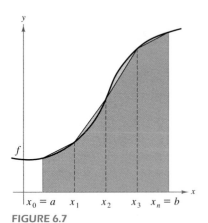

Next, form n trapezoids, as shown in Figure 6.7. As you can see in Figure 6.8, the area of the first trapezoid is

$$\text{Area of first trapezoid} = \frac{b-a}{n}\left[\frac{f(x_0)+f(x_1)}{2}\right].$$

The areas of the other trapezoids follow a similar pattern, and the sum of the n areas is

$$\frac{b-a}{n}\left[\frac{f(x_0)+f(x_1)}{2}+\frac{f(x_1)+f(x_2)}{2}+\cdot\cdot\cdot+\frac{f(x_{n-1})+f(x_n)}{2}\right]$$
$$=\frac{b-a}{2n}[f(x_0)+f(x_1)+f(x_1)+f(x_2)+\cdot\cdot\cdot+f(x_{n-1})+f(x_n)]$$
$$=\frac{b-a}{2n}[f(x_0)+2f(x_1)+2f(x_2)+\cdot\cdot\cdot+2f(x_{n-1})+f(x_n)].$$

Although this development assumes f to be continuous *and* nonnegative on $[a, b]$, the resulting formula is valid as long as f is continuous on $[a, b]$.

The Trapezoidal Rule

If f is continuous on $[a, b]$, then

$$\int_a^b f(x)\,dx \approx \frac{b-a}{2n}[f(x_0)+2f(x_1)+\cdot\cdot\cdot+2f(x_{n-1})+f(x_n)].$$

EXAMPLE 1 *Using the Trapezoidal Rule*

Use the Trapezoidal Rule to approximate the definite integral

$$\int_0^1 e^x \, dx.$$

Compare the results for $n = 4$ and $n = 8$.

Solution

When $n = 4$, the width of each subinterval is $(1 - 0)/4 = \frac{1}{4}$ and the endpoints of the subintervals are

$$x_0 = 0, \quad x_1 = \frac{1}{4}, \quad x_2 = \frac{1}{2}, \quad x_3 = \frac{3}{4}, \quad \text{and} \quad x_4 = 1,$$

as indicated in Figure 6.9. Therefore, by the Trapezoidal Rule,

$$\int_0^1 e^x \, dx = \frac{1}{8}(e^0 + 2e^{0.25} + 2e^{0.5} + 2e^{0.75} + e^1)$$

$$\approx 1.7272. \qquad \qquad \textit{Approximation using } n = 4$$

When $n = 8$, the width of each subinterval is $(1 - 0)/8 = \frac{1}{8}$ and the endpoints of the subintervals are

$$x_0 = 0, \quad x_1 = \frac{1}{8}, \quad x_2 = \frac{1}{4}, \quad x_3 = \frac{3}{8}, \quad x_4 = \frac{1}{2}$$

$$x_5 = \frac{5}{8}, \quad x_6 = \frac{3}{4}, \quad x_7 = \frac{7}{8}, \quad \text{and} \quad x_8 = 1,$$

as indicated in Figure 6.10. Therefore, by the Trapezoidal Rule,

$$\int_0^1 e^x \, dx = \frac{1}{16}(e^0 + 2e^{0.125} + 2e^{0.25} + \cdots + 2e^{0.875} + e^1)$$

$$\approx 1.7205. \qquad \qquad \textit{Approximation using } n = 8$$

Of course, for *this particular* integral, you could have found an antiderivative and used the Fundamental Theorem to find the exact value of the definite integral. The exact value is

$$\int_0^1 e^x \, dx = e - 1 \approx 1.7183. \qquad \qquad \textit{Exact value}$$

■

There are two important points that should be made concerning the Trapezoidal Rule. First, the approximation tends to become more accurate as n increases. For instance, in Example 1, if $n = 16$, the Trapezoidal Rule yields an approximation of 1.7188. Second, although you could have used the Fundamental Theorem of Calculus to evaluate the integral in Example 1, this theorem cannot be used to evaluate an integral as simple as $\int_0^1 e^{x^2} \, dx$, because e^{x^2} has no elementary function as an antiderivative. Yet, the Trapezoidal Rule can be easily applied to this integral.

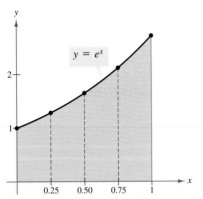

FIGURE 6.9 Four Subintervals

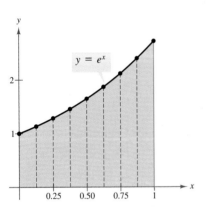

FIGURE 6.10 Eight Subintervals

Simpson's Rule

One way to view the Trapezoidal Rule is to say that on each subinterval, f is approximated by a first-degree polynomial. In Simpson's Rule, f is approximated by a second-degree polynomial on each subinterval.

To develop Simpson's Rule, partition the interval $[a, b]$ into an *even number n* of subintervals, each of width $(b - a)/n$. On the subinterval $[x_0, x_2]$, approximate the function f by the second-degree polynomial $p(x)$ that passes through the points

$$(x_0, f(x_0)), \quad (x_1, f(x_1)), \quad \text{and} \quad (x_2, f(x_2)),$$

as shown in Figure 6.11. The Fundamental Theorem of Calculus can be used to show that

$$\int_{x_0}^{x_2} f(x)\, dx \approx \int_{x_0}^{x_2} p(x)\, dx$$

$$= \frac{x_2 - x_0}{6} \left[p(x_0) + 4p\left(\frac{x_0 + x_2}{2}\right) + p(x_2) \right]$$

$$= \frac{2[(b - a)/n]}{6} [p(x_0) + 4p(x_1) + p(x_2)]$$

$$= \frac{b - a}{3n} [f(x_0) + 4f(x_1) + f(x_2)].$$

Repeating this process on the subintervals $[x_{i-2}, x_i]$ produces

$$\int_a^b f(x)\, dx \approx \frac{b - a}{3n} \Big[f(x_0) + 4f(x_1) + f(x_2) + f(x_2) + 4f(x_3)$$

$$+ f(x_4) + \cdots + f(x_{n-2}) + 4f(x_{n-1}) + f(x_n) \Big].$$

By grouping like terms, you can obtain the following approximation, which is known as Simpson's Rule. This rule is named after the English mathematician Thomas Simpson (1710–1761).

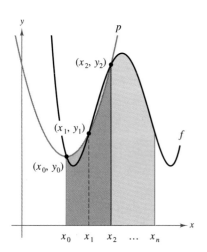

$$\int_{x_0}^{x_2} p(x)\, dx \approx \int_{x_0}^{x_2} f(x)\, dx$$

FIGURE 6.11

Simpson's Rule (n is Even)

If f is continuous on $[a, b]$, then

$$\int_a^b f(x)\, dx \approx \frac{b - a}{3n} [f(x_0) + 4f(x_1) + 2f(x_2) + 4f(x_3) +$$

$$\cdots + 4f(x_{n-1}) + f(x_n)].$$

Note   The coefficients in Simpson's Rule have the pattern

$$1 \quad 4 \quad 2 \quad 4 \quad 2 \quad 4 \quad \cdots \quad 4 \quad 2 \quad 4 \quad 1.$$

In Example 1, the Trapezoidal Rule was used to estimate the value of

$$\int_0^1 e^x \, dx.$$

The next example uses Simpson's Rule to approximate the same integral.

EXAMPLE 2 *Using Simpson's Rule*

Use Simpson's Rule to approximate the definite integral

$$\int_0^1 e^x \, dx.$$

Compare the results for $n = 4$ and $n = 8$.

Solution

When $n = 4$, the width of each subinterval is $(1 - 0)/4 = \frac{1}{4}$ and the endpoints of the subintervals are

$$x_0 = 0, \quad x_1 = \frac{1}{4}, \quad x_2 = \frac{1}{2}, \quad x_3 = \frac{3}{4}, \quad \text{and} \quad x_4 = 1.$$

Therefore, by Simpson's Rule,

$$\int_0^1 e^x \, dx = \frac{1}{12}(e^0 + 4e^{0.25} + 2e^{0.5} + 4e^{0.75} + e^1)$$

$$\approx 1.718319. \qquad \textit{Approximation using } n = 4$$

When $n = 8$, the width of each subinterval is $(1 - 0)/8 = \frac{1}{8}$ and the endpoints of the subintervals are

$$x_0 = 0, \quad x_1 = \frac{1}{8}, \quad x_2 = \frac{1}{4}, \quad x_3 = \frac{3}{8}, \quad x_4 = \frac{1}{2}$$

$$x_5 = \frac{5}{8}, \quad x_6 = \frac{3}{4}, \quad x_7 = \frac{7}{8}, \quad \text{and} \quad x_8 = 1.$$

Therefore, by Simpson's Rule,

$$\int_0^1 e^x \, dx = \frac{1}{24}(e^0 + 4e^{0.125} + 2e^{0.25} + \cdots + 4e^{0.875} + e^1)$$

$$\approx 1.718284. \qquad \textit{Approximation using } n = 8$$

Recall that the exact value of this integral is

$$\int_0^1 e^x \, dx = e - 1 \approx 1.718282.$$

Thus, with only eight subintervals, you obtained an approximation that is correct to the nearest 0.00002—an impressive result! ■

Note Comparing the results of Examples 1 and 2, you can see that for a given value of n, Simpson's Rule tends to be more accurate than the Trapezoidal Rule.

TECHNOLOGY

Programming Simpson's Rule

In Section 5.6, you saw how to program the Midpoint Rule into a computer or programmable calculator. The program below will evaluate Simpson's Rule on a *TI-81* or *TI-82* calculator. See Appendix D for other models.

```
Prgm1:SIMPSONS
:Disp ''LOWER LIMIT''          Prompt for value of a.
:Input A                       Input value of a.
:Disp ''UPPER LIMIT''          Prompt for value of b.
:Input B                       Input value of b.
:Disp ''N/2 DIVISIONS''        Prompt for value of n/2.
:Input D                       Input value of n/2.
:0→S                           Initialize sum of areas.
:(B−A)/(2D)→W                  Calculate width of subinterval.
:1→J                           Initialize counter.
:Lbl 1                         Begin loop.
:A+2(J−1)W→L                   Calculate left endpoint.
:A+2JW→R                       Calculate right endpoint.
:(L+R)/2→M                     Calculate midpoint of subinterval.
:L→X
:Y₁→L                          Evaluate f(x) at left endpoint.
:M→X
:Y₁→M                          Evaluate f(x) at midpoint.
:R→X
:Y₁→R                          Evaluate f(x) at right endpoint.
:W(L+4M+R)/3+S→S               Simpson's Rule
:IS>(J,D)                      Check value of index, J.
:Goto 1                        End loop.
:Disp ''APPROXIMATION''
:Disp S                        Display approximation.
```

Before executing the program, enter the function as Y_1. When the program is executed, you will be prompted to enter the lower and upper limits of integration, and *half* the number of subintervals you want to use.

To check your steps, try running the program with the integral in Example 2. When $n = 8$ (input 4), you should obtain an approximation of

$$\int_0^1 e^x \, dx \approx 1.718284155.$$

Error Analysis

In Examples 1 and 2, you were able to calculate the exact value of the integral and compare that value with the approximations to see how good they were. In practice, you need to have a different way of telling how good an approximation is: such a way is provided in the next result.

Error in Trapezoidal and Simpson's Rules

The error E in approximating $\int_a^b f(x)\,dx$ is as follows.

Trapezoidal Rule: $|E| \leq \dfrac{(b-a)^3}{12n^2}\left[\max|f''(x)|\right]$, $a \leq x \leq b$

Simpson's Rule: $|E| \leq \dfrac{(b-a)^5}{180n^4}\left[\max|f^{(4)}(x)|\right]$, $a \leq x \leq b$

This result indicates that the errors generated by the Trapezoidal Rule and Simpson's Rule have upper bounds dependent on the extreme values of $f''(x)$ and $f^{(4)}(x)$ in the interval $[a, b]$. Furthermore, the bounds for the errors can be made arbitrarily small by *increasing n*. To determine what value of n to choose, consider the following steps.

Trapezoidal Rule

1. Find $f''(x)$.
2. Find the maximum of $|f''(x)|$ on the interval $[a, b]$.
3. Set up the inequality $|E| \leq \dfrac{(b-a)^3}{12n^2}\left[\max|f''(x)|\right]$.
4. For an error less than ϵ, solve for n in the inequality
$$\frac{(b-a)^3}{12n^2}\left[\max|f''(x)|\right] < \epsilon.$$
5. Partition $[a, b]$ into n subintervals and apply the Trapezoidal Rule.

Simpson's Rule

1. Find $f^{(4)}(x)$.
2. Find the maximum of $|f^{(4)}(x)|$ on the interval $[a, b]$.
3. Set up the inequality $|E| \leq \dfrac{(b-a)^5}{180n^4}\left[\max|f^{(4)}(x)|\right]$.
4. For an error less than ϵ, solve for n in the inequality
$$\frac{(b-a)^5}{180n^4}\left[\max|f^{(4)}(x)|\right] < \epsilon.$$
5. Partition $[a, b]$ into n subintervals and apply Simpson's Rule.

Discovery

How does the error in the Trapezoidal Rule decrease as you increase n, the number of subintervals? In Example 1 on page 425, you approximated the integral $\int_0^1 e^x\,dx$ using the Trapezoidal Rule with four and eight subintervals. Because the exact value of this integral is known to be $e - 1$, the errors are

$n = 4$ error: 0.008918

and

$n = 8$ error: 0.002218.

Notice that the error diminishes by a factor of approximately 4 when the number of subintervals is doubled. Explain how this is consistent with the formula for the Trapezoidal Rule error. What do you think the error would be if you used 16 subintervals in Example 1? (The actual error is 0.0005593, which is approximately 0.002218/4.) How does the error in Simpson's Rule diminish as you increase the number of subintervals?

EXAMPLE 3 *Using the Trapezoidal Rule*

Use the Trapezoidal Rule to estimate the value of the integral

$$\int_0^1 e^{-x^2}\, dx$$

so that the approximation error is less than 0.01.

Solution

1. Begin by finding the second derivative of $f(x) = e^{-x^2}$.

$$f(x) = e^{-x^2}$$
$$f'(x) = -2xe^{-x^2}$$
$$f''(x) = 4x^2 e^{-x^2} - 2e^{-x^2} = 2e^{-x^2}(2x^2 - 1)$$

2. Using calculus, you can determine that f'' has only one critical number in the interval [0, 1] and that the maximum value of $|f''(x)|$ on this interval is $|f''(0)| = 2$.

3. The error E using the Trapezoidal Rule is bounded by

$$|E| \le \frac{(b-a)^3}{12n^2}(2) = \frac{1}{12n^2}(2) = \frac{1}{6n^2}.$$

4. To ensure that the approximation has an error of less than 0.01, you should choose n so that

$$\frac{1}{6n^2} < 0.01.$$

Solving for n, you can determine that n must be 5 or more.

5. Partition [0, 1] into five subintervals, as shown in Figure 6.12. Then apply the Trapezoidal Rule to obtain

$$\int_0^1 e^{-x^2}\, dx = \frac{1}{10}\left(\frac{1}{e^0} + \frac{2}{e^{0.04}} + \frac{2}{e^{0.16}} + \frac{2}{e^{0.36}} + \frac{2}{e^{0.64}} + \frac{1}{e^1}\right)$$

$$\approx 0.744.$$

Therefore, with an error of less than 0.01, you know that

$$0.734 \le \int_0^1 e^{-x^2}\, dx \le 0.754.$$

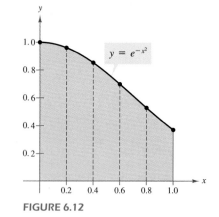

$y = e^{-x^2}$

FIGURE 6.12

Discussion Problem   **Simpson's Rule Error**

Rework Example 3 using Simpson's Rule instead of the Trapezoidal Rule.

Warm Up

The following warm-up exercises involve skills that were covered in earlier sections. You will use these skills in the exercise set for this section.

In Exercises 1–6, find the indicated derivative.

1. $f(x) = \dfrac{1}{x},\ f''(x)$

2. $f(x) = \ln(2x + 1),\ f^{(4)}(x)$

3. $f(x) = 2\ln x,\ f^{(4)}(x)$

4. $f(x) = x^3 - 2x^2 + 7x - 12,$
$f''(x)$

5. $f(x) = e^{2x},\ f^{(4)}(x)$

6. $f(x) = e^{x^2},\ f''(x)$

In Exercises 7 and 8, find the absolute maximum of *f* on the interval.

7. $f(x) = -x^2 + 6x + 9,\ [0,\ 4]$

8. $f(x) = \dfrac{8}{x^3},\ [1,\ 2]$

In Exercises 9 and 10, solve for *n*.

9. $\dfrac{1}{4n^2} < 0.001$

10. $\dfrac{1}{16n^4} < 0.0001$

E X E R C I S E S   6.5   means that technology can help you solve or check the exercise(s).

In Exercises 1–10, use the Trapezoidal Rule and Simpson's Rule to approximate the value of the definite integral for the indicated value of *n*. Compare these results with the exact value of the definite integral. Round your answers to four decimal places.

1. $\displaystyle\int_0^2 x^2\,dx,\ n = 4$

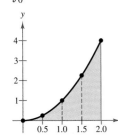

2. $\displaystyle\int_0^1 \left(\dfrac{x^2}{2} + 1\right)dx,\ n = 4$

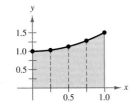

3. $\displaystyle\int_0^2 (x^4 + 1)dx,\ n = 4$

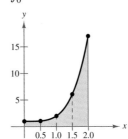

4. $\displaystyle\int_1^2 \dfrac{1}{x}\,dx,\ n = 4$

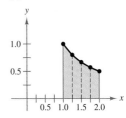

5. $\displaystyle\int_0^2 x^3\,dx,\ n = 8$

6. $\displaystyle\int_1^2 \dfrac{1}{x}\,dx,\ n = 8$

7. $\displaystyle\int_1^2 \dfrac{1}{x^2}\,dx,\ n = 4$

8. $\displaystyle\int_0^4 \sqrt{x}\,dx,\ n = 8$

9. $\displaystyle\int_0^1 \dfrac{1}{1+x}\,dx,\ n = 4$

10. $\displaystyle\int_0^2 x\sqrt{x^2 + 1}\,dx,\ n = 4$

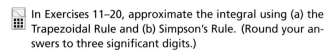

In Exercises 11–20, approximate the integral using (a) the Trapezoidal Rule and (b) Simpson's Rule. (Round your answers to three significant digits.)

Definite Integral	Subdivisions

11. $\displaystyle\int_0^2 (1 + x^3)\, dx$ $n = 4$

12. $\displaystyle\int_0^2 \frac{1}{\sqrt{1 + x^3}}\, dx$ $n = 4$

13. $\displaystyle\int_0^1 \sqrt{x}\, \sqrt{1 - x}\, dx$ $n = 4$

14. $\displaystyle\int_0^1 \frac{1}{1 + x^2}\, dx$ $n = 4$

15. $\displaystyle\int_0^1 \sqrt{1 - x^2}\, dx$ $n = 4$

16. $\displaystyle\int_0^2 e^{-x^2}\, dx$ $n = 2$

17. $\displaystyle\int_0^1 \sqrt{1 - x^2}\, dx$ $n = 8$

18. $\displaystyle\int_0^2 e^{-x^2}\, dx$ $n = 4$

19. $\displaystyle\int_0^3 \frac{1}{2 - 2x + x^2}\, dx$ $n = 6$

20. $\displaystyle\int_0^3 \frac{x}{2 + x + x^2}\, dx$ $n = 6$

Present Value In Exercises 21 and 22, use Simpson's Rule (with $n = 8$) to approximate the present value of the income $c(t)$ over t_1 years at the given annual interest rate r. (Present value is defined in Section 6.2.)

21. $c(t) = 6000 + 200\sqrt{t}$, $r = 7\%$, $t_1 = 4$

22. $c(t) = 200{,}000 + 15{,}000\sqrt[3]{t}$, $r = 10\%$, $t_1 = 8$

Marginal Analysis In Exercises 23 and 24, use Simpson's Rule (with $n = 4$) to approximate the change in revenue from the marginal revenue function dR/dx. In each case, assume that the number of units sold x increases from 14 to 16.

23. $\dfrac{dR}{dx} = 5\sqrt{8000 - x^3}$ **24.** $\dfrac{dR}{dx} = 50\sqrt{x}\, \sqrt{20 - x}$

Probability In Exercises 25 and 26, use Simpson's Rule (with $n = 6$) to approximate the indicated normal probability. The standard normal probability density function is

$$f(x) = \frac{1}{\sqrt{2\pi}}\, e^{-x^2/2}.$$

If x is chosen at random from a population with this density, then the probability that x lies in the interval $[a, b]$ is $P(a \le x \le b) = \int_a^b f(x)\, dx$.

25. $P(0 \le x \le 1)$ **26.** $P(0 \le x \le 2)$

Surveying In Exercises 27 and 28, use Simpson's Rule to estimate the number of square feet of land in the lot where x and y are measured in feet, as shown in the accompanying figures. In each case the land is bounded by a stream and two straight roads.

27.

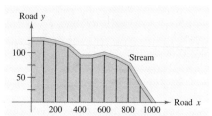

x	0	100	200	300	400	500
y	125	125	120	112	90	90

x	600	700	800	900	1000
y	95	88	75	35	0

28.

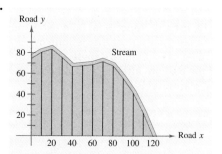

x	0	10	20	30	40	50	60
y	75	81	84	76	67	68	69

x	70	80	90	100	110	120
y	72	68	56	42	23	0

In Exercises 29–32, use the error formula to find bounds for the error in approximating the integral using (a) the Trapezoidal Rule and (b) Simpson's Rule. (Let $n = 4$.)

29. $\int_0^2 x^4 \, dx$

30. $\int_0^1 \frac{1}{x+1} \, dx$

31. $\int_0^1 e^{x^3} \, dx$

32. $\int_0^1 e^{-x^2} \, dx$

In Exercises 33–36, use the error formula to find n so that the error in the approximation of the definite integral is less than 0.0001 using (a) the Trapezoidal Rule and (b) Simpson's Rule.

33. $\int_0^1 x^4 \, dx$

34. $\int_1^3 \frac{1}{x} \, dx$

35. $\int_1^3 e^{2x} \, dx$

36. $\int_3^5 \ln x \, dx$

 In Exercises 37–40, use the program for Simpson's Rule given on page 428 to approximate the integral.

37. $\int_1^4 x\sqrt{x+4} \, dx$

38. $\int_1^4 x^2\sqrt{x+4} \, dx$

39. $\int_2^5 10xe^{-x} \, dx$

40. $\int_2^5 10x^2 e^{-x} \, dx$

41. Prove that Simpson's Rule is exact when used to approximate the integral of a cubic polynomial function, and demonstrate the result for

$$\int_0^1 x^3 \, dx, \quad n = 2.$$

 42. Use Simpson's Rule (with $n = 4$) to find the volume of the solid generated by revolving the region bounded by the graphs of

$$y = x\sqrt[3]{x+4}, \quad y = 0, \quad \text{and} \quad x = 4$$

about the x-axis.

In Exercises 43 and 44, use the following definite integral to find the required arc length. If f has a continuous derivative, then the arc length of f between the points $(a, f(a))$ and $(b, f(b))$ is

$$\int_a^b \sqrt{1 + [f'(x)]^2} \, dx.$$

43. *Arc Length* The suspension cable on a bridge that is 400 feet long is in the shape of a parabola whose equation is $y = x^2/800$ (see figure). Use Simpson's Rule (with $n = 12$) to approximate the length of the cable. Compare this result with the length obtained by using the table in Section 6.4 to perform the integration.

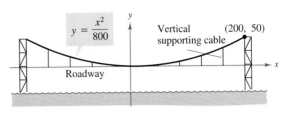

FIGURE FOR 43

44. *Arc Length* A fleeing object leaves the origin and moves up the y-axis (see figure). At the same time, a pursuer leaves the point $(1, 0)$ and always moves toward the fleeing object. If the pursuer's speed is twice that of the fleeing object, the equation of the pursuer's path is

$$y = \frac{1}{3}(x^{3/2} - 3x^{1/2} + 2).$$

Find the distance traveled by the pursuer by integrating over the interval $[0, 1]$.

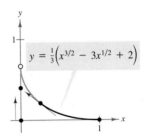

45. *Medicine* A body assimilates a 12-hour cold tablet at a rate modeled by

$$\frac{dC}{dt} = 8 - \ln(t^2 - 2t + 4), \quad 0 \le t \le 12,$$

where dC/dt is measured in grams per hour and t is the time (in hours). Find the total amount of the drug absorbed into the body during the 12 hours.

46. *Medicine* The concentration M (in grams per liter) of a 6-hour allergy medicine in a body is modeled by

$$M = 12 - 4\ln(t^2 - 4t + 6), \quad 0 \le t \le 6,$$

where t is the time in hours since the allergy medication was taken. Find the average level of concentration in the body over the 6-hour period.

47. *Magazine Subscribers* The rate of change S of subscribers to a newly introduced magazine is modeled by

$$\frac{dS}{dt} = 1000t^2 e^{-t}, \quad 0 \le t \le 6,$$

where t is the time in years. Find the total increase in the number of subscribers during the first 6 years.

Improper Integrals

Improper Integrals ■ *Integrals with Infinite Limits of Integration* ■
Integrals with Infinite Integrands ■ *Applications*

Improper Integrals

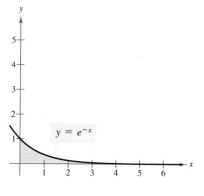

The definition of the definite integral

$$\int_a^b f(x)\,dx$$

includes the requirements that the interval $[a, b]$ be finite and that f be bounded on $[a, b]$. In this section, you will study integrals that do not satisfy these requirements because of one of the following.

1. One or both of the limits of integration are infinite.
2. f has an infinite discontinuity on the interval $[a, b]$.

Integrals possessing either of these characteristics are called **improper integrals.** For instance, the integrals

$$\int_0^\infty e^{-x}\,dx \quad \text{and} \quad \int_{-\infty}^\infty \frac{1}{x^2+1}\,dx$$

are improper because one or both limits of integration are infinite, as indicated in Figure 6.13. Similarly, the integrals

$$\int_1^5 \frac{1}{\sqrt{x-1}}\,dx \quad \text{and} \quad \int_{-2}^2 \frac{1}{(x+1)^2}\,dx$$

are improper because their integrands approach infinity somewhere in the interval of integration, as indicated in Figure 6.14.

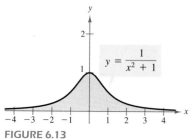

FIGURE 6.13

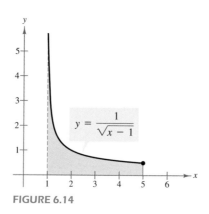
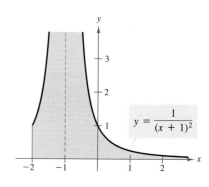

FIGURE 6.14

Integrals with Infinite Limits of Integration

To see how to evaluate an improper integral, consider the integral shown in Figure 6.15. As long as b is a real number that is greater than 1 (no matter how large), this is a definite integral whose value is

$$\int_1^b \frac{1}{x^2}\,dx = \left[-\frac{1}{x}\right]_1^b = -\frac{1}{b} + 1 = 1 - \frac{1}{b}.$$

Table 6.2 shows the values of this integral for several values of b.

TABLE 6.2

b	2	5	10	100	1000	10,000
$\int_1^b \frac{1}{x^2}\,dx = 1 - \frac{1}{b}$	0.5000	0.8000	0.9000	0.9900	0.9990	0.9999

From this table, it appears that the value of the integral is approaching a limit as b increases without bound. This limit is denoted by the following *improper integral.*

$$\int_1^\infty \frac{1}{x^2}\,dx = \lim_{b\to\infty} \int_1^b \frac{1}{x^2}\,dx$$
$$= \lim_{b\to\infty}\left(1 - \frac{1}{b}\right)$$
$$= 1$$

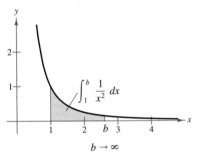

FIGURE 6.15

Improper Integrals (Infinite Limits of Integration)

1. If f is continuous on the interval $[a, \infty)$, then

$$\int_a^\infty f(x)\,dx = \lim_{b\to\infty} \int_a^b f(x)\,dx.$$

2. If f is continuous on the interval $(-\infty, b]$, then

$$\int_{-\infty}^b f(x)\,dx = \lim_{a\to-\infty} \int_a^b f(x)\,dx.$$

3. If f is continuous on the interval $(-\infty, \infty)$, then

$$\int_{-\infty}^\infty f(x)\,dx = \int_{-\infty}^c f(x)\,dx + \int_c^\infty f(x)\,dx,$$

where c is any real number.

In each case, if the limit exists, then the improper integral **converges;** otherwise the improper integral **diverges.** Thus, in the third case, the integral will diverge if either one of the integrals on the right diverges.

EXAMPLE 1 *Evaluating an Improper Integral*

Determine the convergence or divergence of the improper integral

$$\int_1^\infty \frac{1}{x}\,dx.$$

Solution

Begin by applying the definition of an improper integral.

$$\int_1^\infty \frac{1}{x}\,dx = \lim_{b\to\infty} \int_1^b \frac{1}{x}\,dx \qquad \textit{Definition of improper integral}$$

$$= \lim_{b\to\infty} \left[\ln x\right]_1^b \qquad \textit{Find antiderivative.}$$

$$= \lim_{b\to\infty} (\ln b - 0) \qquad \textit{Apply Fundamental Theorem.}$$

$$= \infty \qquad \textit{Evaluate limit.}$$

Because the limit is infinite, the improper integral diverges. ▬

As you begin to work with improper integrals, you will find that integrals that appear to be similar can have very different values. For instance, consider the two improper integrals

$$\int_1^\infty \frac{1}{x}\,dx = \infty \qquad \textit{Divergent integral}$$

and

$$\int_1^\infty \frac{1}{x^2}\,dx = 1. \qquad \textit{Convergent integral}$$

The first integral diverges and the second converges to 1. Graphically, this means that the areas shown in Figure 6.16 are very different. The region lying between the graph of $y = 1/x$ and the x-axis (for $x \geq 1$) has an *infinite* area, and the region lying between the graph of $y = 1/x^2$ and the x-axis (for $x \geq 1$) has a *finite* area.

▦ **TECHNOLOGY**

Symbolic integration utilities, such as *Derive*, *Maple*, or *Mathmematica*, evaluate improper integrals in much the same way they evaluate definite integrals. For instance, the steps below show how to use *Derive* to evaluate the improper integral $\int_1^\infty (1/x^2)\,dx$.

1: $\dfrac{1}{x^2}$

2: $\displaystyle\int_1^\infty \frac{1}{x^2}\mathbf{dx}$

3: 1

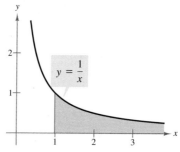

Diverges (infinite area)

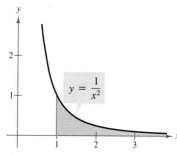

Converges (finite area)

FIGURE 6.16

EXAMPLE 2 *Evaluating an Improper Integral*

Evaluate the improper integral

$$\int_{-\infty}^{0} \frac{1}{(1-2x)^{3/2}}\,dx.$$

Solution
Begin by applying the definition of an improper integral.

$$\int_{-\infty}^{0} \frac{1}{(1-2x)^{3/2}}\,dx = \lim_{a\to-\infty} \int_{a}^{0} \frac{1}{(1-2x)^{3/2}}\,dx \qquad \text{\textit{Definition of improper integral}}$$

$$= \lim_{a\to-\infty} \left[\frac{1}{\sqrt{1-2x}} \right]_{a}^{0} \qquad \text{\textit{Find antiderivative.}}$$

$$= \lim_{a\to-\infty} \left(1 - \frac{1}{\sqrt{1-2a}} \right) \qquad \text{\textit{Apply Fundamental Theorem.}}$$

$$= 1 - 0 \qquad \text{\textit{Evaluate limit.}}$$

$$= 1 \qquad \text{\textit{Simplify.}}$$

Thus, the improper integral converges to 1. As shown in Figure 6.17, this implies that the region lying between the graph of $y = 1/(1-2x)^{3/2}$ and the x-axis (for $x \le 0$) has an area of 1. ■

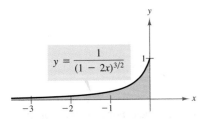

$$y = \frac{1}{(1-2x)^{3/2}}$$

FIGURE 6.17

EXAMPLE 3 *Evaluating an Improper Integral*

Evaluate the improper integral

$$\int_{0}^{\infty} 2xe^{-x^2}\,dx.$$

Solution
Begin by applying the definition of an improper integral.

$$\int_{0}^{\infty} 2xe^{-x^2}\,dx = \lim_{b\to\infty} \int_{0}^{b} 2xe^{-x^2}\,dx \qquad \text{\textit{Definition of improper integral}}$$

$$= \lim_{b\to\infty} \left[-e^{-x^2} \right]_{0}^{b} \qquad \text{\textit{Find antiderivative.}}$$

$$= \lim_{b\to\infty} \left(-e^{-b^2} + 1 \right) \qquad \text{\textit{Apply Fundamental Theorem.}}$$

$$= 0 + 1 \qquad \text{\textit{Evaluate limit.}}$$

$$= 1 \qquad \text{\textit{Simplify.}}$$

Thus, the improper integral converges to 1. As shown in Figure 6.18, this implies that the region lying between the graph of $y = 2xe^{-x^2}$ and the x-axis (for $x \ge 0$) has an area of 1. ■

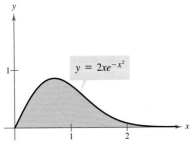

$$y = 2xe^{-x^2}$$

FIGURE 6.18

Integrals with Infinite Integrands

Improper Integrals (Infinite Integrands)

1. If f is continuous on the interval $[a, b)$ and approaches infinity at b, then

$$\int_a^b f(x)\,dx = \lim_{c \to b^-} \int_a^c f(x)\,dx.$$

2. If f is continuous on the interval $(a, b]$ and approaches infinity at a, then

$$\int_a^b f(x)\,dx = \lim_{c \to a^+} \int_c^b f(x)\,dx.$$

3. If f is continuous on the interval $[a, b]$, except for some c in (a, b) at which f approaches infinity, then

$$\int_a^b f(x)\,dx = \int_a^c f(x)\,dx + \int_c^b f(x)\,dx.$$

In each case, if the limit exists, then the improper integral **converges;** otherwise the improper integral **diverges.**

EXAMPLE 4 *Evaluating an Improper Integral*

Evaluate the improper integral

$$\int_1^2 \frac{1}{\sqrt[3]{x-1}}\,dx.$$

Solution

$$
\begin{aligned}
\int_1^2 \frac{1}{\sqrt[3]{x-1}}\,dx &= \lim_{b \to 1^+} \int_b^2 \frac{1}{\sqrt[3]{x-1}}\,dx && \text{\textit{Definition of improper integral}} \\[2mm]
&= \lim_{b \to 1^+} \left[\frac{3}{2}(x-1)^{2/3} \right]_b^2 && \text{\textit{Find antiderivative.}} \\[2mm]
&= \lim_{b \to 1^+} \left[\frac{3}{2} - \frac{3}{2}(b-1)^{2/3} \right] && \text{\textit{Apply Fundamental Theorem.}} \\[2mm]
&= \frac{3}{2} - 0 && \text{\textit{Evaluate limit.}} \\[2mm]
&= \frac{3}{2} && \text{\textit{Simplify.}}
\end{aligned}
$$

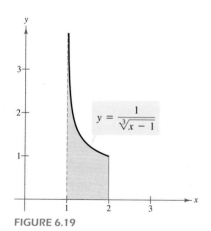

FIGURE 6.19

$$y = \frac{1}{\sqrt[3]{x-1}}$$

Thus, the integral converges to $\frac{3}{2}$. This implies that the region shown in Figure 6.19 has an area of $\frac{3}{2}$.

EXAMPLE 5 *Evaluating an Improper Integral*

Evaluate the improper integral

$$\int_1^2 \frac{2}{x^2 - 2x}\,dx.$$

Solution

$$\int_1^2 \frac{2}{x^2 - 2x}\,dx = \int_1^2 \left(\frac{1}{x - 2} - \frac{1}{x}\right)dx \qquad \textit{Use partial fractions.}$$

$$= \lim_{b \to 2^-} \int_1^b \left(\frac{1}{x - 2} - \frac{1}{x}\right)dx \qquad \begin{matrix}\textit{Definition of}\\\textit{improper integral}\end{matrix}$$

$$= \lim_{b \to 2^-} \left[\ln|x - 2| - \ln|x|\right]_1^b \qquad \textit{Find antiderivative.}$$

$$= -\infty \qquad \textit{Evaluate limit.}$$

Thus, you can conclude that the integral diverges. This implies that the region shown in Figure 6.20 has an infinite area.

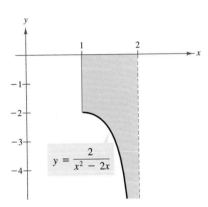

$$y = \frac{2}{x^2 - 2x}$$

FIGURE 6.20

EXAMPLE 6 *Evaluating an Improper Integral*

Evaluate the improper integral

$$\int_{-1}^2 \frac{1}{x^3}\,dx.$$

Solution

This integral is improper because the integrand has an infinite discontinuity at the interior value $x = 0$, as shown in Figure 6.21. Thus, you can write

$$\int_{-1}^2 \frac{1}{x^3}\,dx = \int_{-1}^0 \frac{1}{x^3}\,dx + \int_0^2 \frac{1}{x^3}\,dx.$$

By applying the definition of an improper integral, you can show that each of these integrals diverges. Therefore, the original improper integral also diverges.

Note Had you not recognized that the integral in Example 6 was improper, you would have obtained the incorrect result

$$\int_{-1}^2 \frac{1}{x^3}\,dx = \left[-\frac{1}{2x^2}\right]_{-1}^2 = -\frac{1}{8} + \frac{1}{2} = \frac{3}{8}. \qquad \textit{Incorrect}$$

Improper integrals in which the integrand has an infinite discontinuity *between* the limits of integration are often overlooked, so keep alert for such possibilities. Even symbolic integrators can have trouble with this type of integral—when we used *Derive* to evaluate this integral, it gave us the same incorrect result.

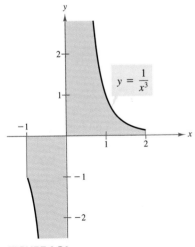

$$y = \frac{1}{x^3}$$

FIGURE 6.21

Applications

In Section 4.2, you studied the graph of the *normal probability density function*

$$f(x) = \frac{1}{\sigma\sqrt{2\pi}} e^{-(x-\mu)^2/2\sigma^2}.$$

This function is used in statistics to represent a population that is normally distributed with a mean of μ and a standard deviation of σ. Specifically, if an outcome x is chosen at random from the population, the probability that x will have a value between a and b is

$$P(a \leq x \leq b) = \int_a^b \frac{1}{\sigma\sqrt{2\pi}} e^{-(x-\mu)^2/2\sigma^2} \, dx.$$

As shown in Figure 6.22, the probability $P(-\infty < x < \infty)$ is

$$P(-\infty < x < \infty) = \int_{-\infty}^{\infty} \frac{1}{\sigma\sqrt{2\pi}} e^{-(x-\mu)^2/2\sigma^2} \, dx = 1.$$

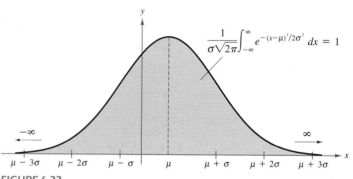

$$\frac{1}{\sigma\sqrt{2\pi}} \int_{-\infty}^{\infty} e^{-(x-\mu)^2/2\sigma^2} \, dx = 1$$

$-\infty$ ∞

$\mu - 3\sigma$ $\mu - 2\sigma$ $\mu - \sigma$ μ $\mu + \sigma$ $\mu + 2\sigma$ $\mu + 3\sigma$

FIGURE 6.22

FPG International/© J. Zimmerman

Many professional basketball players are over $6\frac{1}{2}$ feet tall. If a man is chosen at random from the population, the probability that he is $6\frac{1}{2}$ feet tall or taller is less than half of one percent.

Real Life

EXAMPLE 7 *Finding a Probability*

In 1992, the mean height of American men (between 18 and 24 years old) was 70 inches, and the standard deviation was 3 inches. If an 18- to 24-year-old man was chosen at random from the population, what is the probability that he was 6 feet tall or taller? *(Source: National Center for Health Statistics)*

Solution

Using a mean of $\mu = 70$ and a standard deviation of $\sigma = 3$, the probability $P(72 \leq x < \infty)$ is given by the improper integral

$$P(72 \leq x < \infty) = \int_{72}^{\infty} \frac{1}{3\sqrt{2\pi}} e^{-(x-70)^2/18} \, dx.$$

Using a symbolic integrator, you can approximate the value of this integral to be 0.252. Thus, the probability that the man was 6 feet tall or taller is about 25.2%.

Real
Life

Discovery

Example 8 shows that the volume of Gabriel's horn is finite. What is the *area* under the curve $f(x) = 1/x$ and above the x-axis for $x \geq 1$? It can be shown that the surface area of Gabriel's horn is given by the integral

$$2\pi \int_1^\infty \frac{1}{x} \sqrt{1 + \frac{1}{x^4}}\, dx.$$

Use a symbolic integration utility to show that this improper integral diverges. Hence, Gabriel's horn has finite volume but infinite surface area! How could you paint Gabriel's horn?

EXAMPLE 8 *Finding Volume*

The solid formed by revolving the graph of

$$f(x) = \frac{1}{x}, \qquad 1 \leq x < \infty$$

about the x-axis is called **Gabriel's horn.** (See Figure 6.23.) Find the volume of Gabriel's horn.

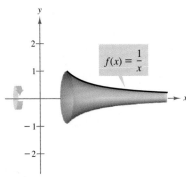

$f(x) = \dfrac{1}{x}$

FIGURE 6.23

Solution

You can find the volume of the horn using the Disc Method (see Section 5.7).

$$\begin{aligned}
\text{Volume} &= \int_1^\infty \pi \left(\frac{1}{x} \right)^2 dx && \textit{Disc Method} \\
&= \lim_{b \to \infty} \int_1^b \frac{\pi}{x^2}\, dx && \textit{Definition of improper integral} \\
&= \lim_{b \to \infty} \left[-\frac{\pi}{x} \right]_1^b && \textit{Find antiderivative.} \\
&= \lim_{b \to \infty} \left(\pi - \frac{\pi}{b} \right) && \textit{Apply Fundamental Theorem.} \\
&= \pi && \textit{Evaluate limit.}
\end{aligned}$$

Thus, Gabriel's horn has a volume of π cubic units. ■

Discussion Problem *Finite or Infinite?*

Which of the following improper integrals converge? What can you conclude?

a. $\displaystyle\int_1^\infty \frac{1}{x^{0.9}}\, dx$ **b.** $\displaystyle\int_1^\infty \frac{1}{x^{1.0}}\, dx$ **c.** $\displaystyle\int_1^\infty \frac{1}{x^{1.1}}\, dx$

Warm Up

The following warm-up exercises involve skills that were covered in earlier sections. You will use these skills in the exercise set for this section.

In Exercises 1–6, find the limit.

1. $\lim\limits_{x \to 2} (2x + 5)$

2. $\lim\limits_{x \to 1} \left(\dfrac{1}{x} + 2x^2\right)$

3. $\lim\limits_{x \to -4} \dfrac{x + 4}{x^2 - 16}$

4. $\lim\limits_{x \to 0} \dfrac{x^2 - 2x}{x^3 + 3x^2}$

5. $\lim\limits_{x \to 1} \dfrac{1}{\sqrt{x - 1}}$

6. $\lim\limits_{x \to -3} \dfrac{x^2 + 2x - 3}{x + 3}$

In Exercises 7–10, evaluate the expression when $x = b$ and when $x = 0$.

7. $\dfrac{4}{3}(2x - 1)^3$

8. $\dfrac{1}{x - 5} + \dfrac{3}{(x - 2)^2}$

9. $\ln(5 - 3x^2) - \ln(x + 1)$

10. $e^{3x^2} + e^{-3x^2}$

EXERCISES 6.6

In Exercises 1–14, determine whether the improper integral converges. If it does, evaluate the integral.

1. $\displaystyle\int_{0}^{\infty} e^{-x}\, dx$

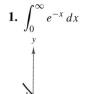

2. $\displaystyle\int_{-\infty}^{0} e^{2x}\, dx$

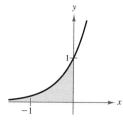

3. $\displaystyle\int_{1}^{\infty} \dfrac{1}{x^2}\, dx$

4. $\displaystyle\int_{1}^{\infty} \dfrac{1}{\sqrt{x}}\, dx$

5. $\displaystyle\int_{0}^{\infty} e^{x/3}\, dx$

6. $\displaystyle\int_{0}^{\infty} \dfrac{5}{e^{2x}}\, dx$

7. $\displaystyle\int_{5}^{\infty} \dfrac{x}{\sqrt{x^2 - 16}}\, dx$

8. $\displaystyle\int_{1/2}^{\infty} \dfrac{1}{\sqrt{2x - 1}}\, dx$

9. $\displaystyle\int_{-\infty}^{0} e^{-x}\, dx$

10. $\displaystyle\int_{-\infty}^{-1} \dfrac{1}{x^2}\, dx$

11. $\displaystyle\int_{-\infty}^{\infty} e^{|x|}\, dx$

12. $\displaystyle\int_{-\infty}^{\infty} \dfrac{|x|}{x^2 + 1}\, dx$

13. $\displaystyle\int_{-\infty}^{\infty} 2xe^{-3x^2}\, dx$

14. $\displaystyle\int_{-\infty}^{\infty} x^2 e^{-x^3}\, dx$

In Exercises 15–18, determine the divergence or convergence of the given improper integral. Evaluate the integral if it converges.

15. $\displaystyle\int_{0}^{4} \dfrac{1}{\sqrt{x}}\, dx$

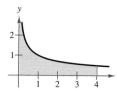

16. $\displaystyle\int_{3}^{4} \dfrac{1}{\sqrt{x - 3}}\, dx$

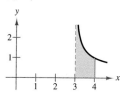

17. $\int_0^2 \frac{1}{(x-1)^{2/3}}\, dx$ **18.** $\int_0^2 \frac{1}{(x-1)^2}\, dx$

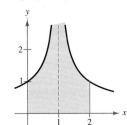 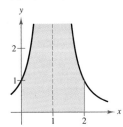

In Exercises 19–28, evaluate the improper integral.

19. $\int_0^1 \frac{1}{1-x}\, dx$ **20.** $\int_0^{27} \frac{5}{\sqrt[3]{x}}\, dx$

21. $\int_0^9 \frac{1}{\sqrt{9-x}}\, dx$ **22.** $\int_0^2 \frac{x}{\sqrt{4-x^2}}\, dx$

23. $\int_0^1 \frac{1}{x^2}\, dx$ **24.** $\int_0^1 \frac{1}{x}\, dx$

25. $\int_0^2 \frac{1}{\sqrt[3]{x-1}}\, dx$ **26.** $\int_0^2 \frac{1}{(x-1)^{4/3}}\, dx$

27. $\int_3^4 \frac{1}{\sqrt{x^2-9}}\, dx$ **28.** $\int_3^5 \frac{1}{x^2\sqrt{x^2-9}}\, dx$

In Exercises 29–32, complete the table for the specified values of a and n to demonstrate that

$$\lim_{x\to\infty} x^n e^{-ax} = 0, \quad a > 0,\, n > 0.$$

x	1	10	25	50
$x^n e^{-ax}$				

29. $a = 1,\, n = 1$ **30.** $a = 2,\, n = 4$
31. $a = \frac{1}{2},\, n = 2$ **32.** $a = \frac{1}{2},\, n = 5$

In Exercises 33–36, use the results of Exercises 29–32 to evaluate the improper integral.

33. $\int_0^\infty x^2 e^{-x}\, dx$ **34.** $\int_0^\infty (x-1)e^{-x}\, dx$

35. $\int_0^\infty x e^{-2x}\, dx$ **36.** $\int_0^\infty x e^{-x}\, dx$

37. *Present Value* You are asked to calculate the pay for a business that is forecast to yield a continuous flow of profit at the rate of $500,000 per year. If money will earn interest at the nominal rate of 9% per year compounded continuously, what is the present value of the business (a) for 20 years and (b) forever?

38. *Present Value* Repeat Exercise 37 for a farm that is expected to produce a profit of $75,000 per year. Assume that money will earn interest at the nominal rate of 8% compounded continuously.

Capitalized Cost In Exercises 39 and 40, find the capitalized cost C of an asset (a) for $n = 5$ years, (b) for $n = 10$ years, and (c) forever. The **capitalized cost** is given by

$$C = C_0 + \int_0^n c(t)e^{-rt}\, dt,$$

where C_0 is the original investment, t is the time in years, r is the annual interest rate compounded continuously, and $c(t)$ is the annual cost of maintenance. [*Hint:* For part (c), see Exercises 29–32.]

39. $C_0 = \$650{,}000,\ c(t) = \$25{,}000,\ r = 10\%$
40. $C_0 = \$650{,}000,\ c(t) = \$25{,}000(1 + 0.08t),\ r = 12\%$

In Exercises 41 and 42, (a) find the area of the region bounded by the graphs of the given equations and (b) find the volume of the solid generated by revolving the region about the x-axis.

41. $y = \frac{1}{x^2},\ y = 0,\ x \geq 1$
42. $y = e^{-x},\ y = 0,\ x \geq 0$

43. *Household Size* In 1991, the number of people per household in the United States was normally distributed with a mean of 3.07 and a standard deviation of 1.65. Find the probability that a person chosen at random lives

(a) alone.
(b) with 2 people.
(c) with 6 people.

(Source: U.S. Bureau of the Census)

44. *Quality Control* A company manufactures wooden yardsticks. The lengths of the yardsticks are normally distributed, with a mean of 36 inches and a standard deviation of 0.2 inch. Find the probability that a yardstick is (a) longer than 35.5 inches and (b) longer than 35.9 inches.

Chapter Summary and Study Tips

After studying this chapter, you should have acquired the following skills. The exercise numbers are keyed to the Review Exercises that begin on page 446. Answers to odd-numbered Review Exercises are given in the back of the text.

■ Use the basic integration formulas to find antiderivatives. (*Section 6.1*)   Review Exercises 1–12

Constant Rule: $\int k\, dx = kx + C$

Power Rules: $\int x^n\, dx = \dfrac{x^{n+1}}{n+1} + C,$ $\int u^n u'\, dx = \dfrac{u^{n+1}}{n+1} + C,\ n \neq -1$

Exponential Rules: $\int e^x\, dx = e^x + C,$ $\int e^u u'\, dx = \int e^u\, du = e^u + C$

Log Rules: $\int \dfrac{1}{x}\, dx = \ln|x| + C,$ $\int \dfrac{u'}{u}\, dx = \int \dfrac{1}{u}\, du = \ln|u| + C$

■ Use substitution to find antiderivatives. (*Section 6.1*)   Review Exercises 13–20

■ Use substitution to evaluate definite integrals. (*Section 6.1*)   Review Exercises 21–24

■ Use integration to solve real-life problems. (*Section 6.1*)   Review Exercises 25, 26

■ Use integration by parts to find antiderivatives. (*Section 6.2*)   Review Exercises 27–30

$\int u\, dv = uv - \int v\, du$

■ Use integration by parts repeatedly to find antiderivatives. (*Section 6.2*)   Review Exercises 31, 32

■ Find the present value of future income. (*Section 6.2*)   Review Exercises 33–36

■ Use partial fractions to find antiderivatives. (*Section 6.3*)   Review Exercises 37–42

■ Use logistics growth functions to model real-life situations. (*Section 6.3*)   Review Exercises 43, 44

■ Use integration tables to find antiderivatives. (*Section 6.4*)   Review Exercises 45–50

■ Use reduction formulas to find antiderivatives. (*Section 6.4*)   Review Exercises 51–54

■ Use completing the square to find antiderivatives. (*Section 6.4*)   Review Exercises 55–58

■ Use the Trapezoidal Rule to approximate definite integrals. (*Section 6.5*)   Review Exercises 59–62

$\int_a^b f(x)\, dx \approx \dfrac{b-a}{2n}[f(x_0) + 2f(x_1) + \cdots + 2f(x_{n-1}) + f(x_n)]$

■ Use Simpson's Rule to approximate definite integrals. (*Section 6.5*)   Review Exercises 63–66

$\int_a^b f(x)\, dx \approx \dfrac{b-a}{3n}[f(x_0) + 4f(x_1) + 2f(x_2) + 4f(x_3) + \cdots + 4f(x_{n-1}) + f(x_n)]$

■ Analyze the sizes of the errors when approximating definite integrals with the Trapezoidal Rule. (*Section 6.5*)   Review Exercises 67, 68

$|E| \leq \dfrac{(b-a)^3}{12n^2}\Big[\max|f''(x)|\Big], \qquad a \leq x \leq b$

*Several student study aids are available with this text. The *Student Solutions Guide* includes detailed solutions to all odd-numbered exercises, as well as practice chapter tests with answers. The *Graphics Calculator Guide* offers instructions on the use of a variety of graphing calculators and computer graphing software. The *Brief Calculus TUTOR* includes additional examples for selected exercises in the text.

■ Analyze the sizes of the errors when approximating definite integrals with Simpson's Rule. (*Section 6.5*)

Review Exercises 69, 70

$$|E| \le \frac{(b-a)^5}{180n^4}\left[\max|f^{(4)}(x)|\right], \qquad a \le x \le b$$

■ Evaluate improper integrals with infinite limits of integration. (*Section 6.6*)

Review Exercises 71–74

$$\int_a^\infty f(x)\,dx = \lim_{b\to\infty}\int_a^b f(x)\,dx, \qquad \int_{-\infty}^b f(x)\,dx = \lim_{a\to-\infty}\int_a^b f(x)\,dx.$$

■ Evaluate improper integrals with infinite integrands. (*Section 6.6*)

Review Exercises 75–78

$$\int_a^b f(x)\,dx = \lim_{c\to b^-}\int_a^c f(x)\,dx, \qquad \int_a^b f(x)\,dx = \lim_{c\to a^+}\int_c^b f(x)\,dx$$

■ Use improper integrals to solve real-life problems. (*Section 6.6*)

Review Exercises 79–82

■ **Use a Variety of Approaches**   To be efficient at finding antiderivatives, you need to use a variety of approaches.

1. Check to see whether the integral fits one of the basic integration formulas—you should have these formulas memorized.
2. Try an integration technique such as substitution, integration by parts, partial fractions, or completing the square to rewrite the integral in a form that fits one of the basic integration formulas.
3. Use a table of integrals.
4. Use a symbolic integration utility such as *Derive*, *Maple*, or *Mathematica*.

■ **Use Numerical Integration**   When solving a definite integral, remember that you cannot apply the Fundamental Theorem of Calculus unless you can find an antiderivative of the integrand. This is not always possible—even with a symbolic integration utility. In such cases, you can use a numerical technique such as the Midpoint Rule, the Trapezoidal Rule, or Simpson's Rule to approximate the value of the integral.

■ **Improper Integrals**   When solving integration problems, remember that the symbols used to denote definite integrals are the same as those used to denote improper integrals. Evaluating an improper integral as a definite integral can lead to an incorrect value. For instance, if you evaluated the integral

$$\int_{-2}^1 \frac{1}{x^2}\,dx$$

as though it were a definite integral, you would obtain a value of $-\frac{3}{2}$. This is not, however, correct. This integral is actually a divergent improper integral. If you have access to a symbolic integration utility, try using it to evaluate this integral—it will probably make the same mistake.

Review Exercises

In Exercises 1–12, use a basic integration formula to find the indefinite integral.

1. $\int dt$

2. $\int (x^2 + 2x - 1)\, dx$

3. $\int (x + 5)^3\, dx$

4. $\int \dfrac{2}{(x - 1)^2}\, dx$

5. $\int e^{10x}\, dx$

6. $\int 3xe^{-x^2}\, dx$

7. $\int \dfrac{1}{5x}\, dx$

8. $\int \dfrac{2x^3 - x}{x^4 - x^2 + 1}\, dx$

9. $\int x\sqrt{x^2 + 4}\, dx$

10. $\int \dfrac{1}{\sqrt{2x - 9}}\, dx$

11. $\int \dfrac{2e^x}{3 + e^x}\, dx$

12. $\int (x^2 - 1)e^{x^3 - 3x}\, dx$

In Exercises 13–20, use substitution to find the indefinite integral.

13. $\int x(x - 2)^3\, dx$

14. $\int x(1 - x)^2\, dx$

15. $\int x\sqrt{x + 1}\, dx$

16. $\int x^2\sqrt{x + 1}\, dx$

17. $\int 2x\sqrt{x - 3}\, dx$

18. $\int \dfrac{\sqrt{x}}{1 + \sqrt{x}}\, dx$

19. $\int (x + 1)\sqrt{1 - x}\, dx$

20. $\int \dfrac{x}{x - 1}\, dx$

In Exercises 21–24, use substitution to evaluate the definite integral.

21. $\int_{2}^{3} x\sqrt{x - 2}\, dx$

22. $\int_{2}^{3} x^2\sqrt{x - 2}\, dx$

23. $\int_{1}^{3} x^2(x - 1)^3\, dx$

24. $\int_{-3}^{0} x(x + 3)^4\, dx$

25. *Probability* The probability of recall in an experiment is found to be

$$P(a \le x \le b) = \int_{a}^{b} \frac{105}{16} x^2\sqrt{1 - x}\, dx,$$

where x represents the percent of recall (see figure).

(a) Find the probability that a randomly chosen individual will recall 80% of the material.

(b) What is the median percent recall? That is, for what value of b is it true that $P(0 \le x \le b) = 0.5$?

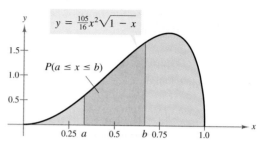

FIGURE FOR 25

26. *Probability* The probability of locating between a and b percent of oil and gas deposits in a region is

$$P(a \le x \le b) = \int_{a}^{b} \frac{140}{27} x^2(1 - x)^{1/3}\, dx$$

(see figure).

(a) Find the probability that between 40% and 60% of the deposits will be found.

(b) Find the probability that between 0% and 50% of the deposits will be found.

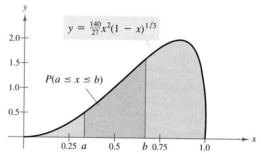

In Exercises 27–30, use integration by parts to find the indefinite integral.

27. $\int \dfrac{\ln x}{\sqrt{x}}\, dx$

28. $\int \sqrt{x}\, \ln x\, dx$

29. $\int (x - 1)e^x\, dx$

30. $\int \ln\left(\dfrac{x}{x + 1}\right)\, dx$

In Exercises 31 and 32, use integration by parts repeatedly to find the indefinite integral.

31. $\int 2x^2 e^{2x}\, dx$

32. $\int (\ln x)^3\, dx$

In Exercises 33–36, find the present value of the income given by $c(t)$ (measured in dollars) over t_1 years at the given annual inflation rate.

33. $c(t) = 10,000, \ r = 4\%, \ t_1 = 5$ years

34. $c(t) = 20,000 + 1500t, \ r = 6\%, \ t_1 = 10$ years

35. $c(t) = 12,000t, \ r = 5\%, \ t_1 = 10$ years

36. $c(t) = 10,000 + 100e^{t/2}, \ r = 5\%, \ t_1 = 5$ years

In Exercises 37–42, use partial fractions to find the indefinite integral.

37. $\displaystyle \int \frac{1}{x(x+5)}\,dx$

38. $\displaystyle \int \frac{4x-2}{3(x-1)^2}\,dx$

39. $\displaystyle \int \frac{4x-13}{x^2-3x-10}\,dx$

40. $\displaystyle \int \frac{4x^2-x-5}{x^2(x+5)}\,dx$

41. $\displaystyle \int \frac{x^2}{x^2+2x-15}\,dx$

42. $\displaystyle \int \frac{x^2+2x-12}{x(x+3)}\,dx$

43. *Sales Growth* When it is introduced to the market, a new product initially sells 1250 units per week. After 6 months, the number of sales increases to 6500. The sales can be modeled by logistics growth with a limit of 10,000 units per week.

(a) Find a logistics growth model for the number of units.

(b) Use the model to complete the table.

Time, t	0	3	6	12	24
Sales, y					

(c) Use the graph below to approximate the time t that sales will be 7500.

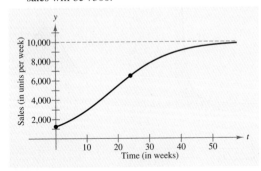

44. *Population Growth* A conservation society has introduced a population of 300 ring-neck pheasants into a new area. After 5 years the population has increased to 966. The population can be modeled by logistics growth with a limit of 2700 pheasants.

(a) Find a logistics growth model for the population of ring-neck pheasants.

(b) How many pheasants were present after 4 years?

(c) How long will it take to establish a population of 1750 pheasants?

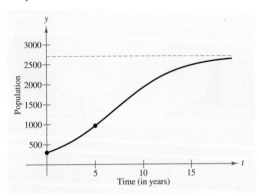

FIGURE FOR 44

In Exercises 45–50, use the table of integrals in Section 6.4 to evaluate the integral.

45. $\displaystyle \int \frac{\sqrt{x^2+25}}{x}\,dx$

46. $\displaystyle \int \frac{1}{x(4+3x)}\,dx$

47. $\displaystyle \int \frac{1}{(x^2-4)}\,dx$

48. $\displaystyle \int x(\ln x^2)^2\,dx$

49. $\displaystyle \int_0^3 \frac{x}{\sqrt{1+x}}\,dx$

50. $\displaystyle \int_1^3 \frac{1}{x^2\sqrt{16-x^2}}\,dx$

In Exercises 51–54, use a reduction formula from the table of integrals in Section 6.4 to find the indefinite integral.

51. $\displaystyle \int \frac{\sqrt{1+x}}{x}\,dx$

52. $\displaystyle \int \frac{1}{(x^2-9)^2}\,dx$

53. $\displaystyle \int (x-5)^3 e^{x-5}\,dx$

54. $\displaystyle \int (\ln x)^4\,dx$

In Exercises 55–58, complete the square and then use the table of integrals given in Section 6.4 to find the indefinite integral.

55. $\displaystyle \int \frac{1}{x^2+4x-21}\,dx$

56. $\displaystyle \int \frac{1}{x^2-8x-52}\,dx$

57. $\displaystyle \int \sqrt{x^2-10x}\,dx$

58. $\displaystyle \int \frac{x}{\sqrt{x^4+6x^2+10}}\,dx$

In Exercises 59–62, use the Trapezoidal Rule to approximate the definite integral.

59. $\int_1^3 \frac{1}{x^2}\,dx$, $n=4$

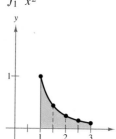

60. $\int_0^2 (x^2+1)\,dx$, $n=4$

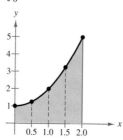

61. $\int_1^2 \frac{1}{1+\ln x}\,dx$, $n=4$

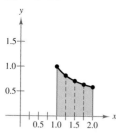

62. $\int_0^2 \frac{1}{\sqrt{1+x^3}}\,dx$, $n=8$

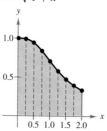

In Exercises 63–66, use Simpson's Rule to approximate the definite integral.

63. $\int_1^2 \frac{1}{x^3}\,dx$, $n=4$

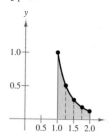

64. $\int_1^2 x^3\,dx$, $n=4$

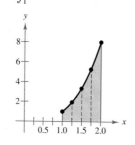

65. $\int_0^1 \frac{x^{3/2}}{2-x^2}\,dx$, $n=4$

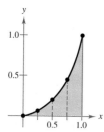

66. $\int_0^1 e^{x^2}\,dx$, $n=6$

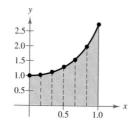

In Exercises 67 and 68, use the error formula to find bounds for the error in approximating the integral using the Trapezoidal Rule.

67. $\int_0^2 e^{2x}\,dx$, $n=4$

68. $\int_0^2 e^{2x}\,dx$, $n=8$

In Exercises 69 and 70, use the error formula to find bounds for the error in approximating the integral using Simpson's Rule.

69. $\int_2^4 \frac{1}{x-1}\,dx$, $n=4$

70. $\int_2^4 \frac{1}{x-1}\,dx$, $n=8$

In Exercises 71–78, evaluate the improper integral.

71. $\int_0^\infty 4xe^{-2x^2}\,dx$

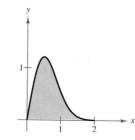

72. $\int_{-\infty}^0 \frac{3}{(1-3x)^{2/3}}\,dx$

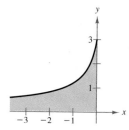
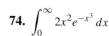

73. $\int_{-\infty}^0 \frac{1}{3x^2}\,dx$

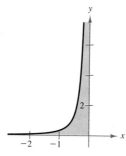

74. $\int_0^\infty 2x^2e^{-x^3}\,dx$

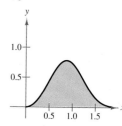

75. $\int_0^4 \frac{1}{\sqrt{4x}}\,dx$

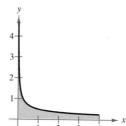

76. $\int_1^2 \frac{x}{16(x-1)^2}\,dx$

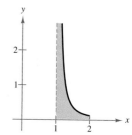

77. $\displaystyle\int_{2}^{3} \frac{1}{\sqrt{x-2}}\, dx$

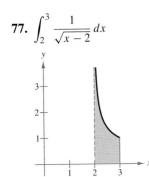

78. $\displaystyle\int_{0}^{2} \frac{x+2}{(x-1)^2}\, dx$

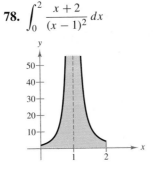

79. *Present Value* You are considering buying a franchise that yields a continuous income stream of $50,000 per year. Find the present value of the franchise (a) for 15 years and (b) forever. Assume that money earns 6% interest per year, compounded continuously.

80. *Capitalized Cost* A company invests $1.5 million in a new manufacturing plant that will cost $50,000 per year in maintenance. Find the capitalized cost for (a) 20 years and (b) forever. Assume that money earns 6% interest, compounded continuously.

81. *Per Capita Income* In 1991, the per capita income per state was approximately normally distributed with a mean of $18,188.38 and a standard deviation of $2736.16 (see figure). Find the probability that a state has a per capita income of (a) $20,000 or greater and (b) $30,000 or greater. *(Source: U.S. Bureau of Economic Analysis)*

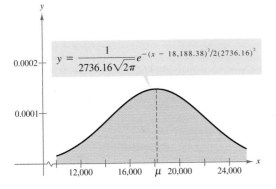

82. *Eastern Warblers* The average length (from beak to tail) of different species of warblers in the eastern United States is approximately normally distributed with a mean of 12.9 centimeters and a standard deviation of 0.95 centimeters (see figure). Find the probability that a species of warbler has a length that is (a) 13 centimeters or greater and (b) 15 centimeters or greater. *(Source: Peterson Field Guide: Eastern Birds)*

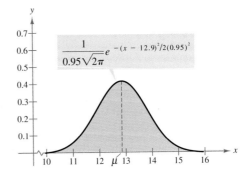

Sample Post-Graduation Exam Questions

The following questions were taken from certified public accountant (CPA) exams, graduate management admission tests (GMAT), graduate records exams (GRE), actuarial exams, or college-level academic skills tests (CLAST). The answers to the questions are given in the back of the book.

For Questions 1–4, the total 1988 population was 12,335,000; the total 1992 population was 13,488,000. Also use the data given in the graphs. *(Source: U.S. Bureau of the Census)*

1. Find the number of people aged 75 and over for the year 1992.
   (a) 986,800   (b) 944,160   (c) 1,079,840   (d) 863,450

2. Find the increase in population of 35 to 54 year olds from 1988 to 1992.
   (a) 411,600   (b) 18,490   (c) 153,370   (d) 288,250

3. In 1988, how many people were 54 years old or younger?
   (a) 8,881,200   (b) 8,757,850   (c) 9,576,480   (d) 11,471,550

4. In what age group did the population and the population percent decrease between 1988 and 1992?
   (a) 0–17   (b) 75 and over   (c) 18–34   (d) 55–74

5. $\displaystyle\int_{1}^{6} \frac{x}{\sqrt{x+3}} =$

   (a) $\frac{10}{3}$ (b) $-\frac{20}{3}$ (c) -4 (d) $\frac{20}{3}$

6. $\displaystyle\lim_{x \to \infty} \frac{2x^2 + 5x - 16}{5x^2 - 15x + 48} =$

   (a) 0   (b) $\frac{2}{5}$ (c) ∞ (d) $\frac{1}{3}$

7. The city council is planning to construct a new street as shown in the figure. Construction costs of the new street are $110 per linear foot. What is the projected cost for constructing the new street?
   (a) $2,904,000   (b) $4,065,600   (c) $1,904,000
   (d) $3,864,000

8. The following information pertains to Varn Co.:
   Sales $1,000,000,   Variable Costs $200,000,
   Fixed Costs $50,000

   What is Varn's break-even point in sales dollars?
   (a) $40,000   (b) $250,000   (c) $62,500   (d) $200,000

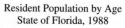
Resident Population by Age
State of Florida, 1988

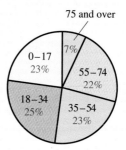

Resident Population by Age
State of Florida, 1992

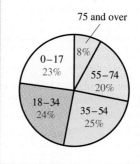

FIGURE FOR 1–4

FIGURE FOR 7

$$\frac{\partial z}{\partial x} = \lim_{\Delta x \to 0} \frac{f(x + \Delta x, y) - f(x, y)}{\Delta x}$$

7

Functions of Several Variables

Calculus was originally developed to study functions of a single variable. It was later discovered that the two basic operations of calculus—differentiation and integration—can be meaningfully extended to functions of two or more variables.

Graphs of functions of *one* variable can be represented by curves in the plane. Graphs of functions of *two* variables can be represented by surfaces in space. A technique for computer generation of such "three-dimensional" graphs was developed by the African-American mathematician David Hedgley, Jr. (1937–).

$$\frac{\partial z}{\partial x} = f_x(x, y)$$

Hedgley

Courtesy of NASA

451

7.1

The Three-Dimensional Coordinate System

The Three-Dimensional Coordinate System ■ *The Distance and Midpoint Formulas* ■ *The Equation of a Sphere* ■ *Traces of Surfaces*

The Three-Dimensional Coordinate System

Recall that the Cartesian plane is determined by two perpendicular number lines called the x-axis and the y-axis. These axes together with their point of intersection (the origin) allow you to develop a two-dimensional coordinate system for identifying points in a plane. To identify a point in space, you must introduce a third dimension to the model. The geometry of this three-dimensional model is called **solid analytic geometry.**

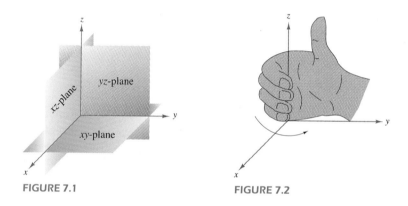

FIGURE 7.1   FIGURE 7.2

You can construct a **three-dimensional coordinate system** by passing a z-axis perpendicular to both the x- and y-axes at the origin. Figure 7.1 shows the positive portion of each coordinate axis. Taken as pairs, the axes determine three **coordinate planes:** the **xy-plane,** the **xz-plane,** and the **yz-plane.** These three coordinate planes separate the three-dimensional coordinate system into eight **octants.** The first octant is the one for which all three coordinates are positive. In this three-dimensional system, a point P in space is determined by an ordered triple (x, y, z), where x, y, and z are as follows.

$x = $ directed distance from yz-plane to P

$y = $ directed distance from xz-plane to P

$z = $ directed distance from xy-plane to P

A three-dimensional coordinate system can have either a **left-handed** or a **right-handed** orientation. In this text, we work exclusively with right-handed systems, as shown in Figure 7.2.

EXAMPLE 1 *Plotting Points in Space*

Plot the following points in space.

a. $(2, -3, 3)$ **b.** $(-2, 6, 2)$ **c.** $(1, 4, 0)$ **d.** $(2, 2, -3)$

Solution

To plot the point $(2, -3, 3)$, notice that $x = 2$, $y = -3$, and $z = 3$. To help visualize the point (see Figure 7.3), locate the point $(2, -3)$ in the xy-plane (denoted by a cross). The point $(2, -3, 3)$ lies three units above the cross. The other three points are also shown in the figure.

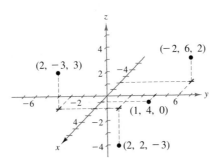

FIGURE 7.3

The Distance and Midpoint Formulas

Many of the formulas established for the two-dimensional coordinate system can be extended to three dimensions. For example, to find the distance between two points in space, you can use the Pythagorean Theorem twice, as shown in Figure 7.4. By doing this, you will obtain the formula for the distance between two points in space.

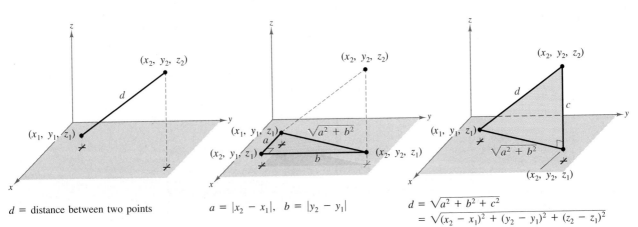

d = distance between two points

$a = |x_2 - x_1|$, $b = |y_2 - y_1|$

$d = \sqrt{a^2 + b^2 + c^2}$
$ = \sqrt{(x_2 - x_1)^2 + (y_2 - y_1)^2 + (z_2 - z_1)^2}$

FIGURE 7.4

Distance Formula in Space

The distance between the points (x_1, y_1, z_1) and (x_2, y_2, z_2) is
$$d = \sqrt{(x_2 - x_1)^2 + (y_2 - y_1)^2 + (z_2 - z_1)^2}.$$

EXAMPLE 2 *Finding the Distance Between Two Points*

Find the distance between $(1, 0, 2)$ and $(2, 4, -3)$.

Solution

$$
\begin{aligned}
d &= \sqrt{(x_2 - x_1)^2 + (y_2 - y_1)^2 + (z_2 - z_1)^2} && \textit{Distance Formula} \\
&= \sqrt{(2 - 1)^2 + (4 - 0)^2 + (-3 - 2)^2} && \textit{Substitute.} \\
&= \sqrt{1 + 16 + 25} && \textit{Simplify.} \\
&= \sqrt{42}. && \textit{Simplify.}
\end{aligned}
$$

Notice the similarity between the Distance Formula in the plane and the Distance Formula in space. The Midpoint Formulas in the plane and in space are also similar.

Midpoint Formula in Space

The midpoint of the line segment joining the points (x_1, y_1, z_1) and (x_2, y_2, z_2) is
$$\text{Midpoint} = \left(\frac{x_1 + x_2}{2}, \frac{y_1 + y_2}{2}, \frac{z_1 + z_2}{2} \right).$$

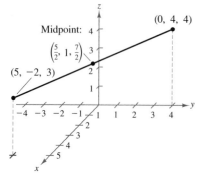

FIGURE 7.5

EXAMPLE 3 *Using the Midpoint Formula*

Find the midpoint of the line segment joining $(5, -2, 3)$ and $(0, 4, 4)$.

Solution

Using the Midpoint Formula, the midpoint is
$$\left(\frac{5 + 0}{2}, \frac{-2 + 4}{2}, \frac{3 + 4}{2} \right) = \left(\frac{5}{2}, 1, \frac{7}{2} \right),$$
as shown in Figure 7.5.

The Equation of a Sphere

A **sphere** with center at (h, k, l) and radius r is defined to be the set of all points (x, y, z) such that the distance between (x, y, z) and (h, k, l) is r, as shown in Figure 7.6. Using the Distance Formula, this condition can be written as

$$\sqrt{(x - h)^2 + (y - k)^2 + (z - l)^2} = r.$$

By squaring both sides of this equation, you obtain the standard equation of a sphere.

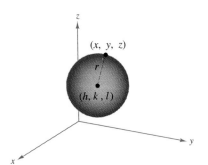

FIGURE 7.6   Sphere: Radius r, Center (h, k, l)

Standard Equation of a Sphere

The **standard equation of a sphere** whose center is (h, k, l) and whose radius is r is

$$(x - h)^2 + (y - k)^2 + (z - l)^2 = r^2.$$

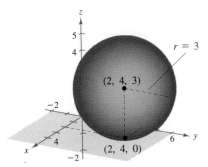

FIGURE 7.7

EXAMPLE 4 *Finding the Equation of a Sphere*

Find the standard equation for the sphere whose center is $(2, 4, 3)$ and whose radius is 3. Does this sphere intersect the xy-plane?

Solution

$$(x - h)^2 + (y - k)^2 + (z - l)^2 = r^2 \qquad \textit{Standard equation}$$

$$(x - 2)^2 + (y - 4)^2 + (z - 3)^2 = 3^2 \qquad \textit{Substitute.}$$

From the graph shown in Figure 7.7, you can see that the center of the sphere lies three units above the xy-plane. Because the sphere has a radius of 3, you can conclude that it does intersect the xy-plane—at the point $(2, 4, 0)$. ▪

EXAMPLE 5 *Finding the Equation of a Sphere*

Find the equation of the sphere that has the points $(3, -2, 6)$ and $(-1, 4, 2)$ as endpoints of a diameter.

Solution

By the Midpoint Formula, the center of the sphere is

$$(h, k, l) = \left(\frac{3-1}{2}, \frac{-2+4}{2}, \frac{6+2}{2} \right) \qquad \textit{Midpoint Formula}$$

$$= (1, 1, 4). \qquad \textit{Simplify.}$$

By the Distance Formula, the radius is

$$r = \sqrt{(3-1)^2 + (-2-1)^2 + (6-4)^2} \qquad \textit{Distance Formula}$$

$$= \sqrt{17}. \qquad \textit{Simplify.}$$

Therefore, the standard equation of the sphere is

$$(x - h)^2 + (y - k)^2 + (z - l)^2 = r^2 \qquad \textit{Formula for a sphere}$$

$$(x - 1)^2 + (y - 1)^2 + (z - 4)^2 = 17. \qquad \textit{Substitute.} \qquad ■$$

EXAMPLE 6 *Finding the Center and Radius of a Sphere*

Find the center and radius of the sphere whose equation is

$$x^2 + y^2 + z^2 - 2x + 4y - 6z + 8 = 0.$$

Solution

You can obtain the standard equation of the sphere by completing the square. To do this, begin by grouping terms with the same variable. Then add "the square of half the coefficient of each linear term" to both sides of the equation. For instance, to complete the square of $(x^2 - 2x)$, add $[\frac{1}{2}(-2)]^2 = 1$ to both sides.

$$x^2 + y^2 + z^2 - 2x + 4y - 6z + 8 = 0$$

$$(x^2 - 2x + \quad) + (y^2 + 4y + \quad) + (z^2 - 6z + \quad) = -8$$

$$(x^2 - 2x + 1) + (y^2 + 4y + 4) + (z^2 - 6z + 9) = -8 + 1 + 4 + 9$$

$$(x - 1)^2 + (y + 2)^2 + (z - 3)^2 = 6$$

Therefore, the center of the sphere is $(1, -2, 3)$, and its radius is $\sqrt{6}$, as shown in Figure 7.8. ◼

Note in Example 6 that the points satisfying the equation of the sphere are "surface points," not "interior points." In general, the collection of points satisfying an equation involving x, y, and z is called a **surface in space.**

Sphere:
$(x - 1)^2 + (y + 2)^2 + (z - 3)^2 = 6$

Center: $(1, -2, 3)$

FIGURE 7.8

Traces of Surfaces

Finding the intersection of a surface with one of the three coordinate planes (or with a plane parallel to one of the three coordinate planes) helps visualize the surface. Such an intersection is called a **trace** of the surface. For example, the xy-trace of a surface consists of all points that are common to both the surface *and* the xy-plane. Similarly, the xz-trace of a surface consists of all points that are common to both the surface and the xz-plane.

EXAMPLE 7 *Finding a Trace of a Surface*

Sketch the xy-trace of the sphere whose equation is

$$(x - 3)^2 + (y - 2)^2 + (z + 4)^2 = 5^2.$$

Solution

To find the xy-trace of this surface, use the fact that every point in the xy-plane has a z-coordinate of zero. This means that if you substitute $z = 0$ into the given equation, the resulting equation will represent the intersection of the surface with the xy-plane.

$$(x - 3)^2 + (y - 2)^2 + (z + 4)^2 = 5^2 \qquad \text{\textit{Equation of sphere}}$$
$$(x - 3)^2 + (y - 2)^2 + (0 + 4)^2 = 25 \qquad \text{\textit{Let } z = 0 \text{ to find } xy\text{-trace.}}$$
$$(x - 3)^2 + (y - 2)^2 + 16 = 25$$
$$(x - 3)^2 + (y - 2)^2 = 9$$
$$(x - 3)^2 + (y - 2)^2 = 3^2 \qquad \text{\textit{Equation of circle}}$$

From this equation, you can see that the xy-trace is a circle of radius 3, as shown in Figure 7.9. ■

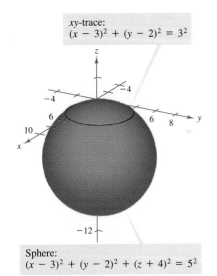

xy-trace:
$(x - 3)^2 + (y - 2)^2 = 3^2$

Sphere:
$(x - 3)^2 + (y - 2)^2 + (z + 4)^2 = 5^2$

FIGURE 7.9

Discussion Problem   Comparing Two and Three Dimensions

In this section, you saw similarities between formulas in two-dimensional coordinate geometry and three-dimensional coordinate geometry. In two-dimensional coordinate geometry, the graph of the equation

$$ax + by + c = 0$$

is a line. In three-dimensional coordinate geometry, what is the graph of the equation

$$ax + by + c = 0?$$

Is it a line? Explain your reasoning.

Warm Up

The following warm-up exercises involve skills that were covered in earlier sections. You will use these skills in the exercise set for this section.

In Exercises 1–4, find the distance between the points.

1. $(5, 1), (3, 5)$

2. $(2, 3), (-1, -1)$

3. $(-5, 4), (-5, -4)$

4. $(-3, 6), (-3, -2)$

In Exercises 5–8, find the midpoint of the line segment connecting the points.

5. $(2, 5), (6, 9)$

6. $(-1, -2), (3, 2)$

7. $(-6, 0), (6, 6)$

8. $(-4, 3), (2, -1)$

In Exercises 9 and 10, write the standard equation of the circle.

9. Center: $(2, 3)$; Radius: 2

10. Diameter endpoints: $(4, 0), (-2, 8)$

EXERCISES 7.1

 means that technology can help you solve or check the exercise(s).

In Exercises 1–4, plot the points on the same three-dimensional coordinate system.

1. (a) $(2, 1, 3)$
   (b) $(-1, 2, 1)$

2. (a) $(3, -2, 5)$
   (b) $\left(\frac{3}{2}, 4, -2\right)$

3. (a) $(5, -2, 2)$
   (b) $(5, -2, -2)$

4. (a) $(0, 4, -5)$
   (b) $(4, 0, 5)$

In Exercises 5–8, find the distance between the two points.

5. $(4, 1, 5), (8, 2, 6)$

6. $(-4, -1, 1), (2, -1, 5)$

7. $(-1, -5, 7), (-3, 4, -4)$

8. $(8, -2, 2), (8, -2, 4)$

In Exercises 9–12, find the coordinates of the midpoint of the line segment joining the given points.

9. $(6, -9, 1), (-2, -1, 5)$

10. $(4, 0, -6), (8, 8, 20)$

11. $(-5, -2, 5), (6, 3, -7)$

12. $(0, -2, 5), (4, 2, 7)$

In Exercises 13–16, find (x, y, z).

13.

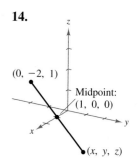

14.

15.

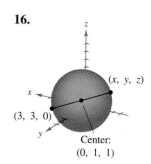

16.

In Exercises 17–20, find the lengths of the sides of the triangle with the indicated vertices, and determine whether the triangle is a right triangle, an isosceles triangle, or neither of these.

17. $(0, 0, 0)$, $(2, 2, 1)$, $(2, -4, 4)$

18. $(5, 3, 4)$, $(7, 1, 3)$, $(3, 5, 3)$

19. $(-2, 2, 4)$, $(-2, 2, 6)$, $(-2, 4, 8)$

20. $(5, 0, 0)$, $(0, 2, 0)$, $(0, 0, -3)$

In Exercises 21–28, find the standard form of the equation of the sphere.

21.

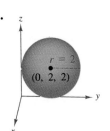

22.

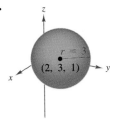

23.

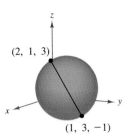

24.

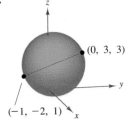

25. Center: $(1, 1, 5)$; Radius: 3

26. Center: $(4, -1, 1)$; Radius: 5

27. Diameter endpoints: $(2, 0, 0)$, $(0, 6, 0)$

28. Center: $(-2, 1, 1)$; Tangent to the xy-coordinate plane

In Exercises 29–36, find the sphere's center and radius.

29. $x^2 + y^2 + z^2 - 2x + 6y + 8z + 1 = 0$

30. $x^2 + y^2 + z^2 - 5x = 0$

31. $x^2 + y^2 + z^2 - 8y = 0$

32. $x^2 + y^2 + z^2 - 4y + 6z + 4 = 0$

33. $2x^2 + 2y^2 + 2z^2 - 4x - 12y - 8z + 3 = 0$

34. $x^2 + y^2 + z^2 = 36$

35. $9x^2 + 9y^2 + 9z^2 - 6x + 18y + 1 = 0$

36. $4x^2 + 4y^2 + 4z^2 - 4x - 32y + 8z + 33 = 0$

In Exercises 37 and 38, sketch the xy-trace of the sphere.

37. $(x - 1)^2 + (y - 3)^2 + (z - 2)^2 = 25$

38. $x^2 + y^2 + z^2 - 6x - 10y + 6z + 30 = 0$

In Exercises 39 and 40, sketch the yz-trace of the sphere.

39. $x^2 + y^2 + z^2 - 4x - 4y - 6z - 12 = 0$

40. $x^2 + y^2 + z^2 - 6x - 10y + 6z + 30 = 0$

In Exercises 41–44, sketch the trace of the intersection of each plane with the given sphere.

41. $x^2 + y^2 + z^2 = 25$; (a) $z = 3$, (b) $x = 4$

42. $x^2 + y^2 + z^2 = 169$; (a) $x = 5$, (b) $y = 12$

43. $x^2 + y^2 + z^2 - 4x - 6y + 9 = 0$; (a) $x = 2$, (b) $y = 3$

44. $x^2 + y^2 + z^2 - 8x - 6z + 16 = 0$; (a) $x = 4$, (b) $z = 3$

45. *Crystals* Crystals are classified according to their symmetry. Crystals shaped like cubes are classified as isometric. Suppose you have mapped the vertices of a crystal onto a three-dimensional coordinate system. Determine (x, y, z) if the crystal is isometric.

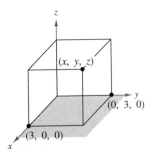

© Breck P. Kent

Halite crystals (rock salt) are classified as isometric.

Equations of Planes in Space

In Section 7.1, you studied one type of surface in space—a sphere. In this section, you will study a second type—a plane in space. The **general equation of a plane** in space is

$$ax + by + cz = d.$$ *General equation of a plane*

Note the similarity of this equation to the general equation of a line in the plane. In fact, if you intersect the plane represented by this equation with each of the three coordinate planes, you will obtain traces that are lines, as shown in Figure 7.10.

In Figure 7.10, the points where the plane intersects the three coordinate axes are the x-, y-, and z-intercepts of the plane. By connecting these three points, you can form a triangular region, which helps you visualize the plane in space.

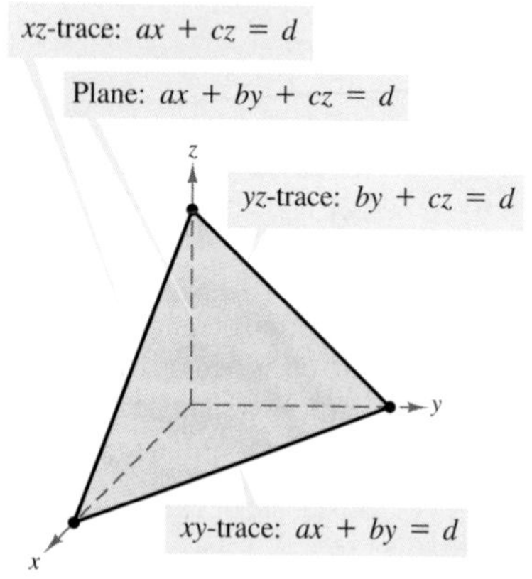

xz-trace: $ax + cz = d$

Plane: $ax + by + cz = d$

yz-trace: $by + cz = d$

xy-trace: $ax + by = d$

FIGURE 7.10

EXAMPLE 1 *Sketching a Plane in Space*

Find the x-, y-, and z-intercepts of the plane given by

$$3x + 2y + 4z = 12.$$

Then sketch the plane.

Solution
To find the x-intercept, let both y and z be zero.

$$3x + 2(0) + 4(0) = 12$$ *Substitute 0 for y and z.*
$$3x = 12$$ *Simplify.*
$$x = 4$$ *Solve for x.*

Thus, the x-intercept is $(4, 0, 0)$. To find the y-intercept, let x and z be zero and conclude that $y = 6$. Thus, the y-intercept is $(0, 6, 0)$. Similarly, by letting x and y be zero, you can determine that $z = 3$ and that the z-intercept is $(0, 0, 3)$. Figure 7.11 shows the triangular portion of the plane formed by connecting the three intercepts.

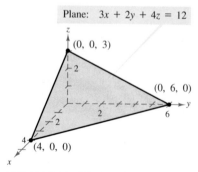

Plane: $3x + 2y + 4z = 12$

$(0, 0, 3)$

$(0, 6, 0)$

$(4, 0, 0)$

FIGURE 7.11   Sketch made by connecting intercepts: $(4, 0, 0)$, $(0, 6, 0)$, $(0, 0, 3)$

Drawing Planes in Space

The planes shown in Figures 7.10 and 7.11 each have three intercepts. When this occurs, we suggest that you draw the plane by sketching the triangular region formed by connecting the three intercepts.

It is possible for a plane in space to have fewer than three intercepts. This occurs when one or more of the coefficients in the equation $ax + by + cz = d$ is zero. Figure 7.12 shows some planes in space that have only one intercept, and Figure 7.13 shows some that have only two intercepts. In each figure, note the use of dashed lines and shading to give the illusion of three dimensions.

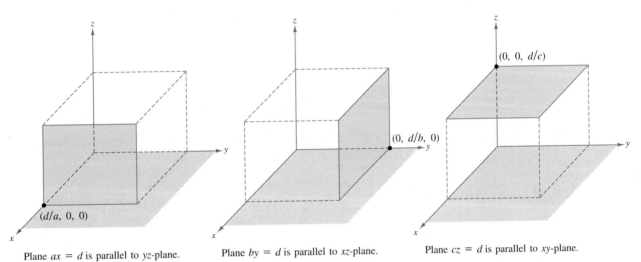

Plane $ax = d$ is parallel to yz-plane.   Plane $by = d$ is parallel to xz-plane.   Plane $cz = d$ is parallel to xy-plane.

FIGURE 7.12   Planes Parallel to Coordinate Planes

Plane $ax + by = d$ is parallel to z-axis.   Plane $ax + cz = d$ is parallel to y-axis.   Plane $by + cz = d$ is parallel to x-axis.

FIGURE 7.13   Planes Parallel to Coordinate Axes

Quadric Surfaces

A third common surface is a **quadric surface.** Every quadric surface has an equation of the form

$$Ax^2 + By^2 + Cz^2 + Dx + Ey + Fz + G = 0. \qquad \textit{2nd-degree equation}$$

There are six basic quadric surfaces.

1. Elliptic cone   **2.** Elliptic paraboloid

3. Hyperbolic paraboloid   **4.** Ellipsoid

5. Hyperboloid of one sheet   **6.** Hyperboloid of two sheets

The six types are summarized on pages 464 and 465. Notice that each surface is pictured with two types of three-dimensional sketches. The computer-generated sketches use traces with hidden lines to give the illusion of three dimensions. The artist-rendered sketches use shading to create the same illusion.

All of the quadric surfaces on pages 464 and 465 are centered at the origin and have axes along the coordinate axes. Moreover, only one of several possible orientations of each surface is shown. If the surface has a different center or is oriented along a different axis, then its standard equation will change accordingly. For instance, the ellipsoid

$$\frac{x^2}{1^2} + \frac{y^2}{3^2} + \frac{z^2}{2^2} = 1$$

has $(0, 0, 0)$ as its center, but the ellipsoid

$$\frac{(x-2)^2}{1^2} + \frac{(y+1)^2}{3^2} + \frac{(z-4)^2}{2^2} = 1$$

has $(2, -1, 4)$ as its center. A computer-generated graph of the first ellipsoid is shown in Figure 7.14.

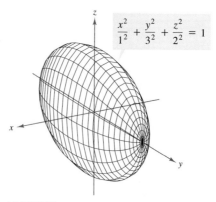

FIGURE 7.14

▦ TECHNOLOGY

Using a Three-Dimensional Graphing Utility

Most three-dimensional graphing utilities represent surfaces by sketching several traces of the surface. The traces are usually taken in equally spaced parallel planes. Depending on the graphing utility, the sketch can be made with one set, two sets, or three sets of traces. For instance, the two sketches shown below use two sets of traces: in one set the traces are parallel to the xz-plane and in the other set the traces are parallel to the yz-plane.

To sketch the graph of an equation involving x, y, and z with a three-dimensional "function grapher," you must first solve the equation for z. After entering the equation, you need to specify a rectangular viewing box (the three-dimensional analog of a viewing rectangle).

The two sketches shown below were generated by *Derive*. If you have access to a three-dimensional graphing utility, try using it to duplicate these graphs.

Equation: $z = (x^2 + y^2)e^{1-x^2-y^2}$

Grid: 20 traces by 20 traces

Viewing Box: $-5 \le x \le 5$
$-5 \le y \le 5$
$-5 \le z \le 5$

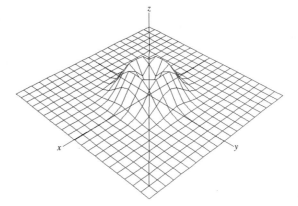

Generated by Derive

Equation: $z = \frac{1}{4}x^2 - \frac{1}{4}y^2$

Grid: 10 traces by 10 traces

Viewing Box: $-4 \le x \le 4$
$-4 \le y \le 4$
$-4 \le z \le 4$

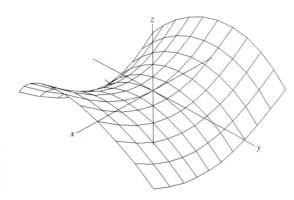

Generated by Derive

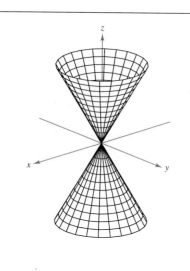

Elliptic Cone

$$\frac{x^2}{a^2} + \frac{y^2}{b^2} - \frac{z^2}{c^2} = 0$$

Trace	*Plane*
Ellipse	Parallel to xy-plane
Hyperbola	Parallel to xz-plane
Hyperbola	Parallel to yz-plane

The axis of the cone corresponds to the variable whose coefficient is negative. The traces in the coordinate planes parallel to this axis are intersecting lines.

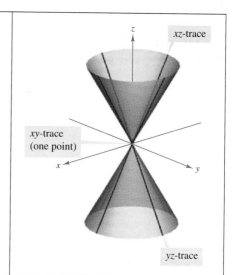

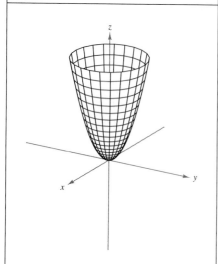

Elliptic Paraboloid

$$z = \frac{x^2}{a^2} + \frac{y^2}{b^2}$$

Trace	*Plane*
Ellipse	Parallel to xy-plane
Parabola	Parallel to xz-plane
Parabola	Parallel to yz-plane

The axis of the paraboloid corresponds to the variable raised to the first power.

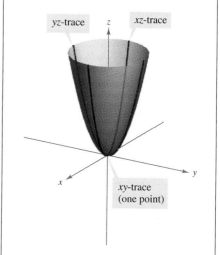

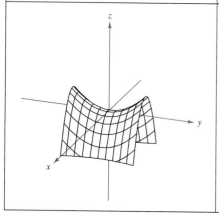

Hyperbolic Paraboloid

$$z = \frac{y^2}{b^2} - \frac{x^2}{a^2}$$

Trace	*Plane*
Hyperbola	Parallel to xy-plane
Parabola	Parallel to xz-plane
Parabola	Parallel to yz-plane

The axis of the paraboloid corresponds to the variable raised to the first power.

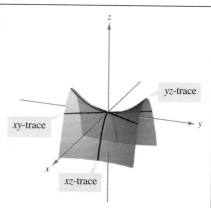

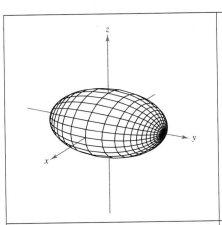

Ellipsoid

$$\frac{x^2}{a^2} + \frac{y^2}{b^2} + \frac{z^2}{c^2} = 1$$

Trace	Plane
Ellipse	Parallel to xy-plane
Ellipse	Parallel to xz-plane
Ellipse	Parallel to yz-plane

The surface is a sphere if the coefficients a, b, and c are equal and nonzero.

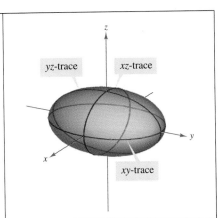

Hyperboloid of One Sheet

$$\frac{x^2}{a^2} + \frac{y^2}{b^2} - \frac{z^2}{c^2} = 1$$

Trace	Plane
Ellipse	Parallel to xy-plane
Hyperbola	Parallel to xz-plane
Hyperbola	Parallel to yz-plane

The axis of the hyperboloid corresponds to the variable whose coefficient is negative.

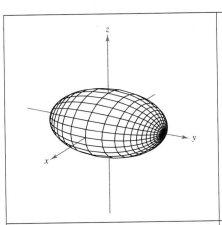

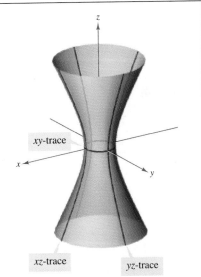

Hyperboloid of Two Sheets

$$\frac{z^2}{c^2} - \frac{x^2}{a^2} - \frac{y^2}{b^2} = 1$$

Trace	Plane
Ellipse	Parallel to xy-plane
Hyperbola	Parallel to xz-plane
Hyperbola	Parallel to yz-plane

The axis of the hyperboloid corresponds to the variable whose coefficient is positive. There is no trace in the coordinate plane perpendicular to this axis.

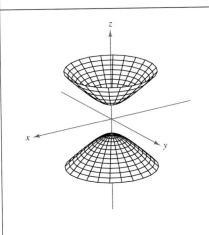

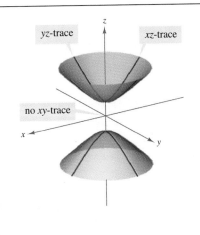

When classifying quadric surfaces, note that the two types of paraboloids have one variable raised to the first power. The other four types of quadric surfaces have equations that are of second degree in *all* three variables.

EXAMPLE 2 *Classifying a Quadric Surface*

Classify the surface given by

$$x - y^2 - z^2 = 0.$$

Describe the traces of the surface in the xy-plane, the xz-plane, and the plane given by $x = 1$.

Solution

Because x is raised only to the first power, the surface is a paraboloid whose axis is the x-axis, as shown in Figure 7.15. In standard form, the equation is

$$x = y^2 + z^2.$$

The traces in the xy-plane, the xz-plane, and the plane given by $x = 1$ are as follows.

Trace in xy-plane ($z = 0$):	$x = y^2$	*Parabola*
Trace in xz-plane ($y = 0$):	$x = z^2$	*Parabola*
Trace in plane $x = 1$:	$y^2 + z^2 = 1$	*Circle*

These three traces are shown in Figure 7.16. From the traces, you can see that the surface is an elliptic (or circular) paraboloid. If you have access to a three-dimensional graphing utility, try using it to graph this surface. If you do this, you will discover that sketching surfaces in space is not a simple task—even with a graphing utility.

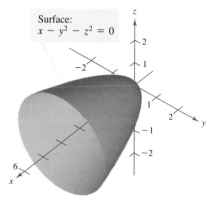

Surface:
$x - y^2 - z^2 = 0$

FIGURE 7.15 Elliptic Paraboloid

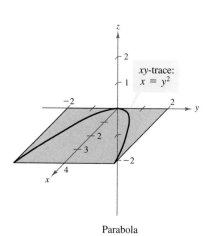

Parabola

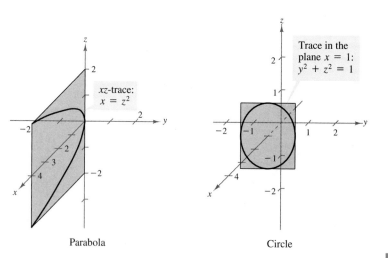

Parabola

Circle

FIGURE 7.16

EXAMPLE 3 *Classifying Quadric Surfaces*

a. The equation

$$x^2 - 4y^2 - 4z^2 - 4 = 0 \qquad \textit{Original equation}$$

can be written in standard form as

$$\frac{x^2}{4} - y^2 - z^2 = 1. \qquad \textit{Standard form}$$

From the standard form, you can see that the graph is a hyperboloid of two sheets, with the x-axis as its axis, as shown in Figure 7.17(a).

b. The equation

$$x^2 + 4y^2 + z^2 - 4 = 0 \qquad \textit{Original equation}$$

can be written in standard form as

$$\frac{x^2}{4} + y^2 + \frac{z^2}{4} = 1. \qquad \textit{Standard form}$$

From the standard form, you can see that the graph is an ellipsoid, as shown in Figure 7.17(b).

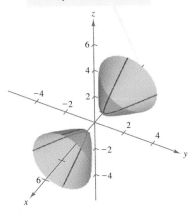

Surface:
$x^2 - 4y^2 - 4z^2 - 4 = 0$

(a)

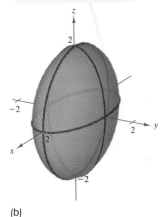

Surface:
$x^2 + 4y^2 + z^2 - 4 = 0$

(b)

FIGURE 7.17

Discussion Problem Classifying Quadric Surfaces

Classify the following quadric surfaces.

a. $\dfrac{x^2}{2^2} - \dfrac{y^2}{4^2} + \dfrac{z^2}{3^2} = 0$ **b.** $\dfrac{x^2}{2^2} + \dfrac{y^2}{4^2} + \dfrac{z^2}{3^2} = 1$ **c.** $\dfrac{x^2}{2^2} - \dfrac{y^2}{4^2} + \dfrac{z^2}{3^2} = 1$

Warm Up   The following warm-up exercises involve skills that were covered in earlier sections. You will use these skills in the exercise set for this section.

In Exercises 1–4, find the x- and y-intercepts of the function.

1. $3x + 4y = 12$ **2.** $6x + y = -8$ **3.** $-2x + y = -2$ **4.** $-x - y = 5$

In Exercises 5–8, rewrite the expression by completing the square.

5. $x^2 + y^2 + z^2 - 2x - 4y - 6z + 15 = 0$ **6.** $x^2 + y^2 - z^2 - 8x + 4y - 6z + 11 = 0$

7. $z - 2 = x^2 + y^2 + 2x - 2y$ **8.** $x^2 + y^2 + z^2 - 6x + 10y + 26z = -202$

In Exercises 9 and 10, write the expression in standard form.

9. $16x^2 - 16y^2 + 16z^2 = 4$ **10.** $9x^2 - 9y^2 + 9z^2 = 36$

EXERCISES 7.2   means that technology can help you solve or check the exercise(s).

In Exercises 1–16, find the intercepts and sketch the graph of the plane.

1. $4x + 2y + 6z = 12$ **2.** $3x + 6y + 2z = 6$

3. $3x + 3y + 5z = 15$ **4.** $x + y + z = 3$

5. $2x - y + 3z = 8$ **6.** $2x - y + z = 4$

7. $z = 3$ **8.** $y = -4$

9. $y + z = 5$ **10.** $x + 2y = 8$

11. $x + y - z = 0$ **12.** $x - 3z = 3$

13. $-x + 5y - 3z = 10$ **14.** $2y + 4z = 1$

15. $2y - 5z = 5$ **16.** $-x - y = 6$

In Exercises 17–24, determine whether the planes $a_1x + b_1y + c_1z = d_1$ and $a_2x + b_2y + c_2z = d_2$ are parallel, perpendicular, or neither. The planes are parallel if there exists a nonzero constant k such that $a_1 = ka_2$, $b_1 = kb_2$, $c_1 = kc_2$, and perpendicular if $a_1a_2 + b_1b_2 + c_1c_2 = 0$.

17. $5x - 3y + z = 4, x + 4y + 7z = 1$

18. $3x + y - 4z = 3, -9x - 3y + 12z = 4$

19. $x - 5y - z = 1, 5x - 25y - 5z = -3$

20. $x = 6, y = -1$

21. $x + 2y = 3, 4x + 8y = 5$

22. $x + 3y + z = 7, x - 5z = 0$

23. $2x + y = 3, 3x - 5z = 0$

24. $2x - z = 1, 4x + y + 8z = 10$

In Exercises 25–28, find the distance between the point and the plane (see figure). The distance D between a point (x_0, y_0, z_0) and the plane $ax + by + cz + d = 0$ is

$$D = \frac{|ax_0 + by_0 + cz_0 + d|}{\sqrt{a^2 + b^2 + c^2}}.$$

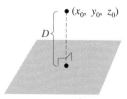

Plane:
$ax + by + cz + d = 0$

FIGURE FOR 25–28

25. $(0, 0, 0), 2x + 3y + z = 12$

26. $(1, 5, -4), 3x - y + 2z = 6$

27. $(1, 2, 3), 2x - y + z = 4$

28. $(10, 0, 0), x - 3y + 4z = 6$

In Exercises 29–36, match the given equation with the correct graph. [The graphs are labeled (a)–(h).]

29. $\dfrac{x^2}{9} + \dfrac{y^2}{16} + \dfrac{z^2}{9} = 1$

30. $x^2 - \dfrac{4y^2}{15} + z^2 = -\dfrac{4}{15}$

31. $4x^2 + 4y^2 - z^2 = 4$

32. $y^2 = 4x^2 + 9z^2$

33. $4x^2 - 4y + z^2 = 0$

34. $12z = -3y^2 + 4x^2$

35. $4x^2 - y^2 + 4z = 0$

36. $x^2 + y^2 + z^2 = 9$

(a)

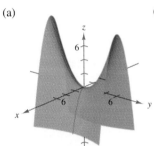

(b)

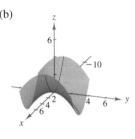

(c)

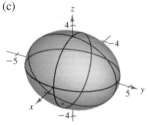

(d)

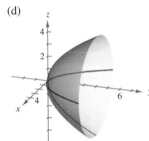

(e)

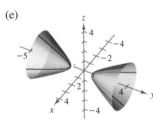

(f)

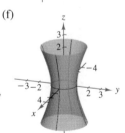

(g)

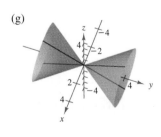

(h)

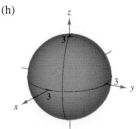

In Exercises 37–40, describe the trace of the surface in the indicated planes.

Surface	*Planes*
37. $x^2 - y - z^2 = 0$	xy-plane, $y = 1$, yz-plane
38. $y = x^2 + z^2$	xy-plane, $y = 1$, yz-plane
39. $\dfrac{x^2}{4} + y^2 + z^2 = 1$	xy-plane, xz-plane, yz-plane
40. $y^2 + z^2 - x^2 = 1$	xy-plane, xz-plane, yz-plane

In Exercises 41–56, identify the quadric surface.

41. $x^2 + \dfrac{y^2}{4} + z^2 = 1$

42. $\dfrac{x^2}{9} + \dfrac{y^2}{16} + \dfrac{z^2}{16} = 1$

43. $25x^2 + 25y^2 - z^2 = 5$

44. $9x^2 + 4y^2 - 8z^2 = 72$

45. $x^2 - y + z^2 = 0$

46. $z = 4x^2 + y^2$

47. $x^2 - y^2 + z = 0$

48. $z^2 - x^2 - \dfrac{y^2}{4} = 1$

49. $4x^2 - y^2 + 4z^2 = -16$

50. $z^2 = x^2 + \dfrac{y^2}{4}$

51. $z^2 = 9x^2 + y^2$

52. $4y = x^2 + z^2$

53. $3z = -y^2 + x^2$

54. $z^2 = 2x^2 + 2y^2$

55. $2x^2 + 2y^2 + 2z^2 - 3x + 4z = 10$

56. $4x^2 + y^2 - 4z^2 - 16x - 6y - 16z + 9 = 0$

In Exercises 57–60, use a three-dimensional graphing utility to graph the function.

57. $z = y^2 - x^2 + 1$

58. $z = x^2 + y^2 + 1$

59. $z = \dfrac{x^2}{2} + \dfrac{y^2}{4}$

60. $z = \frac{1}{12}\sqrt{144 - 16x^2 - 9y^2}$

Functions of Several Variables

Functions of Several Variables ■ *The Graph of a Function of Two Variables* ■
Level Curves and Contour Maps ■ *Applications*

Functions of Several Variables

In the first six chapters of this text, you studied functions of a single independent variable. Many quantities in science, business, and technology, however, are functions not of one, but of two or more variables. For instance, the demand function for a product is often dependent on the price *and* the advertising, rather than on the price alone.

The notation for functions of two or more variables is similar to that used for functions of a single variable. For example,

$$f(\underbrace{x, y}) = x^2 + xy \quad \text{and} \quad g(\underbrace{x, y}) = e^{x+y}$$

$$\quad\; \text{2 variables} \qquad\qquad\qquad \text{2 variables}$$

are functions of two variables, and

$$f(\underbrace{x, y, z}) = x + 2y - 3z$$

$$\quad\;\; \text{3 variables}$$

is a function of three variables.

Definition of a Function of Two Variables

If to each ordered pair (x, y) in some set D there corresponds a unique real number $z = f(x, y)$, then f is called a **function of x and y.** The set D is the **domain** of f, and the corresponding set of z-values is the **range** of f. Functions of three, four, or more variables are defined similarly.

EXAMPLE 1 *Evaluating Functions of Several Variables*

a. For $f(x, y) = 2x^2 - y^2$, you can evaluate $f(2, 3)$ as follows.

$$f(2, 3) = 2(2)^2 - (3)^2 = 8 - 9 = -1$$

b. For $f(x, y, z) = e^x(y + z)$, you can evaluate $f(0, -1, 4)$ as follows.

$$f(0, -1, 4) = e^0(-1 + 4) = (1)(3) = 3$$

The Graph of a Function of Two Variables

A function of two variables can be represented graphically as a surface in space by letting $z = f(x, y)$. When sketching the graph of a function of x and y, remember that even though the graph is three-dimensional, the domain of the function is two-dimensional—it consists of the points in the xy-plane for which the function is defined. As with functions of a single variable, unless specifically restricted, the domain of a function of two variables is assumed to be the set of all points (x, y) for which the defining equation has meaning.

EXAMPLE 2 *Finding the Domain and Range of a Function*

Find the domain and range of the function

$$f(x, y) = \sqrt{64 - x^2 - y^2}.$$

Solution

Because no restrictions are given, the domain is assumed to be the set of all points for which the defining equation makes sense.

$$64 - x^2 - y^2 \geq 0 \qquad \textit{Quantity inside radical must be nonnegative.}$$
$$x^2 + y^2 \leq 64 \qquad \textit{Domain of the function}$$

Thus, the domain is the set of all points that lie on or inside the circle given by $x^2 + y^2 = 8^2$. The range of f is the set

$$0 \leq z \leq 8. \qquad \textit{Range of the function}$$

As shown in Figure 7.18, the graph of the function is a hemisphere.   ▬

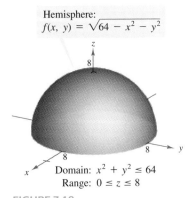

Hemisphere:
$f(x, y) = \sqrt{64 - x^2 - y^2}$

Domain: $x^2 + y^2 \leq 64$
Range: $0 \leq z \leq 8$

FIGURE 7.18

▦ TECHNOLOGY

Some three-dimensional graphing utilities can graph equations in x, y, and z. Others are programmed to graph only functions of x and y. A surface in space represents the graph of a function of x and y only if each vertical line intersects the surface at most once. For instance, the surface shown in Figure 7.18 does pass this vertical line test, but the surface at the right (drawn by *Mathematica*) does not represent the graph of a function of x and y.

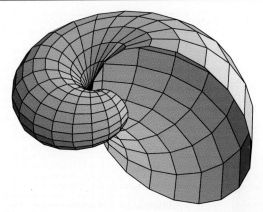

Some vertical lines intersect this surface more than once. Thus, the surface does not pass the vertical line test and is not a function of x and y.

Level Curves and Contour Maps

A **contour map** of a surface is created by *projecting* traces, taken in evenly spaced planes that are parallel to the xy-plane, onto the xy-plane. Each projection is a **level curve** of the surface.

Contour maps are used to create weather maps, topographical maps, and population density maps. For instance, Figure 7.19(a) shows a graph of a "mountain and valley" surface given by $z = f(x, y)$. Each of the level curves in Figure 7.19(b) represents the intersection of the surface $z = f(x, y)$ with a plane $z = c$, where $c = 828, 830, \ldots, 854$.

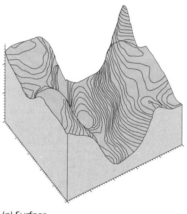
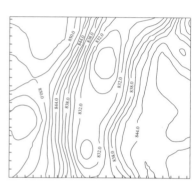

(a) Surface   (b) Contour Map

FIGURE 7.19

Real Life

EXAMPLE 3 *Reading a Contour Map*

The "contour map" in Figure 7.20 was computer generated using data collected by satellite instrumentation on August 25, 1992. The map uses color to represent levels of chlorine nitrate in the atmosphere. Chlorine nitrate contributes to the ozone depletion in the earth's atmosphere. The red areas represent the highest level of chlorine nitrate and the dark blue areas represent the lowest level. Describe the areas that have the highest levels of chlorine nitrate. *(Source: Lockheed Missiles and Space Company)*

Solution

The highest levels of chlorine nitrate are in the Antarctic Ocean, surrounding Antarctica. Although chlorine nitrate is not itself harmful to ozone, it has a tendency to convert to chlorine monoxide, which *is* harmful to ozone. Once the chlorine nitrate is converted to chlorine monoxide, it no longer shows on the contour map. Thus, Antarctica itself shows little chlorine nitrate—the nitrate has been converted to monoxide. If you have seen maps showing the "ozone hole" in the earth's atmosphere, you know that the hole occurs over Antarctica.

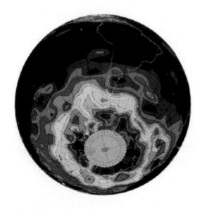

FIGURE 7.20

Real Life

EXAMPLE 4 *Reading a Contour Map*

The contour map shown in Figure 7.21 represents the population density of the United States in 1990. Discuss the use of color to represent the level curves. *(Source: U.S. Bureau of Census)*

Solution

You can see from the key that the light yellow regions have population densities of less than 10 people per square mile. As the color darkens, the population density increases, so that the brown regions have population densities of more than 500 people per square mile. The most densely populated regions are urban areas and are represented by three different types of dots.

One advantage of such a map is that it allows you to "see" the population density of the country at a glance. From the map, it is clear that the Rocky Mountain region of the country has a population density that is much less than the region between Chicago and New York City.

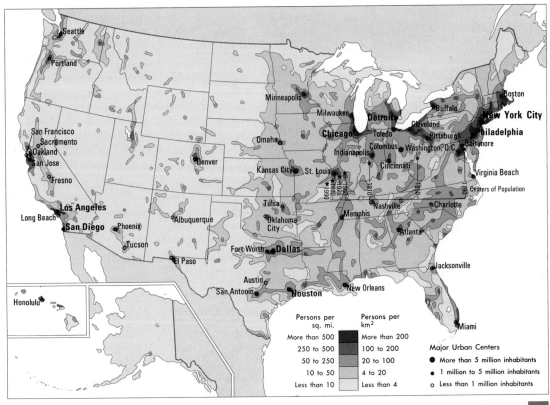

Persons per sq. mi.	Persons per km²
More than 500	More than 200
250 to 500	100 to 200
50 to 250	20 to 100
10 to 50	4 to 20
Less than 10	Less than 4

Major Urban Centers
● More than 5 million inhabitants
• 1 million to 5 million inhabitants
○ Less than 1 million inhabitants

FIGURE 7.21   From *The World Book Encyclopedia.* © World Book, Inc. By permission of the publisher.

Note In Figure 7.21, the level curves do not correspond to equally spaced population density levels. If, for example, you wanted to represent population density levels of 0–50, 50–100, . . . , and 450–500, how would the map change? Do you think that the change would better illustrate the population density levels in the United States?

Applications

The **Cobb-Douglas production function** is used in economics to represent the number of units produced by varying amounts of labor and capital. Let x represent the number of units of labor and let y represent the number of units of capital. Then, the number of units produced is modeled by

$$f(x, y) = Cx^a y^{1-a},$$

where C is a constant and $0 < a < 1$.

Real
Life

EXAMPLE 5 *Using a Production Function*

A manufacturer estimates that its production (measured in units of a product) can be modeled by

$$f(x, y) = 100x^{0.6}y^{0.4},$$

where the labor x is measured in person-hours and the capital y is measured in thousands of dollars.

a. What is the production level when $x = 1000$ and $y = 500$?

b. What is the production level when $x = 2000$ and $y = 1000$?

c. How does doubling the amounts of labor and capital from part a to part b affect the production?

Solution

a. When $x = 1000$ and $y = 500$, the production level is

$$f(1000, 500) = 100(1000)^{0.6}(500)^{0.4}$$
$$\approx 75{,}786 \text{ units.}$$

b. When $x = 2000$ and $y = 1000$, the production level is

$$f(2000, 1000) = 100(2000)^{0.6}(1000)^{0.4}$$
$$\approx 151{,}572 \text{ units.}$$

c. When the amounts of labor and capital are doubled, the production level also doubles. In Exercise 42, you are asked to show that this is characteristic of the Cobb-Douglas production function.

A contour graph of this function is shown in Figure 7.22.

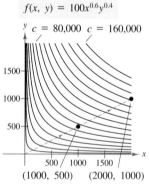

$f(x, y) = 100x^{0.6}y^{0.4}$

$c = 80{,}000 \quad c = 160{,}000$

(1000, 500) (2000, 1000)

FIGURE 7.22 Level Curves (at increments of 10,000)

Note In Figure 7.22, note that the level curves of the function

$$f(x, y) = 100x^{0.6}y^{0.4}$$

occur at increments of 10,000.

Real
Life

EXAMPLE 6 *Finding Monthly Payments*

For many Americans, buying a house is the largest single purchase they will ever make. During the 1970s, 1980s, and 1990s, the annual interest rate on home mortgages varied drastically. It was as high as 18% and as low as 5%. Such variations can change monthly payments by hundreds of dollars.

The monthly payment M for an installment loan of P dollars taken out over t years at an annual rate of r is

$$M = f(P, r, t) = \frac{\dfrac{Pr}{12}}{1 - \left[\dfrac{1}{1 + (r/12)}\right]^{12t}}.$$

a. Find the monthly payment for a home mortgage of $95,000 taken out for 30 years at an annual interest rate of 9%.

b. Find the monthly payment for a car loan of $14,000 taken out for 5 years at an annual interest rate of 11%.

Solution

a. If $P = \$95,000$, $r = 0.09$, and $t = 30$, then the monthly payment is

$$M = f(95,000, 0.09, 30)$$

$$= \frac{\dfrac{(95,000)(0.09)}{12}}{1 - \left[\dfrac{1}{1 + (0.09/12)}\right]^{12(30)}}$$

$$= \$764.39.$$

b. If $P = \$14,000$, $r = 0.11$, and $t = 5$, then the monthly payment is

$$M = f(14,000, 0.11, 5)$$

$$= \frac{\dfrac{(14,000)(0.11)}{12}}{1 - \left[\dfrac{1}{1 + (0.11/12)}\right]^{12(5)}}$$

$$= \$304.39.$$

Discussion Problem   **Monthly Payments**

You are taking out a home mortgage for $100,000, and you are given the following options. Which option would you choose? Explain your reasoning.

a. A fixed annual rate of 10%, over a term of 20 years.

b. A fixed annual rate of 9%, over a term of 30 years.

c. An adjustable annual rate of 9%, over a term of 20 years. The annual rate can fluctuate—each year it is set at 1% above the prime rate.

d. A fixed annual rate of 9%, over a term of 15 years.

Warm Up   The following warm-up exercises involve skills that were covered in earlier sections. You will use these skills in the exercise set for this section.

In Exercises 1–4, evaluate the function when $x = -3$.

1. $f(x) = 5 - 2x$

2. $f(x) = -x^2 + 4x + 5$

3. $y = \sqrt{4x^2 - 3x + 4}$

4. $y = \sqrt[3]{34 - 4x + 2x^2}$

In Exercises 5–8, find the domain of the function.

5. $f(x) = 5x^2 + 3x - 2$

6. $g(x) = \dfrac{1}{2x} - \dfrac{2}{x+3}$

7. $h(y) = \sqrt{y - 5}$

8. $f(y) = \sqrt{y^2 - 5}$

In Exercises 9 and 10, evaluate the expression.

9. $(476)^{0.65}$

10. $(251)^{0.35}$

E X E R C I S E S   7.3   means that technology can help you solve or check the exercise(s).

In Exercises 1–14, find the function values.

1. $f(x, y) = \dfrac{x}{y}$

(a) $f(3, 2)$ (b) $f(-1, 4)$ (c) $f(30, 5)$
(d) $f(5, y)$ (e) $f(x, 2)$ (f) $f(5, t)$

2. $f(x, y) = 4 - x^2 - 4y^2$

(a) $f(0, 0)$ (b) $f(0, 1)$ (c) $f(2, 3)$
(d) $f(1, y)$ (e) $f(x, 0)$ (f) $f(t, 1)$

3. $f(x, y) = xe^y$

(a) $f(5, 0)$ (b) $f(3, 2)$ (c) $f(2, -1)$
(d) $f(5, y)$ (e) $f(x, 2)$ (f) $f(t, t)$

4. $g(x, y) = \ln|x + y|$

(a) $g(2, 3)$ (b) $g(5, 6)$ (c) $g(e, 0)$
(d) $g(0, 1)$ (e) $g(2, -3)$ (f) $g(e, e)$

5. $h(x, y, z) = \dfrac{xy}{z}$

(a) $h(2, 3, 9)$ (b) $h(1, 0, 1)$

6. $f(x, y, z) = \sqrt{x + y + z}$

(a) $f(0, 5, 4)$ (b) $f(6, 8, -3)$

7. $V(r, h) = \pi r^2 h$

(a) $V(3, 10)$ (b) $V(5, 2)$

8. $F(r, N) = 500\left[1 + \dfrac{r}{12}\right]^N$

(a) $F(0.09, 60)$ (b) $F(0.14, 240)$

9. $A(P, r, t) = P\left[\left(1 + \dfrac{r}{12}\right)^{12t} - 1\right]\left(1 + \dfrac{12}{r}\right)$

(a) $A(100, 0.10, 10)$ (b) $A(275, 0.0925, 40)$

10. $A(P, r, t) = Pe^{rt}$

(a) $A(500, 0.10, 5)$ (b) $A(1500, 0.12, 20)$

11. $f(x, y) = \displaystyle\int_x^y (2t - 3)\,dt$

(a) $f(1, 2)$ (b) $f(1, 4)$

12. $g(x, y) = \displaystyle\int_x^y \dfrac{1}{t}\,dt$

(a) $g(4, 1)$ (b) $g(6, 3)$

13. $f(x, y) = x^2 - 2y$

(a) $f(x + \Delta x, y)$ (b) $\dfrac{f(x, y + \Delta y) - f(x, y)}{\Delta y}$

14. $f(x, y) = 3xy + y^2$

(a) $f(x + \Delta x, y)$ (b) $\dfrac{f(x, y + \Delta y) - f(x, y)}{\Delta y}$

In Exercises 15–18, describe the region R in the xy-coordinate plane that corresponds to the domain of the function, and find the range of the function.

15. $f(x, y) = \sqrt{16 - x^2 - y^2}$

16. $f(x, y) = x^2 + y^2 - 1$

17. $f(x, y) = e^{x/y}$

18. $f(x, y) = \ln(x + y)$

In Exercises 19–28, describe the region R in the xy-coordinate plane that corresponds to the domain of the function.

19. $f(x, y) = \sqrt{9 - 9x^2 - y^2}$

20. $f(x, y) = \sqrt{x^2 + y^2 - 1}$

21. $f(x, y) = \dfrac{x}{y}$

22. $f(x, y) = \dfrac{4y}{x - 1}$

23. $f(x, y) = \dfrac{1}{xy}$

24. $g(x, y) = \dfrac{1}{x - y}$

25. $h(x, y) = x\sqrt{y}$

26. $f(x, y) = \sqrt{xy}$

27. $g(x, y) = \ln(4 - x - y)$

28. $f(x, y) = ye^{1/x}$

In Exercises 29–32, match the graph of the surface with one of the contour maps. [The contour maps are labeled (a)–(d).]

29. $f(x, y) = x^2 + \dfrac{y^2}{4}$

30. $f(x, y) = e^{1-x^2+y^2}$

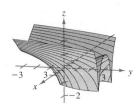

31. $f(x, y) = e^{1-x^2-y^2}$

32. $f(x, y) = \ln|y - x^2|$

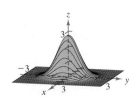

(a)

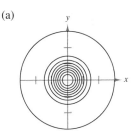

(b)

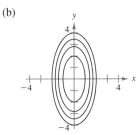

(c)

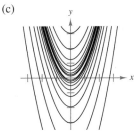

(d)

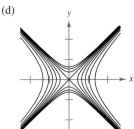

In Exercises 33–40, describe the level curves of the function. Sketch the level curves for the given c-values.

Function	c-Values
33. $z = x + y$	$c = -1, 0, 2, 4$
34. $z = 6 - 2x - 3y$	$c = 0, 2, 4, 6, 8, 10$
35. $z = \sqrt{16 - x^2 - y^2}$	$c = 0, 1, 2, 3, 4$
36. $f(x, y) = x^2 + y^2$	$c = 0, 2, 4, 6, 8$
37. $f(x, y) = xy$	$c = \pm1, \pm2, \ldots, \pm6$
38. $z = e^{xy}$	$c = 1, 2, 3, 4, \frac{1}{2}, \frac{1}{3}, \frac{1}{4}$
39. $f(x, y) = \dfrac{x}{x^2 + y^2}$	$c = \pm\frac{1}{2}, \pm 1, \pm\frac{3}{2}, \pm2$
40. $f(x, y) = \ln(x - y)$	$c = 0, \pm\frac{1}{2}, \pm1, \pm\frac{3}{2}, \pm2$

41. *Cobb-Douglas Production Function* A manufacturer estimates the Cobb-Douglas production function to be

$$f(x, y) = 100x^{0.75}y^{0.25}.$$

Estimate the production level when $x = 1500$ and $y = 1000$.

42. *Cobb-Douglas Production Function* Use the Cobb-Douglas production function (Example 5) to show that if both the number of units of labor and the number of units of capital are doubled, the production level is also doubled.

43. *Cost* A company manufactures two types of woodburning stoves: a free-standing model and a fireplace-insert model. The cost function for producing x freestanding stoves and y fireplace-insert stoves is

$$C(x, y) = 27\sqrt{xy} + 195x + 215y + 980.$$

Find the cost when $x = 80$ and $y = 20$.

44. *Forestry* The **Doyle Log Rule** is one of several methods used to determine the lumber yield of a log in board feet in terms of its diameter d in inches and its length L in feet. The number of board feet is given by

$$N(d, L) = \left(\frac{d - 4}{4}\right)^2 L.$$

(a) Find the number of board feet of lumber in a log with a diameter of 22 inches and a length of 12 feet.
(b) Find $N(30, 12)$.

45. *Profit* A corporation manufactures a product at two locations. The costs of producing x_1 units at location 1 and x_2 units at location 2 are

$$C_1(x_1) = 0.02x_1^2 + 4x_1 + 500$$

and

$$C_2(x_2) = 0.05x_2^2 + 4x_2 + 275,$$

respectively. If the product sells for $15 per unit, then the profit function for the product is

$$P(x_1, x_2) = 15(x_1 + x_2) - C_1(x_1) - C_2(x_2).$$

Find (a) $P(250, 150)$, and (b) $P(300, 200)$.

46. *Queuing Model* The average amount of time that a customer waits in line for service is given by

$$W(x, y) = \frac{1}{x - y}, \qquad y < x,$$

where y is the average arrival rate and x is the average service rate (x and y are measured in the number of customers per hour). Evaluate W at the following points.

(a) $(15, 10)$ (b) $(12, 9)$ (c) $(12, 6)$ (d) $(4, 2)$

47. *Investment* In 1994, an investment of $1000 was made in a bond earning 10% compounded annually. The investor pays tax at rate R and the annual rate of inflation is I. In the year 2004, the value V of the bond in constant 1994 dollars is

$$V(I, R) = 1000\left[\frac{1 + 0.10(1 - R)}{1 + I}\right]^{10}.$$

Use this function of two variables to complete the table.

	Inflation Rate, I		
	0.00	0.03	0.05
Tax Rate, R 0.00			
0.28			
0.35			

48. *Investment* The sum of $1000 is deposited in a savings account earning r percent interest compounded continuously. The amount $A(r, t)$ after t years is given by

$$A(r, t) = 1000e^{rt}.$$

Use this function of two variables to complete the table.

	Number of Years, t			
	5	10	15	20
Rate, r 0.06				
0.08				
0.10				

49. *Equity* The equity per share for Phillips Petroleum Company from 1986 through 1991 can be modeled by

$$z = 1.085x + 0.779y - 10.778,$$

where x is the total revenue (in billions of dollars) and y is the total assets (in billions of dollars). *(Source: Phillips Petroleum Company, 1992 Annual Report)*

(a) Find the equity per share when $x = 15$ and $y = 10$.
(b) Which of the two variables in this model has the greater influence on the equity per share of common stock? Explain your reasoning.

50. *A Contour Map* The contour map shown below represents the estimated annual hours of air conditioner use for an average household in 1991. *(Source: Association of Home Appliance Manufacturers)*

(a) Discuss the use of color to represent the level curves.
(b) Do the level curves correspond to equally spaced annual usage hours? Explain.
(c) Describe how to obtain a more detailed contour map.

Partial Derivatives

Functions of Two Variables ■ *Graphical Interpretation of Partial Derivatives* ■
Functions of Three Variables ■ *Higher-Order Partial Derivatives*

Functions of Two Variables

Real-life applications of functions of several variables are often concerned with how changes in one of the variables will affect the values of the functions. For instance, an economist who wants to determine the effect of a tax increase on the economy might make calculations using different tax rates while holding all other variables, such as unemployment, constant.

You can follow a similar procedure to find the rate of change of a function f with respect to one of its independent variables. That is, you find the derivative of f with respect to one independent variable, while holding the other variables constant. This process is called **partial differentiation**, and each derivative is called a **partial derivative.** A function of several variables has as many partial derivatives as it has independent variables.

Note Note that this definition indicates that partial derivatives of a function of two variables are determined by temporarily considering one variable to be fixed. For instance, if $z = f(x, y)$, then to find $\partial z/\partial x$, you consider y to be constant and differentiate with respect to x. Similarly, to find $\partial z/\partial y$, you consider x to be constant and differentiate with respect to y.

Partial Derivatives of a Function of Two Variables

If $z = f(x, y)$, then the **first partial derivatives of f with respect to x and y** are the functions $\partial z/\partial x$ and $\partial z/\partial y$, defined as follows.

$$\frac{\partial z}{\partial x} = \lim_{\Delta x \to 0} \frac{f(x + \Delta x, y) - f(x, y)}{\Delta x} \qquad \text{y is held constant.}$$

$$\frac{\partial z}{\partial y} = \lim_{\Delta y \to 0} \frac{f(x, y + \Delta y) - f(x, y)}{\Delta y} \qquad \text{x is held constant.}$$

EXAMPLE 1 *Finding Partial Derivatives*

Find $\partial z/\partial x$ and $\partial z/\partial y$ for the function $z = 3x - x^2y^2 + 2x^3y$.

Solution

$$\frac{\partial z}{\partial x} = 3 - 2xy^2 + 6x^2y \qquad \text{\textit{Hold y constant and differentiate with respect to x.}}$$

$$\frac{\partial z}{\partial y} = -2x^2y + 2x^3 \qquad \text{\textit{Hold x constant and differentiate with respect to y.}}$$

Notation for First Partial Derivatives

The first partial derivatives of $z = f(x, y)$ are denoted by

$$\frac{\partial z}{\partial x} = f_x(x, y) = z_x = \frac{\partial}{\partial x}[f(x, y)]$$

and

$$\frac{\partial z}{\partial y} = f_y(x, y) = z_y = \frac{\partial}{\partial y}[f(x, y)].$$

The values of the first partial derivatives at the point (a, b) are denoted by

$$\left.\frac{\partial z}{\partial x}\right|_{(a,b)} = f_x(a, b) \quad \text{and} \quad \left.\frac{\partial z}{\partial y}\right|_{(a,b)} = f_y(a, b).$$

TECHNOLOGY

Symbolic differentiation utilities, such as *Derive*, *Maple*, and *Mathematica*, can be used to find partial derivatives of a function of two variables. For instance, when *Derive* is used to find $f_x(x, y)$ for the function in Example 2, you obtain the following.

1: $x\hat{e}^{x^2 y}$

2: $\dfrac{d}{dx}\left[x\hat{e}^{x^2 y}\right]$

3: $\hat{e}^{x^2 y}(2x^2 y + 1)$

EXAMPLE 2 *Finding and Evaluating Partial Derivatives*

Find the first partial derivatives of $f(x, y) = xe^{x^2 y}$ and evaluate each at the point $(1, \ln 2)$.

Solution

To find the first partial derivative with respect to x, hold y constant and differentiate using the Product Rule.

$$f_x(x, y) = x\frac{\partial}{\partial x}[e^{x^2 y}] + e^{x^2 y}\frac{\partial}{\partial x}[x] \qquad \textit{Product Rule}$$

$$= x(2xy)e^{x^2 y} + e^{x^2 y} \qquad \textit{y is held constant.}$$

$$= e^{x^2 y}(2x^2 y + 1) \qquad \textit{Simplify.}$$

At the point $(1, \ln 2)$, the value of this derivative is

$$f_x(1, \ln 2) = e^{(1)^2(\ln 2)}[2(1)^2(\ln 2) + 1] \qquad \textit{Substitute for x and y.}$$

$$= 2(2\ln 2 + 1) \qquad \textit{Simplify.}$$

$$\approx 4.773. \qquad \textit{Use a calculator.}$$

To find the first partial derivative with respect to y, hold x constant and differentiate to obtain

$$f_y(x, y) = x(x^2)e^{x^2 y} \qquad \textit{Constant Multiple Rule}$$

$$= x^3 e^{x^2 y}. \qquad \textit{Simplify.}$$

At the point $(1, \ln 2)$, the value of this derivative is

$$f_y(1, \ln 2) = (1)^3 e^{(1)^2(\ln 2)} \qquad \textit{Substitute for x and y.}$$

$$= 2 \qquad \textit{Simplify.}$$

Graphical Interpretation of Partial Derivatives

At the beginning of this course, you spent a lot of time studying graphical interpretations of the derivative of a function of a single variable. There, you found
that $f'(x_0)$ represents the slope of the tangent line to the graph of $y = f(x)$ at
the point (x_0, y_0). The partial derivatives of a function of two variables also
have useful graphical interpretations. Consider the function

$z = f(x, y).$ *Function of two variables*

As shown in Figure 7.23(a), the graph of this function is a surface in space. If
the variable y is fixed, say $y = y_0$, then

$z = f(x, y_0)$ *Function of one variable*

is a function of one variable. The graph of this function is the curve that is the
intersection of the plane $y = y_0$ and the surface $z = f(x, y)$. On this curve,
the partial derivative

$f_x(x, y_0)$ *Slope in x-direction*

represents the slope in the plane $y = y_0$, as shown in Figure 7.23(a). In a
similar way, if the variable x is fixed, say $x = x_0$, then

$z = f(x_0, y)$ *Function of one variable*

is a function of one variable. Its graph is the intersection of the plane $x = x_0$
and the surface $z = f(x, y)$. On this curve, the partial derivative

$f_y(x_0, y)$ *Slope in y-direction*

represents the slope in the plane $x = x_0$, as shown in Figure 7.23(b).

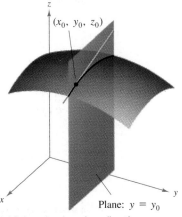

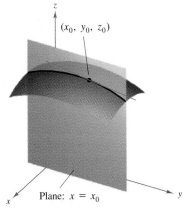

(a) $f_x(x, y_0)$ = slope in x-direction   (b) $f_y(x_0, y)$ = slope in y-direction

FIGURE 7.23

EXAMPLE 3 *Finding Slopes in the x- and y-Directions*

Find the slope of the surface given by

$$f(x, y) = -\frac{x^2}{2} - y^2 + \frac{25}{8}$$

at the point $\left(\frac{1}{2}, 1, 2\right)$ in (a) the x-direction and (b) the y-direction.

Solution

a. To find the slope in the x-direction, hold y constant and differentiate with respect to x to obtain

$$f_x(x, y) = -x. \qquad \textit{Partial derivative with respect to x}$$

At the point $\left(\frac{1}{2}, 1, 2\right)$, the slope in the x-direction is

$$f_x\left(\frac{1}{2}, 1\right) = -\frac{1}{2}, \qquad \textit{Slope in x-direction}$$

as shown in Figure 7.24(a).

b. To find the slope in the y-direction, hold x constant and differentiate with respect to y to obtain

$$f_y(x, y) = -2y. \qquad \textit{Partial derivative with respect to y}$$

At the point $\left(\frac{1}{2}, 1, 2\right)$, the slope in the y-direction is

$$f_y\left(\frac{1}{2}, 1\right) = -2, \qquad \textit{Slope in y-direction}$$

as shown in Figure 7.24(b).

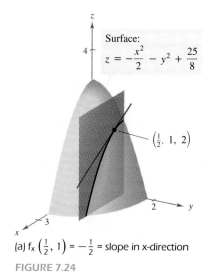

Surface:
$$z = -\frac{x^2}{2} - y^2 + \frac{25}{8}$$

$\left(\frac{1}{2}, 1, 2\right)$

(a) $f_x\left(\frac{1}{2}, 1\right) = -\frac{1}{2}$ = slope in x-direction

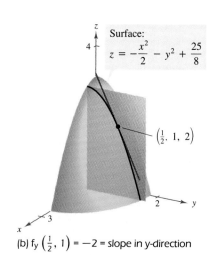

Surface:
$$z = -\frac{x^2}{2} - y^2 + \frac{25}{8}$$

$\left(\frac{1}{2}, 1, 2\right)$

(b) $f_y\left(\frac{1}{2}, 1\right) = -2$ = slope in y-direction

FIGURE 7.24

Consumer products in the same or related markets can be classified as **complementary** or **substitute products.** If two products have a complementary relationship, an increase in the sale of one product will be accompanied by an increase in the sale of the other product. For instance, videocassette recorders and videocassettes have a complementary relationship.

If two products have a substitute relationship, an increase in the sale of one product will be accompanied by a decrease in the sale of the other product. For instance, videocassette recorders and videodisc recorders both compete in the same home entertainment market and you would expect a drop in the price of one to be a deterrent to the sale of the other.

Real Life

EXAMPLE 4 *Examining a Demand Function*

The demand functions for two products are represented by

$$x_1 = f(p_1, p_2) \quad \text{and} \quad x_2 = g(p_1, p_2),$$

where p_1 and p_2 are the prices per unit for the two products, and x_1 and x_2 are the numbers of units sold. The graphs of two different demand functions for x_1 are shown below. Use the graphs to classify the products as complementary or substitute.

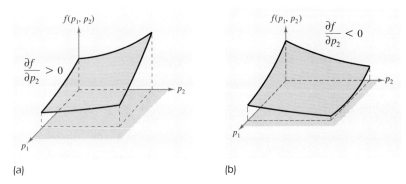

(a)   (b)

Bill Keefrey Studios

In 1993, Subway was chosen as the number one franchise by *Entrepreneur* magazine. At that time, Subway was opening more than 100 new stores each month. What type of product is complementary to a Subway sandwich? What type of product is a substitute?

Solution

a. Notice that graph (a) represents the demand for the *first product*. From the graph of this function, you can see that for a fixed price p_1, an increase in p_2 results in an increase in the demand for the first product. This means that $\partial f / \partial p_2 > 0$ and the two products have a substitute relationship.

b. Notice that graph (b) represents a different demand for the *first product*. From the graph of this function, you can see that for a fixed price p_1, an increase in p_2 results in a decrease in the demand for the first product. This means that $\partial f / \partial p_2 < 0$ and the two products have a complementary relationship.

Functions of Three Variables

The concept of a partial derivative can be extended in a natural way to functions of three or more variables. For instance, the function $w = f(x, y, z)$ has three partial derivatives, each of which is formed by considering two of the variables to be constant. That is, to define the partial derivative of w with respect to x, consider y *and* z to be constant and write

$$\frac{\partial w}{\partial x} = f_x(x, y, z) = \lim_{\Delta x \to 0} \frac{f(x + \Delta x, y, z) - f(x, y, z)}{\Delta x}.$$

To define the partial derivative of w with respect to y, consider x *and* z to be constant and write

$$\frac{\partial w}{\partial y} = f_y(x, y, z) = \lim_{\Delta y \to 0} \frac{f(x, y + \Delta y, z) - f(x, y, z)}{\Delta y}.$$

To define the partial derivative of w with respect to z, consider x *and* y to be constant and write

$$\frac{\partial w}{\partial z} = f_z(x, y, z) = \lim_{\Delta z \to 0} \frac{f(x, y, z + \Delta z) - f(x, y, z)}{\Delta z}.$$

TECHNOLOGY

Symbolic differentiation utilities, such as *Derive*, *Maple*, and *Mathematica*, can be used to find partial derivatives of a function of three or more variables. For instance, when *Derive* is used to find $f_x(x, y, z)$ for the function in Example 5, you obtain the following.

1: $x\hat{e}^{xy+2z}$

2: $\dfrac{d}{dx}\left[x\hat{e}^{xy+2z}\right]$

3: $\hat{e}^{xy+2z}(xy + 1)$

EXAMPLE 5 *Finding Partial Derivatives of a Function*

Find the three partial derivatives of the function

$$w = xe^{xy+2z}.$$

Solution

Holding y and z constant, you obtain

$$\frac{\partial w}{\partial x} = x\frac{\partial}{\partial x}[e^{xy+2z}] + e^{xy+2z}\frac{\partial}{\partial x}[x] \qquad \text{\textit{Product Rule}}$$

$$= x(ye^{xy+2z}) + e^{xy+2z}(1) \qquad \text{\textit{Hold y and z constant.}}$$

$$= (xy + 1)e^{xy+2z}. \qquad \text{\textit{Simplify.}}$$

Holding x and z constant, you obtain

$$\frac{\partial w}{\partial y} = x(x)e^{xy+2z} \qquad \text{\textit{Hold x and z constant.}}$$

$$= x^2 e^{xy+2z}. \qquad \text{\textit{Simplify.}}$$

Holding x and y constant, you obtain

$$\frac{\partial w}{\partial z} = x(2)e^{xy+2z} \qquad \text{\textit{Hold x and y constant.}}$$

$$= 2xe^{xy+2z}. \qquad \text{\textit{Simplify.}}$$

Note that the Product Rule was used only when finding the partial derivative with respect to x. Can you see why? ■

Higher-Order Partial Derivatives

As with ordinary derivatives, it is possible to find partial derivatives of second, third, or higher order. For instance, there are four different ways to find a second partial derivative of $z = f(x, y)$.

$$\frac{\partial}{\partial x}\left(\frac{\partial f}{\partial x}\right) = \frac{\partial^2 f}{\partial x^2} = f_{xx} \qquad \text{\textit{Differentiate twice with respect to x.}}$$

$$\frac{\partial}{\partial y}\left(\frac{\partial f}{\partial y}\right) = \frac{\partial^2 f}{\partial y^2} = f_{yy} \qquad \text{\textit{Differentiate twice with respect to y.}}$$

$$\frac{\partial}{\partial y}\left(\frac{\partial f}{\partial x}\right) = \frac{\partial^2 f}{\partial y \partial x} = f_{xy} \qquad \begin{array}{l}\textit{Differentiate first with respect to x}\\ \textit{and then with respect to y.}\end{array}$$

$$\frac{\partial}{\partial x}\left(\frac{\partial f}{\partial y}\right) = \frac{\partial^2 f}{\partial x \partial y} = f_{yx} \qquad \begin{array}{l}\textit{Differentiate first with respect to y}\\ \textit{and then with respect to x.}\end{array}$$

The third and fourth cases are *mixed* partial derivatives. Notice that with the two types of notation for mixed partials, different conventions are used for indicating the order of differentiation. For instance, the partial derivative

$$\frac{\partial}{\partial y}\left(\frac{\partial f}{\partial x}\right) = \frac{\partial^2 f}{\partial y \partial x} \qquad \text{\textit{Right-to-left order}}$$

indicates differentiation with respect to x first, but the partial

$$(f_y)_x = f_{yx} \qquad \text{\textit{Left-to-right order}}$$

indicates differentiation with respect to y first. To remember this, note that in each case you differentiate first with respect to the variable "nearest" f.

EXAMPLE 6 *Finding Second Partial Derivatives*

Find the second partial derivatives of

$$f(x, y) = 3xy^2 - 2y + 5x^2y^2$$

and determine the value of $f_{xy}(-1, 2)$.

Solution

Note Notice in Example 6 that the two mixed partials are equal. This is often the case. In fact, it can be shown that if a function has continuous second partial derivatives, then the order in which the partial derivatives are taken is irrelevant.

Begin by finding the first partial derivatives

$$f_x(x, y) = 3y^2 + 10xy^2 \quad \text{and} \quad f_y(x, y) = 6xy - 2 + 10x^2y.$$

Then, differentiating with respect to x and y produces

$$f_{xx}(x, y) = 10y^2, \qquad f_{yy}(x, y) = 6x + 10x^2$$
$$f_{xy}(x, y) = 6y + 20xy, \qquad f_{yx}(x, y) = 6y + 20xy.$$

Finally, the value of $f_{xy}(x, y)$ at the point $(-1, 2)$ is

$$f_{xy}(-1, 2) = 6(2) + 20(-1)(2) = 12 - 40 = -28.$$

A function of two variables has two first partial derivatives and four second partial derivatives. For a function of three variables, there are three first partials,

$$f_x, f_y, \text{ and } f_z$$

and nine second partials,

$$f_{xx}, f_{xy}, f_{xz}, f_{yx}, f_{yy}, f_{yz}, f_{zx}, f_{zy}, \text{ and } f_{zz},$$

six of which are mixed partials. To find partial derivatives of order three and higher, follow the same pattern used to find second partial derivatives. For instance, if $z = f(x, y)$, then

$$z_{xxx} = \frac{\partial}{\partial x}\left(\frac{\partial^2 f}{\partial x^2}\right) = \frac{\partial^3 f}{\partial x^3} \quad \text{and} \quad z_{xxy} = \frac{\partial}{\partial y}\left(\frac{\partial^2 f}{\partial x^2}\right) = \frac{\partial^3 f}{\partial y \partial x^2}.$$

EXAMPLE 7 *Finding Second Partial Derivatives*

Find the second partial derivatives of

$$f(x, y, z) = ye^x + x \ln z.$$

Solution

Begin by finding the first partial derivatives.

$$f_x(x, y, z) = ye^x + \ln z, \qquad f_y(x, y, z) = e^x, \qquad f_z(x, y, z) = \frac{x}{z}$$

Then, differentiate with respect to x, y, and z to find the nine second partial derivatives.

$$f_{xx}(x, y, z) = ye^x, \qquad f_{xy}(x, y, z) = e^x, \qquad f_{xz}(x, y, z) = \frac{1}{z}$$

$$f_{yx}(x, y, z) = e^x, \qquad f_{yy}(x, y, z) = 0, \qquad f_{yz}(x, y, z) = 0$$

$$f_{zx}(x, y, z) = \frac{1}{z}, \qquad f_{zy}(x, y, z) = 0, \qquad f_{zz}(x, y, z) = -\frac{x}{z^2}$$

Discussion Problem **Finding Partial Derivatives**

Find the indicated partial derivatives.

Function	*Partial Derivative*
a. $f(x, y) = 2x^2 + 3y^2$	$f_x(x, y)$
b. $g(r, t) = e^{rt^2}$	$g_t(r, t)$
c. $h(x, y, z) = 3xy^2 + 4yz^2$	$h_y(x, y, z)$
d. $f(x, y) = 4xy^2 + 2x^2 y$	$f_{xy}(x, y)$

Warm Up

The following warm-up exercises involve skills that were covered in earlier sections. You will use these skills in the exercise set for this section.

In Exercises 1–8, find the derivative of the function.

1. $f(x) = \sqrt{x^2 + 3}$

2. $g(x) = (3 - x^2)^3$

3. $g(t) = te^{2t+1}$

4. $f(x) = e^{2x}\sqrt{1 - e^{2x}}$

5. $f(x) = \ln(3 - 2x)$

6. $u(t) = \ln\sqrt{t^3 - 6t}$

7. $g(x) = \dfrac{5x^2}{(4x - 1)^2}$

8. $f(x) = \dfrac{(x + 2)^3}{(x^2 - 9)^2}$

In Exercises 9 and 10, evaluate the derivative at the point $(2, 6)$.

9. $f(x) = x^2 e^{x-2}$

10. $g(x) = x\sqrt{x^2 - x + 2}$

E X E R C I S E S 7.4   means that technology can help you solve or check the exercise(s).

In Exercises 1–18, find the first partial derivatives with respect to x and with respect to y.

1. $f(x, y) = 2x - 3y + 5$

2. $f(x, y) = x^2 - 3y^2 + 7$

3. $f(x, y) = 5\sqrt{x} - 6y^2$

4. $f(x, y) = x^{-1/2} + 4y^{3/2}$

5. $f(x, y) = xy$

6. $f(x, y) = \dfrac{x}{y}$

7. $z = x\sqrt{y}$

8. $z = x^2 - 3xy + y^2$

9. $f(x, y) = \sqrt{x^2 + y^2}$

10. $f(x, y) = \dfrac{xy}{x^2 + y^2}$

11. $z = x^2 e^{2y}$

12. $z = xe^{x+y}$

13. $h(x, y) = e^{-(x^2+y^2)}$

14. $g(x, y) = e^{x/y}$

15. $z = \ln(x^2 + y^2)$

16. $z = \ln\sqrt{xy}$

17. $z = \ln\dfrac{x - y}{(x + y)^2}$

18. $g(x, y) = \ln\sqrt{x^2 + y^2}$

In Exercises 19–24, let $f(x, y) = 3x^2 ye^{x-y}$ and $g(x, y) = 3xy^2 e^{y-x}$. Find each of the following.

19. $f_x(x, y)$

20. $f_y(x, y)$

21. $g_x(x, y)$

22. $g_y(x, y)$

23. $f_x(1, 1)$

24. $g_x(-2, -2)$

In Exercises 25–28, evaluate f_x and f_y at the point.

	Function	*Point*
25.	$f(x, y) = 3x^2 + xy - y^2$	$(2, 1)$
26.	$f(x, y) = e^{3xy}$	$(0, 4)$
27.	$f(x, y) = \dfrac{xy}{x - y}$	$(2, -2)$
28.	$f(x, y) = \dfrac{4xy}{\sqrt{x^2 + y^2}}$	$(1, 0)$

In Exercises 29–36, evaluate w_x, w_y, and w_z at the point.

	Function	*Point*
29.	$w = \sqrt{x^2 + y^2 + z^2}$	$(2, -1, 2)$
30.	$w = \dfrac{xy}{x + y + z}$	$(1, 2, 0)$
31.	$w = \ln\sqrt{x^2 + y^2 + z^2}$	$(3, 0, 4)$
32.	$w = \dfrac{1}{\sqrt{1 - x^2 - y^2 - z^2}}$	$(0, 0, 0)$
33.	$w = 2xz^2 + 3xyz - 6y^2z$	$(1, -1, 2)$
34.	$w = xye^{z^2}$	$(2, 1, 0)$
35.	$w = \dfrac{x + 2z}{(x + y + z)^2}$	$(2, 1, 2)$
36.	$w = \dfrac{x + y + z}{(xyz + 1)^2}$	$(1, 1, 1)$

In Exercises 37–40, find values of x and y such that $f_x(x, y) = 0$ and $f_y(x, y) = 0$ simultaneously.

37. $f(x, y) = x^2 + 4xy + y^2 - 4x + 16y + 3$

38. $f(x, y) = 3x^3 - 12xy + y^3$

39. $f(x, y) = \dfrac{1}{x} + \dfrac{1}{y} + xy$

40. $f(x, y) = \ln(x^2 + y^2 + 1)$

In Exercises 41–48, find the slope of the surface at the indicated point in (a) the x-direction and (b) the y-direction.

Function	Point
41. $z = 2x - 3y + 5$	$(2, 1, 6)$
42. $z = xy$	$(1, 2, 2)$
43. $z = x^2 - 9y^2$	$(3, 1, 0)$
44. $z = x^2 + 4y^2$	$(2, 1, 8)$
45. $z = \sqrt{25 - x^2 - y^2}$	$(3, 0, 4)$
46. $z = \dfrac{x}{y}$	$(3, 1, 3)$
47. $z = 4 - x^2 - y^2$	$(1, 1, 2)$

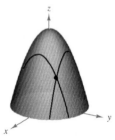

48. $z = x^2 - y^2$ $(-2, 1, 3)$

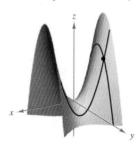

In Exercises 49–52, show that $\partial^2 z/\partial x \partial y = \partial^2 z/\partial y \partial x$.

49. $z = x^2 - 2xy + 3y^2$

50. $z = x^4 - 3x^2 y^2 + y^4$

51. $z = \dfrac{e^{2xy}}{4x}$

52. $z = \dfrac{x^2 - y^2}{2xy}$

In Exercises 53–64, find the second partial derivatives
$$\frac{\partial^2 z}{\partial x^2}, \quad \frac{\partial^2 z}{\partial y^2}, \quad \frac{\partial^2 z}{\partial y \partial x}, \quad \text{and} \quad \frac{\partial^2 z}{\partial x \partial y}.$$

53. $z = x^3 - 4y^2$

54. $z = 3x^2 - xy + 2y^3$

55. $z = 4x^3 + 3xy^2 - 4y^3$

56. $z = x^4 - 3x^2 y^2 + y^4$

57. $z = 9 + 4x - 6y - x^2 - y^2$

58. $z = \sqrt{9 - x^2 - y^2}$

59. $z = \dfrac{xy}{x - y}$

60. $z = \dfrac{x}{x + y}$

61. $z = \sqrt{x^2 + y^2}$

62. $z = \ln(x - y)$

63. $z = xe^{-y^2}$

64. $z = xe^y + ye^x$

In Exercises 65–68, find the first partial derivatives with respect to x, y, and z.

65. $w = 3x^2 y - 5xyz + 10yz^2$

66. $w = \sqrt{x^2 + y^2 + z^2}$

67. $w = \dfrac{xy}{x + y + z}$

68. $w = \dfrac{1}{\sqrt{1 - x^2 - y^2 - z^2}}$

69. *Marginal Cost* A company manufactures two models of bicycles: a mountain bike and a racing bike. The cost function for producing x mountain bikes and y racing bikes is
$$C = 10\sqrt{xy} + 149x + 189y + 675.$$
Find the marginal costs ($\partial C/\partial x$ and $\partial C/\partial y$) when $x = 120$ and $y = 160$.

70. *Marginal Revenue* A corporation has two plants that produce the same product. If x_1 and x_2 are the numbers of units produced at plant 1 and plant 2, respectively, then the total revenue for the product is given by
$$R = 200x_1 + 200x_2 - 4x_1^2 - 8x_1 x_2 - 4x_2^2.$$
If $x_1 = 4$ and $x_2 = 12$, find the following.
(a) The marginal revenue for plant 1, $\partial R/\partial x_1$
(b) The marginal revenue for plant 2, $\partial R/\partial x_2$

71. Marginal Productivity Let $x = 1000$ and $y = 500$ in the Cobb-Douglas production function

$$f(x, y) = 100x^{0.6}y^{0.4}.$$

(a) Find the marginal productivity of labor, $\partial f/\partial x$.
(b) Find the marginal productivity of capital, $\partial f/\partial y$.

72. Marginal Productivity Repeat Exercise 71 for the production function

$$f(x, y) = 100x^{0.75}y^{0.25}.$$

73. Complementary and Substitute Products Using the notation of Example 4 for this section, we let x_1 and x_2 be the demands for products 1 and 2, respectively, and p_1 and p_2 the prices of products 1 and 2, respectively. Determine if the following demand functions describe complementary or substitute product relationships.

(a) $x_1 = 150 - 2p_1 - \frac{5}{2}p_2$, $\quad x_2 = 350 - \frac{3}{2}p_1 - 3p_2$

(b) $x_1 = 150 - 2p_1 + 1.8p_2$, $\quad x_2 = 350 + \frac{3}{4}p_1 - 1.9p_2$

(c) $x_1 = \dfrac{1000}{\sqrt{p_1 p_2}}$, $\qquad x_2 = \dfrac{750}{p_2\sqrt{p_1}}$

74. Investment The value of an investment of $1000 earning 10% compounded annually is given by

$$V(I, R) = 1000\left[\frac{1 + 0.10(1 - R)}{1 + I}\right]^{10},$$

where I is the annual rate of inflation and R is the tax rate for the person making the investment. Calculate $V_I(0.03, 0.28)$ and $V_R(0.03, 0.28)$ and determine whether the tax rate or rate of inflation is the greater negative factor in the growth of the investment.

75. University Admissions Let N be the number of applicants to a university, p the charge for food and housing at the university, and t the tuition. Suppose that N is a function of p and t such that $\partial N/\partial p < 0$ and $\partial N/\partial t < 0$. How would you interpret the fact that both partials are negative?

76. Apparent Temperature A measure of what hot weather feels like to two average persons is the Apparent Temperature index. A model for this index is

$$A = 0.885t - 78.7h + 1.20th + 2.70,$$

where A is the apparent temperature, t is the air temperature, and h is the relative humidity in decimal form. *(Source: The UMAP Journal)*

(a) Find $\partial A/\partial t$ and $\partial A/\partial h$ when $t = 90°C$ and $h = 0.80$.
(b) Which has a greater effect on A, air temperature or humidity? Explain.

77. Marginal Utility The utility function $U = f(x, y)$ is a measure of the utility (or satisfaction) derived by a person from the consumption of two goods x and y. Suppose the utility function is

$$U = -5x^2 + xy - 3y^2.$$

(a) Determine the marginal utility of good x.
(b) Determine the marginal utility of good y.
(c) When $x = 2$ and $y = 3$, should a person consume one more unit of good x or one more unit of good y? Explain your reasoning.
(d) Use a three-dimensional graphing utility to graph the function. Interpret the marginal utilities of goods x and y graphically.

78. Research Project Use your school's library or some other reference source to research a company that increased the demand for its product by creative advertising. Write a paper about the company. Use graphs to show how a change in demand is a change in a product's marginal utility.

Business Capsule

Courtesy The Volant Ski Corporation, Colorado

Hank Kashiwa, co-founder of Volant Ski Corp., used a video brochure to educate customers about their stainless steel skis. Sales increased in 1993 to 20,000 pairs of skis.

7.5 Extrema of Functions of Two Variables

Relative Extrema ■ *The First-Partials Test for Relative Extrema* ■
The Second-Partials Test for Relative Extrema ■ *Applications of Extrema*

Relative Extrema

Earlier in the text, you learned how to use derivatives to find the relative minimum and relative maximum values of a function of a single variable. In this section, you will learn how to use partial derivatives to find the relative minimum and relative maximum values of a function of two variables.

Relative Extrema of a Function of Two Variables

Let f be a function defined on a region containing (x_0, y_0). The number $f(x_0, y_0)$ is a **relative maximum** of f if there is a circular region R centered at (x_0, y_0) such that

$$f(x, y) \leq f(x_0, y_0) \qquad f(x_0, y_0) \text{ is a relative maximum.}$$

for all (x, y) in R. The number $f(x_0, y_0)$ is a **relative minimum** of f if there is a circular region R centered at (x_0, y_0) such that

$$f(x, y) \geq f(x_0, y_0) \qquad f(x_0, y_0) \text{ is a relative minimum.}$$

for all (x, y) in R.

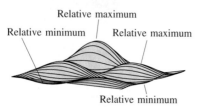

Relative maximum

Relative minimum · Relative maximum

Relative minimum

FIGURE 7.25 Relative Extrema

Graphically, you can think of a point on a surface as a relative maximum if it is at least as high as all "nearby" points on the surface, and you can think of a point on a surface as a relative minimum if it is at least as low as all "nearby" points on the surface. Several relative maxima and relative minima are shown in Figure 7.25.

As in single-variable calculus, you need to distinguish between relative extrema and absolute extrema of a function of two variables. The number $f(x_0, y_0)$ is an absolute maximum of f in the region R if it is greater than or equal to all other function values in the region. For instance, the function

$$f(x, y) = -(x^2 + y^2)$$

graphs as a paraboloid, opening downward, with vertex at $(0, 0, 0)$. The number $f(0, 0) = 0$ is an absolute maximum of the function over the entire xy-plane. An absolute minimum of f in a region is defined similarly.

The First-Partials Test for Relative Extrema

To locate the relative extrema of a function of two variables, you can use a procedure that is similar to the First-Derivative Test used for functions of a single variable.

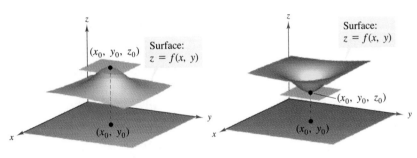

Relative maximum   Relative minimum

FIGURE 7.26

First-Partials Test for Relative Extrema

If $f(x_0, y_0)$ is a relative extremum of f on an open region R in the xy-plane, and the first partial derivatives of f exist in R, then

$$f_x(x_0, y_0) = 0 \quad \text{and} \quad f_y(x_0, y_0) = 0,$$

as shown in Figure 7.26.

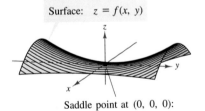

Surface: $z = f(x, y)$

Saddle point at $(0, 0, 0)$:
$f_x(0, 0) = f_y(0, 0) = 0$

FIGURE 7.27

Note   An *open* region in the xy-plane is similar to an open interval on the real number line. For instance, the region R consisting of the interior of the circle $x^2 + y^2 = 1$ is an open region. If the region R consists of the interior of the circle *and* the points on the circle, then it would be a *closed* region.

A point (x_0, y_0) is a **critical point** of f if $f_x(x_0, y_0)$ or $f_y(x_0, y_0)$ is undefined or if

$$f_x(x_0, y_0) = 0 \quad \text{and} \quad f_y(x_0, y_0) = 0. \qquad \textit{Critical point}$$

The First-Partials Test states that if the first partial derivatives exist, then you need only examine values of $f(x, y)$ at critical points to find the relative extrema. As is true for a function of a single variable, however, the critical points of a function of two variables do not always yield relative extrema. For instance, the point $(0, 0)$ is a critical point of the surface shown in Figure 7.27, but $f(0, 0)$ is not a relative extremum of the function. Such points are called **saddle points** of the function.

EXAMPLE 1 *Finding Relative Extrema*

Find the relative extrema of

$$f(x, y) = 2x^2 + y^2 + 8x - 6y + 20.$$

Solution

Begin by finding the first partial derivatives of f.

$$f_x(x, y) = 4x + 8 \quad \text{and} \quad f_y(x, y) = 2y - 6$$

Because these partial derivatives are defined for all points in the xy-plane, the only critical points are those for which both first partial derivatives are 0. To locate these points, set $f_x(x, y)$ and $f_y(x, y)$ equal to 0, and solve the resulting system of equations.

$$4x + 8 = 0 \qquad\qquad \textit{Set } f_x(x, y) \textit{ equal to 0.}$$
$$2y - 6 = 0 \qquad\qquad \textit{Set } f_y(x, y) \textit{ equal to 0.}$$

The solution of this system is $x = -2$ and $y = 3$. Therefore, the point $(-2, 3)$ is the only critical number of f. From the graph of the function, as shown in Figure 7.28, you can see that this critical point yields a relative minimum of the function. Thus, the function has only one relative extremum, which is

$$f(-2, 3) = 3. \qquad\qquad \textit{Relative minimum} \qquad ■$$

Example 1 shows a relative minimum occurring at one type of critical point—the type for which both $f_x(x, y)$ and $f_y(x, y)$ are zero. The next example shows a relative maximum that occurs at the other type of critical point—the type for which either $f_x(x, y)$ or $f_y(x, y)$ is undefined.

EXAMPLE 2 *Finding Relative Extrema*

Find the relative extrema of

$$f(x, y) = 1 - (x^2 + y^2)^{1/3}.$$

Solution

Begin by finding the first partial derivatives of f.

$$f_x(x, y) = -\frac{2x}{3(x^2 + y^2)^{2/3}} \quad \text{and} \quad f_y(x, y) = -\frac{2y}{3(x^2 + y^2)^{2/3}}$$

These partial derivatives are defined for all points in the xy-plane *except* the point $(0, 0)$. Thus, $(0, 0)$ is a critical point of f. Moreover, this is the only critical point because there are no other values of x and y for which either partial is undefined or for which both partials are 0. From the graph of the function, as shown in Figure 7.29, you can see that this critical point yields a relative maximum of the function. Thus, the function has only one relative extremum, which is

$$f(0, 0) = 1. \qquad\qquad \textit{Relative maximum} \qquad ■$$

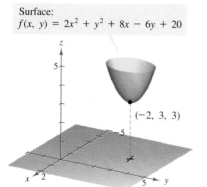

Surface:
$f(x, y) = 2x^2 + y^2 + 8x - 6y + 20$

$(-2, 3, 3)$

FIGURE 7.28

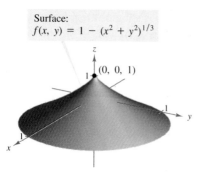

Surface:
$f(x, y) = 1 - (x^2 + y^2)^{1/3}$

$(0, 0, 1)$

FIGURE 7.29 $f_x(x, y)$ and $f_y(x, y)$ are undefined at $(0, 0)$.

▦ TECHNOLOGY

Classifying Extrema with a Graphing Utility

The First-Partials Test tells you that the relative extrema of a function $z = f(x, y)$ can occur only at the critical points of the function. The test does not, however, give you a method for determining whether a critical point actually yields a relative minimum, a relative maximum, or neither. Once you have located the critical points of a function, you can use technology to find the relative extrema graphically. Four examples of surfaces are shown below. Try using the graphs and the given critical numbers to classify the relative extrema.

Surface: $f(x, y) = x^2 - 2x + y^2 + 4y + 5$

Critical point: $(1, -2)$

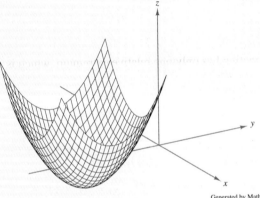

Generated by Mathematica

Surface: $f(x, y) = \sqrt{x^2 + y^2}$

Critical point: $(0, 0)$

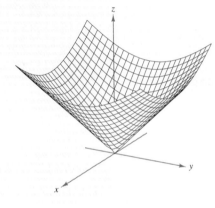

Generated by Mathematica

Surface: $f(x, y) = \dfrac{1}{1 + x^2 + y^2}$

Critical point: $(0, 0)$

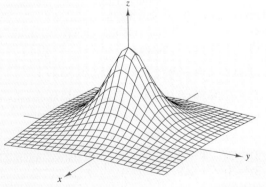

Generated by Mathematica

Surface: $f(x, y) = 8xye^{-x^2 - y^2}$

Critical points: $(0, 0),\ \left(\pm\dfrac{1}{\sqrt{2}}, \pm\dfrac{1}{\sqrt{2}}\right),\ \left(\pm\dfrac{1}{\sqrt{2}}, \mp\dfrac{1}{\sqrt{2}}\right)$

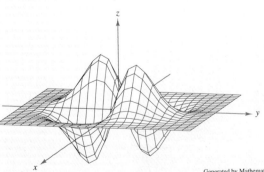

Generated by Mathematica

The Second-Partials Test for Relative Extrema

For functions such as those in Examples 1 and 2, you can determine the *type* of extrema at the critical points by sketching the graph of the function. For more complicated functions, a graphical approach is not so easy to use. The **Second-Partials Test** is an analytical test that can be used to determine whether a critical number yields a relative minimum, a relative maximum, or neither.

Second-Partials Test

If f has continuous first and second partial derivatives in an open region and there exists a point (a, b) in the region such that $f_x(a, b) = 0$ and $f_y(a, b) = 0$, then the quantity

$$d = f_{xx}(a, b) f_{yy}(a, b) - [f_{xy}(a, b)]^2$$

can be used as follows.

1. $f(a, b)$ is a relative minimum if $d > 0$ and $f_{xx}(a, b) > 0$.
2. $f(a, b)$ is a relative maximum if $d > 0$ and $f_{xx}(a, b) < 0$.
3. $(a, b, f(a, b))$ is a saddle point if $d < 0$.
4. The test gives no information if $d = 0$.

Note If $d > 0$, then $f_{xx}(a, b)$ and $f_{yy}(a, b)$ must have the same sign. Thus, you can replace $f_{xx}(a, b)$ with $f_{yy}(a, b)$ in the first two parts of the test.

EXAMPLE 3 *Applying the Second-Partials Test*

Find the relative extrema and saddle points of $f(x, y) = xy - \frac{1}{4}x^4 - \frac{1}{4}y^4$.

Solution
Begin by finding the critical points of f. Because

$$f_x(x, y) = y - x^3 \quad \text{and} \quad f_y(x, y) = x - y^3$$

are defined for all points in the xy-plane, the only critical points are those for which both first partial derivatives are 0. By solving the equations $y - x^3 = 0$ and $x - y^3 = 0$ simultaneously, you can determine that the critical points are $(1, 1)$, $(-1, -1)$, and $(0, 0)$. Furthermore, because

$$f_{xx}(x, y) = -3x^2, \quad f_{yy}(x, y) = -3y^2, \quad \text{and} \quad f_{xy}(x, y) = 1,$$

you can use the quantity $d = f_{xx}(a, b) f_{yy}(a, b) - [f_{xy}(a, b)]^2$ to classify the critical points as follows.

Critical Point	d	$f_{xx}(x, y)$	Conclusion
$(1, 1)$	$(-3)(-3) - 1 = 8$	-3	Relative maximum
$(-1, -1)$	$(-3)(-3) - 1 = 8$	-3	Relative maximum
$(0, 0)$	$(0)(0) - 1 = -1$	0	Saddle point

The graph of f is shown in Figure 7.30.

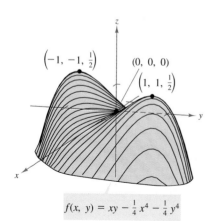

$f(x, y) = xy - \frac{1}{4}x^4 - \frac{1}{4}y^4$

FIGURE 7.30

Applications of Extrema

Real Life

EXAMPLE 4 *Finding a Maximum Profit*

Note In Example 4, you can check that the two products are substitutes by observing that x_1 increases as p_2 increases and x_2 increases as p_1 increases.

A company makes two substitute products whose demand functions are

$$x_1 = 200(p_2 - p_1) \qquad \textit{Demand for Product 1}$$
$$x_2 = 500 + 100p_1 - 180p_2, \qquad \textit{Demand for Product 2}$$

where p_1 and p_2 are the prices per unit (in dollars) and x_1 and x_2 are the numbers of units sold. The costs of producing the two products are \$0.50 and \$0.75 per unit, respectively. Find the prices that will yield a maximum profit.

Solution
The cost and revenue functions are as follows.

$$C = 0.5x_1 + 0.75x_2 \qquad \textit{Cost}$$
$$= 0.5(200)(p_2 - p_1) + 0.75(500 + 100p_1 - 180p_2) \qquad \textit{Substitute.}$$
$$= 375 - 25p_1 - 35p_2 \qquad \textit{Simplify.}$$
$$R = p_1x_1 + p_2x_2 \qquad \textit{Revenue}$$
$$= p_1(200)(p_2 - p_1) + p_2(500 + 100p_1 - 180p_2) \qquad \textit{Substitute.}$$
$$= -200p_1^2 - 180p_2^2 + 300p_1p_2 + 500p_2 \qquad \textit{Simplify.}$$

This implies that the profit function is

$$P = R - C \qquad \textit{Profit}$$
$$= -200p_1^2 - 180p_2^2 + 300p_1p_2 + 500p_2 - (375 - 25p_1 - 35p_2)$$
$$= -200p_1^2 - 180p_2^2 + 300p_1p_2 + 25p_1 + 535p_2 - 375.$$

The maximum profit occurs when the two first partial derivatives are zero.

$$\frac{\partial P}{\partial p_1} = -400p_1 + 300p_2 + 25 = 0$$

$$\frac{\partial P}{\partial p_2} = 300p_1 - 360p_2 + 535 = 0$$

By solving this system simultaneously, you can conclude that the solution is $p_1 = \$3.14$ and $p_2 = \$4.10$. From the graph of the function shown in Figure 7.31, you can see that this critical number yields a maximum. Thus, the maximum profit is

$$P = P(3.14, 4.10) = \$761.48.$$

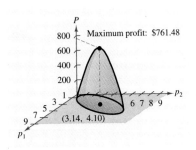

FIGURE 7.31

Note In Example 4, to convince yourself that the maximum profit is \$761.48, try substituting other prices into the profit function. For each pair of prices, you will obtain a profit that is less than \$761.48. For instance, if $p_1 = \$2$ and $p_2 = \$3$, then the profit is $P = P(2, 3) = \$660.00$.

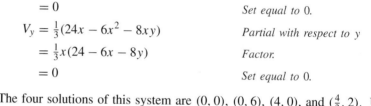

EXAMPLE 5 *Finding a Maximum Volume*

Real Life

Consider all possible rectangular boxes that are resting on the xy-plane with one vertex at the origin and the opposite vertex in the plane $6x + 4y + 3z = 24$, as shown in Figure 7.32. Of all such boxes, which has the greatest volume?

Solution

Because one vertex of the box lies in the plane given by $6x + 4y + 3z = 24$ or $z = \frac{1}{3}(24 - 6x - 4y)$, you can write the volume of the box as

$$V = xyz \qquad\qquad \textit{Volume = (width)(length)(height)}$$
$$= xy\left(\tfrac{1}{3}\right)(24 - 6x - 4y) \qquad\qquad \textit{Substitute for z.}$$
$$= \tfrac{1}{3}(24xy - 6x^2y - 4xy^2). \qquad\qquad \textit{Simplify.}$$

To find the critical numbers, set the first partial derivatives equal to zero.

$$V_x = \tfrac{1}{3}(24y - 12xy - 4y^2) \qquad\qquad \textit{Partial with respect to x}$$
$$= \tfrac{1}{3}y(24 - 12x - 4y) \qquad\qquad \textit{Factor.}$$
$$= 0 \qquad\qquad \textit{Set equal to 0.}$$
$$V_y = \tfrac{1}{3}(24x - 6x^2 - 8xy) \qquad\qquad \textit{Partial with respect to y}$$
$$= \tfrac{1}{3}x(24 - 6x - 8y) \qquad\qquad \textit{Factor.}$$
$$= 0 \qquad\qquad \textit{Set equal to 0.}$$

The four solutions of this system are $(0, 0)$, $(0, 6)$, $(4, 0)$, and $(\frac{4}{3}, 2)$. Using the Second-Partials Test, you can determine that the maximum volume occurs when the width is $x = \frac{4}{3}$ and the length is $y = 2$. For these values, the height of the box is

$$z = \tfrac{1}{3}\left[24 - 6\left(\tfrac{4}{3}\right) - 4(2)\right] = \tfrac{8}{3}.$$

Thus, the maximum volume is

$$V = xyz = \left(\tfrac{4}{3}\right)(2)\left(\tfrac{8}{3}\right) = \tfrac{64}{9} \text{ cubic units.}$$

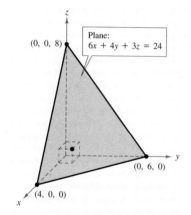

Plane:
$6x + 4y + 3z = 24$

$(0, 0, 8)$

$(0, 6, 0)$

$(4, 0, 0)$

FIGURE 7.32

Discussion Problem *Using the Second-Partials Test*

In Example 5, explain why you can disregard the three critical numbers $(0, 0)$, $(0, 6)$, and $(4, 0)$. Then use the second partial derivatives

$$V_{xx}(x, y) = -4y$$
$$V_{yy}(x, y) = -\tfrac{8}{3}x$$
$$V_{xy}(x, y) = \tfrac{1}{3}(24 - 12x - 8y)$$

to explain why the critical number $\left(\frac{4}{3}, 2\right)$ yields a maximum.

$\mathcal{Warm\ Up}$

The following warm-up exercises involve skills that were covered in earlier sections. You will use these skills in the exercise set for this section.

In Exercises 1–6, solve the system of equations.

1. $\begin{cases} 5x = 15 \\ 3x - 2y = 5 \end{cases}$

2. $\begin{cases} \frac{1}{2}y = 3 \\ -x + 5y = 19 \end{cases}$

3. $\begin{cases} x + y = 5 \\ x - y = -3 \end{cases}$

4. $\begin{cases} 2x - 4y = 14 \\ 3x + y = 7 \end{cases}$

5. $\begin{cases} x^2 + x = 0 \\ 2yx + y = 0 \end{cases}$

6. $\begin{cases} 3y^2 + 6y = 0 \\ xy + x + 2 = 0 \end{cases}$

In Exercises 7–10, find all first and second partial derivatives of the function.

7. $z = 4x^3 - 3y^2$

8. $z = 2x^2 - 3xy + y^2$

9. $z = x^4 - \sqrt{xy} + 2y$

10. $z = xe^{xy}$

E X E R C I S E S   7.5   means that technology can help you solve or check the exercise(s).

In Exercises 1–4, find any critical points and relative extrema of the function.

1. $f(x, y) = x^2 - y^2 + 4x - 8y - 11$

2. $f(x, y) = x^2 + y^2 + 2x - 6y + 6$

3. $f(x, y) = \sqrt{x^2 + y^2 + 1}$

4. $f(x, y) = \sqrt{25 - (x - 2)^2 - y^2}$

In Exercises 5–20, examine each function for relative extrema and saddle points.

5. $f(x, y) = (x - 1)^2 + (y - 3)^2$

6. $f(x, y) = 9 - (x - 3)^2 - (y + 2)^2$

7. $f(x, y) = 2x^2 + 2xy + y^2 + 2x - 3$

8. $f(x, y) = -x^2 - 5y^2 + 8x - 10y - 13$

9. $f(x, y) = -5x^2 + 4xy - y^2 + 16x + 10$

10. $f(x, y) = x^2 + 6xy + 10y^2 - 4y + 4$

11. $f(x, y) = 3x^2 + 2y^2 - 12x - 4y + 7$

12. $f(x, y) = -3x^2 - 2y^2 + 3x - 4y + 5$

13. $f(x, y) = x^2 - y^2 + 4x - 4y - 8$

14. $f(x, y) = x^2 - 3xy - y^2$

15. $f(x, y) = xy$

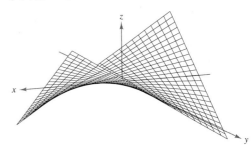

16. $f(x, y) = 12x + 12y - xy - x^2 - y^2$

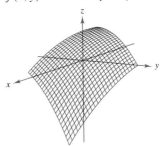

17. $f(x, y) = (x^2 + 4y^2)e^{1-x^2-y^2}$

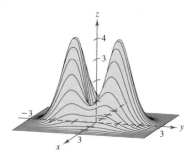

18. $f(x, y) = e^{-(x^2+y^2)}$

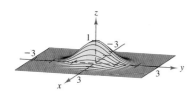

19. $f(x, y) = e^{xy}$

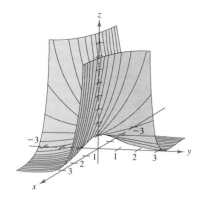

20. $f(x, y) = -\dfrac{4x}{x^2 + y^2 + 1}$

In Exercises 21–24, determine whether there is a relative maximum, a relative minimum, a saddle point, or insufficient information to determine the nature of the function $f(x, y)$ at the critical point (x_0, y_0).

21. $f_{xx}(x_0, y_0) = 16$
$f_{yy}(x_0, y_0) = 4$
$f_{xy}(x_0, y_0) = 8$

22. $f_{xx}(x_0, y_0) = -4$
$f_{yy}(x_0, y_0) = -6$
$f_{xy}(x_0, y_0) = 3$

23. $f_{xx}(x_0, y_0) = -7$
$f_{yy}(x_0, y_0) = 4$
$f_{xy}(x_0, y_0) = 9$

24. $f_{xx}(x_0, y_0) = 20$
$f_{yy}(x_0, y_0) = 8$
$f_{xy}(x_0, y_0) = 9$

In Exercises 25–30, find the critical points and test for relative extrema. List the critical points for which the Second-Partials Test fails.

25. $f(x, y) = (xy)^2$

26. $f(x, y) = \sqrt{x^2 + y^2}$

27. $f(x, y) = x^3 + y^3$

28. $f(x, y) = x^3 + y^3 - 3x^2 + 6y^2 + 3x + 12y + 7$

29. $f(x, y) = x^{2/3} + y^{2/3}$ **30.** $f(x, y) = (x^2 + y^2)^{2/3}$

In Exercises 31 and 32, find the critical points of the function and from the form of the function determine whether each critical point is a relative maximum or a relative minimum.

31. $f(x, y, z) = (x - 1)^2 + (y + 3)^2 + z^2$

32. $f(x, y, z) = 6 - [x(y + 2)(z - 1)]^2$

True or False In Exercises 33–36, determine whether the statement is true or false. If it is false, explain why or give an example that shows it is false.

33. If $d > 0$ and $f_x(a, b) < 0$, then $f(a, b)$ is a relative minimum.

34. A saddle point always occurs at a critical point.

35. If $f(x, y)$ has a relative maximum at (x_0, y_0, z_0), then $f_x(x_0, y_0) = f_y(x_0, y_0) = 0$.

36. The function $f(x, y) = \sqrt[3]{x^2 + y^2}$ has a relative maximum at the origin.

In Exercises 37–40, find three positive numbers *x*, *y*, and *z* that satisfy the given conditions.

37. The sum is 30 and the product is maximum.

38. The sum is 32 and $P = xy^2z$ is maximum.

39. The sum is 30 and the sum of the squares is minimum.

40. The sum is 1 and the sum of the squares is minimum.

41. *Revenue* A company manufactures two products. The total revenue from x_1 units of product 1 and x_2 units of product 2 is

$$R = -5x_1^2 - 8x_2^2 - 2x_1x_2 + 42x_1 + 102x_2.$$

Find x_1 and x_2 so as to maximize the revenue.

42. *Revenue* A retail outlet sells two competitive products, the prices of which are p_1 and p_2. Find p_1 and p_2 so as to maximize the total revenue

$$R = 500p_1 + 800p_2 + 1.5p_1p_2 - 1.5p_1^2 - p_2^2.$$

Revenue In Exercises 43 and 44, find p_1 and p_2 so as to maximize the total revenue $R = x_1p_1 + x_2p_2$ for a retail outlet that sells two competitive products with the given demand functions.

43. $x_1 = 1000 - 2p_1 + p_2, \ x_2 = 1500 + 2p_1 - 1.5p_2$

44. $x_1 = 1000 - 4p_1 + 2p_2, \ x_2 = 900 + 4p_1 - 3p_2$

45. *Profit* A corporation manufactures a product at two locations. The costs of producing x_1 units at location 1 and x_2 units at location 2 are $C_1 = 0.04x_1^2 + 5x_1 + 375$ and $C_2 = 0.07x_2^2 + 3.1x_2 + 295$, respectively. If the product sells for $15 per unit, find the quantity that must be produced at each location to maximize the profit

$$P = 15(x_1 + x_2) - C_1 - C_2.$$

46. *Profit* Repeat Exercise 45 for a product that sells for $50 per unit with cost functions $C_1 = 0.1x_1^3 + 20x_1 + 150$ and $C_2 = 0.05x_2^3 + 20.6x_2 + 125$.

47. *Profit* A corporation manufactures a product at two locations. The cost functions for producing x_1 units at location 1 and x_2 units at location 2 are

$$C_1 = 0.05x_1^2 + 15x_1 + 5400$$

and

$$C_2 = 0.03x_2^2 + 15x_2 + 6100,$$

respectively. The demand function for the product is given by

$$p = 225 - 0.4(x_1 + x_2),$$

and therefore the total revenue function is

$$R = [225 - 0.4(x_1 + x_2)](x_1 + x_2).$$

Find the production levels at the two locations that will maximize the profit

$$P = R - C_1 - C_2.$$

48. *Cost* The material for constructing the base of an open box costs 1.5 times as much as the material for constructing the sides. Find the dimensions of the box of largest volume that can be made for a fixed amount of money C.

49. *Volume* Find the dimensions of a rectangular package of largest volume that may be sent by parcel post assuming that the sum of the length and the girth (perimeter of a cross section) cannot exceed 108 inches.

50. *Volume* Show that the rectangular box of given volume and minimum surface area is a cube.

51. *Hardy-Weinberg Law* Common blood types are determined genetically by three alleles A, B, and O. (An allele is any of a group of possible mutational forms of a gene.) A person whose blood type is AA, BB, or OO is homozygous. A person whose blood type is AB, AO, or BO is heterozygous. The Hardy-Weinberg Law states that the proportion P of heterozygous individuals in any given population is

$$P(p, q, r) = 2pq + 2pr + 2qr,$$

where p represents the percent of allele A in the population, q represents the percent of allele B in the population, and r represents the percent of allele O in the population. Use the fact that $p + q + r = 1$ (the sum of the three must equal 100%) to show that the maximum proportion of heterozygous individuals in any population is $\frac{2}{3}$.

52. *Stocking a Lake* A lake is to be stocked with smallmouth and largemouth bass. Let x represent the number of smallmouth bass and y represent the number of largemouth bass in the lake. The weight of each fish is dependent on the population densities. After a 6-month period, the weight of a single smallmouth bass is

$$W_1 = 3 - 0.002x - 0.005y,$$

and the weight of a single largemouth bass is

$$W_2 = 4.5 - 0.003x - 0.004y.$$

Assuming no fish die during the 6-month period, how many smallmouth and largemouth bass should be stocked in the lake so that the *total* weight T of bass in the lake is a maximum?

Lagrange Multipliers

Lagrange Multipliers with One Constraint

In Example 5 in Section 7.5, you were asked to find the dimensions of the rectangular box of maximum volume that would fit in the first octant beneath the plane $6x + 4y + 3z = 24$, as shown again in Figure 7.33. Another way of stating this problem is to say that you are asked to find the maximum of

$$V = xyz \qquad \textit{Objective function}$$

subject to the constraint

$$6x + 4y + 3z - 24 = 0. \qquad \textit{Constraint}$$

This type of problem is called a **constrained optimization** problem. In Section 7.5, you answered this question by solving for z in the constraint equation and then rewriting V as a function of two variables.

In this section, you will study a different (and often better) way to solve constrained optimization problems. This involves the use of variables called **Lagrange multipliers,** named after the French mathematician Joseph Louis Lagrange (1736–1813).

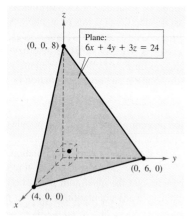

Plane:
$6x + 4y + 3z = 24$

$(0, 0, 8)$
$(0, 6, 0)$
$(4, 0, 0)$

FIGURE 7.33

Method of Lagrange Multipliers

If $f(x, y)$ has a maximum or minimum subject to the constraint $g(x, y) = 0$, then it will occur at one of the critical numbers of the function F defined by

$$F(x, y, \lambda) = f(x, y) - \lambda g(x, y).$$

The variable λ (the lowercase Greek letter lambda) is called a **Lagrange multiplier.** For functions of three variables, F has the form

$$F(x, y, z, \lambda) = f(x, y, z) - \lambda g(x, y, z).$$

The method of Lagrange multipliers gives you a way of finding critical points but does not tell you whether these points yield minimums, maximums, or neither. To make this distinction, you must rely on the context of the problem.

EXAMPLE 1 *Using Lagrange Multipliers: One Constraint*

Find the maximum of

$$V = xyz \qquad \text{\textit{Objective function}}$$

subject to the constraint

$$6x + 4y + 3z - 24 = 0. \qquad \text{\textit{Constraint}}$$

Solution

First, let $f(x, y, z) = xyz$ and $g(x, y, z) = 6x + 4y + 3z - 24$. Then, define a new function F by

$$F(x, y, z, \lambda) = f(x, y, z) - \lambda g(x, y, z)$$
$$= xyz - \lambda(6x + 4y + 3z - 24).$$

To find the critical numbers of F, set the partial derivatives of F with respect to x, y, z, and λ equal to zero and obtain

$$F_x(x, y, z, \lambda) = yz - 6\lambda = 0$$
$$F_y(x, y, z, \lambda) = xz - 4\lambda = 0$$
$$F_z(x, y, z, \lambda) = xy - 3\lambda = 0$$
$$F_\lambda(x, y, z, \lambda) = -6x - 4y - 3z + 24 = 0.$$

Solving for λ in the first equation and substituting into the second and third equations produces the following.

$$xz - 4\left(\frac{yz}{6}\right) = 0 \quad \longrightarrow \quad y = \frac{3}{2}x$$
$$xy - 3\left(\frac{yz}{6}\right) = 0 \quad \longrightarrow \quad z = 2x$$

Next, substitute these values for y and z into the equation $F_\lambda(x, y, z, \lambda) = 0$ and solve for x.

$$F_\lambda(x, y, z, \lambda) = 0$$
$$-6x - 4\left(\tfrac{3}{2}x\right) - 3(2x) + 24 = 0$$
$$-18x = -24$$
$$x = \tfrac{4}{3}$$

Using this x-value, you can conclude that the critical values are $x = \frac{4}{3}$, $y = 2$, and $z = \frac{8}{3}$, which implies that the maximum is

$$V = xyz \qquad \text{\textit{Objective function}}$$
$$= \left(\frac{4}{3}\right)(2)\left(\frac{8}{3}\right) \qquad \text{\textit{Substitute values of x, y, and z.}}$$
$$= \frac{64}{9} \text{ cubic units.} \qquad \text{\textit{Maximum volume}}$$

Note Example 1 shows how Lagrange multipliers can be used to solve the same problem that was solved in Example 5 in Section 7.5.

Real Life

EXAMPLE 2 *Finding a Maximum Production*

For many industrial applications, a simple robot costs as much as a year's wages and benefits for three workers. Thus, manufacturers must carefully balance the amount of money spent on labor and capital.

A manufacturer's production is modeled by the Cobb-Douglas function

$$f(x, y) = 100x^{3/4}y^{1/4}, \qquad \text{\textit{Objective function}}$$

where x represents the units of labor and y represents the units of capital. Each labor unit costs $150 and each capital unit costs $250. The total expenses for labor and capital cannot exceed $50,000. Find the maximum production level.

Solution

As total labor and capital expenses cannot exceed $50,000, the constraint is

$$150x + 250y = 50,000 \qquad \text{\textit{Constraint}}$$
$$150x + 250y - 50,000 = 0. \qquad \text{\textit{Standard form}}$$

To find the maximum production level, begin by writing the function

$$F(x, y, \lambda) = 100x^{3/4}y^{1/4} - \lambda(150x + 250y - 50,000).$$

Next, set the partial derivatives of this function equal to zero.

$$F_x(x, y, \lambda) = 75x^{-1/4}y^{1/4} - 150\lambda = 0 \qquad \text{\textit{Equation 1}}$$
$$F_y(x, y, \lambda) = 25x^{3/4}y^{-3/4} - 250\lambda = 0 \qquad \text{\textit{Equation 2}}$$
$$F_\lambda(x, y, \lambda) = -150x - 250y + 50,000 = 0 \qquad \text{\textit{Equation 3}}$$

The strategy for solving such a system must be customized to the particular system. In this case, you can solve for λ in the first equation, substitute into the second, solve for x, substitute into the third equation, and solve for y.

$$75x^{-1/4}y^{1/4} - 150\lambda = 0 \qquad \text{\textit{Equation 1}}$$
$$\lambda = \tfrac{1}{2}x^{-1/4}y^{1/4} \qquad \text{\textit{Solve for }} \lambda.$$
$$25x^{3/4}y^{-3/4} - 250(\tfrac{1}{2})x^{-1/4}y^{1/4} = 0 \qquad \text{\textit{Substitute in Equation 2.}}$$
$$25x - 125y = 0 \qquad \text{\textit{Multiply by }} x^{1/4}y^{3/4}.$$
$$x = 5y \qquad \text{\textit{Solve for }} x.$$
$$-150(5y) - 250y + 50,000 = 0 \qquad \text{\textit{Substitute in Equation 3.}}$$
$$-1000y = -50,000 \qquad \text{\textit{Simplify.}}$$
$$y = 50 \qquad \text{\textit{Solve for }} y.$$

Using this value for y, it follows that $x = 5(50) = 250$. Thus, the maximum production level of

$$f(250, 50) = 100(250)^{3/4}(50)^{1/4} \qquad \text{\textit{Substitute for x and y.}}$$
$$\approx 16,719 \text{ units} \qquad \text{\textit{Maximum production}}$$

occurs when $x = 250$ units of labor and $y = 50$ units of capital.

Economists call the Lagrange multiplier obtained in a production function the **marginal productivity of money.** For instance, in Example 2, the marginal productivity of money when $x = 250$ and $y = 50$ is

$$\lambda = \tfrac{1}{2}x^{-1/4}y^{1/4} = \tfrac{1}{2}(250)^{-1/4}(50)^{1/4} \approx 0.334.$$

This means that if one additional dollar is spent on production, approximately 0.334 additional units of the product can be produced.

Real Life

EXAMPLE 3 *Finding a Maximum Production*

In Example 2, suppose that $70,000 is available for labor and capital. What is the maximum number of units that can be produced?

Solution
You could rework the entire problem, as demonstrated in Example 2. However, because the only change in the problem is the availability of additional money to spend on labor and capital, you can use the fact that the marginal productivity of money is

$$\lambda \approx 0.334.$$

Because an additional $20,000 is available and the maximum production in Example 2 was 16,719 units, you can conclude that the maximum production is now

$$16,719 + (0.334)(20,000) \approx 23,400 \text{ units.}$$

Try using the procedure demonstrated in Example 2 to confirm this result.

⊞ TECHNOLOGY

You can use a three-dimensional graphing utility to confirm graphically the results of Examples 2 and 3. Begin by graphing the surface $f(x, y) = 100x^{3/4}y^{1/4}$. Then graph the vertical plane given by $150x + 250y = 50,000$. As shown at the right, the maximum production level corresponds to the highest point on the intersection of the surface and the plane.

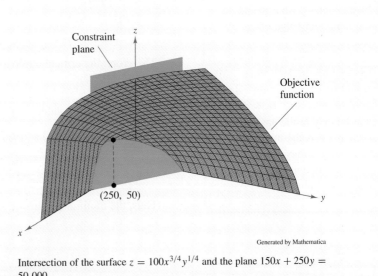

Generated by Mathematica

Intersection of the surface $z = 100x^{3/4}y^{1/4}$ and the plane $150x + 250y = 50,000$.

EXAMPLE 4 *Optimizing with Four Variables*

Find the minimum of the function

$$f(x, y, z, w) = x^2 + y^2 + z^2 + w^2 \qquad \text{\textit{Objective function}}$$

subject to the constraint

$$3x + 2y - 4z + w - 3 = 0. \qquad \text{\textit{Constraint}}$$

Solution

Begin by forming the function

$$F(x, y, z, w, \lambda) = x^2 + y^2 + z^2 + w^2 - \lambda(3x + 2y - 4z + w - 3).$$

Next, set the five partial derivatives of F equal to zero.

$$F_x(x, y, z, w, \lambda) = 2x - 3\lambda = 0 \qquad \text{\textit{Equation 1}}$$
$$F_y(x, y, z, w, \lambda) = 2y - 2\lambda = 0 \qquad \text{\textit{Equation 2}}$$
$$F_z(x, y, z, w, \lambda) = 2z + 4\lambda = 0 \qquad \text{\textit{Equation 3}}$$
$$F_w(x, y, z, w, \lambda) = 2w - \lambda = 0 \qquad \text{\textit{Equation 4}}$$
$$F_\lambda(x, y, z, w, \lambda) = -3x - 2y + 4z - w + 3 = 0 \qquad \text{\textit{Equation 5}}$$

For this system, you can solve the first four equations in terms of λ.

$$2x - 3\lambda = 0 \longrightarrow x = \tfrac{3}{2}\lambda$$
$$2y - 2\lambda = 0 \longrightarrow y = \lambda$$
$$2z + 4\lambda = 0 \longrightarrow z = -2\lambda$$
$$2w - \lambda = 0 \longrightarrow w = \tfrac{1}{2}\lambda$$

Substituting these values into the fifth equation allows you to solve for λ.

$$-3x - 2y + 4z - w + 3 = 0$$
$$-3\left(\tfrac{3}{2}\lambda\right) - 2(\lambda) + 4(-2\lambda) - \left(\tfrac{1}{2}\lambda\right) = -3$$
$$-15\lambda = -3$$
$$\lambda = \tfrac{1}{5}$$

Therefore, the critical numbers are

$$x = \tfrac{3}{10}, \quad y = \tfrac{1}{5}, \quad z = -\tfrac{2}{5}, \quad \text{and} \quad w = \tfrac{1}{10},$$

and the minimum value of f subject to the given constraint is

$$f\left(\tfrac{3}{10}, \tfrac{1}{5}, -\tfrac{2}{5}, \tfrac{1}{10}\right) = \left(\tfrac{3}{10}\right)^2 + \left(\tfrac{1}{5}\right)^2 + \left(-\tfrac{2}{5}\right)^2 + \left(\tfrac{1}{10}\right)^2 \quad \text{\textit{Substitute.}}$$
$$= \tfrac{9}{100} + \tfrac{1}{25} + \tfrac{4}{25} + \tfrac{1}{100} \quad \text{\textit{Simplify.}}$$
$$= \tfrac{3}{10}. \quad \text{\textit{Minimum value}}$$

Thus, the minimum value of $f(x, y, z, w)$ is $\tfrac{3}{10}$.

Discovery

In Example 4, what would be the minimum value of the objective function $f(x, y, z, w) = x^2 + y^2 + z^2 + w^2$ if there were no constraint? What would be the maximum value of f in this case? How do you know that the answer to Example 4 is the minimum value of f subject to the given constraint, and not the maximum value?

In Example 4 in Section 7.5, you found the maximum profit for two substitute products whose demand functions are

$$x_1 = 200(p_2 - p_1)$$ *Demand for Product 1*
$$x_2 = 500 + 100p_1 - 180p_2.$$ *Demand for Product 2*

With this model, the total demand, $x_1 + x_2$, is completely determined by the prices p_1 and p_2. In many real-life situations, this assumption is too simplistic: regardless of the prices of the substitute brands, the annual total demands for some products, such as toothpaste, are relatively constant. In such situations, the total demand is **limited,** and variations in price do not affect the total demand as much as they affect the market share of the substitute brands.

Real Life

EXAMPLE 5 *Finding the Maximum Profit*

A company makes two substitute products whose demand functions are

$$x_1 = 200(p_2 - p_1)$$ *Demand for Product 1*
$$x_2 = 500 + 100p_1 - 180p_2,$$ *Demand for Product 2*

where p_1 and p_2 are the prices per unit (in dollars) and x_1 and x_2 are the numbers of units sold. The costs of producing the two products are $0.50 and $0.75 per unit, respectively. The total demand is limited to 200 units per year. Find the prices that will yield a maximum profit.

Solution

From Example 4 in Section 7.5, the profit function is modeled by

$$P = -200p_1{}^2 - 180p_2{}^2 + 300p_1p_2 + 25p_1 + 535p_2 - 375.$$

The total demand for the two products is

$$x_1 + x_2 = 200(p_2 - p_1) + 500 + 100p_1 - 180p_2$$
$$= -100p_1 + 20p_2 + 500.$$

Because the total demand is limited to 200 units,

$$-100p_1 + 20p_2 + 500 = 200.$$ *Constraint*

Using Lagrange multipliers, you can determine that the maximum profit occurs when $p_1 = \$3.94$ and $p_2 = \$4.69$. This corresponds to an annual profit of $712.21. ▬

Note   The constrained optimization problem in Example 5 is represented graphically in Figure 7.34. The graph of the objective function is a paraboloid and the graph of the constraint is a vertical plane. In the "unconstrained" optimization problem on page 495, the maximum profit occurred at the vertex of the paraboloid. In this "constrained" problem, however, the maximum profit corresponds to the highest point on the curve that is the intersection of the paraboloid and the vertical "constraint" plane.

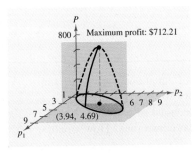

FIGURE 7.34

Lagrange Multipliers with Two Constraints

In Examples 1 through 5, each of the optimization problems contained only one constraint. When an optimization problem has two constraints, you need to introduce a second Lagrange multiplier. The customary symbol for this second multiplier is μ, the Greek letter mu.

EXAMPLE 6   *Using Lagrange Multipliers: Two Constraints*

Find the minimum value of

$$f(x, y, z) = x^2 + y^2 + z^2 \qquad \textit{Objective function}$$

subject to the constraints

$$x + y - 3 = 0 \qquad \textit{Constraint 1}$$
$$x + z - 5 = 0. \qquad \textit{Constraint 2}$$

Solution

Begin by forming the function

$$F(x, y, z, \lambda, \mu) = x^2 + y^2 + z^2 - \lambda(x + y - 3) - \mu(x + z - 5).$$

Next, set the five partial derivatives equal to zero, and solve the resulting system of equations for x, y, and z.

$$F_x(x, y, z, \lambda, \mu) = 2x - \lambda - \mu = 0 \qquad \textit{Equation 1}$$
$$F_y(x, y, z, \lambda, \mu) = 2y - \lambda = 0 \qquad \textit{Equation 2}$$
$$F_z(x, y, z, \lambda, \mu) = 2z - \mu = 0 \qquad \textit{Equation 3}$$
$$F_\lambda(x, y, z, \lambda, \mu) = -x - y + 3 = 0 \qquad \textit{Equation 4}$$
$$F_\mu(x, y, z, \lambda, \mu) = -x - z + 5 = 0 \qquad \textit{Equation 5}$$

Solving this system of equations produces $x = \frac{8}{3}$, $y = \frac{1}{3}$, and $z = \frac{7}{3}$. Thus, the minimum value of $f(x, y, z)$ is

$$f\left(\frac{8}{3}, \frac{1}{3}, \frac{7}{3}\right) = \left(\frac{8}{3}\right)^2 + \left(\frac{1}{3}\right)^2 + \left(\frac{7}{3}\right)^2 = \frac{38}{3}.$$

Discussion Problem   Solving a System of Equations

Explain how you would solve the system of equations shown in Example 6. Illustrate your description by actually solving the system.

Suppose that in Example 6 you had been asked to find the maximum value of $f(x, y, z)$ subject to the two constraints. How would your answer have been different?

Warm Up

The following warm-up exercises involve skills that were covered in earlier sections. You will use these skills in the exercise set for this section.

In Exercises 1–6, solve the system of linear equations.

1. $\begin{cases} 4x - 6y = 3 \\ 2x + 3y = 2 \end{cases}$

2. $\begin{cases} 6x - 6y = 5 \\ -3x - y = 1 \end{cases}$

3. $\begin{cases} 5x - y = 25 \\ x - 5y = 15 \end{cases}$

4. $\begin{cases} 4x - 9y = 5 \\ -x + 8y = -2 \end{cases}$

5. $\begin{cases} 2x - y + z = 3 \\ 2x + 2y + z = 4 \\ -x + 2y + 3z = -1 \end{cases}$

6. $\begin{cases} -x - 4y + 6z = -2 \\ x - 3y - 3z = 4 \\ 3x + y + 3z = 0 \end{cases}$

In Exercises 7–10, find all first partial derivatives.

7. $f(x, y) = x^2 y + xy^2$

8. $f(x, y) = 25(xy + y^2)^2$

9. $f(x, y, z) = x(x^2 - 2xy + yz)$

10. $f(x, y, z) = z(xy + xz + yz)$

EXERCISES 7.6

 means that technology can help you solve or check the exercise(s).

In Exercises 1–12, use Lagrange multipliers to find the indicated extrema. In each case, assume that x and y are positive.

Objective Function	*Constraint*
1. Maximize $f(x, y) = xy$	$x + y = 10$
2. Maximize $f(x, y) = xy$	$2x + y = 4$
3. Minimize $f(x, y) = x^2 + y^2$	$x + y - 4 = 0$
4. Minimize $f(x, y) = x^2 + y^2$	$-2x - 4y + 5 = 0$
5. Maximize $f(x, y) = x^2 - y^2$	$y - x^2 = 0$
6. Maximize $f(x, y) = x^2 - y^2$	$x - 2y + 6 = 0$
7. Maximize $f(x, y) = 3x + xy + 3y$	$x + y = 25$
8. Maximize $f(x, y) = 3x + y + 10$	$x^2 y = 6$
9. Maximize $f(x, y) = \sqrt{6 - x^2 - y^2}$	$x + y - 2 = 0$
10. Minimize $f(x, y) = \sqrt{x^2 + y^2}$	$2x + 4y - 15 = 0$
11. Maximize $f(x, y) = e^{xy}$	$x^2 + y^2 - 8 = 0$
12. Minimize $f(x, y) = 2x + y$	$xy = 32$

In Exercises 13–18, use Lagrange multipliers to find the indicated extrema. In each case, assume that x, y, and z are positive.

13. Minimize $f(x, y, z) = 2x^2 + 3y^2 + 2z^2$

Constraint: $x + y + z - 24 = 0$

14. Maximize $f(x, y, z) = xyz$

Constraint: $x + y + z - 6 = 0$

15. Minimize $f(x, y, z) = x^2 + y^2 + z^2$

Constraint: $x + y + z = 1$

16. Minimize $f(x, y) = x^2 - 8x + y^2 - 12y + 48$

Constraint: $x + y = 8$

17. Maximize $f(x, y, z) = x + y + z$

Constraint: $x^2 + y^2 + z^2 = 1$

18. Maximize $f(x, y, z) = x^2 y^2 z^2$

Constraint: $x^2 + y^2 + z^2 = 1$

In Exercises 19 and 20, use Lagrange multipliers with the objective function

$$f(x, y, z, w) = 2x^2 + y^2 + z^2 + 2w^2$$

with the given constraints to find the indicated extrema. In each case, assume that $x, y, z,$ and w are nonnegative.

19. Maximize $f(x, y, z, w)$

Constraint: $2x + 2y + z + w = 2$

20. Maximize $f(x, y, z, w)$

Constraint: $x + y + 2z + 2w = 4$

In Exercises 21–24, use Lagrange multipliers to find the indicated extrema of f subject to two constraints. In each case, assume $x, y,$ and z are nonnegative.

21. Maximize $f(x, y, z) = xyz$

Constraints: $x + y + z = 24, \ x - y + z = 12$

22. Minimize $f(x, y, z) = x^2 + y^2 + z^2$

Constraints: $x + 2z = 4, \ x + y = 8$

23. Maximize $f(x, y, z) = xyz$

Constraints: $x^2 + z^2 = 5, \ x - 2y = 0$

24. Maximize $f(x, y, z) = xy + yz$

Constraints: $x + 2y = 6, \ x - 3z = 0$

In Exercises 25–28, find three positive numbers $x, y,$ and z that satisfy the given conditions.

25. The sum is 120 and the product is maximum.

26. The sum is 120 and the sum of the squares is minimum.

27. The sum is S and the product is maximum.

28. The sum is S and the sum of the squares is minimum.

In Exercises 29–32, find the minimum distance from the curve or surface to the specified point. (*Hint:* Start by minimizing the square of the distance.)

29. Line: $x + 2y = 5, \ (0, 0)$

Minimize $d^2 = x^2 + y^2$.

30. Circle: $(x - 4)^2 + y^2 = 4, \ (0, 10)$

Minimize $d^2 = x^2 + (y - 10)^2$.

31. Plane: $x + y + z = 1, \ (2, 1, 1)$

Minimize $d^2 = (x - 2)^2 + (y - 1)^2 + (z - 1)^2$.

32. Cone: $z = \sqrt{x^2 + y^2}, \ (4, 0, 0)$

Minimize $d^2 = (x - 4)^2 + y^2 + z^2$.

33. *Volume* Find the dimensions of the rectangular package of largest volume subject to the constraint that the sum of the length and the girth cannot exceed 108 inches (see figure). (*Hint:* Maximize $V = xyz$ subject to the constraint $x + 2y + 2z = 108$.)

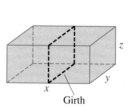

FIGURE FOR 33   FIGURE FOR 34

34. *Cost* In redecorating an office, the cost for new carpeting is five times the cost of wallpapering a wall. Find the dimensions of the largest office that can be redecorated for a fixed cost C (see figure). (*Hint:* Maximize $V = xyz$ subject to $5xy + 2xz + 2yz = C$.)

35. *Cost* A cargo container (in the shape of a rectangular solid) must have a volume of 480 cubic feet. Use Lagrange multipliers to find the dimensions of the container of this size that has a minimum cost, if the bottom will cost $5 per square foot to construct and the sides and top will cost $3 per square foot to construct.

36. *Cost* A manufacturer has an order for 1000 units that can be produced at two locations. Let x_1 and x_2 be the numbers of units produced at the two plants. Find the number of units that should be produced at each plant to minimize the cost if the cost function is given by

$$C = 0.25x_1^2 + 25x_1 + .05x_2^2 + 12x_2.$$

37. *Cost* A manufacturer has an order for 2000 units that can be produced at two locations. Let x_1 and x_2 be the numbers of units produced at the two plants. The cost function is modeled by

$$C = 0.25x_1^2 + 10x_1 + 0.15x_2^2 + 12x_2.$$

Find the number of units that should be produced at each plant to minimize the cost.

38. *Hardy-Weinberg Law* Repeat Exercise 51 in Section 7.5 using Lagrange multipliers—that is, maximize

$$P(p, q, r) = 2pq + 2pr + 2qr$$

subject to the constraint

$$p + q + r = 1.$$

39. *Least-Cost Rule* The production function for a company is

$$f(x, y) = 100x^{0.25}y^{0.75},$$

where x is the number of units of labor and y is the number of units of capital. Suppose that labor costs \$48 per unit, capital costs \$36 per unit, and management sets a production goal of 20,000 units.

(a) Find the numbers of units of labor and capital needed to meet the production goal while minimizing the cost.

(b) Show that the conditions of part (a) are met when

$$\frac{\text{Marginal productivity of labor}}{\text{Marginal productivity of capital}} = \frac{\text{unit price of labor}}{\text{unit price of capital}}.$$

This proportion is called the **Least-Cost Rule** (or Equimarginal Rule).

40. *Least-Cost Rule* Repeat Exercise 39 for the production function

$$f(x, y) = 100x^{0.6}y^{0.4}.$$

41. *Production* The production function for a company is

$$f(x, y) = 100x^{0.25}y^{0.75},$$

where x is the number of units of labor and y is the number of units of capital. Suppose that labor costs \$48 per unit and capital costs \$36 per unit. The total cost of labor and capital is limited to \$100,000.

(a) Find the maximum production level for this manufacturer.

(b) Find the marginal productivity of money.

(c) Use the marginal productivity of money to find the maximum number of units that can be produced if \$125,000 is available for labor and capital.

42. *Production* Repeat Exercise 41 for the production function

$$f(x, y) = 100x^{0.06}y^{0.04}.$$

43. *Bacteria Culture* A microbiologist must prepare a culture medium in which to grow a certain type of bacteria. The percent of salt contained in this medium is $S = 12xyz$, where x, y, and z are the nutrient solutions to be mixed in the medium. For the bacteria to grow, the medium must be 13% salt. The nutrient solutions cost \$1, \$2, and \$3 per liter. How much of each nutrient solution should be used to minimize the cost of the culture medium?

44. *Bacteria Culture* Repeat Exercise 43 for a salt-content model of

$$S = 0.01x^2y^2z^2.$$

45. *Investment Strategy* An investor is considering three different stocks in which to invest \$300,000. The average annual dividends for the stocks are

Dow Chemicals (D)	3.9%
CIGNA Corp. (C)	4.5%
Pennzoil Company (P)	4.0%.

The amount invested in Pennzoil must follow the equation

$$3000D - 3000C + P^2 = 0.$$

How much should be invested in each stock to yield a maximum of dividends?

46. *Research Project* Use your school's library or some other reference source to write a paper about two different types of investment options available. Find examples of each type and find data about their dividends for the past 10 years. What are the similarities and differences between the two types?

Least Squares Regression Analysis

Measuring the Accuracy of a Mathematical Model ■
Least Squares Regression Line ■ *Least Squares Regression Quadratic*

Measuring the Accuracy of a Mathematical Model

When seeking a mathematical model to fit real-life data, you should try to find a model that is both as *simple* and as *accurate* as possible. For instance, a simple linear model for the points shown in Figure 7.35 is

$$f(x) = 1.8566x - 5.0246. \qquad \textit{Linear model}$$

However, Figure 7.36 shows that by choosing a slightly more complicated quadratic model,

$$g(x) = 0.1996x^2 - 0.7281x + 1.3749, \qquad \textit{Quadratic model}$$

you can obtain significantly greater accuracy.

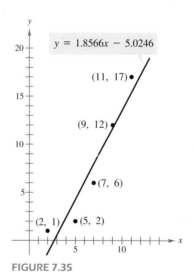

FIGURE 7.35

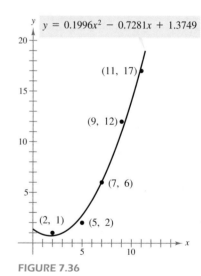

FIGURE 7.36

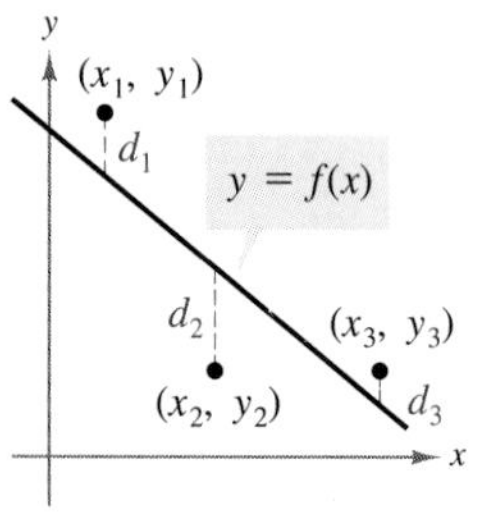

$$S = d_1^2 + d_2^2 + d_3^2$$

FIGURE 7.37

To measure how well the model $y = f(x)$ fits a collection of points, sum the squares of the differences between the actual y-values and the model's y-values. This sum is called the **sum of the squared errors** and is denoted by S. Graphically, S is the sum of the squares of the vertical distances between the graph of f and the given points in the plane, as shown in Figure 7.37.

Definition of the Sum of the Squared Errors

The **sum of the squared errors** for the model $y = f(x)$ with respect to the points $(x_1, y_1), (x_2, y_2), \ldots, (x_n, y_n)$ is given by

$$S = [f(x_1) - y_1]^2 + [f(x_2) - y_2]^2 + \cdots + [f(x_n) - y_n]^2.$$

EXAMPLE 1 *Finding the Sum of the Squared Errors*

Find the sum of the squared errors for the linear and quadratic models

$$f(x) = 1.8566x - 5.0246. \qquad \textit{Linear model}$$
$$g(x) = 0.1996x^2 - 0.7281x + 1.3749 \qquad \textit{Quadratic model}$$

(see Figures 7.35 and 7.36) with respect to the points

$$(2, 1), (5, 2), (7, 6), (9, 12), (11, 17).$$

Solution

Begin by evaluating each model at the given x-values, as shown in Table 7.1.

TABLE 7.1

x	2	5	7	9	11
Actual y-values	1	2	6	12	17
Linear model, $f(x)$	−1.3114	4.2584	7.9716	11.6848	15.3980
Quadratic model, $g(x)$	0.7171	2.7244	6.0586	10.9896	17.5174

For the linear model f, the sum of the squared errors is

$$S = (-1.3114 - 1)^2 + (4.2584 - 2)^2 + (7.9716 - 6)^2$$
$$+ (11.6848 - 12)^2 + (15.3980 - 17)^2$$
$$\approx 16.9959.$$

Similarly, the sum of the squared errors for the quadratic model g is

$$S = (0.7171 - 1)^2 + (2.7244 - 2)^2 + (6.0586 - 6)^2$$
$$+ (10.9896 - 12)^2 + (17.5174 - 17)^2$$
$$\approx 1.8968.$$ ▬

Note   In Example 1, note that the sum of the squared errors for the quadratic model is less than the sum of the squared errors for the linear model, which confirms that the quadratic model is a better fit.

Least Squares Regression Line

The sum of the squared errors can be used to determine which of several models is the best fit for a collection of data. In general, if the sum of the squared errors of f is less than the sum of the squared errors of g, then f is said to be a better fit for the data than g. In regression analysis, you consider all possible models of a certain type. The one that is defined to be the best-fitting model is the one with the least sum of the squared errors. Example 2 shows how to use the optimization techniques described in Section 7.5 to find the best-fitting linear model for a collection of data.

EXAMPLE 2   *Finding the Best Linear Model*

Find the values of a and b so that the linear model

$$f(x) = ax + b$$

has a minimum sum of the squared errors for the points

$$(-3, 0), (-1, 1), (0, 2), (2, 3).$$

Solution

The sum of the squared errors is

$$S = [f(x_1) - y_1]^2 + [f(x_2) - y_2]^2 + [f(x_3) - y_3]^2 + [f(x_4) - y_4]^2$$
$$= (-3a + b - 0)^2 + (-a + b - 1)^2 + (b - 2)^2 + (2a + b - 3)^2$$
$$= 14a^2 - 4ab + 4b^2 - 10a - 12b + 14.$$

To find the values of a and b for which S is a minimum, you can use the techniques described in Section 7.5. That is, find the partial derivatives of S.

$$\frac{\partial S}{\partial a} = 28a - 4b - 10 \qquad\qquad \textit{Differentiate with respect to a.}$$

$$\frac{\partial S}{\partial b} = -4a + 8b - 12 \qquad\qquad \textit{Differentiate with respect to b.}$$

Next, set each partial derivative equal to zero.

$$28a - 4b - 10 = 0 \qquad\qquad \textit{Set } \partial S/\partial a \textit{ equal to 0.}$$

$$-4a + 8b - 12 = 0 \qquad\qquad \textit{Set } \partial S/\partial b \textit{ equal to 0.}$$

The solution of this system of linear equations is

$$a = \frac{8}{13} \quad \text{and} \quad b = \frac{47}{26}.$$

Thus, the best-fitting linear model for the given points is

$$f(x) = \frac{8}{13}x + \frac{47}{26}.$$

The graph of this model is shown in Figure 7.38.

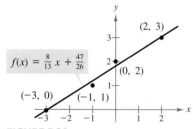

$f(x) = \frac{8}{13}x + \frac{47}{26}$

FIGURE 7.38

The line in Example 2 is called the **least squares regression line** for the given data. The solution shown in Example 2 can be generalized to find a formula for the least squares regression line, as follows. Consider the linear model

$$f(x) = ax + b$$

and the points $(x_1, y_1), (x_2, y_2), \ldots, (x_n, y_n)$. The sum of the squared errors is

$$S = \sum_{i=1}^{n} [f(x_i) - y_i]^2 = \sum_{i=1}^{n} (ax_i + b - y_i)^2.$$

To minimize S, set the partial derivatives $\partial S / \partial a$ and $\partial S / \partial b$ equal to zero and solve for a and b. The results are summarized below.

The Least Squares Regression Line

The **least squares regression line** for the points

$$(x_1, y_1), (x_2, y_2), \ldots, (x_n, y_n)$$

is $y = ax + b$, where

$$a = \frac{n \sum_{i=1}^{n} x_i y_i - \sum_{i=1}^{n} x_i \sum_{i=1}^{n} y_i}{n \sum_{i=1}^{n} x_i^2 - \left(\sum_{i=1}^{n} x_i \right)^2} \quad \text{and} \quad b = \frac{1}{n} \left(\sum_{i=1}^{n} y_i - a \sum_{i=1}^{n} x_i \right).$$

In the formula for the least squares regression line, note that if the x-values are symmetrically spaced about zero, then

$$\sum_{i=1}^{n} x_i = 0$$

and the formulas for a and b simplify to

$$a = \frac{n \sum_{i=1}^{n} x_i y_i}{n \sum_{i=1}^{n} x_i^2} \quad \text{and} \quad b = \frac{1}{n} \sum_{i=1}^{n} y_i.$$

Note also that only the *development* of the least squares regression line involves partial derivatives. The *application* of this formula is simply a matter of computing the values of a and b—a task that is performed much more simply on a calculator or a computer than by hand.

Discovery

Graph the three points (2, 2), (2, 1), and (2.1, 1.5) and visually estimate the least squares regression line for this data. Now use the formulas on this page or a graphing utility to show that the equation of the line is actually $y = 1.5$. In general, the least squares regression line for "nearly vertical" data can be quite unusual. Show that by interchanging the roles of x and y, you can obtain a better linear approximation.

Real
Life

EXAMPLE 3 *Modeling Median Income*

The median income y (in thousands of dollars) for American households from 1981 through 1990 is listed in Table 7.2. Find the least squares regression line for the data and use the result to predict the median income for an American household in 1995. *(Source: U.S. Bureau of Census)*

TABLE 7.2

Year	1981	1982	1983	1984	1985	1986	1987	1988	1989	1990
y	19.1	20.2	21.0	22.4	23.6	24.9	26.1	27.2	28.9	29.9

Solution

Let t represent the year, with $t = 1$ corresponding to 1981. Then, you need to find the linear model that best fits the points

$$(1, 19.1), (2, 20.2), (3, 21.0), (4, 22.4), (5, 23.6),$$
$$(6, 24.9), (7, 26.1), (8, 27.2), (9, 28.9), \text{ and } (10, 29.9).$$

Using a calculator with a built-in least squares program, you can determine that the best-fitting line is

$$y = 1.22t + 17.6.$$

With this model, you can predict the 1995 median income to be

$$y = 1.22(15) + 17.6 = 35.9,$$

or \$35,900 per year. This result is shown graphically in Figure 7.39. ■

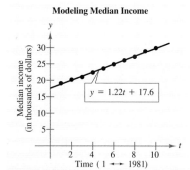

Modeling Median Income

$y = 1.22t + 17.6$

Time (1 ↔ 1981)

FIGURE 7.39

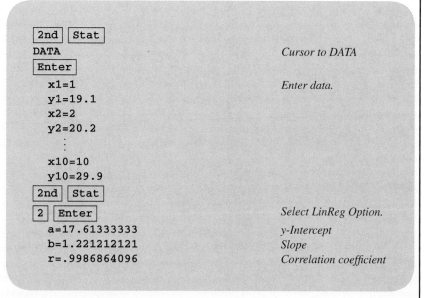

TECHNOLOGY

Most graphing utilities have a built-in linear regression program. For instance, the steps for finding the linear regression model in Example 3 on a *TI-81* are shown at the right. Steps for other calculators are given in Appendix D. When you run such a program, the "r-value" gives a measure of how well the model fits the data. The closer the value of $|r|$ is to 1, the better the fit. For the data in Example 3, $r \approx 0.9987$, which implies that the model is a very good fit.

2nd	Stat	
DATA		*Cursor to DATA*
Enter		
x1=1		*Enter data.*
y1=19.1		
x2=2		
y2=20.2		
⋮		
x10=10		
y10=29.9		
2nd	Stat	
2	Enter	*Select LinReg Option.*
a=17.61333333		*y-Intercept*
b=1.221212121		*Slope*
r=.9986864096		*Correlation coefficient*

Least Squares Regression Quadratic

When using regression analysis to model data, remember that the least squares regression line provides only the best *linear* model for a set of data. It does not necessarily provide the best possible model. For instance, in Example 1, you saw that the quadratic model was a better fit than the linear model.

Regression analysis can be performed with many different types of models, such as exponential or logarithmic models. The following development shows how to find the best-fitting quadratic model for a collection of data points. Consider a quadratic model of the form

$$f(x) = ax^2 + bx + c.$$

The sum of the squared errors for this model is

$$S = \sum_{i=1}^{n} [f(x_i) - y_i]^2 = \sum_{i=1}^{n} (ax_i^2 + bx_i + c - y_i)^2.$$

To find the values of a, b, and c that minimize S, set the three partial derivatives, $\partial S/\partial a$, $\partial S/\partial b$, and $\partial S/\partial c$, equal to zero.

$$\frac{\partial S}{\partial a} = \sum_{i=1}^{n} 2x_i^2 (ax_i^2 + bx_i + c - y_i) = 0$$

$$\frac{\partial S}{\partial b} = \sum_{i=1}^{n} 2x_i (ax_i^2 + bx_i + c - y_i) = 0$$

$$\frac{\partial S}{\partial c} = \sum_{i=1}^{n} 2(ax_i^2 + bx_i + c - y_i) = 0$$

By expanding this system, you obtain the result given in the following summary.

Least Squares Regression Quadratic

The **least squares regression quadratic** for the points

$$(x_1, y_1), (x_2, y_2), \ldots, (x_n, y_n)$$

is $y = ax^2 + bx + c$, where a, b, and c are the solutions of the following system of equations.

$$a\sum_{i=1}^{n} x_i^4 + b\sum_{i=1}^{n} x_i^3 + c\sum_{i=1}^{n} x_i^2 = \sum_{i=1}^{n} x_i^2 y_i$$

$$a\sum_{i=1}^{n} x_i^3 + b\sum_{i=1}^{n} x_i^2 + c\sum_{i=1}^{n} x_i = \sum_{i=1}^{n} x_i y_i$$

$$a\sum_{i=1}^{n} x_i^2 + b\sum_{i=1}^{n} x_i + cn = \sum_{i=1}^{n} y_i$$

EXAMPLE 4 *Modeling the Average Price of a Home* Real Life

The average price y (in thousands of dollars) of a new one-family house in the United States from 1965 to 1991 is listed in Table 7.3. Find the least squares regression quadratic for this data and use the result to predict the average home price for an American household in 1998. *(Source: U.S. Bureau of Census)*

TABLE 7.3

Year	1965	1970	1975	1980	1985	1988	1990	1991
y	21.5	26.6	42.6	76.4	100.8	138.3	149.8	147.2

Solution

Let t represent the year, with $t = 65$ corresponding to 1965. Then, you need to find the quadratic model that best fits the points

$(65, 21.5)$, $(70, 26.6)$, $(75, 42.6)$, $(80, 76.4)$,
$(85, 100.8)$, $(88, 138.3)$, $(90, 149.8)$, and $(91, 147.2)$

Using a computer or calculator with a built-in least squares program, you can determine that the best-fitting quadratic model is

$$y = 601.9 - 19.12t + 0.156t^2.$$

With this model, you can predict the 1998 average home price to be

$$y = 601.9 - 19.12(98) + 0.156(98)^2 \approx 226.4,$$

or $226,400 per year. This result is shown graphically in Figure 7.40. ■

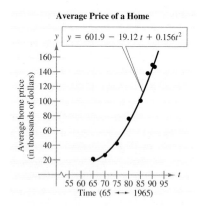

Average Price of a Home

$y = 601.9 - 19.12\,t + 0.156t^2$

FIGURE 7.40

Discussion Problem **Least Squares Regression Analysis**

Most graphing utilities, such as a *TI-81* or *Casio fx-7700g*, have several built-in least squares regression programs. Try entering the points

$(65, 21.5)$, $(70, 26.6)$, $(75, 42.6)$, $(80, 76.4)$,
$(85, 100.8)$, $(88, 138.3)$, $(90, 149.8)$, and $(91, 147.2)$

from Example 4 in a graphing utility. Then run one of the regression programs for the points. The quadratic model in Example 4 has a correlation coefficient of $r \approx 0.994$. Can you find a model, such as an exponential model or power model

$y = ab^x$	*Exponential model*
$y = ax^b$,	*Power model*

that has a better correlation coefficient?

Warm Up

The following warm-up exercises involve skills that were covered in earlier sections. You will use these skills in the exercise set for this section.

In Exercises 1 and 2, evaluate the expression.

1. $(2.5 - 1)^2 + (3.25 - 2)^2 + (4.1 - 3)^2$ **2.** $(1.1 - 1)^2 + (2.08 - 2)^2 + (2.95 - 3)^2$

In Exercises 3 and 4, find the partial derivatives of S.

3. $S = a^2 + 6b^2 - 4a - 8b - 4ab + 6$ **4.** $S = 4a^2 + 9b^2 - 6a - 4b - 2ab + 8$

In Exercises 5–10, evaluate the summation.

5. $\sum_{i=1}^{5} i$ **6.** $\sum_{i=1}^{6} 2i$ **7.** $\sum_{i=1}^{4} \frac{1}{i}$ **8.** $\sum_{i=1}^{3} i^2$ **9.** $\sum_{i=1}^{6} (2 - i)^2$ **10.** $\sum_{i=1}^{5} (30 - i^2)$

E X E R C I S E S   7.7

 means that technology can help you solve or check the exercise(s).

In Exercises 1–4, (a) use the method of least squares to find the least squares regression line, and (b) calculate the sum of the squared errors.

1.
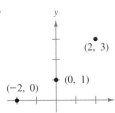

2.
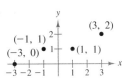

3.
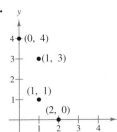

4.
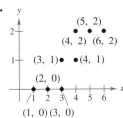

In Exercises 5–14, find the least squares regression line for the given points.

5. $(-2, 0), (-1, 1), (0, 1), (1, 2), (2, 3)$

6. $(-4, -1), (-2, 0), (2, 4), (4, 5)$

7. $(-2, 2), (2, 6), (3, 7)$

8. $(-5, 1), (1, 3), (2, 3), (2, 5)$

9. $(-3, 4), (-1, 2), (1, 1), (3, 0)$

10. $(-10, 10), (-5, 8), (3, 6), (7, 4), (5, 0)$

11. $(0, 0), (1, 1), (3, 4), (4, 2), (5, 5)$

12. $(1, 0), (3, 3), (5, 6)$

13. $(0, 6), (4, 3), (5, 0), (8, -4), (10, -5)$

14. $(5, 2), (0, 0), (2, 1), (7, 4), (10, 6), (12, 6)$

In Exercises 15–18, use partial derivatives to find the values of a and b such that the linear model $f(x) = ax + b$ has a minimum sum of the squared errors for the points.

15. $(-2, -1), (0, 0), (2, 3)$

16. $(-3, 0), (-1, 1), (1, 1), (3, 2)$

17. $(-2, 4), (-1, 1), (0, -1), (1, -3)$

18. $(-5, -3), (-4, -2), (-2, -1), (-1, 1)$

In Exercises 19–22, find the least squares regression quadratic for the given points. Then plot the points and sketch the graph of the least squares quadratic.

19. $(-2, 0), (-1, 0), (0, 1), (1, 2), (2, 5)$

20. $(-4, 5), (-2, 6), (2, 6), (4, 2)$

21. $(1, 0), (2, 1), (3, 7), (4, 13)$

22. $(0, 10), (1, 9), (2, 6), (3, 0)$

In Exercises 23–26, use a computer with a built-in least squares program to find a model (linear and quadratic) for the data. State which model best fits the data.

23. $(-4, 1), (-3, 2), (-2, 2), (-1, 4), (0, 6), (1, 8), (2, 9)$

24. $(-1, -4), (0, -3), (1, -3), (2, 0), (4, 5), (6, 9), (9, 3)$

25. $(0, 769), (1, 677), (2, 601), (3, 543), (4, 489), (5, 411)$

26. $(1, 10.3), (2, 14.2), (3, 18.9), (4, 23.7), (5, 29.1), (6, 35)$

27. *Demand* A store manager wants to know the demand for a product as a function of price. The daily sales for three different prices of the product are listed in the table.

Price, x	$1.00	$1.25	$1.50
Demand, y	450	375	330

(a) Find the least squares regression line for the data.
(b) Estimate the demand when the price is $1.40.
(c) What price will create a demand of 500 products?

28. *Demand* A hardware retailer wants to know the demand for a tool as a function of price. The monthly sales for four different prices of the tool are listed in the table.

Price, x	$25	$30	$35	$40
Demand, y	82	75	67	55

(a) Find the least squares regression line for the data.
(b) Estimate the demand when the price is $32.95.
(c) What price will yield a demand of 83 tools?

29. *Crop Yield* A farmer used four test plots to determine the relationship between wheat yield in bushels per acre and the amount of fertilizer in hundreds of pounds per acre. The results are given in the table.

Fertilizer, x	1.0	1.5	2.0	2.5
Yield, y	35	44	50	56

(a) Find the least squares regression line for the data.
(b) Estimate the yield for a fertilizer application of 160 pounds per acre.

30. *Carcinogen Contamination* After contamination by a carcinogen, people in different geographic regions were assigned an exposure index that represented the degree of contamination. Using the following data, find a least squares regression line to estimate the mortality per 100,000 people for a given exposure.

Exposure, x	1.35	2.67	3.93	5.14	7.43
Mortality, y	118.5	135.2	167.3	197.6	204.7

31. *Infant Mortality* To study the number of infant deaths per 1000 live births in the United States, a medical researcher obtains the following data. *(Source: Department of Health and Human Services)*

Year	1950	1960	1970	1980	1988	1991
Deaths, y	29.2	26.0	20.0	12.6	10.0	9.0

(a) Find the least squares regression line for the data and use it to estimate the number of infant deaths in 1998. Let $t = 0$ represent 1970.
(b) Find the least squares regression quadratic for the data and use it to estimate the number of infant deaths in 1998.

32. *Population Growth* The table gives the approximate world population (in billions) for five different years.

Year	1800	1850	1900	1950	1990
t	-2	-1	0	1	1.8
Population, y	0.9	1.1	1.5	2.4	5.2

(a) During the 1800s, population growth was almost linear. Find a least squares regression line for those years and use it to estimate the population in 1875.
(b) Find a least squares regression quadratic for the data from 1850 through 1990 and use it to estimate the population in the year 2000.
(c) Even though the rate of growth of the population has begun to decline, most demographers believe the population size will pass the eight billion mark sometime in the next 50 years. What do you think?

33. *Engine Design* After developing a new turbocharger for an automobile engine, the following experimental data was obtained for speed in miles per hour at 2-second intervals.

Time, x	0	2	4	6	8	10
Speed, y	0	15	30	50	65	70

(a) Find a least squares regression quadratic for the data.
(b) Use the model to estimate the speed after 5 seconds.

34. *New York Stock Exchange* The daily average of shares traded (in thousands) on the NYSE for selected years is listed in the table. *(Source: New York Stock Exchange)*

Year	1970	1975	1980	1985	1990
Shares	11,564	18,551	44,871	109,169	156,777

(a) Find a least squares regression quadratic that models the data.

(b) Use the model to estimate the number of shares that will be traded in 2000.

 In Exercises 35–38, use a graphing utility to find any model that best fits the data points.

35. $(1, 13)$, $(2, 16.5)$, $(4, 24)$, $(5, 28)$, $(8, 39)$, $(11, 50.25)$, $(17, 72)$, $(20, 85)$

36. $(1, 5.5)$, $(3, 7.75)$, $(6, 15.2)$, $(8, 23.5)$, $(11, 46)$, $(15, 110)$

37. $(1, 1.5)$, $(2.5, 8.5)$, $(5, 13.5)$, $(8, 16.7)$, $(9, 18)$, $(20, 22)$

38. $(0, 0.5)$, $(1, 7.6)$, $(3, 60)$, $(4.2, 117)$, $(5, 170)$, $(7.9, 380)$

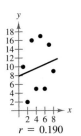 In Exercises 39–42, plot the points and determine whether the data has positive, negative, or no correlation. Then use a graphing utility to find the value of *r* and confirm your result. The number *r* is called the **correlation coefficient.** It is a measure of how well the model fits the data. Correlation coefficients vary between −1 and 1, and the closer |*r*| is to 1, the better the model.

39. $(1, 4)$, $(2, 6)$, $(3, 8)$, $(4, 11)$, $(5, 13)$, $(6, 15)$

40. $(1, 7.5)$, $(2, 7)$, $(3, 7)$, $(4, 6)$, $(5, 5)$, $(6, 4.9)$

41. $(1, 3)$, $(2, 6)$, $(3, 2)$, $(4, 3)$, $(5, 9)$, $(6, 1)$

42. $(0.5, 2)$, $(0.75, 1.75)$, $(1, 3)$, $(1.5, 3.2)$, $(2, 3.7)$, $(2.6, 4)$

True or False In Exercises 43–46, determine whether the statement is true or false. Explain your reasoning.

43. Data that is modeled by $y = 3.29x - 4.17$ has a negative correlation.

44. Data that is modeled by $y = -0.238x + 25$ has a negative correlation.

45. If the correlation coefficient $r \approx -0.98781$, the model is a good fit.

46. A correlation coefficient of $r \approx 0.201$ implies that the data has no correlation.

$r = 0.981$

Positive Correlation

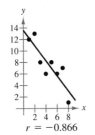

$r = -0.866$

Negative Correlation

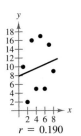

$r = 0.190$

No Correlation

Case Study: Projecting Retail Sales

PROFILE

COMPANY
Regional Economic Analysis Divison, Bureau of Economic Analysis, U.S. Department of Commerce

■

YEARS IN BUSINESS
47

■

LOCATION
Washington, D.C.

■

TYPE OF BUSINESS
Government agency responsible for interpreting regional economics

■

EMPLOYEES
Approx. 475

■

The Regional Economic Analysis Division (READ) of the Bureau of Economic Analysis is a unit of the U.S. Department of Commerce. READ is responsible for developing and interpreting the regional economic accounts of the United States. READ develops methods for estimating gross state product; maintains and improves regional economic projections; and develops analytical techniques for regional economic impact studies.

READ performs regional economic impact analyses, which estimate the effect of a particular event or change in policy on a state's economy. In these analyses, one of the techniques READ economists use to estimate the mathematical relationships among economic variables is regression analysis.

For instance, READ analysts estimate that the total amount spent in eating and drinking establishments in Ohio depends primarily on the number of adults in the state and their level of disposable (after tax) personal income. When either of these variables increases, there are more dollars spent on eating and drinking in Ohio restaurants and bars.

Using annual data on these three variables for the 1970 to 1991 period, READ has used multiple least squares regression methods, a variation of the least squares regression line discussed in Section 7.7, to determine that the actual relationship is:

$$\text{Sales} = -8092.1 + 0.0385 \cdot \text{Income} + 1.1815 \cdot \text{Population}$$

where:

■ Sales is retail sales of Ohio eating and drinking establishments, in millions of dollars,

■ Income is disposable personal income of Ohioans, in millions of dollars, and

■ Population is the adult (18+) population of Ohio, in thousands.

This regression equation shows that an increase of $1 million in Ohio income will result in $38,500 more spent in Ohio's Red Lobsters, TGIFridays, and other restaurants and bars (0.0385 · $1,000,000). Similarly, if the state's adult population increases by a thousand, there will be $1,181,500 more spent in Ohio's eating and drinking places.

In the business world, it is useful to have an idea of the impact that forces outside a firm's control will have on its business. If projections of Ohio income and population are available, this equation will give an expected value for retail sales in eating and drinking establishments for future years. These types of equations could be quite useful to firms in their planning process. The relationship estimated above could give guidance to those in the restaurant business, as well as to firms that supply restaurants with food, drink, and equipment.

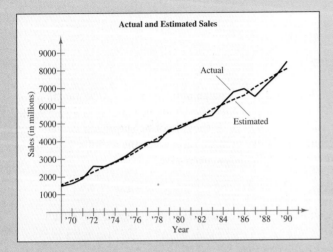

Actual and Estimated Sales

Courtesy of Clifford H. Woodruff

Clifford H. Woodruff is a regional economist for the Regional Economic Analysis Division of the Bureau of Economic Analysis ■

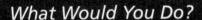

What Would You Do?

1. If you were in the business of manufacturing and selling restaurant equipment such as commercial refrigerators, stoves, tables, and chairs in Ohio, what would you do if economists were predicting a recession for the coming year? A recession typically means higher unemployment and lower income levels. Suppose a predicted recession is expected to lead to a 7% overall reduction in income of Ohioans for a certain year while population remains constant. Use the regression equation to show how this decrease will affect the sales figure that would normally be predicted for this year.

2. If Ohio universities plan to substantially increase their admissions of out-of-state students, such that another 5000 students would be arriving in the state next fall, what effect would you expect this to have on the restaurant business? If you ran a restaurant near a college campus, which of the following scenarios would you prefer (if all other factors remained the same) and why? Make a mathematical argument and explain all calculations.

(A) An increase in out-of-state admissions of 8000 students

OR

(B) An increase in out-of-state admissions of 4500 where each student is given a $50 grant (considered as income) as a reward for enrolling in Ohio.

3. Why are both the population and income variables needed in this equation? Why couldn't we simply use one or the other to generate the estimates of sales? Can you think of other variables that affect restaurant sales, which might be added to the regression equation to improve its ability to predict?

Double Integrals and Area in the Plane

Double Integrals ■ *Finding Area with a Double Integral*

Double Integrals

In Section 7.4, you learned that it is meaningful to differentiate functions of several variables by differentiating with respect to one variable at a time while holding the other variables constant. It should not be surprising to learn that you can *integrate* functions of two or more variables using a similar procedure. For instance, if you are given the partial derivative

$$f_x(x, y) = 2xy, \qquad \text{Partial with respect to } x$$

then, by holding y constant, you can integrate with respect to x to obtain

$$\int f_x(x, y)\, dx = f(x, y) = x^2 y + C(y).$$

This procedure is called **partial integration with respect to x.** Note that the "constant of integration" $C(y)$ is assumed to be a function of y, because y is fixed during integration with respect to x. Similarly, if you are given the partial derivative

$$f_y(x, y) = x^2 + 2, \qquad \text{Partial with respect to } y$$

then, by holding x constant, you can integrate with respect to y to obtain

$$\int f_y(x, y)\, dy = f(x, y) = x^2 y + 2y + C(x).$$

In this case, the "constant of integration" $C(x)$ is assumed to be a function of x, because x is fixed during integration with respect to y.

To evaluate a definite integral of a function of two or more variables, you can apply the Fundamental Theorem of Calculus to one variable while holding the others constant, as follows.

$$\int_1^{2y} 2xy\, dx = x^2 y \Big]_1^{2y} = (2y)^2 y - (1)^2 y = 4y^3 - y.$$

x is the variable of integration and y is fixed.

Replace x by the limits of integration.

The result is a function of y.

Note that you omit the constant of integration, just as you do for a definite integral of a function of one variable.

EXAMPLE 1 *Finding Partial Integrals*

a. $$\int_1^x (2x^2y^{-2} + 2y)\,dy = \left[\frac{-2x^2}{y} + y^2\right]_1^x \qquad \textit{Hold } x \textit{ constant.}$$
$$= \left(\frac{-2x^2}{x} + x^2\right) - \left(\frac{-2x^2}{1} + 1\right)$$
$$= 3x^2 - 2x - 1$$

b. $$\int_y^{5y} \sqrt{x-y}\,dx = \left[\frac{2}{3}(x-y)^{3/2}\right]_y^{5y} \qquad \textit{Hold } y \textit{ constant.}$$
$$= \frac{2}{3}[(5y-y)^{3/2} - (y-y)^{3/2}]$$
$$= \frac{16}{3}y^{3/2}$$

TECHNOLOGY

A symbolic integration utility, such as *Derive*, *Maple*, or *Mathematica*, can be used to evaluate double integrals. To do this, you need to enter the integrand, then integrate twice—once with respect to one of the variables and then with respect to the other variable. For instance, when *Derive* is used to evaluate the integral in Example 2, you obtain the following.

1: $2xy + 3$

2: $\int_0^x (2xy + 3)\,dy$

3: $\int_1^2 \int_0^x (2xy + 3)\,dy\,dx$

4: $\frac{33}{4}$

In Example 1a, note that the definite integral defines a function of x and can *itself* be integrated. An "integral of an integral" is called a **double integral.** With a function of two variables, there are two types of double integrals.

$$\int_a^b \int_{g_1(x)}^{g_2(x)} f(x,y)\,dy\,dx = \int_a^b \left[\int_{g_1(x)}^{g_2(x)} f(x,y)\,dy\right]dx$$

$$\int_a^b \int_{g_1(y)}^{g_2(y)} f(x,y)\,dx\,dy = \int_a^b \left[\int_{g_1(y)}^{g_2(y)} f(x,y)\,dx\right]dy$$

Notice that the difference between these two types is the order in which the integration is performed, $dy\,dx$ or $dx\,dy$.

EXAMPLE 2 *Evaluating a Double Integral*

$$\int_1^2 \int_0^x (2xy+3)\,dy\,dx = \int_1^2 \left[\int_0^x (2xy+3)\,dy\right]dx$$
$$= \int_1^2 \left[xy^2 + 3y\right]_0^x dx$$
$$= \int_1^2 (x^3 + 3x)\,dx$$
$$= \left[\frac{x^4}{4} + \frac{3x^2}{2}\right]_1^2$$
$$= \left(\frac{2^4}{4} + \frac{3(2^2)}{2}\right) - \left(\frac{1^4}{4} + \frac{3(1^2)}{2}\right)$$
$$= \frac{33}{4}$$

Finding Area with a Double Integral

One of the simplest applications of a double integral is finding the area of a plane region. For instance, consider the region R that is bounded by

$$a \leq x \leq b \quad \text{and} \quad g_1(x) \leq y \leq g_2(x).$$

Using the techniques described in Section 5.5, you know that the area of R is

$$\int_a^b [g_2(x) - g_1(x)]\, dx.$$

This same area is also given by the double integral

$$\int_a^b \int_{g_1(x)}^{g_2(x)} dy\, dx$$

because

$$\int_a^b \int_{g_1(x)}^{g_2(x)} dy\, dx = \int_a^b \left[\, y \,\right]_{g_1(x)}^{g_2(x)} dx$$

$$= \int_a^b [g_2(x) - g_1(x)]\, dx.$$

Figure 7.41 shows the two basic types of plane regions whose areas can be determined by a double integral.

Note In Figure 7.41, note that the horizontal or vertical orientation of the narrow rectangle indicates the order of integration. The "outer" variable of integration always corresponds to the width of the rectangle. Notice also that the outer limits of integration for a double integral are constant, whereas the inner limits may be functions of the outer variable.

Determining Area in the Plane by Double Integrals

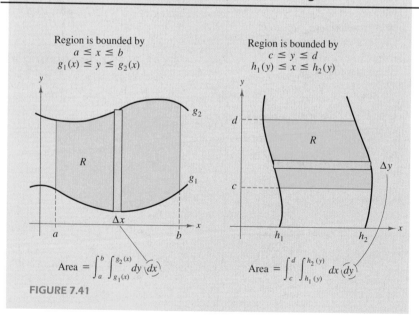

Region is bounded by
$$a \leq x \leq b$$
$$g_1(x) \leq y \leq g_2(x)$$

$$\text{Area} = \int_a^b \int_{g_1(x)}^{g_2(x)} dy\, dx$$

Region is bounded by
$$c \leq y \leq d$$
$$h_1(y) \leq x \leq h_2(y)$$

$$\text{Area} = \int_c^d \int_{h_1(y)}^{h_2(y)} dx\, dy$$

FIGURE 7.41

EXAMPLE 3 *Finding Area by a Double Integral*

Use a double integral to find the area of the rectangular region shown in Figure 7.42.

Solution
The bounds for x are $1 \le x \le 5$ and the bounds for y are $2 \le y \le 4$. Therefore, the area of the region is

$$\int_1^5 \int_2^4 dy\,dx = \int_1^5 \left[y \right]_2^4 dx \qquad \text{\textit{Integrate with respect to y.}}$$

$$= \int_1^5 (4-2)\,dx \qquad \text{\textit{Apply Fundamental Theorem.}}$$

$$= \int_1^5 2\,dx \qquad \text{\textit{Simplify.}}$$

$$= \left[2x \right]_1^5 \qquad \text{\textit{Integrate with respect to x.}}$$

$$= 10 - 2 \qquad \text{\textit{Apply Fundamental Theorem.}}$$

$$= 8. \qquad \text{\textit{Simplify.}}$$

You can confirm this by noting that the rectangle is two units by four units.

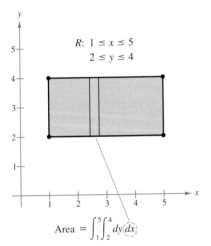

R: $1 \le x \le 5$
    $2 \le y \le 4$

$$\text{Area} = \int_1^5 \int_2^4 dy\,dx$$

FIGURE 7.42

EXAMPLE 4 *Finding Area by a Double Integral*

Use a double integral to find the area of the region bounded by the graphs of $y = x^2$ and $y = x^3$.

Solution
As shown in Figure 7.43, the two graphs intersect when $x = 0$ and $x = 1$. Choosing x to be the outer variable, the bounds for x are $0 \le x \le 1$ and the bounds for y are $x^3 \le y \le x^2$. This implies that the area of the region is

$$\int_0^1 \int_{x^3}^{x^2} dy\,dx = \int_0^1 \left[y \right]_{x^3}^{x^2} dx \qquad \text{\textit{Integrate with respect to y.}}$$

$$= \int_0^1 (x^2 - x^3)\,dx \qquad \text{\textit{Apply Fundamental Theorem.}}$$

$$= \left[\frac{x^3}{3} - \frac{x^4}{4} \right]_0^1 \qquad \text{\textit{Integrate with respect to x.}}$$

$$= \frac{1}{3} - \frac{1}{4} \qquad \text{\textit{Apply Fundamental Theorem.}}$$

$$= \frac{1}{12}. \qquad \text{\textit{Simplify.}}$$

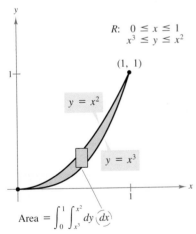

R: $0 \le x \le 1$
    $x^3 \le y \le x^2$

(1, 1)

$y = x^2$

$y = x^3$

$$\text{Area} = \int_0^1 \int_{x^3}^{x^2} dy\,dx$$

FIGURE 7.43

In setting up double integrals, the most difficult task is likely to be determining the correct limits of integration. This can be simplified by making a sketch of the region R and identifying the appropriate bounds for x and y.

EXAMPLE 5 *Changing the Order of Integration*

For the double integral

$$\int_0^2 \int_{y^2}^4 dx\, dy,$$

a. sketch the region R whose area is represented by the integral,
b. rewrite the integral so that x is the outer variable, and
c. show that both orders of integration yield the same value.

Solution

a. From the limits of integration, you know that

$$y^2 \le x \le 4, \qquad\qquad \textit{Variable bounds for } x$$

which means that the region R is bounded on the left by the parabola $x = y^2$ and on the right by the line $x = 4$. Furthermore, because

$$0 \le y \le 2, \qquad\qquad \textit{Constant bounds for } y$$

you know that the region lies above the x-axis, as shown in Figure 7.44.

b. If you interchange the order of integration so that x is the outer variable, then x will have constant bounds of integration given by $0 \le x \le 4$. Solving for y in the equation $x = y^2$ implies that the bounds for y are $0 \le y \le \sqrt{x}$, as shown in Figure 7.45. Thus, with x as the outer variable, the integral can be written as

$$\int_0^4 \int_0^{\sqrt{x}} dy\, dx.$$

c. Both integrals yield the same value.

$$\int_0^2 \int_{y^2}^4 dx\, dy = \int_0^2 \Big[x \Big]_{y^2}^4 dy = \int_0^2 (4 - y^2)\, dy = \left[4y - \frac{y^3}{3} \right]_0^2 = \frac{16}{3}$$

$$\int_0^4 \int_0^{\sqrt{x}} dy\, dx = \int_0^4 \Big[y \Big]_0^{\sqrt{x}} dx = \int_0^4 \sqrt{x}\, dx = \left[\frac{2}{3} x^{3/2} \right]_0^4 = \frac{16}{3}$$

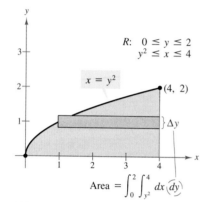

R: $0 \le y \le 2$
    $y^2 \le x \le 4$

$x = y^2$

$(4, 2)$

Δy

$\text{Area} = \int_0^2 \int_{y^2}^4 dx\, dy$

FIGURE 7.44

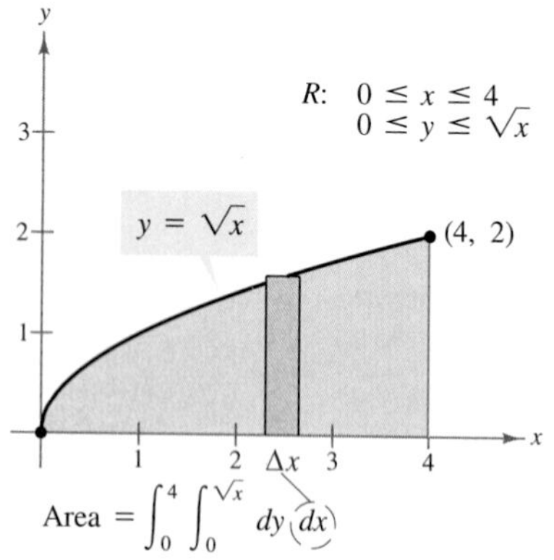

R: $0 \le x \le 4$
    $0 \le y \le \sqrt{x}$

$y = \sqrt{x}$

$(4, 2)$

Δx

$\text{Area} = \int_0^4 \int_0^{\sqrt{x}} dy\, dx$

FIGURE 7.45

To designate a double integral or an area of a region without specifying a particular order of integration, you can use the symbol

$$\int_R \int dA,$$

where $dA = dx\, dy$ or $dA = dy\, dx$.

EXAMPLE 6 *Finding Area by a Double Integral*

Use a double integral to calculate the area denoted by

$$\int_R \int dA,$$

where R is the region bounded by $y = x$ and $y = x^2 - x$.

Solution

Begin by sketching the region R, as shown in Figure 7.46. From the sketch, you can see that vertical rectangles of width dx are more convenient than horizontal ones. Therefore, x is the outer variable of integration and its constant bounds are $0 \le x \le 2$. This implies that the bounds for y are $x^2 - x \le y \le x$, and the area is given by

$$\int_R \int dA = \int_0^2 \int_{x^2-x}^x dy \, dx$$

$$= \int_0^2 \Big[y \Big]_{x^2-x}^x dx$$

$$= \int_0^2 [x - (x^2 - x)] \, dx$$

$$= \int_0^2 (2x - x^2) \, dx$$

$$= \Big[x^2 - \frac{x^3}{3} \Big]_0^2$$

$$= 4 - \frac{8}{3}$$

$$= \frac{4}{3}.$$

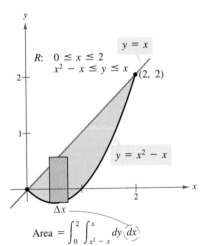

R: $0 \le x \le 2$
$x^2 - x \le y \le x$

$y = x$

$(2, 2)$

$y = x^2 - x$

Δx

Area $= \int_0^2 \int_{x^2-x}^x dy \, dx$

FIGURE 7.46

As you are working the exercises for this section, you should be aware that the primary uses of double integrals will be discussed in Section 7.9. We have introduced double integrals by way of areas in the plane so that you can gain practice in finding the limits of integration. When setting up a double integral, remember that your first step should be to sketch the region R. After doing this, you have two choices of integration orders: $dx \, dy$ or $dy \, dx$.

Discussion Problem *Sketching Regions in the Plane*

Sketch the region represented by the double integral.

a. $\displaystyle\int_0^4 \int_0^y dx \, dy$ **b.** $\displaystyle\int_0^4 \int_0^x dy \, dx$ **c.** $\displaystyle\int_0^4 \int_{x^2}^x dy \, dx$

Warm Up   The following warm-up exercises involve skills that were covered in earlier sections. You will use these skills in the exercise set for this section.

In Exercises 1–6, evaluate the definite integral.

1. $\displaystyle\int_0^1 dx$

2. $\displaystyle\int_1^4 2x^2\,dx$

3. $\displaystyle\int_1^2 (x^3 - 2x + 4)\,dx$

4. $\displaystyle\int_1^2 \frac{2}{7x^2}\,dx$

5. $\displaystyle\int_0^2 \frac{2x}{x^2 + 1}\,dx$

6. $\displaystyle\int_0^2 xe^{x^2+1}\,dx$

In Exercises 7–10, sketch the region bounded by the graphs of the equations.

7. $y = x,\ y = 0,\ x = 0,\ x = 3$

8. $y = x,\ y = 3,\ y = 0,\ x = 0$

9. $y = 4 - x^2,\ y = 0,\ x = 0,\ x = 2$

10. $y = x^2,\ y = 4x,\ x = 0,\ x = 4$

EXERCISES 7.8

   means that technology can help you solve or check the exercise(s).

In Exercises 1–10, evaluate the partial integral.

1. $\displaystyle\int_0^x (2x - y)\,dy$

2. $\displaystyle\int_x^{x^2} \frac{y}{x}\,dy$

3. $\displaystyle\int_1^{2y} \frac{y}{x}\,dx$

4. $\displaystyle\int_0^{e^y} y\,dx$

5. $\displaystyle\int_0^{\sqrt{9 - x^2}} x^2 y\,dy$

6. $\displaystyle\int_{x^2}^{\sqrt{x}} (x^2 + y^2)\,dy$

7. $\displaystyle\int_{e^y}^{y} \frac{y \ln x}{x}\,dx$

8. $\displaystyle\int_{-\sqrt{1 - y^2}}^{\sqrt{1 - y^2}} (x^2 + y^2)\,dx$

9. $\displaystyle\int_0^{x^3} ye^{-y/x}\,dy$

10. $\displaystyle\int_y^3 \frac{xy}{\sqrt{x^2 + 1}}\,dx$

In Exercises 11–24, evaluate the double integral.

11. $\displaystyle\int_0^2 \int_0^1 (x - y)\,dy\,dx$

12. $\displaystyle\int_0^2 \int_0^2 (6 - x^2)\,dy\,dx$

13. $\displaystyle\int_0^4 \int_0^3 xy\,dy\,dx$

14. $\displaystyle\int_0^1 \int_0^x \sqrt{1 - x^2}\,dy\,dx$

15. $\displaystyle\int_0^1 \int_0^{\sqrt{1 - y^2}} (x + y)\,dx\,dy$

16. $\displaystyle\int_0^2 \int_{3y^2 - 6y}^{2y - y^2} 3y\,dx\,dy$

17. $\displaystyle\int_1^2 \int_0^4 (x^2 - 2y^2 + 1)\,dx\,dy$

18. $\displaystyle\int_0^1 \int_y^{2y} (1 + 2x^2 + 2y^2)\,dx\,dy$

19. $\displaystyle\int_0^2 \int_0^{\sqrt{1 - y^2}} - 5xy\,dx\,dy$

20. $\displaystyle\int_0^4 \int_0^x \frac{2}{(x + 1)(y + 1)}\,dy\,dx$

21. $\displaystyle\int_0^2 \int_0^{4 - x^2} x^3\,dy\,dx$

22. $\displaystyle\int_0^a \int_0^{a-x} (x^2 + y^2)\,dy\,dx$

23. $\displaystyle\int_0^{\infty} \int_0^{\infty} e^{-(x+y)/2}\,dy\,dx$

24. $\displaystyle\int_0^{\infty} \int_0^{\infty} xye^{-(x^2+y^2)}\,dx\,dy$

In Exercises 25–32, sketch the region R whose area is given by the double integral. Then change the order of integration and show that both orders yield the same area.

25. $\int_0^1 \int_0^2 dy\,dx$

26. $\int_1^2 \int_2^4 dx\,dy$

27. $\int_0^1 \int_{2y}^2 dx\,dy$

28. $\int_0^4 \int_0^{\sqrt{x}} dy\,dx$

29. $\int_0^2 \int_{x/2}^1 dy\,dx$

30. $\int_0^4 \int_{\sqrt{x}}^2 dy\,dx$

31. $\int_0^1 \int_{y^2}^{\sqrt[3]{y}} dx\,dy$

32. $\int_{-2}^2 \int_0^{4-y^2} dx\,dy$

In Exercises 33 and 34, evaluate the double integral. Note that it is necessary to change the order of integration.

33. $\int_0^3 \int_y^3 e^{x^2}\,dx\,dy$

34. $\int_0^2 \int_x^2 e^{-y^2}\,dy\,dx$

In Exercises 35–40, use a double integral to find the area of the specified region.

35.

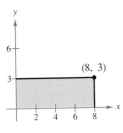

36.

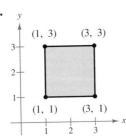

37.

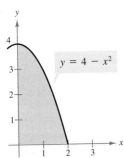

38.

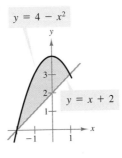

39.

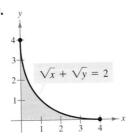

40.

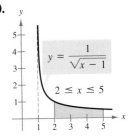

In Exercises 41–46, use a double integral to find the area of the region bounded by the graphs of the equations.

41. $y = 25 - x^2, y = 0$

42. $y = x^{3/2}, y = x$

43. $5x - 2y = 0, x + y = 3, y = 0$

44. $xy = 9, y = x, y = 0, x = 9$

45. $y = x, y = 2x, x = 2$

46. $y = x^2 + 2x + 1, y = 3(x + 1)$

In Exercises 47–52, use a symbolic integration utility to evaluate the double integral.

47. $\int_0^1 \int_0^2 e^{-x^2-y^2}\,dx\,dy$

48. $\int_0^3 \int_0^{x^2} \sqrt{x}\sqrt{1+x}\,dy\,dx$

49. $\int_1^2 \int_0^x e^{xy}\,dy\,dx$

50. $\int_1^2 \int_y^{2y} \ln(x+y)\,dx\,dy$

51. $\int_0^1 \int_x^1 \sqrt{1-x^2}\,dy\,dx$

52. $\int_0^4 \int_0^y \frac{2}{(x+1)(y+1)}\,dx\,dy$

True or False In Exercises 53 and 54, determine whether the statement is true or false. Explain your reasoning.

53. Changing the order of integration will sometimes change the value of a double integral.

54. $\int_2^5 \int_1^6 x\,dy\,dx = \int_1^6 \int_2^5 x\,dx\,dy$

Applications of Double Integrals

Volume of a Solid Region ∎
Average Value of a Function over a Region

Volume of a Solid Region

In Section 7.8, you used double integrals as an alternative way to find the area of a plane region. In this section, you will study the primary uses of double integrals—to find the volume of a solid region and to find the average value of a function.

Consider a function $z = f(x, y)$ that is continuous and nonnegative over a region R. Let S be the solid region that lies between the xy-plane and the surface

$$z = f(x, y) \qquad \textit{Surface lying above the } xy\textit{-plane}$$

directly above the region R, as shown in Figure 7.47. You can find the volume of S by integrating $f(x, y)$ over the region R.

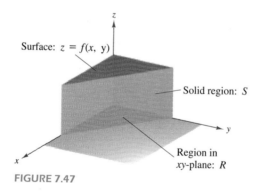

FIGURE 7.47

Determining Volume by Double Integrals

If R is a bounded region in the xy-plane and f is continuous and nonnegative over R, then the **volume of the solid** region between the surface $z = f(x, y)$ and R is given by the double integral

$$\int_R \int f(x, y)\, dA,$$

where $dA = dx\, dy$ or $dA = dy\, dx$.

EXAMPLE 1 *Finding the Volume of a Solid*

Find the volume of the solid region bounded in the first octant by the plane

$$z = 2 - x - 2y.$$

Solution

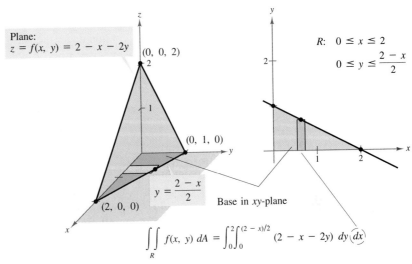

FIGURE 7.48

Note Example 1 uses $dy\,dx$ as the order of integration. Try using the other order, $dx\,dy$, as indicated in Figure 7.49, to find the volume of the region.

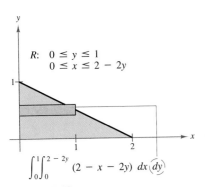

FIGURE 7.49

To set up the double integral for the volume, it is helpful to sketch both the solid and the region R in the xy-plane. In Figure 7.48, you can see that the region R is bounded by the lines $x = 0$, $y = 0$, and $y = \frac{1}{2}(2 - x)$. One way to set up the double integral is to choose x as the outer variable. With that choice, the constant bounds for x are $0 \le x \le 2$ and the variable bounds for y are $0 \le y \le \frac{1}{2}(2 - x)$. Thus, the volume of the solid region is

$$
\begin{aligned}
V &= \int_0^2 \int_0^{(2-x)/2} (2 - x - 2y)\,dy\,dx \\
&= \int_0^2 \left[(2 - x)y - y^2 \right]_0^{(2-x)/2} dx \\
&= \int_0^2 \left[(2 - x)\left(\frac{1}{2}\right)(2 - x) - \left[\frac{1}{2}(2 - x)\right]^2 \right] dx \\
&= \frac{1}{4} \int_0^2 (2 - x)^2\,dx \\
&= \left[-\frac{1}{12}(2 - x)^3 \right]_0^2 \\
&= \frac{2}{3}.
\end{aligned}
$$

In Example 1, the order of integration was arbitrary. Although the example used x as the outer variable, you could just as easily have used y as the outer variable. The next example describes a situation in which one order of integration is more convenient than the other.

EXAMPLE 2 *Comparing Different Orders of Integration*

Find the volume under the surface $f(x, y) = e^{-x^2}$, bounded by the xz-plane and the planes $y = x$ and $x = 1$, as shown in Figure 7.50.

Solution

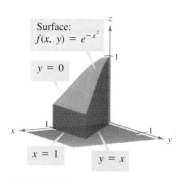

Surface:
$f(x, y) = e^{-x^2}$

$y = 0$

$x = 1$

$y = x$

FIGURE 7.50

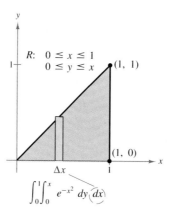

R: $0 \le x \le 1$
     $0 \le y \le x$ (1, 1)

(1, 0)

Δx

$\int_0^1 \int_0^x e^{-x^2} \, dy \, dx$

FIGURE 7.51

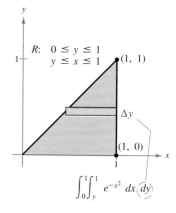

R: $0 \le y \le 1$
     $y \le x \le 1$ (1, 1)

Δy

(1, 0)

$\int_0^1 \int_y^1 e^{-x^2} \, dx \, dy$

In the xy-plane, the bounds of region R are the lines $y = 0$, $x = 1$, and $y = x$. The two possible orders of integration are indicated in Figure 7.51. If you attempt to evaluate the two double integrals shown in the figure, you will discover that the one on the right involves finding the antiderivative of e^{-x^2}, which you know is not an elementary function. The integral on the left, however, can be easily evaluated, as follows.

$$V = \int_0^1 \int_0^x e^{-x^2} \, dy \, dx$$
$$= \int_0^1 \left[e^{-x^2} y \right]_0^x dx$$
$$= \int_0^1 x e^{-x^2} \, dx$$
$$= \left[-\frac{1}{2} e^{-x^2} \right]_0^1$$
$$= -\frac{1}{2} \left(\frac{1}{e} - 1 \right)$$

Guidelines for Finding the Volume of a Solid

1. Write the equation of the surface in the form $z = f(x, y)$ and sketch the solid region.
2. Sketch the region R in the xy-plane and determine the order and limits of integration.
3. Evaluate the double integral

$$\int_R \int f(x, y)\, dA$$

using the order and limits determined in the second step.

The first step above suggests that you sketch the three-dimensional solid region. This is a good suggestion, but, it is not always feasible and is not as important as making a sketch of the two-dimensional region R.

EXAMPLE 3 *Finding the Volume of a Solid*

Find the volume of the solid bounded above by the surface

$$f(x, y) = 6x^2 - 2xy$$

and below by the plane region R shown in Figure 7.52.

Solution
Because the region R is bounded by the parabola $y = 3x - x^2$ and the line $y = x$, the limits for y are $x \le y \le 3x - x^2$. The limits for x are $0 \le x \le 2$, and the volume of the solid is

$$V = \int_0^2 \int_x^{3x-x^2} (6x^2 - 2xy)\, dy\, dx$$

$$= \int_0^2 \left[6x^2 y - xy^2 \right]_x^{3x-x^2} dx$$

$$= \int_0^2 [(18x^3 - 6x^4 - 9x^3 + 6x^4 - x^5) - (6x^3 - x^3)]\, dx$$

$$= \int_0^2 (4x^3 - x^5)\, dx$$

$$= \left[x^4 - \frac{x^6}{6} \right]_0^2$$

$$= \frac{16}{3}.$$

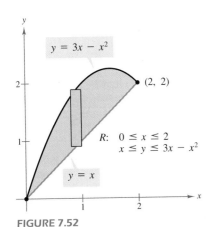

$y = 3x - x^2$

(2, 2)

$R:\ \ 0 \le x \le 2$
$\quad\ \ x \le y \le 3x - x^2$

$y = x$

FIGURE 7.52

A *population density function* $p = f(x, y)$ is a model that describes density (in people per square unit) of a region. To find the population of a region R, evaluate the double integral

$$\int_R \int f(x, y)\, dA.$$

Real
Life

EXAMPLE 4 *Finding a City's Population*

The population density (in people per square mile) of the city shown in Figure 7.53 can be modeled by

$$f(x, y) = \frac{50,000}{x + |y| + 1},$$

where x and y are measured in miles. Approximate the city's population. What is the city's average population density?

Solution

Because the model involves the absolute value of y, it follows that the population density is symmetrical about the x-axis. Thus, the population in the first quadrant is equal to the population in the fourth quadrant. This means that you can find the total population by doubling the population in the first quadrant.

$$\text{Population} = 2\int_0^4 \int_0^5 \frac{50,000}{x + y + 1}\, dy\, dx$$

$$= 100,000 \int_0^4 \left[\ln(x + y + 1) \right]_0^5 dx$$

$$= 100,000 \int_0^4 [\ln(x + 6) - \ln(x + 1)]\, dx$$

$$= 100,000 \bigg[(x + 6)\ln(x + 6) - (x + 6)$$

$$- (x + 1)\ln(x + 1) + (x + 1) \bigg]_0^4$$

$$= 100,000 \bigg[(x + 6)\ln(x + 6) - (x + 1)\ln(x + 1) - 5 \bigg]_0^4$$

$$= 100,000[10\ln(10) - 5\ln(5) - 5 - 6\ln(6) + 5]$$

$$\approx 422,810 \text{ people}$$

Thus, the city's population is about 422,810. Because the city covers a region 4 miles wide and 10 miles long, its area is 40 square miles. Thus, the average population density is

$$\text{Average population density} = \frac{422,810}{40}$$

$$\approx 10,570 \text{ people per square mile.}$$

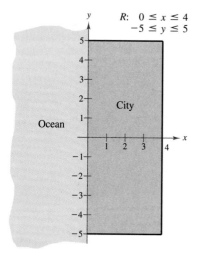

$R: \ 0 \le x \le 4$
$\quad\ -5 \le y \le 5$

FIGURE 7.53

Average Value of a Function over a Region

Average Value of a Function over a Region

If f is integrable over the plane region R with area A, then its **average value** over R is

$$\text{Average value} = \frac{1}{A} \int_{R} \int f(x, y)\, dA.$$

Real Life

EXAMPLE 5 *Finding the Average Profit*

A manufacturer determines that the profit for selling x units of one product and y units of a second product is

$$P = -(x - 200)^2 - (y - 100)^2 + 5000.$$

The weekly sales for product 1 vary between 150 and 200 units and the weekly sales for product 2 vary between 80 and 100 units. Estimate the average weekly profit for the two products.

Solution

Because $150 \leq x \leq 200$ and $80 \leq y \leq 100$, you can estimate the weekly profit to be the average of the profit function over the rectangular region shown in Figure 7.54. Because the area of this rectangular region is $(50)(20) = 1000$, it follows that the average profit V is

$$V = \frac{1}{1000} \int_{150}^{200} \int_{80}^{100} [-(x - 200)^2 - (y - 100)^2 + 5000]\, dy\, dx$$

$$= \frac{1}{1000} \int_{150}^{200} \left[-(x - 200)^2 y - \frac{(y - 100)^3}{3} + 5000y \right]_{80}^{100} dx$$

$$= \frac{1}{1000} \int_{150}^{200} \left[-20(x - 200)^2 - \frac{292{,}000}{3} \right] dx$$

$$= \frac{1}{3000} \left[-20(x - 200)^3 + 292{,}000x \right]_{150}^{200}$$

$$\approx \$4033.$$

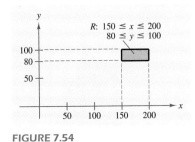

FIGURE 7.54

Discussion Problem   Finding an Average Value

What is the average value of the constant function $f(x, y) = 100$ over a region R? Explain your reasoning.

Warm Up

The following warm-up exercises involve skills that were covered in earlier sections. You will use these skills in the exercise set for this section.

In Exercises 1–4, sketch the region that is described.

1. $0 \le x \le 2,\ 0 \le y \le 1$ 2. $1 \le x \le 3,\ 2 \le y \le 3$

3. $0 \le x \le 4,\ 0 \le y \le 2x - 1$ 4. $0 \le x \le 2,\ 0 \le y \le x^2$

In Exercises 5–10, evaluate the double integral.

5. $\displaystyle\int_0^1 \int_1^2 dy\,dx$ 6. $\displaystyle\int_0^3 \int_1^3 dx\,dy$ 7. $\displaystyle\int_0^1 \int_0^x x\,dy\,dx$

8. $\displaystyle\int_0^4 \int_1^y y\,dx\,dy$ 9. $\displaystyle\int_1^3 \int_x^{x^2} 2\,dy\,dx$ 10. $\displaystyle\int_0^1 \int_x^{-x^2+2} dy\,dx$

EXERCISES 7.9

 means that technology can help you solve or check the exercise(s).

In Exercises 1–6, sketch the region of integration and evaluate the double integral.

1. $\displaystyle\int_0^2 \int_0^1 (3x + 4y)\,dy\,dx$ 2. $\displaystyle\int_{-a}^a \int_{-\sqrt{a^2-x^2}}^{\sqrt{a^2-x^2}} dy\,dx$

3. $\displaystyle\int_0^1 \int_y^{\sqrt{y}} x^2 y^2\,dx\,dy$ 4. $\displaystyle\int_0^6 \int_{y/2}^3 (x + y)\,dx\,dy$

5. $\displaystyle\int_0^1 \int_0^{\sqrt{1-x^2}} y\,dy\,dx$ 6. $\displaystyle\int_0^2 \int_0^{4-x^2} xy^2\,dy\,dx$

In Exercises 7–10, set up the integral for both orders of integration and use the more convenient order to evaluate the integral over the region R.

Integral   *Region R*

7. $\displaystyle\int_R \int xy\,dA$ Rectangle with vertices at $(0, 0)$, $(0, 5)$, $(3, 5)$, $(3, 0)$

8. $\displaystyle\int_R \int \frac{y}{x^2 + y^2}\,dA$ Triangle bounded by $y = x$, $y = 2x$, $x = 2$

9. $\displaystyle\int_R \int \frac{y}{1 + x^2}\,dA$ Region bounded by $y = 0$, $y = \sqrt{x}$, $x = 4$

10. $\displaystyle\int_R \int x\,dA$ Semicircle bounded by $y = \sqrt{25 - x^2}$ and $y = 0$

In Exercises 11 and 12, evaluate the double integral. Note that it is necessary to change the order of integration.

11. $\displaystyle\int_0^1 \int_{y/2}^{1/2} e^{-x^2}\,dx\,dy$ 12. $\displaystyle\int_0^{\ln 10} \int_{e^x}^{10} \frac{1}{\ln y}\,dy\,dx$

In Exercises 13–24, use a double integral to find the volume of the specified solid.

13.   14.

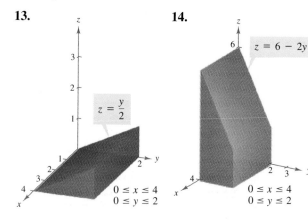

15.

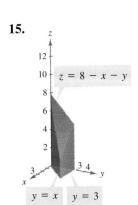

$z = 8 - x - y$

$y = x$ $y = 3$

16.

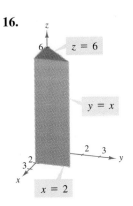

$z = 6$

$y = x$

$x = 2$

17.

$2x + 3y + 4z = 12$

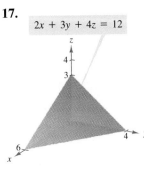

18.

$x + y + z = 1$

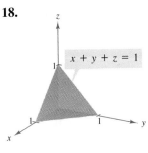

19.

$z = 1 - xy$

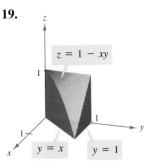

$y = x$ $y = 1$

20.

$z = 4 - y^2$

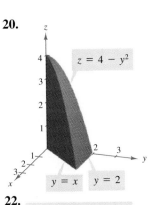

$y = x$ $y = 2$

21.

$z = 4 - x^2 - y^2$

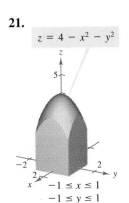

$-1 \le x \le 1$
$-1 \le y \le 1$

22.

$z = \dfrac{1}{(x + 1)^2(y + 1)^2}$

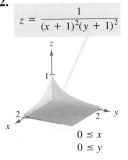

$0 \le x$
$0 \le y$

23.

$x^2 + z^2 = 1$

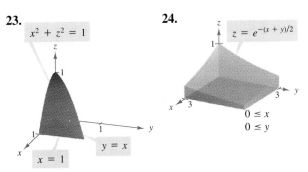

$y = x$

$x = 1$

24.

$z = e^{-(x + y)/2}$

$0 \le x$
$0 \le y$

In Exercises 25–28, use a double integral to find the volume of the solid bounded by the graphs of the equations.

25. $z = xy, z = 0, y = 0, y = 4, x = 0, x = 1$

26. $z = x, z = 0, y = x, y = 0, x = 0, x = 4$

27. $z = x^2, z = 0, x = 0, x = 2, y = 0, y = 4$

28. $z = x + y, x^2 + y^2 = 4$ (first octant)

In Exercises 29–32, find the average value of $f(x, y)$ over the region R.

Integral	*Region R*
29. $f(x, y) = x$	Rectangle with vertices $(0, 0)$, $(4, 0)$, $(4, 2)$, $(0, 2)$
30. $f(x, y) = xy$	Rectangle with vertices $(0, 0)$, $(4, 0)$, $(4, 2)$, $(0, 2)$
31. $f(x, y) = x^2 + y^2$	Square with vertices $(0, 0)$, $(2, 0)$, $(2, 2)$, $(0, 2)$
32. $f(x, y) = e^{x+y}$	Triangle with vertices $(0, 0)$, $(0, 1)$, $(1, 1)$

33. *Average Revenue*   A company sells two products whose demand functions are

$$x_1 = 500 - 3p_1 \quad \text{and} \quad x_2 = 750 - 2.4p_2.$$

Therefore, the total revenue is given by

$$R = x_1 p_1 + x_2 p_2.$$

Estimate the average revenue if the price p_1 varies between $50 and $75 and the price p_2 varies between $100 and $150.

34. *Average Weekly Profit*   A firm's weekly profit in marketing two products is given by

$$P = 192x_1 + 576x_2 - x_1^2 - 5x_2^2 - 2x_1x_2 - 5000,$$

where x_1 and x_2 represent the numbers of units of each product sold weekly. Estimate the average weekly profit if x_1 varies between 40 and 50 units and x_2 varies between 45 and 50 units.

Chapter Summary and Study Tips

After studying this chapter, you should have acquired the following skills. The exercise numbers are keyed to the Review Exercises that begin on page 540. Answers to odd–numbered Review Exercises are given in the back of the text.

- ▪ Plot points in space. (*Section 7.1*)   Review Exercises 1, 2
- ▪ Find the distance between two points in space. (*Section 7.1*)   Review Exercises 3, 4

$$d = \sqrt{(x_2 - x_1)^2 + (y_2 - y_1)^2 + (z_2 - z_1)^2}$$

- ▪ Find the midpoints of line segments in space. (*Section 7.1*)   Review Exercises 5, 6

$$\text{Midpoint} = \left(\frac{x_1 + x_2}{2}, \frac{y_1 + y_2}{2}, \frac{z_1 + z_2}{2} \right)$$

- ▪ Write the standard forms of the equations of spheres. (*Section 7.1*)   Review Exercises 7–10

$$(x - h)^2 + (y - k)^2 + (z - l)^2 = r^2$$

- ▪ Find the centers and radii of spheres. (*Section 7.1*)   Review Exercises 11, 12
- ▪ Sketch the coordinate plane traces of spheres. (*Section 7.1*)   Review Exercises 13, 14
- ▪ Sketch planes in space. (*Section 7.2*)   Review Exercises 15–18
- ▪ Classify quadric surfaces in space. (*Section 7.2*)   Review Exercises 19–26
- ▪ Evaluate functions of several variables. (*Section 7.3*)   Review Exercises 27, 28
- ▪ Find the domains and ranges of functions of several variables. (*Section 7.3*)   Review Exercises 29, 30
- ▪ Sketch the level curves of functions of two variables. (*Section 7.3*)   Review Exercises 31–34
- ▪ Use functions of several variables to answer questions about real life. (*Section 7.3*)   Review Exercises 35–38
- ▪ Find the first partial derivatives of functions of several variables. (*Section 7.4*)   Review Exercises 39–48

$$\frac{\partial z}{\partial x} = \lim_{\Delta x \to 0} \frac{f(x + \Delta x, y) - f(x, y)}{\Delta x}$$

$$\frac{\partial z}{\partial y} = \lim_{\Delta y \to 0} \frac{f(x, y + \Delta y) - f(x, y)}{\Delta y}$$

- ▪ Find the slopes of surfaces in the *x*- or *y*-direction. (*Section 7.4*)   Review Exercises 49–52
- ▪ Find the second partial derivatives of functions of several variables. (*Section 7.4*)   Review Exercises 53–56
- ▪ Use partial derivatives to answer questions about real life. (*Section 7.4*)   Review Exercises 57, 58
- ▪ Find the relative extrema of functions of two variables. (*Section 7.5*)   Review Exercises 59–66
- ▪ Use relative extrema to answer questions about real life. (*Section 7.5*)   Review Exercises 67, 68
- ▪ Use Lagrange multipliers to find extrema of functions of several variables. (*Section 7.6*)   Review Exercises 69–74
- ▪ Use Lagrange multipliers to answer questions about real life. (*Section 7.6*)   Review Exercises 75, 76

*Several student study aids are available with this text. The *Student Solutions Guide* includes detailed solutions to all odd-numbered exercises, as well as practice chapter tests with answers. The *Graphics Calculator Guide* offers instructions on the use of a variety of graphing calculators and computer graphing software. The *Brief Calculus TUTOR* includes additional examples for selected exercises in the text.

■   Find the least squares regression lines, $y = ax + b$, for data. (*Section 7.7*)   Review Exercises 77, 78

$$a = \frac{n\sum_{i=1}^{n} x_i y_i - \sum_{i=1}^{n} x_i \sum_{i=1}^{n} y_i}{n\sum_{i=1}^{n} x_i^2 - \left(\sum_{i=1}^{n} x_i\right)^2}, \qquad b = \frac{1}{n}\left(\sum_{i=1}^{n} y_i - a\sum_{i=1}^{n} x_i\right)$$

■   Use least squares regression lines to model real-life data. (*Section 7.7*)   Review Exercises 79, 80

■   Find the least squares regression quadratics for data. (*Section 7.7*)   Review Exercises 81, 82

■   Evaluate double integrals. (*Section 7.8*)   Review Exercises 83–86

■   Use double integrals to find the areas of regions. (*Section 7.8*)   Review Exercises 87–90

■   Use double integrals to find the volumes of solids. (*Section 7.9*)   Review Exercises 91, 92

$$\text{Volume} = \int_R \int f(x, y)\, dA$$

■   Use double integrals to find the average values of real-life models. (*Section 7.9*)   Review Exercises 93, 94

$$\text{Average value} = \frac{1}{A}\int_R \int f(x, y)\, dA$$

■   Comparing Two Dimensions with Three Dimensions   Many of the formulas and techniques in this chapter are generalizations of formulas and techniques used in earlier chapters in the text. Here are several examples.

Two-Dimensional Coordinate System	*Three-Dimensional Coordinate System*
Distance Formula $d = \sqrt{(x_2 - x_1)^2 + (y_2 - y_1)^2}$	*Distance Formula* $d = \sqrt{(x_2 - x_1)^2 + (y_2 - y_1)^2 + (z_2 - z_1)^2}$
Midpoint Formula $\text{Midpoint} = \left(\dfrac{x_1 + x_2}{2}, \dfrac{y_1 + y_2}{2}\right)$	*Midpoint Formula* $\text{Midpoint} = \left(\dfrac{x_1 + x_2}{2}, \dfrac{y_1 + y_2}{2}, \dfrac{z_1 + z_2}{2}\right)$
Equation of Circle $(x - h)^2 + (y - k)^2 = r^2$	*Equation of Sphere* $(x - h)^2 + (y - k)^2 + (z - l)^2 = r^2$
Equation of Line $ax + by = c$	*Equation of Plane* $ax + by + cz = d$
Derivative of $y = f(x)$ $\dfrac{dy}{dx} = \lim\limits_{\Delta x \to 0} \dfrac{f(x + \Delta x) - f(x)}{\Delta x}$	*Partial derivative of* $z = f(x, y)$ $\dfrac{\partial z}{\partial x} = \lim\limits_{\Delta x \to 0} \dfrac{f(x + \Delta x, y) - f(x, y)}{\Delta x}$
Area of Region $A = \displaystyle\int_a^b f(x)\, dx$	*Volume of Region* $V = \displaystyle\int_R \int f(x, y)\, dA$

Review Exercises

 means that technology can help you solve or check the exercise(s).

In Exercises 1 and 2, plot the points.

1. $(2, -1, 4), (-1, 3, -3)$ **2.** $(1, -2, -3), (-4, -3, 5)$

In Exercises 3 and 4, find the distance between the two points.

3. $(0, 0, 0), (2, 5, 9)$ **4.** $(-4, 1, 5), (1, 3, 7)$

In Exercises 5 and 6, find the midpoint of the line segment joining the two points.

5. $(2, 6, 4), (-4, 2, 8)$ **6.** $(5, 0, 7), (-1, -2, 9)$

In Exercises 7–10, find the standard form of the equation of the sphere.

7. Center: $(0, 1, 0)$; Radius: 5

8. Center: $(4, -5, 3)$; Radius: 10

9. Diameter endpoints: $(3, 4, 0), (5, 8, 2)$

10. Diameter endpoints: $(-2, 5, 1), (4, -3, 3)$

In Exercises 11 and 12, find the center and radius of the sphere.

11. $x^2 + y^2 + z^2 + 4x - 2y - 8z + 5 = 0$

12. $x^2 + y^2 + z^2 + 4y - 10z - 7 = 0$

In Exercises 13 and 14, sketch the *xy*-trace of the sphere.

13. $(x + 2)^2 + (y - 1)^2 + (z - 3)^2 = 25$

14. $(x - 1)^2 + (y + 3)^2 + (z - 6)^2 = 72$

In Exercises 15–18, find the intercepts and sketch the graph of the plane.

15. $x + 2y + 3z = 6$ **16.** $2y + z = 4$

17. $6x + 3y - 6z = 12$ **18.** $4x - y + 2z = 8$

In Exercises 19–26, identify the surface.

19. $x^2 + y^2 + z^2 - 2x + 4y - 6z + 5 = 0$

20. $16x^2 + 16y^2 - 9z^2 = 0$

21. $x^2 + \dfrac{y^2}{16} + \dfrac{z^2}{9} = 1$ **22.** $-x^2 + \dfrac{y^2}{16} + \dfrac{z^2}{9} = 1$

23. $z = \dfrac{x^2}{9} + y^2$ **24.** $-4x^2 + y^2 + z^2 = 4$

25. $z = \sqrt{x^2 + y^2}$ **26.** $z = 9x + 3y - 5$

In Exercises 27 and 28, find the function values.

27. $f(x, y) = xy^2$

(a) $f(2, 3)$ (b) $f(0, 1)$
(c) $f(-5, 7)$ (d) $f(-2, -4)$

28. $f(x, y) = \dfrac{x^2}{y}$

(a) $f(6, 9)$ (b) $f(8, 4)$
(c) $f(t, 2)$ (d) $f(r, r)$

In Exercises 29 and 30, describe the region *R* in the *xy*-plane that corresponds to the domain of the function. Then find the range of the function.

29. $f(x, y) = \sqrt{1 - x^2 - y^2}$ **30.** $f(x, y) = \dfrac{1}{x + y}$

 In Exercises 31–34, describe the level curves of the function. Sketch the level curves for the given *c*-values.

31. $z = 10 - 2x - 5y, \ c = 0, 2, 4, 5, 10$

32. $z = \sqrt{9 - x^2 - y^2}, \ c = 0, 1, 2, 3$

33. $z = (xy)^2, \ c = 1, 4, 9, 12, 16$

34. $z = 2e^{xy}, \ c = 1, 2, 3, 4, 5$

35. *Average Precipitation* The contour map shown below represents the average yearly precipitation for Iowa. *(Source: U.S. National Oceanic and Atmospheric Administration)*

(a) Discuss the use of color to represent the level curves.
(b) Which part of Iowa receives the most precipitation?
(c) Which part of Iowa receives the least precipitation?

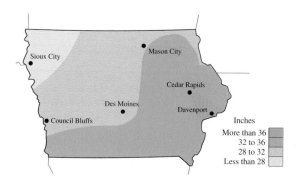

36. *Population Density* The contour map below represents the population density of New York. *(Source: U.S. Bureau of Census)*

(a) Discuss the use of color to represent the level curves.

(b) Do the level curves correspond to equally spaced population densities?

(c) Describe how to obtain a more detailed contour map.

37. *Equation of Exchange* Economists use an equation of exchange to express the relation among money, prices, and business transactions. This equation can be written as

$$P = \frac{MV}{T},$$

where M is the money supply, V is the velocity of circulation, T is the total number of transactions, and P is the price level. Find P when $M = \$2500$, $V = 6$, and $T = 6000$.

38. *Biomechanics* The Froude number F, defined as

$$F = \frac{v^2}{gl},$$

where v represents velocity, g represents gravitational acceleration, and l represents stride length, is an example of a "similarity criterion." Find the Froude number of a rabbit whose velocity is 2 meters/sec, gravitational acceleration is 3 m/sec^2, and stride length is 0.75 meters.

In Exercises 39–48, find the first partial derivatives.

39. $f(x, y) = x^2y + 3xy + 2x - 5y$

40. $f(x, y) = 5x^2 + 4xy - 3y^2$

41. $z = 6x^2\sqrt{y} + 3\sqrt{xy} - 7xy$

42. $z = (xy + 2x + 4y)^2$

43. $f(x, y) = \ln(2x + 3y)$ **44.** $f(x, y) = \ln\sqrt{2x + 3y}$

45. $f(x, y) = x^2e^y - y^2e^x$ **46.** $f(x, y) = x^2e^{-2y}$

47. $w = xyz^2$

48. $w = xyz + 2xy - 9xz + 4yz - y^2z + 4z^2$

In Exercises 49–52, find the slope of the surface at the indicated point in (a) the *x*-direction and (b) the *y*-direction.

49. $z = 3x - 4y + 9$, $(3, 2, 10)$

50. $z = 4x^2 - y^2$, $(2, 4, 0)$

51. $z = 8 - x^2 - y^2$, $(1, 2, 3)$

52. $z = x^2 - y^2$, $(5, -4, 9)$

In Exercises 53–56, find all second partial derivatives.

53. $f(x, y) = x^3 - 4xy^2 + y^3$ **54.** $f(x, y) = \dfrac{y}{x + y}$

55. $f(x, y) = \sqrt{64 - x^2 - y^2}$ **56.** $f(x, y) = x^2e^{-y^2}$

57. *Marginal Cost* A company manufactures two models of skis: cross-country skis and downhill skis. The cost function for producing x cross-country skis and y downhill skis is

$$C = 15(xy)^{1/3} + 99x + 139y + 2293.$$

Find the marginal costs when $x = 250$ and $y = 175$.

58. *Marginal Revenue* At a baseball stadium, souvenir caps are sold at two locations. If x_1 and x_2 are the numbers of baseball caps sold at location 1 and location 2, respectively, then the total revenue for the caps is

$$R = 15x_1 + 16x_2 - \frac{1}{10}x_1^2 - \frac{1}{10}x_2^2 - \frac{1}{100}x_1x_2.$$

Given that $x_1 = 50$ and $x_2 = 40$, find the marginal revenue at location 1 and at location 2.

In Exercises 59–66, find any critical points and relative extrema of the function.

59. $f(x, y) = x^2 + 2xy + y^2$

60. $f(x, y) = x^3 - 3xy + y^2$

61. $f(x, y) = x^2 + 6xy + 3y^2 + 6x + 8$

62. $f(x, y) = x + y^2 - e^x$

63. $f(x, y) = x^3 + y^2 - xy$

64. $f(x, y) = y^2 + xy + 3y - 2x + 5$

65. $f(x, y) = x^3 + y^3 - 3x - 3y + 2$

66. $f(x, y) = y^2 - x^2$

67. *Revenue*   A company manufactures and sells two products. The demand functions for the products are

$$p_1 = 100 - x_1 \quad \text{and} \quad p_2 = 200 - 0.5x_2.$$

(a) Find the total revenue functions for x_1 and x_2.
(b) Find x_1 and x_2 so that the revenue is maximized.
(c) What is the maximum revenue?

68. *Profit*   A company manufactures a product at two locations. The costs of manufacturing x_1 units at plant 1 and x_2 units at plant 2 are

$$C_1 = 0.03x_1{}^2 + 4x_1 + 300$$
$$C_2 = 0.05x_2{}^2 + 7x_2 + 175.$$

If the product sells for \$10 per unit, find x_1 and x_2 so that the profit, $P = 10(x_1 + x_2) - C_1 - C_2$, is maximized.

In Exercises 69–74, locate any extrema of the functions by using Lagrange multipliers.

69. $f(x, y) = x^2 y$,   Constraint: $x + 2y = 2$

70. $f(x, y) = x^2 + y^2$,   Constraint: $x + y = 4$

71. $f(x, y, z) = xyz$,   Constraint: $x + 2y + z - 4 = 0$

72. $f(x, y, z) = xz + yz$,   Constraint: $x + y + z = 6$

73. $f(x, y, z) = x^2 + y^2 + z^2$,   Constraints: $x + z = 6$,
$$y + z = 8$$

74. $f(x, y, z) = xyz$,   Constraints: $x + y + z = 32$,
$$x - y + z = 0$$

75. *Maximum Production Level*   The production function for a manufacturer is

$$f(x, y) = 4x + xy + 2y.$$

Assume that the total amount available for labor x and capital y is \$2000 and that units of labor and capital cost \$20 and \$4, respectively. Find the maximum production level for this manufacturer.

76. *Minimum Cost*   A manufacturer has an order for 1500 units that can be produced at two locations. Let x_1 and x_2 be the numbers of units produced at the two locations. Find the number that should be produced at each to meet the order and minimize cost if the cost function is

$$C = 0.20x_1{}^2 + 10x_1 + 0.15x_2{}^2 + 12x_2.$$

 In Exercises 77 and 78, (a) use the method of least squares to find the least squares regression line and (b) calculate the sum of the squared errors.

77. $(-2, -3), (-1, -1), (1, 2), (3, 2)$

78. $(-3, -1), (-2, -1), (0, 0), (1, 1), (2, 1)$

 79. *Biotechnology*   The biotechnology industry, which includes genetic engineering, has produced over 27 new wonder drugs. The biotechnology industry product sales (in billions of dollars) from 1990 to 1993 are listed in the table. *(Source: Ernst and Young)*

Year	1990	1991	1992	1993
Sales	2.9	4.0	5.9	7.0

(a) Use a graphing utility with a built-in least squares regression program to find a linear model for the data.
(b) Use the model to estimate the product sales of the biotechnology industry in the year 2000.

 80. *Federal Employees*   The per capita federal government employment decreased steadily from 1966 to 1992. The table lists the number of federal employees per 10,000 people for selected years. *(Source: U.S. Bureau of Census)*

Year	1966	1971	1976	1981	1986	1992
Employees	111	105	102	96	92	89

(a) Find a least squares regression line to model the data.
(b) Use the model to estimate when the number of federal employees per 10,000 will be 80.

 In Exercises 81 and 82, find the least squares regression quadratic for the given points. Plot the points and sketch the least squares regression quadratic.

81. $(-1, 9), (0, 7), (1, 5), (2, 6), (4, 23)$

82. $(0, 10), (2, 9), (3, 7), (4, 4), (5, 0)$

In Exercises 83–86, evaluate the double integral.

83. $\displaystyle\int_0^1 \int_0^{1+x} (3x + 2y)\, dy\, dx$

84. $\displaystyle\int_{-2}^2 \int_0^4 (x - y^2)\, dx\, dy$

85. $\int_{1}^{2} \int_{1}^{2y} \frac{x}{y^2} \, dx \, dy$

86. $\int_{0}^{4} \int_{0}^{\sqrt{16-x^2}} 2x \, dy \, dx$

In Exercises 87–90, use a double integral to find the area of the region.

87.

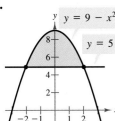

$y = 9 - x^2$
$y = 5$

88.

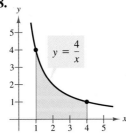

$y = \frac{4}{x}$

89.

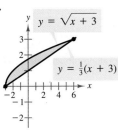

$y = \sqrt{x + 3}$
$y = \frac{1}{3}(x + 3)$

90.

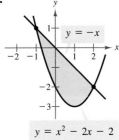

$y = -x$
$y = x^2 - 2x - 2$

91. Find the volume of the solid bounded by the graphs of $z = (xy)^2$, $z = 0$, $y = 0$, $y = 4$, $x = 0$, and $x = 4$.

92. Find the volume of the solid bounded by the graphs of $z = x + y$, $z = 0$, $x = 0$, $x = 3$, $y = x$, and $y = 0$.

93. *Average Elevation* In a triangular coastal area, the elevation in miles above sea level at the point (x, y) is $f(x, y) = 0.25 - 0.025x - 0.01y$, where x and y are measured in miles (see figure). Find the average elevation of the triangular area.

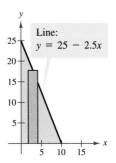

Line:
$y = 25 - 2.5x$

94. *Real Estate Value* The average value of real estate (in dollars per square foot) for a rectangular section of a city is given by $f(x, y) = 2.5x^{3/2}y^{3/4}$. The section of the city is shown in the figure below. Find the average value of real estate for this section.

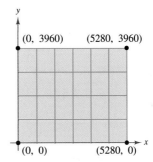

(0, 3960)   (5280, 3960)

(0, 0)   (5280, 0)

Sample Post-Graduation Exam Questions

The following questions were taken from certified public accountant (CPA) exams, graduate management admission tests (GMAT), graduate records exams (GRE), actuarial exams, or college-level academic skills tests (CLAST). The answers to the questions are given in the back of the book.

1. What is the derivative of $f(x, y) = y^2(x + y)^3$ with respect to y?
(a) $6y(x + y)^2$ (b) $y(x + y)^2(2x + 5y)$
(c) $3y^2(x + y)^2$ (d) $y(x + y)^2(5x + 2y)$

2. Let $f(x, y) = x^2 + y^2 + 6x - 2y + 4$. At which point does f have a relative minimum?
(a) $(-3, 1, -13)$ (b) $(-3, 1, -6)$
(c) $(3, 1, -2)$ (d) $(-3, -1, -2)$

For Questions 3–4, use the following excerpts from the 1993 Tax Rate Schedules.

Schedule X–Use if your filing status is **Single**

If the amount on Form 1040, line 37, is over—	But not over—	Enter on Form 1040, line 38	of the amount over—
$0	$22,100	........15%	$0
22,100	53,500	3,315.00 + 28%	22,100
53,500	115,000	12,107.00 + 31%	53,500
115,100	250,000	31,172.00 + 36%	115,000
250,000	....	79,772.00 + 39.6%	250,000

Schedule Y-1–Use if your filing status is **Married filing jointly** or **Qualifying widow(er)**

If the amount on Form 1040, line 37, is over—	But not over—	Enter on Form 1040, line 38	of the amount over—
$0	$36,900	........15%	$0
36,900	89,150	5,535.00 + 28%	36,900
89,150	140,000	20,165.00 + 31%	89,150
140,000	250,000	35,928.50 + 36%	140,000
250,000	....	75,528.50 + 39.6%	250,000

3. The tax for a married couple filing jointly whose amount on Form 1040, line 37 is $125,480 is
(a) $11,262.30   (b) $47,801.50
(c) $31,427.30   (d) $34,944.80

4. The tax for a single whose amount on Form 1040, line 37 is $1,000,000 is
(a) $372,528.50   (b) $297,000
(c) $368,028.50   (d) $376,772

5. If $x, y,$ and z are chosen from the three numbers $\frac{1}{3}, -2,$ and 4, what is the largest possible value of the expression $[(x^2 + z)]/(y^2)$?
(a) 126   (b) 42   (c) 24   (d) 72

6. If $xz = 4y$, then $(x^3)/2$ equals
(a) $\frac{2y^3}{z^3}$ (b) $\frac{16y^2}{z^3}$ (c) $\frac{32y^3}{z^3}$ (d) $\frac{64y^3}{z^3}$

7. Mave Co. calculated the following ratios for one of its profit centers:
Gross margin 31%, Return on Sales 26%, Capital Turnover .5 times
What is Mave's return on investment for this profit center?
(a) 8.5%   (b) 13%   (c) 15.5%   (d) 26%

$$\int_{-\infty}^{\infty} \frac{1}{\sigma\sqrt{2\pi}} \; e^{-(x-\mu)^2/2\sigma^2} \, dx = 1$$

P

Probability and Calculus

P.1 Discrete Probability
P.2 Continuous Random Variables
P.3 Expected Value and Variance
Chapter Summary and Study Tips
Review Exercises

The study of modern probability is often traced to the French mathematician Blaise Pascal. In 1654, Pascal was asked why it was unprofitable to bet even money that double sixes would turn up at least once in 24 tosses of two dice. Pascal answered the question by calculating the probability of double sixes to be 1/36.

At first, probability dealt with sample spaces that contained only finitely many outcomes, such as the number of ways two dice can be tossed. When combined with calculus, however, probability theory was able to analyze problems in which there are infinitely many outcomes. These two outlooks still divide the study of probability into two fields—discrete probability and continuous probability.

Since Pascal's time, the applications of probability have expanded greatly—to fields as diverse as management, quality control, psychology, and biology. Modern work in probability is exemplified by the Japanese quality control expert, Kaoru Ishikawa (1915–1989).

$$P(a \leq x \leq b) = \int_{a}^{b} f(x) \, dx$$

Ishikawa

Courtesy Japanese Union of Scientists and Engineers

Discrete Probability

Sample Spaces ■ *Discrete Random Variables* ■ *Discrete Probability* ■
Expected Value ■ *Variance and Standard Deviation*

Sample Spaces

When assigning measurements to the uncertainties of everyday life, people often use ambiguous terminology such as "fairly certain," "probable," and "highly unlikely." Probability theory allows you to remove this ambiguity by assigning a number to the likelihood of the occurrence of an event. This number is called the **probability** that the event will occur. For example, if you toss a fair coin, the probability that it will land heads up is one-half or 0.5.

In probability theory, any happening whose result is uncertain is called an **experiment.** The possible results of the experiment are the **outcomes,** a collection of outcomes is an **event,** and the collection of all possible outcomes is the **sample space** of the experiment. For instance, consider an experiment in which a coin is tossed. The sample space of this experiment consists of two outcomes: either the coin will land heads up (denoted by H) or it will land tails up (denoted by T). Thus, the sample space S is

$S = \{H, T\}.$ *Sample space*

In this text, all outcomes of a sample space are assumed to be equally likely. For instance, when a coin is tossed, H and T are assumed to be equally likely.

The Stock Market/© 92 Tom Ives

When a weather forecaster states that there is a 50% chance of thunderstorms, it means that thunderstorms have occurred on half of all days that have had similar weather conditions.

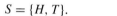

EXAMPLE 1 *Finding a Sample Space*   *Real Life*

An experiment consists of tossing a six-sided die.

a. What is the sample space?
b. Describe the event corresponding to a number greater than 2 turning up.

Solution

a. The sample space consists of six outcomes, which can be represented by the numbers 1 through 6. That is,

$S = \{1, 2, 3, 4, 5, 6\}.$ *Sample space*

Note that each of the outcomes in the sample space is equally likely.

b. The event corresponding to a number greater than 2 turning up is the following subset of S.

$E = \{3, 4, 5, 6\}.$ *Event*

Discrete Random Variables

A function that assigns a numerical value to each of the outcomes in a sample space is called a **random variable.** For instance, in the sample space $S = \{HH, HT, TH, TT\}$, the outcomes could be assigned the numbers 2, 1, and 0, depending on the number of heads in the outcome.

Definition of a Discrete Random Variable

Let S be a sample space. A **random variable** is a function x that assigns a numerical value to each outcome in S. If the set of values taken on by the random variable is finite, then the random variable is **discrete.** The number of times a specific value of x occurs is the **frequency** of x and is denoted by $n(x)$.

Real Life

EXAMPLE 2 *Finding Frequencies*

Three coins are tossed. A random variable assigns the number 0, 1, 2, or 3 to each possible outcome, depending on the number of heads that turn up.

$$S = \{HHH, HHT, HTH, HTT, THH, THT, TTH, TTT\}$$

$$\downarrow \quad \downarrow \quad \downarrow \quad \downarrow \quad \downarrow \quad \downarrow \quad \downarrow \quad \downarrow$$

$$3 \quad\quad 2 \quad\quad 2 \quad\quad 1 \quad\quad 2 \quad\quad 1 \quad\quad 1 \quad\quad 0$$

Find the frequency of 0, 1, 2, and 3. Then use a bar graph to represent the result.

Solution
To find the frequencies, simply count the number of occurrences of each value of the random variable, as shown in Table P.1.

TABLE P.1

Random variable, x	0	1	2	3
Frequency of x, $n(x)$	1	3	3	1

This table is called a **frequency distribution** of the random variable. The result is shown graphically by the bar graph in Figure P.1.

Note In Example 2, note that the sample space consists of *eight* outcomes, each of which is *equally likely.* The sample space does *not* consist of the outcomes "0 heads," "1 head," "2 heads," and "3 heads." You cannot consider these events to be outcomes because they are not equally likely.

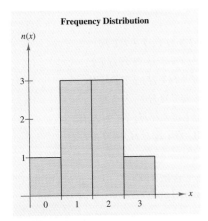

Frequency Distribution

FIGURE P.1

Discrete Probability

The probability of a random variable x is

$$P(x) = \frac{\text{Frequency of } x}{\text{Number of outcomes in } S} = \frac{n(x)}{n(S)}, \qquad \textit{Probability}$$

where $n(S)$ is the number of outcomes in the sample space. By this definition, it follows that the probability of an event must be a number between 0 and 1. That is, $0 \le P(x) \le 1$.

The collection of probabilities corresponding to the values of the random variable is called the **probability distribution** of the random variable. If the range of a discrete random variable consists of m different values $\{x_1, x_2, x_3, \dots, x_m\}$, then the sum of the probabilities of x_i is 1. This can be written as

$$P(x_1) + P(x_2) + P(x_3) + \cdots + P(x_m) = 1.$$

Real Life

EXAMPLE 3 *Finding a Probability Distribution*

Five coins are tossed. Graph the probability distribution for the random variable giving the number of heads that turn up.

Solution

x	Event	$n(x)$
0	$TTTTT$	1
1	$HTTTT, THTTT, TTHTT, TTTHT, TTTTH$	5
2	$HHTTT, HTHTT, HTTHT, HTTTH, THHTT$ $THTHT, THTTH, TTHHT, TTHTH, TTTHH$	10
3	$HHHTT, HHTHT, HHTTH, HTHHT, HTHTH$ $HTTHH, THHHT, THHTH, THTHH, TTHHH$	10
4	$HHHHT, HHHTH, HHTHH, HTHHH, THHHH$	5
5	$HHHHH$	1

The number of outcomes in the sample space is $n(S) = 32$. The probability of each value of the random variable is shown in Table P.2.

TABLE P.2

Random variable, x	0	1	2	3	4	5
Probability, $P(x)$	$\frac{1}{32}$	$\frac{5}{32}$	$\frac{10}{32}$	$\frac{10}{32}$	$\frac{5}{32}$	$\frac{1}{32}$

A graph of this probability distribution is shown in Figure P.2. Note that values of the random variable are represented by intervals on the x-axis.

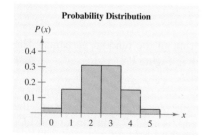

Probability Distribution

FIGURE P.2

Expected Value

Suppose you repeated the coin-tossing experiment in Example 3 several times. On the average, how many heads would you expect to turn up? From Figure P.2, it seems reasonable that the average number of heads would be $2\frac{1}{2}$. This "average" is the **expected value** of the random variable.

Note Although the expected value of x is denoted by $E(x)$, the mean of x is usually denoted by μ. Because the mean often occurs near the center of the values in the range of the random variable, it is called a **measure of central tendency.**

Definition of Expected Value

If the range of a discrete random variable consists of m different values $\{x_1, x_2, x_3, \ldots, x_m\}$, then the **expected value** of the random variable is

$$E(x) = x_1 P(x_1) + x_2 P(x_2) + x_3 P(x_3) + \cdots + x_m P(x_m).$$

The expected value is also called the **mean** of the random variable.

Real Life

EXAMPLE 4 *Finding an Expected Value*

Five coins are tossed. Find the expected value for the number of heads that will turn up.

Solution
Using the results of Example 3, you have the following.

$$E(x) = \overbrace{(0)\left(\tfrac{1}{32}\right)}^{0\ Heads} + \overbrace{(1)\left(\tfrac{5}{32}\right)}^{1\ Head} + \overbrace{(2)\left(\tfrac{10}{32}\right)}^{2\ Heads} + \overbrace{(3)\left(\tfrac{10}{32}\right)}^{3\ Heads} + \overbrace{(4)\left(\tfrac{5}{32}\right)}^{4\ Heads} + \overbrace{(5)\left(\tfrac{1}{32}\right)}^{5\ Heads}$$
$$= \tfrac{80}{32} = 2.5$$

Real Life

EXAMPLE 5 *Finding an Expected Value*

Over a period of 1 year (225 selling days), a sales representative sold between 0 and 8 units per day, as shown in Figure P.3. From this data, what is the average number of units per day the sales representative should expect to sell?

Solution
One way to answer this question is to calculate the expected value for the number of units.

$$E(x) = (0)\left(\tfrac{33}{225}\right) + (1)\left(\tfrac{45}{225}\right) + (2)\left(\tfrac{52}{225}\right) + (3)\left(\tfrac{46}{225}\right) + (4)\left(\tfrac{24}{225}\right)$$
$$+ (5)\left(\tfrac{11}{225}\right) + (6)\left(\tfrac{8}{225}\right) + (7)\left(\tfrac{5}{225}\right) + (8)\left(\tfrac{1}{225}\right)$$
$$= \tfrac{529}{225} \approx 2.35 \text{ units per day}$$

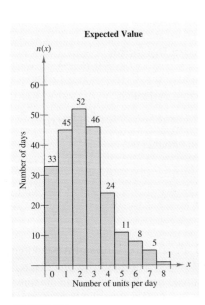

Expected Value

FIGURE P.3

Variance and Standard Deviation

Discovery

The average grade on the calculus final in a class of 20 students was 80 out of 100 possible points. Describe a distribution of grades for which ten students scored above 95 points. Describe another distribution of grades for which only one student scored above 85. In general, how does the standard deviation influence the grade distribution in a course?

The expected value or mean gives a measure of the average value assigned by a random variable. But the mean doesn't tell the whole story. For instance, all three of the following distributions have a mean of 2.

Distribution 1

Random variable, x	0	1	2	3	4
Frequency of x, $n(x)$	2	2	2	2	2

Distribution 2

Random variable, x	0	1	2	3	4
Frequency of x, $n(x)$	0	3	4	3	0

Distribution 3

Random variable, x	0	1	2	3	4
Frequency of x, $n(x)$	5	0	0	0	5

Even though each distribution has the same mean, the patterns of the distributions are quite different. In the first distribution, each value has the same frequency. In the second, the values are clustered about the mean. In the third distribution, the values are far from the mean. To measure how much the distribution varies from the mean, you can use the concept of **standard deviation.**

Definition of Variance and Standard Deviation

Consider a random variable whose range is $\{x_1, x_2, x_3, \ldots, x_m\}$ with a mean of μ. The **variance** of the random variable is

$$V(x) = (x_1 - \mu)^2 P(x_1) + (x_2 - \mu)^2 P(x_2) + \cdots + (x_m - \mu)^2 P(x_m).$$

The **standard deviation** of the random variable is

$$\sigma = \sqrt{V(x)}$$

(σ is the lowercase Greek letter sigma).

When the standard deviation is small, most of the values of the random variable are clustered near the mean. As the standard deviation becomes larger, the distribution becomes more and more spread out. For instance, in the three distributions above, you would expect the second to have the smallest standard deviation and the third to have the largest. This is confirmed in Example 6.

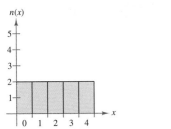

(a) Mean $= 2$; standard deviation $= \sqrt{1.41}$

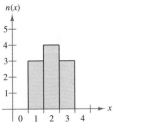

(b) Mean $= 2$; standard deviation $= 0.77$

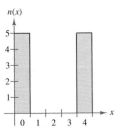

(c) Mean $= 2$; standard deviation $= 2$

FIGURE P.4

EXAMPLE 6 *Finding Variance and Standard Deviation*

Find the variance and standard deviation of each of the three distributions shown on page P-6.

Solution

a. For distribution 1, the mean is $\mu = 2$, the variance is

$$V(x) = (0-2)^2 \left(\tfrac{2}{10}\right) + (1-2)^2 \left(\tfrac{2}{10}\right) + (2-2)^2 \left(\tfrac{2}{10}\right)$$
$$+ (3-2)^2 \left(\tfrac{2}{10}\right) + (4-2)^2 \left(\tfrac{2}{10}\right)$$
$$= 2, \qquad\qquad\qquad\qquad\qquad\qquad \textit{Variance}$$

and the standard deviation is

$$\sigma = \sqrt{2} \approx 1.41. \qquad\qquad\qquad \textit{Standard deviation}$$

b. For distribution 2, the mean is $\mu = 2$, the variance is

$$V(x) = (0-2)^2 \left(\tfrac{0}{10}\right) + (1-2)^2 \left(\tfrac{3}{10}\right) + (2-2)^2 \left(\tfrac{4}{10}\right)$$
$$+ (3-2)^2 \left(\tfrac{3}{10}\right) + (4-2)^2 \left(\tfrac{0}{10}\right)$$
$$= 0.6, \qquad\qquad\qquad\qquad\qquad\qquad \textit{Variance}$$

and the standard deviation is

$$\sigma = \sqrt{0.6} \approx 0.77. \qquad\qquad\qquad \textit{Standard deviation}$$

c. For distribution 3, the mean is $\mu = 2$, the variance is

$$V(x) = (0-2)^2 \left(\tfrac{5}{10}\right) + (1-2)^2 \left(\tfrac{0}{10}\right) + (2-2)^2 \left(\tfrac{0}{10}\right)$$
$$+ (3-2)^2 \left(\tfrac{0}{10}\right) + (4-2)^2 \left(\tfrac{5}{10}\right)$$
$$= 4, \qquad\qquad\qquad\qquad\qquad\qquad \textit{Variance}$$

and the standard deviation is

$$\sigma = \sqrt{4} = 2. \qquad\qquad\qquad\qquad \textit{Standard deviation}$$

The graphs of the three distributions are shown in Figure P.4. As expected, note that the second distribution has the smallest standard deviation and the third distribution has the largest. ■

Discussion Problem   **Creating Distributions**

a. Create a distribution of 10 numbers whose mean is 2 and whose standard deviation is smaller than any of the standard deviations in Example 6. Is it possible to create a distribution whose standard deviation is 0?

b. Create a distribution of 10 numbers whose mean is 2 and whose standard deviation is larger than any of the standard deviations in Example 6.

Warm Up   The following warm-up exercises involve skills that were covered in earlier sections. You will use these skills in the exercise set for this section.

In Exercises 1 and 2, solve for x.

1. $\dfrac{1}{9} + \dfrac{2}{3} + \dfrac{2}{9} = x$

2. $\dfrac{1}{3} + \dfrac{5}{12} + \dfrac{1}{8} + \dfrac{1}{12} + \dfrac{x}{24} = 1$

In Exercises 3–6, evaluate the expression.

3. $0\left(\frac{1}{16}\right) + 1\left(\frac{3}{16}\right) + 2\left(\frac{8}{16}\right) + 3\left(\frac{3}{16}\right) + 4\left(\frac{1}{16}\right)$

4. $0\left(\frac{1}{12}\right) + 1\left(\frac{2}{12}\right) + 2\left(\frac{6}{12}\right) + 3\left(\frac{2}{12}\right) + 4\left(\frac{1}{12}\right)$

5. $(0-1)^2\left(\frac{1}{4}\right) + (1-1)^2\left(\frac{1}{2}\right) + (2-1)^2\left(\frac{1}{4}\right)$

6. $(0-2)^2\left(\frac{1}{12}\right) + (1-2)^2\left(\frac{2}{12}\right) + (2-2)^2\left(\frac{6}{12}\right) + (3-2)^2\left(\frac{2}{12}\right) + (4-2)^2\left(\frac{1}{12}\right)$

In Exercises 7–10, write the fraction as a percent.

7. $\frac{3}{8}$

8. $\frac{9}{11}$

9. $\frac{13}{24}$

10. $\frac{112}{256}$

E X E R C I S E S   P.1   means that technology can help you solve or check the exercise(s).

In Exercises 1–4, list the elements in the specified set.

1. *Tossing a Coin*   A coin is tossed three times.

(a) The sample space S
(b) The event A that at least two heads occur
(c) The event B that no more than one head occurs

2. *Tossing a Coin*   A coin is tossed. If a head occurs, the coin is tossed again; otherwise, a die is tossed.

(a) The sample space S
(b) The event A that a number greater than 3 occurs on the die
(c) The event B that two heads occur

3. *Selecting an Integer*   An integer is selected from the set of all integers between 1 and 50 that are divisible by 3.

(a) The sample space S
(b) The event A that the integer is divisible by 12
(c) The event B that the integer is a perfect square

4. *Taking a Poll*   Three people are asked their opinions on a political issue. They can answer "In favor" (I), "Opposed" (O), or "Undecided" (U).

(a) The sample space S
(b) The event A that at least 2 are in favor
(c) The event B that no more than 1 is opposed

5. *Election Poll*   Three people have been nominated for president of a college class. From a small poll it is estimated that Jane has a probability of 0.29 of winning and Larry has a probability of 0.47. What is the probability of the third candidate winning the election?

6. *Random Selection*   In a class of 72 students, 44 are girls and of these, 12 are going to college. Of the 28 boys in the class, 9 are going to college. If a student is selected at random from the class, what is the probability that the person chosen is (a) going to college? (b) not going to college? (c) a girl who is not going to college?

7. *Quality Control* A component of a spacecraft has both a main system and a backup system. The probability of at least one of the systems performing satisfactorily throughout the duration of the flight is 0.9855. What is the probability of both of them failing?

8. *Choosing a Card* A card is chosen at random from a standard 52-card deck of playing cards. What is the probability that the card will be black and a face card?

9. *Tossing Coins* Two coins are tossed. A random variable assigns the number 0, 1, or 2 to each possible outcome, depending on the number of heads that turn up. Find the frequencies of 0, 1, and 2.

10. *Tossing Coins* Four coins are tossed. A random variable assigns the number 0, 1, 2, 3, or 4 to each possible outcome, depending on the number of heads that turn up. Find the frequencies of 0, 1, 2, 3, and 4.

11. *True-False Exam* Three students answer a true-false question on an examination. A random variable assigns the number 0, 1, 2, or 3 to each outcome, depending on the number of answers of *true* among the three students. Find the frequencies of 0, 1, 2, and 3.

12. *True-False Exam* Four students answer a true-false question on an examination. A random variable assigns the number 0, 1, 2, 3, or 4 to each outcome, depending on the number of answers of *true* among the four students. Find the frequencies of 0, 1, 2, 3, and 4.

In Exercises 13–16, sketch a graph of the probability distribution and find the required probabilities.

13.

x	0	1	2	3	4
$P(x)$	$\frac{1}{20}$	$\frac{3}{20}$	$\frac{6}{20}$	$\frac{6}{20}$	$\frac{4}{20}$

(a) $P(1 \leq x \leq 3)$ (b) $P(x \geq 2)$

14.

x	0	1	2	3	4
$P(x)$	$\frac{8}{20}$	$\frac{6}{20}$	$\frac{3}{20}$	$\frac{2}{20}$	$\frac{1}{20}$

(a) $P(x \leq 2)$ (b) $P(x > 2)$

15.

x	0	1	2	3	4	5
$P(x)$	0.041	0.189	0.247	0.326	0.159	0.038

(a) $P(x \leq 3)$ (b) $P(x > 3)$

16.

x	0	1	2	3
$P(x)$	0.027	0.189	0.441	0.343

(a) $P(1 \leq x \leq 2)$ (b) $P(x < 2)$

17. *Boy or Girl?* Consider a couple that has four children. Assume that it is equally likely that each child is a girl or a boy.

(a) Complete the set to form the sample space consisting of 16 elements.

$S = \{gggg, gggb, ggbg, \ldots\}$

(b) Complete the table, in which the random variable x is the number of girls in the family.

x	0	1	2	3	4
$P(x)$					

(c) Use the table in part (b) to sketch the graph of the probability distribution.

(d) Use the table in part (b) to find the probability of having at least one boy.

18. *Tossing Dice* Consider the experiment of tossing a six-sided die twice.

(a) Complete the set to form the sample space of 36 elements. Note that each element is an ordered pair in which the entries are the numbers of points on the first and second tosses, respectively.

$S = \{(1, 1), (1, 2), \ldots, (2, 1), (2, 2), \ldots\}$

(b) Complete the table, in which the random variable x is the sum of the number of points.

x	2	3	4	5	6	7	8	9	10	11	12
$P(x)$											

(c) Use the table in part (b) to sketch the graph of the probability distribution.

(d) Use the table in part (b) to find $P(5 \leq x \leq 9)$.

In Exercises 19–22, find $E(x)$, $V(x)$, and σ for the given probability distribution.

19.

x	1	2	3	4	5
$P(x)$	$\frac{1}{16}$	$\frac{3}{16}$	$\frac{8}{16}$	$\frac{3}{16}$	$\frac{1}{16}$

20.

x	1	2	3	4	5
$P(x)$	$\frac{4}{10}$	$\frac{2}{10}$	$\frac{2}{10}$	$\frac{1}{10}$	$\frac{1}{10}$

21.

x	-3	-1	0	3	5
$P(x)$	$\frac{1}{5}$	$\frac{1}{5}$	$\frac{1}{5}$	$\frac{1}{5}$	$\frac{1}{5}$

22.

x	$-\$5000$	$-\$2500$	$\$300$
$P(x)$	0.008	0.052	0.940

Tossing Dice   In Exercises 23 and 24, find the mean and variance of the discrete random variable *x*.

23. *x* is (a) the number of points when a four-sided die is tossed once and (b) the sum of the points when the four-sided die is tossed twice.

24. *x* is the number of heads when a coin is tossed four times.

25. *Revenue*   A publishing company introduces a new weekly magazine that sells for $1.50 on the newsstand. The marketing group of the company estimates that sales *x* in thousands will be approximated by the following probability function.

x	10	15	20	30	40
$P(x)$	0.25	0.30	0.25	0.15	0.05

(a) Find $E(x)$ and σ.
(b) Find the expected revenue.

26. *Personal Income*   In a certain section of a large city, the probability distribution of the random variable *x*, the annual income of a family in thousands of dollars, is given in the following table.

x	30	40	50	60	80
$P(x)$	0.10	0.20	0.50	0.15	0.05

Find $E(x)$ and σ.

27. *Insurance*   An insurance company needs to determine the annual premium required to break even on fire protection policies with a face value of $90,000. If *x* is the claim size on these policies and the analysis is restricted to the losses $30,000, $60,000, and $90,000, then the probability distribution of *x* is given by the following table.

x	0	30,000	60,000	90,000
$P(x)$	0.995	0.0036	0.0011	0.0003

What premium should customers be charged for the company to break even?

28. *Insurance*   An insurance company needs to determine the annual premium required to break even for collision protection for cars with a value of $10,000. If *x* is the claim size on these policies and the analysis is restricted to the losses $1,000, $5,000, and $10,000, then the probability distribution of *x* is given by the following table.

x	0	1,000	5,000	10,000
$P(x)$	0.936	0.040	0.020	0.004

What premium should customers be charged for the company to break even?

Games of Chance   If *x* is the net gain to a player in a game of chance, then $E(x)$ is usually negative. This value gives the average amount per game the player can expect to lose over the long run. In Exercises 29 and 30, find the expected net gain to the player for one play of the specified game.

29. In roulette, the wheel has the 38 numbers 00, 0, 1, 2, . . . , 34, 35, and 36, marked on equally spaced slots. If a player bets $1 on a number and wins, then the player keeps the dollar and receives an additional 35 dollars. Otherwise, the dollar is lost.

30. A service organization is selling $2 raffle tickets as part of a fund-raising program. The first prize is a boat valued at $2950, and the second prize is a camping tent valued at $400. In addition to the first and second prizes, there are 25 $20 gift certificates to be awarded. The number of tickets sold is 3000.

31. *Market Analysis*   After considerable market study, a sporting goods company has decided on two possible cities in which to open a new store. Management estimates that city 1 will yield $20 million in revenues if successful and will lose $4 million if not, whereas city 2 will yield $50 million in revenues if successful and lose $9 million if not. City 1 has a 0.3 probability of being successful and city 2 has a 0.2 probability of being successful. In which city should the sporting goods company open the new store with respect to the expected return from each store?

32. Repeat Exercise 31 if the probabilities of city 1 and city 2 being successful are 0.4 and 0.25, respectively.

33. Use a graphing utility to find the mean and standard deviation of the following data, which gives the ages of 50 residents of a long-term-care nursing facility.

62, 64, 66, 66, 68, 69, 69, 70, 70, 70,
71, 72, 72, 72, 72, 73, 73, 74, 74, 75,
75, 75, 75, 76, 76, 77, 77, 77, 77, 77,
77, 78, 78, 79, 79, 79, 80, 81, 81, 83,
84, 84, 87, 88, 89, 91, 94, 94, 97, 98

34. Use a graphing utility to find the mean and standard deviation of the following data, which gives the number of hours of overtime for 60 employees of a particular company during a given week.

0.0, 0.0, 0.0, 0.0, 0.0, 0.5, 0.5, 1.0, 1.0, 1.0,
2.0, 2.0, 2.0, 2.0, 2.0, 2.0, 2.0, 2.5, 2.5, 2.5,
3.0, 3.0, 3.0, 3.0, 3.0, 3.0, 3.0, 4.0, 4.0, 4.0,
4.0, 4.5, 4.5, 5.0, 5.0, 5.0, 5.0, 5.0, 6.0, 6.0,
6.5, 6.5, 6.5, 7.0, 7.0, 7.0, 7.0, 8.0, 8.0, 8.0,
8.0, 8.0, 8.0, 8.0, 8.0, 8.5, 8.5, 8.5, 9.0, 9.5

Continuous Random Variables

Continuous Random Variables

In many applications of probability, it is useful to consider a random variable whose range is an interval on the real line. Such a random variable is called **continuous.** For instance, the random variable that measures the height of a person in a population is continuous.

To define the probability of an event involving a continuous random variable, you cannot simply count the number of ways the event can occur (as you can with a discrete random variable). Rather, you need to define a function f called a **probability density function.**

Definition of a Probability Density Function

Consider a function f of a continuous random variable x whose range is the interval $[a, b]$. The function is a **probability density function** if it is nonnegative and continuous on the interval $[a, b]$ and if

$$\int_a^b f(x)\, dx = 1.$$

The probability that x lies in the interval $[c, d]$ is

$$P(c \le x \le d) = \int_c^d f(x)\, dx,$$

as shown in Figure P.5. If the range of the continuous random variable is an infinite interval, then the integrals are improper integrals.

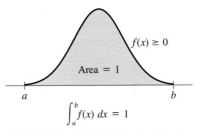

$$\int_a^b f(x)\, dx = 1$$

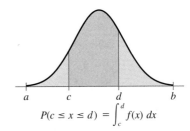

$$P(c \le x \le d) = \int_c^d f(x)\, dx$$

FIGURE P.5 Probability Density Function

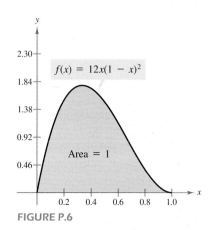

FIGURE P.6

EXAMPLE 1 *Verifying a Probability Density Function*

Show that $f(x) = 12x(1 - x)^2$ is a probability density function over the interval $[0, 1]$.

Solution

Begin by observing that f is continuous and nonnegative on the interval $[0, 1]$.

$$f(x) = 12x(1 - x)^2 \geq 0, \qquad 0 \leq x \leq 1 \qquad \textit{Nonnegative}$$

Next, evaluate the following integral.

$$\int_0^1 12x(1 - x)^2 \, dx = 12 \int_0^1 (x^3 - 2x^2 + x) \, dx$$

$$= 12 \left[\frac{x^4}{4} - \frac{2x^3}{3} + \frac{x^2}{2} \right]_0^1$$

$$= 12 \left(\frac{1}{4} - \frac{2}{3} + \frac{1}{2} \right)$$

$$= 1$$

Because this value is 1, you can conclude that f is a probability density function over the interval $[0, 1]$. The graph of f is shown in Figure P.6. ▬

The next example deals with an infinite interval and its corresponding improper integral.

EXAMPLE 2 *Verifying a Probability Density Function*

Show that $f(t) = 0.1e^{-0.1t}$ is a probability density function over the infinite interval $[0, \infty)$.

Solution

Begin by observing that f is continuous and nonnegative on the interval $[0, \infty)$.

$$f(t) = 0.1e^{-0.1t} \geq 0, \qquad 0 \leq t \qquad \textit{Nonnegative}$$

Next, evaluate the following integral.

$$\int_0^\infty 0.1e^{-0.1t} \, dt = \lim_{b \to \infty} \left[-e^{-0.1t} \right]_0^b$$

$$= \lim_{b \to \infty} (-e^{-0.1b} + 1)$$

$$= 1$$

Because this value is 1, you can conclude that f is a probability density function over the interval $[0, \infty)$. The graph of f is shown in Figure P.7. ▬

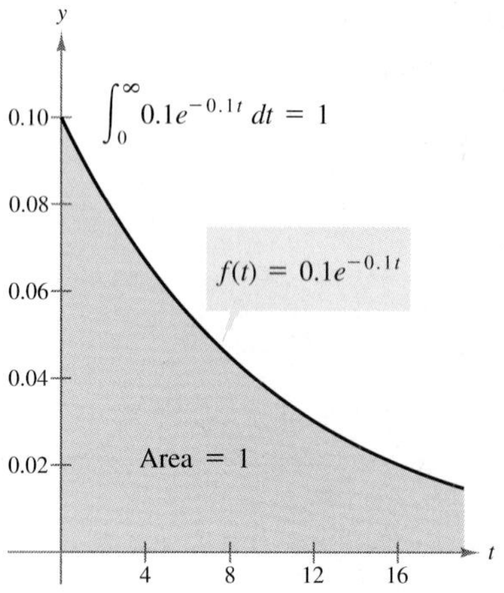

FIGURE P.7

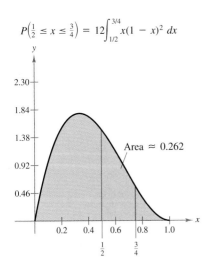

$$P\left(\tfrac{1}{2} \le x \le \tfrac{3}{4}\right) = 12\int_{1/2}^{3/4} x(1-x)^2\,dx$$

FIGURE P.8

EXAMPLE 3 *Finding a Probability*

For the probability density function in Example 1,

$$f(x) = 12x(1-x)^2,$$

find the probability that x lies in the interval $\tfrac{1}{2} \le x \le \tfrac{3}{4}$.

Solution

$$P\left(\tfrac{1}{2} \le x \le \tfrac{3}{4}\right) = 12\int_{1/2}^{3/4} x(1-x)^2\,dx$$

$$= 12\int_{1/2}^{3/4} (x^3 - 2x^2 + x)\,dx$$

$$= 12\left[\frac{x^4}{4} - \frac{2x^3}{3} + \frac{x^2}{2}\right]_{1/2}^{3/4}$$

$$= 12\left[\frac{\left(\tfrac{3}{4}\right)^4}{4} - \frac{2\left(\tfrac{3}{4}\right)^3}{3} + \frac{\left(\tfrac{3}{4}\right)^2}{2} - \frac{\left(\tfrac{1}{2}\right)^4}{4} + \frac{2\left(\tfrac{1}{2}\right)^3}{3} - \frac{\left(\tfrac{1}{2}\right)^2}{2}\right]$$

$$\approx 0.262$$

Therefore, the probability that x lies in the interval $\left[\tfrac{1}{2}, \tfrac{3}{4}\right]$ is approximately 0.262 or 26.2%, as indicated in Figure P.8. ▪

In Example 3, note that if you had been asked to find the probability that x lies in any of the intervals $\tfrac{1}{2} < x < \tfrac{3}{4}$, $\tfrac{1}{2} \le x < \tfrac{3}{4}$, or $\tfrac{1}{2} < x \le \tfrac{3}{4}$, you would have obtained the same solution as that given in the example. In other words, the inclusion of either endpoint adds nothing to the probability. This demonstrates an important difference between discrete and continuous random variables. For a continuous random variable, the probability that x will be precisely one value (such as 0.5) is considered to be zero, because

$$P(0.5 \le x \le 0.5) = \int_{0.5}^{0.5} f(x)\,dx = 0.$$

You should not interpret this result to mean that it is impossible for the continuous random variable x to have the value 0.5. It simply means that the probability that x will have this *exact* value is insignificant.

EXAMPLE 4 *Finding a Probability*

Consider a probability density function defined over the interval $[0, 5]$. If the probability that x lies in the interval $[0, 2]$ is 0.7, what is the probability that x lies in the interval $[2, 5]$?

Solution

Because the probability that x lies in the interval $[0, 5]$ is 1, you can conclude that the probability that x lies in the interval $[2, 5]$ is $1 - 0.7 = 0.3$. ▪

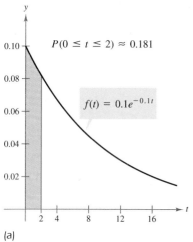

(a)

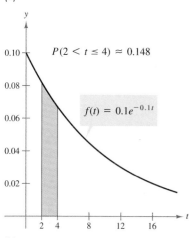

(b)

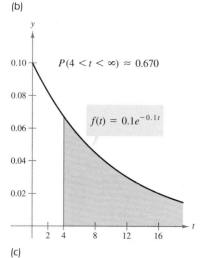

(c)

FIGURE P.9

Applications

Real Life

EXAMPLE 5 *Modeling the Lifetime of a Product*

The usable lifetime (in years) of a product is modeled by the probability density function

$$f(t) = 0.1e^{-0.1t}, \qquad 0 \le t < \infty.$$

Find the probability that a randomly selected unit will have a lifetime falling in the given interval.

a. No more than 2 years
b. More than 2 years, but no more than 4 years
c. More than 4 years

Solution

a. The probability that the unit will last no more than 2 years is

$$P(0 \le t \le 2) = 0.1 \int_0^2 e^{-0.1t}\, dt$$

$$= \left[-e^{-0.1t} \right]_0^2$$

$$= -e^{-0.2} + 1$$

$$\approx 0.181.$$

b. The probability that the unit will last more than 2 years, but no more than 4 years, is

$$P(2 < t \le 4) = 0.1 \int_2^4 e^{-0.1t}\, dt$$

$$= \left[-e^{-0.1t} \right]_2^4$$

$$= -e^{-0.4} + e^{-0.2}$$

$$\approx 0.148.$$

c. The probability that the unit will last more than 4 years is

$$P(4 < t < \infty) = 0.1 \int_4^\infty e^{-0.1t}\, dt$$

$$= \lim_{b \to \infty} \left[-e^{-0.1t} \right]_4^b$$

$$= \lim_{b \to \infty} (-e^{-0.1b} + e^{-0.4})$$

$$= e^{-0.4}$$

$$\approx 0.670.$$

These three probabilities are illustrated graphically in Figure P.9. Note that the sum of the three probabilities is 1.

Real
Life

EXAMPLE 6 *Modeling Weekly Demand*

The weekly demand for a product is modeled by the probability density function

$$f(x) = \frac{1}{36}(-x^2 + 6x), \qquad 0 \le x \le 6,$$

where x is the number of units sold (in thousands). What are the minimum and maximum weekly sales? Find the probability that the sales for a randomly chosen week will be between 2000 and 4000 units.

Solution

Because $0 \le x \le 6$, the weekly sales vary between a minimum of 0 and a maximum of 6000 units. The probability is given by the integral

$$P(2 \le x \le 4) = \frac{1}{36}\int_{2}^{4}(-x^2 + 6x)\,dx$$

$$= \frac{1}{36}\left[-\frac{x^3}{3} + 3x^2\right]_{2}^{4}$$

$$= \frac{1}{36}\left(-\frac{64}{3} + 48 + \frac{8}{3} - 12\right)$$

$$= \frac{13}{27}$$

$$\approx 0.481.$$

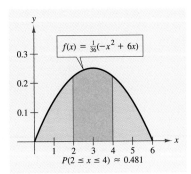

$P(2 \le x \le 4) \approx 0.481$

FIGURE P.10

Thus, the probability that the weekly sales will be between 2000 and 4000 units is about 0.481 or 48.1%, as indicated in Figure P.10.

Discussion Problem   ***Modeling Population Heights***

Which of the following probability density functions do you think is a better model of the heights of American women (ages 18–24)? Explain your reasoning.

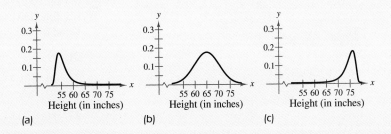

Use the model to estimate the average height of American women.

Warm Up   The following warm-up exercises involve skills that were covered in earlier sections. You will use these skills in the exercise set for this section.

In Exercises 1–4, determine whether f is continuous and nonnegative on the given interval.

1. $f(x) = \dfrac{1}{x}$, $[1, 4]$ **2.** $f(x) = x^2 - 1$, $[0, 1]$

3. $f(x) = 3 - x$, $[1, 5]$ **4.** $f(x) = e^{-x}$, $[0, 1]$

In Exercises 5–10, evaluate the definite integral.

5. $\displaystyle\int_0^4 \frac{1}{4}\, dx$ **6.** $\displaystyle\int_1^3 \frac{1}{4}\, dx$

7. $\displaystyle\int_0^2 \frac{2-x}{2}\, dx$ **8.** $\displaystyle\int_1^2 \frac{2-x}{2}\, dx$

9. $\displaystyle\int_0^\infty 0.4e^{-0.4t}\, dt$ **10.** $\displaystyle\int_0^\infty 3e^{-3t}\, dt$

E X E R C I S E S   P.2   means that technology can help you solve or check the exercise(s).

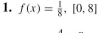 In Exercises 1–14, use a graphing utility to graph the function. Then verify that f is a probability density function over the given interval.

1. $f(x) = \frac{1}{8}$, $[0, 8]$ **2.** $f(x) = \frac{1}{5}$, $[0, 5]$

3. $f(x) = \dfrac{4-x}{8}$, $[0, 4]$ **4.** $f(x) = \dfrac{x}{18}$, $[0, 6]$

5. $f(x) = 6x(1-x)$, $[0, 1]$ **6.** $f(x) = \dfrac{x(6-x)}{36}$, $[0, 6]$

7. $f(x) = \frac{1}{5}e^{-x/5}$, $[0, \infty)$ **8.** $f(x) = \frac{1}{6}e^{-x/6}$, $[0, \infty)$

9. $f(x) = \dfrac{3x}{8}\sqrt{4 - x^2}$, $[0, 2]$

10. $f(x) = 12x^2(1 - x)$, $[0, 1]$

11. $f(x) = \frac{4}{27}x^2(3 - x)$, $[0, 3]$

12. $f(x) = \frac{2}{9}x(3 - x)$, $[0, 3]$

13. $f(x) = \frac{1}{3}e^{-x/3}$, $[0, \infty)$ **14.** $f(x) = \frac{1}{4}$, $[8, 12]$

In Exercises 15–20, find the constant k so that the function f is a probability density function over the given interval.

15. $f(x) = kx$, $[1, 4]$ **16.** $f(x) = kx^3$, $[0, 4]$

17. $f(x) = k(4 - x^2)$, $[-2, 2]$

18. $f(x) = k\sqrt{x}(1 - x)$, $[0, 1]$

19. $f(x) = ke^{-x/2}$, $[0, \infty)$ **20.** $f(x) = \dfrac{k}{b-a}$, $[a, b]$

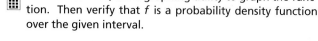 In Exercises 21–26, sketch the graph of the probability density function over the indicated interval and find the indicated probabilities.

21. $f(x) = \frac{1}{10}$, $[0, 10]$

   (a) $P(0 < x < 6)$ (b) $P(4 < x < 6)$
   (c) $P(8 < x < 10)$ (d) $P(x \geq 2)$

22. $f(x) = \dfrac{x}{50}$, $[0, 10]$

   (a) $P(0 < x < 6)$ (b) $P(4 < x < 6)$
   (c) $P(8 < x < 10)$ (d) $P(x \geq 2)$

23. $f(x) = \frac{3}{16}\sqrt{x}$, $[0, 4]$

(a) $P(0 < x < 2)$ (b) $P(2 < x < 4)$
(c) $P(1 < x < 3)$ (d) $P(x \le 3)$

24. $f(x) = \dfrac{5}{4(x+1)^2}$, $[0, 4]$

(a) $P(0 < x < 2)$ (b) $P(2 < x < 4)$
(c) $P(1 < x < 3)$ (d) $P(x \le 3)$

25. $f(t) = \frac{1}{3}e^{-t/3}$, $[0, \infty)$

(a) $P(t < 2)$ (b) $P(t \ge 2)$
(c) $P(1 < t < 4)$ (d) $P(t = 3)$

26. $f(t) = \frac{3}{256}(16 - t^2)$, $[-4, 4]$

(a) $P(t < -2)$ (b) $P(t > 2)$
(c) $P(-1 < t < 1)$ (d) $P(t > -2)$

27. *Transportation*   Buses arrive and depart from a college every 30 minutes. The probability density function for the waiting time t (in minutes) for a person arriving at the bus stop is

$$f(t) = \frac{1}{30}, \quad [0, 30].$$

Find the probability that the person will wait (a) no more than 5 minutes, and (b) at least 18 minutes.

28. *Learning Theory*   The time t (in hours) required for a new employee to successfully learn to operate a machine in a manufacturing process is described by the probability density function

$$f(t) = \frac{5}{324}t\sqrt{9 - t}, \quad [0, 9].$$

Find the probability that a new employee will learn to operate the machine (a) in less than 3 hours, and (b) in more than 4 hours but less than 8 hours.

In Exercises 29–32, find the required probabilities using the **exponential density function**

$$f(t) = \frac{1}{\lambda}e^{-t/\lambda}, \quad [0, \infty).$$

29. *Waiting Time*   The waiting time (in minutes) for service at the checkout at a grocery store is exponentially distributed with $\lambda = 3$. Find the probability of waiting (a) less than 2 minutes, (b) more than 2 minutes but less than 4 minutes, and (c) at least 2 minutes.

30. *Useful Life*   The lifetime (in years) of a battery is exponentially distributed with $\lambda = 5$. Find the probability that the lifetime of a given battery will be (a) less than 6 years, (b) more than 2 years but less than 6 years, and (c) more than 8 years.

31. *Waiting Time*   The length of time (in hours) required to unload trucks at a depot is exponentially distributed with $\lambda = \frac{3}{4}$. What proportion of the trucks can be unloaded in less than 1 hour?

32. *Useful Life*   The time (in years) until failure of a component in a machine is exponentially distributed with $\lambda = 3.5$. A manufacturer has a large number of these machines and plans to replace the components in all the machines during regularly scheduled maintenance periods. How much time should elapse between maintenance periods if at least 90% of the components are to remain working throughout the period?

33. *Demand*   The weekly demand x (in tons) for a certain product is a continuous random variable with the density function

$$f(x) = \frac{1}{36}xe^{-x/6}, \quad [0, \infty).$$

Find the following.

(a) $P(x < 6)$ (b) $P(6 < x < 12)$
(c) $P(x > 12) = 1 - P(x \le 12)$

34. *Demand*   Given the conditions of Exercise 33, determine the number of tons that should be ordered each week so that the demand can be met for 90% of the weeks.

35. *Learning Theory*   The probability density function for the percentage of recall in a certain learning experiment is found to be

$$f(x) = \frac{15}{4}x\sqrt{1 - x}, \quad [0, 1].$$

What is the probability that a randomly chosen individual in the experiment will recall (a) between 0% and 25%, and (b) between 50% and 75%, of the material?

36. *Metallurgy*   The probability density function for the percent of iron in ore samples taken from a certain region is

$$f(x) = \frac{1155}{32}x^3(1 - x)^{3/2}, \quad [0, 1].$$

Find the probability that a sample will contain (a) from 0% to 25% iron and (b) from 50% to 100% iron.

37. The probability of obtaining 49, 50, or 51 heads when a fair coin is tossed 100 times is

$$P(49 \le x \le 51) \approx \int_{48.5}^{51.5} \frac{1}{5\sqrt{2\pi}}e^{-(x-50)^2/50}\,dx.$$

Use a computer or graphing utility and Simpson's Rule (with $n = 12$) to approximate this integral.

P.3 Expected Value and Variance

Expected Value ▪ *Variance and Standard Deviation* ▪ *Median* ▪
Special Probability Density Functions

Expected Value

In Section P.1, you studied the concepts of expected value, mean, variance, and standard deviation for *discrete* random variables. In this section, you will extend these concepts to *continuous* random variables.

Definition of Expected Value

If f is a probability density function for a continuous random variable x over the interval $[a, b]$, then the **expected value** or **mean** of x is

$$\mu = E(x) = \int_a^b xf(x)\,dx.$$

EXAMPLE 1 *Finding Average Weekly Demand*

Real Life

In Example 6 in Section P.2, the weekly demand for a product was modeled by the probability density function

$$f(x) = \frac{1}{36}(-x^2 + 6x), \qquad 0 \le x \le 6.$$

Find the expected weekly demand for this product.

Solution

$$\begin{aligned}
\mu = E(x) &= \frac{1}{36}\int_0^6 x(-x^2 + 6x)\,dx \\
&= \frac{1}{36}\int_0^6 (-x^3 + 6x^2)\,dx \\
&= \frac{1}{36}\left[\frac{-x^4}{4} + 2x^3\right]_0^6 \\
&= 3
\end{aligned}$$

From Figure P.11, you can see that an expected value of 3 seems reasonable because the region is symmetric about the line $x = 3$.

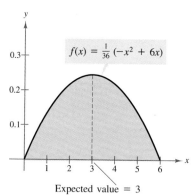

Expected value = 3

FIGURE P.11

Variance and Standard Deviation

> ### Definition of Variance and Standard Deviation
>
> If f is a probability density function for a continuous random variable x over the interval $[a, b]$, then the **variance** of x is
>
> $$V(x) = \int_a^b (x - \mu)^2 f(x)\, dx,$$
>
> where μ is the mean of x. The **standard deviation** of x is
>
> $$\sigma = \sqrt{V(x)}.$$

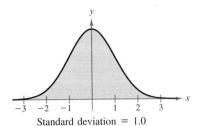

Standard deviation = 1.0

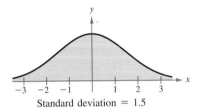

Standard deviation = 1.5

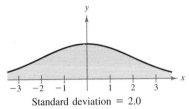

Standard deviation = 2.0

FIGURE P.12

Recall from Section P.1 that distributions that are clustered about the mean tend to have smaller standard deviations than distributions that are more dispersed. For instance, all three of the probability density distributions shown in Figure P.12 have a mean of $\mu = 0$, but they have different standard deviations. Because the first distribution is clustered more toward the mean, its standard deviation is the smallest of the three.

EXAMPLE 2 *Finding Variance and Standard Deviation*

Find the variance and standard deviation for the probability density function

$$f(x) = 2 - 2x, \qquad 0 \le x \le 1.$$

Solution
Begin by finding the mean.

$$\mu = \int_0^1 x(2 - 2x)\, dx = \frac{1}{3} \qquad\qquad \textit{Mean}$$

Next, apply the formula for variance.

$$
\begin{aligned}
V(x) &= \int_0^1 \left(x - \frac{1}{3}\right)^2 (2 - 2x)\, dx \\
&= \int_0^1 \left(-2x^3 + \frac{10x^2}{3} - \frac{14x}{9} + \frac{2}{9}\right) dx \\
&= \left[-\frac{x^4}{2} + \frac{10x^3}{9} - \frac{7x^2}{9} + \frac{2x}{9}\right]_0^1 \\
&= \frac{1}{18} \qquad\qquad\qquad \textit{Variance}
\end{aligned}
$$

Finally, you can conclude that the standard deviation is

$$\sigma = \sqrt{\frac{1}{18}} \approx 0.236. \qquad\qquad \textit{Standard deviation}$$

The integral for variance can be difficult to evaluate. The following alternative formula is often simpler.

Alternative Formula for Variance

If f is a probability density function for a continuous random variable x over the interval $[a, b]$, then the **variance** of x is

$$V(x) = \int_a^b x^2 f(x)\, dx - \mu^2,$$

where μ is the mean of x.

EXAMPLE 3 *Using the Alternative Formula*

Find the standard deviation of the probability density function

$$f(x) = \frac{2}{\pi(x^2 - 2x + 2)}, \qquad 0 \le x \le 2.$$

What percent of the distribution lies within one standard deviation of the mean?

Solution

Begin by using a symbolic integrator to find the mean.

$$\mu = \int_0^2 \left(\frac{2}{\pi(x^2 - 2x + 2)} \right)(x)\, dx$$

$$= 1 \qquad\qquad\qquad\qquad\qquad\qquad\qquad\text{Mean}$$

Next, use a symbolic integrator to find the variance.

$$V(x) = \int_0^2 \left(\frac{2}{\pi(x^2 - 2x + 2)} \right)(x^2)\, dx - 1^2$$

$$\approx 0.273 \qquad\qquad\qquad\qquad\qquad\qquad\text{Variance}$$

This implies that the standard deviation is

$$\sigma = \sqrt{0.273} \approx 0.522. \qquad\qquad\qquad\text{Standard deviation}$$

To find the percent of the distribution that lies within one standard deviation of the mean, integrate the probability density function between $\mu - \sigma = 0.478$ and $\mu + \sigma = 1.522$.

$$\int_{0.478}^{1.522} \frac{2}{\pi(x^2 - 2x + 2)}\, dx \approx 0.613$$

Thus, about 61.3% of the distribution lies within one standard deviation of the mean. This result is illustrated in Figure P.13.

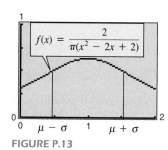

$$f(x) = \frac{2}{\pi(x^2 - 2x + 2)}$$

FIGURE P.13

Median

The mean of a probability density function is an example of a **measure of central tendency.** Another useful measure of central tendency is the **median.**

Definition of Median

If f is a probability density function for a continuous random variable x over the interval $[a, b]$, then the **median** of x is the number m such that

$$\int_{a}^{m} f(x)\, dx = 0.5.$$

Real Life

EXAMPLE 4 *Comparing Mean and Median*

In Example 5 in Section P.2, the probability density function

$$f(t) = 0.1e^{-0.1t}, \qquad 0 \le t < \infty$$

was used to model the usable lifetime of a product. Find the mean and median usable lifetime.

Solution
Using integration by parts or a symbolic integrator, you can find the mean to be

$$\mu = \int_{0}^{\infty} 0.1te^{-0.1t}\, dt$$
$$= 10 \text{ years.} \qquad\qquad \textit{Mean}$$

The median is given by

$$\int_{0}^{m} 0.1e^{-0.1t}\, dt = 0.5$$
$$-e^{-0.1m} + 1 = 0.5$$
$$e^{-0.1m} = 0.5$$
$$-0.1m = \ln 0.5$$
$$m = -10\ln 0.5$$
$$m \approx 6.93 \text{ years.} \qquad\qquad \textit{Median}$$

From this, you can see that the mean and median of a probability distribution can be quite different. Using the mean, the "average" lifetime of a product is 10 years, but using the median, the "average" lifetime is 6.93 years. In Figure P.14, note that half of the products have usable lifetimes of 6.93 years or less. ■

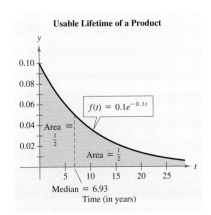

Usable Lifetime of a Product

$f(t) = 0.1e^{-0.1t}$

Area $= \frac{1}{2}$

Area $= \frac{1}{2}$

Median ≈ 6.93

Time (in years)

FIGURE P.14

Special Probability Density Functions

The remainder of this section describes three common types of probability density functions: uniform, exponential, and normal. The **uniform probability density function** is defined by

$$f(x) = \frac{1}{b - a}, \qquad a \leq x \leq b. \qquad \textit{Uniform probability density function}$$

This probability density function represents a continuous random variable for which each outcome is equally likely.

EXAMPLE 5 *Analyzing a Probability Density Function*

Find the expected value and standard deviation for the uniform probability density function

$$f(x) = \frac{1}{8}, \qquad 0 \leq x \leq 8.$$

Solution

The expected value (or mean) is

$$\mu = \int_0^8 \frac{1}{8} x \, dx$$

$$= \left[\frac{x^2}{16} \right]_0^8$$

$$= 4. \qquad \textit{Expected value}$$

The variance is

$$V(x) = \int_0^8 \frac{1}{8} x^2 \, dx - 4^2$$

$$= \left[\frac{x^3}{24} \right]_0^8 - 16$$

$$\approx 5.333. \qquad \textit{Variance}$$

The standard deviation is

$$\sigma = \sqrt{5.333}$$

$$\approx 2.309. \qquad \textit{Standard deviation}$$

The graph of f is shown in Figure P.15.

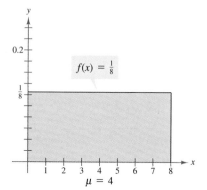

FIGURE P.15 Uniform Probability Density Function

Note   Try showing that the mean and variance of the general uniform probability density function $f(x) = 1/(b - a)$ is $\mu = \frac{1}{2}(a + b)$ and $V(x) = \frac{1}{12}(b - a)^2$.

The second special type of probability density function is an **exponential probability density function** and has the form

$$f(x) = ae^{-ax}, \qquad 0 \le x < \infty.$$

Exponential probability density function, a > 0

The probability density function in Example 4 is of this type. Try showing that this function has a mean of $1/a$ and a variance of $1/a^2$.

The third special type of probability density function (and the most widely used) is the **normal probability density function** given by

$$f(x) = \frac{1}{\sigma\sqrt{2\pi}}e^{-(x-\mu)^2/2\sigma^2}, \qquad -\infty < x < \infty.$$

Normal probability density function

The expected value for this function is μ and the standard deviation is σ. Figure P.16 shows the graph of a typical normal probability density function.

A normal probability density function for which $\mu = 0$ and $\sigma = 1$ is called a **standard normal probability density function.**

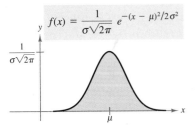

$$y \quad f(x) = \frac{1}{\sigma\sqrt{2\pi}}e^{-(x-\mu)^2/2\sigma^2}$$

FIGURE P.16   Normal Probability Density Function

The SAT and GMAT standardized tests are published by the College Entrance Examination Board. On each test, scores vary between 200 and 800. The mean tends to be around 500 and the standard deviation tends to be around 100.

Real Life

EXAMPLE 6 *Finding a Probability*

In 1990, the scores for the *Graduate Management Admission Test* (GMAT) could be modeled by a normal probability density function with a mean of $\mu = 490$ and a standard deviation of $\sigma = 110$. If you select a person who took the GMAT in 1990, what is the probability that the person scored between 600 and 700? What is the probability that the person scored between 700 and 800?

Solution

Using a calculator or computer, you can find the first probability to be

$$P(600 \le x \le 700) = \int_{600}^{700} \frac{1}{110\sqrt{2\pi}}e^{-(x-490)^2/2(110)^2}\,dx$$

$$\approx 0.13.$$

Thus, the probability of choosing a person who scored between 600 and 700 is about 13%. In a similar way, you can find the probability of choosing a person who scored between 700 and 800 to be

$$P(700 \le x \le 800) = \int_{700}^{800} \frac{1}{110\sqrt{2\pi}}e^{-(x-490)^2/2(110)^2}\,dx$$

$$\approx 0.026,$$

or about 2.6%.

Note In Example 6, note that probabilities connected with normal distributions must be evaluated with a table of values or with a symbolic integration utility.

TECHNOLOGY   Using the Normal Probability Density Function

The *standard* normal probability density function is

$$f(x) = \frac{1}{\sqrt{2\pi}} e^{-x^2/2}.$$

For this function, the mean is $\mu = 0$ and the standard deviation is $\sigma = 1$. This function does not have an antiderivative that is an elementary function. It does, however, have an antiderivative, which is denoted by $\frac{1}{2}\text{ERF}(x/\sqrt{2})$. Symbolic integration utilities such as *Derive*, *Mathematica*, and *Maple* use built-in programs that can evaluate this function. Here are some examples.

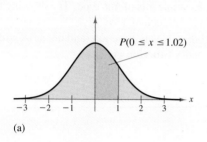

(a)

a. The probability that a random value is between 0 and 1.02 is

$$P(0 \le x \le 1.02) = \int_0^{1.02} \frac{1}{\sqrt{2\pi}} e^{-x^2/2}\, dx$$

$$= \frac{1}{2}\text{ERF}\left(\frac{1.02}{\sqrt{2}}\right) - \frac{1}{2}\text{ERF}(0)$$

$$\approx 0.3461.$$

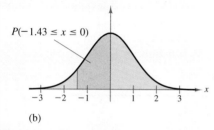

(b)

b. The probability that a random value is between -1.43 and 0 is

$$P(-1.43 \le x \le 0) = \int_{-1.43}^{0} \frac{1}{\sqrt{2\pi}} e^{-x^2/2}\, dx$$

$$= \frac{1}{2}\text{ERF}(0) - \frac{1}{2}\text{ERF}\left(\frac{-1.43}{\sqrt{2}}\right)$$

$$\approx 0.4236.$$

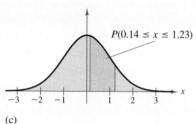

(c)

c. The probability that a random value is between 0.14 and 1.23 is

$$P(0.14 \le x \le 1.23) = \int_{0.14}^{1.23} \frac{1}{\sqrt{2\pi}} e^{-x^2/2}\, dx$$

$$= \frac{1}{2}\text{ERF}\left(\frac{1.23}{\sqrt{2}}\right) - \frac{1}{2}\text{ERF}\left(\frac{0.14}{\sqrt{2}}\right)$$

$$\approx 0.3350.$$

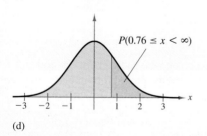

(d)

d. The probability that a random value is greater than or equal to 0.76 is

$$P(0.76 \le x < \infty) = \int_{0.76}^{\infty} \frac{1}{\sqrt{2\pi}} e^{-x^2/2}\, dx$$

$$= \frac{1}{2} - \frac{1}{2}\text{ERF}\left(\frac{0.76}{\sqrt{2}}\right)$$

$$\approx 0.2236.$$

Graphical representations of these four probabilities are shown in the figure at left.

Allstock/© G. Robert Bishop/HMS Group

In addition to being popular zoo animals, rhesus monkeys are commonly used in medical and behavioral research. Research on rhesus monkeys led to the discovery of the *Rh factor* in human red blood cells.

EXAMPLE 7 *Modeling Body Weight*

The weight of adult male rhesus monkeys is normally distributed with a mean of 15 pounds and a standard deviation of 3 pounds. In a typical population of adult male rhesus monkeys, what percent of the monkeys would have a weight that is within one standard deviation of the mean?

Solution

For this population, the normal probability density function is

$$f(x) = \frac{1}{3\sqrt{2\pi}} e^{-(x-15)^2/18}.$$

The probability that a randomly chosen adult male monkey will weigh between 12 and 18 pounds (that is, within 3 pounds of 15 pounds) is

$$P(12 \le x \le 18) = \int_{12}^{18} \frac{1}{3\sqrt{2\pi}} e^{-(x-15)^2/18} \, dx$$

$$\approx 0.683.$$

Thus, about 68% of the adult male monkeys have weights that lie within one standard deviation of the mean, as shown in Figure P.17.

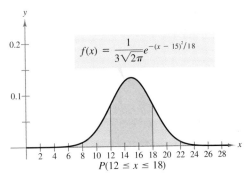

$$f(x) = \frac{1}{3\sqrt{2\pi}} e^{-(x-15)^2/18}$$

$P(12 \le x \le 18)$

FIGURE P.17

The result that is described in Example 7 can be generalized to all normal distributions. That is, in any normal distribution, the probability that x lies within one standard deviation of the mean is about 68%. For normal distributions, 95.4% of the x-values lie within two standard deviations of the mean, and almost all (99.7%) of the x-values lie within three standard deviations of the mean.

Discussion Problem   **Exploring Normal Distributions**

In a normal distribution, what percent of the x-values lie within 1.5 standard deviations of the mean? What percent lie within 0.5 standard deviation of the mean?

Warm Up   The following warm-up exercises involve skills that were covered in earlier sections. You will use these skills in the exercise set for this section.

In Exercises 1–4, solve for m.

1. $\displaystyle\int_0^m \frac{1}{10}\,dx = 0.5$ **2.** $\displaystyle\int_0^m \frac{1}{16}\,dx = 0.5$

3. $\displaystyle\int_0^m \frac{1}{3}e^{-t/3}\,dt = 0.5$ **4.** $\displaystyle\int_0^m \frac{1}{9}e^{-t/9}\,dt = 0.5$

In Exercises 5–8, evaluate the definite integral.

5. $\displaystyle\int_0^2 \frac{x^2}{2}\,dx$ **6.** $\displaystyle\int_1^2 x(4-2x)\,dx$

7. $\displaystyle\int_2^5 x^2\left(\frac{1}{3}\right)dx - \left(\frac{7}{2}\right)^2$ **8.** $\displaystyle\int_2^4 x^2\left(\frac{4-x}{2}\right)dx - \left(\frac{8}{3}\right)^2$

In Exercises 9 and 10, find the indicated probability using the given probability density function.

9. $f(x) = \frac{1}{8}$, $[0, 8]$; (a) $P(x \le 2)$, (b) $P(3 < x < 7)$

10. $f(x) = 6x - 6x^2$, $[0, 1]$; (a) $P\left(x \le \frac{1}{2}\right)$, (b) $P\left(\frac{1}{4} \le x \le \frac{3}{4}\right)$

EXERCISES P.3   means that technology can help you solve or check the exercise(s).

In Exercises 1–10, use the given probability density function over the indicated interval to find (a) the mean, (b) the variance, and (c) the standard deviation of the random variable. Sketch the graph of the density function, and then locate the mean on the graph.

1. $f(x) = \frac{1}{8}$, $[0, 8]$ **2.** $f(x) = \frac{1}{4}$, $[0, 4]$

3. $f(t) = \dfrac{t}{18}$, $[0, 6]$ **4.** $f(x) = \dfrac{4}{3x^2}$, $[1, 4]$

5. $f(x) = 6x(1-x)$, $[0, 1]$

6. $f(x) = \frac{3}{32}x(4-x)$, $[0, 4]$

7. $f(x) = \frac{5}{2}x^{3/2}$, $[0, 1]$ **8.** $f(x) = \frac{3}{16}\sqrt{4-x}$, $[0, 4]$

9. $f(x) = \dfrac{4}{3(x+1)^2}$, $[0, 3]$ **10.** $f(x) = \dfrac{1}{18}\sqrt{9-x}$, $[0, 9]$

In Exercises 11 and 12, find the median of the exponential probability density function.

11. $f(t) = \frac{1}{9}e^{-t/9}$, $[0, \infty)$ **12.** $f(t) = \frac{2}{5}e^{-2t/5}$, $[0, \infty)$

In Exercises 13–18, identify the probability density function. Then find the mean, variance, and standard deviation without integrating.

13. $f(x) = \frac{1}{10}$, $[0, 10]$ **14.** $f(x) = \frac{1}{9}$, $[0, 9]$

15. $f(x) = \frac{1}{8}e^{-x/8}$, $[0, \infty)$ **16.** $f(x) = \frac{5}{3}e^{-5x/3}$, $[0, \infty)$

17. $f(x) = \dfrac{1}{11\sqrt{2\pi}}e^{-(x-100)^2/242}$, $(-\infty, \infty)$

18. $f(x) = \dfrac{1}{6\sqrt{2\pi}}e^{-(x-30)^2/72}$, $(-\infty, \infty)$

In Exercises 19–24, use a symbolic differentiation utility to find the mean, standard deviation, and indicated probability.

Function	*Probability*
19. $f(x) = \dfrac{1}{\sqrt{2\pi}} e^{-x^2/2}$	$P(0 \le x \le 0.85)$
20. $f(x) = \dfrac{1}{\sqrt{2\pi}} e^{-x^2/2}$	$P(-1.21 \le x \le 1.21)$
21. $f(x) = \frac{1}{6} e^{-x/6}$	$P(x \ge 2.23)$
22. $f(x) = \frac{3}{4} e^{-3x/4}$	$P(x \ge 0.27)$
23. $f(x) = \dfrac{1}{2\sqrt{2\pi}} e^{-(x-8)^2/8}$	$P(3 \le x \le 13)$
24. $f(x) = \dfrac{1}{1.5\sqrt{2\pi}} e^{-(x-2)^2/4.5}$	$P(-2.5 \le x \le 2.5)$

 25. Let x be a random variable that is normally distributed with a mean of 60 and a standard deviation of 12. Find the required probabilities using a symbolic integration utility.
(a) $P(x > 64)$ (b) $P(x > 70)$
(c) $P(x < 70)$ (d) $P(33 < x < 65)$

 26. Let x be a random variable that is normally distributed with a mean of 75 and a standard deviation of 12. Find the required probabilities using a symbolic integration utility.
(a) $P(x > 69)$ (b) $P(x < 99)$
(c) $P(x < 57)$ (d) $P(63 < x < 78)$

27. *Arrival Time* The arrival time t of a bus at a bus stop is uniformly distributed between the times of 10:00 A.M. and 10:10 A.M.
(a) Find the mean and standard deviation of the random variable t.
(b) What is the probability that you will miss the bus if you arrive at the bus stop at 10:03 A.M.?

28. *Arrival Time* Repeat Exercise 27 for a bus that arrives between the times of 10:00 A.M. and 10:05 A.M.

29. *Useful Life* The time t until failure of an appliance is exponentially distributed with a mean of 2 years.
(a) Find the probability density function for the random variable t.
(b) Find the probability that the appliance will fail in less than 1 year.

30. *Arrival Time* The time t between arrivals of customers at a gasoline station is exponentially distributed with a mean of 2 minutes.
(a) Find the probability density function for the random variable t.
(b) If a customer has just arrived, determine the probability that it will be more than 3 minutes before the next arrival.

31. *Waiting Time* The waiting time t for service at a customer service desk in a department store is exponentially distributed with a mean of 5 minutes.
(a) Find the probability density function for the random variable t.
(b) Find the probability that t is within one standard deviation of the mean.

32. *Service Time* The service time t for a customer at the service desk in a department store is exponentially distributed with a mean of 3.5 minutes.
(a) Find the probability density function for the random variable t.
(b) Find the probability that t is within one standard deviation of the mean.

33. *Standardized Test Scores* The scores on a national exam are normally distributed with a mean of 150 and a standard deviation of 16. You scored 174 on the exam.
(a) How far, in standard deviations, did your score exceed the national mean?
(b) What percent of those who took the exam had scores lower than yours?

34. *Useful Life* The length of life of a battery is normally distributed with a mean of 400 hours and a standard deviation of 24 hours. You purchased one of the batteries and its useful life was 340 hours.
(a) How far, in standard deviations, did the useful life of your battery fall short of the expected life?
(b) What percent of all other batteries of this type have useful lives that exceed yours?

35. *Demand* The daily demand x for a certain product (in hundreds of pounds) is a random variable with the probability density function
$f(x) = \frac{1}{36} x(6 - x), \;\; [0, 6].$
(a) Determine the expected value and the standard deviation of demand.
(b) Determine the median of the random variable.
(c) Find the probability that x is within one standard deviation of the mean.

36. *Demand* Repeat Exercise 35 for a probability density function of
$f(x) = \frac{3}{256} (x - 2)(10 - x), \;\; [2, 10].$

37. *Learning Theory* The percent recall x in a learning experiment is a random variable with the probability density function
$f(x) = \frac{15}{4} x \sqrt{1 - x}, \;\; [0, 1].$
Determine the mean and variance of random variable x.

38. *Metallurgy* The percent of iron x in samples of ore is a random variable with the probability density function

$$f(x) = \frac{1155}{32}x^3(1 - x)^{3/2}, \quad [0, 1].$$

Determine the expected percent of iron in each ore sample.

39. *Demand* The daily demand x for a certain product (in thousands of units) is a random variable with the probability density function

$$f(x) = \frac{1}{25}xe^{-x/5}, \quad [0, \infty).$$

(a) Determine the expected daily demand.
(b) Find $P(x \leq 4)$.

40. *Recovery Time* The time t (in days) until recovery after a certain medical procedure is a random variable with the probability density function

$$f(t) = \frac{1}{2\sqrt{t-2}}, \quad [3, 6].$$

(a) Find the probability that a patient selected at random will take more than 4 days to recover.
(b) Determine the expected time for recovery.

In Exercises 41–46, find the mean and median.

41. $f(x) = \frac{1}{11}$ $[0, 11]$

42. $f(x) = 0.04$ $[0, 25]$

43. $f(x) = 4(1 - 2x)$ $\left[0, \frac{1}{2}\right]$

44. $f(x) = \frac{4}{3} - \frac{2}{3}x$ $[0, 3]$

45. $f(x) = \frac{1}{5}e^{-x/5}$ $[0, \infty)$

46. $f(x) = \frac{2}{3}e^{-2x/3}$ $[0, \infty)$

47. *Cost of Electricity* The daily cost of electricity x in a city is a random variable with the probability density function

$$f(x) = 0.28e^{-0.28x}, \quad 0 \leq x < \infty.$$

Find the median daily cost of electricity.

48. *Coupon Sales* The number of coupons used by a customer in a grocery store is a random variable with the probability density function

$$f(x) = \frac{2x + 1}{10}, \quad 1 \leq x \leq 2.$$

Find the expected number of coupons a customer will use.

49. *Demand* The daily demand x for water (in millions of gallons) in a town is a random variable with the probability density function

$$f(x) = \frac{1}{9}xe^{-x/3}, \quad [0, \infty).$$

(a) Determine the expected value and the standard deviation of the demand.
(b) Find the probability that the demand is greater than 4 million gallons on a given day.

50. *Useful Life* The length of life of an electrical component is normally distributed with a mean of 5.3 years and a standard deviation of 0.8 year. How long should this component be guaranteed if the producer does not want to replace any more than 10% of the components during the time covered by the guarantee?

51. *Machine Precision* An automatic filling machine fills cans so that the weights are normally distributed with a mean of μ and a standard deviation of σ. The value of μ can be controlled by settings on the machine, but σ depends on the precision and design of the machine. For a particular substance, $\sigma = 0.15$ ounce. If 12-ounce cans are being filled, determine the setting for μ so that no more than 5% of the cans weigh less than the stated weight.

52. *Useful Life* A storage battery has an expected lifetime of 4.5 years with a standard deviation of 0.5 year. Assume that the useful lives of these batteries are normally distributed. Use a computer or calculator and Simpson's Rule (with $n = 12$) to approximate the probability that a given battery will last for 4 to 5 years.

53. *Wages* The employees of a large corporation are paid an average wage of $12.30 per hour with a standard deviation of $1.50. Assume that these wages are normally distributed. Use a computer or calculator and Simpson's Rule (with $n = 10$) to approximate the percent of employees that earn hourly wages of $9.00 to $12.00.

54. *Profit* The weekly profit x (in thousands of dollars) of a store is normally distributed with $\mu = 8$ and $\sigma = 3$. Use a symbolic integration utility to find $P(8 < x < 11)$.

55. *Intelligence Quotient* The IQs of students in a school are normally distributed with a mean of 110 and a standard deviation of 10. Use a symbolic integration utility to find the probability that a student selected at random will have an IQ within one standard deviation of the mean.

Chapter Summary and Study Tips

After studying this chapter, you should have acquired the following skills. The exercise numbers are keyed to the Review Exercises that begin on page P–30. Answers to odd-numbered Review Exercises are given in the back of the text.

■ Describe sample spaces for experiments. (*Section P.1*)   Review Exercises 1–4

■ Assign values to discrete random variables. (*Section P.1*)   Review Exercises 5, 6

■ Form frequency distributions for discrete random variables. (*Section P.1*)   Review Exercises 7, 8

■ Find the probabilities of events for discrete random variables. (*Section P.1*)   Review Exercises 9–12

$$P(x) = \frac{\text{Frequency of } x}{\text{Number of outcomes in } S} = \frac{n(x)}{n(S)}$$

■ Find the expected values or means of discrete random variables. (*Section P.1*)   Review Exercises 13–16

$$\mu = E(x) = x_1 P(x_1) + x_2 P(x_2) + x_3 P(x_3) + \cdots + x_m P(x_m).$$

■ Find the variances and standard deviations of discrete random variables. (*Section P.1*)   Review Exercises 17–20

$$V(x) = (x_1 - \mu)^2 P(x_1) + \cdots + (x_m - \mu)^2 P(x_m), \qquad \sigma = \sqrt{V(x)}$$

■ Verify continuous probability density functions. (*Section P.2*)   Review Exercises 21–24

■ Use continuous probability density functions to find probabilities. (*Section P.2*)   Review Exercises 25–28

$$P(c \leq x \leq d) = \int_c^d f(x)\, dx$$

■ Use continuous probability density functions to answer questions about real life. (*Section P.2*)   Review Exercises 29–32

■ Find the means of continuous probability density functions. (*Section P.3*)   Review Exercises 33–36

$$\mu = E(x) = \int_a^b x f(x)\, dx$$

■ Find the variances and standard deviations of continuous probability density functions. (*Section P.3*)   Review Exercises 37–40

$$V(x) = \int_a^b (x - \mu)^2 f(x)\, dx, \qquad \sigma = \sqrt{V(x)}$$

■ Find the medians of continuous probability density functions. (*Section P.3*)   Review Exercises 41–44

■ Use special probability density functions to answer questions about real life. (*Section P.3*)   Review Exercises 45–50

■ Using Technology   Integrals that arise with continuous probability density functions tend to be difficult to evaluate by hand. When evaluating such integrals, we suggest that you use a symbolic integration utility such as *Derive*, *Mathematica*, or *Maple*, or that you use a numerical integration technique such as Simpson's Rule with a programmable calculator.

*Several student study aids are available with this text. The *Student Solutions Guide* includes detailed solutions to all odd-numbered exercises, as well as practice chapter tests with answers. The *Graphics Calculator Guide* offers instructions on the use of a variety of graphing calculators and computer graphing software. The *Brief Calculus TUTOR* includes additional examples for selected exercises in the text.

Review Exercises

 means that technology can help you solve or check the exercise(s).

In Exercises 1–4, describe the sample space of the experiment.

1. A month of the year is chosen for a vacation.

2. A letter from the word *calculus* is selected.

3. A student must answer three questions from a selection of four essay questions.

4. A winner in a game show must choose two out of five prizes.

5. Three numbers are drawn in a lottery. Each number is a digit from 0 to 9. Find the sample space giving the number of 7's drawn.

6. As cans of soft drink are filled on the production line, four are randomly selected and labeled with an "S" if the weight is satisfactory and a "U" if the weight is unsatisfactory. Find the sample space giving the satisfactory/unsatisfactory classification of each can.

In Exercises 7 and 8, complete the table to form the frequency distribution of the random variable x. Then construct a bar graph to represent the result.

7. A computer randomly selects a three-digit bar code. Each digit can be 0 or 1, and x is the number of 1's in the bar code.

x	0	1	2	3
$n(x)$				

8. A cat has a litter of four kittens. Let x represent the number of male kittens.

x	0	1	2	3	4
$n(x)$					

In Exercises 9 and 10, sketch a graph of the given probability distribution and find the required probabilities.

9.

x	1	2	3	4	5
$P(x)$	$\frac{1}{18}$	$\frac{7}{18}$	$\frac{5}{18}$	$\frac{3}{18}$	$\frac{2}{18}$

(a) $P(2 \leq x \leq 4)$ (b) $P(x \geq 3)$

10.

x	-2	-1	1	3	5
$P(x)$	$\frac{1}{11}$	$\frac{2}{11}$	$\frac{4}{11}$	$\frac{3}{11}$	$\frac{1}{11}$

(a) $P(x < 0)$ (b) $P(x > 1)$

11. *Tossing Dice* Consider an experiment in which two six-sided dice are tossed. Find the indicated probability.

(a) The probability that the total is 8.
(b) The probability that the total is greater than 4.
(c) The probability that doubles are thrown.
(d) The probability of getting double sixes.

12. *Selecting a Card* Consider an experiment in which one card is randomly selected from a standard deck of 52 playing cards. Find the probability of

(a) selecting a face card.
(b) selecting a card that is not a face card.
(c) selecting a black card that is not a face card.
(d) selecting a card whose value is 6 or less.

13. *Quiz Scores* An instructor gave a 25-point quiz to 52 students. Use the following frequency distribution of quiz scores to find the mean quiz score.

Score	9	10	11	12	13	14	15	16	17
Frequency	1	0	1	0	0	0	3	4	7

Score	18	19	20	21	22	23	24	25
Frequency	3	0	9	11	6	3	0	4

14. *Cost Increases* A small business uses three different materials, A, B, and C, to create a particular product. The following table gives the cost and the percent increase of the cost of each of the three materials. Find the mean percent increase of the three products.

Material	Percent Increase	Cost of Supplies
A	8%	$650
B	23%	$375
C	16%	$800

15. *Revenue* A publishing company introduces a new weekly newspaper that sells for 75 cents. The marketing group of the company estimates that sales x (in thousands) will be approximated by the following probability function.

x	10	15	20	30	40
$P(x)$	0.10	0.20	0.50	0.15	0.05

(a) Find $E(x)$.   (b) Find the expected revenue.

16. *Game of Chance* A service organization is selling $3 raffle tickets as part of a fund-raising program. The first and second prizes are $2000 and $1000, respectively. In addition to the first and second prizes, there are 50 $20 gift certificates to be awarded. The number of tickets sold is 2000. Find the expected net gain to the player when one ticket is purchased.

17. *Sales Volume* A company sells five different models of personal computers. During one month the sales for the five models were as follows.

    Model 1   24 sold at $1200 each
    Model 2   12 sold at $1500 each
    Model 3   35 sold at $2000 each
    Model 4   5 sold at $2200 each
    Model 5   4 sold at $3000 each

    Find the variance and standard deviation of the prices.

18. *Textbook Prices* A school buys multiple copies of five different mathematics textbooks. The quantities and prices per book are as follows.

    Geometry   12 copies at $54 each
    Algebra   45 copies at $45 each
    Precalculus   25 copies at $52 each
    Calculus   20 copies at $60 each
    Statistics   15 copies at $65 each

    Find the variance and standard deviation of the prices.

19. *Car Ownership* A random survey of households recorded the number of cars per household. The results of the survey are shown in the table.

x	0	1	2	3	4	5
$P(x)$	0.10	0.28	0.39	0.17	0.04	0.02

    Find the variance and standard deviation of x.

20. *Number of Children* The probability distribution for the number of children in a sample of families is given in the following table.

x	0	1	2	3	4
$P(x)$	0.12	0.31	0.43	0.12	0.02

    Find the variance and standard deviation of x.

In Exercises 21–24, use a graphing utility to graph the function. Then verify that f is a probability density function.

21. $f(x) = \frac{1}{8}(4 - x)$, $[0, 4]$

22. $f(x) = \frac{3}{4}x^2(2 - x)$, $[0, 2]$

23. $f(x) = \dfrac{1}{4\sqrt{x}}$, $[1, 9]$

24. $f(x) = 8.75x^{3/2}(1 - x)$, $[0, 1]$

In Exercises 25–28, find the indicated probability for the probability density function.

25. $f(x) = \frac{1}{50}(10 - x)$, $[0, 10]$, $P(0 < x < 2)$

26. $f(x) = \frac{1}{36}(9 - x^2)$, $[-3, 3]$, $P(-1 < x < 2)$

27. $f(x) = \dfrac{2}{(x + 1)^2}$, $[0, 1]$, $P\left(0 < x < \dfrac{1}{2}\right)$

28. $f(x) = \dfrac{3}{128}\sqrt{x}$, $[0, 16]$, $P(4 < x < 9)$

29. *Transportation* Buses arrive and depart from a college every 20 minutes. The probability density function for the waiting time t (in minutes) for a person arriving at the bus stop is

    $$f(t) = \frac{1}{20}, \ [0, 20].$$

    Find the probability that the person will wait (a) no more than 10 minutes, and (b) at least 15 minutes.

30. *Learning Theory* The time t (in hours) required for a new employee to successfully learn to operate a machine in a manufacturing process is described by the probability density function

    $$f(t) = \frac{5}{324}t\sqrt{9 - t}, \ [0, 9].$$

    Find the probability that a new employee will learn to operate the machine in less than 4 hours.

31. *Recovery Time* The time t (in days) until recovery after a certain medical procedure is a random variable with the probability density function

    $$f(t) = \frac{1}{4\sqrt{t - 4}}, \ [5, 13].$$

    Find the probability that a patient selected at random will take more than 8 days to recover.

32. *Demand* The monthly demand x (in kilograms) for a certain product is a random variable with density function

$$f(x) = \tfrac{1}{64}xe^{-x/8}.$$

Find the probability that the demand is at least 9 kilograms for a randomly selected month.

In Exercises 33–36, find the mean of the continuous probability density function.

33. $f(x) = \dfrac{1}{5}$, $[0, 5]$

34. $f(x) = \dfrac{8 - x}{32}$, $[0, 8]$

35. $f(x) = \tfrac{1}{6}e^{-x/6}$, $[0, \infty)$

36. $f(x) = 0.3e^{-0.3x}$, $[0, \infty)$

In Exercises 37–40, find the variance and standard deviation of the continuous probability density function.

37. $f(x) = \tfrac{2}{9}x(3 - x)$, $[0, 3]$

38. $f(x) = \tfrac{3}{16}\sqrt{x}$, $[0, 4]$

39. $f(x) = \tfrac{1}{2}e^{-x/2}$, $[0, \infty)$

40. $f(x) = 0.8e^{-0.8x}$, $[0, \infty)$

In Exercises 41–44, find the median of the continuous probability density function.

41. $f(x) = 6x(1 - x)$, $[0, 1]$

42. $f(x) = 12x^2(1 - x)$, $[0, 1]$

43. $f(x) = 0.25e^{-x/4}$, $[0, \infty)$

44. $f(x) = \tfrac{5}{6}e^{-5x/6}$, $[0, \infty)$

45. *Waiting Time* The waiting time t (in minutes) for service at the checkout at a grocery store is exponentially distributed with the probability density function

$$f(t) = \tfrac{1}{3}e^{-t/3}, \quad [0, \infty).$$

Find the probability of waiting (a) less than 2 minutes, and (b) more than 2 minutes but less than 4 minutes.

46. *Useful Life* The lifetime t (in hours) of a mechanical unit is exponentially distributed with the following density function.

$$f(t) = \tfrac{1}{350}e^{-t/350}, \quad [0, \infty).$$

Find the probability that a given unit chosen at random will perform satisfactorily for more than 400 hours.

47. *Botany* In a botany experiment, plants are grown in a nutrient solution. The heights of the plants are found to be normally distributed with a mean of 42 centimeters and a standard deviation of 3 centimeters. Find the probability that a plant in the experiment is at least 50 centimeters tall.

48. *Wages* The distribution of hourly wages for the workers in a certain company is normally distributed with a mean of $14.50 and a standard deviation of $1.40. (a) What percent of the workers receive an hourly wage between $13 and $15, inclusive? (b) The highest 10% of the hourly wages is greater than what amount?

49. *Meteorology* The monthly rainfall x in a certain state is normally distributed with a mean of 3.75 inches and a standard deviation of 0.5 inch. Use Simpson's Rule (with $n = 12$) to approximate the probability that in a randomly selected month the rainfall is between 3.5 and 4 inches.

50. *Demand* The weekly demand x (in tons) for a certain product is a continuous random variable with the density function

$$f(x) = \tfrac{1}{25}xe^{-x/5}, \quad [0, \infty).$$

Use a symbolic integration utility to find (a) $P(x < 5)$, (b) $P(5 < x < 10)$, and (c) $P(x > 10) = 1 - P(x \le 10)$.

Appendix A

Alternate Introduction to the Fundamental Theorem of Calculus

In this appendix we use a summation process to provide an alternate development of the definite integral. It is intended that this supplement follow Section 5.3 in the text. If used, this appendix should replace the material preceding Example 2 in Section 5.4.

Area of a Region in the Plane

We begin by showing how the area of a region in the plane can be approximated by the use of rectangles.

EXAMPLE 1 Using Rectangles to Approximate the Area of a Region

Use the four rectangles indicated in Figure A.1 to approximate the area of the region lying between the graph of

$$f(x) = \frac{x^2}{2}$$

and the x-axis, between $x = 0$ and $x = 4$.

SOLUTION

We can find the heights of the rectangles by evaluating the function f at each of the midpoints of the subintervals

$$[0, 1], \quad [1, 2], \quad [2, 3], \quad [3, 4].$$

Since the width of each rectangle is 1, the sum of the areas of the four rectangles is

$$S = \overset{\text{width}}{\overbrace{(1)}} \overset{\text{height}}{\overbrace{f\left(\frac{1}{2}\right)}} + \overset{\text{width}}{\overbrace{(1)}} \overset{\text{height}}{\overbrace{f\left(\frac{3}{2}\right)}} + \overset{\text{width}}{\overbrace{(1)}} \overset{\text{height}}{\overbrace{f\left(\frac{5}{2}\right)}} + \overset{\text{width}}{\overbrace{(1)}} \overset{\text{height}}{\overbrace{f\left(\frac{7}{2}\right)}}$$

$$= \frac{1}{8} + \frac{9}{8} + \frac{25}{8} + \frac{49}{8}$$

$$= \frac{84}{8} = 10.5.$$

Thus, we approximate the area of the given region to be 10.5.

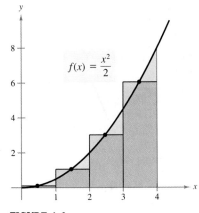

FIGURE A.1

$f(x) = \dfrac{x^2}{2}$

REMARK   The approximation technique used in Example 1 is called the **Midpoint Rule.** We will say more about the Midpoint Rule in Section 5.6.

We can generalize the procedure illustrated in Example 1. Let f be a continuous function defined on the closed interval $[a, b]$. To begin, we partition the interval into n subintervals, each of width $\Delta x = (b - a)/n$, as follows.

$$a = x_0 < x_1 < x_2 < \cdots < x_{n-1} < x_n = b$$

In each subinterval $[x_{i-1}, x_i]$ we choose an arbitrary point c_i and form the sum

$$S = f(c_1)\Delta x + f(c_2)\Delta x + \cdots + f(c_{n-1})\Delta x + f(c_n)\Delta x.$$

This type of summation is called a **Riemann sum,** and is often written using summation notation as follows.

$$S = \sum_{i=1}^{n} f(c_i)\Delta x$$

For the Riemann sum in Example 1, the interval is $[a, b] = [0, 4]$, the number of subintervals is $n = 4$, the width of each interval is $\Delta x = 1$, and the point c_i in each subinterval is its midpoint. Thus, we can write the approximation in Example 1 as

$$S = \sum_{i=1}^{n} f(c_i)\Delta x = \sum_{i=1}^{4} f(c_i)(1) = \frac{1}{8} + \frac{9}{8} + \frac{25}{8} + \frac{49}{8} = \frac{84}{8}.$$

EXAMPLE 2   Using a Riemann Sum to Approximate Area

Use a Riemann sum to approximate the area of the region bounded by the graph of $f(x) = -x^2 + 2x$ and the x-axis, for $0 \le x \le 2$. In the Riemann sum, let $n = 6$ and choose c_i to be the left endpoint of each subinterval.

SOLUTION

We subdivide the interval $[0, 2]$ into six subintervals, each of width

$$\Delta x = \frac{2 - 0}{6} = \frac{1}{3},$$

as shown in Figure A.2. Since c_i is the left endpoint of each subinterval, the Riemann sum is given by

$$S = \sum_{i=1}^{n} f(c_i)\Delta x$$

$$= \left[f(0) + f\left(\frac{1}{3}\right) + f\left(\frac{2}{3}\right) + f(1) + f\left(\frac{4}{3}\right) + f\left(\frac{5}{3}\right) \right]\left(\frac{1}{3}\right)$$

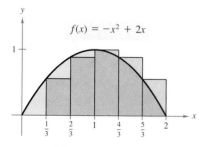

$f(x) = -x^2 + 2x$

FIGURE A.2

$$= \left[0 + \frac{5}{9} + \frac{8}{9} + 1 + \frac{8}{9} + \frac{5}{9} \right] \left(\frac{1}{3} \right)$$

$$= \frac{35}{27}.$$

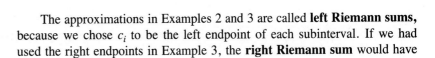

Example 2 illustrates an important point. If a function f is continuous and nonnegative over the interval $[a, b]$, then the Riemann sum

$$S = \sum_{i=1}^{n} f(c_i) \Delta x$$

can be used to approximate the area of the region bounded by the graph of f and the x-axis, between $x = a$ and $x = b$. Moreover, for a given interval, as the number of subintervals increases, the approximation to the actual area will improve. We illustrate this in the next two examples by using Riemann sums to approximate the area of a triangle.

EXAMPLE 3   Approximating the Area of a Triangle

Use a Riemann sum to approximate the area of the triangular region bounded by the graph of $f(x) = 2x$ and the x-axis, $0 \leq x \leq 3$. Use a partition of six subintervals and choose c_i to be the left endpoint of each subinterval.

SOLUTION

We subdivide the interval $[0, 3]$ into six subintervals, each of width

$$\Delta x = \frac{3 - 0}{6} = \frac{1}{2},$$

as shown in Figure A.3. Since c_i is the left endpoint of each subinterval, the Riemann sum is given by

$$S = \sum_{i=1}^{n} f(c_i) \Delta x$$

$$= \left[f(0) + f\left(\frac{1}{2} \right) + f(1) + f\left(\frac{3}{2} \right) + f(2) + f\left(\frac{5}{2} \right) \right] \left(\frac{1}{2} \right)$$

$$= [0 + 1 + 2 + 3 + 4 + 5]\left(\frac{1}{2} \right)$$

$$= \frac{15}{2}.$$

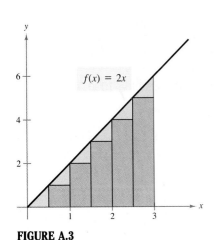

FIGURE A.3

The approximations in Examples 2 and 3 are called **left Riemann sums,** because we chose c_i to be the left endpoint of each subinterval. If we had used the right endpoints in Example 3, the **right Riemann sum** would have been $\frac{21}{2}$. (Try verifying this.) Note that the exact area of the triangular region in Example 3 is

$$\text{Area} = \frac{1}{2}(\text{base})(\text{height}) = \frac{1}{2}(3)(6) = 9.$$

Thus, the left Riemann sum gives us an approximation that is less than the actual area, and the right Riemann sum gives us an approximation that is greater than the actual area.

In Example 4, we will show that the approximation improves as the number of subintervals increases.

EXAMPLE 4   Increasing the Number of Subintervals

Let $f(x) = 2x$ for $0 \le x \le 3$. Use a computer to determine the left and right Riemann sums for $n = 10$, $n = 100$, and $n = 1000$ subintervals.

SOLUTION

A basic computer program for this problem is as follows.

```
 10   INPUT; N
 20   DELTA=3/N
 30   LSUM=0: RSUM=0
 40   FOR I=1 TO N
 50   LC=(I-1)*DELTA: RC=I*DELTA
 60   LSUM=LSUM+2*LC*DELTA: RSUM=RSUM+2*RC*DELTA
 70   NEXT
 80   PRINT "LEFT RIEMANN SUM: "; LSUM
 90   PRINT "RIGHT RIEMANN SUM: "; RSUM
100   END
```

After running this program for $n = 10$, $n = 100$, and $n = 1000$, we obtained the results shown in Table A.1.

TABLE A.1

n	Left Riemann Sum	Right Riemann Sum
10	8.100	9.900
100	8.910	9.090
1000	8.991	9.009

From the results of Example 4, it appears that the Riemann sums are approaching the limit 9 as n approaches infinity. It is this observation that motivates the following definition of a **definite integral.** In this definition, we consider the partition of $[a, b]$ into n subintervals of equal width $\Delta x = (b - a)/n$ as follows.

$$a = x_0 < x_1 < x_2 < \cdots < x_{n-1} < x_n = b$$

Moreover, we consider c_i to be an arbitrary point in the ith subinterval $[x_{i-1}, x_i]$. To say that the number of subintervals n tends to infinity is equivalent to saying that the width, Δx, of the subintervals tends to zero.

Definition of Definite Integral

If f is a continuous function defined on the closed interval $[a, b]$, then the **definite integral of f on $[a, b]$** is

$$\int_a^b f(x)\, dx = \lim_{\Delta x \to 0} \sum_{i=1}^{n} f(c_i)\Delta x = \lim_{n \to \infty} \sum_{i=1}^{n} f(c_i)\Delta x.$$

If f is continuous and nonnegative on the interval $[a, b]$, then the definite integral of f on $[a, b]$ gives the area of the region bounded by the graph of f, the x-axis, and the vertical lines $x = a$ and $x = b$.

Evaluation of a definite integral by its limit definition can be difficult. However, there are times when we can solve a definite integral by recognizing that it represents the area of a common type of geometric figure.

EXAMPLE 5   The Areas of Common Geometric Figures

Sketch the region corresponding to each of the following definite integrals. Then evaluate each definite integral using a geometric formula.

a. $\displaystyle\int_1^3 4\, dx$ **b.** $\displaystyle\int_0^3 (x + 2)\, dx$ **c.** $\displaystyle\int_{-2}^2 \sqrt{4 - x^2}\, dx$

SOLUTION

A sketch of each region is shown in Figure A.4 (page A6).

a. The region associated with this definite integral is a rectangle of height 4 and width 2. Moreover, since the function $f(x) = 4$ is continuous and nonnegative on the interval $[1, 3]$, we can conclude that the area of the rectangle is given by the definite integral. Thus, the value of the definite integral is

$$\int_1^3 4\, dx = 4(2) = 8.$$

b. The region associated with this definite integral is a trapezoid with an altitude of 3 and parallel bases of lengths 2 and 5. The formula for the area of a trapezoid is $\frac{1}{2}h(b_1 + b_2)$, and so we have

$$\int_0^3 (x + 2)\, dx = \frac{1}{2}(3)(2 + 5) = \frac{21}{2}.$$

c. The region associated with this definite integral is a semicircle of radius 2. Thus, the area is $\frac{1}{2}\pi r^2$, and we have

$$\int_{-2}^2 \sqrt{4 - x^2}\, dx = \frac{1}{2}\pi(2^2) = 2\pi.$$

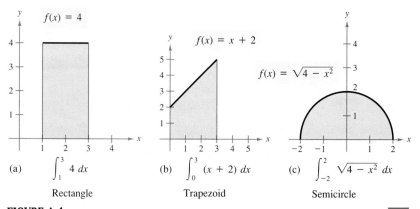

(a) $\displaystyle\int_1^3 4\, dx$ (b) $\displaystyle\int_0^3 (x + 2)\, dx$ (c) $\displaystyle\int_{-2}^2 \sqrt{4 - x^2}\, dx$

    Rectangle   Trapezoid   Semicircle

FIGURE A.4

For some simple functions it is possible to evaluate definite integrals by the Riemann sum definition. In the next example, we will use the fact that the sum of the first n integers is given by the formula

$$1 + 2 + \cdots + n = \sum_{i=1}^n i = \frac{n(n + 1)}{2} \qquad \textit{See Exercise 27.}$$

to compute the area of the triangular region of Examples 3 and 4.

EXAMPLE 6   Evaluating a Definite Integral by Its Definition

Evaluate the definite integral $\displaystyle\int_0^3 2x\, dx$.

SOLUTION

Let $\Delta x = (b - a)/n = 3/n$, and choose c_i to be the right endpoint of each subinterval, $c_i = 3i/n$. Then we have

$$\int_0^3 2x \, dx = \lim_{\Delta x \to 0} \sum_{i=1}^{n} f(c_i)\Delta x$$

$$= \lim_{n \to \infty} \sum_{i=1}^{n} 2\left(i\frac{3}{n}\right)\left(\frac{3}{n}\right)$$

$$= \lim_{n \to \infty} \frac{18}{n^2} \sum_{i=1}^{n} i$$

$$= \lim_{n \to \infty} \left(\frac{18}{n^2}\right)\left(\frac{n(n+1)}{2}\right)$$

$$= \lim_{n \to \infty} \left(9 + \frac{9}{n}\right).$$

This limit can be evaluated in the same way that we calculated horizontal asymptotes in Section 3.6. In particular, as n approaches infinity, we see that $9/n$ approaches 0, and the above limit is 9. Thus, we conclude that

$$\int_0^3 2x \, dx = 9.$$

From Example 6, you can see that it can be difficult to evaluate the definite integral of even a simple function by using Riemann sums. A computer can help in calculating these sums for large values of n, but this procedure would only give an approximation of the definite integral. Fortunately, the **Fundamental Theorem of Calculus** provides a technique for evaluating definite integrals using antiderivatives, and for this reason it is often thought to be the most important theorem in calculus. In the remainder of this appendix, we will show how derivatives and integrals are related via the Fundamental Theorem of Calculus.

To simplify the discussion, we will assume that f is a continuous nonnegative function defined on the interval $[a, b]$. Let $A(x)$ be the area of the region under the graph of f from a to x, as indicated in Figure A.5. The area under the shaded region in Figure A.6 is thus $A(x + \Delta x) - A(x)$.

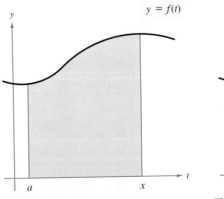

FIGURE A.5   **FIGURE A.6**

If Δx is small, then this area is approximately given by the area of the rectangle of height $f(x)$ and width Δx. Thus, we have $A(x + \Delta x) - A(x) \approx f(x)\Delta x$. Dividing by Δx produces

$$f(x) \approx \frac{A(x + \Delta x) - A(x)}{\Delta x}.$$

By taking the limit as Δx approaches 0, we see that

$$f(x) = \lim_{\Delta x \to 0} \frac{A(x + \Delta x) - A(x)}{\Delta x} = A'(x)$$

and we have established the fact that the area function $A(x)$ is an antiderivative of f. Although we assumed that f is continuous and nonnegative, this development is valid if the function f is simply continuous on the closed interval $[a, b]$. We use this result in the proof of the Fundamental Theorem of Calculus.

Fundamental Theorem of Calculus

If f is a continuous function on the closed interval $[a, b]$, then

$$\int_a^b f(x)\, dx = F(b) - F(a),$$

where F is any function such that $F'(x) = f(x)$.

Proof

From the previous discussion, we know that

$$\int_a^x f(x)\, dx = A(x)$$

and in particular

$$A(a) = \int_a^a f(x)\, dx = 0 \quad \text{and} \quad A(b) = \int_a^b f(x)\, dx.$$

If F is *any* antiderivative of f, then we know that F differs from A by a constant. That is, $A(x) = F(x) + C$. Hence,

$$\int_a^b f(x)\, dx = A(b) - A(a) = [F(b) + C] - [F(a) + C] = F(b) - F(a).$$

You are now ready to continue Section 5.4, on page 342, just after the statement of the Fundamental Theorem of Calculus.

APPENDIX A ▪ EXERCISES

In Exercises 1–6, use the left Riemann sum and the right Riemann sum to approximate the area of the given region using the indicated number of subintervals.

1. $y = \sqrt{x}$

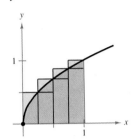

2. $y = \sqrt{x} + 1$

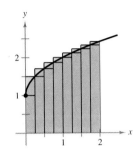

3. $y = \dfrac{1}{x}$

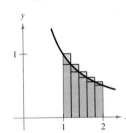

4. $y = \dfrac{1}{x - 2}$

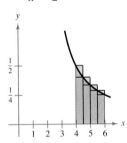

5. $y = \sqrt{1 - x^2}$

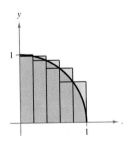

6. $y = \sqrt{x + 1}$

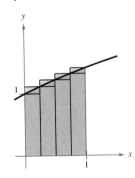

7. Repeat Exercise 1 using the midpoint Riemann sum.

8. Repeat Exercise 2 using the midpoint Riemann sum.

9. Consider a triangle of area 2 bounded by the graphs of $y = x$, $y = 0$, and $x = 2$.
   (a) Sketch the graph of the region.

(b) Divide the interval $[0, 2]$ into n equal subintervals and show that the endpoints are

$$0 < 1\left(\frac{2}{n}\right) < \cdots < (n - 1)\left(\frac{2}{n}\right) < n\left(\frac{2}{n}\right).$$

(c) Show that the left Riemann sum is

$$S_L = \sum_{i=1}^{n} \left[(i - 1)\left(\frac{2}{n}\right) \right]\left(\frac{2}{n}\right).$$

(d) Show that the right Riemann sum is

$$S_R = \sum_{i=1}^{n} \left[i\left(\frac{2}{n}\right) \right]\left(\frac{2}{n}\right).$$

(e) Complete the following table.

n	5	10	50	100
Left sum S_L				
Right sum S_R				

(f) Show that $\lim\limits_{n \to \infty} S_L = \lim\limits_{n \to \infty} S_R = 2.$

10. Consider a trapezoid of area 4 bounded by the graphs of $y = x$, $y = 0$, $x = 1$, and $x = 3$.
   (a) Sketch the graph of the region.
   (b) Divide the interval $[1, 3]$ into n equal subintervals and show that the endpoints are

$$1 < 1 + 1\left(\frac{2}{n}\right) < \cdots < 1 + (n - 1)\left(\frac{2}{n}\right) < 1 + n\left(\frac{2}{n}\right).$$

(c) Show that the left Riemann sum is

$$S_L = \sum_{i=1}^{n} \left[1 + (i - 1)\left(\frac{2}{n}\right) \right]\left(\frac{2}{n}\right).$$

(d) Show that the right Riemann sum is

$$S_R = \sum_{i=1}^{n} \left[1 + i\left(\frac{2}{n}\right) \right]\left(\frac{2}{n}\right).$$

(e) Complete the following table.

n	5	10	50	100
Left sum S_L				
Right sum S_R				

(f) Show that $\lim\limits_{n \to \infty} S_L = \lim\limits_{n \to \infty} S_R = 4.$

In Exercises 11–16, set up a definite integral that yields the area of the given region. (Do not evaluate the integral.)

11. $f(x) = 3$

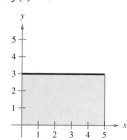

12. $f(x) = 4 - 2x$

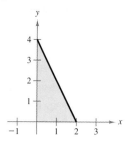

13. $f(x) = 4 - |x|$

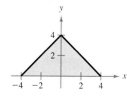

14. $f(x) = x^2$

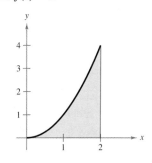

15. $f(x) = 4 - x^2$

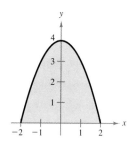

16. $f(x) = \dfrac{1}{x^2 + 1}$

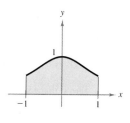

In Exercises 17–26, sketch the region whose area is indicated by the given definite integral. Then use a geometric formula to evaluate the integral.

17. $\displaystyle\int_0^3 4\, dx$

18. $\displaystyle\int_{-a}^{a} 4\, dx$

19. $\displaystyle\int_0^4 x\, dx$

20. $\displaystyle\int_0^4 \frac{x}{2}\, dx$

21. $\displaystyle\int_0^2 (2x + 5)\, dx$

22. $\displaystyle\int_0^5 (5 - x)\, dx$

23. $\displaystyle\int_{-1}^{1} (1 - |x|)\, dx$

24. $\displaystyle\int_{-a}^{a} (a - |x|)\, dx$

25. $\displaystyle\int_{-3}^{3} \sqrt{9 - x^2}\, dx$

26. $\displaystyle\int_{-r}^{r} \sqrt{r^2 - x^2}\, dx$

27. Show that $\displaystyle\sum_{i=1}^{n} i = \frac{n(n + 1)}{n}$. Hint: Add the following two sums.

$$S = 1 + 2 + 3 + \cdots + (n - 2) + (n - 1) + n$$
$$S = n + (n - 1) + (n - 2) + \cdots + 3 + 2 + 1$$

28. Use the Riemann sum definition of the definite integral and the result of Exercise 27 to evaluate $\int_1^2 x\, dx$.

Appendix B

Reference Tables

REFERENCE TABLE 1   The Greek Alphabet

Letter		Name	Letter		Name	Letter		Name
A	α	Alpha	I	ι	Iota	P	ρ	Rho
B	β	Beta	K	κ	Kappa	Σ	σ	Sigma
Γ	γ	Gamma	Λ	λ	Lambda	T	τ	Tau
Δ	δ	Delta	M	μ	Mu	Υ	υ	Upsilon
E	ε	Epsilon	N	ν	Nu	Φ	ϕ	Phi
Z	ζ	Zeta	Ξ	ξ	Xi	X	χ	Chi
H	η	Eta	O	o	Omicron	Ψ	ψ	Psi
Θ	θ	Theta	Π	π	Pi	Ω	ω	Omega

REFERENCE TABLE 2   The Number of Each Day in the Year

Day of Month	Jan.	Feb.	Mar.	Apr.	May	June	July	Aug.	Sept.	Oct.	Nov.	Dec.	Day of Month
1	1	32	60	91	121	152	182	213	244	274	305	335	**1**
2	2	33	61	92	122	153	183	214	245	275	306	336	**2**
3	3	34	62	93	123	154	184	215	246	276	307	337	**3**
4	4	35	63	94	124	155	185	216	247	277	308	338	**4**
5	5	36	64	95	125	156	186	217	248	278	309	339	**5**
6	6	37	65	96	126	157	187	218	249	279	310	340	**6**
7	7	38	66	97	127	158	188	219	250	280	311	341	**7**
8	8	39	67	98	128	159	189	220	251	281	312	342	**8**
9	9	40	68	99	129	160	190	221	252	282	313	343	**9**
10	10	41	69	100	130	161	191	222	253	283	314	344	**10**
11	11	42	70	101	131	162	192	223	254	284	315	345	**11**
12	12	43	71	102	132	163	193	224	255	285	316	346	**12**
13	13	44	72	103	133	164	194	225	256	286	317	347	**13**
14	14	45	73	104	134	165	195	226	257	287	318	348	**14**
15	15	46	74	105	135	166	196	227	258	288	319	349	**15**
16	16	47	75	106	136	167	197	228	259	289	320	350	**16**
17	17	48	76	107	137	168	198	229	260	290	321	351	**17**
18	18	49	77	108	138	169	199	230	261	291	322	352	**18**
19	19	50	78	109	139	170	200	231	262	292	323	353	**19**
20	20	51	79	110	140	171	201	232	263	293	324	354	**20**
21	21	52	80	111	141	172	202	233	264	294	325	355	**21**
22	22	53	81	112	142	173	203	234	265	295	326	356	**22**
23	23	54	82	113	143	174	204	235	266	296	327	357	**23**
24	24	55	83	114	144	175	205	236	267	297	328	358	**24**
25	25	56	84	115	145	176	206	237	268	298	329	359	**25**
26	26	57	85	116	146	177	207	238	269	299	330	360	**26**
27	27	58	86	117	147	178	208	239	270	300	331	361	**27**
28	28	59	87	118	148	179	209	240	271	301	332	362	**28**
29	29	*	88	119	149	180	210	241	272	302	333	363	**29**
30	30	—	89	120	150	181	211	242	273	303	334	364	**30**
31	31	—	90	—	151	—	212	243	—	304	—	365	**31**

*On leap year, Feb. 29 is day 60 and the number of each day after Feb. 29 is increased by one.

REFERENCE TABLE 3   Units of Measurement of Length

English System: Inch (in), Foot (ft), Yard (yd), Mile (mi)

1 mi = 5280 ft   1 mi = 1760 yd
1 yd = 3 ft   1 ft = 12 in

Metric System: Millimeter (mm), Centimeter (cm), Meter (m), Kilometer (km)

1 km = 1000 m   1 m = 100 cm
1 m = 1000 mm   1 cm = 10 mm

Conversion Factors (six significant figures)

Metric to English   **English to Metric**

1 mm = 0.0393701 in   1 in = 25.4000 mm
1 cm = 0.393701 in   1 in = 2.54000 cm
1 m = 3.28084 ft   1 ft = 0.304800 m
1 m = 1.09361 yd   1 yd = 0.914400 m
1 km = 0.621371 mi   1 mi = 1.60934 km

Miscellaneous

1 fathom = 6 ft
1 astronomical unit = 93,000,000 mi (average distance between earth and sun)
1 light year = 5,800,000,000,000 mi (distance traveled by light in one year)

REFERENCE TABLE 4   Units of Measurement of Area

English System: Square Inch (in^2), Square Foot (ft^2), Square Yard (yd^2), Acre, Square Mile (mi^2)

1 mi^2 = 640 acres   1 acre = 43,560 ft^2
1 yd^2 = 9 ft^2 1 ft^2 = 144 in^2

Metric System: Square Centimeter (cm²), Square Meter (m²), Square Kilometer (km²)

$1 \text{ km}^2 = 1{,}000{,}000 \text{ m}^2$ $\qquad\qquad$ $1 \text{ m}^2 = 10{,}000 \text{ cm}^2$

Conversion Factors (six significant figures)

Metric to English	English to Metric
$1 \text{ cm}^2 = 0.155000 \text{ in}^2$	$1 \text{ in}^2 = 6.45160 \text{ cm}^2$
$1 \text{ m}^2 = 10.7640 \text{ ft}^2$	$1 \text{ ft}^2 = 0.0929030 \text{ m}^2$
$1 \text{ m}^2 = 1.19599 \text{ yd}^2$	$1 \text{ yd}^2 = 0.836127 \text{ m}^2$
$1 \text{ km}^2 = 0.386102 \text{ mi}^2$	$1 \text{ mi}^2 = 2.58999 \text{ km}^2$

Miscellaneous

1 square mile = 1 section = 4 quarters

REFERENCE TABLE 5   Units of Measurement of Volume

English System: Cubic Inch (in³), Cubic Foot (ft³), Cubic Yard (yd³), Fluid Ounce (fl oz), Pint (pt), Quart (qt), Gallon (gal)

$1 \text{ yd}^3 = 27 \text{ ft}^3$	$1 \text{ ft}^3 = 1728 \text{ in}^3$
$1 \text{ gal} = 4 \text{ qt}$	$1 \text{ qt} = 2 \text{ pt}$
$1 \text{ qt} = 32 \text{ fl oz}$	$1 \text{ pt} = 16 \text{ fl oz}$
$1 \text{ gal} = 231 \text{ in}^3$	$1 \text{ ft}^3 = 7.48052 \text{ gal}$

Metric System: Cubic Centimeter (cm³), Cubic Meter (m³), Milliliter (cc), Liter

$1 \text{ m}^3 = 1000 \text{ liters}$	$1 \text{ liter} = 1000 \text{ cc}$
$1 \text{ cm}^3 = 1 \text{ cc}$	

Conversion Factors (six significant figures)

Metric to English	English to Metric
$1 \text{ cm}^3 = 0.0610237 \text{ in}^3$	$1 \text{ in}^3 = 16.3871 \text{ cm}^3$
$1 \text{ m}^3 = 35.3147 \text{ ft}^3$	$1 \text{ ft}^3 = 0.0283168 \text{ m}^3$
$1 \text{ m}^3 = 1.30795 \text{ yd}^3$	$1 \text{ yd}^3 = 0.764555 \text{ m}^3$

REFERENCE TABLE 5   (Continued)

Miscellaneous

1 gallon = 5 fifths	1 fifth = 0.757 liter
1 quart = 4 cups	1 cup = 16 tablespoons
1 tablespoon = 3 teaspoons (cooking)	
1 tablespoon = 4 teaspoons (medical)	
1 barrel (bl) = 42 gallons (petroleum)	

REFERENCE TABLE 6   Units of Measurement of Mass and Force

English System (Force or Weight): Ounce (oz), Pound (lb), Ton

1 ton = 2000 lb   1 lb = 16 oz

Metric System (Mass): Gram (g), Kilogram (kg), Metric Ton

1 metric ton = 1000 kg   1 kg = 1000 g

Conversion Factors (six significant figures at sea level)

Metric to English	**English to Metric**
1 g = 0.0352740 oz	1 oz = 28.3495 g
1 kg = 2.20462 lb	1 lb = 0.453592 kg
1 metric ton = 1.10231 ton	1 ton = 0.907185 metric ton

REFERENCE TABLE 7   Units of Measurement of Time

Second (sec), Minute (min), Hour (hr)

1 hr = 60 min   1 min = 60 sec

Calendar Units: Day, Week, Month, Year (yr)

1 yr = 365 days (366 days on leap year)
1 yr = 365.256 mean solar days
1 yr = 12 months
1 week = 7 days
1 day = 24 hr

REFERENCE TABLE 8   Units of Measurement of Temperature

Fahrenheit (F), Celsius (C), Kelvin or Absolute (K)

Celsius to Fahrenheit

$$°F = \frac{9}{5}°C + 32$$

Fahrenheit to Celsius

$$°C = \frac{5}{9}°F - 32$$

Celsius to Kelvin
$$°K = °C + 273.15$$

Kelvin to Celsius
$$°C = °K - 273.15$$

Freezing temperature for water = 32° F = 0° C
Boiling temperature for water = 212° F = 100° C
Absolute zero temperature = 0° K = −273.15° C = −459.67° F
  (0° K is, by definition, the coldest possible temperature, at which there is no molecular activity.)

REFERENCE TABLE 9   Miscellaneous Units and Number Constants

1 dozen (doz) = 12 units   1 gross = 12 dozen units
1 score = 20 units

$\pi \approx 3.1415926535$ $e \approx 2.7182818284$

Equatorial radius of the earth = 3963.34 mi = 6378.388 km
Polar radius of the earth = 3949.99 mi = 6356.912 km
Acceleration due to gravity at sea level = 32.1726 ft/sec^2
Speed of sound at sea level (standard atmosphere) = 1116.45 ft/sec
Speed of light in vacuum = 186,284 mi/sec
Density of water: 1 ft^3 = 62.425 lb

REFERENCE TABLE 10   Algebra

Operations with Exponents

1. $x^n x^m = x^{n+m}$

2. $\dfrac{x^n}{x^m} = x^{n-m}$

3. $(xy)^n = x^n y^n$

4. $\left(\dfrac{x}{y}\right)^n = \dfrac{x^n}{y^n}$

5. $(x^n)^m = x^{nm}$

6. $-x^n = -(x^n)$

7. $cx^n = c(x^n)$

8. $x^{n^m} = x^{(n^m)}$

Exponents and Radicals (*n* and *m* are positive integers)

1. $x^n = \underbrace{x \cdot x \cdot x \cdots x}_{n \text{ factors}}$

2. $x^0 = 1, \; x \neq 0$

3. $x^{-n} = \dfrac{1}{x^n}, \; x \neq 0$

*4. $\sqrt[n]{x} = a \;\;\Longrightarrow\;\; x = a^n$

5. $x^{1/n} = \sqrt[n]{x}$

6. $x^{m/n} = (x^{1/n})^m = (\sqrt[n]{x})^m$
   $x^{m/n} = (x^m)^{1/n} = \sqrt[n]{x^m}$

7. $\sqrt[2]{x} = \sqrt{x}$

Operations with Fractions

1. $\dfrac{a}{b} + \dfrac{c}{d} = \dfrac{a}{b}\left(\dfrac{d}{d}\right) + \dfrac{c}{d}\left(\dfrac{b}{b}\right) = \dfrac{ad}{bd} + \dfrac{bc}{bd} = \dfrac{ad + bc}{bd}$

2. $\dfrac{a}{b} - \dfrac{c}{d} = \dfrac{a}{b}\left(\dfrac{d}{d}\right) - \dfrac{c}{d}\left(\dfrac{b}{b}\right) = \dfrac{ad}{bd} - \dfrac{bc}{bd} = \dfrac{ad - bc}{bd}$

3. $\left(\dfrac{a}{b}\right)\left(\dfrac{c}{d}\right) = \dfrac{ac}{bd}$

4. $\dfrac{a/b}{c/d} = \left(\dfrac{a}{b}\right)\left(\dfrac{d}{c}\right) = \dfrac{ad}{bc}$

   $\dfrac{a/b}{c} = \dfrac{a/b}{c/1} = \left(\dfrac{a}{b}\right)\left(\dfrac{1}{c}\right) = \dfrac{a}{bc}$

5. $\dfrac{\cancel{a}b}{\cancel{a}c} = \dfrac{b}{c}$

   $\dfrac{ab + ac}{ad} = \dfrac{\cancel{a}(b + c)}{\cancel{a}d} = \dfrac{b + c}{d}$

*If *n* is even, the principal *n*th root is defined to be positive.

Quadratic Formula

$$ax^2 + bx + c = 0 \implies x = \frac{-b \pm \sqrt{b^2 - 4ac}}{2a}$$

Factors and Special Products

1. $x^2 - a^2 = (x - a)(x + a)$
2. $x^3 - a^3 = (x - a)(x^2 + ax + a^2)$
3. $x^3 + a^3 = (x + a)(x^2 - ax + a^2)$
4. $x^4 - a^4 = (x - a)(x + a)(x^2 + a^2)$

Factoring by Grouping

$$acx^3 + adx^2 + bcx + bd = ax^2(cx + d) + b(cx + d) = (ax^2 + b)(cx + d)$$

Binomial Theorem

1. $(x + a)^2 = x^2 + 2ax + a^2$
2. $(x - a)^2 = x^2 - 2ax + a^2$
3. $(x + a)^3 = x^3 + 3ax^2 + 3a^2x + a^3$
4. $(x - a)^3 = x^3 - 3ax^2 + 3a^2x - a^3$
5. $(x + a)^4 = x^4 + 4ax^3 + 6a^2x^2 + 4a^3x + a^4$
6. $(x - a)^4 = x^4 - 4ax^3 + 6a^2x^2 - 4a^3x + a^4$
7. $(x + a)^n = x^n + nax^{n-1} + \dfrac{n(n-1)}{2!}a^2x^{n-2} + \dfrac{n(n-1)(n-2)}{3!}a^3x^{n-3} +$
$\cdots + na^{n-1}x + a^n$
8. $(x - a)^n = x^n - nax^{n-1} + \dfrac{n(n-1)}{2!}a^2x^{n-2} - \dfrac{n(n-1)(n-2)}{3!}a^3x^{n-3} +$
$\cdots \pm na^{n-1}x \mp a^n$

Miscellaneous

1. If $ab = 0$, then $a = 0$ or $b = 0$.
2. If $ac = bc$ and $c \neq 0$, then $a = b$.
3. Factorial: $0! = 1$, $1! = 1$, $2! = 2 \cdot 1$, $3! = 3 \cdot 2 \cdot 1$, $4! = 4 \cdot 3 \cdot 2 \cdot 1$, etc.

REFERENCE TABLE 10 (Continued)

Sequences

1. Arithmetic: $a,\ a + b,\ a + 2b,\ a + 3b,\ a + 4b,\ a + 5b,\ \ldots$

2. Geometric: $ar^0,\ ar^1,\ ar^2,\ ar^3,\ ar^4,\ ar^5,\ \ldots$

$$ar^0 + ar^1 + ar^2 + ar^3 + \cdots + ar^n = \frac{a(1 - r^{n+1})}{1 - r}$$

3. General Harmonic: $\dfrac{1}{a},\ \dfrac{1}{a + b},\ \dfrac{1}{a + 2b},\ \dfrac{1}{a + 3b},\ \dfrac{1}{a + 4b},\ \dfrac{1}{a + 5b},\ \ldots$

4. Harmonic: $\dfrac{1}{1},\ \dfrac{1}{2},\ \dfrac{1}{3},\ \dfrac{1}{4},\ \dfrac{1}{5},\ \ldots$

5. p-Sequence: $\dfrac{1}{1^p},\ \dfrac{1}{2^p},\ \dfrac{1}{3^p},\ \dfrac{1}{4^p},\ \dfrac{1}{5^p},\ \ldots$

Series

$$\frac{1}{x} = 1 - (x - 1) + (x - 1)^2 - (x - 1)^3 + (x - 1)^4 - \cdots + (-1)^n(x - 1)^n + \cdots, \quad 0 < x < 2$$

$$\frac{1}{1 + x} = 1 - x + x^2 - x^3 + x^4 - x^5 + \cdots + (-1)^n x^n + \cdots, \quad -1 < x < 1$$

$$\ln x = (x - 1) - \frac{(x - 1)^2}{2} + \frac{(x - 1)^3}{3} - \frac{(x - 1)^4}{4} + \cdots + \frac{(-1)^{n-1}(x - 1)^n}{n} + \cdots, \quad 0 < x \leq 2$$

$$e^x = 1 + x + \frac{x^2}{2!} + \frac{x^3}{3!} + \frac{x^4}{4!} + \frac{x^5}{5!} + \cdots + \frac{x^n}{n!} + \cdots, \quad -\infty < x < \infty$$

$$\sin x = x - \frac{x^3}{3!} + \frac{x^5}{5!} - \frac{x^7}{7!} + \cdots, \quad -\infty < x < \infty$$

$$\cos x = 1 - \frac{x^2}{2!} + \frac{x^4}{4!} - \frac{x^6}{6!} + \cdots, \quad -\infty < x < \infty$$

$$(1 + x)^k = 1 + kx + \frac{k(k - 1)x^2}{2!} + \frac{k(k - 1)(k - 2)x^3}{3!} + \frac{k(k - 1)(k - 2)(k - 3)x^4}{4!} + \cdots, \quad -1 < x < 1^*$$

$$(1 + x)^{-k} = 1 - kx + \frac{k(k + 1)x^2}{2!} - \frac{k(k + 1)(k + 2)x^3}{3!} + \frac{k(k + 1)(k + 2)(k + 3)x^4}{4!} - \cdots, \quad -1 < x < 1^*$$

*The convergence at $x = \pm 1$ depends on the value k.

REFERENCE TABLE 11   Table of Square Roots and Cube Roots

n	$\sqrt{n}$	$\sqrt[3]{n}$	n	$\sqrt{n}$	$\sqrt[3]{n}$	n	$\sqrt{n}$	$\sqrt[3]{n}$
1	1.00000	1.00000	51	7.14143	3.70843	101	10.0499	4.65701
2	1.41421	1.25992	52	7.21110	3.73251	102	10.0995	4.67233
3	1.73205	1.44225	53	7.28011	3.75629	103	10.1489	4.68755
4	2.00000	1.58740	54	7.34847	3.77976	104	10.1980	4.70267
5	2.23607	1.70998	55	7.41620	3.80295	105	10.2470	4.71769
6	2.44949	1.81712	56	7.48331	3.82586	106	10.2956	4.73262
7	2.64575	1.91293	57	7.54983	3.84850	107	10.3441	4.74746
8	2.82843	2.00000	58	7.61577	3.87088	108	10.3923	4.76220
9	3.00000	2.08008	59	7.68115	3.89300	109	10.4403	4.77686
10	3.16228	2.15443	60	7.74597	3.91487	110	10.4881	4.79142
11	3.31662	2.22398	61	7.81025	3.93650	111	10.5357	4.80590
12	3.46410	2.28943	62	7.87401	3.95789	112	10.5830	4.82028
13	3.60555	2.35133	63	7.93725	3.97906	113	10.6301	4.83459
14	3.74166	2.41014	64	8.00000	4.00000	114	10.6771	4.84881
15	3.87298	2.46621	65	8.06226	4.02073	115	10.7238	4.86294
16	4.00000	2.51984	66	8.12404	4.04124	116	10.7703	4.87700
17	4.12311	2.57128	67	8.18535	4.06155	117	10.8167	4.89097
18	4.24264	2.62074	68	8.24621	4.08166	118	10.8628	4.90487
19	4.35890	2.66840	69	8.30662	4.10157	119	10.9087	4.91868
20	4.47214	2.71442	70	8.36660	4.12129	120	10.9545	4.93242
21	4.58258	2.75892	71	8.42615	4.14082	121	11.0000	4.94609
22	4.69042	2.80204	72	8.48528	4.16017	122	11.0454	4.95968
23	4.79583	2.84387	73	8.54400	4.17934	123	11.0905	4.97319
24	4.89898	2.88450	74	8.60233	4.19834	124	11.1355	4.98663
25	5.00000	2.92402	75	8.66025	4.21716	125	11.1803	5.00000
26	5.09902	2.96250	76	8.71780	4.23582	126	11.2250	5.01330
27	5.19615	3.00000	77	8.77496	4.25432	127	11.2694	5.02653
28	5.29150	3.03659	78	8.83176	4.27266	128	11.3137	5.03968
29	5.38516	3.07232	79	8.88819	4.29084	129	11.3578	5.05277
30	5.47723	3.10723	80	8.94427	4.30887	130	11.4018	5.06580
31	5.56776	3.14138	81	9.00000	4.32675	131	11.4455	5.07875
32	5.65685	3.17480	82	9.05539	4.34448	132	11.4891	5.09164
33	5.74456	3.20753	83	9.11043	4.36207	133	11.5326	5.10447
34	5.83095	3.23961	84	9.16515	4.37952	134	11.5758	5.11723
35	5.91608	3.27107	85	9.21954	4.39683	135	11.6190	5.12993
36	6.00000	3.30193	86	9.27362	4.41400	136	11.6619	5.14256
37	6.08276	3.33222	87	9.32738	4.43105	137	11.7047	5.15514
38	6.16441	3.36198	88	9.38083	4.44796	138	11.7473	5.16765
39	6.24500	3.39121	89	9.43398	4.46475	139	11.7898	5.18010
40	6.32456	3.41995	90	9.48683	4.48140	140	11.8322	5.19249
41	6.40312	3.44822	91	9.53939	4.49794	141	11.8743	5.20483
42	6.48074	3.47603	92	9.59166	4.51436	142	11.9164	5.21710
43	6.55744	3.50340	93	9.64365	4.53065	143	11.9583	5.22932
44	6.63325	3.53035	94	9.69536	4.54684	144	12.0000	5.24148
45	6.70820	3.55689	95	9.74679	4.56290	145	12.0416	5.25359
46	6.78233	3.58305	96	9.79796	4.57886	146	12.0830	5.26564
47	6.85565	3.60883	97	9.84886	4.59470	147	12.1244	5.27763
48	6.92820	3.63424	98	9.89949	4.61044	148	12.1655	5.28957
49	7.00000	3.65931	99	9.94987	4.62606	149	12.2066	5.30146
50	7.07107	3.68403	100	10.0000	4.64159	150	12.2474	5.31329

REFERENCE TABLE 11   (Continued)

n	$\sqrt{n}$	$\sqrt[3]{n}$	n	$\sqrt{n}$	$\sqrt[3]{n}$	n	$\sqrt{n}$	$\sqrt[3]{n}$
151	12.2882	5.32507	168	12.9615	5.51785	185	13.6015	5.69802
152	12.3288	5.33680	169	13.0000	5.52877	186	13.6382	5.70827
153	12.3693	5.34848	170	13.0384	5.53966	187	13.6748	5.71848
154	12.4097	5.36011	171	13.0767	5.55050	188	13.7113	5.72865
155	12.4499	5.37169	172	13.1149	5.56130	189	13.7477	5.73879
156	12.4900	5.38321	173	13.1529	5.57205	190	13.7840	5.74890
157	12.5300	5.39469	174	13.1909	5.58277	191	13.8203	5.75897
158	12.5698	5.40612	175	13.2288	5.59344	192	13.8564	5.76900
159	12.6095	5.41750	176	13.2665	5.60408	193	13.8924	5.77900
160	12.6491	5.42884	177	13.3041	5.61467	194	13.9284	5.78896
161	12.6886	5.44012	178	13.3417	5.62523	195	13.9642	5.79889
162	12.7279	5.45136	179	13.3791	5.63574	196	14.0000	5.80879
163	12.7671	5.46256	180	13.4164	5.64622	197	14.0357	5.81865
164	12.8062	5.47370	181	13.4536	5.65665	198	14.0712	5.82848
165	12.8452	5.48481	182	13.4907	5.66705	199	14.1067	5.83827
166	12.8841	5.49586	183	13.5277	5.67741	200	14.1421	5.84804
167	12.9228	5.50688	184	13.5647	5.68773			

REFERENCE TABLE 12   Exponential Tables

x	e^x	e^{-x}	x	e^x	e^{-x}	x	e^x	e^{-x}
0.0	1.0000	1.0000	3.5	33.115	0.0302	7.0	1096.63	0.0009
0.1	1.1052	0.9048	3.6	36.598	0.0273	7.1	1211.97	0.0008
0.2	1.2214	0.8187	3.7	40.447	0.0247	7.2	1339.43	0.0007
0.3	1.3499	0.7408	3.8	44.701	0.0224	7.3	1480.30	0.0007
0.4	1.4918	0.6703	3.9	49.402	0.0202	7.4	1635.98	0.0006
0.5	1.6487	0.6065	4.0	54.598	0.0183	7.5	1808.04	0.0006
0.6	1.8221	0.5488	4.1	60.340	0.0166	7.6	1998.20	0.0005
0.7	2.0138	0.4966	4.2	66.686	0.0150	7.7	2208.35	0.0005
0.8	2.2255	0.4493	4.3	73.700	0.0136	7.8	2440.60	0.0004
0.9	2.4596	0.4066	4.4	81.451	0.0123	7.9	2697.28	0.0004
1.0	2.7183	0.3679	4.5	90.017	0.0111	8.0	2980.96	0.0003
1.1	3.0042	0.3329	4.6	99.484	0.0101	8.1	3294.47	0.0003
1.2	3.3201	0.3012	4.7	109.95	0.0091	8.2	3640.95	0.0003
1.3	3.6693	0.2725	4.8	121.51	0.0082	8.3	4023.87	0.0002
1.4	4.0552	0.2466	4.9	134.29	0.0074	8.4	4447.07	0.0002
1.5	4.4817	0.2231	5.0	148.41	0.0067	8.5	4914.77	0.0002
1.6	4.9530	0.2019	5.1	164.02	0.0061	8.6	5431.66	0.0002
1.7	5.4739	0.1827	5.2	181.27	0.0055	8.7	6002.91	0.0002
1.8	6.0496	0.1653	5.3	200.34	0.0050	8.8	6634.24	0.0002
1.9	6.6859	0.1496	5.4	221.41	0.0045	8.9	7331.97	0.0001
2.0	7.3891	0.1353	5.5	244.69	0.0041	9.0	8103.08	0.0001
2.1	8.1662	0.1225	5.6	270.43	0.0037	9.1	8955.29	0.0001
2.2	9.0250	0.1108	5.7	298.87	0.0033	9.2	9897.13	0.0001
2.3	9.9742	0.1003	5.8	330.30	0.0030	9.3	10938.02	0.0001
2.4	11.023	0.0907	5.9	365.04	0.0027	9.4	12088.38	0.0001
2.5	12.182	0.0821	6.0	403.43	0.0025	9.5	13359.73	0.0001
2.6	13.464	0.0743	6.1	445.86	0.0022	9.6	14764.78	0.0001
2.7	14.880	0.0672	6.2	492.75	0.0020	9.7	16317.61	0.0001
2.8	16.445	0.0608	6.3	544.57	0.0018	9.8	18033.74	0.0001
2.9	18.174	0.0550	6.4	601.85	0.0017	9.9	19930.37	0.0001
3.0	20.086	0.0498	6.5	665.14	0.0015	10.0	22026.47	0.0000
3.1	22.198	0.0450	6.6	735.10	0.0014			
3.2	24.533	0.0408	6.7	812.41	0.0012			
3.3	27.113	0.0369	6.8	897.85	0.0011			
3.4	29.964	0.0334	6.9	992.27	0.0010			

REFERENCE TABLE 13 Natural Logarithmic Tables

	0.00	0.01	0.02	0.03	0.04	0.05	0.06	0.07	0.08	0.09
1.0	0.0000	0.0100	0.0198	0.0296	0.0392	0.0488	0.0583	0.0677	0.0770	0.0862
1.1	0.0953	0.1044	0.1133	0.1222	0.1310	0.1398	0.1484	0.1570	0.1655	0.1740
1.2	0.1823	0.1906	0.1989	0.2070	0.2151	0.2231	0.2311	0.2390	0.2469	0.2546
1.3	0.2624	0.2700	0.2776	0.2852	0.2927	0.3001	0.3075	0.3148	0.3221	0.3293
1.4	0.3365	0.3436	0.3507	0.3577	0.3646	0.3716	0.3784	0.3853	0.3920	0.3988
1.5	0.4055	0.4121	0.4187	0.4253	0.4318	0.4383	0.4447	0.4511	0.4574	0.4637
1.6	0.4700	0.4762	0.4824	0.4886	0.4947	0.5008	0.5068	0.5128	0.5188	0.5247
1.7	0.5306	0.5365	0.5423	0.5481	0.5539	0.5596	0.5653	0.5710	0.5766	0.5822
1.8	0.5878	0.5933	0.5988	0.6043	0.6098	0.6152	0.6206	0.6259	0.6313	0.6366
1.9	0.6419	0.6471	0.6523	0.6575	0.6627	0.6678	0.6729	0.6780	0.6831	0.6881
2.0	0.6931	0.6981	0.7031	0.7080	0.7129	0.7178	0.7227	0.7275	0.7324	0.7372
2.1	0.7419	0.7467	0.7514	0.7561	0.7608	0.7655	0.7701	0.7747	0.7793	0.7839
2.2	0.7885	0.7930	0.7975	0.8020	0.8065	0.8109	0.8154	0.8198	0.8242	0.8286
2.3	0.8329	0.8372	0.8416	0.8459	0.8502	0.8544	0.8587	0.8629	0.8671	0.8713
2.4	0.8755	0.8796	0.8838	0.8879	0.8920	0.8961	0.9002	0.9042	0.9083	0.9123
2.5	0.9163	0.9203	0.9243	0.9282	0.9322	0.9361	0.9400	0.9439	0.9478	0.9517
2.6	0.9555	0.9594	0.9632	0.9670	0.9708	0.9746	0.9783	0.9821	0.9858	0.9895
2.7	0.9933	0.9969	1.0006	1.0043	1.0080	1.0116	1.0152	1.0188	1.0225	1.0260
2.8	1.0296	1.0332	1.0367	1.0403	1.0438	1.0473	1.0508	1.0543	1.0578	1.0613
2.9	1.0647	1.0682	1.0716	1.0750	1.0784	1.0818	1.0852	1.0886	1.0919	1.0953
3.0	1.0986	1.1019	1.1053	1.1086	1.1119	1.1151	1.1184	1.1217	1.1249	1.1282
3.1	1.1314	1.1346	1.1378	1.1410	1.1442	1.1474	1.1506	1.1537	1.1569	1.1600
3.2	1.1632	1.1663	1.1694	1.1725	1.1756	1.1787	1.1817	1.1848	1.1878	1.1909
3.3	1.1939	1.1969	1.2000	1.2030	1.2060	1.2090	1.2119	1.2149	1.2179	1.2208
3.4	1.2238	1.2267	1.2296	1.2326	1.2355	1.2384	1.2413	1.2442	1.2470	1.2499
3.5	1.2528	1.2556	1.2585	1.2613	1.2641	1.2669	1.2698	1.2726	1.2754	1.2782
3.6	1.2809	1.2837	1.2865	1.2892	1.2920	1.2947	1.2975	1.3002	1.3029	1.3056
3.7	1.3083	1.3110	1.3137	1.3164	1.3191	1.3218	1.3244	1.3271	1.3297	1.3324
3.8	1.3350	1.3376	1.3403	1.3429	1.3455	1.3481	1.3507	1.3533	1.3558	1.3584
3.9	1.3610	1.3635	1.3661	1.3686	1.3712	1.3737	1.3762	1.3788	1.3813	1.3838
4.0	1.3863	1.3888	1.3913	1.3938	1.3962	1.3987	1.4012	1.4036	1.4061	1.4085
4.1	1.4110	1.4134	1.4159	1.4183	1.4207	1.4231	1.4255	1.4279	1.4303	1.4327
4.2	1.4351	1.4375	1.4398	1.4422	1.4446	1.4469	1.4493	1.4516	1.4540	1.4563
4.3	1.4586	1.4609	1.4633	1.4656	1.4679	1.4702	1.4725	1.4748	1.4770	1.4793
4.4	1.4816	1.4839	1.4861	1.4884	1.4907	1.4929	1.4951	1.4974	1.4996	1.5019
4.5	1.5041	1.5063	1.5085	1.5107	1.5129	1.5151	1.5173	1.5195	1.5217	1.5239
4.6	1.5261	1.5282	1.5304	1.5326	1.5347	1.5369	1.5390	1.5412	1.5433	1.5454
4.7	1.5476	1.5497	1.5518	1.5539	1.5560	1.5581	1.5602	1.5623	1.5644	1.5665
4.8	1.5686	1.5707	1.5728	1.5748	1.5769	1.5790	1.5810	1.5831	1.5851	1.5872
4.9	1.5892	1.5913	1.5933	1.5953	1.5974	1.5994	1.6014	1.6034	1.6054	1.6074
5.0	1.6094	1.6114	1.6134	1.6154	1.6174	1.6194	1.6214	1.6233	1.6253	1.6273
5.1	1.6292	1.6312	1.6332	1.6351	1.6371	1.6390	1.6409	1.6429	1.6448	1.6467
5.2	1.6487	1.6506	1.6525	1.6544	1.6563	1.6582	1.6601	1.6620	1.6639	1.6658
5.3	1.6677	1.6696	1.6715	1.6734	1.6752	1.6771	1.6790	1.6808	1.6827	1.6845
5.4	1.6864	1.6882	1.6901	1.6919	1.6938	1.6956	1.6974	1.6993	1.7011	1.7029

REFERENCE TABLE 13 (Continued)

	0.00	0.01	0.02	0.03	0.04	0.05	0.06	0.07	0.08	0.09
5.5	1.7047	1.7066	1.7084	1.7102	1.7120	1.7138	1.7156	1.7174	1.7192	1.7210
5.6	1.7228	1.7246	1.7263	1.7281	1.7299	1.7317	1.7334	1.7352	1.7370	1.7387
5.7	1.7405	1.7422	1.7440	1.7457	1.7475	1.7492	1.7509	1.7527	1.7544	1.7561
5.8	1.7579	1.7596	1.7613	1.7630	1.7647	1.7664	1.7681	1.7699	1.7716	1.7733
5.9	1.7750	1.7766	1.7783	1.7800	1.7817	1.7834	1.7851	1.7867	1.7884	1.7901
6.0	1.7918	1.7934	1.7951	1.7967	1.7984	1.8001	1.8017	1.8034	1.8050	1.8066
6.1	1.8083	1.8099	1.8116	1.8132	1.8148	1.8165	1.8181	1.8197	1.8213	1.8229
6.2	1.8245	1.8262	1.8278	1.8294	1.8310	1.8326	1.8342	1.8358	1.8374	1.8390
6.3	1.8405	1.8421	1.8437	1.8453	1.8469	1.8485	1.8500	1.8516	1.8532	1.8547
6.4	1.8563	1.8579	1.8594	1.8610	1.8625	1.8641	1.8656	1.8672	1.8687	1.8703
6.5	1.8718	1.8733	1.8749	1.8764	1.8779	1.8795	1.8810	1.8825	1.8840	1.8856
6.6	1.8871	1.8886	1.8901	1.8916	1.8931	1.8946	1.8961	1.8976	1.8991	1.9006
6.7	1.9021	1.9036	1.9051	1.9066	1.9081	1.9095	1.9110	1.9125	1.9140	1.9155
6.8	1.9169	1.9184	1.9199	1.9213	1.9228	1.9242	1.9257	1.9272	1.9286	1.9301
6.9	1.9315	1.9330	1.9344	1.9359	1.9373	1.9387	1.9402	1.9416	1.9430	1.9445
7.0	1.9459	1.9473	1.9488	1.9502	1.9516	1.9530	1.9544	1.9559	1.9573	1.9587
7.1	1.9601	1.9615	1.9629	1.9643	1.9657	1.9671	1.9685	1.9699	1.9713	1.9727
7.2	1.9741	1.9755	1.9769	1.9782	1.9796	1.9810	1.9824	1.9838	1.9851	1.9865
7.3	1.9879	1.9892	1.9906	1.9920	1.9933	1.9947	1.9961	1.9974	1.9988	2.0001
7.4	2.0015	2.0028	2.0042	2.0055	2.0069	2.0082	2.0096	2.0109	2.0122	2.0136
7.5	2.0149	2.0162	2.0176	2.0189	2.0202	2.0215	2.0229	2.0242	2.0255	2.0268
7.6	2.0281	2.0295	2.0308	2.0321	2.0334	2.0347	2.0360	2.0373	2.0386	2.0399
7.7	2.0412	2.0425	2.0438	2.0451	2.0464	2.0477	2.0490	2.0503	2.0516	2.0528
7.8	2.0541	2.0554	2.0567	2.0580	2.0592	2.0605	2.0618	2.0631	2.0643	2.0656
7.9	2.0669	2.0681	2.0694	2.0707	2.0719	2.0732	2.0744	2.0757	2.0769	2.0782
8.0	2.0794	2.0807	2.0819	2.0832	2.0844	2.0857	2.0869	2.0882	2.0894	2.0906
8.1	2.0919	2.0931	2.0943	2.0956	2.0968	2.0980	2.0992	2.1005	2.1017	2.1029
8.2	2.1041	2.1054	2.1066	2.1078	2.1090	2.1102	2.1114	2.1126	2.1138	2.1150
8.3	2.1163	2.1175	2.1187	2.1199	2.1211	2.1223	2.1235	2.1247	2.1258	2.1270
8.4	2.1282	2.1294	2.1306	2.1318	2.1330	2.1342	2.1353	2.1365	2.1377	2.1389
8.5	2.1401	2.1412	2.1424	2.1436	2.1448	2.1459	2.1471	2.1483	2.1494	2.1506
8.6	2.1518	2.1529	2.1541	2.1552	2.1564	2.1576	2.1587	2.1599	2.1610	2.1622
8.7	2.1633	2.1645	2.1656	2.1668	2.1679	2.1691	2.1702	2.1713	2.1725	2.1736
8.8	2.1748	2.1759	2.1770	2.1782	2.1793	2.1804	2.1815	2.1827	2.1838	2.1849
8.9	2.1861	2.1872	2.1883	2.1894	2.1905	2.1917	2.1928	2.1939	2.1950	2.1961
9.0	2.1972	2.1983	2.1994	2.2006	2.2017	2.2028	2.2039	2.2050	2.2061	2.2072
9.1	2.2083	2.2094	2.2105	2.2116	2.2127	2.2138	2.2148	2.2159	2.2170	2.2181
9.2	2.2192	2.2203	2.2214	2.2225	2.2235	2.2246	2.2257	2.2268	2.2279	2.2289
9.3	2.2300	2.2311	2.2322	2.2332	2.2343	2.2354	2.2364	2.2375	2.2386	2.2396
9.4	2.2407	2.2418	2.2428	2.2439	2.2450	2.2460	2.2471	2.2481	2.2492	2.2502
9.5	2.2513	2.2523	2.2534	2.2544	2.2555	2.2565	2.2576	2.2586	2.2597	2.2607
9.6	2.2618	2.2628	2.2638	2.2649	2.2659	2.2670	2.2680	2.2690	2.2701	2.2711
9.7	2.2721	2.2732	2.2742	2.2752	2.2762	2.2773	2.2783	2.2793	2.2803	2.2814
9.8	2.2824	2.2834	2.2844	2.2854	2.2865	2.2875	2.2885	2.2895	2.2905	2.2915
9.9	2.2925	2.2935	2.2946	2.2956	2.2966	2.2976	2.2986	2.2996	2.3006	2.3016

REFERENCE TABLE 14   Properties of Logarithms

Inverse Properties

1. $\ln e^x = x$

2. $e^{\ln x} = x$

Properties of Logarithms

1. $\ln 1 = 0$

2. $\ln e = 1$

3. $\ln xy = \ln x + \ln y$

4. $\ln \dfrac{x}{y} = \ln x - \ln y$

5. $\ln x^y = y \ln x$

6. $\log_a x = \dfrac{\ln x}{\ln a}$

REFERENCE TABLE 15   Geometry

Triangles

1. General triangle

   Sum of triangles $\alpha + \beta + \theta = 180°$

   Area $= \dfrac{1}{2}(\text{base})(\text{height}) = \dfrac{1}{2}bh$

2. Similar triangles

   $\dfrac{a}{b} = \dfrac{A}{B}$

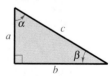

3. Right triangle

   $c^2 = a^2 + b^2$ (Pythagorean Theorem)

   Sum of acute angles $\alpha + \beta = 90°$

4. Equilateral triangle

   Height $= h = \dfrac{\sqrt{3}\,s}{2}$

   Area $= \dfrac{\sqrt{3}\,s^2}{4}$

5. Isosceles right triangle

$$\text{Area} = \frac{s^2}{2}$$

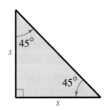

Quadrilaterals (Four-Sided Figures)

1. Rectangle

$$\text{Area} = (\text{length})(\text{height}) = bh$$

2. Square

$$\text{Area} = (\text{side})^2 = s^2$$

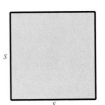

3. Parallelogram

$$\text{Area} = bh$$

4. Trapezoid

$$\text{Area} = \frac{1}{2}h(a + b)$$

Circles and Ellipses

1. Circle

$$\text{Area} = \pi r^2$$

$$\text{Circumference} = 2\pi r$$

2. Sector of circle (θ in radians)

$$\text{Area} = \frac{\theta r^2}{2}$$

$$s = r\theta$$

REFERENCE TABLE 15 (Continued)

3. Circular ring

$$\text{Area} = \pi(R^2 - r^2)$$

4. Ellipse

$$\text{Area} = \pi ab$$

$$\text{Circumference} \approx 2\pi\sqrt{\frac{a^2 + b^2}{2}}$$

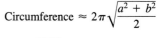

Solid Figures

1. Cone (A = area of base)

$$\text{Volume} = \frac{Ah}{3}$$

2. Right circular cone

$$\text{Volume} = \frac{\pi r^2 h}{3}$$

$$\text{Lateral surface area} = \pi r\sqrt{r^2 + h^2}$$

3. Frustum of right circular cone

$$\text{Volume} = \frac{\pi(r^2 + rR + R^2)h}{3}$$

$$\text{Lateral surface area} = \pi s(R + r)$$

4. Right circular cylinder

$$\text{Volume} = \pi r^2 h$$

$$\text{Lateral surface area} = 2\pi rh$$

5. Sphere

$$\text{Volume} = \frac{4}{3}\pi r^3$$

$$\text{Surface area} = 4\pi r^2$$

REFERENCE TABLE 16   Trigonometry

Definitions of the Six Trigonometric Functions

Right triangle definition: $0 < \theta < \pi/2$

$$\sin \theta = \frac{\text{opp.}}{\text{hyp.}} \qquad \csc \theta = \frac{\text{hyp.}}{\text{opp.}}$$

$$\cos \theta = \frac{\text{adj.}}{\text{hyp.}} \qquad \sec \theta = \frac{\text{hyp.}}{\text{adj.}}$$

$$\tan \theta = \frac{\text{opp.}}{\text{adj.}} \qquad \cot \theta = \frac{\text{adj.}}{\text{opp.}}$$

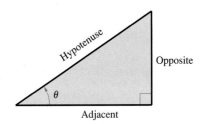

Circular function definition: θ is any angle and (x, y) is a point on the terminal ray of the angle.

$$\sin \theta = \frac{y}{r} \qquad \csc \theta = \frac{r}{y}$$

$$\cos \theta = \frac{x}{r} \qquad \sec \theta = \frac{r}{x}$$

$$\tan \theta = \frac{y}{x} \qquad \cot \theta = \frac{x}{y}$$

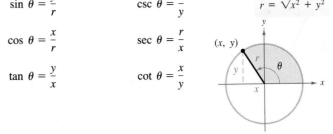

Signs of the Trigonometric Functions by Quadrant

Quadrant	sin	cos	tan	cot	sec	csc
I	+	+	+	+	+	+
II	+	−	−	−	−	+
III	−	−	+	+	−	−
IV	−	+	−	−	+	−

REFERENCE TABLE 16   (Continued)

Values of Trigonometric Functions at Common Angles

Degrees	0°	30°	45°	60°	90°
Radians	0	$\dfrac{\pi}{6}$	$\dfrac{\pi}{4}$	$\dfrac{\pi}{3}$	$\dfrac{\pi}{2}$
$\sin \theta$	0	$\dfrac{1}{2}$	$\dfrac{\sqrt{2}}{2}$	$\dfrac{\sqrt{3}}{2}$	1
$\cos \theta$	1	$\dfrac{\sqrt{3}}{2}$	$\dfrac{\sqrt{2}}{2}$	$\dfrac{1}{2}$	0
$\tan \theta$	0	$\dfrac{1}{\sqrt{3}}$	1	$\sqrt{3}$	undefined

Trigonometric Identities

Reciprocal identities

$$\sin \theta = \frac{1}{\csc \theta} \qquad \cos \theta = \frac{1}{\sec \theta} \qquad \tan \theta = \frac{1}{\cot \theta}$$

$$\csc \theta = \frac{1}{\sin \theta} \qquad \sec \theta = \frac{1}{\cos \theta} \qquad \cot \theta = \frac{1}{\tan \theta}$$

$$\tan \theta = \frac{\sin \theta}{\cos \theta} \qquad \cot \theta = \frac{\cos \theta}{\sin \theta}$$

Pythagorean identities

$$\sin^2 \theta + \cos^2 \theta = 1 \qquad \tan^2 \theta + 1 = \sec^2 \theta \qquad \cot^2 \theta + 1 = \csc^2 \theta$$

Reduction formulas

$$\sin(-\theta) = -\sin \theta \qquad \cos(-\theta) = \cos \theta \qquad \tan(-\theta) = -\tan \theta$$
$$\sin \theta = -\sin(\theta - \pi) \qquad \cos \theta = -\cos(\theta - \pi) \qquad \tan \theta = \tan(\theta - \pi)$$

Sum or difference of two angles

$$\sin(\theta \pm \phi) = \sin \theta \cos \phi \pm \cos \theta \sin \phi$$
$$\cos(\theta \pm \phi) = \cos \theta \cos \phi \mp \sin \theta \sin \phi$$
$$\tan(\theta \pm \phi) = \frac{\tan \theta \pm \tan \phi}{1 \mp \tan \theta \tan \phi}$$
$$\sin(\theta + \phi) \sin(\theta - \phi) = \sin^2 \theta - \sin^2 \phi$$
$$\cos(\theta + \phi) \cos(\theta - \phi) = \cos^2 \theta - \sin^2 \phi$$

Double-angle identities

$$\sin 2\theta = 2 \sin \theta \cos \theta$$
$$\cos 2\theta = 2 \cos^2 \theta - 1 = 1 - 2 \sin^2 \theta$$
$$\tan 2\theta = \frac{2 \tan \theta}{1 - \tan^2 \theta}$$

Multiple-angle identities

$$\sin 3\theta = 3 \sin \theta - 4 \sin^3 \theta$$
$$\cos 3\theta = -3 \cos \theta + 4 \cos^3 \theta$$
$$\tan 3\theta = \frac{3 \tan \theta - \tan^2 \theta}{1 - 3 \tan^2 \theta}$$
$$\sin 4\theta = 4 \sin \theta \cos \theta - 8 \sin^3 \theta \cos \theta$$
$$\cos 4\theta = 8 \cos^4 \theta - 8 \cos^2 \theta + 1$$
$$\tan 4\theta = \frac{4 \tan \theta - 4 \tan^3 \theta}{1 - 6 \tan^2 \theta + \tan^4 \theta}$$

Half-angle identities

$$\sin^2 \theta = \frac{1}{2}(1 - \cos 2\theta)$$
$$\cos^2 \theta = \frac{1}{2}(1 + \cos 2\theta)$$

Product identities

$$\sin \theta \sin \phi = \frac{1}{2} \cos(\theta - \phi) - \frac{1}{2} \cos(\theta + \phi)$$
$$\cos \theta \cos \phi = \frac{1}{2} \cos(\theta - \phi) + \frac{1}{2} \cos(\theta + \phi)$$
$$\cos \theta \sin \phi = \frac{1}{2} \sin(\theta - \phi) + \frac{1}{2} \sin(\theta + \phi)$$

REFERENCE TABLE 17 Tables of Trigonometric Functions

1 degree ≈ 0.01745 radians
1 radian ≈ 57.29578 degrees

For $0 \leqslant \theta \leqslant 45$, read from upper left.
For $45 \leqslant \theta \leqslant 90$, read from lower right.
For $90 \leqslant \theta \leqslant 360$, use the identities:

θ	Quadrant II	Quadrant III	Quadrant IV
$\sin\theta$	$\sin(180 - \theta)$	$-\sin(\theta - 180)$	$-\sin(360 - \theta)$
$\cos\theta$	$-\cos(180 - \theta)$	$-\cos(\theta - 180)$	$\cos(360 - \theta)$
$\tan\theta$	$-\tan(180 - \theta)$	$\tan(\theta - 180)$	$-\tan(360 - \theta)$
$\cot\theta$	$-\cot(180 - \theta)$	$\cot(\theta - 180)$	$-\cot(360 - \theta)$

Degrees	Radians	sin	cos	tan	cot		
0° 00′	.0000	.0000	1.0000	.0000	–	1.5708	90° 00′
10	.0029	.0029	1.0000	.0029	343.774	1.5679	50
20	.0058	.0058	1.0000	.0058	171.885	1.5650	40
30	.0087	.0087	1.0000	.0087	114.589	1.5621	30
40	.0116	.0116	.9999	.0116	85.940	1.5592	20
50	.0145	.0145	.9999	.0145	68.750	1.5563	10
1° 00′	.0175	.0175	.9998	.0175	57.290	1.5533	89° 00′
10	.0204	.0204	.9998	.0204	49.104	1.5504	50
20	.0233	.0233	.9997	.0233	42.964	1.5475	40
30	.0262	.0262	.9997	.0262	38.188	1.5446	30
40	.0291	.0291	.9996	.0291	34.368	1.5417	20
50	.0320	.0320	.9995	.0320	31.242	1.5388	10
2° 00′	.0349	.0349	.9994	.0349	28.636	1.5359	88° 00′
10	.0378	.0378	.9993	.0378	26.432	1.5330	50
20	.0407	.0407	.9992	.0407	24.542	1.5301	40
30	.0436	.0436	.9990	.0437	22.904	1.5272	30
40	.0465	.0465	.9989	.0466	21.470	1.5243	20
50	.0495	.0494	.9988	.0495	20.206	1.5213	10
3° 00′	.0524	.0523	.9986	.0524	19.081	1.5184	87° 00′
10	.0553	.0552	.9985	.0553	18.075	1.5155	50
20	.0582	.0581	.9983	.0582	17.169	1.5126	40
30	.0611	.0610	.9981	.0612	16.350	1.5097	30
40	.0640	.0640	.9980	.0641	15.605	1.5068	20
50	.0669	.0669	.9978	.0670	14.924	1.5039	10
4° 00′	.0698	.0698	.9976	.0699	14.301	1.5010	86° 00′
10	.0727	.0727	.9974	.0729	13.727	1.4981	50
20	.0756	.0756	.9971	.0758	13.197	1.4952	40
30	.0785	.0785	.9969	.0787	12.706	1.4923	30
40	.0814	.0814	.9967	.0816	12.251	1.4893	20
50	.0844	.0843	.9964	.0846	11.826	1.4864	10
5° 00′	.0873	.0872	.9962	.0875	11.430	1.4835	85° 00′
10	.0902	.0901	.9959	.0904	11.059	1.4806	50
20	.0931	.0929	.9957	.0934	10.712	1.4777	40
30	.0960	.0958	.9954	.0963	10.385	1.4748	30
40	.0989	.0987	.9951	.0992	10.078	1.4719	20
50	.1018	.1016	.9948	.1022	9.788	1.4690	10
6° 00′	.1047	.1045	.9945	.1051	9.514	1.4661	84° 00′
10	.1076	.1074	.9942	.1080	9.255	1.4632	50
20	.1105	.1103	.9939	.1110	9.010	1.4603	40
30	.1134	.1132	.9936	.1139	8.777	1.4573	30
40	.1164	.1161	.9932	.1169	8.556	1.4544	20
50	.1193	.1190	.9929	.1198	8.345	1.4515	10
7° 00′	.1222	.1219	.9925	.1228	8.144	1.4486	83° 00′
10	.1251	.1248	.9922	.1257	7.953	1.4457	50
20	.1280	.1276	.9918	.1287	7.770	1.4428	40
30	.1309	.1305	.9914	.1317	7.596	1.4399	30
40	.1338	.1334	.9911	.1346	7.429	1.4370	20
50	.1367	.1363	.9907	.1376	7.269	1.4341	10
		cos	sin	cot	tan	Radians	Degrees

Degrees	Radians	sin	cos	tan	cot		
8°00′	.1396	.1392	.9903	.1405	7.115	1.4312	82°00′
10	.1425	.1421	.9899	.1435	6.968	1.4283	50
20	.1454	.1449	.9894	.1465	6.827	1.4254	40
30	.1484	.1478	.9890	.1495	6.691	1.4224	30
40	.1513	.1507	.9886	.1524	6.561	1.4195	20
50	.1542	.1536	.9881	.1554	6.435	1.4166	10
9°00′	.1571	.1564	.9877	.1584	6.314	1.4137	81°00′
10	.1600	.1593	.9872	.1614	6.197	1.4108	50
20	.1629	.1622	.9868	.1644	6.084	1.4079	40
30	.1658	.1650	.9863	.1673	5.976	1.4050	30
40	.1687	.1679	.9858	.1703	5.871	1.4021	20
50	.1716	.1708	.9853	.1733	5.769	1.3992	10
10°00′	.1745	.1736	.9848	.1763	5.671	1.3963	80°00′
10	.1774	.1765	.9843	.1793	5.576	1.3934	50
20	.1804	.1794	.9838	.1823	5.485	1.3904	40
30	.1833	.1822	.9833	.1853	5.396	1.3875	30
40	.1862	.1851	.9827	.1883	5.309	1.3846	20
50	.1891	.1880	.9822	.1914	5.226	1.3817	10
11°00′	.1920	.1908	.9816	.1944	5.145	1.3788	79°00′
10	.1949	.1937	.9811	.1974	5.066	1.3759	50
20	.1978	.1965	.9805	.2004	4.989	1.3730	40
30	.2007	.1994	.9799	.2035	4.915	1.3701	30
40	.2036	.2022	.9793	.2065	4.843	1.3672	20
50	.2065	.2051	.9787	.2095	4.773	1.3643	10
12°00′	.2094	.2079	.9781	.2126	4.705	1.3614	78°00′
10	.2123	.2108	.9775	.2156	4.638	1.3584	50
20	.2153	.2136	.9769	.2186	4.574	1.3555	40
30	.2182	.2164	.9763	.2217	4.511	1.3526	30
40	.2211	.2193	.9757	.2247	4.449	1.3497	20
50	.2240	.2221	.9750	.2278	4.390	1.3468	10
13°00′	.2269	.2250	.9744	.2309	4.331	1.3439	77°00′
10	.2298	.2278	.9737	.2339	4.275	1.3410	50
20	.2327	.2306	.9730	.2370	4.219	1.3381	40
30	.2356	.2334	.9724	.2401	4.165	1.3352	30
40	.2385	.2363	.9717	.2432	4.113	1.3323	20
50	.2414	.2391	.9710	.2462	4.061	1.3294	10
14°00′	.2443	.2419	.9703	.2493	4.011	1.3265	76°00′
10	.2473	.2447	.9696	.2524	3.962	1.3235	50
20	.2502	.2476	.9689	.2555	3.914	1.3206	40
30	.2531	.2504	.9681	.2586	3.867	1.3177	30
40	.2560	.2532	.9674	.2617	3.821	1.3148	20
50	.2589	.2560	.9667	.2648	3.776	1.3119	10
15°00′	.2618	.2588	.9659	.2679	3.732	1.3090	75°00′
10	.2647	.2616	.9652	.2711	3.689	1.3061	50
20	.2676	.2644	.9644	.2742	3.647	1.3032	40
30	.2705	.2672	.9636	.2773	3.606	1.3003	30
40	.2734	.2700	.9628	.2805	3.566	1.2974	20
50	.2763	.2728	.9621	.2836	3.526	1.2945	10
16°00′	.2793	.2756	.9613	.2867	3.487	1.2915	74°00′
10	.2822	.2784	.9605	.2899	3.450	1.2886	50
20	.2851	.2812	.9596	.2931	3.412	1.2857	40
30	.2880	.2840	.9588	.2962	3.376	1.2828	30
40	.2909	.2868	.9580	.2994	3.340	1.2799	20
50	.2938	.2896	.9572	.3026	3.305	1.2770	10
17°00′	.2967	.2924	.9563	.3057	3.271	1.2741	73°00′
10	.2996	.2952	.9555	.3089	3.237	1.2712	50
20	.3025	.2979	.9546	.3121	3.204	1.2683	40
30	.3054	.3007	.9537	.3153	3.172	1.2654	30
40	.3083	.3035	.9528	.3185	3.140	1.2625	20
50	.3113	.3062	.9520	.3217	3.108	1.2595	10
	cos	sin	cot	tan	Radians	Degrees	

Degrees	Radians	sin	cos	tan	cot		
18°00′	.3142	.3090	.9511	.3249	3.078	1.2566	72°00′
10	.3171	.3118	.9502	.3281	3.047	1.2537	50
20	.3200	.3145	.9492	.3314	3.018	1.2508	40
30	.3229	.3173	.9483	.3346	2.989	1.2479	30
40	.3258	.3201	.9474	.3378	2.960	1.2450	20
50	.3287	.3228	.9465	.3411	2.932	1.2421	10
19°00′	.3316	.3256	.9455	.3443	2.904	1.2392	71°00′
10	.3345	.3283	.9446	.3476	2.877	1.2363	50
20	.3374	.3311	.9436	.3508	2.850	1.2334	40
30	.3403	.3338	.9426	.3541	2.824	1.2305	30
40	.3432	.3365	.9417	.3574	2.798	1.2275	20
50	.3462	.3393	.9407	.3607	2.773	1.2246	10
20°00′	.3491	.3420	.9397	.3640	2.747	1.2217	70°00′
10	.3520	.3448	.9387	.3673	2.723	1.2188	50
20	.3549	.3475	.9377	.3706	2.699	1.2159	40
30	.3578	.3502	.9367	.3739	2.675	1.2130	30
40	.3607	.3529	.9356	.3772	2.651	1.2101	20
50	.3636	.3557	.9346	.3805	2.628	1.2072	10
21°00′	.3665	.3584	.9336	.3839	2.605	1.2043	69°00′
10	.3694	.3611	.9325	.3872	2.583	1.2014	50
20	.3723	.3638	.9315	.3906	2.560	1.1985	40
30	.3752	.3665	.9304	.3939	2.539	1.1956	30
40	.3782	.3692	.9293	.3973	2.517	1.1926	20
50	.3811	.3719	.9283	.4006	2.496	1.1897	10
22°00′	.3840	.3746	.9272	.4040	2.475	1.1868	68°00′
10	.3869	.3773	.9261	.4074	2.455	1.1839	50
20	.3898	.3800	.9250	.4108	2.434	1.1810	40
30	.3927	.3827	.9239	.4142	2.414	1.1781	30
40	.3956	.3854	.9228	.4176	2.394	1.1752	20
50	.3985	.3881	.9216	.4210	2.375	1.1723	10
23°00′	.4014	.3907	.9205	.4245	2.356	1.1694	67°00′
10	.4043	.3934	.9194	.4279	2.337	1.1665	50
20	.4072	.3961	.9182	.4314	2.318	1.1636	40
30	.4102	.3987	.9171	.4348	2.300	1.1606	30
40	.4131	.4014	.9159	.4383	2.282	1.1577	20
50	.4160	.4041	.9147	.4417	2.264	1.1548	10
24°00′	.4189	.4067	.9135	.4452	2.246	1.1519	66°00′
10	.4218	.4094	.9124	.4487	2.229	1.1490	50
20	.4247	.4120	.9112	.4522	2.211	1.1461	40
30	.4276	.4147	.9100	.4557	2.194	1.1432	30
40	.4305	.4173	.9088	.4592	2.177	1.1403	20
50	.4334	.4200	.9075	.4628	2.161	1.1374	10
25°00′	.4363	.4226	.9063	.4663	2.145	1.1345	65°00′
10	.4392	.4253	.9051	.4699	2.128	1.1316	50
20	.4422	.4279	.9038	.4734	2.112	1.1286	40
30	.4451	.4305	.9026	.4770	2.097	1.1257	30
40	.4480	.4331	.9013	.4806	2.081	1.1228	20
50	.4509	.4358	.9001	.4841	2.066	1.1199	10
26°00′	.4538	.4384	.8988	.4877	2.050	1.1170	64°00′
10	.4567	.4410	.8975	.4913	2.035	1.1141	50
20	.4596	.4436	.8962	.4950	2.020	1.1112	40
30	.4625	.4462	.8949	.4986	2.006	1.1083	30
40	.4654	.4488	.8936	.5022	1.991	1.1054	20
50	.4683	.4514	.8923	.5059	1.977	1.1025	10
27°00′	.4712	.4540	.8910	.5095	1.963	1.0996	63°00′
10	.4741	.4566	.8897	.5132	1.949	1.0966	50
20	.4771	.4592	.8884	.5169	1.935	1.0937	40
30	.4800	.4617	.8870	.5206	1.921	1.0908	30
40	.4829	.4643	.8857	.5243	1.907	1.0879	20
50	.4858	.4669	.8843	.5280	1.894	1.0850	10
	cos	sin	cot	tan	Radians	Degrees	

REFERENCE TABLE 17   (Continued)

Degrees	Radians	sin	cos	tan	cot		
28°00′	.4887	.4695	.8829	.5317	1.881	1.0821	62°00′
10	.4916	.4720	.8816	.5354	1.868	1.0792	50
20	.4945	.4746	.8802	.5392	1.855	1.0763	40
30	.4974	.4772	.8788	.5430	1.842	1.0734	30
40	.5003	.4797	.8774	.5467	1.829	1.0705	20
50	.5032	.4823	.8760	.5505	1.816	1.0676	10
29°00′	.5061	.4848	.8746	.5543	1.804	1.0647	61°00′
10	.5091	.4874	.8732	.5581	1.792	1.0617	50
20	.5120	.4899	.8718	.5619	1.780	1.0588	40
30	.5149	.4924	.8704	.5658	1.767	1.0559	30
40	.5178	.4950	.8689	.5696	1.756	1.0530	20
50	.5207	.4975	.8675	.5735	1.744	1.0501	10
30°00′	.5236	.5000	.8660	.5774	1.732	1.0472	60°00′
10	.5265	.5025	.8646	.5812	1.720	1.0443	50
20	.5294	.5050	.8631	.5851	1.709	1.0414	40
30	.5323	.5075	.8616	.5890	1.698	1.0385	30
40	.5352	.5100	.8601	.5930	1.686	1.0356	20
50	.5381	.5125	.8587	.5969	1.675	1.0327	10
31°00′	.5411	.5150	.8572	.6009	1.664	1.0297	59°00′
10	.5440	.5175	.8557	.6048	1.653	1.0268	50
20	.5469	.5200	.8542	.6088	1.643	1.0239	40
30	.5498	.5225	.8526	.6128	1.632	1.0210	30
40	.5527	.5250	.8511	.6168	1.621	1.0181	20
50	.5556	.5275	.8496	.6208	1.611	1.0152	10
32°00′	.5585	.5299	.8480	.6249	1.600	1.0123	58°00′
10	.5614	.5324	.8465	.6289	1.590	1.0094	50
20	.5643	.5348	.8450	.6330	1.580	1.0065	40
30	.5672	.5373	.8434	.6371	1.570	1.0036	30
40	.5701	.5398	.8418	.6412	1.560	1.0007	20
50	.5730	.5422	.8403	.6453	1.550	.9977	10
33°00′	.5760	.5446	.8387	.6494	1.540	.9948	57°00′
10	.5789	.5471	.8371	.6536	1.530	.9919	50
20	.5818	.5495	.8355	.6577	1.520	.9890	40
30	.5847	.5519	.8339	.6619	1.511	.9861	30
40	.5876	.5544	.8323	.6661	1.501	.9832	20
50	.5905	.5568	.8307	.6703	1.492	.9803	10
34°00′	.5934	.5592	.8290	.6745	1.483	.9774	56°00′
10	.5963	.5616	.8274	.6787	1.473	.9745	50
20	.5992	.5640	.8258	.6830	1.464	.9716	40
30	.6021	.5664	.8241	.6873	1.455	.9687	30
40	.6050	.5688	.8225	.6916	1.446	.9657	20
50	.6080	.5712	.8208	.6959	1.437	.9628	10
35°00′	.6109	.5736	.8192	.7002	1.428	.9599	55°00′
10	.6138	.5760	.8175	.7046	1.419	.9570	50
20	.6167	.5783	.8158	.7089	1.411	.9541	40
30	.6196	.5807	.8141	.7133	1.402	.9512	30
40	.6225	.5831	.8124	.7177	1.393	.9483	20
50	.6254	.5854	.8107	.7221	1.385	.9454	10
36°00′	.6283	.5878	.8090	.7265	1.376	.9425	54°00′
10	.6312	.5901	.8073	.7310	1.368	.9396	50
20	.6341	.5925	.8056	.7355	1.360	.9367	40
30	.6370	.5948	.8039	.7400	1.351	.9338	30
40	.6400	.5972	.8021	.7445	1.343	.9308	20
50	.6429	.5995	.8004	.7490	1.335	.9279	10
		cos	sin	cot	tan	Radians	Degrees

Degrees	Radians	sin	cos	tan	cot		
37°00′	.6458	.6018	.7986	.7536	1.327	.9250	53°00′
10	.6487	.6041	.7969	.7581	1.319	.9221	50
20	.6516	.6065	.7951	.7627	1.311	.9192	40
30	.6545	.6088	.7934	.7673	1.303	.9163	30
40	.6574	.6111	.7916	.7720	1.295	.9134	20
50	.6603	.6134	.7898	.7766	1.288	.9105	10
38°00′	.6632	.6157	.7880	.7813	1.280	.9076	52°00′
10	.6661	.6180	.7862	.7860	1.272	.9047	50
20	.6690	.6202	.7844	.7907	1.265	.9018	40
30	.6720	.6225	.7826	.7954	1.257	.8988	30
40	.6749	.6248	.7808	.8002	1.250	.8959	20
50	.6778	.6271	.7790	.8050	1.242	.8930	10
39°00′	.6807	.6293	.7771	.8098	1.235	.8901	51°00′
10	.6836	.6316	.7753	.8146	1.228	.8872	50
20	.6865	.6338	.7735	.8195	1.220	.8843	40
30	.6894	.6361	.7716	.8243	1.213	.8814	30
40	.6923	.6383	.7698	.8292	1.206	.8785	20
50	.6952	.6406	.7679	.8342	1.199	.8756	10
40°00′	.6981	.6428	.7660	.8391	1.192	.8727	50°00′
10	.7010	.6450	.7642	.8441	1.185	.8698	50
20	.7039	.6472	.7623	.8491	1.178	.8668	40
30	.7069	.6494	.7604	.8541	1.171	.8639	30
40	.7098	.6517	.7585	.8591	1.164	.8610	20
50	.7127	.6539	.7566	.8642	1.157	.8581	10
41°00′	.7156	.6561	.7547	.8693	1.150	.8552	49°00′
10	.7185	.6583	.7528	.8744	1.144	.8523	50
20	.7214	.6604	.7509	.8796	1.137	.8494	40
30	.7243	.6626	.7490	.8847	1.130	.8465	30
40	.7272	.6648	.7470	.8899	1.124	.8436	20
50	.7301	.6670	.7451	.8952	1.117	.8407	10
42°00′	.7330	.6691	.7431	.9004	1.111	.8378	48°00′
10	.7359	.6713	.7412	.9057	1.104	.8348	50
20	.7389	.6734	.7392	.9110	1.098	.8319	40
30	.7418	.6756	.7373	.9163	1.091	.8290	30
40	.7447	.6777	.7353	.9217	1.085	.8261	20
50	.7476	.6799	.7333	.9271	1.079	.8232	10
43°00′	.7505	.6820	.7314	.9325	1.072	.8203	47°00′
10	.7534	.6841	.7294	.9380	1.066	.8174	50
20	.7563	.6862	.7274	.9435	1.060	.8145	40
30	.7592	.6884	.7254	.9490	1.054	.8116	30
40	.7621	.6905	.7234	.9545	1.048	.8087	20
50	.7650	.6926	.7214	.9601	1.042	.8058	10
44°00′	.7679	.6947	.7193	.9657	1.036	.8029	46°00′
10	.7709	.6967	.7173	.9713	1.030	.7999	50
20	.7738	.6988	.7153	.9770	1.024	.7970	40
30	.7767	.7009	.7133	.9827	1.018	.7941	30
40	.7796	.7030	.7112	.9884	1.012	.7912	20
50	.7825	.7050	.7092	.9942	1.006	.7883	10
45°00′	.7854	.7071	.7071	1.0000	1.000	.7854	45°00′
		cos	sin	cot	tan	Radians	Degrees

REFERENCE TABLE 18   *Plane Analytic Geometry*

Distance Between (x_1, y_1) and (x_2, y_2)

$$d = \sqrt{(x_2 - x_1)^2 + (y_2 - y_1)^2}$$

Midpoint Between (x_1, y_1) and (x_2, y_2)

$$\text{Midpoint} = \left(\frac{x_1 + x_2}{2}, \frac{y_1 + y_2}{2} \right)$$

Slope of Line Passing Through (x_1, y_1) and (x_2, y_2)

$$m = \frac{y_2 - y_1}{x_2 - x_1}$$

Slopes of Parallel Lines

$$m_1 = m_2$$

Slopes of Perpendicular Lines

$$m_1 = -\frac{1}{m_2}$$

Equations of Lines

Point-slope form: $y - y_1 = m(x - x_1)$ General form: $Ax + By + C = 0$
Vertical line: $x = a$ Horizontal line: $y = b$

Equations of Circles [Center: (h, k), Radius: r]

Standard form: $(x - h)^2 + (y - k)^2 = r^2$
General form: $Ax^2 + Ay^2 + Dx + Ey + F = 0$

REFERENCE TABLE 18   (Continued)

Equations of Parabolas [Vertex: (h, k)]

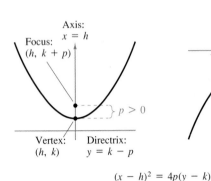

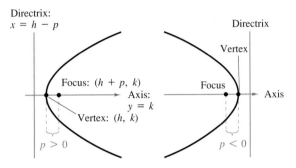

$$(x - h)^2 = 4p(y - k)$$

$$(y - k)^2 = 4p(x - h)$$

(a) Vertical axis: $p > 0$ (b) Vertical axis: $p < 0$ (c) Horizontal axis: $p > 0$ (d) Horizontal axis: $p < 0$

Equations of Ellipses [Center: (h, k)]

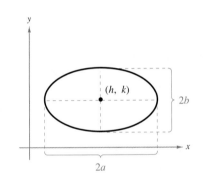

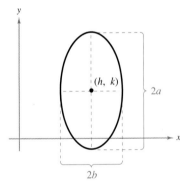

$$\frac{(x - h)^2}{a^2} + \frac{(y - k)^2}{b^2} = 1$$

$$\frac{(x - h)^2}{b^2} + \frac{(y - k)^2}{a^2} = 1$$

Equations of Hyperbolas [Center: (h, k)]

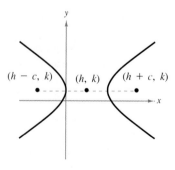

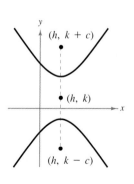

$$\frac{(x - h)^2}{a^2} - \frac{(y - k)^2}{b^2} = 1$$

$$\frac{(y - k)^2}{a^2} - \frac{(x - h)^2}{b^2} = 1$$

REFERENCE TABLE 19   Solid Analytic Geometry

Distance Between (x_1, y_1, z_1) and (x_2, y_2, z_2)

$$d = \sqrt{(x_2 - x_1)^2 + (y_2 - y_1)^2 + (z_2 - z_1)^2}$$

Midpoint Between (x_1, y_1, z_1) and (x_2, y_2, z_2)

$$\text{Midpoint} = \left(\frac{x_1 + x_2}{2}, \frac{y_1 + y_2}{2}, \frac{z_1 + z_2}{2} \right)$$

Equation of Plane

$$Ax + By + Cz + D = 0$$

Equation of Sphere [Center: (h, k, l), Radius: r]

$$(x - h)^2 + (y - k)^2 + (z - l)^2 = r^2$$

REFERENCE TABLE 20   Differentiation Formulas

1. $\dfrac{d}{dx}[cu] = cu'$

2. $\dfrac{d}{dx}[u \pm v] = u' \pm v'$

3. $\dfrac{d}{dx}[uv] = uv' + vu'$

4. $\dfrac{d}{dx}\left[\dfrac{u}{v}\right] = \dfrac{vu' - uv'}{v^2}$

5. $\dfrac{d}{dx}[c] = 0$

6. $\dfrac{d}{dx}[u^n] = nu^{n-1}u'$

7. $\dfrac{d}{dx}[x] = 1$

8. $\dfrac{d}{dx}[|u|] = \dfrac{u}{|u|}(u'), \ u \neq 0$

9. $\dfrac{d}{dx}[\ln u] = \dfrac{u'}{u}$

10. $\dfrac{d}{dx}[e^u] = e^u u'$

11. $\dfrac{d}{dx}[\sin u] = (\cos u)u'$

12. $\dfrac{d}{dx}[\cos u] = -(\sin u)u'$

13. $\dfrac{d}{dx}[\tan u] = (\sec^2 u)u'$

14. $\dfrac{d}{dx}[\cot u] = -(\csc^2 u)u'$

15. $\dfrac{d}{dx}[\sec u] = (\sec u \tan u)u'$

16. $\dfrac{d}{dx}[\csc u] = -(\csc u \cot u)u'$

REFERENCE TABLE 21   Integration Formulas

Forms Involving u^n

1. $\displaystyle \int u^n \, du = \frac{u^{n+1}}{n+1} + C, \; n \neq -1$

2. $\displaystyle \int \frac{1}{u} \, du = \ln |u| + C$

Forms Involving $a + bu$

3. $\displaystyle \int \frac{u}{a+bu} \, du = \frac{1}{b^2}[bu - a \ln |a+bu|] + C$

4. $\displaystyle \int \frac{u}{(a+bu)^2} \, du = \frac{1}{b^2}\left[\frac{a}{a+bu} + \ln |a+bu|\right] + C$

5. $\displaystyle \int \frac{u}{(a+bu)^n} \, du = \frac{1}{b^2}\left[\frac{-1}{(n-2)(a+bu)^{n-2}} + \frac{a}{(n-1)(a+bu)^{n-1}}\right] + C, \; n \neq 1, 2$

6. $\displaystyle \int \frac{u^2}{a+bu} \, du = \frac{1}{b^3}\left[-\frac{bu}{2}(2a - bu) + a^2 \ln |a+bu|\right] + C$

7. $\displaystyle \int \frac{u^2}{(a+bu)^2} \, du = \frac{1}{b^3}\left[bu - \frac{a^2}{a+bu} - 2a \ln |a+bu|\right] + C$

8. $\displaystyle \int \frac{u^2}{(a+bu)^3} \, du = \frac{1}{b^3}\left[\frac{2a}{a+bu} - \frac{a^2}{2(a+bu)^2} + \ln |a+bu|\right] + C$

9. $\displaystyle \int \frac{u^2}{(a+bu)^n} \, du = \frac{1}{b^3}\left[\frac{-1}{(n-3)(a+bu)^{n-3}} + \frac{2a}{(n-2)(a+bu)^{n-2}} - \frac{a^2}{(n-1)(a+bu)^{n-1}}\right] + C, \; n \neq 1, 2, 3$

10. $\displaystyle \int \frac{1}{u(a+bu)} \, du = \frac{1}{a} \ln \left|\frac{u}{a+bu}\right| + C$

11. $\displaystyle \int \frac{1}{u(a+bu)^2} \, du = \frac{1}{a}\left[\frac{1}{a+bu} + \frac{1}{a} \ln \left|\frac{u}{a+bu}\right|\right] + C$

12. $\displaystyle \int \frac{1}{u^2(a+bu)} \, du = -\frac{1}{a}\left[\frac{1}{u} + \frac{b}{a} \ln \left|\frac{u}{a+bu}\right|\right] + C$

13. $\displaystyle \int \frac{1}{u^2(a+bu)^2} \, du = -\frac{1}{a^2}\left[\frac{a+2bu}{u(a+bu)} + \frac{2b}{a} \ln \left|\frac{u}{a+bu}\right|\right] + C$

Forms Involving $\sqrt{a + bu}$

14. $\displaystyle\int u^n\sqrt{a + bu}\ du = \frac{2}{b(2n + 3)}\left[u^n(a + bu)^{3/2} - na\int u^{n-1}\sqrt{a + bu}\ du\right]$

15. $\displaystyle\int \frac{1}{u\sqrt{a + bu}}\ du = \frac{1}{\sqrt{a}}\ln\left|\frac{\sqrt{a + bu} - \sqrt{a}}{\sqrt{a + bu} + \sqrt{a}}\right| + C,\ 0 < a$

16. $\displaystyle\int \frac{1}{u^n\sqrt{a + bu}}\ du = \frac{-1}{a(n - 1)}\left[\frac{\sqrt{a + bu}}{u^{n-1}} + \frac{(2n - 3)b}{2}\int \frac{1}{u^{n-1}\sqrt{a + bu}}\ du\right],\ n \neq 1$

17. $\displaystyle\int \frac{\sqrt{a + bu}}{u}\ du = 2\sqrt{a + bu} + a\int \frac{1}{u\sqrt{a + bu}}\ du$

18. $\displaystyle\int \frac{\sqrt{a + bu}}{u^n}\ du = \frac{-1}{a(n - 1)}\left[\frac{(a + bu)^{3/2}}{u^{n-1}} + \frac{(2n - 5)b}{2}\int \frac{\sqrt{a + bu}}{u^{n-1}}\ du\right],\ n \neq 1$

19. $\displaystyle\int \frac{u}{\sqrt{a + bu}}\ du = -\frac{2(2a - bu)}{3b^2}\sqrt{a + bu} + C$

20. $\displaystyle\int \frac{u^n}{\sqrt{a + bu}}\ du = \frac{2}{(2n + 1)b}\left[u^n\sqrt{a + bu} - na\int \frac{u^{n-1}}{\sqrt{a + bu}}\ du\right]$

Forms Involving $u^2 - a^2,\ 0 < a$

21. $\displaystyle\int \frac{1}{u^2 - a^2}\ du = -\int \frac{1}{a^2 - u^2}\ du = \frac{1}{2a}\ln\left|\frac{u - a}{u + a}\right| + C$

22. $\displaystyle\int \frac{1}{(u^2 - a^2)^n}\ du = \frac{-1}{2a^2(n - 1)}\left[\frac{u}{(u^2 - a^2)^{n-1}} + (2n - 3)\int \frac{1}{(u^2 - a^2)^{n-1}}\ du\right],\ n \neq 1$

Forms Involving $\sqrt{u^2 \pm a^2},\ 0 < a$

23. $\displaystyle\int \sqrt{u^2 \pm a^2}\ du = \frac{1}{2}[u\sqrt{u^2 \pm a^2} \pm a^2\ln|u + \sqrt{u^2 \pm a^2}|] + C$

24. $\displaystyle\int u^2\sqrt{u^2 \pm a^2}\ du = \frac{1}{8}[u(2u^2 \pm a^2)\sqrt{u^2 \pm a^2} - a^4\ln|u + \sqrt{u^2 \pm a^2}|] + C$

25. $\displaystyle\int \frac{\sqrt{u^2 + a^2}}{u}\ du = \sqrt{u^2 + a^2} - a\ln\left|\frac{a + \sqrt{u^2 + a^2}}{u}\right| + C$

26. $\displaystyle\int \frac{\sqrt{u^2 \pm a^2}}{u^2}\ du = \frac{-\sqrt{u^2 \pm a^2}}{u} + \ln|u + \sqrt{u^2 \pm a^2}| + C$

REFERENCE TABLE 21   (Continued)

27. $\int \dfrac{1}{\sqrt{u^2 \pm a^2}} \, du = \ln |u + \sqrt{u^2 \pm a^2}| + C$

28. $\int \dfrac{1}{u\sqrt{u^2 + a^2}} \, du = \dfrac{-1}{a} \ln \left| \dfrac{a + \sqrt{u^2 + a^2}}{u} \right| + C$

29. $\int \dfrac{u^2}{\sqrt{u^2 \pm a^2}} \, du = \dfrac{1}{2}[u\sqrt{u^2 \pm a^2} \mp a^2 \ln |u + \sqrt{u^2 \pm a^2}|] + C$

30. $\int \dfrac{1}{u^2\sqrt{u^2 \pm a^2}} \, du = \mp \dfrac{\sqrt{u^2 \pm a^2}}{a^2 u} + C$

31. $\int \dfrac{1}{(u^2 \pm a^2)^{3/2}} \, du = \dfrac{\pm u}{a^2\sqrt{u^2 \pm a^2}} + C$

Forms Involving $\sqrt{a^2 - u^2}$, $0 < a$

32. $\int \dfrac{\sqrt{a^2 - u^2}}{u} \, du = \sqrt{a^2 - u^2} - a \ln \left| \dfrac{a + \sqrt{a^2 - u^2}}{u} \right| + C$

33. $\int \dfrac{1}{u\sqrt{a^2 - u^2}} \, du = \dfrac{-1}{a} \ln \left| \dfrac{a + \sqrt{a^2 - u^2}}{u} \right| + C$

34. $\int \dfrac{1}{u^2\sqrt{a^2 - u^2}} \, du = \dfrac{-\sqrt{a^2 - u^2}}{a^2 u} + C$

35. $\int \dfrac{1}{(a^2 - u^2)^{3/2}} \, du = \dfrac{u}{a^2\sqrt{a^2 - u^2}} + C$

Forms Involving e^u

36. $\int e^u \, du = e^u + C$

37. $\int u e^u \, du = (u - 1)e^u + C$

38. $\int u^n e^u \, du = u^n e^u - n \int u^{n-1} e^u \, du$

39. $\int \dfrac{1}{1 + e^u} \, du = u - \ln (1 + e^u) + C$

40. $\int \dfrac{1}{1 + e^{nu}} \, du = u - \dfrac{1}{n} \ln (1 + e^{nu}) + C$

Forms Involving ln u

41. $\displaystyle\int \ln u \; du = u[-1 + \ln u] + C$

42. $\displaystyle\int u \ln u \; du = \frac{u^2}{4}[-1 + 2 \ln u] + C$

43. $\displaystyle\int u^n \ln u \; du = \frac{u^{n+1}}{(n+1)^2}[-1 + (n+1) \ln u] + C, \; n \neq -1$

44. $\displaystyle\int (\ln u)^2 \; du = u[2 - 2 \ln u + (\ln u)^2] + C$

45. $\displaystyle\int (\ln u)^n \; du = u(\ln u)^n - n \int (\ln u)^{n-1} \; du$

Forms Involving sin u or cos u

46. $\displaystyle\int \sin u \; du = -\cos u + C$

47. $\displaystyle\int \cos u \; du = \sin u + C$

48. $\displaystyle\int \sin^2 u \; du = \frac{1}{2}(u - \sin u \cos u) + C$

49. $\displaystyle\int \cos^2 u \; du = \frac{1}{2}(u + \sin u \cos u) + C$

50. $\displaystyle\int \sin^n u \; du = -\frac{\sin^{n-1} u \cos u}{n} + \frac{n-1}{n} \int \sin^{n-2} u \; du$

51. $\displaystyle\int \cos^n u \; du = \frac{\cos^{n-1} u \sin u}{n} + \frac{n-1}{n} \int \cos^{n-2} u \; du$

52. $\displaystyle\int u \sin u \; du = \sin u - u \cos u + C$

53. $\displaystyle\int u \cos u \; du = \cos u + u \sin u + C$

54. $\displaystyle\int u^n \sin u \; du = -u^n \cos u + n \int u^{n-1} \cos u \; du$

55. $\displaystyle\int u^n \cos u \; du = u^n \sin u - n \int u^{n-1} \sin u \; du$

56. $\displaystyle\int \frac{1}{1 \pm \sin u} \; du = \tan u \mp \sec u + C$

57. $\displaystyle\int \frac{1}{1 \pm \cos u} \; du = -\cot u \pm \csc u + C$

58. $\displaystyle\int \frac{1}{\sin u \cos u} \; du = \ln |\tan u| + C$

REFERENCE TABLE 21   (Continued)

Forms Involving tan u, cot u, sec u, csc u

59. $\int \tan u \, du = -\ln |\cos u| + C$

60. $\int \cot u \, du = \ln |\sin u| + C$

61. $\int \sec u \, du = \ln |\sec u + \tan u| + C$

62. $\int \csc u \, du = \ln |\csc u - \cot u| + C$

63. $\int \tan^2 u \, du = -u + \tan u + C$

64. $\int \cot^2 u \, du = -u - \cot u + C$

65. $\int \sec^2 u \, du = \tan u + C$

66. $\int \csc^2 u \, du = -\cot u + C$

67. $\int \tan^n u \, du = \dfrac{\tan^{n-1} u}{n-1} - \int \tan^{n-2} u \, du, \; n \neq 1$

68. $\int \cot^n u \, du = -\dfrac{\cot^{n-1} u}{n-1} - \int (\cot^{n-2} u) \, du, \; n \neq 1$

69. $\int \sec^n u \, du = \dfrac{\sec^{n-2} u \tan u}{n-1} + \dfrac{n-2}{n-1} \int \sec^{n-2} u \, du, \; n \neq 1$

70. $\int \csc^n u \, du = -\dfrac{\csc^{n-2} u \cot u}{n-1} + \dfrac{n-2}{n-1} \int \csc^{n-2} u \, du, \; n \neq 1$

71. $\int \dfrac{1}{1 \pm \tan u} \, du = \dfrac{1}{2}(u \pm \ln |\cos u \pm \sin u|) + C$

72. $\int \dfrac{1}{1 \pm \cot u} \, du = \dfrac{1}{2}(u \mp \ln |\sin u \pm \cos u|) + C$

73. $\int \dfrac{1}{1 \pm \sec u} \, du = u + \cot u \mp \csc u + C$

74. $\int \dfrac{1}{1 \pm \csc u} \, du = u - \tan u \pm \sec u + C$

REFERENCE TABLE 22   Formulas from Business

Basic Terms

x = number of units produced (or sold)
p = price per unit
R = total revenue from selling x units
C = total cost of producing x units
$\overline{C}$ = average cost per unit
P = total profit from selling x units

Basic Equations

$$R = xp \qquad \overline{C} = \frac{C}{x} \qquad P = R - C$$

Demand Function: $p = f(x)$ = price required to sell x units

$$\eta = \frac{p/x}{dp/dx} = \text{price elasticity of demand}$$

(If $|\eta| < 1$, the demand is inelastic. If $|\eta| > 1$, the demand is elastic.)

Typical Graphs of Revenue, Cost, and Profit Functions

The low prices required to sell more units eventually result in a decreasing revenue.

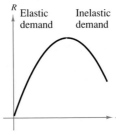

Revenue Function

As x increases, the average cost per unit decreases.

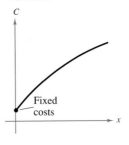

Cost Function

The break-even point occurs when $R = C$.

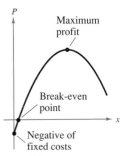

Profit Function

REFERENCE TABLE 22 (Continued)

Marginals

$\dfrac{dR}{dx}$ = marginal revenue ≈ the *extra* revenue for selling one additional unit

$\dfrac{dC}{dx}$ = marginal cost ≈ the *extra* cost of producing one additional unit

$\dfrac{dP}{dx}$ = marginal profit ≈ the *extra* profit for selling one additional unit

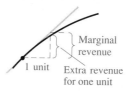

Revenue Function

REFERENCE TABLE 23 Formulas from Finance

Basic Terms

P = amount of deposit
r = interest rate
n = number of times interest is compounded per year
t = number of years
A = balance after t

Compound Interest Formulas

1. Balance when interest is compounded n times per year

$$A = P\left(1 + \frac{r}{n}\right)^{nt}$$

2. Balance when interest is compounded continuously

$$A = Pe^{rt}$$

Balance of an Increasing Annuity after n Deposits per Year of P for t Years

$$A = P\left[\left(1 + \frac{r}{n}\right)^{nt} - 1\right]\left(1 + \frac{n}{r}\right)$$

Initial Deposit for a Decreasing Annuity with n Withdrawals per Year of W for t Years

$$P = W\left(\frac{n}{r}\right)\left[1 - \left(\frac{1}{1 + [r/n]}\right)^{nt}\right]$$

Monthly Installment M for a Loan of P Dollars over t Years at $r\%$ Interest

$$M = P\left[\frac{r/12}{1 - \left(\dfrac{1}{1 + [r/12]}\right)^{12t}}\right]$$

REFERENCE TABLE 24   Standard Normal Distribution (z-scores)

	Hundredths Digit for z-score									
z	0.00	0.01	0.02	0.03	0.04	0.05	0.06	0.07	0.08	0.09
0.0	0.0000	0.0040	0.0080	0.0120	0.0160	0.0199	0.0239	0.0279	0.0319	0.0359
0.1	0.0398	0.0438	0.0478	0.0517	0.0557	0.0596	0.0636	0.0675	0.0714	0.0753
0.2	0.0793	0.0832	0.0871	0.0910	0.0948	0.0987	0.1026	0.1064	0.1103	0.1141
0.3	0.1179	0.1217	0.1255	0.1293	0.1331	0.1368	0.1406	0.1443	0.1480	0.1517
0.4	0.1554	0.1591	0.1628	0.1664	0.1700	0.1736	0.1772	0.1808	0.1844	0.1879
0.5	0.1915	0.1950	0.1985	0.2019	0.2054	0.2088	0.2123	0.2157	0.2190	0.2224
0.6	0.2257	0.2291	0.2324	0.2357	0.2389	0.2422	0.2454	0.2486	0.2517	0.2549
0.7	0.2580	0.2611	0.2642	0.2673	0.2704	0.2734	0.2764	0.2794	0.2823	0.2852
0.8	0.2881	0.2910	0.2939	0.2967	0.2995	0.3023	0.3051	0.3078	0.3106	0.3133
0.9	0.3159	0.3186	0.3212	0.3238	0.3264	0.3289	0.3315	0.3340	0.3365	0.3389
1.0	0.3413	0.3438	0.3461	0.3485	0.3508	0.3531	0.3554	0.3577	0.3599	0.3621
1.1	0.3643	0.3665	0.3686	0.3708	0.3729	0.3749	0.3770	0.3790	0.3810	0.3830
1.2	0.3849	0.3869	0.3888	0.3907	0.3925	0.3944	0.3962	0.3980	0.3997	0.4015
1.3	0.4032	0.4049	0.4066	0.4082	0.4099	0.4115	0.4131	0.4147	0.4162	0.4177
1.4	0.4192	0.4207	0.4222	0.4236	0.4251	0.4265	0.4279	0.4292	0.4306	0.4319
1.5	0.4332	0.4345	0.4357	0.4370	0.4382	0.4394	0.4406	0.4418	0.4429	0.4441
1.6	0.4452	0.4463	0.4474	0.4484	0.4495	0.4505	0.4515	0.4525	0.4535	0.4545
1.7	0.4554	0.4564	0.4573	0.4582	0.4591	0.4599	0.4608	0.4616	0.4625	0.4633
1.8	0.4641	0.4649	0.4656	0.4664	0.4671	0.4678	0.4686	0.4693	0.4699	0.4706
1.9	0.4713	0.4719	0.4726	0.4732	0.4738	0.4744	0.4750	0.4756	0.4761	0.4767
2.0	0.4772	0.4778	0.4783	0.4788	0.4793	0.4798	0.4803	0.4808	0.4812	0.4817
2.1	0.4821	0.4826	0.4830	0.4834	0.4838	0.4842	0.4846	0.4850	0.4854	0.4857
2.2	0.4861	0.4864	0.4868	0.4871	0.4875	0.4878	0.4881	0.4884	0.4887	0.4890
2.3	0.4893	0.4896	0.4898	0.4901	0.4904	0.4906	0.4909	0.4911	0.4913	0.4916
2.4	0.4918	0.4920	0.4922	0.4925	0.4927	0.4929	0.4931	0.4932	0.4934	0.4936
2.5	0.4938	0.4940	0.4941	0.4943	0.4945	0.4946	0.4948	0.4949	0.4951	0.4952
2.6	0.4953	0.4955	0.4956	0.4957	0.4959	0.4960	0.4961	0.4962	0.4963	0.4964
2.7	0.4965	0.4966	0.4967	0.4968	0.4969	0.4970	0.4971	0.4972	0.4973	0.4974
2.8	0.4974	0.4975	0.4976	0.4977	0.4977	0.4978	0.4979	0.4979	0.4980	0.4981
2.9	0.4981	0.4982	0.4982	0.4983	0.4984	0.4984	0.4985	0.4985	0.4986	0.4986
3.0	0.4987	0.4987	0.4987	0.4988	0.4988	0.4989	0.4989	0.4989	0.4990	0.4990

Tenths Digit for z-score

Appendix C

Graphing Utilities

Introduction

In Section 1.2, you studied the point-plotting method for sketching the graph of an equation. One of the disadvantages of the point-plotting method is that in order to get a good idea about the shape of a graph, you need to plot *many* points. With only a few points, you could badly misrepresent the graph. For instance, consider the equation

$$y = \frac{1}{30}x(39 - 10x^2 + x^4).$$

Suppose you plotted only five points: $(-3, -3)$, $(-1, -1)$, $(0, 0)$, $(1, 1)$, and $(3, 3)$, as shown in Figure A.7. From these five points, you might assume that the graph of the equation is a straight line. This, however, is not correct. By plotting several more points you can see that the actual graph is not straight at all! (See Figure A.8.)

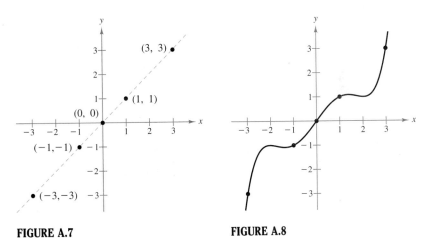

FIGURE A.7   **FIGURE A.8**

Thus, the point-plotting method leaves us with a dilemma. On the one hand, the method can be very inaccurate if only a few points are plotted. But, on the other hand, it is very time consuming to plot a dozen (or more) points. Technology can help us solve this dilemma. Plotting several (even several hundred) points in a rectangular coordinate system is something that a graphing utility can do easily.

The point-plotting method is the method used by *all* graphing packages for computers and *all* graphing calculators. Each computer or calculator screen is made up of a grid of hundreds or thousands of small areas called **pixels.** Screens that have many pixels per inch are said to have a higher **resolution** than screens that don't have as many. For instance, the screen shown in Figure A.9(a) has a higher resolution than the screen shown in Figure A.9(b). Note that the "graph" of the line on the first screen looks more like a line than the "graph" on the second screen.

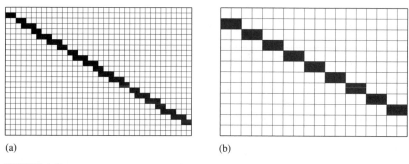

(a)   (b)

FIGURE A.9

Screens on most graphing calculators have 48 pixels per inch. Screens on computer monitors typically have between 32 and 100 pixels per inch.

EXAMPLE 1   Using Pixels to Sketch a Graph

Use the grid shown in Figure A.10 to sketch a graph of $y = \frac{1}{2}x^2$. Each pixel on the grid must be either on (shaded black) or off (unshaded).

SOLUTION

To shade the grid, we use the following rule. If a pixel contains a plotted point of the graph, then it will be "on"; otherwise, the pixel will be "off." Using this rule, the graph of the curve looks like that shown in Figure A.11.

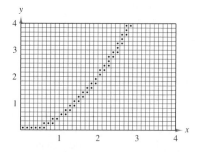

FIGURE A.10

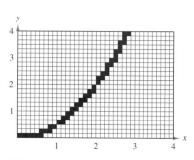

FIGURE A.11

Basic Graphing

There are many different types of graphing utilities—graphing calculators and software packages for computers. The procedures used to draw a graph are similar for most of these utilities.

Basic Graphing Steps for a Graphing Utility

To draw the graph of an equation involving x and y with a graphing utility, use the following steps.

1. Rewrite the equation so that y is isolated on the left side of the equation.
2. Set the boundaries of the viewing rectangle by entering the minimum and maximum x-values and the minimum and maximum y-values.
3. Enter the equation in the form $y =$ (expression involving x). Read the user's guide that accompanies your graphing utility to see how the equation should be entered.
4. Activate the graphing utility.

EXAMPLE 2   Sketching the Graph of an Equation

Sketch the graph of $2y + x^3 = 4x$.

SOLUTION

To begin, solve the given equation for y in terms of x.

$$2y + x^3 = 4x \qquad \textit{Given equation}$$
$$2y = -x^3 + 4x \qquad \textit{Subtract } x^3 \textit{ from both sides.}$$
$$y = -\frac{1}{2}x^3 + 2x \qquad \textit{Divide both sides by 2.}$$

Set the viewing rectangle so that $-10 \le x \le 10$ and $-10 \le y \le 10$. (On some graphing utilities, this is the default setting.) Next, enter the equation into the graphing utility.

$$Y = -X \wedge 3/2 + 2 * X$$

Finally, activate the graphing utility. The display screen should look like that shown in Figure A.12.

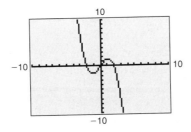

FIGURE A.12

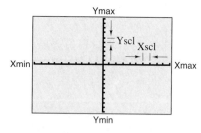

FIGURE A.13

In Figure A.12, notice that the calculator screen does not label the tick marks on the x-axis or the y-axis. To see what the tick marks represent, check the values in the utility's "range."

Range

Xmin $= -10$	*The minimum x-value is -10.*
Xmax $= 10$	*The maximum x-value is 10.*
Xscl $= 1$	*The x-scale is 1 unit per tick mark.*
Ymin $= -10$	*The minimum y-value is -10.*
Ymax $= 10$	*The maximum y-value is 10.*
Yscl $= 1$	*The y-scale is 1 unit per tick mark.*
Xres $= 1$	*The x-resolution is 1 plotted point per pixel.*

These settings are summarized visually in Figure A.13.

EXAMPLE 3   Graphing an Equation Involving Absolute Value

Sketch the graph of $y = |x - 3|$.

SOLUTION

This equation is already written so that y is isolated on the left side of the equation, so you can enter the equation as follows.

$$Y = \text{abs}(X - 3)$$

After activating the graphing utility, its screen should look like the one shown in Figure A.14.

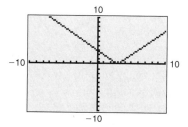

FIGURE A.14

Special Features

In order to be able to use your graphing calculator to its best advantage, you must be able to determine a proper viewing rectangle and use the zoom feature. The next two examples show how this is done.

EXAMPLE 4   Determining a Viewing Rectangle

Sketch the graph of $y = x^2 + 12$.

SOLUTION

Begin as usual by entering the equation.

$$Y = X \wedge 2 + 12$$

Activate the graphing utility. If you used a viewing rectangle in which $-10 \le x \le 10$ and $-10 \le y \le 10$, then no part of the graph will appear on the screen, as shown in Figure A.15(a). The reason for this is that the lowest point on the graph of $y = x^2 + 12$ occurs at the point $(0, 12)$. With the viewing rectangle in Figure A.15(a), the largest y-value is 10. In other words, none of the graph is visible on a screen whose y-values range between -10 and 10.

To be able to see the graph, change Ymax = 10 to Ymax = 30 and Yscl = 1 to Yscl = 5. Now activate the graphing utility and you will obtain the graph shown in Figure A.15(b). On this graph, note that each tick mark on the y-axis represents 5 units because you changed the y-scale to 5. Also note that the highest point on the y-axis is now 30 because you changed the maximum value of y to 30.

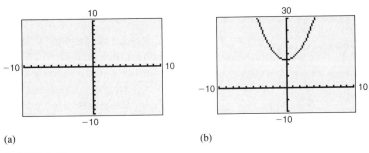

(a)   (b)

FIGURE A.15

EXAMPLE 5   Using the Zoom Feature

Sketch the graph of $y = x^3 - x^2 - x$. How many x-intercepts does this graph have?

SOLUTION

Begin by drawing the graph on a "standard" viewing rectangle, as shown in Figure A.16(a). (See page A52.) From the display screen, it is clear that the graph has at least one intercept (just to the left of $x = 2$), but it is difficult to determine whether the graph has other intercepts. To obtain a better view of the graph near $x = -1$, you can use the zoom feature of the graphing utility. The redrawn screen is shown in Figure A.16(b). From this screen you can tell that the graph has three x-intercepts whose x-coordinates are approximately -0.6, 0, and 1.6.

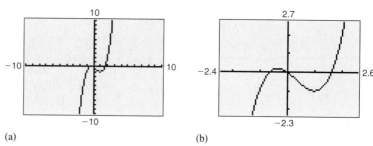

(a)   (b)

FIGURE A.16

EXAMPLE 6   Sketching More Than One Graph on the Same Screen

Sketch the graphs of $y = -\sqrt{36 - x^2}$ and $y = \sqrt{36 - x^2}$ on the same screen.

SOLUTION

To begin, enter both equations into the graphing utility.

$$Y = \sqrt{(36 - X \wedge 2)}$$
$$Y = -\sqrt{(36 - X \wedge 2)}$$

Then, activate the graphing utility to obtain the graph shown in Figure A.17(a). Notice that the graph should be the upper and lower parts of the circle given by $x^2 + y^2 = 6^2$. The reason it doesn't look like a circle is that, with the standard settings, the tick marks on the x-axis are farther apart than the tick marks on the y-axis. To correct this, change the viewing rectangle so that $-15 \le x \le 15$. The redrawn screen is shown in Figure A.17(b). Notice that on this screen the graph appears to be more circular.

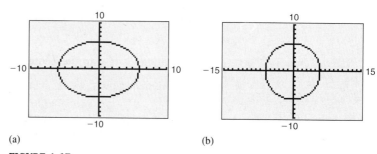

(a)   (b)

FIGURE A.17

Numerical Differentiation

Most graphing utilities have a built-in capability for approximating the derivative of a function $y = f(x)$ at a point x. If your graphing utility does not have this capability, you can use the program provided in Appendix D. On the Texas Instruments calculators (TI-81, TI-82, and TI-85), this numerical approximation is based on the formula

$$f'(x) \approx \frac{f(x + \delta) - f(x - \delta)}{2\delta}$$

where delta (δ) is a small positive number. On some other calculators, such as the Sharp EL 9200 and 9300, this approximation is based on the more familiar formula

$$f'(x) \approx \frac{f(x + \delta) - f(x)}{\delta}.$$

In both cases, the approximation usually becomes more accurate as δ gets closer to zero.

EXAMPLE 7   Approximating the Derivative at a Point

Approximate the derivative of $y = 2x^3 + 1$ at the point $x = 4$. Use $\delta = 0.001$.

SOLUTION

To begin, locate the appropriate command—such as NDerive(, nDeriv(, nDer(, or $\frac{d}{dx}$)—or menu item on your graphing utility. Notice the format of the command and the arguments. The format required for a numerical derivative varies among different graphing calculators and computers. For example, the command format on the TI-81, having two arguments, is NDeriv(expression,δ), and on the TI-82 it is nDeriv(expression,variable, value,δ). Input the appropriate arguments and activate the command. For instance, entering

nDeriv(2X∧3+1,X,4,.001)

on the TI-82 gives 96.000002. For some graphing utilities, such as the TI-81, you will need to store the value 4 as the X variable, and then use the command for the numerical derivative, as shown below.

4 Sto X

NDeriv(2X∧3+1,.001)

Your answer should be close to the exact answer $f'(4) = 96$.   ■

REMARK   For some graphing utilities, the numerical derivative capability can be found in the MATH, CALC, or MATH CALC menu. Consult your user's manual or the Graphing Technology Guide that accompanies this text.

EXAMPLE 8   Using the Derivative of a Function

Graph the function $y = 3x - x^3$ on the standard viewing rectangle.

a. Approximate the derivative at $x = -3$, $x = 0$, $x = 1$ and $x = 3$.
b. What do the signs of the derivative values tell you about the shape of the graph of f at these points?
c. Where does f have horizontal tangent lines?

SOLUTION

The graph of $y = 3x - x^3$ is shown in Figure A.18.

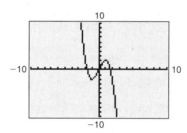

FIGURE A.18

a. Using the numerical derivative key on the graphing utility, you can obtain the following derivative approximations.

$$f'(-3) \approx -24$$
$$f'(0) \approx 3$$
$$f'(1) \approx 0$$
$$f'(3) \approx -24$$

b. The signs of these derivative values indicate that the graph of f is decreasing at $x = -3$, increasing at $x = 0$, and decreasing at $x = 3$. Furthermore, the graph has a horizontal tangent line at $x = 1$.
c. You have already observed that f has a horizontal tangent line at $x = 1$. From the shape of the graph, it appears that there is another horizontal tangent line near $x = -1$. You can use your graphing utility to confirm this by showing that the numerical derivative at $x = -1$ is 0.   ■

Most graphing utilities have a built-in algorithm for approximating definite integrals. If your calculator does not have this capability, you can use the programs for Simpson's Rule provided in Appendix D or Section 6.5.

EXAMPLE 9   Numerical Integration

Use your graphing utility to approximate the area under the curve of $y = \sqrt{16 - x^4}$ between $x = -2$ and $x = 2$.

SOLUTION

The graph of the region is indicated in Figure A.19.

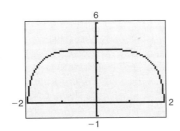

FIGURE A.19

The area is given by the definite integral

$$\int_{-2}^{2} \sqrt{16 - x^4} \, dx.$$

To approximate this integral, you should find the appropriate integration key on your graphing utility. Most commands require that you enter the function, the variable of integration, the lower limit of integration, and the upper limit of integration. For example, on the TI-85, you can use the following command from the CALC menu.

$$\text{fnint}(\sqrt{(16-X\wedge 4)},X,-2,2)$$

The answer should be approximately 13.98.

Discussion Problem

Sketch the graph of $y = x^2 - 12x$, using $-10 \le x \le 10$ and $-10 \le y \le 10$. The graph appears to be a straight line, as shown in the figure. However, this is misleading because the screen doesn't show an important portion of the graph. Can you find a range setting that reveals a better view of this graph?

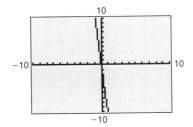

Warm Up

A Misleading Graph

In Exercises 1–10, solve for y in terms of x.

1. $3x + y = 4$ **2.** $x - y = 0$
3. $2x + 3y = 2$ **4.** $4x - 5y = -2$
5. $3x + 4y - 5 = 0$ **6.** $-2x - 3y + 6 = 0$
7. $x^2 + y - 4 = 0$ **8.** $-2x^2 + 3y + 2 = 0$
9. $x^2 + y^2 = 4$ **10.** $x^2 - y^2 = 9$

APPENDIX C ▪ EXERCISES

In Exercises 1–20, use a graphing utility to sketch the graph of the equation. Use a setting on each graph of $-10 \le x \le 10$ and $-10 \le y \le 10$.

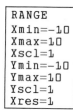

```
RANGE
Xmin=-10
Xmax=10
Xscl=1
Ymin=-10
Ymax=10
Yscl=1
Xres=1
```

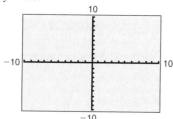

1. $y = x - 5$ **2.** $y = -x + 4$
3. $y = -\frac{1}{2}x + 3$ **4.** $y = \frac{2}{3}x + 1$
5. $2x - 3y = 4$ **6.** $x + 2y = 3$
7. $y = \frac{1}{2}x^2 - 1$ **8.** $y = -x^2 + 6$
9. $y = x^2 - 4x - 5$ **10.** $y = x^2 - 3x + 2$
11. $y = -x^2 + 2x + 1$ **12.** $y = -x^2 + 4x - 1$
13. $2y = x^2 + 2x - 3$ **14.** $3y = -x^2 - 4x + 5$
15. $y = |x + 5|$ **16.** $y = \frac{1}{2}|x - 6|$
17. $y = \sqrt{x^2 + 1}$ **18.** $y = 2\sqrt{x^2 + 2} - 4$
19. $y = \frac{1}{5}(-x^3 + 16x)$ **20.** $y = \frac{1}{8}(x^3 + 8x^2)$

In Exercises 21–30, use a graphing utility to match the equation with its graph. [The graphs are labeled (a), (b), (c), (d), (e), (f), (g), (h), (i), and (j).]

21. $y = x$ **22.** $y = -x$
23. $y = x^2$ **24.** $y = -x^2$
25. $y = x^3$ **26.** $y = -x^3$
27. $y = |x|$ **28.** $y = -|x|$

29. $y = \sqrt{x}$ **30.** $y = -\sqrt{x}$

(a)
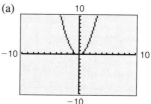

(b)
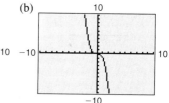

(c)
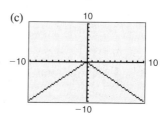

(d)
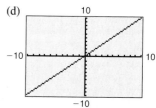

(e)
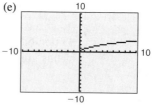

(f)
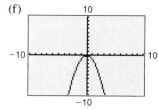

(g)
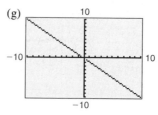

(h)
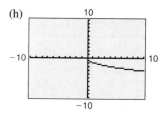

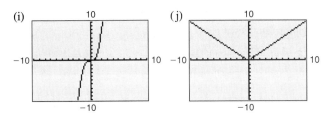

(i)

(j)

In Exercises 31–34, use a graphing utility to sketch the graph of the equation. Use the indicated setting.

31. $y = -2x^2 + 12x + 14$ **32.** $y = -x^2 + 5x + 6$

```
RANGE
Xmin=-5
Xmax=10
Xscl=1
Ymin=-5
Ymax=35
Yscl=5
Xres=1
```

```
RANGE
Xmin=-8
Xmax=4
Xscl=1
Ymin=-5
Ymax=15
Yscl=5
Xres=1
```

33. $y = x^3 + 6x^2$ **34.** $y = -x^3 + 16x$

```
RANGE
Xmin=-10
Xmax=5
Xscl=1
Ymin=-4
Ymax=36
Yscl=3
Xres=1
```

```
RANGE
Xmin=-6
Xmax=6
Xscl=1
Ymin=-25
Ymax=25
Yscl=5
Xres=1
```

In Exercises 35–38, find a setting on a graphing utility such that the graph of the equation agrees with the graph shown.

35. $y = -x^2 - 4x + 20$ **36.** $y = x^2 + 12x - 8$

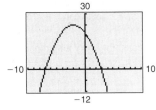

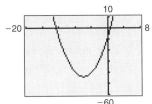

37. $y = -x^3 + x^2 + 2x$ **38.** $y = x^3 + 3x^2 - 2x$

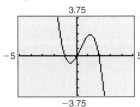

In Exercises 39–42, use a graphing utility to find the number of x-intercepts of the equation.

39. $y = \frac{1}{8}(4x^2 - 32x + 65)$
40. $y = \frac{1}{4}(-4x^2 + 16x - 15)$
41. $y = 4x^3 - 20x^2 - 4x + 61$
42. $y = \frac{1}{4}(2x^3 + 6x^2 - 4x + 1)$

In Exercises 43–46, use a graphing utility to sketch the graphs of the equations on the same screen. Using a "square setting," what geometrical shape is bounded by the graphs?

43. $y = |x| - 4$ **44.** $y = x + |x| - 4$
    $y = -|x| + 4$ $y = x - |x| + 4$
45. $y = -\sqrt{25 - x^2}$ **46.** $y = 6$
    $y = \sqrt{25 - x^2}$ $y = -\sqrt{3}x - 4$
                                 $y = \sqrt{3}x - 4$

Ever Been Married?   In Exercises 47–50, use the following models, which relate ages to the percentages of American males and females who have never been married.

$$y = \frac{0.36 - 0.0056x}{1 - 0.0817x + 0.00226x^2}, \quad \begin{array}{l} \text{Males} \\ 20 \le x \le 50 \end{array}$$

$$y = \frac{100}{8.944 - 0.886x + 0.249x^2}, \quad \begin{array}{l} \text{Females} \\ 20 \le x \le 50 \end{array}$$

In these models, y is the percentage of the population (in decimal form) who have never been married and x is the age of the person. (*Source:* U.S. Bureau of the Census)

47. Use a graphing utility to sketch the graphs of both equations giving the percentages of American males and females who have never been married. Use the following range settings.

```
RANGE
Xmin=20
Xmax=50
Xscl=5
Ymin=0
Ymax=1
Yscl=0.1
Xres=1
```

48. Write a short paragraph describing the relationship between the two graphs that were plotted in Exercise 47.

49. Suppose an American male is chosen at random from the population. If the person is 25 years old, what is the probability that he has never been married?

50. Suppose an American female is chosen at random from the population. If the person is 25 years old, what is the probability that she has never been married?

Earnings and Dividends   In Exercises 51–54, use the following model, which approximates the relationship between dividends per share and earnings per share for the Pall Corporation between 1982 and 1989.

$$y = -0.166 + 0.502x - 0.0953x^2, \qquad 0.25 \le x \le 2$$

In this model, y is the dividends per share (in dollars) and x is the earnings per share (in dollars). (*Source:* Standard ASE Stock Reports)

51. Use a graphing utility to sketch the graph of the model that gives the dividend per share in terms of the earnings per share. Use the following range settings.

```
RANGE
Xmin=0
Xmax=2
Xscl=0.25
Ymin=0
Ymax=0.5
Yscl=0.1
Xres=1
```

52. According to the given model, what size dividend would the Pall Corporation pay if the earnings per share were $1.30?

53. Use the trace feature on your graphing utility to estimate the earnings per share that would produce a dividend per share of $0.25. The choices are labeled (a), (b), (c), and (d). (Find the y-value that is as close to 0.25 as possible. The x-value that is displayed will then be the approximate earnings per share that would produce a dividend per share of $0.25.)

    (a) $1.00   (b) $1.03   (c) $1.06   (d) $1.09

54. The **Payout Ratio** for a stock is the ratio of the dividend per share to earnings per share. Use the model to find the payout ratio for an earnings per share of (a) $0.75, (b) 1.00, and (c) $1.25.

In Exercises 55–58, use a graphing utility to approximate the derivative of the function f at the indicated x-values. What does the value of the derivative tell you about the shape of the graph of f? Verify your answer by graphing the function f on an appropriate viewing rectangle.

55. $f(x) = 4x^4 - x^{1/3} + 1$; $x = -1$, $x = 1$, $x = 3$
56. $f(x) = (x - 2)^3 - 48\sqrt{x}$; $x = 1$, $x = 4$, $x = 6$
57. $f(x) = \dfrac{x^2 - 1}{x^2 + 1}$; $x = -2$, $x = 0$, $x = 2$
58. $f(x) = \dfrac{x^2 + 3}{x - 1}$; $x = -2$, $x = -1$, $x = 2$

In Exercises 59 and 60, use your graphing utility to graph the function f on the indicated domain, and visually estimate the x-values at which the graph of f has a horizontal tangent. Verify your answers by numerically approximating the derivative at these x-values.

59. $f(x) = x\sqrt{x + 2}$, $-2 \le x \le 5$
60. $f(x) = x^3 - 2x^2 + 1$, $-5 \le x \le 5$

61. Use your graphing utility to approximate the derivative of $f(x) = |x|$ at various positive and negative x-values. Does the derivative of f exist at $x = 0$? What is $f'(0)$ according to your graphing utility?

62. Use your graphing utility to approximate the derivative of $f(x) = x^{1/3}$ at various positive and negative x-values. Does the derivative of f exist at $x = 0$? What is $f'(0)$ according to your graphing utility?

In Exercises 63–66, use the numerical integration capability of your graphing utility to approximate the integral.

63. $\displaystyle\int_0^4 \sqrt{25 - x^2}\, dx$ 64. $\displaystyle\int_0^1 \frac{1}{\sqrt{4 - x}}\, dx$

65. $\displaystyle\int_{-2}^2 \frac{4 + x^3}{x^2 - 16}\, dx$ 66. $\displaystyle\int_0^3 \frac{x^{2/3}}{x + 4}\, dx$

67. Find the area under the curve $y = e^{-x^2}$ between $x = 0$ and $x = 5$.

68. Use the numerical integration capability of your graphing utility to approximate the value of the constant b that satisfies $\displaystyle\int_1^b \frac{1}{x}\, dx = 1$.

Appendix D

Programs

Programs for the Texas Instruments TI-81 and TI-82 graphing calculators are given in several sections in the text. This appendix contains translations of these programs for the TI-85, Casio *fx*-7700, Casio *fx*-9700, and Sharp EL 9200/9300, arranged by calculator model. Similar programs can be written for other brands and models of graphics calculators.

Enter a program in your calculator, then refer to the text discussion and apply the program as appropriate. The following section references are provided to help you locate the text discussions of the programs and their uses:

Evaluating Functions	Section 1.4
Midpoint Rule	Section 5.6
Simpson's Rule	Section 6.5
Linear Regression	Section 7.7
Newton's Method	Section S.6

Also included at the beginning of this appendix are the steps for finding a linear regression model using the TI-82. Within the Casio *fx*-7700 section is an additional program that performs numerical differentiation at a point.

Texas Instruments TI-82:   **Linear Regression**

Here are the steps for finding a linear regression model in the Modeling Median Income example.

STAT 4 L₁ , , L₂	*Clear lists L_1 and L_2.*
STAT 1	*Reach the list editor.*
1, 2, ..., 10	*Enter data in list L_1.*
19.1, 20.2, ..., 29.9	*Enter data in list L_2.*
STAT CALC 3	*Set 2-variable statistics with Xlist L_1, Ylist L_2, and frequency 1.*
STAT CALC 5	*Select linear regression option.*
a = 1.221212121	*Slope.*
b = 17.61333333	*y-Intercept.*
r = .9986864096	*Correlation coefficient.*

Texas Instruments TI-85:   **Evaluating Functions**
_____

This program evaluates a function that has been stored as y1.

PROGRAM:Evaluate
:Repeat 1<0   *Start a continuous loop (press ON to break).*
:Input x   *Ask for input.*
:Disp y1   *Calculate the value of the function.*
:End   *End loop.*

Texas Instruments TI-85:   **Midpoint Rule**
_____

This program uses the Midpoint Rule to approximate the definite integral $\int_a^b f(x)dx$. You must store the expression $f(x)$ as y1 before executing the program. The program itself will prompt you for the limits a and b and for the number of subintervals n.

PROGRAM:Midpoint
:Disp "Lower Limit"   *Prompt for value of a.*
:Input A   *Input value of a.*
:Disp "Upper Limit"   *Prompt for value of b.*
:Input B   *Input value of b.*
:Disp "n divisions"   *Prompt for value of n.*
:Input N   *Input value of n.*
:0→S   *Initialize sum of areas.*
:(B-A)/N→W   *Calculate width of subinterval.*
:1→J   *Initialize counter.*
:While J≤N   *Begin conditional loop.*
:A+(J-1)*W→L   *Calculate left endpoint.*
:A+J*W→R   *Calculate right endpoint.*
:(L+R)/2→x   *Calculate midpoint of subinterval.*
:S+W*y1→S   *Add area to sum.*
:J+1→J   *Increase counter.*
:End   *End loop.*
:Disp "Approximation"
:Disp S   *Display approximation.*

Texas Instruments TI-85:   **Simpson's Rule**

This program uses Simpson's Rule to approximate the definite integral $\int_a^b f(x)dx$. You must store the expression $f(x)$ as y1 before executing the program. The program itself will prompt you for the limits a and b and for *half* the number of subintervals you want to use.

PROGRAM:Simpson	
:Disp "Lower Limit"	*Prompt for value of a.*
:Input A	*Input value of a.*
:Disp "Upper Limit"	*Prompt for value of b.*
:Input B	*Input value of b.*
:Disp "n/2 divisions"	*Prompt for value of n/2.*
:Input D	*Input value of n/2.*
:0→S	*Initialize sum of areas.*
:(B-A)/(2D)→W	*Calculate width of subinterval.*
:1→J	*Initialize counter.*
:While J≤D	*Begin conditional loop.*
:A+2(J-1)∗W→L	*Calculate left endpoint.*
:A+2J∗W→R	*Calculate right endpoint.*
:(L+R)/2→M	*Calculate midpoint of subinterval.*
:L→x	
:y1→L	*Evaluate f(x) at left endpoint.*
:M→x	
:y1→M	*Evaluate f(x) at midpoint.*
:R→x	
:y1→R	*Evaluate f(x) at right endpoint.*
:W∗(L+4M+R)/3+S→S	*Add area by Simpson's Rule to sum.*
:J+1→J	*Increase counter.*
:End	*End loop.*
:Disp "Approximation"	
:Disp S	*Display approximation.*

Texas Instruments TI-85: **Linear Regression**

Here are the steps for finding a linear regression model in the Modeling Median Income example.

STAT EDIT	*Set X as xlist Name and Y as ylist Name.*
CLRxy	*Clear any previous data.*
x1 = 1 y1 = 19.1	*Enter data.*
x2 = 2 y2 = 20.2	
&#8942;	
x10 = 10 y10 = 29.9	
STAT CALC LINR	*Calculate linear regression model.*
a = 17.6133333333	*y-Intercept.*
b = 1.22121212121	*Slope.*
corr = .998686409604	*Correlation coefficient.*

Texas Instruments TI-85: **Newton's Method**

This program uses Newton's Method to approximate the zeros of a function. You must store the expression $f(x)$ as y1 before executing the program. Then graph the function to estimate one of its zeros. The program will prompt you for this estimate.

PROGRAM:Newton	
:Disp "Enter Approximation"	*Prompt user for approximation.*
:Input x	*Input approximation.*
:1→N	*Initialize counter.*
:x-y1/der1(y1,x)→R	*Compute new root.*
:While abs (x-R)>1E-10	*Begin conditional loop.*
:R→x	*Continue the iteration.*
:x-y1/der1(y1,x)→R	*Compute new root.*
:N+1→N	*Increase counter by 1.*
:End	*End loop.*
:Disp "Root="	
:Disp R	*Display approximate zero.*
:Disp "Iter="	
:Disp N	*Display number of iterations.*

Casio *fx*-7700:   **Evaluating Functions**

This program evaluates a function that has been stored in function memory location f_1.

EVALUATE
Lbl 1   *Begin loop.*
"X="?→X   *Prompt for input.*
"F(X)=":f₁ ◢   *Display the value of the function, then pause.*
Goto 1   *End loop.*

Casio *fx*-7700:   **Midpoint Rule**

This program uses the Midpoint Rule to approximate the definite integral $\int_a^b f(x)\,dx$. You must store the expression $f(x)$ as f_1 before executing the program. The program itself will prompt you for the limits a and b and for the number of subintervals n.

MIDPOINT
"LOWER LIMIT"?→A   *Prompt for value of a.*
"UPPER LIMIT"?→B   *Prompt for value of b.*
"N DIVISIONS"?→N   *Prompt for value of n.*
0→S   *Initialize sum of areas.*
(B-A)÷N→W   *Calculate width of subinterval.*
1→J   *Initialize counter.*
Lbl 1   *Begin loop.*
A+(J-1)W→L   *Calculate left endpoint.*
A+JW→R   *Calculate right endpoint.*
(L+R)÷2→X   *Calculate midpoint of subinterval.*
S+Wf₁→S   *Add area to sum.*
J+1→J   *Increase counter.*
J≤N⇒Goto 1   *Test counter and end loop.*
"APPROXIMATION"
S   *Display approximation.*

Casio *fx*-7700:   **Simpson's Rule**

This program uses Simpson's Rule to approximate the definite integral $\int_a^b f(x)dx$. You must store the expression $f(x)$ as f_1 before executing the program. The program itself will prompt you for the limits a and b and for *half* the number of subintervals you want to use.

SIMPSON	
"LOWER LIMIT"?→A	*Prompt for value of a.*
"UPPER LIMIT"?→B	*Prompt for value of b.*
"N÷2 DIVISIONS"?→D	*Prompt for value of n/2.*
0→S	*Initialize sum of areas.*
(B-A)÷(2D)→W	*Calculate width of subinterval.*
1→J	*Initialize counter.*
Lbl 1	*Begin loop.*
A+2(J-1)W→L	*Calculate left endpoint.*
A+2JW→R	*Calculate right endpoint.*
(L+R)÷2→M	*Calculate midpoint of subinterval.*
L→X	
f_1→L	*Evaluate f(x) at left endpoint.*
M→X	
f_1→M	*Evaluate f(x) at midpoint.*
R→X	
f_1→R	*Evaluate f(x) at right endpoint.*
W(L+4M+R)÷3+S→S	*Add area by Simpson's Rule to sum.*
J+1→J	*Increase counter.*
J≤D⇒Goto 1	*Test counter and end loop.*
"APPROXIMATION"	
S	*Display approximation.*

Casio *fx*-7700:   **Linear Regression**

Here are the steps for finding a linear regression model in the Modeling Median Income example.

MODE ÷ MODE 4	*Set the calculator for linear regression.*
MODE MODE 1	*Prepare to store statistical data.*
EDIT ERS YES	*Erase any previous data.*
1 , 19.1	*Enter data.*
2 , 20.2	
⋮	
10 , 29.9	
CAL REG	*Calculate the regression model.*
A = 17.61333333	*y-Intercept.*
B = 1.221212121	*Slope.*
r = 0.9986864096	*Correlation coefficient.*

Casio *fx*-7700:	Newton's Method

The main program here uses Newton's Method to approximate the zeros of a function. You must store the expression $f(x)$ as f_1 before executing the program. Then graph the function to estimate one of its zeros. The program will prompt you for this estimate.

The Newton's Method program calls on another program to calculate numerical derivatives. This program also requires the expression $f(x)$ to have been stored as f_1. In the code below, program NEWTON believes NDERIV is stored as program 0. You may store NDERIV in any location, just include the correct address within NEWTON.

NEWTON

"ENTER APPROXIMATION"?→X	*Prompt user for approximation.*
1→N	*Initialize counter.*
Prog 0	*Call program 0 (NDERIV) to calculate the numerical derivative.*
X-f₁÷Y→R	*Compute new root.*
Lbl 1	*Begin loop.*
R→X	*Continue the iteration.*
Prog 0	*Call program 0 (NDERIV) to calculate the numerical derivative.*
X-f₁÷Y→R	*Compute new root.*
N+1→N	*Increase counter by 1.*
Abs (X-R)≥1E-10⇒Goto 1	*End loop.*
"ROOT="	
R◢	*Display approximate zero.*
"ITER="	
N	*Display number of iterations.*

NDERIV

.001→D	*Define value of delta.*
X→A	*Store x.*
A+D→X	*Calculate upper value.*
f₁→B	*Evaluate the function.*
A-D→X	*Calculate lower value.*
f₁→C	*Evaluate the function.*
(B-C)÷2D→Y	*Calculate the symmetric difference to approximate the numerical derivative.*
A→X	*Restore x.*

Casio *fx*-7700:   **Numerical Differentiation**

This program performs numerical differentiation at a point. You must store the expression to be differentiated $f(x)$ as f_1 and store the value of x for which the approximation is to be computed before executing the program.

DERIVATIVE

.001$\mapsto$D	*Set $\Delta x = 0.001$ for approximating derivatives.*
X+D$\mapsto$X	*Evaluate $(x + \Delta x)$.*
$f_1\mapsto$B	*Evaluate $f(x + \Delta x)$.*
X-D$\mapsto$X	*Evaluate $(x - \Delta x)$.*
$f_1\mapsto$C	*Evaluate $f(x - \Delta x)$.*
(B-C)/2D$\mapsto$Y	*Evaluate $[f(x + \Delta x) - f(x - \Delta x)] / 2\Delta x$.*
Y	*Display approximate numerical derivative.*

Casio *fx*-9700:   **Evaluating Functions**

This program evaluates a function that has been stored in function memory location f_1.

EVALUATE↵	
Lbl 1↵	*Begin loop.*
"X="?→X↵	*Ask for input.*
"F(X)=":f₁ ◢	*Display the value of the function, then pause.*
Goto 1	*End loop.*

Casio *fx*-9700:   **Midpoint Rule**

This program uses the Midpoint Rule to approximate the definite integral $\int_a^b f(x)\,dx$. You must store the expression $f(x)$ as f_1 before executing the program. The program itself will prompt you for the limits a and b and for the number of subintervals n.

MIDPOINT↵	
"LOWER LIMIT"?→A↵	*Prompt for value of a.*
"UPPER LIMIT"?→B↵	*Prompt for value of b.*
"N DIVISIONS"?→N↵	*Prompt for value of n.*
0→S↵	*Initialize sum of areas.*
(B-A)÷N→W↵	*Calculate width of subinterval.*
1→J↵	*Initialize counter.*
Lbl 1↵	*Begin loop.*
A+(J-1)W→L↵	*Calculate left endpoint.*
A+JW→R↵	*Calculate right endpoint.*
(L+R)÷2→X↵	*Calculate midpoint of subinterval.*
S+Wf₁→S↵	*Add area to sum.*
J+1→J↵	*Increase counter.*
J≤N⇒Goto 1↵	*Test counter and end loop.*
"APPROXIMATION"↵	
S	*Display approximation.*

Casio *fx*-9700:   **Simpson's Rule**

This program uses Simpson's Rule to approximate the definite integral $\int_a^b f(x)dx$. You must store the expression $f(x)$ as f_1 before executing the program. The program itself will prompt you for the limits a and b and for *half* the number of subintervals you want to use.

SIMPSON↵	
"LOWER LIMIT"?→A↵	*Prompt for value of a.*
"UPPER LIMIT"?→B↵	*Prompt for value of b.*
"N÷2 DIVISIONS"?→D↵	*Prompt for value of n/2.*
0→S↵	*Initialize sum of areas.*
(B-A)÷(2D)→W↵	*Calculate width of subinterval.*
1→J↵	*Initialize counter.*
Lbl 1↵	*Begin loop.*
A+2(J-1)W→L↵	*Calculate left endpoint.*
A+2JW→R↵	*Calculate right endpoint.*
(L+R)÷2→M↵	*Calculate midpoint of subinterval.*
L→X↵	
f_1→L↵	*Evaluate f(x) at left endpoint.*
M→X↵	
f_1→M↵	*Evaluate f(x) at midpoint.*
R→X↵	
f_1→R↵	*Evaluate f(x) at right endpoint.*
W(L+4M+R)÷3+S→S↵	*Add area by Simpson's Rule to sum.*
J+1→J↵	*Increase counter.*
J≤D⇒Goto 1↵	*Test counter and end loop.*
"APPROXIMATION"↵	
S	*Display approximation.*

Casio fx-9700: **Linear Regression**

Here are the steps for finding a linear regression model in the Modeling Median Income example.

MENU 4	Set the calculator for regression.
SETUP	Set the model to linear.
EDIT ERS YES	Erase any previous data.
1 , 19.1	Enter data.
2 , 20.2	
⋮	
10 , 29.9	
REG	Calculate the regression model.
A = 17.6133333333	y-Intercept.
B = 1.22121212121	Slope.
r = 0.998686409604	Correlation coefficient.

Casio fx-9700: **Newton's Method**

This program uses Newton's Method to approximate the zeros of a function. You must store the expression $f(x)$ as f_1 before executing the program. Then graph the function to estimate one of its zeros. The program will prompt you for this estimate.

NEWTON↵	
"ENTER APPROXIMATION"?→X↵	Prompt user for approximation.
1→N↵	Initialize counter.
X−f₁÷d/dx(f₁,X)→R↵	Compute new root.
Lbl 1↵	Begin loop.
R→X↵	Continue the iteration.
X−f₁÷d/dx(f₁,X)→R↵	Compute new root.
N+1→N↵	Increase counter by 1.
Abs (X−R)≥1E−10⇒Goto 1↵	End loop.
"ROOT="↵	
R◢	Display approximate zero.
"ITER="↵	
N	Display number of iterations.

Sharp EL 9200/9300:   **Evaluating Functions**

This program evaluates a function that is entered within the program itself.

evaluate	
Label 1	*Begin loop.*
Input x	*Ask for input.*
y= *insert expression here*	*Calculate the value of the function, written in the local variable* x.
Print y	*Display the value of the function.*
Goto 1	*End loop.*

Sharp EL 9200/9300:   **Midpoint Rule**

This program uses the Midpoint Rule to approximate the definite integral $\int_a^b f(x)dx$. Before running it, you must enter an expression $f(x)$ for the integrand just once in the program itself. The program will prompt you for the limits a and b and for the number of subintervals n.

The program includes a subroutine to evaluate the function $f(x)$. It is placed at the beginning of the program for convenient access when you are changing the integrand.

midpoint	
Goto start	*Jump over the next three lines.*
Label eval	*Begin subroutine.*
y= _f(x)_	*Enter the integrand here as an expression in the local variable* x.
Return	*Continue the main program from where this subroutine was called.*
Label start	*Jumping point for* Goto *command above; start of main program.*
Print "Lower Limit	*Prompt for value of a.*
Input a	*Input value of a.*
Print "Upper Limit	*Prompt for value of b.*
Input b	*Input value of b.*
Print "n divisions	*Prompt for value of n.*
Input n	*Input value of n.*
s=0	*Initialize sum of areas.*
w=(b-a)/n	*Calculate width of subinterval.*
j=1	*Initialize counter.*
Label 1	*Begin loop.*
l=a+(j-1)*w	*Calculate left endpoint.*
r=a+j*w	*Calculate right endpoint.*
x=(l+r)/2	*Calculate midpoint of subinterval.*
Gosub eval	*Evaluate the function.*
s=s+w*y	*Add area to sum.*
j=j+1	*Increase counter.*
If j<=n Goto 1	*Test counter and end loop.*
Print "Approximation	
Print s	*Display approximation.*

Sharp EL 9200/9300:   Simpson's Rule

This program uses Simpson's Rule to approximate the definite integral $\int_a^b f(x)dx$. Before running it, you must enter an expression $f(x)$ for the integrand just once in the program itself. The program will prompt you for the limits a and b and for *half* the number of subintervals you want to use.

The program includes a subroutine to evaluate the function $f(x)$. It is placed at the beginning of the program for convenient access when you are changing the integrand.

```
simpson
Goto start              Jump over the next three lines.
Label eval              Begin subroutine.
y= f(x)                 Enter the integrand here as an expression in the local variable x.
Return                  Continue the main program from where this subroutine was called.
Label start             Jumping point for Goto command above; start of main program.
Print "Lower Limit      Prompt for value of a.
Input a                 Input value of a.
Print "Upper Limit      Prompt for value of b.
Input b                 Input value of b.
Print "n/2 divisions    Prompt for value of n/2.
Input d                 Input value of n/2.
s=0                     Initialize sum of areas.
w=(b-a)/(2d)            Calculate width of subinterval.
j=1                     Initialize counter.
Label 1                 Begin loop.
l=a+2(j-1)*w            Calculate left endpoint.
r=a+2j*w                Calculate right endpoint.
m=(l+r)/2               Calculate midpoint of subinterval.
x=l
Gosub eval              Evaluate f(x) at left endpoint.
l=y
x=m
Gosub eval              Evaluate f(x) at midpoint.
m=y
x=r
Gosub eval              Evaluate f(x) at right endpoint.
r=y
s=w*(l+4m+r)/3+s        Add area by Simpson's Rule to sum.
j=j+1                   Increase counter.
If j<=d Goto 1          Test counter and end loop.
Print "Approximation
Print s                 Display approximation.
```

Sharp EL 9200/9300:   **Linear Regression**

Here are the steps for finding a linear regression model in the Modeling Median Income example.

Statistics	*Press the statistics mode key.*
MENU DEL ALL DATA	*Delete any previous data.*
TWO VARIABLE X,Y	*Select the data format for linear regression.*
x = 1 y = 19.1	*Enter data.*
x = 2 y = 20.2	
&#8942;	
x = 10 y = 29.9	
MENU STAT REG	*Calculate linear regression model.*
a = 17.61333333	*y-Intercept.*
b =1.221212121	*Slope.*
r = 0.998686409	*Correlation coefficient.*

Sharp EL 9200/9300:   Newton's Method

This program uses Newton's Method to approximate the zeros of a function. First graph the function to estimate one of its zeros. The program will prompt you for this estimate.

Before running the program, you must enter the expression $f(x)$ twice within the program itself. Take care that you use the correct variable, local x or global X, each time.

The program includes a subroutine to evaluate the function $f(x)$. Next is a second subroutine to calculate the numerical derivative of the function for the current value of x. They are placed at the beginning of the program for convenient access when you are changing the function.

newton	
Goto start	*Jump over the next six lines.*
Label eval	*Begin subroutine.*
y= _f(x)_	*Enter the function here as an expression in the local variable* x.
Return	*Continue the main program from where this subroutine was called.*
Label nderiv	*Begin subroutine.*
d=d/dx(_f(X)_,x)	*Enter the function here as an expression in the global variable* X.
Return	*Continue the main program from where this subroutine was called.*
Label start	*Jumping point for* Goto *command above; start of main program.*
Print "Enter Approximation	*Prompt user for approximation.*
Input x	*Input approximation.*
n=1	*Initialize counter.*
Gosub eval	*Evaluate the function.*
Gosub nderiv	*Calculate the numerical derivative.*
r=x-y/d	*Compute new root.*
Label 1	*Begin loop.*
x=r	*Continue the iteration.*
Gosub eval	*Evaluate the function.*
Gosub nderiv	*Calculate the numerical derivative.*
r=x-y/d	*Compute new root.*
n=n+1	*Increase counter by 1.*
If abs (x-r)>=1E-10 Goto 1	*Compare approximations and try to end loop.*
Print "Root	
Print r	*Display approximate zero.*
Print "Iter	
Print n	*Display number of iterations.*

Appendix E

Computer Algebra Systems

Throughout the text, explanations of how to use Computer Algebra Systems (CAS) to accomplish certain calculus tasks are given in Technology Notes. In instances where commands are presented for illustration, the commands are often given for the CAS package *Derive*. To help you become acquainted with the rudiments of using a CAS, the following pages contain introductory instructions for *Derive*, as well as for two other popular Computer Algebra Systems, *Maple* and *Mathematica*.

Introduction to Derive

The purpose of this introduction is to illustrate some basic *Derive* commands and demonstrate the use of the different windows and modes of the software. You should be working at the computer terminal. To begin, enter the Author mode (all expressions should be entered in this mode), type

    3+4,

and press the Enter (or Return) key. *Derive* will display this line on the screen. To simplify the expression, press S (for Simplify). The program will then prompt you to check the line number of the expression that you are simplifying. Press Enter again, and 7 should appear on the line beneath 3+4. *Derive* will label each expression in the window with a line number. Here are some basic commands in *Derive*. Note that you can use the Escape key to return to a previous command line.

1. Author 2*(4−1)^3 and press Enter. Simplify will give the answer, 54.
2. *Derive* can calculate limits. Author LIM(x+3,x,1) and Simplify will give 4.
3. *Derive* can differentiate and integrate symbolically. For instance, DIF(SIN(x),x) will Simplify to COS(x).
4. You can enter and simplify an expression by pressing Ctrl-Enter after you have typed it in the Author window. For example, Author INT(x^(1/2),x) and press Ctrl-Enter. The answer $\frac{2x^{3/2}}{3}$ will appear in the window.
5. *Derive* can calculate definite integrals as well: INT(x^(1/2)1,4) will Simplify to 14/3.
6. *Derive* can integrate numerically. Author the expression 4/(x^2+1) and press Enter. Then type C (for Calculus), type I (for Integrate), and press

Enter again. Now respond with x as the variable of integration, 0 as the lower limit of integration, and tab to 1 as the upper limit. Entering this expression and pressing X (for approXimation) will give 3.14159 as the numerical approximation of this definite integral.

7. Both two- and three-dimensional graphs are no problem for the Plot command. Author the expression x^5-x, press Enter, and then type P (for Plot). You should now be in the 2D-Plot window. You can plot the function by pressing P again, and you can adjust the domain and range by changing the Scale or using the Zoom function. Pressing A will take you back to the Algebra window. Now Author $4-x^2-2*y^2$, press Enter, and type P. You should be in the 3D-Plot window. You can plot the surface by pressing P again. (It takes *Derive* a little time to prepare a 3D plot.)

8. *Derive* can calculate important mathematical constants to any degree of accuracy. Author PI, press Enter, and then type O (for Options) and P (for Precision). Change the number of digits to 50 and Escape to the Algebra window. Type X (for approXimation) and press Enter. *Derive* will give all 50 digits of the constant pi.

9. You can Quit *Derive* by typing Q, and answering yes.

For additional information, consult either the *Derive User Manual* or the on-line HELP command and menu.

Introduction to Maple

The purpose of this introduction is to illustrate the basic *Maple* commands. You should be working at the computer terminal. To begin, type the sum

> > 3+4;

and press the Enter key. Note that you need to end the line with a semicolon. Your computer should respond with the answer 7. Continue with the commands below, typing them to the right of the > symbol.

1. Let's try some simple arithmetic.

> > 2*(4−1)^3;
>
>   54

2. *Maple* can calculate limits.

> > limit(x+3,x=1);
>
>   4

3. *Maple* can differentiate and integrate symbolically.

> > diff(sin(x),x);
>
>   cos(x)

> int(x^(1/2),x);
$$2/3x^{3/2}$$
> int(x^(1/2),x=1..4);
$$14/3$$

4. *Maple* can integrate numerically as well. The use of the Int command with a capital I avoids any attempt at symbolic integration, and is usually faster.

> evalf(Int(4/(1+x^2),x=0..1));
$$3.141592654$$

5. We can conveniently graph functions with *Maple*. You might need to define the graphics mode using the command: plotsetup(tek,xterm);.

> plot(x^5−x,x=−10..10);
(you will see the graph of the function.)

You can use a letter to define a function. Let's illustrate this shortcut by plotting the above function again with a different range of *y*-values.

> f:=x^5−x;
> plot(f,x=−10..10,y=−10..10);
(you will see the same function with a different *y*-scale.)

6. *Maple* knows all the important mathematical constants to any degree of accuracy.

> evalf(Pi,50);
3.14159. . . (you will see 50 digits of pi.)

7. For more information, consult the *Maple* documentation. There is also extensive on-line help available. For instance, here is how to learn more about the numerical capabilities of *Maple*.

> help(evalf);

Introduction to Mathematica

The purpose of this introduction is to illustrate the basic *Mathematica* commands. You should be working at the computer terminal. To begin, type the sum

3+4

and press the Enter key. Your computer should respond with the answer 7. Continue with the commands below. You should type in whatever you see after the *In*[]:= symbol, press the Enter key, and *Mathematica* will produce the output after the *Out*[]:= symbol. The numbers in the brackets will increase as you continue your session.

1. *Mathematica* can evaluate expressions such as $2(4 - 1)^3$.

   In[2]:= 2 * (4 − 1)^3
   Out[2]:= 54

2. *Mathematica* can calculate limits.
   In[3]:= Limit[x + 3,x− > 1]
   Out[3]:= 4

3. *Mathematica* can differentiate and integrate symbolically.

   In[4]: = D[Sin[x],x]
   Out[4]: = Cos[x]
   In[5]: = Integrate[Sqrt[x],x]
   Out[5]: = 2x$^{3/2}$/3
   In[6]: = Integrate[Sqrt[x], {x, 1, 4}]
   Out[6]: = 14/3

4. *Mathematica* can numerically approximate integrals as well.

   In[7]: = NIntegrate[4/(1 + x^2), {x, 0, 1}]
   Out[7]: = 3.14159

5. *Mathematica* can produce two- and three-dimensional graphs.

   In[8]: = Plot[x^5 − x, {x, −10, 10}]
   Out[8]: = (you will see the graph of the function.)

   Let's graph the previous function again with a different range.

   In[9]: = Show[%,PlotRange − > {−1, 1}]
   Out[9]: = (the range of the preceding graph has been altered.)
   In[10]: = Plot3D[4 − x^2 − 2y^2, {x, −2, 2}, {y, −1, 1}]
   Out[10]: = (you will see a three-dimensional surface.)

6. *Mathematica* knows all the important mathematical constants to any degree of accuracy.

   In[11]: = N[Pi,50]
   Out[11]: = 3.14159. . . (you will see 50 digits of pi.)

7. *Mathematica* can solve equations.

   In[12]: = FindRoot[Cos[x], ==x, {x, 1}]
   Out[12]: = {x− > 0.739085}

8. For more information, consult the *Mathematica* documentation. There is also extensive on-line help available. For instance, here is how to learn more about the logarithm function.

   In[13]: = ? Log

Answers to Odd-Numbered Exercises

CHAPTER 0

SECTION 0.1 (page 0–7)

1. Rational   **3.** Irrational   **5.** Rational

7. Rational   **9.** Irrational

11. (a) Yes   **(b)** No   **(c)** Yes   **(d)** No

13. (a) Yes   **(b)** No   **(c)** No   **(d)** Yes

15. $x \geq 12$

17. $x < -\frac{1}{2}$

19. $x \geq \frac{1}{2}$

21. $x > 1$

23. $-\frac{1}{2} < x < \frac{7}{2}$

25. $-\frac{3}{4} < x < -\frac{1}{4}$

27. $x > 6$

29. $-3 \leq x \leq 1$

31. $-3 \leq x \leq 2$

33. $r > 12.5\%$

35. $x \geq 36$ units

37. Length of side $\geq 10\sqrt{5}$

39. (a) False   **(b)** True   **(c)** True
(d) True if $ab < 0$ and false if $ab > 0$

SECTION 0.2 (page 0–12)

1. (a) -51 **(b)** 51 **(c)** 51

3. (a) -14.99 **(b)** 14.99 **(c)** 14.99 **5.** 14

7. 1.25 **9.** $-5 < x < 5$

11. $x < -6$ and $x > 6$ **13.** $-7 < x < 3$

15. $x \leq -7$ or $x \geq 13$ **17.** $x < 6$ or $x > 14$

19. $4 < x < 5$ **21.** $a - b \leq x \leq a + b$

23. $|x| \leq 2$ **25.** $|x| > 2$ **27.** $|x - 4| \leq 2$

29. $|x - 2| > 2$ **31.** $|x - 4| < 2$

33. $|y - a| \leq 2$ **35.** $65.8 \leq h \leq 71.2$

37. $175,000 \leq x \leq 225,000$

39. (a) $|4750 - a| \leq 500$, $|4750 - a| \leq 237.50$
(b) At variance

41. (a) $|20,000 - a| \leq 500$, $|20,000 - a| \leq 1000$
(b) At variance

SECTION 0.3 (page 0–18)

1. -24 **3.** $\frac{1}{2}$ **5.** 3 **7.** 44

9. 5 **11.** 9

13. $\frac{1}{2}$ **15.** $\frac{1}{4}$ **17.** 908.3483 **19.** -5.3601

21. $5x^6$ **23.** $24y^{10}$ **25.** $10x^4$ **27.** $7x^5$

29. $\frac{4}{3}(x + y)^2$ **31.** $3x$ **33.** $4x^4$

35. (a) $2\sqrt{2}$ **(b)** $3\sqrt{2}$

37. (a) $2x\sqrt[3]{2x^2}$ **(b)** $2|x|z\sqrt[4]{2z}$

39. (a) $\dfrac{5\sqrt{3}|x|}{y^2}$ **(b)** $(x - y)\sqrt{5(x - y)}$

41. $x \geq 1$ **43.** $(-\infty, \infty)$

45. $(-\infty, 1)$ and $(1, \infty)$ **47.** $x > 3$

49. $1 \leq x \leq 5$ **51.** $\$25,760.55$ **53.** $\$4765.56$

SECTION 0.4 (page 0–24)

1. $\frac{1}{2}, -\frac{1}{3}$ **3.** $\frac{3}{2}$ **5.** $-2 \pm \sqrt{3}$

7. $(x-2)^2$ **9.** $(2x+1)^2$ **11.** $(x+2)(x-1)$

13. $(3x-2)(x-1)$ **15.** $(x-2y)^2$

17. $(3+y)(3-y)(9+y^2)$

19. $(x-2)(x^2+2x+4)$

21. $(y+4)(y^2-4y+16)$

23. $(x-3)(x^2+3x+9)$

25. $(x-4)(x-1)(x+1)$ **27.** $(2x-3)(x^2+2)$

29. $(x-2)(2x^2-1)$ **31.** $0, 5$ **33.** ± 3

35. $\pm\sqrt{3}$ **37.** $0, 6$ **39.** $-2, 1$ **41.** $2, 3$

43. -4 **45.** ± 2 **47.** $1, \pm 2$

49. $x \le 3$ or $x \ge 4$ **51.** $-2 \le x \le 2$

53. $(x+2)(x^2-2x+4)$

55. $(x-1)(2x^2+x-1)$ **57.** ± 1

59. $1, 2, 3$ **61.** $1, \pm\frac{1}{2}$ **63.** 4 **65.** 2000 units

SECTION 0.5 (page 0–32)

1. $\dfrac{x+5}{x-1}$ **3.** $\dfrac{5x-1}{x^2+2}$ **5.** $\dfrac{4x-3}{x^2}$

7. $\dfrac{x-6}{x^2-4}$ **9.** $\dfrac{2}{x-3}$

11. $\dfrac{(A+B)x+3(A-2B)}{(x-6)(x+3)}$ **13.** $-\dfrac{(x-1)^2}{x(x^2+1)}$

15. $\dfrac{x+2}{(x+1)^{3/2}}$ **17.** $-\dfrac{3t}{2\sqrt{1+t}}$

19. $-\dfrac{x^2+3}{(x+1)(x-2)(x-3)}$

21. $\dfrac{(A+C)x^2-(A-B-2C)x-(2A+2B-C)}{(x+1)^2(x-2)}$

23. $\dfrac{x(x^2+2)}{(x+1)^{3/2}}$ **25.** $\dfrac{2}{x^2\sqrt{x^2+2}}$

27. $\dfrac{1}{2\sqrt{x}(x+1)^{3/2}}$ **29.** $\dfrac{\sqrt{3}}{3}$ **31.** $\dfrac{2}{3\sqrt{2}}$

33. $\dfrac{x\sqrt{x-4}}{x-4}$ **35.** $\dfrac{y}{6\sqrt{y}}$ **37.** $\dfrac{49\sqrt{x^2-9}}{x+3}$

39. $\dfrac{\sqrt{14}+2}{2}$ **41.** $\dfrac{x(5+\sqrt{3})}{11}$ **43.** $\sqrt{6}-\sqrt{5}$

45. $\dfrac{1}{x(\sqrt{3}+\sqrt{2})}$ **47.** $\dfrac{2x-1}{2x+\sqrt{4x-1}}$

49. $\$232.68$

CHAPTER 1

SECTION 1.1 (page 8)

Warm Up

1. $3\sqrt{5}$ **2.** $2\sqrt{5}$ **3.** $\frac{1}{2}$ **4.** -2

5. $5\sqrt{3}$ **6.** $-\sqrt{2}$ **7.** $x = -3, x = 9$

8. $y = -8, y = 4$ **9.** $x = 19$ **10.** $y = 1$

1. $a = 4$, $b = 3$, $c = 5$
$4^2 + 3^2 = 5^2$

3. $a = 10$, $b = 3$, $c = \sqrt{109}$
$10^2 + 3^2 = (\sqrt{109})^2$

5. $d = 2\sqrt{5}$
Midpoint: $(4, \ 3)$

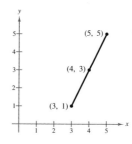

7. $d = 2\sqrt{10}$
Midpoint: $\left(-\frac{1}{2}, \ -2\right)$

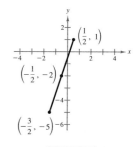

9. $d = 2\sqrt{37}$
Midpoint: $(3, \ 8)$

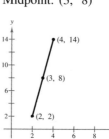

11. $d = \sqrt{8 - 2\sqrt{3}}$
Midpoint: $\left(0, \ \dfrac{\sqrt{3}+1}{2}\right)$

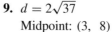
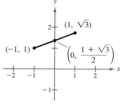
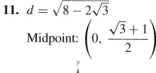

13. $d_1 = \sqrt{45},\, d_2 = \sqrt{20},$
$d_3 = \sqrt{65}$

15. $d_1 = d_2 = d_3 = d_4$
$= \sqrt{5}$

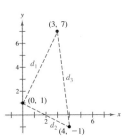

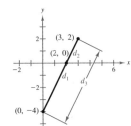

17. Collinear, because
$d_1 + d_2 = d_3$
$2\sqrt{5} + \sqrt{5} = 3\sqrt{5}$

19. Not collinear, because
$d_1 + d_2 \ne d_3$
$d_1 = \sqrt{18},\, d_2 = \sqrt{41}$
$d_3 = \sqrt{113}$

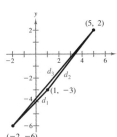

21. $x = 4, -2$　　**23.** $y = \pm\sqrt{55}$

25. $\left(\dfrac{3x_1 + x_2}{4}, \dfrac{3y_1 + y_2}{4}\right), \left(\dfrac{x_1 + x_2}{2}, \dfrac{y_1 + y_2}{2}\right),$
$\left(\dfrac{x_1 + 3x_2}{4}, \dfrac{y_1 + 3y_2}{4}\right)$

27. **(a)** $\left(\frac{7}{4}, -\frac{7}{4}\right), \left(\frac{5}{2}, -\frac{3}{2}\right), \left(\frac{13}{4}, -\frac{5}{4}\right)$

(b) $\left(-\frac{3}{2}, -\frac{9}{4}\right), \left(-1, -\frac{3}{2}\right), \left(-\frac{1}{2}, -\frac{3}{4}\right)$

29. **(a)** 16.76 ft　　**(b)** 1341.04 ft^2

31.

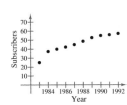

33. **(a)** 2900　　**(b)** 2450　　**(c)** 2970　　**(d)** 3270
35. **(a)** 1.42 million　　**(b)** 1.04 million
　　　(c) 0.85 million　　**(d)** 1.13 million

37. **(a)** Revenue: \$4372.0 million
　　　Profit: \$390.95 million
39. **(a)** $(-1, 2), (1, 1), (2, 3)$
　　　(b)

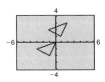

SECTION 1.2　(page 21)

Warm Up

1. $y = \frac{1}{5}(x + 12)$　　**2.** $y = x - 15$

3. $y = \dfrac{1}{x^3 + 2}$

4. $y = \pm\sqrt{x^2 + x - 6} = \pm\sqrt{(x + 3)(x - 2)}$

5. $y = -1 \pm \sqrt{9 - (x - 2)^2}$

6. $y = 5 \pm \sqrt{81 - (x + 6)^2}$

7. $(x - 2)(x - 1)$　　**8.** $(x + 3)(x + 2)$

9. $\left(y - \frac{3}{2}\right)^2$　　**10.** $\left(y - \frac{7}{2}\right)^2$

1. **(a)** Not a solution point　　**(b)** Solution point
　　(c) Solution point
3. **(a)** Not a solution point　　**(b)** Solution point
　　(c) Not a solution point
5. c　　**6.** d　　**7.** b　　**8.** f　　**9.** a　　**10.** e
11. $(0, -3), \left(\frac{3}{2}, 0\right)$　　**13.** $(0, -2), (-2, 0), (1, 0)$
15. $(0, 0), (-3, 0), (3, 0)$　　**17.** $(-2, 0), (0, 2)$
19. $(0, 0)$
21.　　　　　　　　　　　　　**23.**

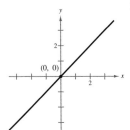

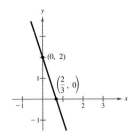

25.

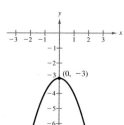

27.

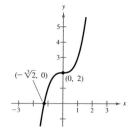

29.

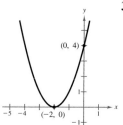

31.

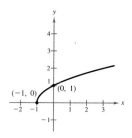

33.

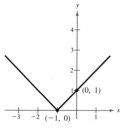

35.

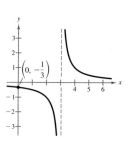

37.

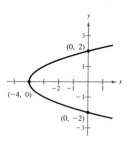

39. $x^2 + y^2 - 9 = 0$

41. $x^2 + y^2 - 4x + 2y - 11 = 0$

43. $x^2 + y^2 + 2x - 4y = 0$

45. $x^2 + y^2 - 6y = 0$

47. $(x - 1)^2 + (y + 3)^2 = 4$

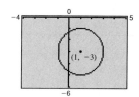

49. $(x + 2)^2 + (y + 3)^2 = 16$

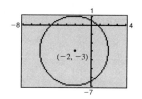

51. $\left(x - \frac{1}{2}\right)^2 + \left(y - \frac{1}{2}\right)^2 = 2$

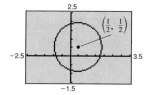

53. $\left(x + \frac{1}{2}\right)^2 + \left(y + \frac{5}{4}\right)^2 = \frac{9}{4}$

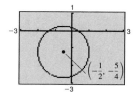

55. $(1, 1)$ **57.** $(5, 2)$ **59.** $(-1, -2), (2, 1)$

61. $(0, 0), (\sqrt{2}, 2\sqrt{2}), (-\sqrt{2}, -2\sqrt{2})$

63. $(-1, 0), (0, 1), (1, 0)$

65. (a) $C = 11.8x + 5000$; $R = 19.3x$
  (b) 667 units
  (c) 680 units
67. 50,000 units   **69.** 193 units
71. (a)

Year	1982	1985	1987
Model	1384.30	1583.14	1819.70

Year	1988	1989	1990
Model	1969.18	2139.46	2330.54

  (b) $3597.94 million
73. (a)

Year	1980	1985	1990
Salary	15,848.32	24,097.00	31,963.44

Year	1991	1994
Salary	33,527.71	38,209.43

  (c) $47,537.98
75. (3, 1)   **77.** (−2, 0), (1, 3)   **79.** (−2, 5), (2, 5)
81. (0, 5.36)

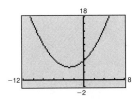

83. (0, −4.3), (1.9749, 0), (−1.7097, 0), (12.7349, 0)

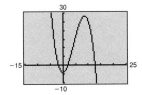

85. (1.4780, 0), (12.8553, 0), (0, 2.3875)

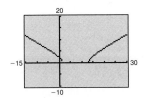

87. (0, 0.4167)

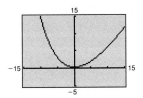

SECTION 1.3   (page 35)

> **Warm Up**
>
> **1.** −1   **2.** $-\frac{7}{3}$ **3.** $\frac{1}{3}$ **4.** $-\frac{7}{6}$
> **5.** $y = 4x + 7$ **6.** $y = 3x - 7$
> **7.** $y = 3x - 10$ **8.** $y = -x - 7$
> **9.** $y = 7x - 17$ **10.** $y = \frac{2}{3}x + \frac{5}{3}$

1. 1   **3.** 0   **5.** −3
7. $m = 3$ **9.** $m = 0$

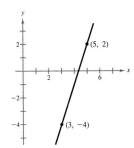

11. m is undefined   **13.** $m = 0$

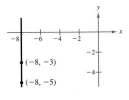

15. (0, 1), (1, 1), (3, 1)   **17.** (7, −6), (5, −2), (4, 0)
19. (0, 10), (2, 4), (3, 1)
21. (−8, 0), (−8, 2), (−8, 3)   **23.** $m = -\frac{1}{5}$, (0, 4)

25. $m = \frac{7}{5}$, $(0, -3)$

27. m is undefined; no y-intercept.

29. $y = 2x - 5$ **31.** $3x + y = 0$

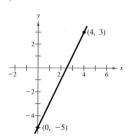
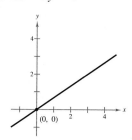

33. $x - 2 = 0$ **35.** $y + 2 = 0$

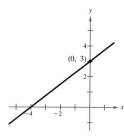
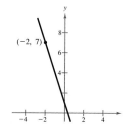

37. $3x - 4y + 12 = 0$ **39.** $2x - 3y = 0$

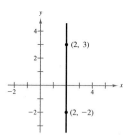
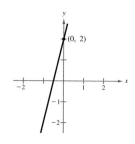

41. $3x + y - 1 = 0$ **43.** $4x - y + 2 = 0$

45. $9x - 12y + 8 = 0$

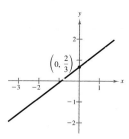

47. The points are not collinear.   **49.** $x - 3 = 0$

51. (a) $x + y + 1 = 0$ (b) $x - y + 5 = 0$

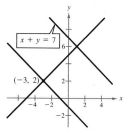

53. (a) $3x + 4y + 2 = 0$ (b) $4x - 3y + 36 = 0$

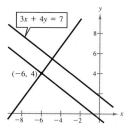

55. (a) $y = 0$ (b) $x + 1 = 0$

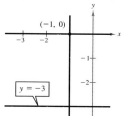

57. (a) $2x - 3y + 1 = 0$ **(b)** $3x + 2y - 5 = 0$

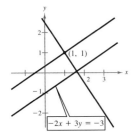

59.

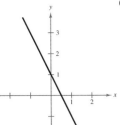

61.

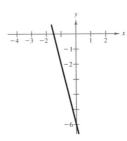

63.

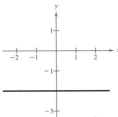

65.

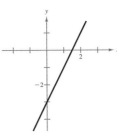

67. $F = \frac{9}{5}C + 32$ **69.** $C = 0.25x + 100$

71. (a) $S = 1700t + 26{,}300$ **(b)** \$34,800
  (c) Salary increases by the same amount (\$1700) each year.
  (d) Salary increases by the same percent each year. Yes, as your salary increases, the amount of your raise increases.

73. (a) $y = -205t + 1025, \; 0 \le t \le 5$
  (b)

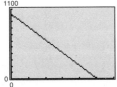

  (c) \$410.00   **(d)** 2.07 years

75. (a) $C = 14.75t + 26{,}500$ **(b)** $R = 25t$
  (c) $P = 10.25t - 26{,}500$ **(d)** 2585.4 hr
77. $x \le 24$ units   **79.** $x \le 70$ units

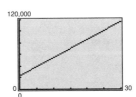
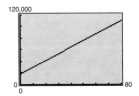

81. $x \le 275$ units

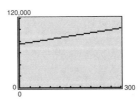

83. (a) $W = 2000 + 0.07S$
  (b) $W = 2300 + 0.05S$
  (c) $(15{,}000, \; 3050)$

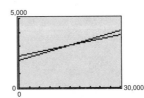

The point of intersection signifies that if your sales were \$15,000, then both sides would yield wages of \$3050.

  (d) No. You will make more money (if sales are \$20,000) at your current job ($W = \$3400$) than in the offered job ($W = \$3300$).

SECTION 1.4 (page 47)

Warm Up

1. 20   **2.** 10   **3.** $x^2 + x - 6$

4. $x^3 + 9x^2 + 26x + 30$ **5.** $\dfrac{1}{x}$ **6.** $\dfrac{2x - 1}{x}$

7. $y = -2x + 17$ **8.** $y = \frac{6}{5}x^2 + \frac{1}{5}$

9. $y = 3 \pm \sqrt{5 + (x + 1)^2}$ **10.** $y = \pm\sqrt{4x^2 + 2}$

1. (a) -3 (b) -9 (c) $2x - 5$ (d) $2\Delta x - 1$

3. (a) 2 (b) 18 (c) $\frac{1}{2}c^2$
  (d) $\frac{1}{2}x^2 + x\Delta x + \frac{1}{2}(\Delta x)^2$

5. (a) 1 (b) -1 (c) 1
  (d) $\dfrac{|x - 1|}{(x - 1)} = \begin{cases} -1, & x < 1 \\ 1, & x > 1 \end{cases}$

7. 3 **9.** $3 + \Delta x$ **11.** $3x^2 + 3x\Delta x + (\Delta x)^2 - 1$

13. y is not a function of x. **15.** y is a function of x.

17. y is a function of x. **19.** y is not a function of x.

21. Domain: $(-\infty, \infty)$ **23.** Domain: $\left[\frac{3}{2}, \infty\right)$
  Range: $(-\infty, \infty)$ Range: $[0, \infty)$

25. Domain: $(-\infty, 0) \cup (0, \infty)$
  Range: $(0, \infty)$

27.

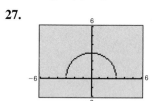

Domain: $[-3, 3]$
Range: $[0, 3]$

29.

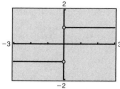

Domain: $(-\infty, 0) \cup (0, \infty)$
Range: $y = -1$ or $y = 1$

31.

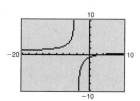

Domain: $(-\infty, -4) \cup (-4, \infty)$
Range: $(-\infty, 1) \cup (1, \infty)$

33. y is a function of x. **35.** y is not a function of x.

37. y is a function of x.

39. (a) $2x$ (b) $(x + 1)(x - 1) = x^2 - 1$
  (c) $\dfrac{x + 1}{x - 1}$ (d) x (e) x

41. (a) $3x^2 - 1$ (b) $2x^4 - 2x^2$ (c) $\dfrac{2x^2}{x^2 - 1}$
  (d) $2x^4 - 4x^2 + 2$ (e) $4x^4 - 1$

43. (a) $x^2 + 5 + \sqrt{1 - x}$ (b) $(x^2 + 5)\sqrt{1 - x}$
  (c) $\dfrac{x^2 + 5}{\sqrt{1 - x}}$ (d) $6 - x, 1 \geq x$
  (e) Not defined

45. (a) $\dfrac{1}{x} + \dfrac{1}{x^2} = \dfrac{x + 1}{x^2}$ (b) $\dfrac{1}{x^3}$ (c) x
  (d) x^2 (e) x^2

47. $f(g(x)) = \left(\sqrt[3]{x}\right)^3 = x$
  $g(f(x)) = \sqrt[3]{x^3} = x$

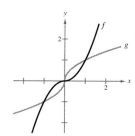

49. $f(g(x)) = 5\left(\dfrac{x - 1}{5}\right) + 1 = x$
  $g(f(x)) = \dfrac{5x + 1 - 1}{5} = x$

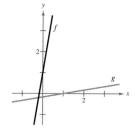

51. $f(g(x)) = \sqrt{x^2 + 4} - 4 = x, x \geq 0$
$g(f(x)) = (\sqrt{x-4})^2 + 4 = x, x \geq 4$

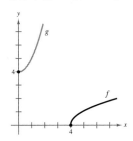

53. $f(x) = 2x - 3, f^{-1}(x) = \dfrac{x+3}{2}$

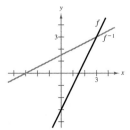

55. $f(x) = x^5, f^{-1}(x) = \sqrt[5]{x}$

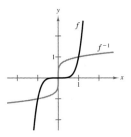

57. $f(x) = \sqrt{9 - x^2}, f'(x) = \sqrt{9 - x^2}, 0 \leq x \leq 3$

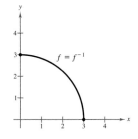

59. $f(x) = x^{2/3}, f^{-1}(x) = x^{3/2}, x \geq 0$

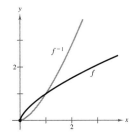

61.

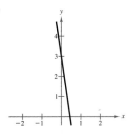

$f(x)$ is one-to-one.
$f^{-1}(x) = \dfrac{3-x}{7}$

63.

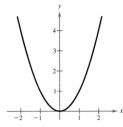

$f(x)$ is not one-to-one.

65.

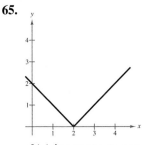

$f(x)$ is not one-to-one.

67. (a) 0   (b) 0   (c) -1 (d) $\sqrt{15}$
(e) $\sqrt{x^2 - 1}$ (f) $x - 1, x \geq 0$

69. The data fits the function
(b) $g(x) = cx^2$ with $c = -2$.

71. The data fits the function
(d) $r(x) = c/x$ with $c = 32$.

73. (a)

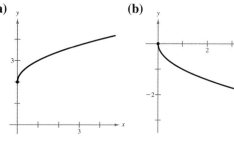

(b)

(c)

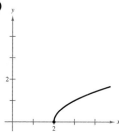

(d)

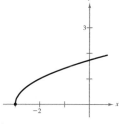

(e)

(f)

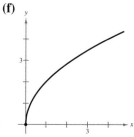

75. $V(x) = 2500x + 750,000$

77. (a) $x = \dfrac{1475}{p} - 100$ **(b)** 47.5 units

79. (a)

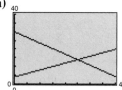

(b) $(25, 14)$
(c) $0 \le x < 25$
(d) $x > 25$

81. (a) $C = 98,000 + 12.30x$
(b) $R = 17.98x$
(c) $P = 5.68x - 98,000$

83.

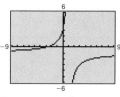

Zeros: $x = 0, \pm 3$
$f(x)$ is not one-to-one.

85.

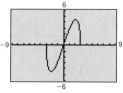

Zeros: $t = -3$
$g(t)$ is one-to-one.

87.

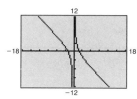

Zeros: $x = \pm 2$
$f(x)$ is not one-to-one.

SECTION 1.5   (page 60)

Warm Up

1. (a) 7   **(b)** $c^2 - 3c + 3$
   (c) $x^2 + 2xh + h^2 - 3x - 3h + 3$
2. (a) -4 **(b)** 10   **(c)** $3t^2 + 4$
3. h **4.** 4
5. Domain: $(-\infty, 0) \cup (0, \infty)$
   Range: $(-\infty, 0) \cup (0, \infty)$

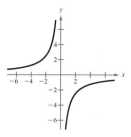

Warm Up *(continued)*

6. Domain: $[-5, 5]$ **7.** Domain: $(-\infty, \infty)$
Range: $[0, 5]$ Range: $[0, \infty)$

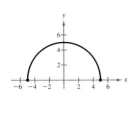   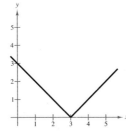

8. Domain: $(-\infty, 0) \cup (0, \infty)$
Range: $-1, 1$

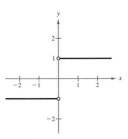

9. y is not a function of x.
10. y is a function of x.

1.

x	1.9	1.99	1.999	2
$f(x)$	13.5	13.95	13.995	?

x	2.001	2.01	2.1
$f(x)$	14.005	14.05	14.5

$\lim\limits_{x \to 2} (5x + 4) = 14$

3.

x	1.9	1.99	1.999	2
$f(x)$	0.2564	0.2506	0.2501	?

x	2.001	2.01	2.1
$f(x)$	0.2499	0.2494	0.2439

$\lim\limits_{x \to 2} \dfrac{x - 2}{x^2 - 4} = \dfrac{1}{4}$

5.

x	-0.1	-0.01	-0.001	0
$f(x)$	0.2911	0.2889	0.2887	?

x	0.001	0.01	0.1
$f(x)$	0.2887	0.2884	0.2863

$\lim\limits_{x \to 0} \dfrac{\sqrt{x + 3} - \sqrt{3}}{x} \approx 0.2887$

$\left(\text{The actual limit is } \dfrac{1}{2\sqrt{3}}. \right)$

7.

x	1.5	1.9	1.99	1.999	2
$f(x)$	0.3780	0.1601	0.0501	0.0158	?

$\lim\limits_{x \to 2} \dfrac{2 - x}{\sqrt{4 - x^2}} = 0$

9. (a) 1   **(b)** 3   **11. (a)** 1   **(b)** 3
13. (a) 12   **(b)** 27   **(c)** $\frac{1}{3}$
15. (a) 1   **(b)** 1   **(c)** 1
17. (a) 0   **(b)** 0   **(c)** 0
19. (a) 3   **(b)** -3 **(c)** Limit does not exist.
21. 4   **23.** -3 **25.** -4 **27.** 2   **29.** -1
31. $\frac{2}{3}$ **33.** -2 **35.** Limit does not exist.   **37.** $\frac{1}{10}$
39. 12   **41.** Limit does not exist.   **43.** -1
45. 2   **47.** 2   **49.** $3x^2$ **51.** $2t - 5$

53. $\dfrac{1}{2\sqrt{2}}$ **55.** $-\infty$ **57.** $-\infty$ **59.** ∞

61. 4   **63.** Limit does not exist.

65. $-\frac{17}{9} \approx -1.8889$ **67.** Yes. 1814.02

SECTION 1.6   (page 71)

Warm Up

1. $\dfrac{x+4}{x-8}$ **2.** $\dfrac{x+1}{x-3}$ **3.** $\dfrac{x+2}{2(x-3)}$

4. $\dfrac{x-4}{x-2}$ **5.** $x = 0, -7$ **6.** $x = -5, 1$

7. $x = -\frac{2}{3}, -2$ **8.** $x = 0, 3, -8$

9. 13   **10.** -1

1. Continuous   **3.** Not continuous ($x \neq \pm 2$)

5. $(-\infty, \infty)$ **7.** $(-\infty, -1) \cup (-1, \infty)$

9. $(-\infty, \infty)$ **11.** $(-\infty, 1) \cup (1, \infty)$

13. $(-\infty, \infty)$ **15.** $(-\infty, 4) \cup (4, 5) \cup (5, \infty)$

17. $(-\infty, 1) \cup (1, \infty)$ **19.** $(-\infty, \infty)$

21. $(-\infty, 2) \cup (2, \infty)$ **23.** $(-\infty, \infty)$

25. $(-\infty, -1) \cup (-1, \infty)$

27. Continuous on all intervals $(c, c+1)$, where c is an integer.

29. Continuous on all intervals $(c, c+1)$, where c is an integer.

31. $(1, \infty)$ **33.** Continuous

35. Nonremovable discontinuity at $x = 2$

37. Removable discontinuity at $x = 4$

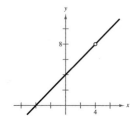

39. Removable discontinuity at $x = 0$

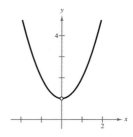

41. Continuous on $(-\infty, 0) \cup (0, \infty)$

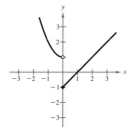

43. $a = 2$

45. Not continuous at $x = 2$ and $x = -1$.

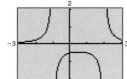

47. $(-\infty, \infty)$

49. Continuous on all intervals $\left(\dfrac{c}{2}, \dfrac{c+1}{2}\right)$, where c is an integer.

51. The graph of $f(x) = \dfrac{x^2 + x}{x}$ appears to be continuous on $[-4, 4]$, but f is not continuous at $x = 0$.

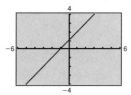

53. (a) Graph has nonremovable discontinuities at
$t = \frac{1}{4}, \frac{1}{2}, \frac{3}{4}, 1, \frac{5}{4}, \ldots$

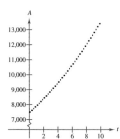

(b) $11,379.17

55. (a)
$$C(t) = \begin{cases} 1.04, & 0 < t \le 2 \\ 1.04 + 0.36 \, [\![t - 1]\!], & t > 2, t \text{ is not an integer} \\ 1.04 + 0.36(t - 2), & t > 2, t \text{ is an integer} \end{cases}$$

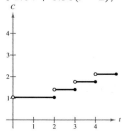

C is not continuous at $t = 2, 3, 4, \ldots$
(b) $3.56

57. (a)

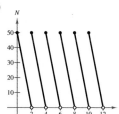

Nonremovable discontinuities at $t = 2, 4, 6, 8, \ldots$
(b) Every two months

CHAPTER 1 REVIEW EXERCISES (page 76)

1. a   **2.** c   **3.** b   **4.** d
5. $\sqrt{29}$ **7.** $3\sqrt{2}$ **9.** $(7, 4)$ **11.** $(-8, 6)$

13. The taller bars in the book represent revenues. The middle bars represent costs. The smaller bars in front represent profits, because $P = R - C$.

15. $(4, 7), (5, 8), (8, 10)$

17. 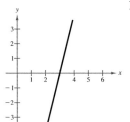   **19.**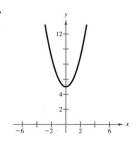

21. 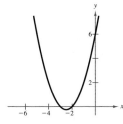   **23.**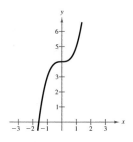

25. $(0, -3), \left(-\frac{3}{4}, 0\right)$ **27.** $x^2 + y^2 = 9$

29. $(x - 3)^2 + (y + 4)^2 = 25$
Center: $(3, -4)$
Radius: 5

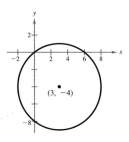

31. $(1, 1)$ **33.** $(0, 0), (1, 1), (-1, -1)$
35. (a) $C = 200 + 10x; R = 14x$
(b) 50 shirts

37. Slope: -3
y-intercept: $(0, -2)$

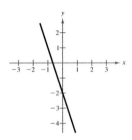

39. Slope: 0 (horizontal line)
y-intercept: $\left(0, -\frac{5}{3}\right)$

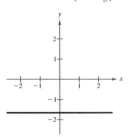

41. Slope: $-\frac{2}{5}$
y-intercept: $(0, -1)$

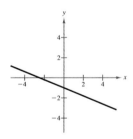

43. $\frac{6}{7}$ **45.** $\frac{20}{21}$
47. $y = -2x + 5$

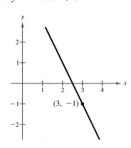

49. (a) $y = \frac{7}{8}x + \frac{69}{8}$ (b) $y = -2x$
(c) $y = -2x$ (d) $y = -\frac{2}{3}x + 4$

51. (a) $x = -10p + 1070$ (b) 725 units
(c) 650 units

53. y is a function of x.   **55.** y is not a function of x.

57. (a) 7   (b) $3x + 7$ (c) $10 + 3\Delta x$

59. Domain: $(-\infty, \infty)$ **61.** Domain: $[-1, \infty)$
Range: $\left[-\frac{1}{4}, \infty\right)$ Range: $[0, \infty)$

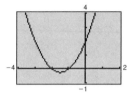
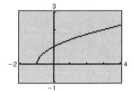

63. Domain: $(-\infty, \infty)$
Range: $(-\infty, 3]$

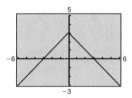

65. (a) $x^2 + 2x$ (b) $x^2 - 2x + 2$
(c) $2x^3 - x^2 + 2x - 1$ (d) $\dfrac{1 + x^2}{2x - 1}$
(e) $4x^2 - 4x + 2$ (f) $2x^2 + 1$

67. $f^{-1}(x) = \frac{2}{3}x$

69. $f(x)$ does not have an inverse.   **71.** 7   **73.** 49

75. $\frac{10}{3}$ **77.** Limit does not exist.   **79.** $-\frac{1}{4}$

81. $-\infty$ **83.** Limit does not exist.

85. $3x^2 - 1$ **87.** 0.5774

89. False   **90.** True   **91.** False

92. True   **93.** False   **94.** True

95. $(-\infty, -4) \cup (-4, \infty)$

97. $(-\infty, -1) \cup (-1, \infty)$

99. Continuous on all intervals of the form $(c, c + 1)$, where c is an integer

101. $(-\infty, 0) \cup (0, \infty)$ **103.** $a = 2$

105. (a)

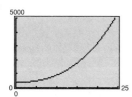

(b)

t	0	5	10
Debt	381	542	909
Model	358.78	494.63	917.48

t	15	20	21
Debt	1817	3206	3599
Model	1777.33	3224.18	3597.19

(c) $5408.03 billion

SAMPLE POST-GRAD EXAM QUESTIONS (page 80)

1. (e)   **2.** (d)   **3.** (b)   **4.** (b)

5. (a)   **6.** (c)   **7.** (e)   **8.** (b)

CHAPTER 2

SECTION 2.1 (page 90)

Warm Up

1. $x = 2$ **2.** $y = 2$ **3.** $2x$ **4.** $3x^2$

5. $\dfrac{1}{x^2}$ **6.** $2x$ **7.** $(-\infty, 1) \cup (1, \infty)$

8. $(-\infty, \infty)$ **9.** $(-\infty, 0) \cup (0, \infty)$

10. $(-\infty, -4) \cup (-4, 3) \cup (3, \infty)$

1.

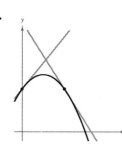

3.

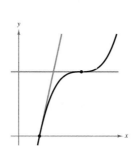

5. $m = 1$ **7.** $m = 0$ **9.** $m = -\frac{1}{3}$

11. 1990: $m \approx 20$

1992: $m \approx 4$

13. $f(x) = 3$

1. $f(x + \Delta x) = 3$

2. $f(x + \Delta x) - f(x) = 0$

3. $\dfrac{f(x + \Delta x) - f(x)}{\Delta x} = 0$

4. $\lim\limits_{\Delta x \to 0} \dfrac{f(x + \Delta x) - f(x)}{\Delta x} = 0$

15. $f(x) = -5x + 3$

1. $f(x + \Delta x) = -5x - 5\Delta x + 3$

2. $f(x + \Delta x) - f(x) = -5\Delta x$

3. $\dfrac{f(x + \Delta x) - f(x)}{\Delta x} = -5$

4. $\lim\limits_{\Delta x \to 0} \dfrac{f(x + \Delta x) - f(x)}{\Delta x} = -5$

17. $f(x) = x^2$

1. $f(x + \Delta x) = x^2 + 2x\Delta x + (\Delta x)^2$

2. $f(x + \Delta x) - f(x) = 2x\Delta x + (\Delta x)^2$

3. $\dfrac{f(x + \Delta x) - f(x)}{\Delta x} = 2x + \Delta x$

4. $\lim\limits_{\Delta x \to 0} \dfrac{f(x + \Delta x) - f(x)}{\Delta x} = 2x$

19. $6x - 5$

21. $h(t) = \sqrt{t - 1}$

1. $h(t + \Delta t) = \sqrt{t + \Delta t - 1}$

2. $h(t + \Delta t) - h(t) = \sqrt{t + \Delta t - 1} - \sqrt{t - 1}$

3. $\dfrac{h(t + \Delta t) - h(t)}{\Delta t} = \dfrac{1}{\sqrt{t + \Delta t - 1} + \sqrt{t - 1}}$

4. $\lim\limits_{\Delta t \to 0} \dfrac{h(t + \Delta t) - h(t)}{\Delta t} = \dfrac{1}{2\sqrt{t - 1}}$

23. $f(t) = t^3 - 12t$

1. $f(t + \Delta t) = t^3 + 3t^2\Delta t + 3t(\Delta t)^2$
$+ (\Delta t)^3 - 12t - 12\Delta t$

2. $f(t + \Delta t) - f(t) = 3t^2\Delta t + 3t(\Delta t)^2$
$+ (\Delta t)^3 - 12\Delta t$

3. $\dfrac{f(t + \Delta t) - f(t)}{\Delta t} = 3t^2 + 3t\Delta t + (\Delta t)^2 - 12$

4. $\lim\limits_{\Delta t \to 0} \dfrac{f(t + \Delta t) - f(t)}{\Delta t} = 3t^2 - 12$

25. $-\dfrac{2}{x^3}$

27. $f'(x) = -2$
$f'(2) = -2$

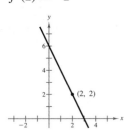

29. $f'(x) = -1$
$f'(0) = -1$

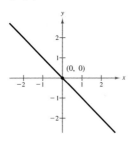

43. $y = \dfrac{x}{4} + 2$

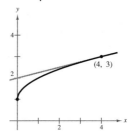

45. $y = -x + 1$ **47.** $y = -6x + 8$
                          $y = -6x - 8$

49. $x \ne -3$ (node)
51. $x \ne -1$ (nonremovable discontinuity)
53. $x \ne 3$ (cusp)   **55.** $x > 1$
57. $x \ne 0$ (nonremovable discontinuity)
59.

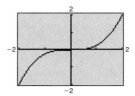

31. $f'(x) = 2x$
$f'(2) = 4$

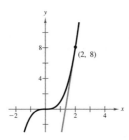

33. $f'(x) = 3x^2$
$f'(2) = 12$

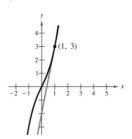

x	-2	$-\frac{3}{2}$	-1	$-\frac{1}{2}$
$f(x)$	-2	-0.8438	-0.25	-0.0313
$f'(x)$	3	1.6875	0.75	0.1875

x	0	$\frac{1}{2}$	1	$\frac{3}{2}$	2
$f(x)$	0	0.0313	0.25	0.8438	2
$f'(x)$	0	0.1875	0.75	1.6875	3

35. $f'(x) = \dfrac{1}{2\sqrt{x+1}}$
$f'(3) = \frac{1}{4}$

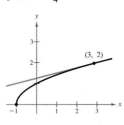

37. $f'(x) = 3x^2 + 2$
$f'(1) = 5$

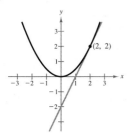

61. True
62. False. $f(x) = |x|$ is continuous, but not differentiable, at $x = 0$.
63. True   **64.** True

39. $y = 2x - 2$

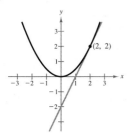

41. $y = -6x - 3$

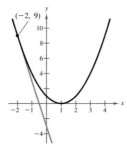

65. $f'(x) = 2x - 4$

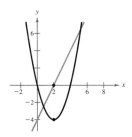

The x-intercept of the derivative indicates a point of horizontal tangency for f.

SECTION 2.2   (page 102)

Warm Up

1. (a) 8   **(b)** 16   **(c)** $\frac{1}{2}$

2. (a) $\frac{1}{36}$ **(b)** $\frac{1}{32}$ **(c)** $\frac{1}{64}$

3. $4x(3x^2 + 1)$ **4.** $\frac{3}{2}x^{1/2}(x^{3/2} - 1)$

5. $\frac{1}{4x^{3/4}}$ **6.** $x^2 - \frac{1}{x^{1/2}} + \frac{1}{3x^{2/3}}$

7. $0, -\frac{2}{3}$ **8.** $0, \pm 1$

9. $-10, 2$ **10.** $-2, 12$

---

1. (a) 2   **(b)** $\frac{1}{2}$ **3. (a)** -1 **(b)** $-\frac{1}{3}$

5. 0   **7.** 1   **9.** $2x$ **11.** $-6t + 2$

13. $3t^2 - 2$ **15.** $\frac{16}{3}t^{1/3}$ **17.** $\frac{2}{\sqrt{x}}$

19. $-\frac{8}{x^3} + 4x$

21. Function: $y = \dfrac{1}{4x^3}$

Rewrite: $y = \dfrac{1}{4}x^{-3}$

Derivative: $y' = \dfrac{-3}{4}x^{-4}$

Simplify: $y' = -\dfrac{3}{4x^4}$

23. Function: $y = \dfrac{1}{(4x)^3}$

Rewrite: $y = \dfrac{1}{64}x^{-3}$

Derivative: $y' = -\dfrac{3}{64}x^{-4}$

Simplify: $y' = -\dfrac{3}{64x^4}$

25. Function: $y = \dfrac{\sqrt{x}}{x}$

Rewrite: $y = x^{-1/2}$

Derivative: $y' = -\dfrac{1}{2}x^{-3/2}$

Simplify: $y' = -\dfrac{1}{2x^{3/2}}$

27. $-\dfrac{1}{x^2}, -1$ **29.** $\dfrac{4}{3t^2}, \dfrac{16}{3}$ **31.** $8x + 4, 4$

33. $\dfrac{2(x^3 + 2)}{x^2}$ **35.** $2x - 2 + \dfrac{8}{x^5}$ **37.** $\dfrac{2x^3 - 6}{x^3}$

39. $3x^2 + 1$ **41.** $\dfrac{4}{5x^{1/5}}$ **43.** $\dfrac{1}{3x^{2/3}} + \dfrac{1}{5x^{4/5}}$

45. $y = 2x - 2$

47. $(0, -1), \left(-\dfrac{\sqrt{6}}{2}, \dfrac{5}{4}\right), \left(\dfrac{\sqrt{6}}{2}, \dfrac{5}{4}\right)$

49. No horizontal tangents

51. (a)   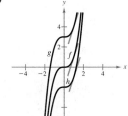   **(b)** $f'(1) = g'(1)$
$= h'(1)$
$= 3$

(c)

53. (a) 3   **(b)** 6   **(c)** −3   **(d)** 6
55. (a) 1990: −4.5   **(c)** Billions of dollars per year
              1992: 6.27
57. $P = 0.40x - 250$
59. $(-0.267,\ 1.577),\ (1.2,\ 0)$

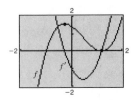

SECTION 2.3   (page 116)

Warm Up

1. 3   **2.** −7   **3.** $y' = 8x - 2$
4. $y' = -9t^2 + 4t$ **5.** $s' = -32t + 24$
6. $y' = -32x + 54$ **7.** $A' = -\frac{3}{5}r^2 + \frac{3}{5}r + \frac{1}{2}$
8. $y' = 2x^2 - 4x + 7$ **9.** $y' = 12 - \dfrac{x}{2500}$
10. $y' = 74 - \dfrac{3x^2}{10{,}000}$

1. (a) $2.5 billion per year
  (b) −$0.714 billion per year
  (c) $5.8 billion per year
  (d) $2.2 billion per year
3. Average rate: 2
  Instantaneous rates: $f'(1) = 2,\ f'(2) = 2$

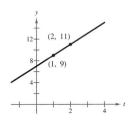

5. Average rate: −1
  Instantaneous rates: $h'(0) = 0,\ h'(1) = -2$

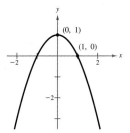

7. Average rate: 2.1
  Instantaneous rates: $f'(1) = 2,\ f'(1.1) = 2.2$

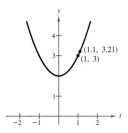

9. Average rate: $-\frac{1}{4}$
  Instantaneous rates: $f'(1) = -1,\ f'(4) = -\frac{1}{16}$

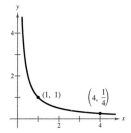

11. Average rate: 1
  Instantaneous rates: $g'(1) = 2,\ g'(9) = \frac{2}{3}$

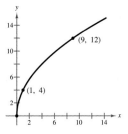

13. (a) Average rate: $\frac{11}{27}$

Instantaneous rates: $E'(0) = \frac{1}{3}$, $E'(1) = \frac{4}{9}$

(b) Average rate: $\frac{11}{27}$

Instantaneous rates: $E'(1) = \frac{4}{9}$, $E'(2) = \frac{1}{3}$

(c) Average rate: $\frac{5}{27}$

Instantaneous rates: $E'(2) = \frac{1}{3}$, $E'(3) = 0$

(d) Average rate: $-\frac{7}{27}$

Instantaneous rates: $E'(3) = 0$, $E'(4) = -\frac{5}{9}$

15. (a) -80 ft/sec

(b) $s'(2) = -64$ ft/sec, $s'(3) = -96$ ft/sec

(c) $\dfrac{\sqrt{555}}{4} \approx 5.89$ sec

(d) $-8\sqrt{555} \approx -188.5$ sec

17. (a) $s'(0) = 0$ ft/sec   **(b)** $s'(1) = 15$ ft/sec
(c) $s'(4) = 30$ ft/sec   **(d)** $s'(9) = 45$ ft/sec

19. 1.47   **21.** $470 - \frac{1}{2}x$ **23.** $50 - x$

25. $-18x^2 + 16x + 200$ **27.** $-4x + 72$

29. $-\frac{1}{2000}x + 12.2$ **31. (a)** 64.998   **(b)** 65

33. (a) 0.584   **(b)** 0.6

35. (a) 37.40   **(b)** 79.80   **(c)** 14.83
(d) -48.40

37. (a) $R = 5x - 0.001x^2$
(b) $P = 3.5x - 0.001x^2 - 35$
(c)

x	600	1200	1800	2400	3000
$\dfrac{dR}{dx}$	3.8	2.6	1.4	0.2	-1.0
$\dfrac{dP}{dx}$	2.3	1.1	-0.1	-1.3	-2.5
P	1705	2725	3025	2605	1465

39. (a) $P = -0.0003x^2 + 17.8x - 85,000$
(b) 5.8 dollars

43. $C(351) - C(350) \approx -1.91$

$\dfrac{dC}{dQ} = -1.93$ when $Q = 350$.

45. $C = \dfrac{16,500}{x}$

$\dfrac{dC}{dx} = -\dfrac{16,500}{x^2}$

x	10	15	20	25
C	\$1650	\$1100	\$825	\$660
$\dfrac{dC}{dx}$	-165	-73.33	-41.25	-26.40

x	30	35	40
C	\$550	\$471.43	\$412.50
$\dfrac{dC}{dx}$	-18.33	-13.47	-10.31

The driver who gets 15 miles per gallon would benefit more from a 1 mile per gallon increase in fuel efficiency. The rate of change is much larger when $x = 15$.

47.

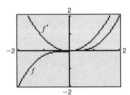

f has a horizontal tangent at $x = 0$.

SECTION 2.4   (page 130)

Warm Up

1. $2(3x^2 + 7x + 1)$ **2.** $4x^2(6 - 5x^2)$

3. $\dfrac{23}{(2x + 7)^2}$ **4.** $-\dfrac{x^2 + 8x + 4}{(x^2 - 4)^2}$

5. $\dfrac{4x^3 - 3x^2 + 3}{x^2}$ **6.** $\dfrac{x^2 - 2x + 4}{(x - 1)^2}$

7. 11   **8.** 0   **9.** $-\frac{1}{4}$ **10.** $\frac{17}{4}$

1. $f'(x) = 15x^4 - 2x$ **3.** $f'(x) = 2x^2$
$f'(1) = 13$ $f'(0) = 0$

5. $g'(x) = 3x^2 - 12x + 11$
$g'(4) = 11$

7. $h'(x) = -\dfrac{5}{(x-5)^2}$
$h'(6) = -5$

9. $f'(t) = \dfrac{2t^2 + 3}{3t^2}$
$f'(2) = \frac{11}{12}$

11. Function: $y = \dfrac{x^2 + 2x}{x}$
Rewrite: $y = x + 2$
Derivative: $y' = 1$
Simplify: $y' = 1$

13. Function: $y = \dfrac{7}{3x^3}$
Rewrite: $y = \dfrac{7}{3}x^{-3}$
Derivative: $y' = -7x^{-4}$
Simplify: $y' = -\dfrac{7}{x^4}$

15. Function: $y = \dfrac{4x^2 - 3x}{8\sqrt{x}}$
Rewrite: $y = \dfrac{1}{2}x^{3/2} - \dfrac{3}{8}x^{1/2}$
Derivative: $y' = \dfrac{3}{4}x^{1/2} - \dfrac{3}{16}x^{-1/2}$
Simplify: $y' = \dfrac{3}{4}\sqrt{x} - \dfrac{3}{16\sqrt{x}}$

17. Function: $y = \dfrac{x^2 - 4x + 3}{x - 1}$
Rewrite: $y = x - 3, x \neq 1$
Derivative: $y' = 1, x \neq 1$
Simplify: $y' = 1, x \neq 1$

19. $10x^4 + 12x^3 - 3x^2 - 18x - 15$
21. $28t^6 - 42t^5 - 15t^4 - 8t + 7$
23. $6p^2(p^3 - 2)$
25. $\dfrac{5}{6x^{1/6}} + \dfrac{1}{x^{2/3}}$
27. $-\dfrac{5}{(2x-3)^2}$
29. $\dfrac{2}{(x+1)^2}$
31. $\dfrac{3(x^4 + 1)}{x^2}$
33. $-\dfrac{1}{(t+3)^2}$
35. $\dfrac{x-1}{2x^{3/2}}$
37. $\dfrac{2x^3 + 11x^2 - 8x - 17}{(x+4)^2}$

39. $15x^4 - 48x^3 - 33x^2 - 32x - 20$
41. $y = \frac{1}{4}x + \frac{1}{2}$
43. $y = -x - 2$
45. $(0,\ 0)$ and $(2,\ 4)$
47. $(0,\ 0)$ and $(\sqrt[3]{-4},\ -2.117)$

49.

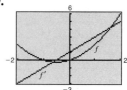

51.

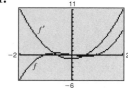

53. -1.87
55. **(a)** -0.480 **(b)** 0.120 **(c)** 0.015
57. 31.55 bacteria/hr

59. **(a)** $p = \dfrac{4000}{\sqrt{x}}$
(b) $C = 250x + 10{,}000$
(c) $P = 4000\sqrt{x} - 250x - 10{,}000$

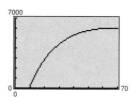

$\$500$ per unit

61. d

63. **(a)** -38.125
(b) -10.37
(c) -3.80
Increasing the order size reduces the cost per item.

SECTION 2.5 *(page 140)*

Warm Up

1. $(1 - 5x)^{2/5}$ **2.** $(2x - 1)^{3/4}$
3. $(4x^2 + 1)^{-1/2}$ **4.** $(x - 6)^{-1/3}$
5. $x^{1/2}(1 - 2x)^{-1/3}$ **6.** $(2x)^{-1}(3 - 7x)^{3/2}$
7. $(x - 2)(3x^2 + 5)$ **8.** $(x - 1)(5\sqrt{x} - 1)$
9. $(x^2 + 1)^2(4 - x - x^3)$
10. $(3 - x^2)(x - 1)(x^2 + x + 1)$

$y = f(g(x))$	$u = g(x)$	$y = f(u)$
1. $y = (6x - 5)^4$	$u = 6x - 5$	$y = u^4$
3. $y = (4 - x^2)^{-1}$	$u = 4 - x^2$	$y = u^{-1}$
5. $y = \sqrt{5x - 2}$	$u = 5x - 2$	$y = \sqrt{u}$
7. $y = \dfrac{1}{3x + 1}$	$u = 3x + 1$	$y = u^{-1}$

9. c **10.** d **11.** b **12.** a

13. $6(2x - 7)^2$ **15.** $12x(x^2 - 1)^2$

17. $-6(4 - 2x)^2$ **19.** $6x(6 - x^2)(2 - x^2)$

21. $\dfrac{4x}{3(x^2 - 9)^{1/3}}$ **23.** $\dfrac{1}{2\sqrt{t + 1}}$

25. $\dfrac{4t + 5}{2\sqrt{2t^2 + 5t + 2}}$ **27.** $\dfrac{6x}{(9x^2 + 4)^{2/3}}$

29. $-\dfrac{2x}{\sqrt{4 - x^2}}$ **31.** $-\dfrac{x}{(25 + x^2)^{3/2}}$

33. $\dfrac{4x^2}{(4 - x^3)^{7/3}}$

35. $y = \frac{8}{3}x - \frac{7}{3}$ or $3y - 8x + 7 = 0$

37. $f'(x) = \dfrac{1 - 3x^2 - 4x^{3/2}}{2\sqrt{x}(x^2 + 1)^2}$

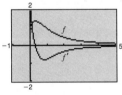

The zero of $f'(x)$ corresponds to the point on the graph of $f(x)$ where the tangent line is horizontal.

39. $f'(x) = -\dfrac{(\sqrt{(x + 1)}/x)}{2x(x + 1)}$

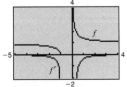

$f'(x)$ has no zeros.

41. $-\dfrac{1}{(x - 2)^2}$ **43.** $\dfrac{72}{(3 - t)^3}$

45. $-\dfrac{9x^2}{(x^3 - 4)^2}$ **47.** $-\dfrac{1}{2(x + 2)^{3/2}}$

49. $-\dfrac{3x^2}{(x^3 - 1)^{4/3}}$ **51.** $\dfrac{3(x + 1)}{\sqrt{2x + 3}}$

53. $\dfrac{t(5t - 8)}{2\sqrt{t - 2}}$ **55.** $x^2(x - 4)(5x - 12)$

57. $\dfrac{5t^2 + 8t - 9}{2\sqrt{t + 2}}$ **59.** $-\dfrac{3}{4\sqrt{3 - 2x}\,x^{3/2}}$

61. $\dfrac{3t(t^2 + 3t - 2)}{(t^2 + 2t - 1)^{3/2}}$ **63.** $-\dfrac{1}{x^2(x^3 + 1)^{2/3}}$

65. $\dfrac{2(6 - 5x)(5x^2 - 12x + 5)}{(x^2 - 1)^3}$

67. **(a)** \$74.00 per 1% **(b)** \$81.59 per 1%

(c) \$89.94 per 1%

69.

t	0	1	2	3	4
$\dfrac{dN}{dt}$	0	177.78	44.44	10.82	3.29

The rate of growth of N is decreasing

71. **(a)** $V = \dfrac{10,000}{\sqrt[3]{t + 1}}$

(b) $-\$1322.83$ per year

(c) $-\$524.97$ per year

SECTION 2.6 (page 147)

Warm Up

1. $t = 0, \frac{3}{2}$ **2.** $t = -2, 7$ **3.** $t = -2, 10$

4. $t = \dfrac{9 \pm 3\sqrt{10,249}}{32}$ **5.** $\dfrac{dy}{dx} = 6x^2 + 14x$

6. $\dfrac{dy}{dx} = 8x^3 + 18x^2 - 10x - 15$

7. $\dfrac{dy}{dx} = \dfrac{2x(x + 7)}{(2x + 7)^2}$ **8.** $\dfrac{dy}{dx} = -\dfrac{6x^2 + 10x + 15}{(2x^2 - 5)^2}$

9. Domain: $(-\infty, \infty)$ **10.** Domain: $[7, \infty)$

Range: $[-4, \infty)$ Range: $[0, \infty)$

1. 0   **3.** 2   **5.** $2t - 8$ **7.** $\dfrac{4}{9t^{7/3}}$

9. $48x^2 - 16$ **11.** $\dfrac{4}{9x^{2/3}}$ **13.** $\dfrac{4}{(x-1)^3}$

15. $12x^2 + 24x + 16$ **17.** $60x^2 - 72x$

19. $120x + 360$ **21.** $-\dfrac{9}{2x^5}$

23. $-\dfrac{3}{8(4-x)^{5/2}}$ **25.** $4x$ **27.** $\dfrac{2}{x^2}$

29. 2   **31.** $f''(x) = 6(x - 3) = 0$ when $x = 3$.

33. $f''(x) = 2(3x + 4) = 0$ when $x = -\frac{4}{3}$.

35. $f''(x) = 36(x^2 - 1) = 0$ when $x = \pm 1$.

37. $f''(x) = \dfrac{2x(x+3)(x-3)}{(x^2+3)^3}$

$\qquad = 0$ when $x = 0$ or $x = \pm 3$.

39. (a) $s(t) = -16t^2 + 144t$

(b) $v(t) = -32t + 144$

$\qquad a(t) = -32$

(c) 4.5 sec, 324 ft

(d) -144 ft/sec, which is the same speed as the initial velocity.

41.

t	0	10	20	30
$\dfrac{ds}{dt}$	0	45	60	67.5
$\dfrac{d^2s}{dt^2}$	9.00	2.25	1.00	0.56

t	40	50	60
$\dfrac{ds}{dt}$	72	75	77.14
$\dfrac{d^2s}{dt^2}$	0.36	0.25	0.18

As t increases, the velocity of the car increases at a diminishing rate (acceleration decreases).

43.

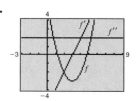

The degrees of the successive derivatives decrease by 1.

45.

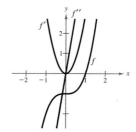

47. False. The product rule is $[f(x)g(x)]'$
$\qquad = f'(x)g(x) + f(x)g'(x)$.

48. True   **49.** True   **50.** True

51. True   **52.** True

53. False. Let $f(x) = x^2$ and $g(x) = x^2 + 1$.

SECTION 2.7   (page 154)

Warm Up

1. $y = x^2 - 2x$ **2.** $y = \dfrac{x - 3}{4}$

3. $y = 1, x \neq -6$ **4.** $y = -4, x \neq \pm\sqrt{3}$

5. $y = \pm\sqrt{5 - x^2}$ **6.** $y = \pm\sqrt{6 - x^2}$

7. $\frac{8}{3}$ **8.** $-\frac{1}{2}$ **9.** $\frac{5}{7}$ **10.** 1

---

1. $-\dfrac{y}{x}$ **3.** $-\dfrac{x}{y}$ **5.** $-\dfrac{1}{5}$ **7.** $-\dfrac{x}{y}, 0$

9. $-\dfrac{y}{x+1}, -\dfrac{1}{4}$ **11.** $\dfrac{2x}{3y^2}, \dfrac{4}{3}$

13. $\dfrac{y - 3x^2}{2y - x}, \dfrac{1}{2}$ **15.** $\dfrac{1 - 3x^2y^3}{3x^3y^2 - 1}, -1$

17. $-\sqrt{\dfrac{y}{x}}, -\dfrac{5}{4}$ **19.** $-\sqrt[3]{\dfrac{y}{x}}, -\dfrac{1}{2}$

21. $\dfrac{2xy - y^2 - x^2}{2xy - x^2}, -\dfrac{1}{8}$ **23.** $3x, 3$

25. $-\dfrac{4x}{9y}, -\dfrac{\sqrt{5}}{3}$ **27.** $-\dfrac{x}{y}, \dfrac{4}{3}$

29. $-\dfrac{9x}{16y}, -\dfrac{\sqrt{3}}{4}$

31. At $(5, 12)$: $5x + 12y - 169 = 0$
At $(-12, 5)$: $12x - 5y + 169 = 0$

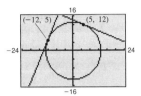

33. At $(1, \sqrt{5})$: $15x - 2\sqrt{5}\,y - 5 = 0$
At $(1, -\sqrt{5})$: $15x + 2\sqrt{5}\,y - 5 = 0$

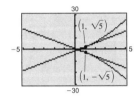

35. $\dfrac{1}{0.024x^3 + 0.04x}$ **37.** $-\dfrac{x^2}{100}$

39. (a) -2
(b)

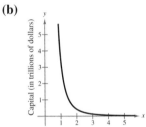

As more labor is used, less capital is available.
As more capital is used, less labor is available.

SECTION 2.8   (page 162)

Warm Up

1. $A = \pi r^2$ **2.** $V = \frac{4}{3}\pi r^3$ **3.** $S = 6s^2$

4. $V = s^3$ **5.** $V = \frac{1}{3}\pi r^2 h$ **6.** $A = \frac{1}{2}bh$

7. $-\dfrac{x}{y}$ **8.** $\dfrac{2x - 3y}{3x}$ **9.** $-\dfrac{2x + y}{x + 2}$

10. $-\dfrac{y^2 - y + 1}{2xy - 2y - x}$

---

1. (a) 62   (b) $\frac{32}{85}$ **3.** (a) $-\frac{5}{8}$ (b) $\frac{3}{2}$

5. (a) 24π in.^{2}/min   (b) 96π in.^{2}/min

7. If $\dfrac{dr}{dt}$ is constant, $\dfrac{dA}{dt} = 2\pi r \dfrac{dr}{dt}$
and thus is proportional to r.

9. (a) $\dfrac{5}{\pi}$ ft/min   (b) $\dfrac{5}{4\pi}$ ft/min

11. $\dfrac{4}{45\pi}$ ft/min   **13.** (a) $9\,\text{cm}^3$/sec   (b) $900\,\text{cm}^3$/sec

15. (a) -12 cm/min   (b) 0 cm/min   (c) 4 cm/min
(d) 12 cm/min

17. (a) $-\frac{7}{12}$ ft/sec   (b) $-\frac{3}{2}$ ft/sec   (c) $-\frac{48}{7}$ ft/sec

19. (a) -750 mi/hr   (b) 20 min

21. -8.33 ft/sec   **23.** \$650 per week

25. (a) \$53.88 million per year
(b) \$165 million per year

CHAPTER 2 REVIEW EXERCISES   (page 166)

1. -2 **3.** 0

5. $t = 7$: slope $\approx$ \$600 million per year (revenues are increasing)
$t = 10$: slope $\approx -\$700$ million per year (revenues are decreasing)

7. -3 **9.** $\dfrac{1}{4}$ **11.** 7 **13.** $-\dfrac{1}{(x - 5)^2}$

15. -5 **17.** $\frac{1}{6}$ **19.** $y = -2x + 6$ **21.** 1

23. 0 **25.** 0 **27.** $5x^4$ **29.** $\dfrac{1}{2\sqrt{x}}$ **31.** $12x^3$

33. $-\dfrac{4}{3t^3}$ **35.** $44x^3 - 10x$ **37.** $\dfrac{x + 1}{2x^{3/2}}$

39. Average rate of change: 4
Instantaneous rate of change when $x = 0$: 3
Instantaneous rate of change when $x = 1$: 5

41. (a) \$82.82 million per year
(b) 1986: $-\$12.18$ million per year
1990: $-\$790.82$ million per year
(c) Revenues were decreasing in 1986 and 1990, but grew during the period 1987–1989.

43. (a) $s(t) = -16t^2 + 276$ **(b)** -32 ft/sec
(c) $t = 2$: -64 ft/sec **(d)** 4.15 sec
  $t = 3$: -96 ft/sec
(e) 132.8 ft/sec

45. $R = 27.50x$ **47.** $\dfrac{dC}{dx} = 320$
  $C = 15x + 2500$
  $P = 12.50 - 2500$

49. $\dfrac{dR}{dx} = \dfrac{35(x-4)}{2(x-2)^{3/2}}$

51. $\dfrac{dP}{dx} = -0.0006x^2 + 12x - 1$ **53.** $15x^2(1-x^2)$

55. $16x^3 - 33x^2 + 12x$ **57.** $-\dfrac{2(3x^2 - 5x - 3)}{(x^2+1)^2}$

59. $\dfrac{4t^5 - 4t^3 + 2t^2 + 1}{t^2(t^3-1)^2}$ **61.** $30x(5x^2+2)^2$

63. $-\dfrac{1}{(x+1)^{3/2}}$ **65.** $\dfrac{2x^2+1}{\sqrt{x^2+1}}$

67. $\dfrac{2t^2+t+1}{\sqrt{t^2+1}}$ **69.** $32x(1-4x^2)$

71. $18x^5(x+1)(2x+3)^2$ **73.** $x(x-1)^4(7x-2)$

75. $\dfrac{3(9t+5)}{2\sqrt{3t+1}(1-3t)^3}$

77. (a) $t = 1$: -6.63 $t = 3$: -6.5
  $t = 5$: -4.33 $t = 10$: -1.36
(b)

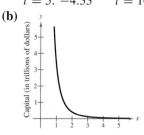

The rate of decrease is approaching zero.

79. 6 **81.** $-\dfrac{120}{x^6}$ **83.** $\dfrac{35x^{3/2}}{2}$ **85.** $\dfrac{2}{x^{2/3}}$

87. (a) $s(t) = -16t^2 + 5t + 30$ **(b)** 1.534 sec
(c) -44.09 ft/sec **(d)** -32 ft/sec^2

89. $-\dfrac{2x+3y}{3(x+y^2)}$ **91.** $\dfrac{x}{y}$ **93.** $\dfrac{3x^2-2xy+y}{x^2-x+4y}$

95. $-\dfrac{4y^{1/2}}{9x^{2/3}}$ **97.** $\dfrac{1}{64}$ ft/min

SAMPLE POST-GRAD EXAM QUESTIONS (page 170)

1. (c) **2.** (b) **3.** (b)
4. (a) **5.** (c) **6.** (a)

CHAPTER 3

SECTION 3.1 (page 179)

Warm Up

1. $x = 0$, $x = 8$ **2.** $x = 0$, $x = 24$
3. $x = \pm 5$ **4.** $x = 0$ **5.** $(-\infty, 3) \cup (3, \infty)$
6. $(-\infty, 1)$ **7.** $(-\infty, -2) \cup (-2, 5) \cup (5, \infty)$
8. $(-\sqrt{3}, \sqrt{3})$
9. $x = -2$: $-\frac{103}{3}$ **10.** $x = -2$: $-\frac{7}{9}$
  $x = 0$: $\frac{1}{3}$ $x = 0$: 3
  $x = 2$: -13 $x = 2$: $-\frac{1}{3}$

1. $f'(-1) = -\frac{8}{25}$ **3.** $f'(-3) = -\frac{2}{3}$
  $f'(0) = 0$ $f'(-2)$ undefined
  $f'(1) = \frac{8}{25}$ $f'(-1) = \frac{2}{3}$

5. Increasing on $(-\infty, -1)$
  Decreasing on $(-1, \infty)$

7. Increasing on $(-1, 0)$ and $(1, \infty)$
  Decreasing on $(-\infty, -1)$ and $(0, 1)$

9. No critical numbers
  Increasing on $(-\infty, \infty)$

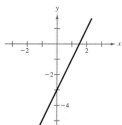

11. Critical number: $x = 1$
  Increasing on $(-\infty, 1)$
  Decreasing on $(1, \infty)$

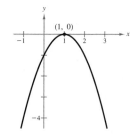

13. Critical number: $x = \frac{5}{2}$
Decreasing on $\left(-\infty, \frac{5}{2}\right)$
Increasing on $\left(\frac{5}{2}, \infty\right)$

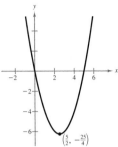

15. Critical numbers: $x = 0, x = 4$
Increasing on $(-\infty, 0)$ and $(4, \infty)$
Decreasing on $(0, 4)$

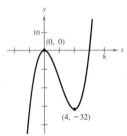

17. Critical number: $x = -1$
Decreasing on $(-\infty, \infty)$

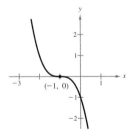

19. Critical number: $x = 1$
Increasing on $(-\infty, 1)$
Decreasing on $(1, \infty)$

21. Critical numbers: $x = -1, x = -\frac{5}{3}$
Increasing on $\left(-\infty, -\frac{5}{3}\right)$ and $(-1, \infty)$
Decreasing on $\left(-\frac{5}{3}, -1\right)$

23. Critical number: $x = 0$
Decreasing on $(-\infty, 0)$
Increasing on $(0, \infty)$

25. Critical numbers: $x = 0, x = \frac{3}{2}$
Decreasing on $\left(-\infty, \frac{3}{2}\right)$
Increasing on $\left(\frac{3}{2}, \infty\right)$

27. Domain: $(-\infty, 3]$
Critical numbers: $x = 2, x = 3$
Increasing on $(-\infty, 2)$
Decreasing on $(2, 3)$

29. Critical numbers: $x = 2, x = -2$
Decreasing on $(-\infty, -2)$ and $(2, \infty)$
Increasing on $(-2, 2)$

31. Critical numbers: $x = -1, x = 1$
Discontinuity: $x = 0$
Increasing on $(-\infty, -1)$ and $(1, \infty)$
Decreasing on $(-1, 0)$ and $(0, 1)$

33. No critical numbers
Discontinuities: $x = \pm 4$
Increasing on $(-\infty, -4), (-4, 4),$ and $(4, \infty)$

35. Critical number: $x = 0$
Increasing on $(-\infty, 0)$
Decreasing on $(0, \infty)$

37. (a) Decreasing on $[1, 4.10)$
Increasing on $(4.10, \infty)$

(b)

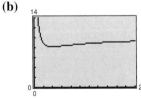

(c) $C = 900$ when $x = 2$ and $x = 15$.
Use an order size of $x = 4$, which will
minimize the cost C.

39. Moving upward on $(0, 3)$
Moving downward on $(3, 6)$

41. (a)

(b) Increasing on $(0, 18.9)$
Decreasing on $(18.9, 21)$

43. (a)

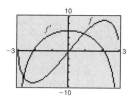

(b) $x = \pm \dfrac{3\sqrt{2}}{2}$

(c) $f' > 0$ on $\left(-\dfrac{3\sqrt{2}}{2}, \dfrac{3\sqrt{2}}{2} \right)$

$f' < 0$ on $\left(-3, -\dfrac{3\sqrt{2}}{2} \right), \left(\dfrac{3\sqrt{2}}{2}, 3 \right)$

f is decreasing when $f' < 0$, and f is increasing when $f' > 0$.

SECTION 3.2   (page 189)

Warm Up

1. $0, \pm\frac{1}{2}$ **2.** $-2, 5$ **3.** 1   **4.** $0, 125$
5. Negative   **6.** Positive
7. Positive   **8.** Negative
9. Increasing   **10.** Decreasing

1. Relative maximum: $(1, 5)$
3. Relative minimum: $(3, -9)$
5. Relative maximum: $\left(\frac{2}{3}, \frac{28}{9} \right)$
   Relative minimum: $(1, 3)$
7. No relative extrema
9. Relative maximum: $(0, 15)$
   Relative minimum: $(4, -17)$
11. Relative minimum: $\left(\frac{3}{2}, -\frac{27}{16} \right)$
13. No relative extrema
15. Relative minimum: $(0, 0)$
17. Relative maximum: $(-1, -2)$
   Relative minimum: $(1, 2)$
19. Relative maximum: $(0, 4)$
21. Minimum: $(2, 2)$ **23.** Maximum: $(0, 5)$
   Maximum: $(-1, 8)$ Minimum: $(3, -13)$

25. Minima: $(-1, -4)$ and $(2, -4)$
   Maxima: $(0, 0)$ and $(3, 0)$
27. Minimum: $(0, 0)$ **29.** Maximum: $(2, 1)$
   Maximum: $(-1, 5)$ Minimum: $(0, \frac{1}{3})$

31.

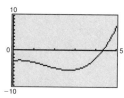

33.

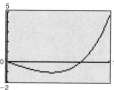

Maximum: $(5, 7)$ Maximum: $(1, 4.7)$
Minimum: $(2.69, -5.55)$ Minimum: $(0.44, -1.06)$
35. Absolute minimum: $(0, 0)$
   Absolute maximum: $(1, 2)$
37. Absolute minimum: $(2, 1)$

39. $\left| f'' \left(\dfrac{\sqrt{3}}{3} \right) \right| = \dfrac{40\sqrt{3}}{3}$

41. $\left| f^{(4)} \left(\frac{1}{2} \right) \right| = 360$ **43.** 82 units   **45.** \$2.10
47. 1981
   There were approximately 95 males for every 100
   females in 1981.

SECTION 3.3   (page 198)

Warm Up

1. $f''(x) = 48x^2 - 54x$
2. $g''(s) = 12s^2 - 18s + 2$
3. $g''(x) = 56x^6 + 120x^4 + 72x^2 + 8$
4. $f''(x) = \dfrac{4}{9(x-3)^{2/3}}$
5. $h''(x) = \dfrac{15}{2x^5}$ **6.** $f''(x) = -\dfrac{42}{(3x+2)^3}$
7. $x = \pm\dfrac{\sqrt{3}}{3}$ **8.** $x = 0, 3$ **9.** $t = \pm 4$
10. $x = 0, \pm 5$

1. Concave upward on $(-\infty, \infty)$
3. Concave upward on $\left(-\infty, -\frac{1}{2} \right)$
   Concave downward on $\left(-\frac{1}{2}, \infty \right)$

5. Concave upward on $(-\infty, -2)$ and $(2, \infty)$
Concave downward on $(-2, 2)$

7. Relative maximum: $(3, 9)$

9. Relative minimum: $(5, 0)$

11. Relative maximum: $(1, 3)$
Relative minimum: $(\frac{7}{3}, 1.\overline{814})$

13. Relative minimum: $(3, -25)$

15. Relative minimum: $(0, -3)$

17. Relative maximum: $(-2, -4)$
Relative minimum: $(2, 4)$

19. Sign of $f'(x)$ on $(0, 2)$ is positive.
Sign of $f''(x)$ on $(0, 2)$ is positive.

21. Sign of $f'(x)$ on $(0, 2)$ is negative.
Sign of $f''(x)$ on $(0, 2)$ is negative.

23. $(3, 0)$ **25.** $(1, 0), (3, -16)$

27. No inflection points   **29.** $\left(\frac{3}{2}, -\frac{1}{16}\right), (2, 0)$

31. Relative maximum: $(-2, 16)$
Relative minimum: $(2, -16)$
Point of inflection: $(0, 0)$

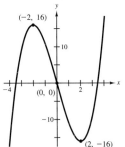

33. Point of inflection: $(2, 8)$

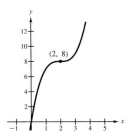

35. Relative maximum: $(0, 0)$
Relative minima: $(\pm 2, -4)$
Points of inflection: $\left(\pm\frac{2}{\sqrt{3}}, -\frac{20}{9}\right)$

35. (*continued*)

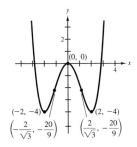

37. Relative maximum: $(-1, 0)$
Relative minimum: $(1, -4)$
Point of inflection: $(0, -2)$

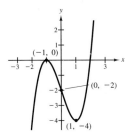

39. Relative minimum: $(-2, -2)$

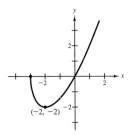

41. Relative maximum: $(0, 4)$
Points of inflection: $\left(\pm\frac{\sqrt{3}}{3}, 3\right)$

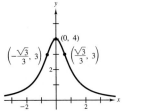

43.

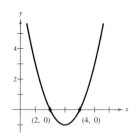

45. (a) f': Positive on $(-\infty, 0)$
f: Increasing on $(-\infty, 0)$
(b) f': Negative on $(0, \infty)$
f: Decreasing on $(0, \infty)$
(c) f': Not increasing
f: Not concave upward
(d) f': Decreasing on $(-\infty, \infty)$
f: Concave downward on $(-\infty, \infty)$

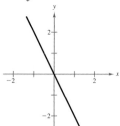

47. $(200, 320)$ **49.** $(3, 36)$ **51.** 100 units
53. $x = 50$ units   **55.** 8:30 P.M.
57. $\sqrt{3} \approx 1.732$ years   **59.**

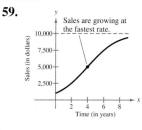

61.

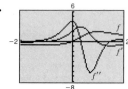

Relative maximum: $(1, 2)$
Relative minimum: $(-1, -\frac{2}{3})$
Points of inflection: $(-1.88, -0.59)$,
$(0.35, 0.90)$, $(1.53, 1.69)$

63.

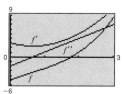
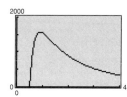

Relative maximum: $(3, 8.5)$
Relative minimum: $(0, -5)$
Point of inflection: $\left(\frac{2}{3}, -3.2963\right)$

65.

SECTION 3.4   (page 207)

Warm Up

1. $x + \frac{1}{2}y = 12$ **2.** $2xy = 24$ **3.** $xy = 24$
4. $\sqrt{(x_2 - x_1)^2 + (y_2 - y_1)^2} = 10$
5. $x = -3$ **6.** $x = -\frac{2}{3}, 1$ **7.** $x = \pm 5$
8. $x = 4$ **9.** $x = \pm 1$ **10.** $x = \pm 3$

1. 55 and 55   **3.** 9, 18   **5.** $\sqrt{192}$ and $\sqrt{192}$
7. 1   **9. (a)** 99 in.^{3} **(b)** 125 in.^{3} **(c)** 117 in.^{3}
11. Length = width = 25 ft   **13.** $x = 25$ ft, $y = \frac{100}{3}$ ft
15. 16 in.^{3} **17.** $x = 5$ m, $y = \frac{10}{3}$ m
19. Length: 50 m, width: $\dfrac{100}{\pi}$ m
21. $x = 3, y = \frac{3}{2}$
23. Length: $2x = \dfrac{10}{\sqrt{2}} \approx 7.07$
       Width: $y = \dfrac{5}{\sqrt{2}} \approx 3.54$
25. $r \approx 1.51$ in., $h = 2r \approx 3.02$ in.
27. $\left(\pm\sqrt{\frac{5}{2}}, \frac{7}{2}\right)$
29. 18 in. $\times$ 18 in. $\times$ 36 in.

31. Radius: $\dfrac{8}{\pi+4}$

Side of square: $\dfrac{16}{\pi+4}$

33. $x = 1$ mi   **35.** $w = 8\sqrt{3}$ in., $h = 8\sqrt{6}$ in.

37. $10\sqrt{2} \times 5\sqrt{2}$

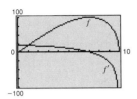

Warm Up

1. 1   **2.** 1.2   **3.** 2   **4.** $\frac{1}{2}$

5. $\dfrac{dC}{dx} = 1.2 + 0.006x$ **6.** $\dfrac{dP}{dx} = 0.02x + 11$

7. $\dfrac{dR}{dx} = 14 - \dfrac{x}{1000}$ **8.** $\dfrac{dR}{dx} = 3.4 - \dfrac{x}{750}$

9. $\dfrac{dP}{dx} = -1.4x + 7$ **10.** $\dfrac{dC}{dx} = 4.2 + 0.003x^2$

---

1. 2000 units   **3.** 200 units   **5.** 80 units

7. 200 units   **9.** $60   **11.** $69.68   **13.** 3 units

15. (a) $80   (b) $45.93

17. Maximum profit: $10,000

Point of diminishing returns: $5833.33

19. 200 units   **21.** $50

23. Line should run from the power station to a point across the river $3/(2\sqrt{7}) \approx 0.57$ mi downstream.

25. 77.46 mph

27. $-\frac{17}{3}$, elastic   **29.** $-\frac{1}{3}$, inelastic

31. $-\frac{3}{2}$, elastic

33. (a) -6.83% (b) -1.37 (c) $-\frac{4}{3}$

(d) $R = 20p - 2p^3$, $x = \frac{40}{3}$, $p = \sqrt{\frac{10}{3}} \approx \$1.83$

35. (a) $-\dfrac{14}{9}$ (b) $x = \dfrac{32}{3}$, $p = \dfrac{4\sqrt{3}}{3}$

37. No, $\eta = -\frac{1}{3}$, demand is inelastic

39. (a) 1987

(b) 1994

(c) 1987: $6006.6 million

1994: $8050.6 million

(d)

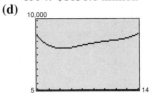

41. (a) Demand function   (b) Cost function

(c) Revenue function   (d) Profit function

Warm Up

1. 3   **2.** 1   **3.** -11 **4.** 4   **5.** 0   **6.** 1

7. $\overline{C} = \dfrac{150}{x} + 3$ **8.** $\overline{C} = \dfrac{1900}{x} + 1.7 + 0.002x$

$\dfrac{dC}{dx} = 3$ $\dfrac{dC}{dx} = 1.7 + 0.004x$

9. $\overline{C} = 0.005x + 0.5 + \dfrac{1375}{x}$

$\dfrac{dC}{dx} = 0.01x + 0.5$

10. $\overline{C} = \dfrac{760}{x} + 0.05$

$\dfrac{dC}{dx} = 0.05$

---

1. Vertical: $x = 0$ **3.** Vertical: $x = -1$, $x = 2$

Horizontal: $y = 1$ Horizontal: $y = 1$

5. Vertical: $x = -1$, $x = 1$ **7.** Vertical: $x = \pm 2$

Horizontal: none   Horizontal: $y = \frac{1}{2}$

9. f   **10.** b   **11.** c   **12.** a   **13.** e

14. d   **15.** ∞ **17.** $-\infty$ **19.** $-\infty$

21. $-\infty$ **23.** $\frac{2}{3}$ **25.** 0   **27.** $-\infty$

29. ∞ **31.** 5

33.

x	10^0	10^1	10^2	10^3
$f(x)$	2.000	0.348	0.101	0.032

x	10^4	10^5	10^6
$f(x)$	0.010	0.003	0.001

$$\lim_{x \to \infty} \frac{x+1}{x\sqrt{x}} = 0$$

35.

x	-10^6	-10^4	-10^2	10^0
$f(x)$	-2	-2	-1.9996	0.8944

x	10^2	10^4	10^6
$f(x)$	1.9996	2	2

$$\lim_{x \to -\infty} \frac{2x}{\sqrt{x^2+4}} = -2, \quad \lim_{x \to \infty} \frac{2x}{\sqrt{x^2+4}} = 2$$

37.
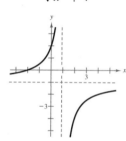

39.
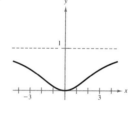

41.
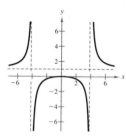

43.
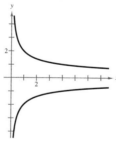

45.
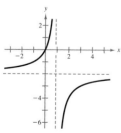

47.
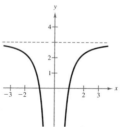

49.
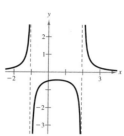

51.
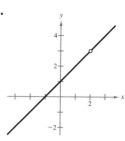

53.
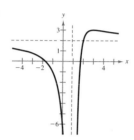

55. (a) \$47.05, \$5.92   (b) \$1.35

57. (a) \$176 million   (b) \$528 million
(c) \$1584 million   (d) ∞

59. a **61.** (a) 5 years: 153   (b) 400
10 years: 215
25 years: 294

63. Horizontal asymptotes: $y = \pm\frac{3}{2}$

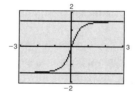

SECTION 3.7   (page 240)

Warm Up

1. Vertical: $x = 0$ **2.** Vertical: $x = 2$
Horizontal: $y = 0$ Horizontal: $y = 0$
3. Vertical: $x = -3$ **4.** Vertical: $x = 1, x = 3$
Horizontal: $y = 40$ Horizontal: $y = 1$
5. Decreasing on $(-\infty, -2)$
Increasing on $(-2, \infty)$
6. Increasing on $(-\infty, -4)$
Decreasing on $(-4, \infty)$

Warm Up *(continued)*

7. Increasing on $(-\infty, -1)$ and $(1, \infty)$
Decreasing on $(-1, 1)$

8. Decreasing on $(-\infty, 0)$ and $\left(\sqrt[3]{2}, \infty\right)$
Increasing on $\left(0, \sqrt[3]{2}\right)$

9. Increasing on $(-\infty, 1)$ and $(1, \infty)$

10. Decreasing on $(-\infty, -3)$ and $\left(\frac{1}{3}, \infty\right)$
Increasing on $\left(-3, \frac{1}{3}\right)$

1.
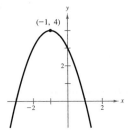

3.
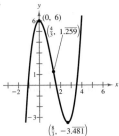

5.
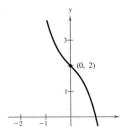

7.
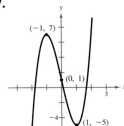

9.
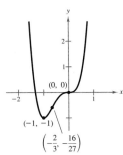

11.
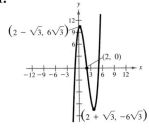

13.
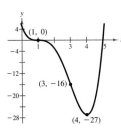

15.
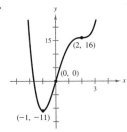

17.
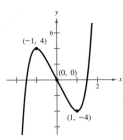

19.
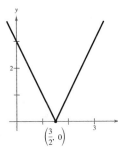

21.
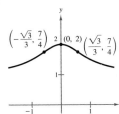

23.
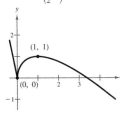

25.
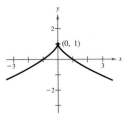

27.
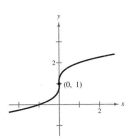

29.
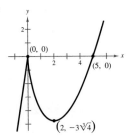

31. Domain: $(-\infty, 2), (2, \infty)$

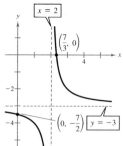

33. Domain: $(-\infty, -1), (-1, 1), (1, \infty)$

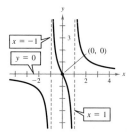

35. Domain: $(-\infty, 4]$ **37.** Domain: $(-\infty, 0), (0, \infty)$

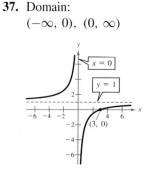

39. Domain: $(-\infty, 1), (1, \infty)$

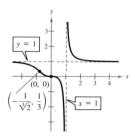

41. $f(x) = -x^3 + x^2 + x + 1$ **43.** $f(x) = x^3 + 1$

45. 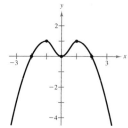   **47.**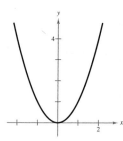

49.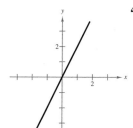

51. (a) $C = 0.24s + \dfrac{900}{s}$, $40 \le s \le 65$

(b) 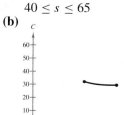   **53.**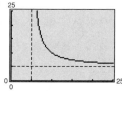

SECTION 3.8   (page 248)

Warm Up

1. $\dfrac{dC}{dx} = 0.18x$ **2.** $\dfrac{dR}{dx} = 1.25 + 0.03\sqrt{x}$

3. $\dfrac{dP}{dx} = -\dfrac{0.01}{\sqrt[3]{x^2}} + 1.4$ **4.** $\dfrac{dA}{dx} = \dfrac{\sqrt{3}}{2}x$

5. $\dfrac{dC}{dr} = 2\pi$ **6.** $\dfrac{dS}{dr} = 8\pi r$

7. $A = \pi r^2$ **8.** $A = x^2$

9. $V = x^3$ **10.** $V = \frac{4}{3}\pi r^3$

1. $6x\,dx$ **3.** $12(4x-1)^2\,dx$ **5.** $\dfrac{x}{\sqrt{x^2+1}}\,dx$

7. $\dfrac{1-2x^2}{\sqrt{1-x^2}}\,dx$ **9.** $\dfrac{-8}{(1-3x)^2}\,dx$

11. 0.1005 **13.** -0.013245

15. $dy = 0.3$ **17.** $dy = -0.04$
$\quad\;\; \Delta y = 0.331$ $\quad\;\; \Delta y \approx -0.0394$

19.

$dx = \Delta x$	dy	Δy	$\Delta y - dy$	$\dfrac{dy}{\Delta x}$
1.000	4.000	5.000	1.000	0.800
0.500	2.000	2.250	0.250	0.889
0.100	0.400	0.410	0.010	0.976
0.010	0.040	0.040	0.000	0.998
0.001	0.004	0.004	0.000	1.000

21.

$dx = \Delta x$	dy	Δy	$\Delta y - dy$	$\dfrac{dy}{\Delta x}$
1.000	80.000	211.000	131.000	0.379
0.500	40.000	65.656	25.656	0.609
0.100	8.000	8.841	0.841	0.905
0.010	0.800	0.808	0.008	0.990
0.001	0.080	0.080	0.000	0.999

23. (a) $\Delta p = -0.25 = dp$ **(b)** $\Delta p = -0.25 = dp$

25. 5.20 **27.** $-\$1250$

29. $R = -\frac{1}{3}x^2 + 100x,\ \$6$

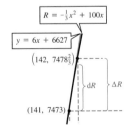

31. $P = -\frac{1}{2000}x^2 + 23x - 275{,}000;\ -\$5$

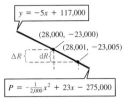

33. (a) $dA = 2x\,\Delta x,\ \Delta A = 2x\,\Delta x + (\Delta x)^2$
(b)

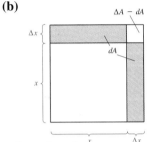

35. $\pm\frac{5}{2}\pi$ in.^{2}, $\pm\frac{1}{40}$

37. $\pm 2.88\pi$ in.^{3}, ± 0.01 **39.** $\$1160$

CHAPTER 3 REVIEW EXERCISES (page 252)

1. $x = 1$ **3.** $x = 0,\ x = 1$
5. Increasing on $\left(-\frac{1}{2},\ \infty\right)$
Decreasing on $\left(-\infty,\ -\frac{1}{2}\right)$
7. Increasing on $(-\infty,\ 3) \cup (3,\ \infty)$
9. (a) $(0.23,\ 6.15)$
(b) $(0,\ 0.23),\ (6.15,\ 12)$
(c) Maximum daily temperature is rising from early January to early June
(d)

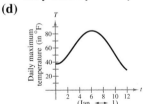

11. Relative maximum: $(0,\ -2)$
Relative minimum: $(1,\ -4)$
13. Relative minimum: $(8,\ -52)$
15. Relative maxima: $(-1,\ 1),\ (1,\ 1)$
Relative minimum: $(0,\ 0)$

17. Relative maximum: $(0, 6)$

19. Relative maximum: $(0, 0)$
Relative minimum: $(4, 8)$

21. Maximum: $(0, 6)$
Minimum: $\left(-\frac{5}{2}, -\frac{1}{4}\right)$

23. Maxima: $(-2, 17), (4, 17)$
Minima: $(-4, -15), (2, -15)$

25. Maximum: $(2, 26)$
Minimum: $(1, -1)$

27. Maximum: $(1, 1)$ **29. (a)** \$253, 1990
Minimum: $(-1, -1)$ **(b)** \$176.08, 1984

31. Concave upward: $(2, \infty)$
Concave downward: $(-\infty, -2)$

33. Concave upward: $\left(-\dfrac{2}{\sqrt{3}}, \dfrac{2}{\sqrt{3}}\right)$
Concave downward: $\left(-\infty, -\dfrac{2}{\sqrt{3}}\right)$ and $\left(\dfrac{2}{\sqrt{3}}, \infty\right)$

35. $(0, 0), (4, -128)$

37. $(0, 0), (1.0652, 4.5244), (2.5348, 3.5246)$

39. Relative maximum: $\left(-\sqrt{3}, 6\sqrt{3}\right)$
Relative minimum: $\left(\sqrt{3}, -6\sqrt{3}\right)$

41. Relative maximum: $(-4, 0)$
Relative minimum: $(-2, -108)$

43. $\left(50, 166\frac{2}{3}\right)$ **45.** $13, 13$

47. (a) \$54,607   **49. (a)** $x = 3$
(b) \$5000   **(b)** $x = \sqrt{3}$

51. $N = 85$ (maximizes revenue)   **53.** 125

55. Elastic for $x > 0$

57. Vertical asymptote: $x = 4$
Horizontal asymptote: $y = 2$

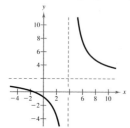

59. Vertical asymptote: $x = 0$
Horizontal asymptote: $y = -3$

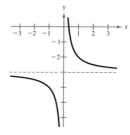

61. Vertical asymptotes: $x = 1, x = 4$
Horizontal asymptote: $y = 0$

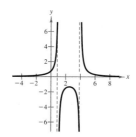

63. $-\infty$ **65.** $\frac{5}{2}$ **67.** $-\infty$

69. (a)

(b) $\lim\limits_{s \to \infty} T = 0.37$

71.

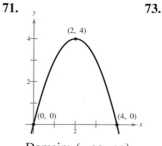

Domain: $(-\infty, \infty)$

73.

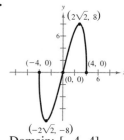

Domain: $[-4, 4]$

75.

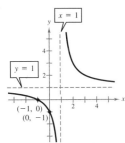

Domain: $(-\infty, 1)\cup(1, \infty)$

77. **79.**

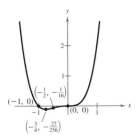 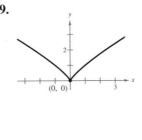

Domain: $(-\infty, \infty)$ Domain: $(-\infty, \infty)$

81. $dy = 12x\,dx$ **83.** $dy = \dfrac{5}{3x^{4/3}}\,dx$ **85.** \$800

87. \$15.25   **89.** $dS = \pm 1.8\pi$ in.^{2}
$dV = \pm 8.1\pi$ in.^{3}

SAMPLE POST-GRAD EXAM QUESTIONS   (page 256)

1. (b)   **2.** (a)   **3.** (c)   **4.** (d)
5. (b)   **6.** (a)

CHAPTER 4

SECTION 4.1   (page 266)

Warm Up

1. Continuous on $(-\infty, \infty)$
2. Removable discontinuity at $x = 4$
3. 0   **4.** 0   **5.** 4   **6.** $\frac{1}{2}$ **7.** $\frac{3}{2}$
8. 6   **9.** 0   **10.** 0

1. (a) 625   (b) 9   (c) $16\sqrt{2}$ (d) 9
 (e) 125   (f) 4
3. (a) 3125   (b) $\frac{1}{5}$ (c) 625   (d) $\frac{1}{125}$

5. (a) $\frac{1}{5}$ (b) 27   (c) 5   (d) 4096

7. (a) e^7 (b) e^{12} (c) $\dfrac{1}{e^6}$ (d) 1

9. 4   **11.** -2 **13.** 2   **15.** 16   **17.** $-\frac{1}{3}$
19. 9   **21.** e **23.** e   **24.** c
25. a   **26.** f   **27.** d   **28.** b
29. $f(x) = 6^x$ **31.** $f(x) = 5^{-x}$

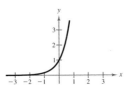 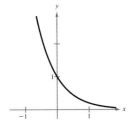

33. $y = 3^{-x^2}$ **35.** $y = 3^{-|x|}$

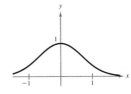 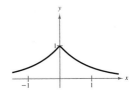

37. $s(t) = \dfrac{3^{-t}}{4}$ **39.** $h(x) = e^{x-2}$

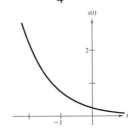 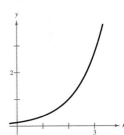

41. $N(t) = 500e^{-0.2t}$ **43.** $g(x) = \dfrac{2}{1 + e^{x^2}}$

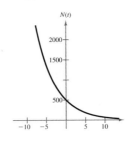 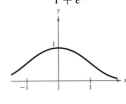

45.

n	1	2	4	12
A	1343.92	1346.86	1348.35	1349.35

n	365	Continuous Compounding
A	1349.84	1349.86

47.

n	1	2	4	12
A	3262.04	3290.66	3305.28	3315.15

n	365	Continuous Compounding
A	3319.95	3320.12

49.

t	1	10	20
P	96,078.94	67,032.00	44,932.90

t	30	40	50
P	30,119.42	20,189.65	13,533.53

51.

t	1	10	20
P	95,132.82	60,716.10	36,864.45

t	30	40	50
P	22,382.66	13,589.88	8251.24

53. (a) \$849.53
  (b) \$421.12
  $\lim_{x \to \infty} p = 0$

55. (a) 0.1535   **(b)** 0.4866   **(c)** 0.8111

57. (a)

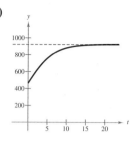

  (b) $\lim\limits_{t \to \infty} \dfrac{925}{1 + e^{-0.3t}} = 925$

59. (a) 0.731   **(b)** 11
  (c) $\lim\limits_{n \to \infty} \dfrac{0.83}{1 + e^{-0.2n}} = 0.83$

61.

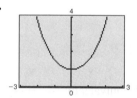

No horizontal asymptotes.
Continuous on the entire real line.

63.

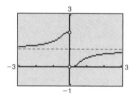

Horizontal asymptote: $y = 1$.
Discontinuous at $x = 0$.

SECTION 4.2   (page 275)

Warm Up

1. $\dfrac{1}{2}e^x(2x^2 - 1)$ **2.** $\dfrac{e^x(x + 1)}{x}$

3. $e^x(x - e^x)$ **4.** $e^{-x}(e^{2x} - x)$

5. $-\dfrac{6}{7x^3}$ **6.** $6x - \dfrac{1}{6}$

7. $6(2x^2 - x + 6)$ **8.** $\dfrac{t + 2}{2t^{3/2}}$

Warm Up *(continued)*

9. Relative maximum: $\left(-\dfrac{4\sqrt{3}}{3}, \dfrac{16\sqrt{3}}{9}\right)$

Relative minimum: $\left(\dfrac{4\sqrt{3}}{3}, -\dfrac{16\sqrt{3}}{9}\right)$

10. Relative maximum: $(0, 5)$
Relative minima: $(-1, 4)$, $(1, 4)$

1. 3   **3.** -1 **5.** $4e^{4x}$

7. $2(x-1)e^{-2x+x^2}$ **9.** $-\dfrac{e^{1/x}}{x^2}$ **11.** $\dfrac{e^{\sqrt{x}}}{2\sqrt{x}}$

13. $e^{4x}(4x^2 + 2x + 4)$

15. $3(e^x - e^{-x})(e^{-x} + e^x)^2$

17. $-\dfrac{2(e^x - e^{-x})}{(e^x + e^{-x})^2}$ **19.** $xe^x + e^x + 4e^{-x}$

21. $\dfrac{dy}{dx} = \dfrac{1}{2}(-x - 1 - 2y)$ $\left(\text{Equivalently, } \dfrac{dy}{dx} = -\dfrac{1}{2}\right)$

23. $6(3e^{3x} + 2e^{-2x})$ **25.** $5e^{-x} - 50e^{-5x}$

27. No relative extrema
No points of inflection
Horizontal asymptote to the right: $y = \frac{1}{2}$
Horizontal asymptote to the left: $y = 0$
Vertical asymptote: $x \approx -0.693$

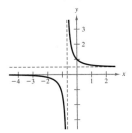

29. Relative minimum: $(0, 0)$
Relative maximum: $(2, 4e^{-2})$
Points of inflection: $(2 - \sqrt{2}, 0.191)$,
$(2 + \sqrt{2}, 0.384)$

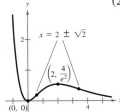

31.

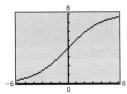

Asymptotes: $y = 0$, $y = 8$

33. (a)

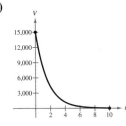

(b) $-\$5028.84$/year
(c) $-\$406.89$/year
(d) In this model, the initial rate of depreciation is greater than in a linear model.

35. (a)

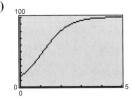

(b) 80.3%
(c) $x \approx 1.1$ or approximately 1100 egg masses

37. (a) $\$433.31$/year   **(b)** $\$890.22$/year
(c) $\$21,839.26$/year

39. $t = 5 : 14.44$
$t = 10 : 3.63$
$t = 25 : 0.58$

41.

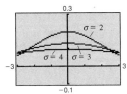

For larger σ, the graph becomes flatter.

SECTION 4.3 (page 283)

Warm Up

1. $\frac{1}{4}$ 2. 64   3. 1   4. $81e^4$
5. $x > -4$ 6. Any real number x
7. $x < -1$ and $x > 1$ 8. $x > 5$
9. \$3462.03   10. \$3374.65

---

1. $e^{0.6931\ldots} = 2$ 3. $e^{-1.6094\ldots} = 0.2$
5. $\ln 1 = 0$ 7. $\ln(0.0498\ldots) = -3$
9. c   10. d   11. b   12. a
13.

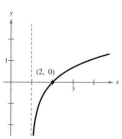

15.

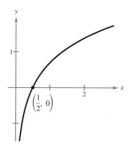

17.

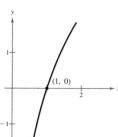

19.

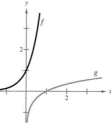

21.

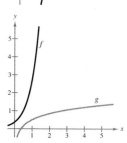

23. x^2 25. $5x + 2$ 27. $\sqrt{x}$
29. (a) 1.7917   (b) 0.4055   (c) 4.3944
   (d) 0.5493
31. $\ln 2 - \ln 3$ 33. $\ln x + \ln y + \ln z$
35. $\frac{1}{2}\ln(x^2 + 1)$

37. $\ln 2 + \ln x - \frac{1}{2}\ln(x+1) - \frac{1}{2}\ln(x-1)$
39. $\ln 3 + \ln x + \ln(x+1) - 2\ln(2x+1)$
41. $\ln \dfrac{x-2}{x+2}$ 43. $\ln \dfrac{x^3 y^2}{z^4}$
45. $\ln \left[\dfrac{x(x+3)}{x+4}\right]^3$ 47. $\ln \left[\dfrac{x(x^2+1)}{x+1}\right]^{3/2}$
49. $\ln \dfrac{(x-1)^3}{x(x+1)^2}$
51. $x = 4$ 53. $x = 1$
55. $x = \ln 4 - 1 \approx 0.3863$
57. $t = \dfrac{\ln 7 - \ln 3}{-0.2} \approx -4.2365$
59. $x = \dfrac{\ln 15}{2 \ln 5} \approx 0.8413$
61. $t = \dfrac{\ln 2}{\ln 1.07} \approx 10.2448$
63. $t = \dfrac{\ln 3}{12 \ln[1 + (0.07/12)]} \approx 15.740$
65. $t = -\ln \frac{4}{5} \approx 0.2231$
67. (a) 14.21 years   (b) 13.88 years
   (c) 13.87 years   (d) 13.86 years
69.

r	2%	4%	6%	8%
t	54.93	27.47	18.31	13.73

r	10%	12%	14%
t	10.99	9.16	7.85

71. 1999   73. 9370 years   75. 12,451 years
77. (a) 80   (b) 57.5   (c) 10 months
79.

x	y	$\dfrac{\ln x}{\ln y}$	$\ln \dfrac{x}{y}$	$\ln x - \ln y$
1	2	0	-0.6931	-0.6931
3	4	0.7925	-0.2877	-0.2877
10	5	1.4307	0.6931	0.6931
4	0.5	-2.0000	2.0794	2.0794

81.

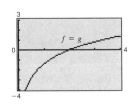

SECTION 4.4　(page 292)

> **Warm Up**
>
> **1.** $2 \ln(x + 1)$　　**2.** $\ln x + \ln(x + 1)$
> **3.** $\ln x - \ln(x + 1)$　　**4.** $3[\ln x - \ln(x - 3)]$
> **5.** $\ln 4 + \ln x + \ln(x - 7) - 2 \ln x$
> **6.** $3 \ln x + \ln(x + 1) - \frac{1}{2} \ln(x - 2)$
> **7.** $-\dfrac{y(y + y^x \ln y)}{x(y + y^x)}$　　**8.** $-\dfrac{-3 + 2xy - y^2}{x(x - 2y)}$
> **9.** $-12x + 2$　　**10.** $-\dfrac{6}{x^4}$

1. 3　　**3.** 2　　**5.** $\dfrac{3}{2}$　　**7.** $\dfrac{2}{x}$　　**9.** $\dfrac{1}{x}$

11. $\dfrac{2(x^3 - 1)}{x(x^3 - 4)}$　　**13.** $\dfrac{3}{x}(\ln x)^5$　　**15.** $1 + \ln x$

17. $\dfrac{2x^2 - 1}{x(x^2 - 1)}$　　**19.** $\dfrac{1 - x^2}{x(x^2 + 1)}$　　**21.** $\dfrac{2}{1 - x^2}$

23. $\dfrac{1}{1 - x^2}$　　**25.** $\dfrac{1 - 2 \ln x}{x^3}$　　**27.** $-\dfrac{4}{x(4 + x^2)}$

29. $\dfrac{2x}{2x^2 - 3}$　　**31.** $e^{-x}\left(\dfrac{1}{x} - \ln x\right)$　　**33.** $2x$

35. $e^{x(\ln 2)}$　　**37.** $\dfrac{1}{\ln 4} \ln x$　　**39.** 2

41. -0.63093　　**43.** 1.49136　　**45.** $(\ln 3)3^x$

47. $\dfrac{1}{x \ln 2}$　　**49.** $(2 \ln 4)4^{2x - 3}$　　**51.** $\dfrac{3}{(3x + 7) \ln 3}$

53. $2x(\ln 10)10^{x^2}$　　**55.** $2^x(1 + x \ln 2)$　　**57.** $\dfrac{2xy}{3 - 2y^2}$

59. $\dfrac{y(1 - 6x^2)}{1 + y}$　　**61.** $\dfrac{1}{2x}$　　**63.** $(\ln 5)^2 5^x$

65. $2, \ y = 2x - 1$　　**67.** $-\frac{8}{5}, \ y = -\frac{8}{5}x - 4$

69. $\dfrac{1}{\ln 2}, \ y = \dfrac{1}{\ln 2}x - \dfrac{1}{\ln 2}$

71. Relative minimum: $(1, 1)$

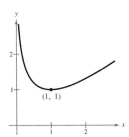

73. Relative maximum: $\left(e, \dfrac{1}{e}\right)$

Point of inflection: $\left(e^{3/2}, \dfrac{3}{2e^{3/2}}\right)$

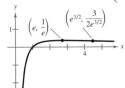

75. Relative minimum: $\left(\dfrac{1}{\sqrt{e}}, -\dfrac{1}{2e}\right)$

Point of inflection: $\left(\dfrac{1}{e^{3/2}}, -\dfrac{3}{2e^3}\right)$

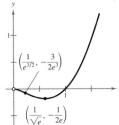

77. $-\dfrac{1}{p}, \ -\dfrac{1}{10}$

79. $-\dfrac{1000p}{(p^2 + 1)[\ln(p^2 + 1)]^2}, \ -4.65$

81. $p = 1000e^{-x}$

$\dfrac{dp}{dx} = -1000e^{-x}$

$p = 10, \ \dfrac{dp}{dx} = -10$

$\dfrac{dp}{dx}$ and $\dfrac{dx}{dp}$ are reciprocals of each other.

83. $e^{8/3} \approx 14.39$

85. (a)

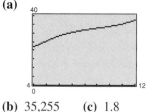

(b) 35,255 **(c)** 1.8

SECTION 4.5 (page 301)

Warm Up

1. $-\frac{1}{4} \ln 2$ **2.** $\frac{1}{5} \ln \frac{10}{3}$ **3.** $\frac{1}{3} \ln \frac{25}{16}$

4. $\frac{1}{20} \ln \frac{11}{16}$ **5.** $7.36e^{0.23t}$ **6.** $-33.6e^{-1.4t}$

7. $1.296e^{0.072t}$ **8.** $-0.025e^{-0.001t}$

9. 4 **10.** 12

---

1. $y = 2e^{0.1014t}$ **3.** $y = 4e^{-0.4159t}$

5. $y = 0.6687e^{0.4024t}$

7. $y = 10e^{2t}$, exponential growth

9. $y = 30e^{-4t}$, exponential decay

11. *Isotope: Ra^{226}*

Half-life (in years): 1620

Initial quantity: 10 grams

Amount after 1000 years: 6.52 grams

Amount after 10,000 years: 0.14 gram

13. *Isotope: C^{14}*

Half-life (in years): 5730

Initial quantity: 6.70 grams

Amount after 1000 years: 5.94 grams

Amount after 10,000 years: 2.00 grams

15. *Isotope: Pu^{239}*

Half-life (in years): 24,360

Initial quantity: 2.16 grams

Amount after 1000 years: 2.10 grams

Amount after 10,000 years: 1.63 grams

17. 68% **19.** $\dfrac{5730(\ln 20 - \ln 3)}{\ln 2} \approx 15,682.8$ years

21. (a) 1350 **(b)** $\dfrac{5 \ln 2}{\ln 3} \approx 3.15$ hr **(c)** No.

23. *Initial investment:* $1000

Annual rate: 12%

Time to double: 5.78 years

Amount after 10 years: $3320.12

Amount after 25 years: $20,085.54

25. *Initial investment:* $750

Annual rate: 8.94%

Time to double: 7.75 years

Amount after 10 years: $1833.67

Amount after 25 years: $7009.86

27. *Initial investment:* $500

Annual rate: 9.50%

Time to double: 7.30 years

Amount after 10 years: $1292.85

Amount after 25 years: $5375.51

29. (b) 6.17%

31.

Number of compoundings/yr	4	12
Effective yield	5.095%	5.116%

Number of compoundings/yr	365	Continuous
Effective yield	5.127%	5.127%

35. (a) $1061.0 million **(b)** $874.9 million

37. (a) $S(t) = 30e^{-1.7918/t}$

(b) $30e^{-0.35836} = 20.9646$ or 20,965 units

(c)

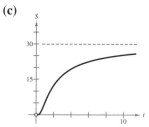

39. (a) $N(t) = 30(1 - e^{-0.0502t})$

(b) $t = \dfrac{\ln 6}{0.0502} \approx 36$ days

41. (a) $C = \dfrac{45}{e^{1000[\ln(45/40)/(-200)]}} \approx 81.0915$

$k = \dfrac{\ln(45/40)}{-200} \approx -0.0005889$

(b) $x = \dfrac{200}{\ln(9/8)} \approx 1698$ units

$p = 45\left(\dfrac{9}{8}\right)^5 \dfrac{1}{e} \approx \$29.83$

43. 2078

45. (a) $I = e^{8.3 \ln 10} \approx 199,526,231.5$

(b) The intensity is squared when R is doubled.

(c) $\dfrac{1}{I \ln 10}$

CHAPTER 4 REVIEW EXERCISES   (page 308)

1. 1024   **3.** $\frac{1}{625}$ **5.** 1

7. $\frac{1}{6}$ **9.** 4   **11.** $\frac{1}{2}$

13. 1950: 2492.4 million shares
1970: 14,677.0 million shares
1990: 86,428.2 million shares

15.
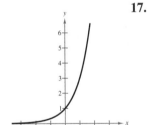

17.
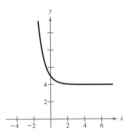

19. $7500

21. $f(2) \approx 5.4366$ **23.** $g(17) \approx 0.4005$

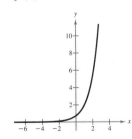
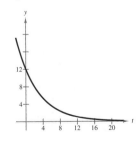

25. (a)
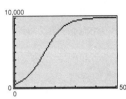

(b) $P \approx 1049$ fish

(c) Yes, P approaches 10,000 fish as t approaches ∞.

(d) The population is increasing most rapidly at the inflection point, which occurs around $t = 15$ months.

27.

n	1	2	4	12
A	\$1216.65	\$1218.99	\$1220.19	\$1221.00

n	365	Continuous Compounding
A	\$1221.39	\$1221.40

29. (b)

31. 1970: $A \approx 22.48$ years
1980: $A \approx 24.96$ years
1990: $A \approx 28.75$ years

33. $8xe^{x^2}$ **35.** $\dfrac{1 - 2x}{e^{2x}}$ **37.** $4e^{2x}$

39. $\dfrac{-10e^{2x}}{(1 + e^{2x})^2}$

41. No relative extrema
No points of inflection
$y = 0$ is a horizontal asymptote

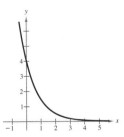

43. (2, 1.847) is a relative minimum.
No points of inflection.
$y = 0$ is a horizontal asymptote.
$x = 0$ is a vertical asymptote.

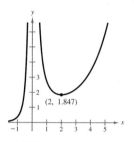

45. $e^{2.4849} \approx 12$ **47.** $\ln 4.4817 \approx 1.5$

49.   **51.**

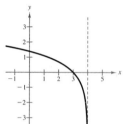   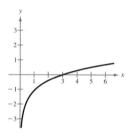

53. $\ln x + \frac{1}{2} \ln(x - 1)$

55. $3[\ln(1 - x) - \ln 3 - \ln x]$

57. 3   **59.** 1   **61.** $\dfrac{3 + \sqrt{13}}{2} \approx 3.3028$ **63.** $\dfrac{1}{4}$

65. (a)

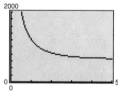

(b) A 30-year term has a smaller monthly payment, but takes more time to pay off, than a 20-year term.

67. $\dfrac{2}{x}$ **69.** $\sqrt{\ln x} + \dfrac{1}{2\sqrt{\ln x}}$

71. $\dfrac{1 - 3 \ln x}{x^4}$ **73.** $-2x$

75. No relative extrema   **77.** No relative extrema
No points of inflection   No points of inflection

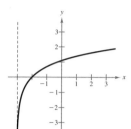

79. 2   **81.** 0   **83.** 1.4307   **85.** 1.5

87. $\dfrac{2}{(2x - 1) \ln 3}$ **89.** $-\dfrac{2}{x \ln 2}$

91. (a) $t = 2$: \$11,250   (b) $t = 1$:
                                  -4315.23 dollars/year
                                  $t = 4$:
                                  -1820.49 dollars/year

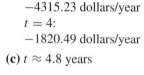

(c) $t \approx 4.8$ years

93. $y = 500e^{-0.01277t}$ **95.** 27.9 yrs

97. \$667.4 million

SAMPLE POST-GRAD EXAM QUESTIONS   (page 312)

1. (b)   **2.** (c)   **3.** (a)   **4.** (c)
5. (d)   **6.** (a)   **7.** (b)

CHAPTER 5

SECTION 5.1   (page 322)

Warm Up

1. $x^{-1/2}$ **2.** $(2x)^{4/3}$ **3.** $5^{1/2}x^{3/2} + x^{5/2}$

4. $x^{-1/2} + x^{-2/3}$ **5.** $(x + 1)^{5/2}$ **6.** $x^{1/6}$

7. -12 **8.** -10 **9.** 14   **10.** 14

9. $6x + C$

$$\frac{d}{dx}[6x + C] = 6$$

11. $t^3 + C$

$$\frac{d}{dt}[t^3 + C] = 3t^2$$

13. $-\dfrac{5}{2x^2} + C$

15. $u + C$

$$\frac{d}{du}[u + C] = 1$$

17. $\frac{2}{5}x^{5/2} + C$

$$\frac{d}{dx}\left[\frac{2}{5}x^{5/2} + C\right] = x^{3/2}$$

	Given	Rewrite	Integrate	Simplify
19.	$\int \sqrt[3]{x}\,dx$	$\int x^{1/3}\,dx$	$\dfrac{x^{4/3}}{4/3} + C$	$\dfrac{3}{4}x^{4/3} + C$
21.	$\int \dfrac{1}{x\sqrt{x}}\,dx$	$\int x^{-3/2}\,dx$	$\dfrac{x^{-1/2}}{-1/2} + C$	$-\dfrac{2}{\sqrt{x}} + C$
23.	$\int \dfrac{1}{2x^3}\,dx$	$\dfrac{1}{2}\int x^{-3}\,dx$	$\dfrac{1}{2}\left(\dfrac{x^{-2}}{-2}\right) + C$	$-\dfrac{1}{4x^2} + C$

25.

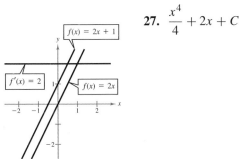

$f(x) = 2x + 1$

$f'(x) = 2$

$f(x) = 2x$

27. $\dfrac{x^4}{4} + 2x + C$

29. $\frac{6}{7}x^{7/3} + \frac{3}{2}x^2 - x + C$

31. $\frac{3}{5}x^{5/3} + C$

33. $-\dfrac{1}{2x^2} + C$

35. $-\dfrac{1}{4x} + C$

37. $t - \dfrac{2}{t} + C$

39. $\frac{3}{4}u^4 + \frac{1}{2}u^2 + C$

41. $2x^3 - \frac{11}{2}x^2 + 5x + C$

43. $\frac{2}{7}y^{7/2} + C$

45. $f(x) = 2x^{3/2} + 3x - 1$

47. $f(x) = 2x^3 - 3x^2$

49. $f(x) = -\dfrac{1}{x^2} + \dfrac{1}{x} + \dfrac{1}{2}$

51. $y = -\frac{5}{2}x^2 - 2x + 2$

53. $f(x) = 4x^{3/2} - 10x + 10$

55. $f(x) = x^2 + x + 4$

57. $f(x) = \frac{9}{4}x^{4/3}$

59. $C = 85x + 5500$

61. $C = \frac{1}{10}\sqrt{x} + 4x + 750$

63. $R = 225x - \frac{3}{2}x^2,\ p = 225 - \frac{3}{2}x$

65. $P = -9x^2 + 1650x$

67. 56.25 feet

69. $v_0 = 40\sqrt{22} \approx 187.617$ ft/sec

71. (a) $C = x^2 - 12x + 125$ **(b)** \$2025

$\overline{C} = x - 12 + \dfrac{125}{x}$ **(c)** \$125 is fixed.

$\qquad\qquad\qquad\qquad$ \$1900 is variable.

73. $S = -\dfrac{0.175}{3}t^3 + 0.2t^2 + 0.81t + 16.7,$

16.16 quadrillion Btu's

SECTION 5.2 (page 331)

Warm Up

1. $\frac{1}{2}x^4 + x + C$ **2.** $\frac{3}{2}x^2 + \frac{2}{3}x^{3/2} - 4x + C$

3. $-\dfrac{1}{x} + C$ **4.** $-\dfrac{1}{6t^2} + C$

5. $\frac{4}{7}t^{7/2} + \frac{2}{5}t^{5/2} + C$ **6.** $\frac{4}{5}x^{5/2} - \frac{2}{3}x^{3/2} + C$

7. $-\dfrac{5(x-2)^4}{16}$ **8.** $-\dfrac{1}{12(x-1)^2}$

9. $9(x^2 + 3)^{2/3}$ **10.** $-\dfrac{5}{(1-x^3)^{1/2}}$

$\displaystyle\int u^n \dfrac{du}{dx}\,dx$	u	$\dfrac{du}{dx}$
1. $\displaystyle\int (5x^2 + 1)^2(10x)\,dx$	$5x^2 + 1$	$10x$
3. $\displaystyle\int \sqrt{1 - x^2}(-2x)\,dx$	$1 - x^2$	$-2x$
5. $\displaystyle\int \left(4 + \dfrac{1}{x^2}\right)\left(\dfrac{-2}{x^3}\right)dx$	$4 + \dfrac{1}{x^2}$	$-\dfrac{2}{x^3}$

7. $\frac{1}{5}(1 + 2x)^5 + C$

9. $\frac{2}{3}(5x^2 - 4)^{3/2} + C$

11. $\frac{1}{5}(x - 1)^5 + C$

13. $\frac{1}{16}(x^2 - 1)^8 + C$

15. $-\dfrac{1}{3(1 + x^3)} + C$

17. $-\dfrac{1}{2(x^2 + 2x - 3)} + C$

19. $\sqrt{x^2 - 4x + 3} + C$

21. $-\frac{15}{8}(1 - x^2)^{4/3} + C$

23. $4\sqrt{1 + x^2} + C$

25. $-3\sqrt{2x + 3} + C$

27. $-\frac{1}{2}\sqrt{1-x^4}+C$ **29.** $\sqrt{2x}+C$

31. $\frac{1}{6}(x^3+3x)^2+C$ **33.** $\frac{1}{48}(6x^2-1)^4+C$

35. $-\frac{2}{45}(2-3x^3)^{5/2}+C$ **37.** $\sqrt{x^2+25}+C$

39. $\frac{2}{3}\sqrt{x^3+3x+4}+C$

41. $\frac{1}{6}(2x-1)^3+C_1=\frac{4}{3}x^3-2x^2+x+C_2$

(Answers differ by a constant: $C_2=C_1-\frac{1}{6}$)

43. $\frac{1}{2}\frac{(x^2-1)^3}{3}+C_1=\frac{1}{6}x^6-\frac{1}{2}x^4+\frac{1}{2}x^2+C_2$

(Answers differ by a constant: $C_2=C_1-\frac{1}{6}$)

45. $\frac{1}{3}[5-(1-x^2)^{3/2}]$

47. (a) $C=8\sqrt{x+1}+18$

(b)

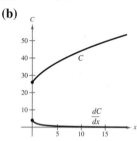

49. $x=\frac{1}{3}(p^2-25)^{3/2}+24$

51. $x=\dfrac{6000}{\sqrt{p^2-16}}+3000$

53. $V=\dfrac{200,000}{t+1}+300,000,\ \$340,000$

55. (a) $Q=(x-19,999)^{0.95}+19,999$

(b)

x	20,000	50,000
Q	20,000	37,916.56
$x-Q$	0	12,083.44

x	100,000	150,000
Q	65,491.59	92,151.16
$x-Q$	34,508.41	57,848.84

55. *(continued)*

(c)

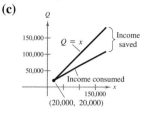

57. $-\frac{2}{3}x^{3/2}+\frac{2}{3}(x+1)^{3/2}+C$

SECTION 5.3   (page 338)

Warm Up

1. $\left(\frac{5}{2},\infty\right)$ **2.** $(-\infty,2)\cup(3,\infty)$

3. $x+2-\dfrac{2}{x+2}$ **4.** $x-2+\dfrac{1}{x-4}$

5. $x+8+\dfrac{2x-4}{x^2-4x}$ **6.** $x^2-x-4+\dfrac{20x+22}{x^2+5}$

7. $\frac{1}{4}x^4-\dfrac{1}{x}+C$ **8.** $\frac{1}{2}x^2+2x+C$

9. $\frac{1}{2}x^2-\dfrac{4}{x}+C$ **10.** $-\dfrac{1}{x}-\dfrac{3}{2x^2}+C$

1. $e^{2x}+C$ **3.** $\frac{1}{4}e^{4x}+C$ **5.** $-\frac{9}{2}e^{-x^2}+C$

7. $\frac{5}{3}e^{x^3}+C$ **9.** $\frac{1}{3}e^{x^3+3x^2-1}+C$ **11.** $-5e^{2-x}+C$

13. $-\frac{1}{2}e^{2/x}+C$ **15.** $2e^{\sqrt{x}}+C$

17. $\ln|x+1|+C$ **19.** $-\frac{1}{2}\ln|3-2x|+C$

21. $\ln\sqrt{x^2+1}+C$ **23.** $\frac{1}{3}\ln|x^3+1|+C$

25. $\frac{1}{2}\ln|x^2+6x+7|+C$ **27.** $\ln|\ln x|+C$

29. $-\ln(1+e^{-x})+C$ **31.** $2\ln|5-e^{2x}|+C$

33. $e^x+2x-e^{-x}+C$ **35.** $-\frac{2}{3}(1-e^x)^{3/2}+C$

37. $-\dfrac{1}{x-1}+C$ **39.** $\dfrac{x^2}{2}-4\ln|x|+C$

41. $\dfrac{x^2}{4}-4\ln|x|+C$ **43.** $2\ln(e^x+1)+C$

45. $\frac{1}{2}x^2 + 3x + 8 \ln |x - 1| + C$

47. $\frac{1}{2}x^2 + 5x + 8 \ln |x - 1| + C$

49. $\ln |e^x + x| + C$

51. (a) $P(t) = 1000[1 + \ln(1 + 0.25t)^{12}]$
 (b) $P(3) \approx 7715$ (c) $t \approx 6$ days

53. (a) $p = -50e^{-x/500} + 45.06$
 (b)

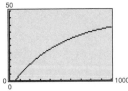

 The price increases as the demand increases.
 (c) 387

55. (a) $T = 9.292t^{5/2} - 2.63t^3 - 44.71e^{-t} + 348.81$
 (b) $T \approx 339$ million transactions

SECTION 5.4 (page 348)

Warm Up

1. $\frac{3}{2}x^2 + 7x + C$ **2.** $\frac{2}{5}x^{5/2} + \frac{4}{3}x^{3/2} + C$

3. $\frac{1}{5} \ln |x| + C$ **4.** $-\frac{1}{6e^{6x}}$ **5.** $-\frac{8}{5}$

6. $-\frac{62}{3}$ **7.** $0.008x^{5/2} + 29,500x + C$

8. $x^2 + 9000x + C$ **9.** $25,000x - 0.005x^2 + C$

10. $0.01x^3 + 4600x + C$

1. Area $= 6$

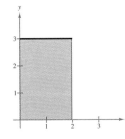

3. Area $= \frac{35}{2}$

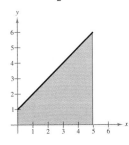

5. $\frac{1}{6}$ **7.** $\frac{8}{5}$ **9.** 6   **11.** 1

13. 0   **15.** $\frac{14}{3}$ **17.** $\frac{1}{3}$ **19.** -4

21. $\frac{22}{3}$ **23.** $-\frac{27}{20}$ **25.** 2

27. $\frac{1}{2}(1 - e^{-2}) \approx 0.432$ **29.** $\frac{e^3 - e}{3} \approx 5.789$

31. 4   **33.** 4   **35.** $\frac{1}{2} \ln 5 - \frac{1}{2} \ln 8 \approx -0.235$

37. $2 \ln(2 + e^3) - 2 \ln 3 \approx 3.993$

39. Area $= 10$ **41.** Area $= \frac{1}{4}$

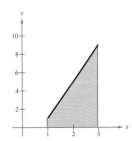
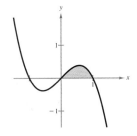

43. Area $= \ln 9$ **45.** 10

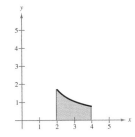

47. $4 + 5 \ln 5 \approx 12.047$

49. (a) 11   (b) 5   (c) -32 (d) -1

51. Average $= \frac{14}{3}$

$$x = \pm \frac{2\sqrt{3}}{3} \approx \pm 1.155$$

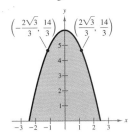

53. Average $= \frac{4}{3}$

$$x = \sqrt{2 + \frac{2\sqrt{5}}{3}} \approx 1.868$$

$$x = \sqrt{2 - \frac{2\sqrt{5}}{3}} \approx 0.714$$

55. Average $= -\frac{2}{3}$

$$x = \frac{4 + 2\sqrt{3}}{3} \approx 2.488$$

$$x = \frac{4 - 2\sqrt{3}}{3} \approx 0.179$$

57. Even   **59.** Neither odd nor even

61. (a) $\frac{8}{3}$ (b) $\frac{16}{3}$ (c) $-\frac{8}{3}$ **63.** \$6.75

65. \$22.50   **67.** \$3.97   **69.** \$2500   **71.** \$1250

73. (a) \$137,000 (b) \$214,720.93 (c) \$338,393.53

75. \$3082.95   **77.** (a) $R = 4.648t^{3/2} - \frac{2}{15}t^3 + 45.2$
    (b) \$61.82 billion

79. $\frac{2}{3}\sqrt{7} - \frac{1}{3}$ **81.** $\frac{39}{200}$

SECTION 5.5   (page 357)

Warm Up

1. $-x^2 + 3x + 2$ **2.** $-2x^2 + 4x + 4$

3. $-x^3 + 2x^2 + 4x - 5$ **4.** $x^3 - 6x - 1$

5. $(0, 4), (4, 4)$ **6.** $(1, -3), (2, -12)$

7. $(-3, 9), (2, 4)$ **8.** $(-2, -4), (0, 0), (2, 4)$

9. $(1, -2), (5, 10)$ **10.** $(1, e)$

1. 36   **3.** 9   **5.** $\frac{3}{2}$ **7.** $e - 2$

9.

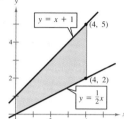

11. Area $= \frac{32}{3}$

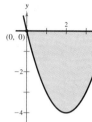

13. Area $= \frac{1}{6}$

15. Area $= 2$

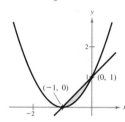
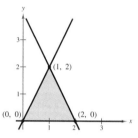

17. Area $= \frac{125}{6}$

19. Area $= 2$

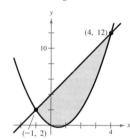
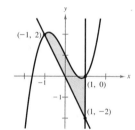

21. Area $= \frac{3}{2}$

23. Area $= \frac{64}{3}$

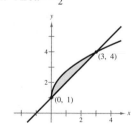
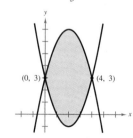

25. Area $= \frac{9}{2}$

27. Area $= 9$

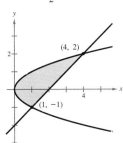

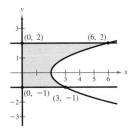

29. Area $= (2e + \ln 2) - 2e^{1/2} \approx 2.832$

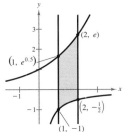

31. Area $= \frac{9}{2}$

33. Area $= -\frac{1}{2}e^{-1} + \frac{1}{2}$
≈ 0.316

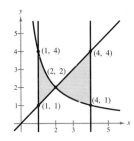

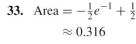

35. 8

37. Point of equilibrium: $(80, 10)$
Consumer surplus $= 1600$
Producer surplus $= 400$

39. Point of equilibrium: $(100, 200)$
Consumer surplus $= 5000$
Producer surplus $= 5000$

41. Point of equilibrium: $(50, 150)$
Consumer surplus ≈ 1666.67
Producer surplus $= 1250$

43. Point of equilibrium: $\left(266\frac{2}{3}, 16\frac{2}{3}\right)$
Consumer surplus ≈ 4444.44
Producer surplus ≈ 888.89

45. Point of equilibrium: $(300, 500)$
Consumer surplus $= 50,000$
Producer surplus $\approx 25,497$

47. A typical demand function is decreasing, whereas a typical supply function is increasing.

49. R_1, \$4.16 billion   **51.** \$501 million

53. Consumer surplus $= \$700,000$
Producer surplus $= \$1,375,000$

55. \$333.33 million   **57.** 648

SECTION 5.6   (page 367)

Warm Up

1. $\frac{1}{6}$ **2.** $\frac{3}{20}$ **3.** $\frac{7}{40}$ **4.** $\frac{13}{12}$ **5.** $\frac{61}{30}$
6. $\frac{53}{18}$ **7.** $\frac{2}{3}$ **8.** $\frac{4}{7}$ **9.** 0   **10.** 5

1. Midpoint Rule: 2
Exact area: 2

3. Midpoint Rule: 0.6730
Exact area: $\frac{2}{3} \approx 0.6667$

5. Midpoint Rule: 4.6250
Exact area:
$\frac{14}{3} \approx 4.6667$

7. Midpoint Rule: 17.2500
Exact area:
$\frac{52}{3} \approx 17.3333$

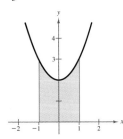

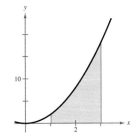

9. Midpoint Rule: 2.7266
Exact area: $\frac{11}{4} = 2.75$

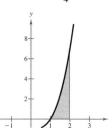

11. Midpoint Rule: 0.5703
Exact area:
$\frac{7}{12} \approx 0.5833$

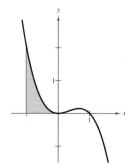

13. Midpoint Rule: 0.0859
Exact area: $\frac{1}{12} \approx 0.0833$

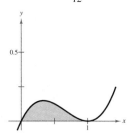

15. Area ≈ 54.6667,
$n = 40$

17. Midpoint Rule: 1.5
Exact area: 1.5

19.

Number of subintervals	2	4	6
Approximations	13.0375	12.7357	12.6591

Number of subintervals	8	10
Approximations	12.6267	12.6096

21.

Number of subintervals	2	4	6
Approximations	3.1523	3.2202	3.2320

Number of subintervals	8	10
Approximations	3.2361	3.2380

23. Exact: 4
Trapezoidal Rule: 4.0625
Midpoint Rule: 3.9688
Midpoint Rule is better in this example.

25. 1.1167   **27.** 1.55

29.

n	Midpoint Rule	Trapezoidal Rule
4	15.3965	15.6055
8	15.4480	15.5010
12	15.4578	15.4814
16	15.4613	15.4745
20	15.4628	15.4713

31. 4.8103   **33.** 916.25 ft

SECTION 5.7   (page 374)

> **Warm Up**
>
> **1.** 0, 2   **2.** 0, 2   **3.** 0, 2, -2 **4.** -1, 2
> **5.** 2, 4   **6.** 1, 5   **7.** 53.5982   **8.** 1.9459
> **9.** 3.3934   **10.** 1.3896

1. $\dfrac{16\pi}{3}$ **3.** $\dfrac{15\pi}{2}$ **5.** $\dfrac{512\pi}{15}$ **7.** $\dfrac{32\pi}{15}$

9. $\dfrac{\pi}{3}$ **11.** $\pi\left(\ln 2 - \dfrac{2}{3}\right) \approx 0.0832$

13. $\dfrac{128\pi}{5}$ **15.** $\dfrac{\pi}{2}(e^2 - 1)$ **17.** 8π

19. $\dfrac{2}{3}\pi$ **21.** $\dfrac{\pi}{4}$ **23.** $\dfrac{256\pi}{15}$ **25.** 18π

27. $V = \pi \displaystyle\int_{-r}^{r} (r^2 - x^2)\,dx = \dfrac{4\pi r^3}{3}$ **29.** 48π

31. (a) 1,256,637 ft^3 **33.** 58.5598
(b) 2513 fish

CHAPTER 5 REVIEW EXERCISES   (page 378)

1. $16x + C$ **3.** $\frac{2}{3}x^3 + \frac{5}{2}x^2 + C$ **5.** $x^{2/3} + C$

7. $\frac{3}{7}x^{7/3} + \frac{3}{2}x^2 + C$ **9.** $f(x) = \frac{3}{2}x^2 + x - 2$

11. $f(x) = \frac{1}{6}x^4 - 8x + \frac{33}{2}$

13. (a) 2.5 sec   **(b)** 100 ft   **(c)** 1.25 sec   **(d)** 75 ft

15. $x + 5x^2 + \frac{25}{3}x^3 + C$ or $\frac{1}{15}(1 + 5x)^3 + C_1$

17. $\frac{2}{5}\sqrt{5x - 1} + C$ **19.** $\frac{1}{2}x^2 - x^4 + C$

21. $\frac{1}{4}(x^4 - 2x)^2 + C$ **23. (a)** 30.5 board-feet
                                            (b) 125.2 board-feet

25. $-e^{-3x} + C$ **27.** $\frac{1}{2}e^{x^2-2x} + C$

29. $-\frac{1}{3}\ln|1 - x^3| + C$ **31.** $\frac{2}{3}x^{3/2} + 2x + 2x^{1/2} + C$

33. $A = 4$ **35.** $A = \frac{8}{3}$ **37.** $A = 2\ln 2$

39. 16   **41.** 0   **43.** 2   **45.** $\frac{1}{8}$ **47.** 3.899

49. 0   **51.** \$700.25   **53.** Average value: $\frac{8}{5}$
                                            $x = \frac{29}{4}$

55. Average value: $\frac{1}{3}(-1 + e^3) \approx 6.362$,
         $x \approx 3.150$

57. \$520.54   **59.** $\int_{-2}^{2} 6x^5\,dx = 0$
                              (Odd function)

61. $\int_{-2}^{-1} \frac{4}{x^2}\,dx = \int_{1}^{2} \frac{4}{x^2}\,dx = 2$
         (Symmetric about y-axis)

63. Area $= \frac{25}{3}$ **65.** Area $= \frac{3}{2}$

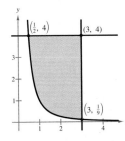

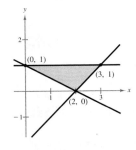

67. Area $= \frac{4}{15}$ **69.** Area $= \frac{64}{3}$

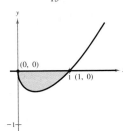

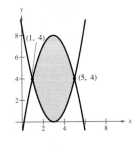

71. Consumer surplus: 11,250   **73.** 3.85 pounds
         Producer surplus: 14,062.5

75. $n = 4$: 13.3203   **77.** $n = 4$: 0.7867
         $n = 20$: 13.7167 $n = 20$: 0.7855

79. $\pi \ln 4 \approx 4.355$ **81.** $\frac{\pi}{2}(e^2 - e^{-2}) \approx 11.394$

83. $\frac{56\pi}{3}$ **85.** $\frac{2\pi}{35}$ **87.** $\frac{5\pi}{16}\sqrt{15}$

SAMPLE POST-GRAD EXAM QUESTIONS   (page 382)

1. (d)   **2.** (b)   **3.** (c)   **4.** (b)
5. (a)   **6.** (b)   **7.** (d)   **8.** (d)

CHAPTER 6

SECTION 6.1   (page 390)

> **Warm Up**
>
> **1.** $5x + C$ **2.** $\frac{2}{5}x^{5/2} + C$
>
> **3.** $\frac{1}{4}(x^2 + 1)^4 + C$ **4.** $e^{6x} + C$
>
> **5.** $\ln|2x + 1| + C$ **6.** $-e^{-x^2} + C$
>
> **7.** $x(x - 1)(2x - 1)$ **8.** $3x(x + 4)^2(x + 8)$
>
> **9.** $(x + 21)(x + 7)^{-1/2}$ **10.** $x(x + 5)^{-2/3}$

1. $\frac{1}{5}(x-2)^5 + C$ **3.** $\frac{2}{9-t} + C$

5. $\frac{2}{3}(1+x)^{3/2} + C$ **7.** $\ln(3x^2 + x)^2 + C$

9. $-\frac{1}{10(5x+1)^2} + C$ **11.** $2(x+1)^{1/2} + C$

13. $-\frac{1}{3}\ln|1-e^{3x}| + C$

15. $\frac{1}{2}x^2 + x + \ln|x-1| + C$

17. $\frac{1}{3}(x^2+4)^{3/2} + C$ **19.** $\frac{1}{5}e^{5x} + C$

21. $\frac{-1}{2(x+1)^2} + \frac{1}{3(x+1)^3} + C$

23. $\frac{1}{9}\left[\ln|3x-1| - \frac{1}{3x-1}\right] + C$

25. $\frac{1}{6}(1-x)^6 - \frac{1}{5}(1-x)^5 + C$

27. $\frac{1}{2}\ln|x^2 - 2x| + C$

29. $\frac{2}{5}(x-3)^{5/2} + 2(x-3)^{3/2} + C$

31. $-\frac{2}{3}(1-x)^{3/2} + \frac{4}{5}(1-x)^{5/2} - \frac{2}{7}(1-x)^{7/2} + C$

33. $\frac{1}{20}(2x-1)^{5/2} + \frac{1}{6}(2x-1)^{3/2} - \frac{3}{4}(2x-1)^{1/2} + C$

35. $\frac{3}{28}(t+1)^{4/3}(4t-3) + C$

37. $2(\sqrt{t}-1) + 2\ln|\sqrt{t}-1| + C$

39. $4\sqrt{t} + \ln t + C$ **41.** $\frac{26}{3}$

43. $\frac{3}{2}(e-1) \approx 2.577$ **45.** $\ln 2 - \frac{1}{2} \approx 0.193$

47. $\frac{13}{320}$ **49.** $\frac{144}{5}$ **51.** $\frac{1209}{28}$ **53.** $\frac{16}{15}\sqrt{2}$

55. $\frac{4}{3}$ **57.** $\frac{4\pi}{15} \approx 0.838$ **59.** $\frac{1}{2}$

61. (a) 0.547 **63.** $\$24{,}520.95$
  (b) 0.586

65. 5.8849

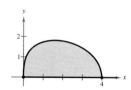

Warm Up

1. $12x^3 + 2$ **2.** $\frac{1}{2}x + \frac{1}{8}$ **3.** $\frac{4x^2+2}{x^3+x}$

4. $3x^2 e^{x^3}$ **5.** $e^x(x^2+2x)$ **6.** $-\frac{4}{(x-3)^2}$

7. $\frac{63}{3}$ **8.** $\frac{4}{3}$ **9.** $\frac{32}{3}$ **10.** 8

1. $\frac{1}{3}xe^{3x} - \frac{1}{9}e^{3x} + C$

3. $-x^2 e^{-x} - 2xe^{-x} - 2e^{-x} + C$

5. $x\ln 2x - x + C$

7. $\frac{1}{4}e^{4x} + C$ **9.** $\frac{e^{4x}}{16}(4x-1) + C$

11. $\frac{1}{2}e^{x^2} + C$ **13.** $x^2 e^x - 2e^x x + 2e^x$

15. $x^3 e^x - 3x^2 e^x + 6xe^x - 6e^x + C$

17. $\frac{1}{4}x^4 \ln x - \frac{1}{16}x^4 + C$

19. $\frac{1}{2}t^2 \ln(t+1) - \frac{1}{2}\ln(t+1) - \frac{1}{4}(t-1)^2 + C$

21. $\frac{x^2}{2}(\ln x)^2 - \frac{x^2}{2}\ln x + \frac{x^2}{4} + C$ **23.** $\frac{1}{3}(\ln x)^3 + C$

25. $\frac{2}{3}x(x-1)^{3/2} - \frac{4}{15}(x-1)^{5/2} + C$

27. $(x^2-1)e^x - 2xe^x + 2e^x + C = (x-1)^2 e^x + C$

29. $\frac{e^{2x}}{4(2x+1)} + C$ **31.** $e - 2 \approx 0.718$

33. $\frac{5}{36}e^6 + \frac{1}{36} \approx 56.060$

35. $\frac{2}{3}x(2x-3)^{3/2} - \frac{2}{15}(2x-3)^{5/2} + C$

37. $\frac{2}{3}x(4+5x)^{1/2} - \frac{4}{75}(4+5x)^{3/2} + C$

39. $\displaystyle \int x^n \ln x \, dx = \frac{x^{n+1}}{n+1}\ln x - \int \frac{x^n}{n+1}\, dx$

$\displaystyle = \frac{x^{n+1}}{n+1}\ln x - \frac{x^{n+1}}{(n+1)^2} + C$

$\displaystyle = \frac{x^{n+1}}{(n+1)^2}[-1 + (n+1)\ln x] + C$

41. $\dfrac{e^{5x}}{125}(25x^2 - 10x + 2) + C$ **43.** $-\dfrac{1}{x} - \dfrac{\ln x}{x} + C$

45. $1 - 5e^{-4} \approx 0.908$ **47. (a)** 2

(b) $4\pi(e - 2) \approx 9.026$

49. $\dfrac{3}{128} - \dfrac{379}{128}e^{-8} \approx 0.022$

51. $\dfrac{1,171,875}{256}\pi \approx 14{,}381.070$

53. (a) Increase   **(b)** 113,212 units   **(c)** 11,321 per year

55. (a) $3.2 \ln 2 - 0.2 \approx 2.018$

(b) $12.8 \ln 4 - 7.2 \ln 3 - 1.8 \approx 8.035$

57. $18,126.92   **59.** $1,332,474.72   **61.** $4103.07

63. (a) $1,200,000   **(b)** $1,094,142.26

65. 4.254   **67.** $45,957.78   **69.** Yes; $81,241.19

SECTION 6.3   (page 411)

Warm Up

1. $(x - 4)(x + 3)$ **2.** $(x - 5)(x + 5)$

3. $x(x - 2)(x + 1)$ **4.** $x(x - 2)^2$

5. $(x - 2)(x - 1)^2$ **6.** $(x - 3)(x - 1)^2$

7. $\dfrac{1}{x - 2} + x$ **8.** $\dfrac{1}{1 - x} + 2x - 2$

9. $\dfrac{2}{2 - x} + x^2 - x - 2$ **10.** $\dfrac{6}{x - 1} + x + 4$

---

1. $\dfrac{5}{x - 5} - \dfrac{3}{x + 5}$ **3.** $\dfrac{9}{x - 3} - \dfrac{1}{x}$

5. $\dfrac{1}{x - 5} + \dfrac{3}{x + 2}$ **7.** $\dfrac{3}{x} - \dfrac{5}{x^2}$

9. $\dfrac{1}{3(x - 2)} + \dfrac{1}{(x - 2)^2}$

11. $\dfrac{8}{x + 1} - \dfrac{1}{(x + 1)^2} + \dfrac{2}{(x + 1)^3}$

13. $\dfrac{1}{2} \ln \left|\dfrac{x - 1}{x + 1}\right| + C$

15. $\frac{1}{4} \ln |4x + 1| - \frac{1}{4} \ln |4 - x| + C$

17. $-\ln |x| + \ln |3x - 1| + C$

19. $\ln |x| - \ln |2x + 1| + C$

21. $\ln |x - 1| - \ln |x + 2| + C$

23. $\frac{3}{2} \ln |2x - 1| - 2 \ln |x + 1| + C$

25. $5 \ln |x - 2| - \ln |x + 2| - 3 \ln |x| + C$

27. $\frac{1}{2}[3 \ln |x - 4| - \ln |x|] + C$

29. $-3 \ln |x - 1| - \dfrac{1}{x - 1} + C$

31. $\dfrac{1}{2}\left[5 \ln |x + 1| - \ln |x| + \dfrac{3}{x + 1}\right] + C$

33. $\frac{1}{6} \ln \frac{4}{7} \approx -0.093$ **35.** $-\frac{4}{5} + 2 \ln \frac{5}{3} \approx 0.222$

37. $\frac{1}{2} - \ln 2 \approx -0.193$ **39.** $\frac{5}{2} - \frac{11}{3} \ln 2 \approx -0.042$

41. $\frac{1}{5} \ln |e^x - 1| - \frac{1}{5} \ln |e^x + 4| + C$

43. $\frac{1}{4} \ln\left(\sqrt{4 + x^2} - 2\right) - \frac{1}{4} \ln\left(\sqrt{4 + x^2} + 2\right) + C$

45. $-\dfrac{2}{3(\sqrt{3x} + 2)} + C$ **47.** $12 - \dfrac{7}{2} \ln 7 \approx 5.1893$

49. $5 \ln 2 - \ln 5 \approx 1.8563$ **51.** $\dfrac{1}{2a}\left(\dfrac{1}{a + x} + \dfrac{1}{a - x}\right)$

53. $\dfrac{1/a}{x} + \dfrac{1/a}{a - x}$ **55.** $\dfrac{\pi}{165}\left[136 - 33 \ln \dfrac{11}{3}\right] \approx 1.7731$

57. $\pi\left[\dfrac{1}{3} + \dfrac{1}{4} \ln 3\right] \approx 1.9100$

59. $y = \dfrac{1000}{1 + 9e^{-0.1656t}}$

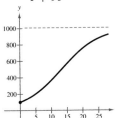

61. $1.077 thousand   **63.** 1467.85 thousand

SECTION 6.4 (page 422)

Warm Up

1. $x^2 + 8x + 16$ **2.** $x^2 + x + \frac{1}{4}$

3. $x^2 - 2x + 1$ **4.** $x^2 - \frac{2}{3}x + \frac{1}{9}$

5. $\frac{2}{x} - \frac{2}{x+2}$ **6.** $-\frac{3}{4x} + \frac{3}{4(x-4)}$

7. $\frac{3}{2(x-2)} - \frac{2}{x^2} - \frac{3}{2x}$

8. $-\frac{3}{x+1} + \frac{2}{x-2} + \frac{4}{x}$ **9.** $2e^x(x-1) + C$

10. $x^3 \ln x - \frac{x^3}{3} + C$

---

1. $\frac{1}{9}\left(\frac{2}{2+3x} + \ln |2+3x|\right) + C$

3. $\frac{2(3x-4)}{27}\sqrt{2+3x} + C$

5. $\ln\left(x^2 + \sqrt{x^4 - 9}\right) + C$

7. $\frac{1}{2}(x^2 - 1)e^{x^2} + C$

9. $\frac{1}{2}(x^2 + 1)[-1 + \ln(x^2 + 1)] + C$

11. $\ln\left|\frac{x}{1+x}\right| + C$ **13.** $-\frac{1}{3}\ln\left|\frac{3 + \sqrt{x^2 + 9}}{x}\right| + C$

15. $-\frac{1}{2}\ln\left|\frac{2 + \sqrt{4 - x^2}}{x}\right| + C$

17. $\frac{1}{4}x^2(-1 + 2\ln x) + C$ **19.** $3x^2 - \ln(1 + e^{3x^2}) + C$

21. $\frac{1}{4}\left(x^2\sqrt{x^4 - 4} - 4\ln\left|x^2 + \sqrt{x^4 - 4}\right|\right) + C$

23. $\frac{1}{27}\left[\frac{4}{2 + 3t} - \frac{4}{2(2 + 3t)^2} + \ln |2 + 3t|\right] + C$

25. $\frac{1}{\sqrt{3}}\ln\left|\frac{\sqrt{3+s} - \sqrt{3}}{\sqrt{3+s} + \sqrt{3}}\right| + C$

27. $\frac{1}{8}\left[\frac{-1}{2(3 + 2x)^2} + \frac{6}{3(3 + 2x)^3} - \frac{9}{4(3 + 2x)^4}\right] + C$

29. $-\frac{\sqrt{1 - x^2}}{x} + C$ **31.** $\frac{1}{9}x^3(-1 + 3\ln x) + C$

33. $\frac{1}{27}\left[3x - \frac{25}{3x - 5} + 10\ln |3x - 5|\right] + C$

35. $\frac{1}{4}\left[x(x^2 + 2)\sqrt{x^2 + 4} - 8\ln\left|x + \sqrt{x^2 + 4}\right|\right] + C$

37. $\frac{1}{2}[4x - \ln(1 + e^{4x})] + C$

39. $\frac{1}{9}[3\ln x - 4\ln|4 + 3\ln x|] + C$

41. $\frac{5\sqrt{5}}{3}$ **43.** $12\ln\left(\frac{2e^2}{1 + e^2}\right) \approx 6.7946$

45. $(x^2 - 2x + 2)e^x + C$ **47.** $-\left[\frac{1}{x} + \ln\left|\frac{x}{x+1}\right|\right] + C$

49. (a) $(x + 3)^2 - 9$ (b) $(x - 4)^2 - 7$
(c) $(x^2 + 1)^2 - 6$ (d) $4 - (x + 1)^2$

51. $\frac{1}{2\sqrt{17}}\ln\left|\frac{(x+3) - \sqrt{17}}{(x+3) + \sqrt{17}}\right| + C$

53. $-\ln\left|\frac{1 + \sqrt{x^2 - 2x + 2}}{x - 1}\right| + C$

55. $\frac{1}{8}\ln\left|\frac{x-3}{x+1}\right| + C$

57. $\frac{1}{2}\ln|x^2 + 1 + \sqrt{x^4 + 2x^2 + 2}| + C$

59. $\frac{40}{3}$ **61.** 42.58   **63.** \$1138.43

SECTION 6.5 (page 431)

Warm Up

1. $\frac{2}{x^3}$ **2.** $-\frac{96}{(2x + 1)^4}$ **3.** $-\frac{12}{x^4}$

4. $6x - 4$ **5.** $16e^{2x}$ **6.** $e^{x^2}(4x^2 + 2)$

7. $(3, 18)$ **8.** $(1, 8)$ **9.** $n < -5\sqrt{10}$
$n > 5\sqrt{10}$

10. $n < -5, n > 5$

	Exact value	*Trapezoidal Rule*	*Simpson's Rule*
1.	2.6667	2.7500	2.6667
3.	8.4000	9.0625	8.4167
5.	4.0000	4.0625	4.0000
7.	0.5000	0.5090	0.5004
9.	0.6931	0.6970	0.6933
11.		6.25	6.00
13.		0.342	0.372
15.		0.749	0.771
17.		0.772	0.780
19.		1.88	1.89

21. $21,831.20   **23.** $678.36

25. $0.3413 = 34.13\%$ **27.** 89,500 ft^2

29. (a) 2

(b) $\dfrac{2^5}{180(4^4)}(24) \approx 0.017$

31. (a) $\dfrac{5e}{64} \approx 0.212$

(b) $\dfrac{13e}{1024} \approx 0.035$

33. (a) $n = 101$ **35.** (a) $n = 3280$
(b) $n = 8$ (b) $n = 60$

37. 19.5215   **39.** 3.6558

41. Exact value: $\displaystyle\int_0^1 x^3\,dx = \dfrac{x^4}{4}\Big]_0^1 = \dfrac{1}{4}$

Simpson's Rule:

$\displaystyle\int_0^1 x^3\,dx = \dfrac{1}{6}\left[0^3 + 4\left(\dfrac{1}{2}\right)^3 + 1^3\right] = \dfrac{1}{4}$

43. 416.1 ft

45. 58.876 grams
(Simpson's Rule with $n = 100$)

47. 1876 subscribers (Simpson's Rule with $n = 100$)

SECTION 6.6   (page 442)

Warm Up

1. 9   **2.** 3   **3.** $-\dfrac{1}{8}$

4. Limit does not exist.

5. Limit does not exist.   **6.** -4

7. (a) $\dfrac{32}{3}b^3 - 16b^2 + 8b - \dfrac{4}{3}$ (b) $-\dfrac{4}{3}$

8. (a) $\dfrac{b^2 - b - 11}{(b-2)^2(b-5)}$ (b) $\dfrac{11}{20}$

9. (a) $\ln\left(\dfrac{5 - 3b^2}{b+1}\right)$ (b) $\ln 5 \approx 1.609$

10. (a) $e^{-3b^2}(e^{6b^2} + 1)$ (b) 2

---

1. 1   **3.** 1   **5.** Diverges   **7.** Diverges

9. Diverges   **11.** Diverges   **13.** 0   **15.** 4

17. 6   **19.** Diverges   **21.** 6   **23.** Diverges

25. 0   **27.** $\ln(4 + \sqrt{7}) - \ln 3 \approx 0.7954$

29.

x	1	10
xe^{-x}	0.3679	0.0005

x	25	50
xe^{-x}	0.0000	0.0000

31.

x	1	10
$x^2 e^{-(1/2)x}$	0.6065	0.6738

x	25	50
$x^2 e^{-(1/2)x}$	0.0023	0.0000

33. 2   **35.** $\dfrac{1}{4}$ **37.** (a) $4,637,228   (b) $5,555,556

39. (a) $748,367.34   **41.** (a) 1   **43.** (a) 0.111
(b) $808,030.14   (b) $\dfrac{\pi}{3}$ (b) 0.194
(c) $900,000.00   (c) 0.052

CHAPTER 6 REVIEW EXERCISES   (page 446)

1. $t + C$ **3.** $\dfrac{(x+5)^4}{4} + C$

5. $\dfrac{1}{10}e^{10x} + C$ **7.** $\dfrac{1}{5}\ln|x| + C$

9. $\dfrac{1}{3}(x^2+4)^{3/2} + C$ **11.** $2\ln(3+e^x) + C$

13. $\dfrac{(x-2)^5}{5} + \dfrac{(x-2)^4}{2} + C$

15. $\dfrac{2}{15}(x+1)^{3/2}(3x-2) + C$

17. $\dfrac{4}{5}(x-3)^{3/2}(x+2) + C$

19. $-\dfrac{2}{15}(1-x)^{3/2}(3x+7) + C$

21. $\dfrac{26}{15}$ **23.** $\dfrac{412}{15}$ **25.** (a) 0.696   (b) 0.693

27. $2\sqrt{x}\ln x - 4\sqrt{x} + C$ **29.** $xe^x - 2e^x + C$

31. $x^2 e^{2x} - xe^{2x} + \dfrac{1}{2}e^{2x} + C$ **33.** \$45,317.31

35. \$432,979.25   **37.** $\dfrac{1}{5}\ln|x| - \dfrac{1}{5}\ln|x+5| + C$

39. $\ln|x-5| + 3\ln|x+2| + C$

41. $x - \dfrac{25}{8}\ln|x+5| + \dfrac{9}{8}\ln|x-3| + C$

43. (a) $y = \dfrac{10,000}{1 + 7e^{-0.098652t}}$

(b)

Time, t	0	3	6
Sales, y	1250	1611	2052

Time, t	12	24
Sales, y	3182	6039

(c) $t \approx 31$ weeks

45. $\sqrt{x^2+25} - 5\ln\left|\dfrac{5+\sqrt{x^2+25}}{x}\right| + C$

47. $\dfrac{1}{4}\ln\left|\dfrac{x-2}{x+2}\right| + C$ **49.** $\dfrac{8}{3}$

51. $2\sqrt{1+x} + \ln\left|\dfrac{\sqrt{1+x}-1}{\sqrt{1+x}+1}\right| + C$

53. $(x-5)^3 e^{x-5} - 3(x-5)^2 e^{x-5} + 6(x-6)e^{x-5} + C$

55. $\dfrac{1}{10}\ln\left|\dfrac{x-3}{x+7}\right| + C$

57. $\dfrac{1}{2}\Big[(x-5)\sqrt{(x-5)^2-5^2}$
$\qquad + 25\ln\left|(x-5) + \sqrt{(x-5)^2-5^2}\right|\Big] + C$

59. 0.705   **61.** 0.741   **63.** 0.376   **65.** 0.289

67. 2.29   **69.** 0.001   **71.** 1   **73.** Diverges

75. 2   **77.** 2   **79.** (a) \$494,525.28
                                      (b) \$833,333.33

81. (a) 0.25   (b) 0.79×10^{-5}

SAMPLE POST-GRAD EXAM QUESTIONS   (page 450)

1. (c)   **2.** (a)   **3.** (b)   **4.** (d)
5. (d)   **6.** (b)   **7.** (a)   **8.** (c)

CHAPTER 7

SECTION 7.1   (page 458)

Warm Up

1. $2\sqrt{5}$ **2.** 5   **3.** 8   **4.** 8   **5.** (4, 7)

6. (1, 0)   **7.** (0, 3)   **8.** (−1, 1)

9. $(x-2)^2 + (y-3)^2 = 4$

10. $(x-1)^2 + (y-4)^2 = 25$

---

1.   **3.**

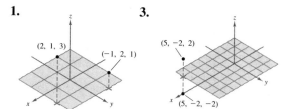

5. $3\sqrt{2}$ **7.** $\sqrt{206}$ **9.** (2, −5, 3)

11. $\left(\dfrac{1}{2}, \dfrac{1}{2}, -1\right)$ **13.** (6, −3, 5)   **15.** (1, 2, 1)

17. $3, 3\sqrt{5}, 6$; right triangle

19. $2, 2\sqrt{5}, 2\sqrt{2}$; the triangle is neither right nor isosceles.

21. $x^2 + (y - 2)^2 + (z - 2)^2 = 4$

23. $\left(x - \frac{3}{2}\right)^2 + (y - 2)^2 + (z - 1)^2 = \frac{21}{4}$

25. $(x - 1)^2 + (y - 1)^2 + (z - 5)^2 = 9$

27. $(x - 1)^2 + (y - 3)^2 + z^2 = 10$

29. Center: $(1, -3, -4)$ **31.** Center: $(0, 4, 0)$
Radius: 5   Radius: 4

33. Center: $(1, 3, 2)$ **35.** Center: $\left(\frac{1}{3}, -1, 0\right)$
Radius: $\dfrac{5}{\sqrt{2}} = \dfrac{5\sqrt{2}}{2}$ Radius: 1

37.

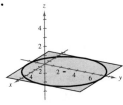

39.

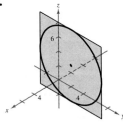

41. (a)

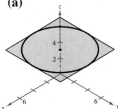

(b)

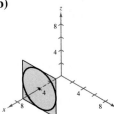

43. (a)

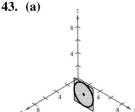

(b)

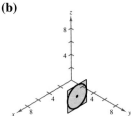

45. $(3, 3, 3)$

SECTION 7.2   (page 468)

Warm Up

1. $(4, 0)$, $(0, 3)$ **2.** $\left(-\frac{4}{3}, 0\right)$, $(0, -8)$

3. $(1, 0)$, $(0, -2)$ **4.** $(-5, 0)$, $(0, -5)$

5. $(x - 1)^2 + (y - 2)^2 + (z - 9)^2 + 1 = 0$

6. $(x - 4)^2 + (y + 2)^2 - (z + 3)^2 = 0$

7. $(x + 1)^2 + (y - 1)^2 - z = 0$

8. $(x - 3)^2 + (y + 5)^2 + (z + 13)^2 = 1$

9. $x^2 - y^2 + z^2 = \frac{1}{4}$ **10.** $x^2 - y^2 + z^2 = 4$

1.

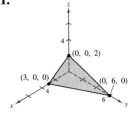

3.

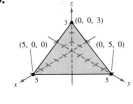

5.

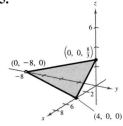

7.

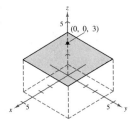

9.

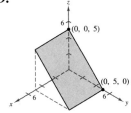

11.

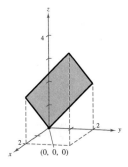

13.

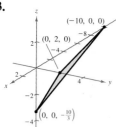

15.

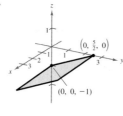

SECTION 7.3   (page 476)

Warm Up

1. 11   **2.** -16 **3.** 7

4. 4   **5.** $(-\infty, \infty)$

6. $(-\infty, -3) \cup (-3, 0) \cup (0, \infty)$

7. $[5, \infty)$ **8.** $\left(-\infty, -\sqrt{5}\,\right] \cup \left[\sqrt{5}, \infty\right)$

9. 55.0104   **10.** 6.9165

17. Perpendicular   **19.** Parallel   **21.** Parallel

23. Neither parallel nor perpendicular   **25.** $\dfrac{6\sqrt{14}}{7}$

27. $\dfrac{\sqrt{6}}{6}$ **29.** c   **30.** e   **31.** f   **32.** g

33. d   **34.** b   **35.** a   **36.** h

37. Trace in xy-plane $(z = 0)$: $y = x^2$ (parabola)
 Trace in plane $y = 1$: $x^2 - z^2 = 1$ (hyperbola)
 Trace in yz-plane $(x = 0)$: $y = -z^2$ (parabola)

39. Trace in xy-plane $(z = 0)$: $\dfrac{x^2}{4} + y^2 = 1$ (ellipse)

 Trace in xz-plane $(y = 0)$: $\dfrac{x^2}{4} + z^2 = 1$ (ellipse)

 Trace in yz-plane $(x = 0)$: $y^2 + z^2 = 1$ (circle)

41. Ellipsoid   **43.** Hyperboloid of one sheet

45. Elliptic paraboloid   **47.** Hyperbolic paraboloid

49. Hyperboloid of two sheets   **51.** Elliptic cone

53. Hyperbolic paraboloid   **55.** Sphere

57.

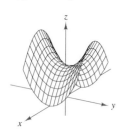

59.

1. (a) $\dfrac{3}{2}$ (b) $-\dfrac{1}{4}$ (c) 6   (d) $\dfrac{5}{y}$ (e) $\dfrac{x}{2}$ (f) $\dfrac{5}{t}$

3. (a) 5   (b) $3e^2$ (c) $2e^{-1}$ (d) $5e^y$ (e) xe^2 (f) te^t

5. (a) $\dfrac{2}{3}$ (b) 0   **7.** (a) 90π (b) 50π

9. (a) 20,655   (b) 1,397,673   **11.** (a) 0   (b) 6

13. (a) $x^2 + 2x\,\Delta x + (\Delta x)^2 - 2y$ (b) -2, $\Delta y \neq 0$

15. Domain: all points (x, y) inside and on the
 circle $x^2 + y^2 = 16$
 Range: $[0, 4]$

17. Domain: all points (x, y) such that $y \neq 0$
 Range: $(0, \infty)$

19. All points inside and on the ellipse $9x^2 + y^2 = 9$

21. All points (x, y) such that $y \neq 0$

23. All points (x, y) such that $x \neq 0$ and $y \neq 0$

25. All points (x, y) such that $y \geq 0$

27. The half-plane below the line $y = -x + 4$

29. (b)   **30.** (d)   **31.** (a)   **32.** (c)

33.

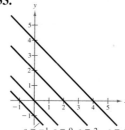

35.

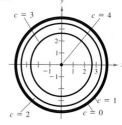

37.

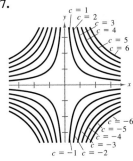

39.

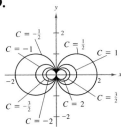

41. 135,540 units **43.** $21,960

45. (a) $1250 **(b)** $925

47.

I R	0	0.03	0.05
0	$2593.74	$1929.99	$1592.33
0.28	$2004.23	$1491.34	$1230.42
0.35	$1877.14	$1396.77	$1152.40

49. (a) $13.287 billion
(b) The variable x has greater influence because its coefficient is larger.

SECTION 7.4 (page 487)

Warm Up

1. $\dfrac{x}{\sqrt{x^2+3}}$ **2.** $-6x(3-x^2)^2$

3. $e^{2t+1}(2t+1)$ **4.** $\dfrac{e^{2x}(2-3e^{2x})}{\sqrt{1-e^{2x}}}$

5. $-\dfrac{2}{3-2x}$ **6.** $\dfrac{3(t^2-2)}{2t(t^2-6)}$

7. $-\dfrac{10x}{(4x-1)^3}$ **8.** $-\dfrac{(x+2)^2(x^2+8x+27)}{(x^2-9)^3}$

9. $f'(2)=8$ **10.** $f'(2)=\frac{7}{2}$

1. $f_x(x,y)=2$
$f_y(x,y)=-3$

3. $f_x(x,y)=\dfrac{5}{2\sqrt{x}}$
$f_y(x,y)=-12y$

5. $f_x(x,y)=y$
$f_y(x,y)=x$

7. $\dfrac{\partial z}{\partial x}=\sqrt{y}$
$\dfrac{\partial z}{\partial y}=\dfrac{x}{2\sqrt{y}}$

9. $f_x(x,y)=\dfrac{x}{\sqrt{x^2+y^2}}$
$f_y(x,y)=\dfrac{y}{\sqrt{x^2+y^2}}$

11. $\dfrac{\partial z}{\partial x}=2xe^{2y}$
$\dfrac{\partial z}{\partial y}=2x^2e^{2y}$

13. $h_x(x,y)=-2xe^{-(x^2+y^2)}$
$h_y(x,y)=-2ye^{-(x^2+y^2)}$

15. $\dfrac{\partial z}{\partial x}=\dfrac{2x}{x^2+y^2}$
$\dfrac{\partial z}{\partial y}=\dfrac{2y}{x^2+y^2}$

17. $\dfrac{\partial z}{\partial x}=\dfrac{3y-x}{x^2-y^2}$
$\dfrac{\partial z}{\partial y}=\dfrac{y-3x}{x^2-y^2}$

19. $f_x(x,y)=3xye^{x-y}(2+x)$

21. $g_x(x,y)=3y^2e^{y-x}(1-x)$ **23.** 9

25. $f_x(x,y)=6x+y,\ 13,\quad f_y(x,y)=x-2y,\ 0$

27. $f_x(x,y)=-\dfrac{y^2}{(x-y)^2},\ -\dfrac{1}{4}$
$f_y(x,y)=\dfrac{x^2}{(x-y)^2},\ \dfrac{1}{4}$

29. $w_x=\dfrac{x}{\sqrt{x^2+y^2+z^2}},\ \dfrac{2}{3}$
$w_y=\dfrac{y}{\sqrt{x^2+y^2+z^2}},\ -\dfrac{1}{3}$
$w_z=\dfrac{z}{\sqrt{x^2+y^2+z^2}},\ \dfrac{2}{3}$

31. $w_x=\dfrac{x}{x^2+y^2+z^2},\ \dfrac{3}{25}$
$w_y=\dfrac{y}{x^2+y^2+z^2},\ 0$
$w_z=\dfrac{z}{x^2+y^2+z^2},\ \dfrac{4}{25}$

33. $w_x = 2z^2 + 3yz$, 2

$w_y = 3xz - 12yz$, 30

$w_z = 4xz + 3xy - 6y^2$, -1

35. $w_x = \dfrac{y - x - 3z}{(x + y + z)^3}$, $-\dfrac{7}{125}$ **37.** $(-6, 4)$

$w_y = -2\dfrac{x + 2z}{(x + y + z)^3}$, $-\dfrac{12}{125}$

$w_z = 2\dfrac{y - z}{(x + y + z)^3}$, $-\dfrac{2}{125}$

39. $(1, 1)$ **41. (a)** 2   **(b)** -3 **43. (a)** 6   **(b)** -18

45. (a) $-\dfrac{3}{4}$ **(b)** 0   **47. (a)** -2 **(b)** -2

49. $\dfrac{\partial^2 z}{\partial y \partial x} = \dfrac{\partial^2 z}{\partial x \partial y}$ **51.** $\dfrac{\partial^2 z}{\partial y \partial x} = \dfrac{\partial^2 z}{\partial x \partial y} = y e^{2xy}$

$\quad = -2$

53. $\dfrac{\partial^2 z}{\partial x^2} = 6x$ **55.** $\dfrac{\partial^2 z}{\partial x^2} = 24x$

$\dfrac{\partial^2 z}{\partial y^2} = -8$ $\dfrac{\partial^2 z}{\partial y^2} = 6x - 24y$

$\dfrac{\partial^2 z}{\partial y \partial x} = \dfrac{\partial^2 z}{\partial x \partial y} = 0$ $\dfrac{\partial^2 z}{\partial y \partial x} = \dfrac{\partial^2 z}{\partial x \partial y} = 6y$

57. $\dfrac{\partial^2 z}{\partial x^2} = -2$ **59.** $\dfrac{\partial^2 z}{\partial x^2} = \dfrac{2y^2}{(x - y)^3}$

$\dfrac{\partial^2 z}{\partial y^2} = -2$ $\dfrac{\partial^2 z}{\partial y^2} = \dfrac{2x^2}{(x - y)^3}$

$\dfrac{\partial^2 z}{\partial y \partial x} = \dfrac{\partial^2 z}{\partial x \partial y} = 0$ $\dfrac{\partial^2 z}{\partial y \partial x} = \dfrac{\partial^2 z}{\partial x \partial y}$

$\quad = -\dfrac{2xy}{(x - y)^3}$

61. $\dfrac{\partial^2 z}{\partial x^2} = \dfrac{y^2}{(x^2 + y^2)^{3/2}}$

$\dfrac{\partial^2 z}{\partial y^2} = \dfrac{x^2}{(x^2 + y^2)^{3/2}}$

$\dfrac{\partial^2 z}{\partial y \partial x} = \dfrac{\partial^2 z}{\partial x \partial y} = -\dfrac{xy}{(x^2 + y^2)^{3/2}}$

63. $\dfrac{\partial^2 z}{\partial x^2} = 0$

$\dfrac{\partial^2 z}{\partial y^2} = 2x e^{-y^2}(2y^2 - 1)$

$\dfrac{\partial^2 z}{\partial y \partial x} = \dfrac{\partial^2 z}{\partial x \partial y} = -2y e^{-y^2}$

65. $w_x = 6xy - 5yz$ **67.** $w_x = \dfrac{y(y + z)}{(x + y + z)^2}$

$w_y = 3x^2 - 5xz + 10z^2$

$w_z = -5xy + 20yz$ $w_y = \dfrac{x(x + z)}{(x + y + z)^2}$

$w_z = -\dfrac{xy}{(x + y + z)^2}$

69. At $(120, 160)$: $\dfrac{\partial C}{\partial x} \approx 154.77$

At $(120, 160)$: $\dfrac{\partial C}{\partial y} \approx 193.33$

71. (a) $f_x(x, y) = 60\left(\dfrac{y}{x}\right)^{0.4}$, $f_x(1000, 500) \approx 45.47$

(b) $f_y(x, y) = 40\left(\dfrac{x}{y}\right)^{0.6}$, $f_x(1000, 500) \approx 60.63$

73. (a) Complementary

(b) Substitute

(c) Complementary

75. An increase in either price will cause a decrease in the number of applicants.

77. (a) $U_x = -10x + y$ **(b)** $U_y = x - 6y$

(c) When $x = 2$ and $y = 3$, $U_x = -17$ and $U_y = -16$. The person should consume one more unit of good y, because the rate of decrease of satisfaction is less for y.

(d)

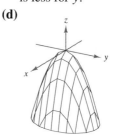

SECTION 7.5   (page 497)

Warm Up

1. $(3, 2)$ **2.** $(11, 6)$ **3.** $(1, 4)$ **4.** $(3, -2)$

5. $(0, 0)$, $(-1, 0)$ **6.** $(-2, 0)$, $(2, -2)$

7. $\dfrac{\partial z}{\partial x} = 12x^2$ $\dfrac{\partial^2 z}{\partial y^2} = -6$

$\dfrac{\partial z}{\partial y} = -6y$ $\dfrac{\partial^2 z}{\partial x \partial y} = 0$

$\dfrac{\partial^2 z}{\partial x^2} = 24x$ $\dfrac{\partial^2 z}{\partial y \partial x} = 0$

8. $\dfrac{\partial z}{\partial x} = 4x - 3y$ $\dfrac{\partial^2 z}{\partial y^2} = 2$

$\dfrac{\partial z}{\partial y} = 2y - 3x$ $\dfrac{\partial^2 z}{\partial x \partial y} = -3$

$\dfrac{\partial^2 z}{\partial x^2} = 4$ $\dfrac{\partial^2 z}{\partial y \partial x} = -3$

9. $\dfrac{\partial z}{\partial x} = 4x^3 - \dfrac{\sqrt{xy}}{2x}$ $\dfrac{\partial^2 z}{\partial y^2} = \dfrac{\sqrt{xy}}{4y^2}$

$\dfrac{\partial z}{\partial y} = -\dfrac{\sqrt{xy}}{2y} + 2$ $\dfrac{\partial^2 z}{\partial x \partial y} = -\dfrac{\sqrt{xy}}{4xy}$

$\dfrac{\partial^2 z}{\partial x^2} = 12x^2 + \dfrac{\sqrt{xy}}{4x^2}$ $\dfrac{\partial^2 z}{\partial y \partial x} = -\dfrac{\sqrt{xy}}{4xy}$

10. $\dfrac{\partial z}{\partial x} = e^{xy}(xy + 1)$ $\dfrac{\partial^2 z}{\partial y^2} = x^3 e^{xy}$

$\dfrac{\partial z}{\partial y} = x^2 e^{xy}$ $\dfrac{\partial^2 z}{\partial x \partial y} = xe^{xy}(xy + 2)$

$\dfrac{\partial^2 z}{\partial x^2} = ye^{xy}(xy + 2)$ $\dfrac{\partial^2 z}{\partial y \partial x} = xe^{xy}(xy + 2)$

---

1. Critical point: $(-2, -4)$
No relative extrema
$(-2, -4, 1)$ is a saddle point.

3. Critical point: $(0, 0)$
Relative minimum: $(0, 0, 1)$

5. Relative minimum: $(1, 3, 0)$

7. Relative minimum: $(-1, 1, -4)$

9. Relative maximum: $(8, 16, 74)$

11. Relative minimum: $(2, 1, -7)$

13. Saddle point: $(-2, -2, -8)$

15. Saddle point: $(0, 0, 0)$

17. Relative maxima: $(0, \pm 1, 4)$
Relative minimum: $(0, 0, 0)$
Saddle points: $(\pm 1, 0, 1)$

19. Saddle point: $(0, 0, 1)$

21. Insufficient information

23. $f(x_0, y_0)$ is a saddle point.

25. Relative minima: $(a, 0, 0)$, $(0, b, 0)$
Second-Partials Test fails at $(a, 0)$ and $(0, b)$.

27. Saddle point: $(0, 0, 0)$
Second-Partials Test fails at $(0, 0)$.

29. Relative minimum: $(0, 0, 0)$
Second-Partials Test fails at $(0, 0)$.

31. Relative minimum: $(1, -3, 0)$

33. False. (a, b) must be a critical point.

34. True   **35.** False. The function need not have
partial derivatives.

36. False. The origin is a minimum.   **37.** 10, 10, 10

39. 10, 10, 10   **41.** $x_1 = 3$, $x_2 = 6$

43. $p_1 = 2500$, $p_2 = 3000$ **45.** $x_1 = 125$, $x_2 = 85$

47. $x_1 \approx 94$, $x_2 \approx 157$ **49.** $36 \times 18 \times 18$ in.   **51.** Proof

SECTION 7.6   (page 507)

Warm Up

1. $\left(\dfrac{7}{8}, \dfrac{1}{12}\right)$ **2.** $\left(-\dfrac{1}{24}, -\dfrac{7}{8}\right)$ **3.** $\left(\dfrac{55}{12}, -\dfrac{25}{12}\right)$

4. $\left(\dfrac{22}{23}, -\dfrac{3}{23}\right)$ **5.** $\left(\dfrac{5}{3}, \dfrac{1}{3}, 0\right)$ **6.** $\left(\dfrac{14}{19}, -\dfrac{10}{19}, -\dfrac{32}{57}\right)$

7. $f_x = 2xy + y^2$ **8.** $f_x = 50y^2(x + y)$

$f_y = x^2 + 2xy$ $f_y = 50y(x + y)(x + 2y)$

9. $f_x = 3x^2 - 4xy + yz$

$f_y = -2x^2 - xz$

$f_z = xy$

10. $f_x = yz + z^2$

$f_y = xz + z^2$

$f_z = xy + 2xz + 2yz$

---

1. $f(5, 5) = 25$ **3.** $f(2, 2) = 8$

5. $f\left(\dfrac{\sqrt{2}}{2}, \dfrac{1}{2}\right) = \dfrac{1}{4}$ **7.** $f\left(\dfrac{25}{2}, \dfrac{25}{2}\right) = 231.25$

9. $f(1, 1) = 2$ **11.** $f(2, 2) = e^4$

13. $f(9, 6, 9) = 432$ **15.** $f\left(\dfrac{1}{3}, \dfrac{1}{3}, \dfrac{1}{3}\right) = \dfrac{1}{3}$

17. $f\left(\dfrac{1}{\sqrt{3}}, \dfrac{1}{\sqrt{3}}, \dfrac{1}{\sqrt{3}}\right) = \sqrt{3}$

19. $f\left(\dfrac{4}{15}, \dfrac{8}{15}, \dfrac{4}{15}, \dfrac{2}{15}\right) = \dfrac{8}{15}$ **21.** $f(9, 6, 9) = 486$

23. $f\left(\sqrt{\dfrac{10}{3}}, \dfrac{1}{2}\sqrt{\dfrac{10}{3}}, \sqrt{\dfrac{5}{3}}\right) = \dfrac{5\sqrt{15}}{9}$

25. $40, 40, 40$ **27.** $\dfrac{S}{3}, \dfrac{S}{3}, \dfrac{S}{3}$

29. $\sqrt{5}$ **31.** $\sqrt{3}$ **33.** $36 \times 18 \times 18$ in.

35. Length = Width = $\sqrt[3]{360} \approx 7.1$ ft

Height = $\dfrac{480}{360^{2/3}} \approx 9.5$ ft

37. $x_1 = 752.5,\; x_2 = 1247.5$
To minimize cost, let $x_1 = 753$ units and
$x_2 = 1247$ units.

39. $x = 50\sqrt{2} \approx 71$

$y = 200\sqrt{2} \approx 283$

41. (a) $f\left(\dfrac{3125}{6}, \dfrac{6250}{3}\right) \approx 147{,}314$
(b) 1.473
(c) $184{,}142$

43. $x = \sqrt[3]{0.065} \approx 0.402$
$y = \dfrac{1}{2}\sqrt[3]{0.065} \approx 0.201$
$z = \dfrac{1}{3}\sqrt[3]{0.065} \approx 0.134$

45. Stock D: $147{,}000
Stock C: $150{,}000
Stock P: $3000

SECTION 7.7 (page 517)

Warm Up

1. 5.0225 **2.** 0.0189

3. $S_a = 2a - 4 - 4b$ **4.** $S_a = 8a - 6 - 2b$
$S_b = 12b - 8 - 4a$ $S_b = 18b - 4 - 2a$

5. 15 **6.** 42 **7.** $\dfrac{25}{12}$

8. 14 **9.** 31 **10.** 95

---

1. (a) $y = \dfrac{3}{4}x + \dfrac{4}{3}$ (b) $\dfrac{1}{6}$

3. (a) $y = -2x + 4$ (b) 2 **5.** $y = \dfrac{7}{10}x + \dfrac{7}{5}$

7. $y = x + 4$ **9.** $y = -\dfrac{13}{20}x + \dfrac{7}{4}$

11. $y = \dfrac{37}{43}x + \dfrac{7}{43}$ **13.** $y = -\dfrac{175}{148}x + \dfrac{945}{148}$

15. $y = x + \dfrac{2}{3}$ **17.** $y = -2.3x - 0.9$

19. $y = \dfrac{3}{7}x^2 + \dfrac{6}{5}x + \dfrac{26}{35}$

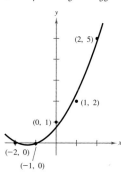

21. $y = 1.25x^2 - 1.75x + 0.25$

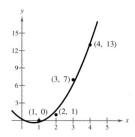

23. Linear: $y = 1.4286x + 6$
Quadratic: $y = 0.1190x^2 + 1.6667x + 5.6429$

25. Linear: $y = -68.9143x + 753.9524$
Quadratic: $y = 2.8214x^2 - 83.0214x + 763.3571$

27. (a) $y = -240x + 685$ **29.** (a) $y = 13.8x + 22.1$
(b) 349 (b) 44.18 bushels/acre
(c) $0.77

31. (a) $y = -0.5267t + 19.4677,\; \approx 4.7$ deaths
(b) $y = -0.000545t^2 - 0.5255t + 19.5881,$
≈ 4.4 deaths

33. (a) $y = -\dfrac{25}{112}x^2 + \dfrac{541}{56}x - \dfrac{25}{14}$ (b) 40.9 mph

35. Linear: $y = 3.7569x + 9.0347$
Quadratic: $y = 0.006316x^2 + 3.6252x + 9.4282$

37. Linear: $y = 0.9374x + 6.2582$
Quadratic: $y = -0.08715x^2 + 2.8159x + 0.3975$

39. Positive correlation,
$r = 0.9981$

41. No correlation, $r = 0$

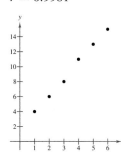

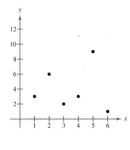

43. False, the slope is positive.
44. True   **45.** True   **46.** True

SECTION 7.8 (page 528)

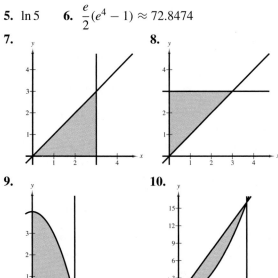

Warm Up

1. 1   **2.** 42   **3.** $\frac{19}{4}$ **4.** $\frac{1}{7}$

5. ln 5   **6.** $\frac{e}{2}(e^4 - 1) \approx 72.8474$

7.

8.

9.

10.

1. $\dfrac{3x^2}{2}$ **3.** $y \ln |2y|$ **5.** $\dfrac{x^2}{2}(9 - x^2)$

7. $\frac{1}{2}y[(\ln y)^2 - y^2]$ **9.** $x^2(1 - e^{-x^2} - x^2 e^{-x^2})$

11. 1   **13.** 36   **15.** $\frac{2}{3}$ **17.** $\frac{20}{3}$ **19.** 5

21. $\frac{16}{3}$ **23.** 4

25. $\displaystyle\int_0^1 \int_0^2 dy\, dx = \int_0^2 \int_0^1 dx\, dy = 2$

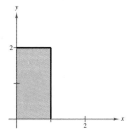

27. $\displaystyle\int_0^1 \int_{2y}^2 dx\, dy = \int_0^2 ds \int_0^{x/2} dy\, dx = 1$

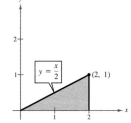

29. $\displaystyle\int_0^2 \int_{x/2}^1 dy\, dx = \int_0^1 \int_0^{2y} dx\, dy = 1$

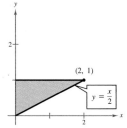

31. $\displaystyle\int_0^1 \int_{y^2}^{\sqrt[3]{y}} dx\, dy = \int_0^1 \int_{x^3}^{\sqrt{x}} dy\, dx = \frac{5}{12}$

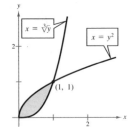

33. $\frac{1}{2}(e^9 - 1) \approx 4051.042$ **35.** 24   **37.** $\frac{16}{3}$

39. $\frac{8}{3}$ **41.** $\frac{500}{3}$ **43.** $\frac{75}{4}$ **45.** 2   **47.** 0.6588

49. 8.1747   **51.** 0.4521   **53.** False

55. True, the region is a rectangle.

SECTION 7.9   (page 536)

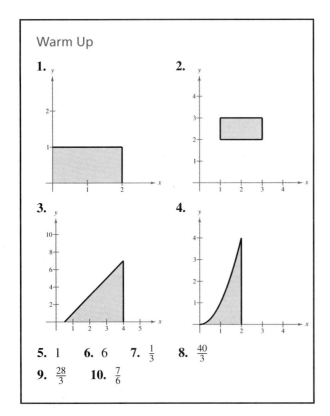

Warm Up

1.

2.

3.

4.

5. 1   **6.** 6   **7.** $\frac{1}{3}$ **8.** $\frac{40}{3}$

9. $\frac{28}{3}$ **10.** $\frac{7}{6}$

1. 10   **3.** $\frac{1}{54}$

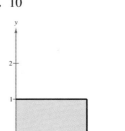

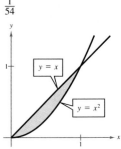

5. $\frac{1}{3}$

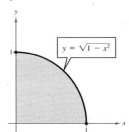

7. $\displaystyle\int_0^3 \int_0^5 xy\, dy\, dx = \int_0^5 \int_0^3 xy\, dx\, dy = \frac{225}{4}$

9. $\displaystyle\int_0^4 \int_0^{\sqrt{x}} \frac{y}{1+x^2}\, dy\, dx = \int_0^2 \int_{y^2}^4 \frac{y}{1+x^2}\, dx\, dy$

$\qquad\qquad = \frac{1}{4}\ln 17 \approx 0.708$

11. $\displaystyle\int_0^{1/2} \int_0^{2x} e^{-x^2}\, dy\, dx = 0.2212$ **13.** 4

15. 22.5   **17.** 12   **19.** $\frac{3}{8}$ **21.** $\frac{40}{3}$ **23.** $\frac{1}{3}$

25. 4   **27.** $\frac{32}{3}$ **29.** 2   **31.** $\frac{8}{3}$ **33.** $75{,}125$

CHAPTER 7 REVIEW EXERCISES   (page 540)

1.   **3.** $\sqrt{110}$

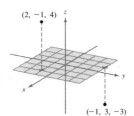

5. $(-1, 4, 6)$ **7.** $x^2 + (y-1)^2 + z^2 = 25$

9. $(x - 4)^2 + (y - 6)^2 + (z - 1)^2 = 6$

11. Center: $(-2, 1, 4)$, radius: 4

13.
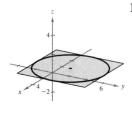

15.
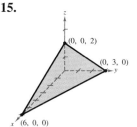

17.
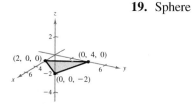

19. Sphere

21. Ellipsoid **23.** The graph is an elliptic paraboloid.

25. Top half of a circular cone

27. (a) 18 **29.** The domain is the set of all
(b) 0 points inside or on the circle
(c) -245 $x^2 + y^2 = 1$, and the range is
(d) -32 $[0, 1]$.

31. The level curves are **33.** The level curves are
lines of slope $-\frac{2}{5}$. hyperbolas.

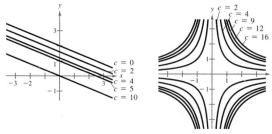

35. (a) The level curves represent lines of equal rainfall
and separate the four colors.
(b) The small eastern portion containing Davenport
(c) the northwestern portion containing Sioux City

37. \$2.50 **39.** $f_x = 2xy + 3y + 2$
$f_y = x^2 + 3x - 5$

41. $z_x = 12x\sqrt{y} + \frac{3}{2}\sqrt{\dfrac{y}{x}} - 7y$ **43.** $f_x = \dfrac{2}{2x + 3y}$

$z_y = \dfrac{3x^2}{\sqrt{y}} + \dfrac{3}{2}\sqrt{\dfrac{x}{y}} - 7x$ $f_y = \dfrac{3}{2x + 3y}$

45. $f_x = 2xe^y - y^2 e^x$ **47.** $w_x = yz^2$
$f_y = x^2 e^y - 2ye^x$ $w_y = xz^2$
$w_z = 2xyz$

49. (a) $z_x = 3$ **51.** (a) $z_x = -2x$
(b) $z_y = -4$ At $(1, 2, 3)$, $z_x = -2$.
(b) $z_y = -2y$
At $(1, 2, 3)$, $z_y = -4$.

53. $f_{xx} = 6x$
$f_{yy} = -8x + 6y$
$f_{xy} = f_{yx} = -8y$

55. $f_{xx} = \dfrac{y^2 - 64}{(64 - x^2 - y^2)^{3/2}}$

$f_{yy} = \dfrac{x^2 - 64}{(64 - x^2 - y^2)^{3/2}}$

$f_{xy} = f_{yx} = \dfrac{-xy}{(64 - x^2 - y^2)^{3/2}}$

57. $C_x(250, 175) \approx 99.70$
$C_y(250, 175) \approx 140.01$

59. Relative minimum: $(x, -x, 0)$

61. Saddle point: $\left(\frac{3}{2}, -\frac{3}{2}, 12.5\right)$

63. Relative minimum: $\left(\frac{1}{6}, \frac{1}{12}, -\frac{1}{432}\right)$
Saddle point: $(0, 0, 0)$

65. Relative minimum: $(1, 1, -2)$
Relative maximum: $(-1, -1, 6)$
Saddle points: $(1, -1, 2)$, $(-1, 1, 2)$

67. (a) $R = -x_1{}^2 - \frac{1}{2}x_2{}^2 + 100x_1 + 200x_2$
(b) $x_1 = 50$, $x_2 = 200$
(c) \$22,500.00

69. $\left(\frac{4}{3}, \frac{1}{3}, \frac{16}{27}\right)$, $(0, 1, 0)$

71. At $\left(\frac{4}{3}, \frac{2}{3}, \frac{4}{3}\right)$, the relative maximum value is $\frac{32}{27}$.

73. At $\left(\frac{4}{3}, \frac{10}{3}, \frac{14}{3}\right)$, the relative minimum is $34\frac{2}{3}$.

75. $f(49.4, 253) \approx 13{,}202$ **77.** (a) $y = \frac{60}{59}x - \frac{15}{59}$
(b) 2.746

79. (a) $S = 1.42t + 2.82$ ($t = 0$ is 1990.)
(b) $17.02 billion

81. $y = 1.71x^2 - 2.57x + 5.56$

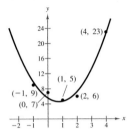

83. $\frac{29}{6}$ **85.** $\frac{7}{4}$

87. $\displaystyle\int_{-2}^{2}\int_{5}^{9-x^2} dy\,dx = \int_{5}^{9}\int_{-\sqrt{9-y}}^{\sqrt{9-y}} dx\,dy = \frac{32}{3}$

89. $\frac{9}{2}$ **91.** $\frac{4096}{9}$ **93.** 0.0833 mile

SAMPLE POST-GRAD EXAM QUESTIONS   (page 544)

1. (b)   **2.** (b)   **3.** (c)   **4.** (d)
5. (a)   **6.** (c)   **7.** (b)

CHAPTER P

SECTION P.1   (page P-8)

> ### Warm Up
>
> **1.** 1   **2.** 1   **3.** 2   **4.** 2   **5.** $\frac{1}{2}$ **6.** 1
> **7.** 37.50%   **8.** $81\frac{9}{11}\%$ **9.** $54\frac{1}{6}\%$
> **10.** 43.75%

1. (a) $S = \{HHH, HHT, HTH, HTT, THH,$
$THT, TTH, TTT\}$
(b) $A = \{HHH, HHT, HTH, THH\}$
(c) $B = \{HTT, THT, TTH, TTT\}$

3. (a) $S = \{3, 6, 9, 12, 15, 18, 21, 24, 27, 30, 33,$
$36, 39, 42, 45, 48\}$
(b) $A = \{12, 24, 36, 48\}$
(c) $B = \{9, 36\}$

5. 0.24   **7.** 0.0145   **9.**

Random variable	0	1	2
Frequency	1	2	1

11.

Random variable	0	1	2	3
Frequency	1	3	3	1

13. (a) $\frac{3}{4}$ **(b)** $\frac{4}{5}$ **15. (a)** 0.803   **(b)** 0.197

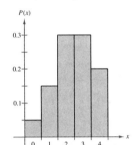   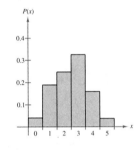

17. (a) $S = \{gggg, gggb, ggbg, gbgg, bggg, ggbb,$
$gbgb, gbbg, bgbg, bbgg, bggb, gbbb,$
$bgbb, bbgb, bbbg, bbbb\}$

(b)

x	0	1	2	3	4
$P(x)$	$\frac{1}{16}$	$\frac{4}{16}$	$\frac{6}{16}$	$\frac{4}{16}$	$\frac{1}{16}$

(c)

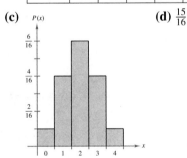

(d) $\frac{15}{16}$

19. $E(x) = 3$
$V(x) = 0.875$
$\sigma = 0.9354$

21. $E(x) = 0.8$
$V(x) = 8.16$
$\sigma = 2.8566$

23. (a) Mean: 2.5
Variance: 1.25
(b) Mean: 5
Variance: 2.5

25. (a) $E(x) = 18.5$
$\sigma = 8.0777$
(b) $27,750

27. \$201   **29.** −\$0.0526   **31.** City 1

33. Mean: 77.3
Standard deviation: 8.198

SECTION P.2   (page P-16)

Warm Up

1. Yes   **2.** No   **3.** No   **4.** Yes   **5.** 1
6. $\frac{1}{2}$ **7.** 1   **8.** $\frac{1}{4}$ **9.** 1   **10.** 1

1. $\displaystyle\int_0^8 \frac{1}{8}\, dx = \left[\frac{1}{8}x\right]_0^8 = 1$

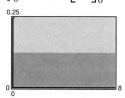

3. $\displaystyle\int_0^4 \frac{4-x}{8}\, dx = \left[\frac{1}{2}x - \frac{1}{16}x^2\right]_0^4 = 1$

5. $\displaystyle\int_0^2 \frac{3}{8}x\sqrt{4-x^2}\, dx = \left[-\frac{(4-x^2)^{3/2}}{8}\right]_0^2 = 1$

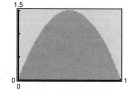

7. $\displaystyle\int_0^\infty \frac{1}{5}e^{-x/5}\, dx = \lim_{b\to\infty}\left[-e^{-x/5}\right]_0^b = 1$

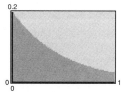

9. $\displaystyle\int_0^1 6x(1-x)\, dx = \left[3x^2 - 2x^3\right]_0^1 = 1$

11. $\displaystyle\int_0^3 \frac{4}{27}x^2(3-x)\, dx = \frac{4}{27}\left[x^3 - \frac{x^4}{4}\right]_0^3 = 1$

13. $\displaystyle\int_0^\infty \frac{1}{3}e^{-x/3}\, dx = \lim_{b\to\infty}\left[-e^{-x/3}\right]_0^b = 1$

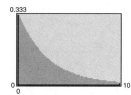

15. $\frac{15}{2}$ **17.** $\frac{3}{32}$ **19.** $\frac{1}{2}$
21. (a) $\frac{3}{5}$ (b) $\frac{1}{5}$ (c) $\frac{1}{5}$ (d) $\frac{4}{5}$

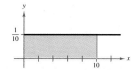

23. (a) $\dfrac{\sqrt{2}}{4} \approx 0.354$

(b) $1 - \dfrac{\sqrt{2}}{4} \approx 0.646$

(c) $\frac{1}{8}\left(3\sqrt{3} - 1\right) \approx 0.525$

(d) $\dfrac{3\sqrt{3}}{8} \approx 0.650$

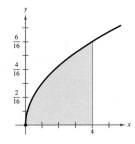

25. (a) $e^{-0/3} - e^{-2/3} \approx 0.4866$

(b) $e^{-2/3} - 0 \approx 0.5134$

(c) $e^{-1/3} - e^{-4/3} \approx 0.4529$

(d) 0

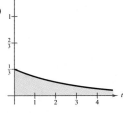

27. (a) $\frac{1}{6}$ **29.** (a) $1 - e^{-2/3} \approx 0.487$

(b) $\frac{2}{5}$ (b) $e^{-2/3} - e^{-4/3} \approx 0.250$

(c) $e^{-2/3} \approx 0.513$

31. $1 - e^{-4/3} \approx 0.736$

33. (a) $-\frac{1}{6}\left(12e^{-1} - 6\right) \approx 0.264$

(b) $2e^{-1} - 3e^{-2} \approx 0.330$

(c) $1 - \left(1 - 3e^{-2}\right) \approx 0.406$

35. (a) 0.107 (b) 0.353 **37.** 0.236

SECTION P.3   (page P-26)

Warm Up

1. 5 **2.** 8 **3.** $3 \ln |2|$ **4.** $9 \ln |2|$

5. $\frac{4}{3}$ **6.** $\frac{4}{3}$ **7.** $\frac{3}{4}$ **8.** $\frac{2}{9}$

9. (a) $\frac{1}{4}$ (b) $\frac{1}{2}$ **10.** (a) $\frac{1}{2}$ (b) $\frac{11}{16}$

1. (a) 4 (b) $\frac{16}{3}$

(c) $\dfrac{4}{\sqrt{3}} = \dfrac{4\sqrt{3}}{3}$

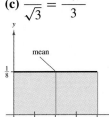

3. (a) 4 (b) 2 (c) $\sqrt{2}$

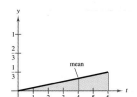

5. (a) $\frac{1}{2}$ (b) $\frac{1}{20}$

(c) $\dfrac{1}{2\sqrt{5}} = \dfrac{\sqrt{5}}{10}$

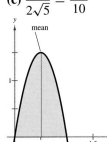

7. (a) $\frac{5}{7}$ (b) $\frac{20}{441}$

(c) $\dfrac{2\sqrt{5}}{21}$

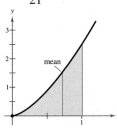

9. (a) $\frac{4}{3} \ln 4 - 1 \approx 0.848$

(b) $4 - \left(\frac{4}{3} \ln 4\right)^2 \approx 0.583$

(c) $\sqrt{4 - \left(\frac{4}{3} \ln 4\right)^2} \approx 0.764$

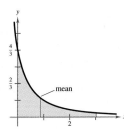

11. $-9 \ln \frac{1}{2} \approx 6.238$

13. Uniform density function

Mean: 5

Variance: $\frac{25}{3}$

Standard deviation: $\dfrac{5\sqrt{12}}{6} \approx 2.887$

15. Exponential density function

Mean: 8

Variance: 64

Standard deviation: 8

17. Normal density function
 Mean: 100
 Variance: 121
 Standard deviation: 11
19. Mean: 0
 Standard deviation: 1
 $P(0 \le x \le 0.85) \approx 0.3023$
21. Mean: 6
 Standard deviation: 6
 $P(x \ge 2.23) \approx 0.6896$
23. Mean: 8
 Standard deviation: 2
 $P(3 \le x \le 13) \approx 0.9876$
25. **(a)** $P(x > 64) \approx 0.3694$
 (b) $P(x > 70) \approx 0.2023$
 (c) $P(x < 70) \approx 0.7977$
 (d) $P(33 < x < 65) \approx 0.6493$
27. **(a)** Mean: 10:05 A.M.

 Standard deviation: $\dfrac{5\sqrt{3}}{3} \approx 2.9$ min

 (b) $\frac{3}{10}$
29. **(a)** $f(t) = \frac{1}{2}e^{-t/2}$
 (b) $P(0 < t < 1) = 1 - e^{-1/2} \approx 0.3935$
31. **(a)** $f(t) = \frac{1}{5}e^{-t/5}$ **(b)** $0.865 = 86.5\%$
33. **(a)** 1.5 standard deviations **(b)** 0.9332

35. **(a)** $\mu = 3$, $\sigma = \dfrac{3\sqrt{5}}{5} \approx 1.342$

 (b) 3 **(c)** $0.626 = 62.6\%$

37. $\mu = \frac{4}{7}$ **39.** **(a)** 10
 $V(x) = \frac{8}{147}$ **(b)** $P(x \le 4) \approx 0.1912$

41. Mean: $\frac{11}{2}$ = median
43. Mean: $\frac{1}{6}$ **45.** Mean: 5
 Median: 0.1465   Median: $5 \ln 2 \approx 3.4657$
47. $\dfrac{1}{-0.28} \ln 0.5 \approx 2.4755$
49. **(a)** Expected value: 6
 Standard deviation: $3\sqrt{2} \approx 4.243$
 (b) 0.615
51. $\mu \approx 12.25$ **53.** 40.68%   **55.** 68.28%

1. $S = \{$January, February, March, April, May, June, July, August, September, October, November, December$\}$
3. If the equations are numbered 1, 2, 3, and 4,
 $S = \{123,\ 124,\ 134,\ 234\}$.
5. $S = \{0,\ 1,\ 2,\ 3\}$

7.

x	0	1	2	3
$n(x)$	1	3	3	1

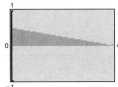

9. **(a)** $\frac{5}{6}$ **(b)** $\frac{5}{9}$

11. **(a)** $\frac{5}{36}$ **(b)** $\frac{5}{6}$ **(c)** $\frac{1}{6}$ **(d)** $\frac{1}{36}$
13. 19.5   **15.** **(a)** 20.5   **(b)** \$15,375
17. $V(x) = 218{,}243.7500$ **19.** $V(x) \approx 1.1611$
 $\sigma \approx 467.1657$ $\sigma \approx 1.0775$
21. $\displaystyle\int_0^4 \frac{1}{8}(4 - x)\,dx = \left[\frac{1}{2}x - \frac{x^2}{16}\right]_0^4 = (2 - 1) = 1$

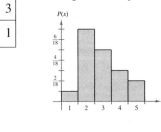

23. $\displaystyle\int_1^9 \frac{1}{4\sqrt{x}}\,dx = \frac{1}{4}\left[2\sqrt{x}\right]_1^9 = 1$

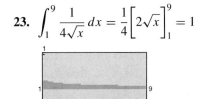

25. $\frac{9}{25}$ **27.** $\frac{2}{3}$ **29.** (a) $\frac{1}{2}$ (b) $\frac{1}{4}$

31. $\frac{1}{2}$ **33.** 2.5   **35.** 6

37. Variance: $\frac{9}{20}$

Standard deviation: $\dfrac{3}{2\sqrt{5}}$

39. Variance: 4
Standard deviation: 2

41. $\frac{1}{2}$ **43.** 2.7726   **45.** (a) 0.4866   (b) 0.2498

47. 0.00383   **49.** 0.3829

CHAPTER D

SECTION D.1   (page D-6)

Warm Up

1. $y' = 6x + 2$
$y'' = 6$

2. $y' = -6x^2 - 8$
$y'' = -12x$

3. $y' = -6e^{2x}$
$y'' = -12e^{2x}$

4. $y' = -6xe^{x^2}$
$y'' = -6e^{x^2}(2x^2 + 1)$

5. $\dfrac{1-x}{y}$ **6.** $\dfrac{2}{3y^2 + 4}$ **7.** $-\dfrac{y}{2x}$

8. $-\dfrac{y}{x}$ **9.** $k = 2\ln 3 - \ln\dfrac{17}{2} \approx 0.0572$

10. $k = \ln 10 - \dfrac{\ln 41}{2} \approx 0.4458$

---

1. $y' = 3x^2$

3. $y' = -2e^{-2x}$ and
$y' + 2y = -2e^{-2x} + 2(e^{-2x}) = 0$

5. $y' = 6x^2$ and
$y' - \dfrac{3}{x}y = 6x^2 - \dfrac{3}{x}(2x^3) = 0$

7. $y'' = 2$ and
$x^2 y'' - 2y = x^2(2) - 2(x^2) = 0$

9. $y' = 4e^{2x}$, $y'' = 8e^{2x}$, and
$y'' - y' - 2y = 8e^{2x} - 4e^{2x} - 2(2e^{2x}) = 0$

11. $\dfrac{dy}{dx} = -\dfrac{1}{x^2}$ **13.** $\dfrac{dy}{dx} = 4Ce^{4x} = 4y$

15. $\dfrac{dy}{dt} = -\dfrac{1}{3}Ce^{-t/3}$ and

$3\dfrac{dy}{dt} + y - 7 = 3\left(-\dfrac{1}{3}Ce^{-t/3}\right) + (Ce^{-t/3} + 7) - 7$
$= 0$

17. $xy' - 3x - 2y = x(2Cx - 3) - 3x - 2(Cx^2 - 3x)$
$= 0$

19. $xy' + y = x\left(2x + 2 - \dfrac{C}{x^2}\right) + \left(x^2 + 2x + \dfrac{C}{x}\right)$
$= x(3x + 4)$

21. $2y'' + 3y' - 2y = 2\left(\frac{1}{4}C_1 e^{x/2} + 4C_2 e^{-2x}\right)$
$+ 3\left(\frac{1}{2}C_1 e^{x/2} - 2C_2 e^{-2x}\right)$
$- 2(C_1 e^{x/2} + C_2 e^{-2x}) = 0$

23.

$y' - \dfrac{ay}{x} = \left(\dfrac{4bx^3}{4-a} + aCx^{a-1}\right) - \dfrac{a}{x}\left(\dfrac{bx^4}{4-a} + Cx^a\right)$
$= bx^3$

25. $y' + 2xy = -\dfrac{4Cxe^{x^2}}{(1 - Ce^{x^2})^2} + 2x\left(\dfrac{2}{1 + Ce^{x^2}}\right) = xy^2$

27. $y' = \ln x + 1 + C$ and
$x(y' - 1) - (y - 4) = x(\ln x + 1 + C - 1)$
$\qquad - (x\ln x + Cx + 4 - 4) = 0$

29. Implicit differentiation: $2x + 2yy' = Cy'$
$y' = \dfrac{2x}{C - 2y} = \dfrac{2xy}{Cy - 2y^2}$
$= \dfrac{2xy}{(x^2 + y^2) - 2y^2} = \dfrac{2xy}{x^2 - y^2}$

31. $x + y = \dfrac{C}{x}$
$y'' = \dfrac{2C}{x^3}$
$x^2 y'' - 2(x + y) = \dfrac{2C}{x} - \dfrac{2C}{x} = 0$

33. Solution **35.** Not a solution

37. Not a solution **39.** Solution

41. $y = 3e^{-2x}$ **43.** $y = 5 + \ln \sqrt{|x|}$

45. $y = 3e^{4x} + 2e^{-3x}$ **47.** $y = \frac{4}{3}(3 - x)e^{2x/3}$

49. $y = Cx^2$ **51.** $y = C(x + 2)^2$

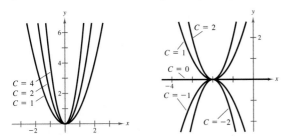

53. $y = x^3 + C$ **55.** $y = x + 3\ln|x| + C$

57. $y = \frac{2}{5}(x - 3)^{3/2}(x + 2) + C$

59. $y^2 = \frac{1}{4}x^3$ **61.** $y = 3e^x$

63. (a) $N = 750 - 650e^{-0.0484t}$

(b)

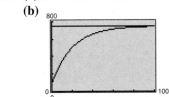

(c) $N \approx 214$

65.

Year, t	2	4	6	8	10
Units, x	3867	7235	10,169	12,725	14,951

67. True

68. False. From Example 1, $y = e^x$ is a solution of $y'' - y = 0$, but $y = e^x + 1$ is not.

71. Because
$$\frac{ds}{dh} = -\frac{13}{\ln 3}\left(\frac{1/2}{h/2}\right) = -\frac{13}{\ln 3}\frac{1}{h} \text{ and } -\frac{13}{\ln 3}$$
is a constant, we can conclude that the equation is a solution to $ds/dh = k/h$ where $k = -13/(\ln 3)$.

SECTION D.2 (page D-14)

> **Warm Up**
>
> **1.** $\frac{2}{5}x^{5/2} + C$ **2.** $\frac{1}{4}t^4 - \frac{3}{4}t^{4/3} + C$
>
> **3.** $2\ln|x - 5| + C$ **4.** $\frac{1}{4}\ln|2y^2 + 1| + C$
>
> **5.** $\frac{1}{2}e^{2y} + C$ **6.** $-\frac{1}{2}e^{1-x^2} + C$ **7.** $C = -10$
>
> **8.** $C = 5$ **9.** $k = \dfrac{\ln 5}{2} \approx 0.8047$
>
> **10.** $k = -2\ln 3 - \ln 2 \approx -2.8904$

1. Yes
$(y + 3)\,dy = x\,dx$

3. Yes
$dy = \left(\dfrac{1}{x} + 1\right)dx$

5. No. The variables cannot be separated.

7. $y = x^2 + C$

9. $y = \ln|x| + C$ **11.** $y = \sqrt[3]{x + C}$

13. $y = Ce^{x^2/2}$ **15.** $y^2 = \frac{1}{2}e^t + C$

17. $y = \ln|t^3 + t + C|$ **19.** $y = C(2 + x)^2$

21. $y = 1 - \left(C - \dfrac{x}{2}\right)^2$ **23.** $y = -3 + Ce^{x^2 - x}$

25. $3y^2 + 2y^3 = 3x^2 + C$ **27.** $y^2 = 2e^x + 14$

29. $y = -4 - e^{-x^2/2}$ **31.** $y = 4e^{x^2} - 1$

33. $5y^2 = 6x^2 - 1$ or $6x^2 - 5y^2 = 1$

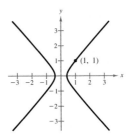

35. $v = 34.56(1 - e^{-0.1t})$ **37.** $T \approx 383.298°$

39. (a) $T \approx 7.277°$ **(b)** $t \approx 5.158$ hr

41. $N = 30 + Ce^{-kt}$ **43.** $y = Cx^{-k}$

SECTION D.3   (page D-20)

Warm Up

1. $e^x + 1$ **2.** $e^{3x} + 1$ **3.** $\dfrac{1}{x^3}$ **4.** $x^2 e^x$

5. $2e^x - e^{-x} + C$ **6.** $e^{3x}\left(\dfrac{x}{3} - \dfrac{1}{9}\right) + \dfrac{1}{2}e^{2x} + C$

7. $\dfrac{1}{2}\ln|2x + 5| + C$

8. $\dfrac{1}{2}\ln|x^2 + 2x + 3| + C$

9. $\dfrac{1}{12}(4x - 3)^3 + C$ **10.** $\dfrac{1}{6}(x^2 - 1)^3 + C$

1. $y' + \dfrac{-3}{2x^2}y = \dfrac{x}{2}$ **3.** $y' + \dfrac{1}{x}y = e^x$

5. $y' + \dfrac{1}{1-x}y = \dfrac{1}{x-1}$ **7.** $y = 2 + Ce^{-3x}$

9. $y = e^{-x}(x + C)$ **11.** $y = x^2 + 2x + \dfrac{C}{x}$

13. $y = \dfrac{1}{5} + Ce^{-(5/2)x^2}$ **15.** $y = \dfrac{x^3 - 3x + C}{3(x - 1)}$

17. $y = \dfrac{x^2}{3} + 1 + \dfrac{C}{x}$ **19.** $y = Ce^{-x} + 4$

21. $y = Ce^{x^2} - 1$ **23.** $y = 3e^x$ **25.** $xy = 4$

27. $y = 5e^{-x} + x - 1$ **29.** $y = x^2(5 - \ln|x|)$

31. $S = t + 95(1 - e^{-t/5})$

t	0	1	2	3	4	5
S	0	18.22	33.32	45.86	56.31	65.05

t	6	7	8	9	10
S	72.39	78.57	83.82	88.30	92.14

33. $p = 400 - 3x$ **35.** $p = 15(4 + e^{-t})$

37. **(a)** $A = \dfrac{P}{r^2}(e^{rt} - rt - 1)$

     (b) $A \approx \$34{,}543{,}402$

SECTION D.4   (page D-4)

Warm Up

1. $y = \dfrac{3}{2}x^2 + C$ **2.** $y^2 = 3x + C$

3. $y = Ce^{x^2}$ **4.** $y^4 = \dfrac{1}{2}(x - 4)^2 + C$

5. $y = 2 + Ce^{-2x}$ **6.** $y = xe^{-2x} + Ce^{-2x}$

7. $y = 1 + Ce^{-x^2/2}$ **8.** $y = \dfrac{1}{4}x^2 + Cx^{-2}$

9. $\dfrac{dy}{dx} = Cx^2$ **10.** $\dfrac{dx}{dt} = C(x - t)$

1. $y = e^{(x\ln 2)/3} \approx e^{0.2310x}$ **3.** $y = 4e^{-(x\ln 4)/4}$

                                    $\approx 4e^{-0.3466x}$

5. $y = \dfrac{1}{2}e^{(\ln 2)x} \approx \dfrac{1}{2}e^{0.6931x}$ **7.** \$4451.08

9. $S = L(1 - e^{-kt})$ **11.** $y = \dfrac{20}{1 + 19e^{-0.5889x}}$

13. $y = \dfrac{5000}{1 + 19e^{-0.10156x}}$ **15.** $N = \dfrac{500}{1 + 4e^{-0.2452t}}$

17. $\dfrac{dP}{dn} = kP(L - P)$, $P = \dfrac{CL}{e^{-Lkn} + C}$

19. $y = \dfrac{360}{8 + 41t}$ **21.** $y = 500e^{-1.6094e^{-0.1451t}}$

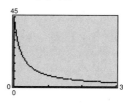 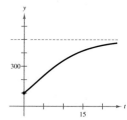

23. 34 beavers   **25.** 92%   **27.** **(a)** $Q = 25e^{-t/20}$

                                      (b) 10.2 min

29. $s = 25 - \dfrac{13\ln(h/2)}{\ln 3}$, $2 \le h \le 15$

31. $y = Cx$ **33.** $A = \dfrac{P}{r}(e^{rt} - 1)$

35. \$7,305,295.15   **37.** **(a)** $C = C_0 e^{-Rt/V}$ **(b)** 0

39. **(a)** $C(t) = \dfrac{Q}{R}(1 - e^{-Rt/V})$ **(b)** $\dfrac{Q}{R}$

CHAPTER D REVIEW EXERCISES (page D-31)

1. $\frac{3}{2}xy' - y = \frac{3}{2}x(2x^{-1/3}) - y = 3x^{2/3} - 3x^{2/3} = 0$

3. $y' - \frac{3y}{x^2} = \frac{3C}{x^2}e^{-3/x} - \frac{3}{x^2}\left(-\frac{1}{3} + Ce^{-3/x}\right) = \frac{1}{x^2}$

5. $y' - \frac{2y}{x} = 2C(x-4) - 2\left[\frac{C(x-4)^2}{x}\right]$

$\qquad = 2C(x-4)\left(1 - \frac{x-4}{x}\right)$

$\qquad = 2C(x-4)\left(\frac{4}{x}\right) = \frac{4}{x}\frac{dy}{dx}$

7. $\frac{dy}{dx} - 3x^2y = e^{x^3} + 3x^2y - 3x^2y = e^{x^3}$

9. $y = 5e^{-x/5}$ **11.** $y = \frac{1}{2}e^x + e^{-x}$

13. $y = x^2 + 2x$ **15.** $y = \dfrac{x^3 - 3x + 10}{3(x-1)}$

17. (a) $y' = Cke^{kt} = ky$ (b) 1389 flies

19. $2x^3 - y^2 = C$ **21.** $y = \dfrac{3(1 + Ce^{2x^3})}{1 - Ce^{2x^3}}$

23. $2(y^2 + 1)(\ln|x| + C) + 1 = 0$

25. (a) $y = Ce^{kt}$ (b) $t = \dfrac{\ln 0.25}{\ln 0.8} \approx 6.2$ hr

27. $y = \frac{3}{2} + Ce^{-6x}$ **29.** $y = -6 + Ce^{x^2/2}$

31. $y = -\frac{1}{3} + Ce^{-3/x}$

33. $y = \frac{1}{5}x^3 - x + C\sqrt{x}$

35. (a) $A = \dfrac{P}{r} + \left(A_0 - \dfrac{P}{r}\right)e^{rt}$
   (b) $A \approx \$1{,}341{,}465.28$
   (c) $A_0 \approx \$399{,}051.74$

37. Mt. St. Helens: $y = 21.796$ in.
    Mt. McKinley: $y = 13.859$ in.

39. (a) $\dfrac{dN}{dt} = kN(L - N)$
   (b) $N \approx 1800$ chickens

CHAPTER S

SECTION S.1 (page S-8)

> **Warm Up**
>
> **1.** 0   **2.** 0   **3.** 2   **4.** ∞ **5.** 0   **6.** 0
>
> **7.** $\dfrac{n-2}{n}$ **8.** $\dfrac{n-3}{n-4}$ **9.** $\dfrac{3n^2+1}{n^3}$
>
> **10.** $\dfrac{2n+1}{(n-1)(n+2)}$

1. 2, 4, 8, 16, 32   **3.** $-\frac{1}{2}, \frac{1}{4}, -\frac{1}{8}, \frac{1}{16}, -\frac{1}{32}$

5. $3, \frac{9}{2}, \frac{27}{6}, \frac{81}{24}, \frac{243}{120}$ **7.** $-1, \frac{1}{4}, -\frac{1}{9}, \frac{1}{16}, -\frac{1}{25}$

9. Converges to 0   **11.** Converges to 1

13. Converges to $\frac{1}{2}$ **15.** Diverges

17. Converges to 0   **19.** Converges to 3

21. Converges to 0   **23.** Diverges   **25.** Diverges

27. Diverges   **29.** $3n - 2$ **31.** $5n - 6$

33. $\dfrac{n+1}{n+2}$ **35.** $\dfrac{(-1)^{n-1}}{2^{n-2}}$ **37.** $\dfrac{n+1}{n}$

39. $2(-1)^n$ **41.** $\dfrac{1}{n!}$ **43.** 2, 5, 8, 11, 14, 17, . . .

45. $1, \frac{5}{3}, \frac{7}{3}, 3, \frac{11}{3}, \frac{13}{3}, \ldots$

47. $3, -\frac{3}{2}, \frac{3}{4}, -\frac{3}{8}, \frac{3}{16}, -\frac{3}{32}, \ldots$

49. 2, 6, 18, 54, 162, 486, . . .

51. Geometric, $20(\frac{1}{2})^{n-1}$ **53.** Arithmetic, $\frac{2}{3}n + 2$

55. $\dfrac{3n+1}{4n}$

57. 9045.00, 9090.23, 9135.68, 9181.35, 9227.26,
    9273.40, 9319.76, 9366.36, 9413.20, 9460.26

59. (a)

Year	1	2	3
Balance	$2200	$4620	$7282

Year	4	5	6
Balance	$10,210.20	$13,431.22	$16,974.34

59. *(continued)*

(b) $126,005.00   (c) $973,703.62

61. $387.97

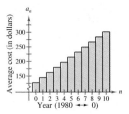

63. (a)

Year	1	2	3
Balance	$100.50	$201.50	$303.01

Year	4	5	6
Balance	$405.03	$507.55	$610.59

(b) $7011.89   (c) $46,435.11

65. (a) $S_1 = 1$ **(b)** $S_{20} = 2870$

$S_2 = 5$

$S_3 = 14$

$S_4 = 30$

$S_5 = 55$

67. (a) $1.3(0.85)^n$

(b)

Year	1	2
Budget amount	$1.105 billion	$0.939 billion

Year	3	4
Budget amount	$0.798 billion	$0.679 billion

(c) Converges to 0

69. $a_1 = 2$ **71.** $37,970.10

$a_{10} = 2.5937$

$a_{100} = 2.7048$

$a_{1000} = 2.7169$

$a_{10,000} = 2.7181$

SECTION S.2   (page S-20)

Warm Up

1. $\frac{77}{60}$ **2.** $\frac{73}{24}$ **3.** $\frac{31}{16}$ **4.** $\frac{40}{9}$ **5.** $\frac{21}{8}$

6. $\frac{31}{32}$ **7.** $\frac{3}{4}$ **8.** 0   **9.** 1   **10.** $\frac{1}{2}$

1. $S_1 = 1$ **3.** $S_1 = 3$

$S_2 = \frac{5}{4} = 1.25$ $S_2 = \frac{9}{2} = 4.5$

$S_3 = \frac{49}{36} \approx 1.361$ $S_3 = \frac{21}{4} = 5.25$

$S_4 = \frac{205}{144} \approx 1.424$ $S_4 = \frac{45}{8} = 5.625$

$S_5 = \frac{5269}{3600} \approx 1.464$ $S_5 = \frac{93}{16} = 5.8125$

5. nth-Term Test: $\displaystyle\lim_{n\to\infty} \frac{n}{n+1} = 1 \neq 0$

7. nth-Term Test: $\displaystyle\lim_{n\to\infty} \frac{n^2}{n^2+1} = 1 \neq 0$

9. Geometric series: $r = \frac{3}{2} > 1$

11. Geometric series: **13.** $r = \frac{3}{4} < 1$
$r = 1.055 > 1$

15. $r = 0.9 < 1$ **17.** 2   **19.** $\frac{2}{3}$

21. $4 + 2\sqrt{2} \approx 6.828$ **23.** $\frac{10}{9}$ **25.** $\frac{3}{2}$ **27.** $\frac{1}{2}$

29. $\frac{17}{6}$ **31.** $\displaystyle\lim_{n\to\infty} \frac{n+10}{10n+1} = \frac{1}{10} \neq 0$; diverges

33. $\displaystyle\lim_{n\to\infty} \frac{n!+1}{n!} = 1 \neq 0$; diverges

35. $\displaystyle\lim_{n\to\infty} \frac{3n-1}{2n+1} = \frac{3}{2} \neq 0$; diverges

37. Geometric series: $r = 1.075 > 1$; diverges

39. Geometric series: $r = \frac{1}{4} < 1$; converges   **41.** $\frac{2}{3}$

43. $\frac{9}{11}$ **45. (a)** $80,000(1 - 0.9^n)$ **47.** 72.89 ft
                          (b) 80,000

49. $7808.24   **51.** $\displaystyle\sum_{n=0}^{\infty} 100(0.75)^n = 400 million

53. $10,485.75   **55.** 2   **57.** 6

59. $e - 1 \approx 1.7183$ **61.** $\dfrac{e^2}{e-1} \approx 4.3003$

SECTION S.3   (page S-29)

Warm Up

1. $\dfrac{1}{n+1}$ **2.** $n+1$ **3.** $\dfrac{n}{n+1}$ **4.** $\dfrac{n+1}{n^2}$

5. 1   **6.** 5   **7.** 1   **8.** $\frac{1}{3}$

9. Geometric series   **10.** Not a geometric series

---

1. p-series   **3.** Not a p-series   **5.** Not a p-series

7. Converges   **9.** Diverges   **11.** Converges

13. Diverges   **15.** Converges   **17.** Converges

19. Diverges   **21.** Converges   **23.** Diverges

25. Converges   **27.** Diverges   **29.** Converges

31. 1.1777, the error is less than $\frac{1}{32}$.

33. 1.9953, the error is less than $\dfrac{2}{\sqrt{10}} \approx 0.6325$.

35. $\displaystyle\lim_{n\to\infty}\left|\dfrac{a_{n+1}}{a_n}\right| = \lim_{n\to\infty}\dfrac{1/[(n+1)^{3/2}]}{1/(n^{3/2})}$

$\qquad = \displaystyle\lim_{n\to\infty}\left(\dfrac{n}{n+1}\right)^{3/2} = 1$

37. Diverges, nth-Term Test

39. Converges, p-series

41. Converges, Geometric Series Test

43. Converges, p-series

45. Diverges, Geometric Series Test

47. Diverges, Ratio Test

49. $\displaystyle\sum_{n=1}^{100}\dfrac{1}{n^2} \approx 1.635,\quad \dfrac{\pi^2}{6} \approx 1.644934$

51. (a) $\displaystyle\sum_{k=1}^{n}[1.00 + 0.06(k-1)]$

   (b) No, $\displaystyle\lim_{k\to\infty}\dfrac{1.00 + 0.06(k)}{1.00 + 0.06(k-1)} = 1$

SECTION S.4   (page S-39)

Warm Up

1. $f(g(x)) = (x-1)^2$ **2.** $f(g(x)) = 6x + 3$

$\quad g(f(x)) = x^2 - 1$ $\qquad g(f(x)) = 6x + 1$

3. $f(g(x)) = \sqrt{x^2 + 4}$

$\quad g(f(x)) = x + 4, \quad x \geq -4$

4. $f(g(x)) = e^{x^2}$ **5.** $\quad f'(x) = 5e^x$

$\quad g(f(x)) = e^{2x}$ $\qquad f''(x) = 5e^x$

$\qquad\qquad\qquad\qquad f'''(x) = 5e^x$

$\qquad\qquad\qquad\qquad f^{(4)}(x) = 5e^x$

6. $\quad f'(x) = \dfrac{1}{x}$ **7.** $\quad f'(x) = 6e^{2x}$

$\quad f''(x) = -\dfrac{1}{x^2}$ $\qquad f''(x) = 12e^{2x}$

$\quad f'''(x) = \dfrac{2}{x^3}$ $\qquad f'''(x) = 24e^{2x}$

$\quad f^{(4)}(x) = -\dfrac{6}{x^4}$ $\qquad f^{(4)}(x) = 48e^{2x}$

8. $\quad f'(x) = \dfrac{1}{x}$ **9.** $\dfrac{n+1}{3}$ **10.** $\dfrac{n+3}{n+1}$

$\quad f''(x) = -\dfrac{1}{x^2}$

$\quad f'''(x) = \dfrac{2}{x^3}$

$\quad f^{(4)}(x) = -\dfrac{6}{x^4}$

---

1. $\displaystyle\sum_{n=0}^{\infty}\left(\dfrac{x}{4}\right)^n = 1 + \dfrac{x}{4} + \left(\dfrac{x}{4}\right)^2 + \left(\dfrac{x}{4}\right)^3 + \left(\dfrac{x}{4}\right)^4 + \cdots$

3. $\displaystyle\sum_{n=0}^{\infty}\dfrac{(-1)^{n+1}(x+1)^n}{n!}$

$\quad = -1 + (x+1) - \dfrac{(x+1)^2}{2} + \dfrac{(x+1)^3}{6}$

$\qquad - \dfrac{(x+1)^4}{24} + \cdots$

5. 2   **7.** 1   **9.** ∞ **11.** 0   **13.** 4

15. 5   **17.** 1   **19.** c **21.** ∞ **23.** ∞

25. $\displaystyle\sum_{n=0}^{\infty} \frac{x^n}{n!}$, $R = \infty$ **27.** $\displaystyle\sum_{n=0}^{\infty} \frac{(2x)^n}{n!}$, $R = \infty$

29. $\displaystyle\sum_{n=0}^{\infty} (-1)^n x^n$, $R = 1$

31. $\sqrt{2} + \frac{1}{4}\sqrt{2}(x - 2)$

$+ \displaystyle\sum_{n=2}^{\infty} \frac{(-1)^{n+1} 1 \cdot 3 \cdots (2n - 3)}{2^{2n} n!} \sqrt{2}\,(x - 2)^n$

$R = 2$

33. $\displaystyle\sum_{n=0}^{\infty} (-1)^n \frac{(n + 2)(n + 1)}{2} x^n$, $R = 1$

35. $1 + \displaystyle\sum_{n=1}^{\infty} \frac{(-1)^n 1 \cdot 3 \cdot 5 \cdots (2n - 1)}{2^n n!} x^n$, $R = 1$

37. $R = 2$ (all parts)   **39.** $R = 1$ (all parts)

41. $\displaystyle\sum_{n=0}^{\infty} \frac{x^{3n}}{n!}$ **43.** $3\displaystyle\sum_{n=0}^{\infty} \frac{x^{3n+2}}{n!}$

45. $\displaystyle\sum_{n=0}^{\infty} (-1)^n x^{4n}$ **47.** $\displaystyle\sum_{n=0}^{\infty} \frac{(-1)^n x^{2n+2}}{n + 1}$

49. $\displaystyle\sum_{n=0}^{\infty} \frac{(-1)^n (x - 1)^{n+1}}{n + 1}$

51. $\displaystyle\sum_{n=1}^{\infty} (-1)^{n+1} n (x - 1)^{n-1}$

53. 1.6487   **55.** -0.6931

SECTION S.5   (page S-49)

Warm Up

1. $\displaystyle\sum_{n=0}^{\infty} \frac{3^n x^n}{n!}$ **2.** $\displaystyle\sum_{n=0}^{\infty} \frac{(-1)^n 3^n x^n}{n!}$

3. $4\displaystyle\sum_{n=0}^{\infty} (-1)^n (x - 1)^n$

Warm Up *(continued)*

4. $\ln 5 + \displaystyle\sum_{n-1}^{\infty} \frac{(-1)^{n-1}(x - 1)^n}{n}$

5. $1 + \dfrac{x}{4} - \dfrac{3x^2}{4^2 2!} + \dfrac{3 \cdot 7 x^3}{4^3 3!} - \dfrac{3 \cdot 7 \cdot 11 x^4}{4^4 4!} + \cdots$

6. $1 + \dfrac{x}{2} - \dfrac{x^2}{2^2 2!} + \dfrac{1 \cdot 3 x^3}{2^3 3!} - \dfrac{1 \cdot 3 \cdot 5 x^4}{2^4 4!} + \cdots$

7. $\frac{47}{60}$ **8.** $\frac{311}{576}$ **9.** $\frac{5}{12}$ **10.** $\frac{77}{192}$

---

1. (a) $S_1(x) = 1 + x$ (b) $S_2(x) = 1 + x + \dfrac{x^2}{2}$

(c) $S_3(x) = 1 + x + \dfrac{x^2}{2} + \dfrac{x^3}{6}$

(d) $S_4(x) = 1 + x + \dfrac{x^2}{2} + \dfrac{x^3}{6} + \dfrac{x^4}{24}$

3. (a) $S_1(x) = 1 + \dfrac{x}{2}$ (b) $S_2(x) = 1 + \dfrac{x}{2} - \dfrac{x^2}{8}$

(c) $S_3(x) = 1 + \dfrac{x}{2} - \dfrac{x^2}{8} + \dfrac{x^3}{16}$

(d) $S_4(x) = 1 + \dfrac{x}{2} - \dfrac{x^2}{8} + \dfrac{x^3}{16} - \dfrac{5x^4}{128}$

5. (a) $S_1(x) = x$

(b) $S_2(x) = x - x^2$

(c) $S_3(x) = x - x^2 + x^3$

(d) $S_4(x) = x - x^2 + x^3 - x^4$

7.

x	0	0.25	0.50	0.75	1.00
$f(x)$	1.0000	1.1331	1.2840	1.4550	1.6487
$S_1(x)$	1.0000	1.1250	1.2500	1.3750	1.5000
$S_2(x)$	1.0000	1.1328	1.2813	1.4453	1.6250
$S_3(x)$	1.0000	1.1331	1.2839	1.4541	1.6458
$S_4(x)$	1.0000	1.1331	1.2840	1.4549	1.6484

9. (a) $S_2(x) = 1 - x^2$
   (b) $S_4(x) = 1 - x^2 + x^4$
   (c) $S_6(x) = 1 - x^2 + x^4 - x^6$
   (d) $S_8(x) = 1 - x^2 + x^4 - x^6 + x^8$

11. $S_4(x) = 1 - x^2 + x^4$ **13.** (d)

14. (c)   **15.** (a)   **16.** (b)   **17.** 0.607

19. 0.4055   **21.** 0.74286   **23.** 0.481

25. 7   **27.** $\dfrac{1}{6!} \approx 0.00139$ **29. (b)** 1   **(c)** \$10

SECTION S.6 (page S-57)

Warm Up

1. $f(2.4) = -0.04$ **2.** $f(-0.6) = 0.064$
   $f'(2.4) = 2.8$ $f'(-0.6) = 3.48$

3. $f(0.35) = 0.01$ **4.** $f(1.4) = 0.30$
   $f'(0.35) = 4.03$ $f'(1.4) = 12.88$

5. $4.9 \leq x \leq 5.1$ **6.** $0.798 \leq x \leq 0.802$

7. $5.97 \leq x \leq 6.03$ **8.** $-3.505 \leq x \leq -3.495$

9. $\left(\dfrac{\sqrt{13} + 3}{2}, \sqrt{13} + 2 \right), \left(\dfrac{3 - \sqrt{13}}{2}, 2 - \sqrt{13} \right)$

10. $\left(\dfrac{1 - \sqrt{5}}{2}, \dfrac{3 - \sqrt{5}}{2} \right), \left(\dfrac{1 + \sqrt{5}}{2}, \dfrac{3 + \sqrt{5}}{2} \right)$

1. 2.2364   **3.** 0.682   **5.** 1.25   **7.** 0.567

9. ± 0.753 **11.** $-4.596, -1.042, 5.638$

13. 2.926   **15.** 2.893   **17.** $f'(x_1) = 0$

19. $1 = x_1 = x_3 = \ldots$ **21.** $x_{n+1} = \dfrac{x_n^2 + a}{2x_n}$
    $0 = x_2 = x_4 = \ldots$

23. 2.646   **25.** 1.565   **29.** $x \approx 1.563$ mi

31. $t \approx 4.486$ hr   **33.** 15.9 yr

35. 11.8033

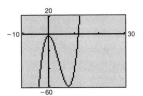

37. $\pm 1.9021, \pm 1.1756$ **39.** 1.3385, 0.2359

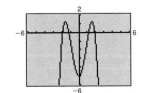 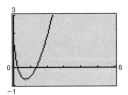

CHAPTER S REVIEW EXERCISES (page S-62)

1. $-\dfrac{1}{3}, \dfrac{1}{9}, -\dfrac{1}{27}, \dfrac{1}{81}, -\dfrac{1}{243}$ **3.** $4, 8, 10\frac{2}{3}, 10\frac{2}{3}, \dfrac{128}{15}$

5. Converges to 0   **7.** Diverges

9. Converges to 5   **11.** Converges to 0

13. $\dfrac{n}{3n}$ or $\dfrac{1}{3}, \quad n \neq 0$

15. $(-1)^n \dfrac{2^n}{3^{n+1}}, \quad n = 0, 1, 2, \ldots$

17. (a) $15{,}000 + 10{,}000(n - 1)$ **(b)** \$175,000

19. $S_0 = 1$ **21.** $S_1 = \frac{1}{2} = 0.5$
   $S_1 = \frac{5}{2} = 2.5$ $S_2 = \frac{11}{24} \approx 0.4583$
   $S_2 = \frac{19}{4} = 4.75$ $S_3 = \frac{331}{720} \approx 0.4597$
   $S_3 = \frac{65}{8} = 8.125$ $S_4 = \frac{18{,}535}{40{,}320} \approx 0.4597$
   $S_4 = \frac{211}{16} = 13.1875$ $S_5 = \frac{1{,}668{,}151}{3{,}628{,}800} \approx 0.4597$

23. Diverges   **25.** Converges

27. $\displaystyle\lim_{n \to \infty} \dfrac{2n}{n + 5} = 2 \neq 0$ **29.** $\displaystyle\lim_{n \to \infty} \left(\dfrac{5}{4} \right)^n = \infty \neq 0$

31. $\dfrac{5}{4} \left[1 - \left(\frac{1}{5} \right)^{n+1} \right]$

33. $2 \left[1 - \left(\frac{1}{2} \right)^{n+1} \right] + \dfrac{4}{3} \left[1 - \left(\frac{1}{4} \right)^{n+1} \right]$

35. Diverges **37.** Converges to $\frac{13}{4}$

39. (a) $D = -8 + 16 + 16(0.7) + 16(0.7)^2 + \cdots$

(b) $\frac{136}{3}$ ft

41. Converges **43.** Converges

45. 1.0172, Error $< \dfrac{1}{(5)4^5} \approx 1.9531 \times 10^{-4}$

47. 2.17857, Error $< \dfrac{1}{(1/4)(7)^{1/4}} < 2.4592$

49. Converges **51.** Diverges **53.** Converges

55. $R = 1$ **57.** $R = 0$ **59.** $\displaystyle\sum_{n-0}^{\infty} \left(-\frac{1}{2}\right)^n \frac{x^n}{n!}$

61. $\displaystyle -\sum_{n=0}^{\infty} (x+1)^n$ **63.** $\displaystyle \ln 2 + \sum_{n=1}^{\infty} (-1)^{n+1} \frac{(x/2)^n}{n}$

65. $1 + 2x^2 + x^4$

67. $\frac{1}{9} - \frac{2}{27}x + \frac{1}{27}x^2 - \frac{4}{243}x^3 + \frac{5}{729}x^4 - \frac{2}{729}x^5 + \frac{7}{6561}x^6$

69. $\ln 3 + \frac{1}{3}(x-1) - \frac{1}{18}(x-1)^2 + \frac{1}{81}(x-1)^3$
$\qquad - \frac{1}{324}(x-1)^4 + \frac{1}{1215}(x-1)^5 - \frac{1}{4374}(x-1)^6$

71. 4.7705 **73.** 0.9163 **75.** $\frac{1}{32}$ **77.** 0.301

79. 0.1233 **81.** 0.25, \$5.75 **83.** 0.313

85. 0.258 **87.** 1.341 **89.** 0.773

CHAPTER T

SECTION T.1 (page T-7)

Warm Up

1. 35 cm^2 **2.** 12 in.^{2} **3.** $c = 13$

4. $b = 4$ **5.** $b = 15$ **6.** $a = 6$

7. Equilateral triangle **8.** Isosceles triangle

9. Right triangle **10.** Isosceles triangle and right triangle

---

1. (a) $405°$, $-315°$
(b) $319°$, $-401°$

3. (a) $660°$, $-60°$
(b) $20°$, $-340°$

5. (a) $\dfrac{19\pi}{9}$, $-\dfrac{17\pi}{9}$

(b) $\dfrac{8\pi}{3}$, $-\dfrac{4\pi}{3}$

7. (a) $\dfrac{7\pi}{4}$, $-\dfrac{\pi}{4}$

(b) $\dfrac{28\pi}{15}$, $-\dfrac{32\pi}{15}$

9. $\dfrac{\pi}{6}$ **11.** $\dfrac{7\pi}{4}$ **13.** $-\dfrac{\pi}{6}$ **15.** $-\dfrac{3\pi}{2}$

17. $270°$ **19.** $-105°$ **21.** $990°$ **23.** $-66°$

25. $c = 10$, $\theta = 60°$ **27.** $a = 4\sqrt{3}$, $\theta = 30°$

29. $\theta = 40°$ **31.** $s = \sqrt{3}, \theta = 60°$

33. $9\sqrt{3}$ in.^{2} **35.** $\dfrac{25\sqrt{2}}{4}$ ft^2

37. False. An obtuse angle is between $90°$ and $180°$.

38. True, $-35° + 360° = 325°$

39. True. The angles would be $90°$, $89°$, and $1°$.

40. True **41.** 18 ft

43.

r	8 ft	15 in.	85 cm	24 in.	$\dfrac{12{,}963}{\pi}$ mi
s	12 ft	24 in.	200.28 cm	96 in.	8642 mi
θ	1.5	1.6	$\dfrac{3\pi}{4}$	4	$\dfrac{2\pi}{3}$

45. (a) $\dfrac{5\pi}{12}$ (b) 7.8125π in. **47.** $4.655°$

SECTION T.2 (page T-18)

Warm Up

1. $\dfrac{3\pi}{4}$ **2.** $-\dfrac{7\pi}{6}$ **3.** $-\dfrac{5\pi}{9}$ **4.** 3π

5. $x = 0$, $x = 1$ **6.** $x = -\frac{1}{2}$, $x = 1$

7. $x = 1$ **8.** $x = 2$, $x = 3$

9. $t = 10$ **10.** $t = \frac{445}{8}$

1. $\sin\theta = \frac{4}{5}$, $\csc\theta = \frac{5}{4}$

$\cos\theta = \frac{3}{5}$, $\sec\theta = \frac{5}{3}$

$\tan\theta = \frac{4}{3}$, $\cot\theta = \frac{3}{4}$

3. $\sin\theta = -\frac{5}{13}$, $\csc\theta = -\frac{13}{5}$

$\cos\theta = -\frac{12}{13}$ $\sec\theta = -\frac{13}{12}$

$\tan\theta = \frac{5}{12}$, $\cot\theta = \frac{12}{5}$

5. $\sin\theta = \frac{1}{2}$, $\csc\theta = 2$

$\cos\theta = -\dfrac{\sqrt{3}}{2}$, $\sec\theta = -\dfrac{2\sqrt{3}}{3}$

$\tan\theta = -\dfrac{\sqrt{3}}{3}$, $\cot\theta = -\sqrt{3}$

7. $\csc\theta = 2$ **9.** $\cot\theta = \frac{4}{3}$ **11.** $\sec\theta = \frac{17}{15}$

13. $\sin\theta = \frac{1}{3}$, $\csc\theta = 3$

$\cos\theta = \dfrac{2\sqrt{2}}{3}$, $\sec\theta = \dfrac{3\sqrt{2}}{4}$

$\tan\theta = \dfrac{\sqrt{2}}{4}$, $\cot\theta = 2\sqrt{2}$

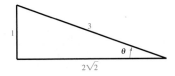

15. $\sin\theta = \dfrac{\sqrt{3}}{2}$, $\csc\theta = \dfrac{2\sqrt{3}}{3}$

$\cos\theta = \frac{1}{2}$, $\sec\theta = 2$

$\tan\theta = \sqrt{3}$, $\cot\theta = \dfrac{\sqrt{3}}{3}$

17. $\sin\theta = \dfrac{3\sqrt{10}}{10}$, $\csc\theta = \dfrac{\sqrt{10}}{3}$

$\cos\theta = \dfrac{\sqrt{10}}{10}$, $\sec\theta = \sqrt{10}$

$\tan\theta = 3$, $\cot\theta = \frac{1}{3}$

19. Quadrant IV   **21.** Quadrant I   **23.** Quadrant II

25. (a) $\sin 60° = \dfrac{\sqrt{3}}{2}$ (b) $\sin\left(-\dfrac{2\pi}{3}\right) = -\dfrac{\sqrt{3}}{2}$

$\cos 60° = \dfrac{1}{2}$ $\cos\left(-\dfrac{2\pi}{3}\right) = -\dfrac{1}{2}$

$\tan 60° = \sqrt{3}$ $\tan\left(-\dfrac{2\pi}{3}\right) = \sqrt{3}$

27. (a) $\sin\left(-\dfrac{\pi}{6}\right) = -\dfrac{1}{2}$ (b) $\sin 150° = \dfrac{1}{2}$

$\cos\left(-\dfrac{\pi}{6}\right) = \dfrac{\sqrt{3}}{2}$ $\cos 150° = -\dfrac{\sqrt{3}}{2}$

$\tan\left(-\dfrac{\pi}{6}\right) = -\dfrac{\sqrt{3}}{3}$ $\tan 150° = -\dfrac{\sqrt{3}}{3}$

29. (a) $\sin 225° = -\dfrac{\sqrt{2}}{2}$ (b) $\sin(-225°) = \dfrac{\sqrt{2}}{2}$

$\cos 225° = -\dfrac{\sqrt{2}}{2}$ $\cos(-225°) = -\dfrac{\sqrt{2}}{2}$

$\tan 225° = 1$ $\tan(-225°) = -1$

31. (a) $\sin 750° = \frac{1}{2}$ (b) $\sin 510° = \frac{1}{2}$

$\cos 750° = \dfrac{\sqrt{3}}{2}$ $\cos 510° = -\dfrac{\sqrt{3}}{2}$

$\tan 750° = \dfrac{\sqrt{3}}{3}$ $\tan 510° = -\dfrac{\sqrt{3}}{3}$

33. (a) 0.2079   **35. (a)** 0.3640   **37. (a)** −0.3420
  (b) 4.8097   **(b)** 0.3640   **(b)** −0.3420

39. (a) 2.0070   **41. (a)** $\dfrac{\pi}{6}, \dfrac{5\pi}{6}$ **43. (a)** $\dfrac{\pi}{3}, \dfrac{2\pi}{3}$
  (b) 2.0000

   (b) $\dfrac{7\pi}{6}, \dfrac{11\pi}{6}$ **(b)** $\dfrac{3\pi}{4}, \dfrac{7\pi}{4}$

45. (a) $\dfrac{3\pi}{4}, \dfrac{7\pi}{4}$ **(b)** $\dfrac{5\pi}{6}, \dfrac{11\pi}{6}$

47. $\dfrac{\pi}{4}, \dfrac{3\pi}{4}, \dfrac{5\pi}{4}, \dfrac{7\pi}{4}$ **49.** $0, \dfrac{\pi}{4}, \pi, \dfrac{5\pi}{4}, 2\pi$

51. $\dfrac{\pi}{6}, \dfrac{\pi}{2}, \dfrac{5\pi}{6}, \dfrac{3\pi}{2}$ **53.** $\dfrac{\pi}{4}, \dfrac{5\pi}{4}$

55. $0, \dfrac{\pi}{2}, \pi, 2\pi$ **57.** $\dfrac{100\sqrt{3}}{3}$ **59.** $\dfrac{25\sqrt{3}}{3}$

61. 15.5572   **63.** $20 \sin 75° \approx 19.32$ ft

65. $150 \cot 3° \approx 2862.2$ ft   **67. (a)** 25.2°F
              (b) 65.1°F

69.

x	0	2	4	6	8	10
$f(x)$	0	2.7021	2.7756	1.2244	1.2979	4

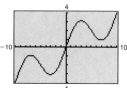

SECTION T.3   (page T-28)

Warm Up

1. 14   **2.** 10   **3.** 0   **4.** 1
5. $-\dfrac{1}{2}$ **6.** $\dfrac{1}{2}$ **7.** 0.9659   **8.** −0.9962
9. 0.9744   **10.** −0.6494

1. Period: π **3.** Period: 4π **5.** Period: 2
  Amplitude: 2   Amplitude: $\dfrac{3}{2}$ Amplitude: $\dfrac{1}{2}$

7. Period: 2π **9.** Period: $\dfrac{\pi}{5}$ **11.** Period: 3π
  Amplitude: 2   Amplitude: 2   Amplitude: $\dfrac{1}{2}$

13. Period: $\dfrac{1}{2}$ **15.** Period: $\dfrac{\pi}{2}$ **17.** Period: $\dfrac{2\pi}{5}$
  Amplitude: 3

19. Period: 6

21. c; Period: π **22.** e; Period: π

23. f; Period: 1   **24.** a; Period: 2π

25. b; Period: 4π **26.** d; Period: 2π

27.   **29.**

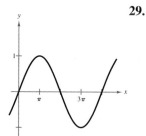   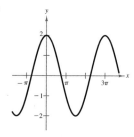

31.   **33.**

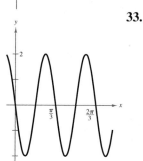   

35.   **37.**

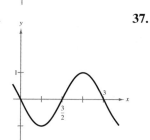   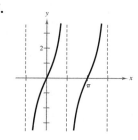

39.

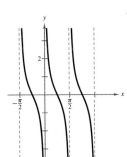

41.

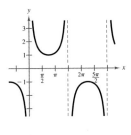

43.

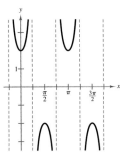

45.

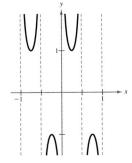

47.

x	-0.1	-0.01	-0.001
$f(x)$	-0.0500	-0.0050	-0.0005

x	0.001	0.01	0.1
$f(x)$	0.0005	0.0050	0.0500

$$\lim_{x\to 0}\frac{1-\cos x}{x}=0$$

49.

x	-0.1	-0.01	-0.001
$f(x)$	0.1997	0.2000	0.2000

x	0.001	0.01	0.1
$f(x)$	0.2000	0.2000	0.1997

$$\lim_{x\to 0}\frac{\sin x}{5x}=\frac{1}{5}$$

51. False, the amplitude is 3 (must be positive).

52. False, the period is $\dfrac{\pi}{4/3}=\dfrac{3\pi}{4}$.

53. True, $\displaystyle\lim_{x\to 0}\frac{\sin 5x}{3x}=\frac{5}{3}\left(\lim_{x\to 0}\frac{\sin 5x}{5x}\right)=\frac{5}{3}(1)=\frac{5}{3}$.

54. False,
$$\tan\left(\frac{5\pi/4}{2}\right)\neq 1.$$

55. **(a)** 6 sec   **(b)** 10
(c)

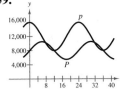

57. **(a)** $\frac{1}{440}$ **(b)** 440
(c)

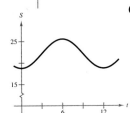

59.

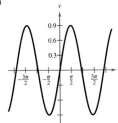

61.

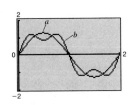

63. $P(13{,}673)\approx 0.1362$
$E(13{,}673)\approx 0.9010$
$I(13{,}673)\approx 0.8660$

65.

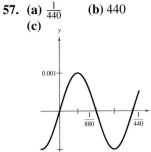

67. $\displaystyle\lim_{x\to 0}\frac{\sin x}{x}=1$

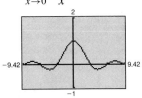

69. January, February, November, December

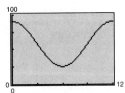

SECTION T.4 (page T-38)

Warm Up

1. $f'(x) = 9x^2 - 4x + 4$
2. $g'(x) = 12x^2(x^3 + 4)^3$
3. $f'(x) = 3x^2 + 2x + 1$
4. $g'(x) = \dfrac{2(5 - x^2)}{(x^2 + 5)^2}$
5. Relative minimum: $(-2, -3)$
6. Relative maximum: $\left(-2, \frac{22}{3}\right)$
   Relative minimum: $\left(2, -\frac{10}{3}\right)$
7. $x = \dfrac{\pi}{3}$, $x = \dfrac{2\pi}{3}$ 8. $x = \dfrac{2\pi}{3}$, $x = \dfrac{4\pi}{3}$
9. $x = \pi$ 10. No solution

1. $2x + \sin x$ 3. $-3\cos x$ 5. $\dfrac{2}{\sqrt{x}} - 3\sin x$
7. $-t^2 \sin t + 2t \cos t$ 9. $-\dfrac{t \sin t + \cos t}{t^2}$
11. $\sec^2 x + 2x$ 13. $5\sec x(x \tan x + 1)$
15. $4\cos 4x$ 17. $-2x \csc x^2 \cot x^2$
19. $-\csc 2x \cot 2x$ 21. $\sin\dfrac{1}{x} - \dfrac{1}{x}\cos\dfrac{1}{x}$
23. $12\sec^2 4x$ 25. $-2\cos x \sin x = -\sin 2x$
27. $-4\cos x \sin x = -2\sin 2x$ 29. $-\csc^2 x$
31. $\dfrac{1}{\sin x}\cos x = \cot x$ 33. $2x \csc x^2$
35. $\sec^2 x - 1 = \tan^2 x$ 37. $\dfrac{\cos x}{2\sqrt{\sin x}}$
39. $\dfrac{1}{2}(x \sec^2 x + \tan x - \sec x \tan x)$ 41. $\dfrac{\cos x}{2\sin 2y}$, 0
43. $y'' + y = (-2\sin x - 3\cos x)$
$\qquad + (2\sin x + 3\cos x) = 0$
45. $y'' + 4y = (-4\cos 2x - 4\sin 2x)$
$\qquad + 4(\cos 2x + \sin 2x) = 0$

47. $\dfrac{5}{4}$ 49. 2   51. 1   53. $y = 2x + \left(\dfrac{\pi}{2} - 1\right)$

55. Relative maximum: $\left(\dfrac{\pi}{3}, \dfrac{3\sqrt{3}}{2}\right)$
    Relative minimum: $\left(\dfrac{5\pi}{3}, -\dfrac{3\sqrt{3}}{2}\right)$

57. Relative minima: $\left(\dfrac{\pi}{3}, \dfrac{\pi}{3} - \sqrt{3}\right)$
    Relative maxima: $\left(\dfrac{5\pi}{3}, \dfrac{5\pi}{3} + \sqrt{3}\right)$

59. Relative maximum: $\left(\dfrac{7\pi}{4}, 0.0029\right)$
    Relative minimum: $\left(\dfrac{3\pi}{4}, -0.0670\right)$

61. (a) Relative minima:
       (3.0369, 168.68), (9.0369, 171.98),
       (15.0369, 175.28), (21.0369, 178.58)
    Relative maxima:
       (0.0369, 194.21), (6.0369, 197.51),
       (12.0369, 200.81), (18.0369, 204.11)
    (b) $197.93
63. (a) $t = 214.5$ (August 2 and 3)
    (b) $t = 32$ (February 1)

65. $\displaystyle\sum_{n=0}^{\infty} \dfrac{(-1)^n x^{2n+1}}{(2n+1)!}$, $R = \infty$

67. $\displaystyle\sum_{n=0}^{\infty} \dfrac{(-1)^n x^{4n+2}}{(2n+1)!}$ 69. $\displaystyle\sum_{n=0}^{\infty} \dfrac{(-1)^n (2x)^{2n}}{(2n)!}$

73. (a)

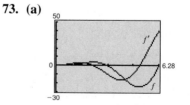

(b) 0, 2.2889, 5.0870
(c) $f' > 0$ on $(0, 2.2889)$, $(5.0870, 2\pi)$
    $f' < 0$ on $(2.2889, 5.0870)$

75. (a)

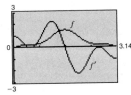

(b) 0.5236, $\pi/2$, 2.6180

(c) $f' > 0$ on $(0, 0.5236)$, $(0.5236, \pi/2)$

$f' < 0$ on $(\pi/2, 2.6180)$, $(2.6180, \pi)$

77. (a)

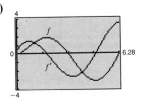

(b) 1.8366, 4.8158

(c) $f' > 0$ on $(0, 1.8366)$, $(4.8158, 2\pi)$

$f' < 0$ on $(1.8366, 4.8158)$

79. 0.6080, 2.8457, 6.4323, 9.3196

81. Relative maximum: 4.49, -4.60

83. Relative maximum: $(1.27, \ 0.07)$

Relative minimum: $(3.38, -1.18)$

85. Relative maximum: $(3.96, \ 1)$

SECTION T.5   (page T-47)

Warm Up

1. $-\dfrac{\sqrt{2}}{2}$ **2.** $-\dfrac{\sqrt{3}}{2}$ **3.** $-\dfrac{\sqrt{3}}{3}$ **4.** -1

5. $\tan x$ **6.** $\sin^2 x$ **7.** 1

8. $\csc x$ **9.** $\dfrac{88}{3}$ **10.** 4

1. $-2\cos x + 3\sin x + C$ **3.** $t + \csc t + C$

5. $-\cot\theta - \sin\theta + C$ **7.** $-\frac{1}{2}\cos 2x + C$

9. $\dfrac{1}{2}\sin x^2 + C$ **11.** $2\tan\dfrac{x}{2} + C$

13. $-\frac{1}{3}\ln|\cos 3x| + C$ **15.** $\frac{1}{4}\tan 4x + C$

17. $\dfrac{1}{\pi}\ln|\sin\pi x| + C$

19. $\frac{1}{2}\ln|\csc 2x - \cot 2x| + C$ **21.** $\frac{1}{2}\ln|\tan 2x| + C$

23. $\ln|\sec x - 1| + C$ **25.** $-\ln|1 + \cos x| + C$

27. $\frac{1}{2}\tan^2 x + C$ **29.** $-\cos e^x + C$

31. $\ln|\cos e^{-x}| + C$ **33.** $x - \frac{1}{4}\cos 4x + C$

35. $x\sin x + \cos x + C$ **37.** $x\tan x + \ln|\cos x| + C$

39. $\dfrac{3\sqrt{3}}{8} \approx 0.6495$ **41.** $2(\sqrt{3} - 1) \approx 1.4641$

43. $\frac{1}{2}$ **45.** $\ln(\cos 0) - \ln(\cos 1) \approx 0.6156$

47. 4   **49.** $\dfrac{\pi^2}{2} + 2 \approx 6.9348$ **51.** 2

53. π **55.** $y = \frac{1}{5}(2\sin x - \cos x) + Ce^{-2x}$

57. 1.3707   **59. (a)** 1.3655   **(b)** 1.3708

61. (a) 225.28 million barrels   **(b)** 225.28 million barrels

(c) 217 million barrels

63. (a) $C \approx \$9.17$ **(b)** $C \approx \$3.14$, savings $\approx \$6.03$

65. 0.5093 liter   **67.** 0.9777   **69.** 3.8202

SECTION T.6   (page T-56)

Warm Up

1. ∞ **2.** 0   **3.** $\frac{2}{3}$ **4.** ∞

5. $-2x\sin x^2$ **6.** $5\cos(5x - 1)$

7. $4\sec(4x)\tan(4x)$ **8.** $2x\sec^2(x - 2)$

9. $-4\sin(2x + 3)$ **10.** $2\sec^2 x\tan x$

1. Yes   **3.** No   **5.** Yes

7.

x	-0.1	-0.01	-0.001
$f(x)$	-0.35	-0.335	-0.3335

x	0.001	0.01	0.1
$f(x)$	-0.3332	-0.332	-0.32

$$\lim_{x \to 0} \frac{e^{-x} - 1}{3x} = -\frac{1}{3}$$

9.

x	-0.1	-0.01	-0.001
$f(x)$	0.1997	0.2	0.2

x	0.001	0.01	0.1
$f(x)$	0.2	0.2	0.1997

$$\lim_{x \to 0} \frac{\sin x}{5x} = \frac{1}{5}$$

11. 0   **13.** -3 **15.** -1 **17.** $\frac{1}{5}$ **19.** 0

21. -3 **23.** 1   **25.** 0   **27.** $\frac{4}{3}$

29. 0   **31.** $\frac{1}{2}$ **33.** $\frac{2}{5}$ **35.** 1

37. ∞ **39.** ∞ **41.** 1   **43.** 0

45. 0   **47.** 0   **49.** 0   **51.** 0   **53.** 0

55.

x	10	10^2	10^3
$\dfrac{(\ln x)^5}{x}$	6.47	20.71	15.73

x	10^4	10^5	10^6
$\dfrac{(\ln x)^5}{x}$	6.63	2.02	0.503

$$\lim_{x \to \infty} \frac{(\ln x)^5}{x} = 0$$

57. The limit of the denominator is not 0.

59. The limit of the numerator is not 0.

61. **(a)** 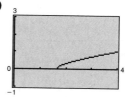   **(b)** $\frac{1}{3}$

63. **(a)** 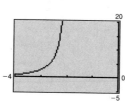   **(b)** Limit does not exist.

65. False, $\lim\limits_{x \to 0} \dfrac{x^2 + 3x - 1}{x + 1} = \dfrac{-1}{1} = -1$

66. False, $\lim\limits_{x \to \infty} \dfrac{x}{1 - x} = -1$ **67.** True

68. False. For example, $\lim\limits_{x \to \infty} \dfrac{x + 1}{x} = 1$, but $x + 1 \neq x$.

69. $\lim\limits_{x \to \infty} \dfrac{x}{\sqrt{x^2 + 1}} = \lim\limits_{x \to \infty} \dfrac{\sqrt{x^2 + 1}}{x} = \lim\limits_{x \to \infty} \dfrac{x}{\sqrt{x^2 + 1}}$

$\lim\limits_{x \to \infty} \dfrac{x}{\sqrt{x^2 + 1}} = \lim\limits_{x \to \infty} \dfrac{1}{\sqrt{1 + (1/x^2)}} = 1$

71. **(a)** Hasbro   **(b)** 1994
(c)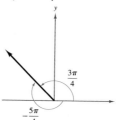

CHAPTER T REVIEW EXERCISES   (page T-61)

1. $\dfrac{3\pi}{4}, \ -\dfrac{5\pi}{4}$ **3.** $470°, \ -250°$

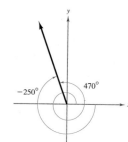

5. $\dfrac{7\pi}{6}$ **7.** $-\dfrac{8\pi}{3}$ **9.** $420°$ **11.** $-108°$

13. $b = 4\sqrt{3}, \ \theta = 60°$

15. $a = \dfrac{5\sqrt{3}}{2}, \ c = 5, \ \theta = 60°$

17. 15.38 ft   **19.** 0 **21.** $72°$ **23.** $\dfrac{\sqrt{2}}{2}$

25. $\sqrt{3}$ **27.** $-\dfrac{\sqrt{3}}{2}$ **29.** -1 **31.** 0.6494

33. -0.3420 **35.** $r \approx 146.19$ **37.** $x \approx 68.69$

39. $\dfrac{2\pi}{3} + 2k\pi, \ \dfrac{4\pi}{3} + 2k\pi$

41. $\dfrac{7\pi}{6} + 2k\pi, \ \dfrac{3\pi}{2} + 2k\pi, \ \dfrac{11\pi}{6} + 2k\pi$

43. $\dfrac{\pi}{3} + 2k\pi, \ \pi + 2k\pi, \ \dfrac{5\pi}{3} + 2k\pi$ **45.** 81.18 ft

47.

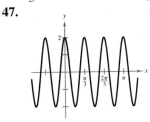

49.

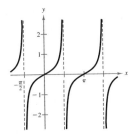

51.

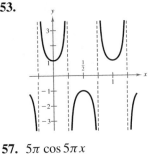

53.

55.

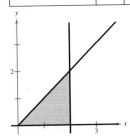

57. $5\pi \cos 5\pi x$

59. $-x \sec^2 x - \tan x$ **61.** $\dfrac{-x \sin x - 2 \cos x}{x^3}$

63. $24 \sin 4x \cos 4x + 1 = 12 \sin 8x + 1$

65. $-6 \csc^3 x \cot x$ **67.** $e^x(\cot x - \csc^2 x)$

69. Relative maximum: $(0.523, \ 1.128)$
Relative minimum: $(2.616, \ 0.443)$

71. Relative maxima: $\left(\dfrac{\pi}{2}, 2\right), \ \left(\dfrac{3\pi}{2}, 0\right)$

Relative minima: $\left(\dfrac{7\pi}{6}, -\dfrac{1}{4}\right), \ \left(\dfrac{11\pi}{6}, -\dfrac{1}{4}\right)$

73. (a) \$114.75 thousand, 92
(b) \$36.25 thousand, 273

75. $-3 \cos x - 2 \sin x + C$ **77.** $\frac{1}{4} \sin^4 x + C$

79. $\pi + 2$ **81.** 0 **83.** 1

85. $\frac{5}{3}$ **87.** 6.9 quads

89. $\frac{1}{5}$ **91.** $-\frac{5}{4}$ **93.** $\frac{1}{2}$ **95.** 0

97. $g(x)$ grows faster than $f(x)$.

APPENDIX A (page A9)

1. Left Riemann sum: 0.518
Right Riemann sum: 0.768

3. Left Riemann sum: 0.746
Right Riemann sum: 0.646

5. Left Riemann sum: 0.859
Right Riemann sum: 0.659

7. Midpoint Rule: 0.673

9.

n	5	10	50	100
Left sum, S_L	1.6	1.8	1.96	1.98
Right sum, S_R	2.4	2.2	2.04	2.02

11. $\int_0^5 3\,dx$

13. $\int_{-4}^4 (4 - |x|)\,dx = \int_{-4}^0 (4 + x)\,dx + \int_0^4 (4 - x)\,dx$

15. $\int_{-2}^2 (4 - x^2)\,dx$

17. $A = 12$

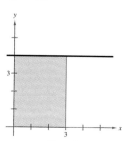

Rectangle

19. $A = 8$

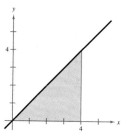

Triangle

21. $A = 14$

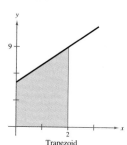

Trapezoid

23. $A = 1$

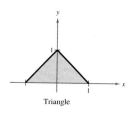

Triangle

25. $A = \dfrac{9\pi}{2}$

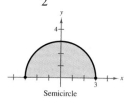

Semicircle

APPENDIX A   (page A55)

Warm Up

1. $y = 4 - 3x$ **2.** $y = x$

3. $y = \frac{2}{3}(1 - x)$ **4.** $y = \frac{2}{3}(2x + 1)$

5. $y = \frac{1}{4}(5 - 3x)$ **6.** $y = \frac{2}{3}(-x + 3)$

7. $y = 4 - x^2$ **8.** $y = \frac{2}{3}(x^2 - 1)$

9. $y = \pm\sqrt{4 - x^2}$ **10.** $y = \pm\sqrt{x^2 - 9}$

1.

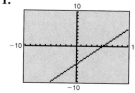

3.

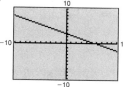

5.

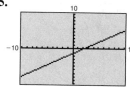

7.

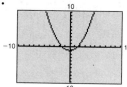

9.

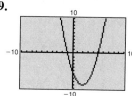

11.

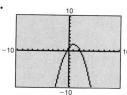

13.

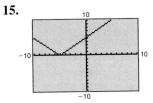

15.

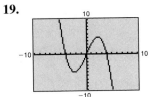

17.

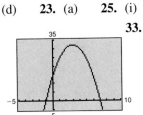

19.

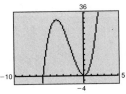

21. (d) **23.** (a) **25.** (i) **27.** (j) **29.** (e)

31.

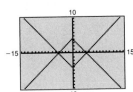

33.

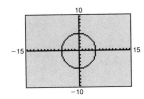

35.

```
RANGE
Xmin=-10
Xmax=10
Xscl=1
Ymin=-12
Ymax=30
Yscl=6
Xres=1
```

37.

```
RANGE
Xmin=-10
Xmax=10
Xscl=1
Ymin=-10
Ymax=10
Yscl=1
Xres=1
```

39. No intercepts **41.** Three x-intercepts

43. Square **45.** Circle

47.

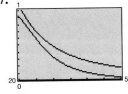

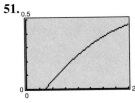

49. 0.59

51.

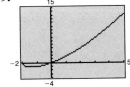

53. (b)

55. $f'(1) \approx -16.333$, decreasing
$f'(1) \approx 15.667$, increasing
$f'(3) \approx 431.840$, increasing

57. $f'(-2) \approx -0.320$, decreasing
$f'(0) \approx 0$, horizontal tangent
$f'(2) \approx 0.320$, increasing

59.

61. $f'(0)$ does not exist.

63. 17.591 **65.** -1.099 **67.** 0.886

Index

L